Fourth *Edition*

Community Nutrition in Action

AN ENTREPRENEURIAL APPROACH

Marie A. Boyle, PhD, RD
College of Saint Elizabeth

David H. Holben, PhD, RD
Ohio University

Australia • Brazil • Canada • Mexico • Singapore • Spain • United Kingdom • United States

Community Nutrition in Action: An Entrepreneurial Approach,
Fourth Edition
Marie A. Boyle and David H. Holben

Publisher: *Peter Marshall*
Assistant Editor: *Elesha Feldman*
Editorial Assistant: *Lauren Vogelbaum*
Technology Project Manager: *Travis Metz*
Marketing Manager: *Jennifer Somerville*
Marketing Assistant: *Michele Colella*
Marketing Communications Manager: *Shemika Britt*
Project Manager, Editorial Production: *Sandra Craig*
Creative Director: *Rob Hugel*
Art Director: *Lee Friedman*
Print Buyer: *Rebecca Cross*
Permissions Editor: *Sarah Harkrader*

Printed in the United States of America
1 2 3 4 5 6 7 09 08 07 06 05

Library of Congress Control Number: 2005925213

ISBN 0-534-46581-1

Development Editor: *Tricia Louvar*
Production Service: *Martha Emry*
Text Designer: *Ellen Pettengell*
Photo Researcher: *Quest Photo Research*
Copy Editor: *Connie Day*
Illustrator: *Atherton Customs*
Cover Designer: *Carolyn Deacy*
Cover Image: ©2005 *Gwendolyn Knight Lawrence/Artists Rights Society (ARS), New York*
Cover Printer: *Phoenix Color Corp*
Compositor: *International Typesetting and Composition*
Printer: *Malloy Incorporated*

Thomson Higher Education
10 Davis Drive
Belmont, CA 94002-3098
USA

For more information about our products, contact us at:
Thomson Learning Academic Resource Center
1-800-423-0563

For permission to use material from this text or product, submit a request online at
http://www.thomsonrights.com.

Any additional questions about permissions can be submitted by e-mail to
thomsonrights@thomson.com.

In memory of my father,
David M. Boyle,
and his love of everything Irish,
especially his daughters,
and my mother,
Marie T. Boyle,
a lifelong educator,
cherished mother,
and treasured friend.

— Marie A. Boyle Struble

This book is dedicated
to the God who made me,
my wife and children who love me,
my family and friends who support me,
and my students who teach,
motivate, and challenge me.

— David H. Holben

About the Authors

MARIE BOYLE STRUBLE, PhD, RD, received her BA in psychology from the University of Southern Maine and her MS and PhD in nutrition from Florida State University. She is coauthor of the basic nutrition textbook *Personal Nutrition* and former professor and director of the Graduate Program in Nutrition at the College of Saint Elizabeth in Morristown, New Jersey. She now teaches online distance courses in such areas as Community Nutrition, Personal and Advanced Nutrition, Nutrition and Aging, and Nutrition Applications of Psychological and Sociological Issues. Her other professional activities include serving as an author and reviewer for the American Dietetic Association and Society for Nutrition Education. She coauthored the current Position Paper of the American Dietetic Association on *Addressing World Hunger, Malnutrition, and Food Insecurity* and serves as editor of the new *Journal of Hunger and Environmental Nutrition* by Haworth Press. She is a member of the American Dietetic Association, the American Public Health Association, and the Society for Nutrition Education.

DAVID HOLBEN, PhD, RD, received his BS in dietetics from Indiana University of Pennsylvania, Indiana, Pennsylvania, and did his dietetic internship at Harper Hospital in Detroit, Michigan. His MA in food science and nutrition was completed at Wayne State University, Detroit, Michigan, and he received both his MS and his PhD in human nutrition from The Ohio State University, Columbus, Ohio. He has broad experience in dietetics and currently is associate professor of Human and Consumer Sciences (Food, Nutrition, and Hospitality) and director of the Didactic Program in Dietetics at Ohio University, Athens, Ohio. Dr. Holben teaches courses in Community Nutrition, Introductory and Advanced Nutrition, Medical Nutrition Therapy, and Research Methods. Professionally, he is active within the American Dietetic Association at the national, state, and local levels and chairs the Ohio Board of Dietetics. Dr. Holben studies food access of individuals and families, especially as it is related to health, and has authored or coauthored many papers and presentations, including the current Position Paper of the American Dietetic Association on *Domestic Food and Nutrition Security*.

Contents in Brief

Contents

Section *Three*

Preface

To succeed in community nutrition today, you must be committed to lifelong learning, because every day brings new research findings, new legislation, new ideas about health promotion, and new technologies, all of which affect the ways in which community nutritionists gather information, solve problems, and reach vulnerable populations. You will probably be an entrepreneur—one who uses innovation and creativity to guide individuals and communities to proper nutrition and good health. You will work well as a member of teams to lobby policy makers, gather information about your community, and design nutrition programs and services. You will be skilled in assessing the activities of "the competition"—the myriad messages about foods, dietary supplements, and research findings that appear in the media.

We spoke, in the first edition of this book, about a sea of change—a shift toward globalization of the workforce and communications, a shift from clinical dietetics to community-based practice. Our second edition reflected the growth of the Internet—a virtual tsunami in communications—and documented the collapse of health care reform, the emergence of managed care, the drive to reform welfare, and the rise of complementary and alternative medicine. Our third edition witnessed the birth of a new millenium, addressed new national health objectives for 2010, applauded the inclusion of medical nutrition therapy as a benefit to certain Medicare recipients, and acknowledged the threat of bioterrorism and the need for nutrition professionals to educate the public on issues related to the safety of our community food and water supplies. Since the third edition was published, our society has been challenged by American eating trends; increasing cultural, ethnic, racial, and generational diversity, the rising tide of obesity; recent advances in genetics and biotechnology; and the need to demonstrate meaningful outcomes for nutrition services.

The Fourth Edition

In this Fourth Edition, we continue to discuss the important issues in community nutrition practice and to present the core information needed by students who are interested in solving nutritional and health problems. The book is organized into three sections. Section One shows the community nutritionist in action within the community. Chapter 1 describes the activities and responsibilities of the community nutritionist and introduces the principles of entrepreneurship and the three arenas of community nutrition practice: people, policy, and programs. Chapter 2 gives a step-by-step analysis of the community needs assessment process and describes the types and sources of data collected about the community. Chapter 3 outlines the questions you'll ask in obtaining information about your target population, including diet assessment methods. Chapter 4 reviews the basic principles of epidemiology. Chapter 5 examines some of the issues surrounding poverty and food insecurity in the domestic arena, considers how these contribute to nutritional risk and malnutrition, and outlines the major domestic food assistance programs designed to help with achieving food security. Chapter 6 focuses on the nuts and bolts of national nutrition policy, including national nutrition monitoring and dietary recommendations. Chapter 7 makes it perfectly clear that if you're a community nutritionist, you're involved in policy making. Chapter 8—new for this edition—discusses the epidemic of obesity, examining some of the epidemic's societal and environmental determinants, along with various policy options for addressing the problem. Chapter 9 discusses today's health care system, the challenge of eliminating health disparities and providing quality health care to all citizens, and the necessity of outcomes assessment in nutrition services.

Section Two describes current federal and nongovernmental programs designed to meet the food and nutritional needs of vulnerable populations. Chapter 10 focuses on programs for pregnant and lactating women and for infants. Chapter 11 describes programs for children and adolescents. Chapter 12 covers a host of programs for adults, including the elderly. Chapter 13 examines the issue of world hunger and food insecurity.

Section Three focuses on the tools used by community nutritionists to design programs to address nutritional and health problems in their communities. Chapter 14 describes the program planning process, covering everything from the factors that trigger program planning to the types of evaluations undertaken to improve program design and delivery. Chapter 15 discusses the reasons why people eat what they eat, what research tells us about how to influence behavior, and the design of program interventions. Chapter 16 addresses the need for cultural competence and explains strategies for providing culturally competent nutrition services. Chapter 17 gets to the heart of any program: the nutrition messages used in community interventions. Chapter 18 introduces the principles of marketing, including social marketing, an important endeavor in community nutrition practice. You are more likely to get good results if your program is marketed successfully! Chapter 19 addresses such important management issues as how to control costs and manage people. Finally, Chapter 20 closes the text with a discussion of grantsmanship—everything you need to know about finding and managing funding for community programs and interventions.

Many of the unique features of the previous editions have been retained:

- **Professional Focus.** This feature is designed to help you develop personal and professional skills and attitudes that will boost your effectiveness and confidence in community settings. The topics range from goal setting and time management to public speaking, working with the media, and leadership.
- **Internet Resources.** Each chapter contains a list of relevant Internet addresses. You'll use these websites to obtain data about your community and to scout for ideas and educational materials. Moreover, you can link with the Internet addresses presented in this book through the publisher's Nutrition Resource Center online (www.wadsworth.com/nutrition). If you still aren't using the Internet regularly, this is the time to begin.
- **Programs in Action.** This feature—found in most chapters—highlights award-winning, innovative, grassroots nutrition programs. It offers a unique perspective on the practice of community nutrition. Our hope is that the insights you gain from these initiatives will inspire you to get involved in learning about your community and its health and nutritional problems and to design similar programs to address the needs you uncover. The feature highlights such programs as "Eat Healthy: Your Kids Are Watching," a program designed to remind parents that they serve as role models for their children; "Feast with the Beasts," a program for disadvantaged elementary school children that teaches children about their own nutritional needs and those of local zoo animals; "Food on the Run," a program to empower teens to make healthful decisions about their nutrition and physical activity patterns; and a "5 a Day Nutrition Education Program," designed to improve the nutrition behaviors and fruit and vegetable intake of firefighters. This feature discusses each program's goals, objectives, and rationale; the practical aspects of its implementation; and its effectiveness in serving the needs of its intended audience.
- **Case Studies.** The book's case studies make use of a transdisciplinary, developmental problem-solving model as a learning framework to enhance students' critical-thinking skills.* They are designed to help students develop competence in applying their knowledge and skills to contemporary nutrition issues with real-life uncertainties—such issues as might be found in the workplace. Each case emphasizes the need to evaluate the information presented, identify and describe uncertainties in the case, locate and distinguish between relevant and irrelevant information, identify assumptions, prioritize alternatives, make decisions, and communicate and evaluate conclusions. Many of the case questions are open-ended.

Finally, we hope that the people, policies, and programs presented in this text inspire you to consider a rewarding career path in community nutrition. We want you to think of yourself as a planner, manager, change agent, thinker, and leader—in short, a nutrition entrepreneur—who has the energy and creativity to open up new vistas for improving the public's health through good nutrition.

Resources for Instructors

Key instructor resources, such as an online Instructor's Manual with Test Bank, accompany the text. (Please consult your local sales representative for a pass code.) In addition, students and instructors will have access to a rich array of teaching and learning tools, such as quizzes and PowerPoint slides with figures, through the book's companion site, as well as Wadsworth's Nutrition Resource Center at nutrition.wadsworth.com. A number of outstanding nutrition resources can also be combined with the text, such as the diet assessment software *Diet Analysis Plus 7.0*. Furthermore, the news-gathering and programming power of CNN can be integrated into the classroom via exclusive videos: CNN® Today: Nutrition Video Series (three volumes). (Available to qualified adopters. Please consult your local sales representative for details.)

Acknowledgments

This book was a community effort. Family and friends provided encouragement and support. Colleagues shared their

*See C. L. Lynch, S. K. Wolcott, and G. E. Huber, *Steps for Better Thinking: A Developmental Problem Solving Process,* May 31, 2002; available at www.WolcottLynch.com.

insights, program materials, and experiences about the practice of community nutrition and the value of focusing on entrepreneurship. We are grateful to this text's contributing authors:

- Kathy Bauer, PhD, RD, associate professor, Montclair State University, Upper Montclair, New Jersey, for Chapter 16: *Gaining Cultural Competence in Community Nutrition.*
- Carol Byrd-Bredbenner, PhD, RD, FADA, professor and extension specialist in nutrition, Rutgers, The State University of New Jersey, New Brunswick, New Jersey, for Chapter 20: *Building Grantsmanship Skills.*
- Deanna M. Hoelscher, PhD, RD, LD, CNS, associate professor and director, Human Nutrition Center, University of Texas School of Public Health, Houston, and Christine McCullum-Gómez, PhD, RD, LD, assistant professor, Human Nutrition Center, University of Texas School of Public Health, Houston, for Chapter 8: *Addressing the Obesity Epidemic: An Issue for Public Health Policy.*
- Mary Kate Harrison, MS, RD, General Manager, Hillsborough County Student Nutrition Services, Tampa, Florida, for her revision of Chapter 11: *Children and Adolescents: Nutrition Issues, Services, and Programs.*
- Kathleen Shimomura, MS, DTR, associate professor, Rutgers University Cooperative Extension, for her contributions to Chapter 17: *Principles of Nutrition Education.*
- Alice Fornari, EdD, RD, assistant director, Medical Education, Department of Family and Social Medicine, Albert Einstein College of Medicine, Bronx, New York, and Alessandra Sarcona, MS, RD, director, Dietetic Internship, C. W. Post Campus of Long Island University, Brookville, New York, for their steadfastness and expertise in developing the case studies that accompany many of this text's chapters.

The text is richer for the contributions made by these authors. Finally, we are grateful for the work that Diane Morris, PhD, RD, contributed to the first two editions of this text as coauthor; her expertise and insights are reflected in this new edition, still. We thank the many people who have prepared the ancillaries for this edition, especially Melanie Burns, Alessandra Sarcona, Alice Fornari, and Carmen Boyd, for their expertise in revising and enhancing the Web-based Instructor's Manual with Test Bank and quizzes that accompany this text.

Special thanks go to our editorial team: Pete Marshall, publisher; Sandra Craig, project production manager; Tricia Louvar, developmental editor; and Jennifer Somerville, marketing manager, for their support and assistance. We are grateful to Elizabeth Howe and Elesha Feldman for their coordination of this book's revision and ancillaries. We appreciate Sarah Harkrader's help in finalizing the permissions. As always, we are most grateful to Martha Emry, our production editor, for her guidance throughout the production of this revision, from organizing the many production activities to her diligent attention to details, all done with her usual grace and creative style. Four other members of the production team also have our thanks: Connie Day, copyeditor; Jim Atherton, artist; Martha Ghent, proofreader; and Pat Quest, photo researcher. The fine quality of this product reflects their hard work and diligence. We are indebted to everyone at International Typesetting and Composition for skillfully producing a text to be proud of. Last, but not least, we owe much to our colleagues who provided articles and course outlines, their favorite Internet addresses, and expert reviews of the manuscript. Their ideas and suggestions are woven into every chapter. We appreciate their time, energy, and enthusiasm, and we hope they take as much pride in this book as all of us with Wadsworth do. Thanks to all of you:

Virginia Bennett, PhD, RD, Central Washington University
Laura Calderon, DrPH, RD, California State University, Los Angeles
Jenell Ciak, PhD, RD, LD, Northwest Missouri State University
Alana D. Cline, University of Northern Colorado
Nancy Cohen, PhD, RD, LDN, University of Massachusetts
Nancy Cotugna, DrPH, RD, University of Delaware
Lynn Duerr, PhD, RD, CD, Indiana State University
Jerald Foote, PhD, RD, University of Arkansas
Lauren Haldeman, University of North Carolina at Greensboro
Nancy Harris, East Carolina University
Terryl J. Hartman, The Pennsylvania State University
M. Jane Heinig, University of California, Davis
Tanya M. Horacek, PhD, RD, Syracuse University
Diana McGuire, MS, RD, CD, CNSD, Brigham Young University
Pamela S. McMahon, University of Florida
Valentina M. Remig, Kansas State University
Padmini Shankar, PhD, RD, LD, Georgia Southern University
Tamara S. Vitale, Utah State University

Marie Boyle Struble
David Holben

Community Nutritionists in Action: Working in the Community

Section *One*

On Saturday morning, Irene H. opens her kitchen cabinet and takes down six small bottles. She lines them up on the countertop and works their caps off. The process takes a few minutes because her fingers are stiff from arthritis. Let's see, there's cod liver oil, chondroitin sulfate, and glucosamine for arthritis; ginkgo biloba and St. John's wort to relieve anxiety and depression; and DHEA to restore youthful vigor. Irene knows her doctor would be surprised—maybe shocked—to learn that she takes these supplements regularly. She knows, too, that her doctor would not approve of her consultations with a naturopath whose office is just a couple of miles from her home.

At 48, Irene figures she is doing all she can to manage the pain from her arthritis and the depression that has afflicted her since her divorce. The supplements and naturopathic counseling are expensive, but she stretches the income from her job as a checkout clerk at a paint supply store to allow for them. After washing down the pills with orange juice, she pops two frozen waffles in the toaster and pours another cup of coffee. She figures she shouldn't eat the waffles—she was diagnosed with Type 2 diabetes just three months ago—but she wants them. After breakfast, she'll enjoy a cigarette with her coffee and then call her oldest daughter. Maybe they can drive out to the mall.

Irene is a typical consumer in many respects. She has chronic health problems for which she has sought traditional medical advice and treatment. Like one in three U.S. adults, she has also sought help from an alternative practitioner. She smokes cigarettes, she is overweight, and about the only exercise she gets is browsing the sale stalls out at the mall. She could do more to improve her health, but she isn't motivated to change her diet or quit smoking. She's looking for the quick fix.

Irene and the thousands of other consumers like her are a challenge for the community nutritionist. To help Irene make changes in her lifestyle—changes that will reduce her demands on the health care system and improve her physical well-being—the community nutritionist must be familiar with a broad spectrum of clinical and epidemiologic research, understand the health care system, and draw on the principles of public health and health promotion. The community nutritionist must know where Irene and people like her live and work, what they eat, and what their attitudes and values are. The community nutritionist must know about the community itself and how it delivers health services to people like Irene. And the community nutritionist must know how to influence policy makers. Perhaps now is the time to call for tighter regulation of dietary supplements and greater government support for health promotion and disease prevention programs.

This section describes the work that community nutritionists do in their communities. It outlines the principles of public health, health promotion, and policy making and reviews the current health care environment. It describes how to conduct a needs assessment in your community and outlines some of the tools you might use to assess the nutritional status of a target population. It focuses on entrepreneurship—the discipline founded on creativity and innovation—and how its principles can be used to reach Irene and other people in the community with health and nutritional problems. The material in this section sets the stage and lays the groundwork for understanding what community nutritionists do: they focus on people, policies, and programs.

Chapter *One*

Opportunities in Community Nutrition

Learning *Objectives*

After you have read and studied this chapter, you will be able to:

- Describe the three arenas of community nutrition practice.
- Explain how community nutrition practice fits into the larger realm of public health.
- Describe the three types of prevention efforts.
- List three major health objectives for the nation and explain why each is important.
- Outline the educational requirements, practice settings, and roles and responsibilities of community nutritionists.
- Explain why entrepreneurship is important to the practice of community nutrition.

Chapter *Outline*

Something To Think About...

Education and health are the two great keys. We must use all public sector institutions, flawed though they may be, to close the gap between rich and poor. We must work with the political sector to convincingly paint the breadth and depth of the problem and the size of the opportunity as well. . . . Above all, we must not abandon the hope of progress.

– SIR GUSTAV NOSSAL, writing on health and the biotechnology revolution in *Public Health Reports,* March/April 1998

Introduction

Community nutritionists face many challenges in the practice of their science and art. There is the challenge of improving the nutritional status of different kinds of people with different education and income levels and different health and nutritional needs: teenagers with anorexia nervosa, pregnant women living in public housing, the homeless, new immigrants from southeast Asia, elderly women alone at home, middle-class adults with high blood cholesterol, professional athletes, children with disabilities. There is the challenge of forming partnerships with colleagues, business leaders, and the public to advocate for change. There is the challenge of influencing lawmakers and other key citizens to enact laws, regulations, and policies that protect and improve the public's health. There is the challenge of studying the scientific literature for new angles on how to help people make good food choices for good health. And there is the challenge of mastering technologies such as the Internet to help meet the needs of clients and communities.

In addition to these challenges, certain social and economic trends also present challenges for community nutritionists. Immigrants from Mexico, Asia, and the Caribbean, many of whom have poor English skills, have streamed into North America in recent years, searching for jobs and improved living conditions.[1] The North American population is aging rapidly, as "baby boomers" mature and life expectancy increases.[2] Financial pressures and increased global competition have forced governments, businesses, and organizations to be creative in the face of scarce resources. Indeed, according to one survey of employers undertaken by the American Dietetic Association, the single greatest challenge for the dietetics professional today is "the need to do more and better with less."[3] Community nutritionists in all practice settings face rising costs, changing consumer expectations about health care services, increased competition in the market, and greater cultural diversity among their clients. They are pressured by downsizing, mergers, cross training, and managed health care.

Community nutritionists who succeed in this changing environment are flexible, innovative, and versatile. They are *focused* on recognizing opportunities for improving people's nutritional status and health and on helping society meet its obligation to alleviate hunger and malnutrition. It is an exciting time for community nutritionists. It is a time for learning new skills and moving into new areas of practice. It is a time of great opportunity and incredible need.

The Concept of Community

"There is no complete agreement as to the nature of community," wrote G. A. Hillery, Jr.[4] Such diverse locales as isolated rural hamlets, mountain villages, prairie towns, state capitals, industrial cities, suburbs or ring cities, resort towns, and major metropolitan areas can all be lumped into a single category called "community."[5] The concept of community is not always circumscribed by a city limits sign or zoning laws. Sometimes the term describes people who share certain interests, beliefs, or values, even though they live in diverse geographical locations; examples include the academic community, the gay community, and the immigrant community. For our purposes in this book, a **community** is a grouping of people who reside in a specific locality and who interact and connect through a definite social structure to fulfill a wide range of daily needs. By this definition, a community has four components: people, a location in space (which can include the realm of cyberspace), social interaction, and shared values.

Communities can be viewed on different scales: global, national, regional, and local. Each of these can be further segmented into specialized communities or groups, such as those individuals who speak Spanish, those who own computers, and those who observe Hanukkah. In the health arena, communities tend to be segmented around particular wellness, disease, or risk factors—for example, adults who exercise regularly, children infected with HIV, black men with high blood pressure, and people with peanut allergy.

Opportunities in Community Nutrition

Founded on the sciences of epidemiology, food, nutrition, and human behavior, **community nutrition** is a discipline that strives to improve the health, nutrition, and well-being of individuals and groups within communities. Its practitioners develop policies and programs that help people improve their eating patterns and health. Indeed, these three arenas—people, policy, and programs—are the focus of community nutrition.

PEOPLE

Individuals who benefit from community nutrition programs and services range from young single mothers on public assistance to senior business executives, from immigrants with poor English skills to college graduates, from pregnant teenagers with iron-deficiency anemia to grandfathers with Alzheimer's disease. They are found in worksites, schools, community centers, health clinics, churches, apartment buildings—virtually any community setting. Through community nutrition programs and services, these individuals and their families have access to food in times of need or learn skills that improve their eating patterns. It is the community nutritionist who identifies a group of people with an unmet nutritional need, gathers information about the group's socioeconomic background, ethnicity, religion, geographical location, and cultural food patterns, and then develops a program or service tailored to the needs of this group.

Community A group of people who are located in a particular space (including cyberspace), have shared values, and interact within a social system.

Community nutrition A discipline that strives to prevent disease and to improve the health, nutrition, and well-being of individuals and groups within communities.

Policy A course of action chosen by public authorities to address a given problem.

POLICY

Policy is a key component of community nutrition practice. **Policy** is a course of action chosen by public authorities to address a given problem.[6] Policy is what governments and organizations *intend* to accomplish through their laws, regulations, and programs.

How does policy apply to the practice of community nutrition? Consider a situation in which a group of community nutritionists address food waste in their community. The impetus for their action came from learning the results of a U.S. Department of Agriculture study that found that one-fourth of all food produced in the United States is wasted[7] and from reading about a successful food assistance program called "gleaning." Gleaning began as a project to deliver an abundance of apples from communities with apple orchards to food banks in neighboring states where apples were scarce.[8] The community nutritionists wanted to try gleaning on a small scale, using farmers' markets in their community. Unfortunately, there was no city bylaw that allowed surplus foods from farmers' markets to be made available to local food banks and soup kitchens. After gaining the support of the farmers' markets, food banks, and soup kitchens, the community nutritionists lobbied the city council to enact a bylaw to allow such transactions. The city council members voted to pass a bylaw to support gleaning projects. In other words, the city council altered its *policy* about recovering and recycling surplus foods.

Community nutritionists are involved in policy when they write letters to their state legislators, lobby Congress to secure expanded Medicare coverage for medical nutrition therapy, advise their municipal governments about food banks and soup kitchens, and use the results of research to influence policy makers. Many aspects of the community nutritionist's job involve policy issues.

PROGRAMS

Programs are the instruments used by community nutritionists to seek behavior changes that improve nutritional status and health. They are wide-ranging and varied. They may target small groups of people—children with developmental disabilities in Nevada schools or teenagers living in a Brooklyn residential home—or they may target large groups such as all adults with high blood cholesterol concentrations. Programs may be as widespread as the U.S. federal Food Stamp Program or as local as a diabetes prevention program for Mohawk people living in the Akwesasne community in northern New York State. They may be tailored to address the specific health and nutritional needs of people with obesity or osteoporosis, or they may be aimed at the general population. Two examples of population-based programs are "Particip*action*," a Canadian program designed to get people moving and fit for health, and "5 a Day for Better Health," a program of the U.S. National Cancer Institute aimed at making Americans more aware of how eating fruits and vegetables can improve their health and may reduce their cancer risk. A variation on the 5 a Day program is the "Gimme 5" school-based nutrition program tested in 12 high schools in the Archdiocese of New Orleans.[9] Regardless of the setting or target audience, community nutrition programs have one desired outcome: behavior change.

Public Health and Community Interventions

Community nutritionists promote good nutrition as one avenue for achieving good health. They develop programs to help people improve their eating habits, and they seek environmental changes (in the form of policy) to support good health habits. But community nutritionists do not work in a vacuum. They work closely with other practitioners, particularly those in public health, to help consumers achieve and maintain behavior change.

Public health can be defined as an effort organized by society to protect, promote, and restore the people's health through the application of science, practical skills, and

Public health Focuses on protecting and promoting people's health through the actions of society.

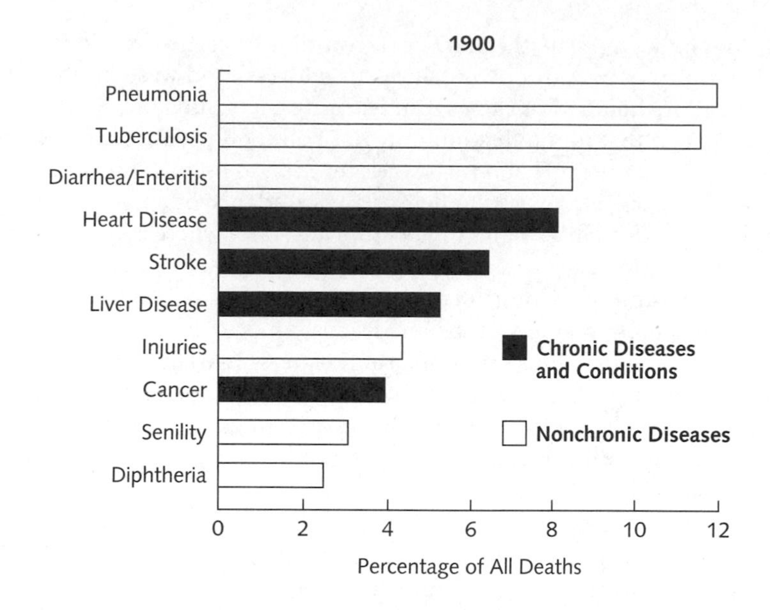

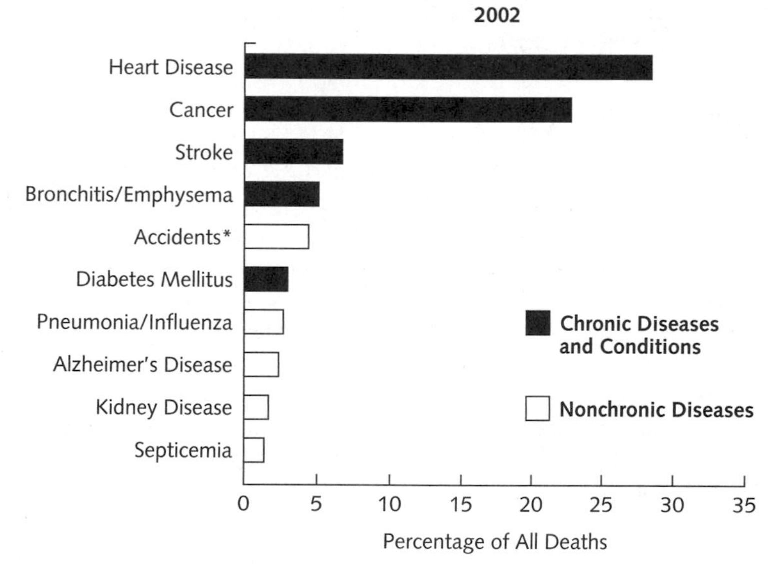

FIGURE 1-1 Leading Causes of Death, United States, 1900 and 2002

Many of the major chronic disease killers, such as heart disease, some types of cancer, stroke, and diabetes, are influenced by a number of factors, including a person's genetic makeup, eating and exercise, and other lifestyle habits.

*The leading cause of death for persons aged 15–24 is motor vehicle and other accidents, followed by homicide, suicide, cancer, and heart disease. About half of all accident fatalities are alcohol-related.

Source: Centers for Disease Control and Prevention, *National Vital Statistics Report, 2004*; available at www.cdc.gov/nchs/data/nvsr/nvsr53/nvsr53_05acc.pdf.

collective actions. "Public health is what we, as a society, do collectively to assure the conditions in which people can be healthy," wrote the authors of a report for the Institute of Medicine.[10] In the nineteenth century, the scope of public health was generally restricted to matters of general sanitation, including building municipal sewer systems, purifying the water supply, and controlling food adulteration. Major public health efforts focused on controlling infectious diseases such as tuberculosis, smallpox, yellow fever, cholera, and typhoid. In 1900, the leading causes of death and disability in the United States were pneumonia, tuberculosis, and diarrhea/enteritis, as shown in Figure 1-1. The morbidity and mortality linked with these disease outbreaks shaped public health practice for many years. Such runaway epidemics, which sometimes killed thousands of people in a single outbreak, are uncommon today because of large-scale public efforts to improve water quality, control the spread of communicable diseases, and enhance personal hygiene and the sanitation of the environment.

The leading causes of morbidity and mortality in the United States today are chronic diseases such as heart disease, cancer, and stroke (refer to Figure 1-1). Seven of every ten Americans, or roughly one and a half million people, die from chronic disease each year. Cardiovascular disease (mainly heart disease and stroke) causes about 38 percent of all deaths, killing 928,000 U.S. adults yearly; almost 2,500 Americans die of cardiovascular disease each day.[11] Although cardiovascular disease is usually assumed to be primarily a disease of men, over half of all deaths resulting from cardiovascular disease occur among women. Cancer kills almost 557,000 Americans each year—about 1,500 people every day. Other serious chronic diseases that reduce the quality of life, disable, or kill include arthritis, diabetes mellitus, osteoporosis, and Alzheimer's disease.[12]

Infectious diseases remain a problem, however. An estimated 38 million people are living with HIV/AIDS worldwide, with approximately 950,000 cases in the United States and about 40,000 new HIV infections occurring in the United States every year.[13] HIV/AIDS is among the top five causes of death for people aged 25–44.[14]

Another infectious disease is tuberculosis, whose incidence had been declining in the general U.S. population for several decades. An estimated 10–15 million Americans are infected with TB bacteria, with the potential to develop active TB disease in the future. About 10 percent of these infected individuals will develop TB at some point in their lives.[15] The AIDS epidemic is partly responsible for the reemerging outbreaks of tuberculosis, although there are other causes, such as increases in homelessness, drug abuse, immigration from other countries where tuberculosis is widespread, and crowded housing among the poor.[16]

The leading causes of death in Canada mirror those of the U.S. population in many respects.[17] The top-ranking cause of death among Canadian men and women is cardiovascular disease, followed by cancer. Because more women smoke today than smoked 20 years ago, lung cancer is now the leading cause of cancer death among Canadian women.[18]

Robb Gregg/PhotoEdit, Inc.

A cooking demonstration is an intervention that promotes awareness of the importance of healthful eating and teaches heart-healthy cooking skills.

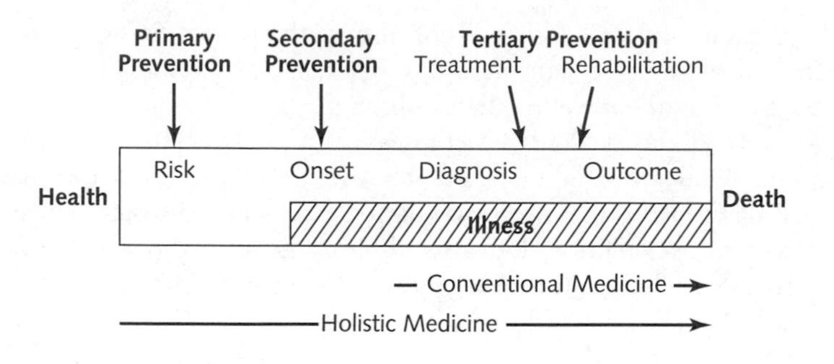

FIGURE 1-2 The Health Continuum and Types of Prevention to Promote Health and Prevent Disease

Source: Adapted from M. P. O'Donnell, Definitions of health promotion, *American Journal of Health Promotion* 1 (1986): 4; and R. H. Fletcher, S. W. Fletcher, and E. H. Wagner, *Clinical Epidemiology: The Essentials* (Baltimore, MD: Williams & Wilkins, 1982), p. 129.

These changes in disease patterns over the last century have spawned changes in public health actions. Because the goals of public health reflect the values and beliefs of society and existing knowledge about disease and health, public health initiatives change as society's perception of health needs changes.

In order to ensure the health of the public in the twenty-first century, public health initiatives have shifted from financing basic population-based measures such as immunization to efforts focused on achieving universal health services, responding rapidly to new infectious diseases such as SARS (sudden acute respiratory syndrome), or responding to new threats from antibiotic-resistant germs or **bioterrorism.** Recognizing the need for increased emphasis on preventive health measures, new efforts are under way to foster better collaboration between public health agencies and other organizations involved in protecting and promoting the public's health.[19]

THE CONCEPT OF HEALTH

Most of us equate health with "feeling good," a concept we understand intuitively but cannot define exactly. The term *health* is a derivative of the old English word for "hale," which means whole, hearty, sound of mind and body.[20] Health can be viewed as the absence of disease and pain, or it can be pictured as a continuum along which the total living experience can be placed, with the presence of disease, impairment, or disability at one end of the spectrum and freedom from disease or injury at the other. These extremes in the health continuum are shown in Figure 1-2.[21]

Health is properly defined from an ecological viewpoint—that is, one that focuses on the interaction of humans among themselves and with their environment. In this sense, **health** is a state characterized by "anatomic integrity; ability to perform personally valued family, work, and community roles; ability to deal with physical, biological, and social stress; a feeling of well-being; and freedom from the risk of disease and untimely death."[22] A healthy individual, then, has the physical, mental, and spiritual capacity to live, work, and interact joyfully with other human beings.

But how is good health achieved? Why does one child in a family become addicted to cocaine, whereas another never touches illicit drugs? Why do people start smoking? Why do some people overeat? Why do some teenagers consume adequate amounts of iron and calcium, whereas others do not? Why is one 80-year-old healthy and vigorous and another 70-year-old infirm? The answers to these questions still elude epidemiologists and other scientists. We know that a constellation of factors, shown in Table 1-1, influence health. Certain individual factors such as age, sex, and race are fixed, inherited traits that influence an individual's health potential. Other factors, such as lifestyle, housing, working conditions, social networks, community services, and even national health policies, represent layers of influence that can theoretically be changed to improve the health of

Bioterrorism The intentional release of disease-causing toxins, microorganisms, or other substances.

Health According to the World Health Organization, a state of complete physical, mental, and social well-being, not merely the absence of disease.

TABLE 1-1 Determinants of Health

BIOLOGY	LIFESTYLE	LIVING, WORKING, AND SOCIAL CONDITIONS	COMMUNITY CONDITIONS	BACKGROUND CONDITIONS
Sex	Physical activity	Housing	Climate and geography	National food and nutrition policy
Race	Diet	Education	Water supply	National minimum wage
Age	Hobbies	Occupation	Type and condition of housing	Cultural beliefs
Other hereditary factors	Leisure-time activities	Income	Number and type of hospitals and clinics	Cultural values
	Use of drugs:	Employment status	Health and medical services	Cultural attitudes
	• Cigarettes, cigars, chewing tobacco	Social networks such as family, friends, coworkers	Social services	Advertising
	• Alcohol	Socioeconomic status/class	Leading industries	Media messages
	• Prescription medications	Economic inequality	Political/government structure	Food distribution system
	• Illicit drugs such as cocaine, marijuana, etc.	Racial and ethnic health disparities	Community health groups and organizations	
	Religion		Number, type, and location of grocery stores, etc.	
	Safety practices such as wearing seatbelts, wearing wrist guards and knee pads		Recreation	
	Medical self-care		Transportation systems	
	Stress management			

Source: Adapted from G. Pickett and J. J. Hanlon, *Public Health: Administration and Practice* (St. Louis: Times Mirror/Mosby College Publishing, 1990), p. 50; M. P. O'Donnell, Definition of health promotion, *American Journal of Health Promotion* 1 (1986): 4–5; and Institute of Medicine, The Future of Public Health in the 21st Century (Washington, D.C.: National Academy Press, 2003).

individuals. In truth, however, less is known about the specific determinants of health than about the factors that contribute to disease, injury, and disability. And understanding the causes of disease and ill health does not necessarily lead to an understanding of the causes of good health.

HEALTH PROMOTION

Some people do things that are not good for their health. They overeat, smoke, refuse to wear a helmet when riding a bicycle, never wear seat belts when driving, fail to take their blood pressure medication—the list is endless. These behaviors reflect personal choices, habits, and customs that are influenced and modified by social forces. Such "lifestyle behaviors" can be changed if the individual is so motivated. Educating people about healthful and unhealthful behaviors is one way to help them adopt positive health behaviors.

Health promotion focuses on changing human behavior, on encouraging people to eat healthful diets, be active, get regular rest, develop leisure-time hobbies for relaxation, strengthen social networks with family and friends, and achieve a balance among family, work, and play.[23] It is "the science and art of helping people change their lifestyle to move toward a state of optimal health."[24] Behavior change is the desired outcome of a health promotion activity—what we call an **intervention**—aimed at a target audience. Interventions focus on promoting health and preventing disease and are designed to change a preexisting condition related to the target audience's behavior.[25]

Health promotion Helping all people achieve their maximum potential for good health.

Intervention A health promotion activity aimed at changing the behavior of a target audience.

There are three types of prevention efforts, as shown in Figure 1-2. Primary prevention is aimed at preventing disease by controlling **risk factors** that are related to injury and disease. Low-fat cooking classes, for example, help people change their eating and cooking patterns to reduce their risk of cardiovascular disease. Secondary prevention focuses on detecting disease early through screening and other forms of risk appraisal.

Risk factors Factors associated with an increased probability of acquiring a disease.

TABLE 1-2 Ways to Promote Good Health

• Safe environment	Control physical, chemical, and biological hazards.
• Enhance immunity	Immunize to protect individuals and communities.
• Sensible behavior	Encourage healthful habits, and discourage harmful habits.
• Good nutrition	Eat a well-balanced diet, containing neither too much nor too little.
• Well-born children	Every child should be a wanted child, and every mother fit and healthy.
• Prudent health care	Cautious skepticism is better than uncritical enthusiasm.

Source: Adapted from J. M. Last, *Public Health and Human Ecology*, 2nd ed. (Stamford, CT: Appleton & Lange, 1998), p. 10.

Public screenings for hypertension at a health fair identify people whose blood pressure is high; these individuals are then referred to a physician or other health professional for follow-up and treatment. Tertiary prevention aims to treat and rehabilitate people who have experienced an illness or injury. Education programs for people recently diagnosed with diabetes help prevent further disability and health problems, such as blindness and end-stage renal disease, from arising from the condition and improve overall health.[26] Prevention has become increasingly important, as the medical community moves away from conventional medicine, which focuses on diagnosing and treating diseases, to a holistic approach that encompasses all aspects of the health spectrum.

Many questions about why people make the choices they make remain unanswered, but the ways to promote good health are widely recognized. Born of decades, if not centuries, of scientific observation and testing, the strategies for promoting good health are outlined in Table 1-2. Although these strategies seem relatively straightforward, putting them into practice is a major challenge for most communities and nations.

HEALTH OBJECTIVES

The challenge of improving the nutrition, health, and quality of life for humans is complex. As outlined by the nations of the world in a 1978 conference on primary health care, convened by the World Health Organization (WHO) and the United Nations Children's Fund (UNICEF), the goal of the world community is to "protect and promote the health of all people of the world."[27] The commitment to the WHO goal of "health for all" through global improvements in health, especially for the most disadvantaged populations, was renewed by nations in 1998.[28] At the 2002 World Summit on Sustainable Development, health was recognized as both a resource for, and an outcome of, sustainable development: "The goals of sustainable development cannot be achieved when there is a high prevalence of debilitating illness and poverty, and the health of a population cannot be maintained without a responsive health system and a healthy environment."[29]

When translating the global goal of "health for all" into action at the local level, one challenge is to understand the many physical, biological, social, and behavioral factors that influence the health of individuals and communities. Another challenge is to change human behavior.

Nations differ in how they formulate health objectives in an effort to help their people achieve behavior change, although there are common themes. Working groups in the European Region of the World Health Organization, for example, outlined the following prerequisites for health:[30]

- Freedom from the fear of war—"the most serious of all threats to health"
- Equal opportunity for all peoples

- The satisfaction of basic needs for food, clean water and sanitation, decent housing, and education
- The right to find meaningful work and perform a useful role in society

Achieving these necessities requires both political will and public support, according to the working groups, which translated these prerequisites into specific targets for health. One such target, for example, called for enhancing life expectancy by reducing infant and maternal mortality. Other targets focused on enhancing social networks and promoting healthful behaviors, controlling water and air pollution, and improving the primary health care system.

In Canada, a new vision for promoting health and preventing disease among Canadians was expressed in documents, released by Health Canada, that aim to promote a balance between individual and societal responsibilities for health. These documents, *Achieving Health for All: A Framework for Health Promotion* and *Toward a Healthy Future,* cite challenges to achieving health for all: reducing inequities in access to and use of the health care system; increasing prevention efforts to change unhealthful behaviors; and enhancing the individual's ability to cope with chronic illnesses and disabilities. A key focus of the proposed implementation strategies is the strengthening of community-based health services, including worksite programs.[31] This vision for health in Canada is a window of opportunity for community nutritionists to promote food and nutrition policies in all Canadian communities.

The health and well-being of individuals and the prosperity of the nation require a well nourished population.

– Joint Steering Committee (Canada), *Nutrition for Health: An Agenda for Action*

The health objectives for the peoples of the United States differ slightly from those of the European and Canadian communities, reflecting the health needs of the U.S. population. A national strategy for improving the health of the nation was laid out in the 1990 publication *Healthy People 2000: National Health Promotion and Disease Prevention Objectives,* which identified health improvement goals and objectives to be reached by the year 2000.[32] The most recent health initiative, issued in the year 2000 and titled *Healthy People 2010: Understanding and Improving Health,* presents a national health promotion and disease prevention agenda for the first decade of the twenty-first century.[33] The *Healthy People* documents represent a national health agenda developed by a consortium of national health organizations, state health departments, the Institute of Medicine, and the U.S. Public Health Service. They grew out of a national health strategy launched in 1979 with the publication of *Healthy People: The Surgeon General's Report on Health Promotion and Disease Prevention* and continued in 1980 with the release of the report *Promoting Health/Preventing Disease: Objectives for the Nation,* which established health objectives leading up to the year 1990.[34]

Healthy People in Healthy Communities

The expert working groups of the *Healthy People 2010* consortium developed two broad goals designed to help Americans achieve their full potential:[35]

- **Goal 1:** *Increase quality and years of healthy life.* This first goal, to help individuals increase life expectancy and improve their quality of life, reflects the change in demographics to an increasingly older nation.
- **Goal 2:** *Eliminate health disparities.* This goal reflects the increasing diversity of the American population and recognizes that gender, race, and ethnicity; income and education; rural or urban location; disability; and sexual orientation are major factors that affect access to health care services and contribute to health disparities.

Access *Healthy People 2010* online at www.health.gov/healthypeople.

Track *Healthy People* data online at www.cdc.gov/nchs/hphome.htm.

These goals represent the nation's hope for the improved health of its citizens, and they can serve as the foundation for all work toward health promotion and disease

TABLE 1-3 ***Healthy People 2010:*** **Goals, Focus Areas, and Leading Health Indicators**

Goals	
1. Increase quality and years of healthy life	2. Eliminate health disparities
Leading Health Indicators	
1. Physical activity	6. Mental health
2. Overweight and obesity	7. Injury and violence
3. Tobacco use	8. Environmental quality
4. Substance abuse	9. Immunization
5. Responsible sexual behavior	10. Access to health care
Focus Areas	
Access to quality health services	Injury and violence prevention
Arthritis, osteoporosis, and chronic back conditions	Maternal, infant, and child health
Cancer	Medical product safety
Chronic kidney disease	Mental health and mental disorders
Diabetes	Nutrition and overweight
Disability and secondary conditions	Occupational safety and health
Educational and community-based programs	Oral health
Environmental health	Physical activity and fitness
Family planning	Public health infrastructure
Food safety	Respiratory diseases
Health communication	Sexually transmitted diseases
Heart disease and stroke	Substance abuse
HIV	Tobacco use
Immunization and infectious diseases	Vision and hearing

Source: U.S. Department of Health and Human Services, *Healthy People 2010.*

Healthy People 2010

LEADING HEALTH INDICATORS

Lifestyle Indicators

- Physical activity
- Overweight and obesity
- Tobacco use
- Substance abuse
- Responsible sexual behavior

Health System Indicators

- Mental health
- Injury and violence
- Environmental quality
- Immunization
- Access to health care

See the *Healthy People Toolkit,* which provides examples of state and national experiences in setting and using objectives, at www.health.gov/healthypeople/state/toolkit.

prevention. As stated, however, they are too broad to implement. Thus the working groups also laid out 467 specific, measurable targets or objectives to be achieved by the year 2010. These objectives are grouped into 28 broad focus areas, such as access to quality health services, cancer, diabetes, food safety, heart disease and stroke, nutrition and overweight, and physical activity and fitness, as shown in Table 1-3. *Healthy People 2010* balances the comprehensive set of health objectives with a list of 10 public health priorities for the decade, known as Leading Health Indicators and listed in the margin. For each of the Leading Health Indicators, specific objectives from *Healthy People 2010* are used to track the progress made in improving the nation's health and to provide periodic "snapshots" of the nation and its communities during the decade (see Table 1-4).

Many nutrition-related activities are considered essential to the overall *Healthy People 2010* initiative, because four of the leading causes of death in the United States are related to dietary imbalance and excess (coronary heart disease, some types of cancer, stroke, and diabetes mellitus). Diet also contributes to the development of other conditions, such as hypertension, osteoporosis, obesity, dental caries, and diseases of the gastrointestinal tract.[36]

Some of the *Healthy People 2010* nutrition-related objectives focus on improving health status. For example, one objective calls for increasing the proportion of adults who are at a healthful weight (see Figure 1-3). Several objectives focus on health risk reduction and specify targets for the intake of nutrients such as fat, saturated fat, and calcium and of foods such as fruits, vegetables, and grain products. Other nutrition-related objectives set targets for the prevalence of iron deficiency and anemia and the proportion

TABLE 1-4 Objectives Used to Track Progress on the *Healthy People 2010* Leading Health Indicators

- **Physical Activity.** Increase the proportion of teens who exercise for 20 minutes a day/3 times a week from 64 percent to 85 percent by 2010. Increase the number of adults who engage in 30 minutes of moderate physical activity every day from 15 percent of the population to 30 percent.
- **Overweight and Obesity.** Lower the proportion of children and teenagers aged 6 to 19 who are overweight or obese from 11 percent to 5 percent by 2010, and reduce the number of adults who are obese from 23 percent of the population to 15 percent.
- **Tobacco Use.** Reduce adolescent cigarette smoking, currently at 36 percent, to 16 percent. Also reduce adult smoking from 24 percent to 12 percent by 2010.
- **Substance Abuse.** Increase the proportion of adolescents not using alcohol or illicit drugs during the past 30 days from 77 percent to 89 percent by 2010. Reduce binge drinking during the past month among adults from 16 percent to 6 percent of the population.
- **Responsible Sexual Behavior.** Increase the proportion of teens who abstain from sexual intercourse or use condoms from 85 percent to 95 percent by 2010. Increase the number of sexually active adults who use condoms from 23 percent of the population to 50 percent.
- **Mental Health.** Increase the proportion of adults with recognized depression who receive treatment from 23 percent to 50 percent by 2010.
- **Injury and Death.** Reduce the homicide rate from 7.2 per 100,000 to 3.2 per 100,000 by 2010, and reduce deaths caused by motor vehicle crashes from 15.8 per 100,000 to 9 per 100,000.
- **Environmental Quality.** Reduce the proportion of persons who are exposed to air that does not meet the Environmental Protection Agency's health-based standards for ozone from 43 percent to 0.
- **Immunization.** Increase the proportion of young children who receive all vaccines that have been recommended for universal administration for at least 5 years from 73 percent to 80 percent by 2010.
- **Access to Health Care.** Increase the number of people with health insurance from 86 percent to 100 percent of the population by 2010, and increase the proportion of pregnant women who begin prenatal care in the first trimester of pregnancy from 83 percent to 90 percent.

of worksites that offer nutrition or weight management classes.[37] Some objectives encompass broad environmental measures, such as reducing foodborne illnesses, improving the workplace to reduce work-related injuries, and lowering people's exposure to lead. Other objectives underscore the premise of *Healthy People 2010*—that the health of the individual is almost inseparable from the health of the larger community. For example, predictors of access to quality health care include having health insurance, a higher income level, and a regular health care provider. Some of the health objectives outlined in the *Healthy People 2010* report address the special needs of various age groups, such as children and adolescents, whereas others focus on special population groups, such as blacks, Hispanics, and Asians.

Each *Healthy People 2010* objective has a target for specific improvements to be achieved by the year 2010. The **surveillance** and data-tracking systems of *Healthy People 2010* systematically collect, analyze, interpret, disseminate, and make use of health data to understand the nation's health status and plan prevention programs.[38]

The ultimate objective of public health is to lower the risk of disease and disability and to discourage risky behaviors in the first place.[39] Thus, many of the educational programs and services developed by public health practitioners to meet the objectives of *Healthy People 2010* focus on people in groups, whether they be families, schools, workplaces, cities, or nations. Such strategies target people of all ages and segments within the community. Refer to the Internet Resources at the end of this chapter for

Surveillance An approach to collecting data on a population's health and nutritional status in which data collection occurs regularly and repeatedly.

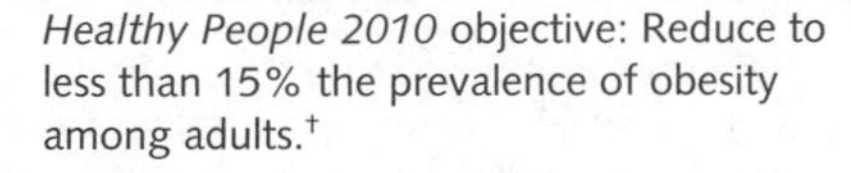

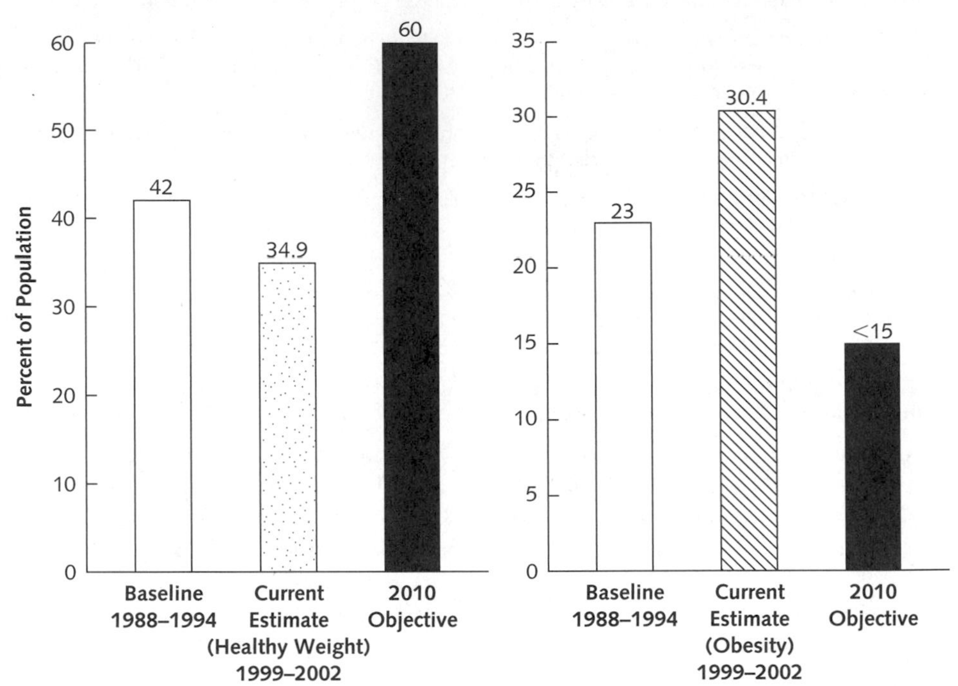

FIGURE 1-3 Healthy Weight Goals

During the baseline period (1988–1994), only 42 percent of adults were at a healthy weight, whereas 23 percent of adults were identified as obese. Current estimates (1999–2002) suggest that even fewer people (less than 35 percent) are at a healthy weight and that some 30 percent of adults are obese.

*Healthy weight is defined as having a body mass index (BMI) from 18.5 through 24.9.

†Obesity is defined as having a BMI at or above 30.

Data Source: Baseline data for healthy weight and obesity among adults is from National Health and Nutrition Examination Survey (NHANES III), 1988–1994. Current estimates are from NHANES 1999–2002. Public Health Service, U.S. Department of Health and Human Services, *HP 2010 Progress Review: Nutrition and Overweight,* January 21, 2004.

Internet addresses related to *Healthy People* initiatives and other sites of interest to community nutritionists.

HEALTHY PEOPLE 2010: PROGRESS REPORT

How are Americans doing in terms of meeting the *Healthy People 2010* goals? When *Healthy People 2010* was released in 2000, life expectancy was 76 years. Today, the average life expectancy at birth is 77 years, and death rates for heart disease, stroke, and certain types of cancer have declined.[40] However, health disparities remain evident among Americans, with significant differences between whites and minorities in mortality, morbidity, health insurance coverage, and the use of health services.[41]

Progress toward meeting the *Healthy People 2010* objectives is also discussed in Chapters 8, 10, 11, and 12.

Since the 1980s the prevalence of overweight has soared (see Chapter 8), and the trend toward obesity continues (see Figure 1-3). In fact, overweight increased among all ethnic and age subgroups of the population. One contributing factor is that people seem to be taking fewer steps to control their weight by adopting sound dietary patterns and being physically active. Some 40 percent of adults engage in no leisure-time physical activity. Little or no improvements are noted for dietary fat intake and consumption of fruits, vegetables, and whole grains. Additional health promotion efforts are also needed to reduce the prevalence of iron deficiency and anemia and to increase the number of women who breastfeed their infants early in the postpartum period. Table 1-5 provides the current status of a sampling of the *Healthy People 2010* objectives for the focus area on nutrition and overweight.

TABLE 1-5 A Sampling of *Healthy People 2010* Objectives for Nutrition and Overweight

	BASELINE	CURRENT ESTIMATE (1999–2002)	TARGET
Increase the proportion of adults who are at a healthy weight (BMI* between 18.5 and 24.9)	42%	34.9%	60%
Reduce the proportion of adults who are obese (BMI $\geq$ 30)	23%	30.4%	15%
Reduce the proportion of children (ages 6 to 19) who are overweight or obese.	11%	16%	5%
Increase the proportion of persons age 2 years and older who consume at least two daily servings of fruit.	28%	NC	75%
Increase the proportion of persons age 2 years and older who consume at least three daily servings of vegetables, with at least one-third being dark green or deep yellow vegetables.	3%	NC	50%
Increase the proportion of persons age 2 years and older who consume at least six daily servings of grain products, with at least three being whole grains (e.g., whole wheat bread and oatmeal).	7%	NC	50%
Increase the proportion of persons age 2 years and older who consume less than 10 percent of calories from saturated fat.	36%	NC	75%
Increase the proportion of persons age 2 years and older who consume no more than 30 percent of calories from fat.	33%	NC	75%
Increase the proportion of persons age 2 years and older who consume 2,400 mg or less of sodium daily.	21%	—	65%
Increase the proportion of persons age 2 years and older who meet dietary recommendations for calcium.	46%	—	75%

*BMI (Body Mass Index) is calculated as weight in kilograms (kg) divided by the square of height in meters (m^2). [BMI = weight (kg)/height (m^2)] To estimate BMI using pounds (lb) and inches (in.), divide weight in pounds by the square of height in inches. Then multiply the resulting number by 703 [BMI = weight (lb)/height $(in.)^2 \times 703$].

NC = Little or no change.

— = Cannot assess; limited data.

Community Nutrition Practice

Earlier in the chapter, we defined community nutrition as a discipline that strives to improve the nutrition and health of individuals and groups within communities. How do community nutritionists do this? What skills are needed to accomplish this goal? What job responsibilities do community nutritionists have? This book answers these questions and introduces you to the challenges of working in communities today. Imagine for a moment that *you* are a community nutritionist in each of the following situations:

- An article in the *New York Times* describes the high rates of substance abuse, teen pregnancy, HIV infection, sexually transmitted diseases, smoking, and eating disorders among

U.S. adolescents. Long concerned about this issue, your public health department plans an assessment of the health and nutritional status of teenagers in your county. Your job is to coordinate and lead the community assessment. Where do you start? What is the purpose of your assessment? What types of data do you collect? What information already exists about this population? Should your department work with other agencies to collect data? How will the results of your assessment be used to improve the health of teenagers in your community?

▶ As the director of health promotion for a large nonprofit health organization, you are responsible for developing and implementing programs to reduce the risk of cardiovascular diseases among people living in your state. Your organization's board of directors has called for an assessment of the effectiveness of all programs in your area. How do you evaluate program effectiveness? What types of data should be collected to show that each program reaches an appropriate number of people at a reasonable cost and helps them make behavioral changes to reduce their risk of heart attack and stroke? How will you present your findings to the board?

▶ You are attracted to the challenge of building a business and believe that your training in nutrition and exercise physiology can help people in your community get fit and improve their lifestyles. What is an attractive name for your business? Where should it be located? What services will you offer and to whom? Who are your competitors? How will you market your services? Can you use the Internet to enhance your business?

▶ You are employed by the Special Supplemental Nutrition Program for Women, Infants, and Children (WIC) in your state, and you have noticed that Spanish is the first language for an increasing number of your clients. You and your colleagues want to offer these clients more materials and services in Spanish. Should you adapt existing English-language materials for these clients or develop new materials from scratch? Are the existing English-language materials culturally appropriate for your Hispanic clients? What are other state WIC programs doing to address this issue?

Common themes are apparent in these situations. All refer to gathering information about the community itself or about people who use or implement community-based programs and services. Although it may not be clear to you now, all involve issues of policy, program management, and cost. All entail making decisions about how to use scarce resources. All are concerned with determining whether nutrition programs and services are reaching the right audience with the right messages and having the desired effect. All describe challenges of a trained professional—the community nutritionist—who identifies a nutritional need in the community and then puts into place a program or service designed to meet that need.

Community versus Public Health Nutrition

Community nutrition and public health nutrition are sometimes considered to be synonymous. In this book, *community nutrition* is the broader of the two terms and encompasses any nutrition program whose target is the community, whether the program is funded by the federal government, as are the WIC program and the National School Lunch program, or sponsored by a private group, such as a worksite weight-management program. *Public health nutrition* refers to those community-based programs conducted by a government agency (federal, state, provincial, territorial, county, or municipal) whose official mandate is the delivery of health services to individuals living in a particular area.

The confusion over these terms stems partly from the traditional practice settings of community dietitians and public health nutritionists, as shown in Figure 1-4.

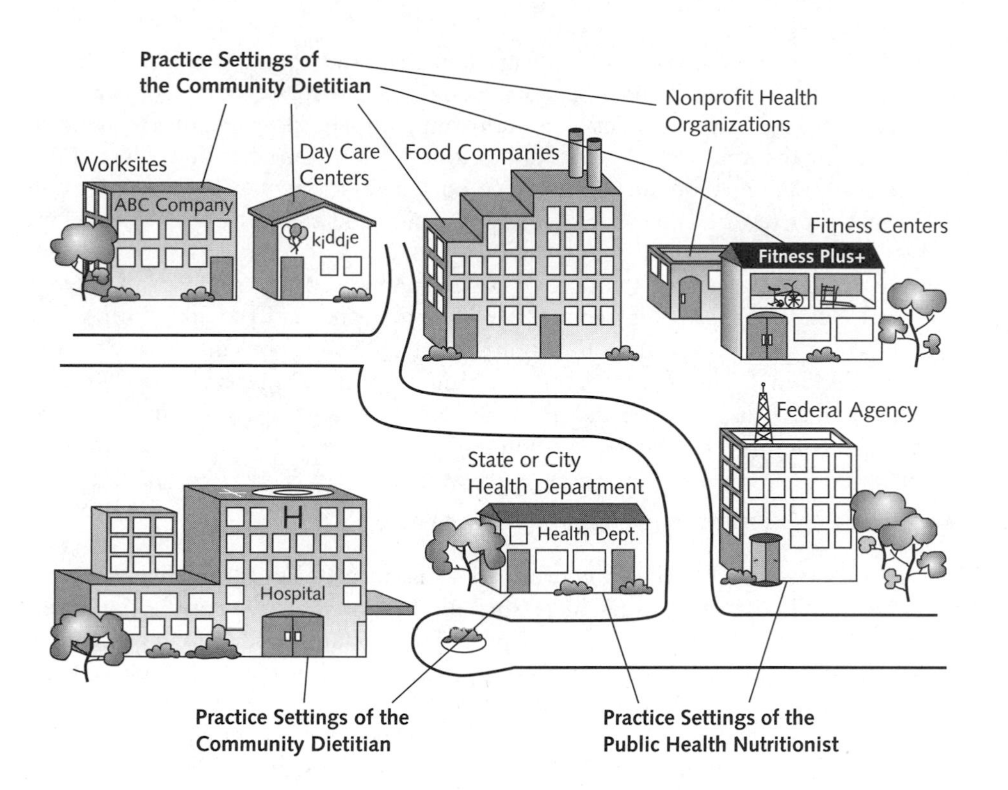

FIGURE 1-4 **Practice Settings of Community Dietitians and Public Health Nutritionists**

Community dietitians, who are always registered dietitians (RDs) or licensed dietitians (LDs), tend to be situated in hospitals, clinics, health maintenance organizations, voluntary health organizations, worksites, and other nongovernment settings. Some community dietitians work in federal, state, and municipal health agencies. Public health nutritionists, some of whom are RDs or LDs, provide nutrition services through government agencies.

In today's practice environment, these two designations overlap considerably, and practitioners in both areas share many goals, responsibilities, target groups, and practice settings. The community dietitian plans, coordinates, directs, manages, and evaluates the nutrition component of his or her organization's programs and services. The public health nutritionist carries out similar activities in a government agency.[43] For our purposes, all nutritionists whose major orientation is community-based programming will be called community nutritionists, whether their official title is community dietitian, public health nutritionist, nutrition education specialist, or some other designation.

EDUCATIONAL REQUIREMENTS

Community nutritionists have a solid background in the nutrition sciences. They have competencies in such areas as nutritional biochemistry, nutrition requirements, nutritional assessment, medical nutrition therapy, nutrition throughout the life cycle, food composition, food safety, and food habits and customs. They are knowledgeable about the theories and principles of health education, epidemiology, community organization, management, and marketing. Marketing skills are especially important, because it is no longer sufficient merely to know *which* nutrition messages to deliver. It is also necessary to know *how* to deliver them effectively in a variety of media formats to a variety of audiences.

The minimum educational requirements for a community nutritionist include a bachelor's degree in community nutrition, foods and nutrition, or dietetics from an accredited college or university. Most community nutrition positions require registration as a dietitian by the American Dietetic Association (ADA). Some positions also require graduate-level training to obtain additional competencies in areas such as quality assurance, biostatistics, research methodology, survey design and analysis, and the behavioral sciences.

Although dietetic technicians, registered (DTRs) are most often employed in the food service sector and clinical settings, some do work in the community arena. Community-based DTRs assist the community nutritionist in determining the community's nutritional needs and in delivering community nutrition programs and services. DTRs must have at least an associate's degree and must pass the registration examination developed by the Commission on Dietetic Registration (CDR).

LICENSURE OF NUTRITION PROFESSIONALS

Licensure of physicians has been the principal social mechanism for quality control in health care. In the past two decades, increasing numbers of *licensed*, nonphysician health care providers, such as nurse practitioners, physician assistants, midwives, physical therapists, dietitians, and dental hygienists, have appeared. They provide at a lower cost many services formerly performed only by doctors.[44]

In medicine, the educational standards for the medical degree (MD) are governed by law. Unfortunately, in some states the term *nutritionist* is not legally defined, and as a result, the public can be the hapless prey of anyone who wishes to use this title. Some nutritionists obtain their diplomas and titles without the rigorous training required for a legitimate nutrition degree. Because of lax state laws, it is even possible for an irresponsible "correspondence school"—a diploma mill—to pass out degrees to anyone who pays a fee.

Licensure of qualified nutritionists protects consumers from unqualified practitioners, particularly those who have no training in nutrition but nevertheless refer to themselves as "nutritionists." Licensure is designed to protect the public, control malpractice, and ensure minimum standards of practice. Its aim is legal recognition of health care professionals with the training and experience to deliver nutrition services.

Licensure Dietetics practitioners are licensed to ensure that only qualified, trained professionals provide nutrition services or advice to individuals requiring or seeking nutrition counseling or information; nonlicensed practitioners may be subject to prosecution for practicing without a license.

Certification Limits the use of particular titles (for example, dietitian or nutritionist) to persons meeting predetermined requirements, but persons not certified can still practice the occupation or profession. For a list of states that have laws that regulate dietitians or nutritionists through licensure or certification, go to www.cdrnet.org/certifications/licensure/index.htm.

Today, better consumer protection with respect to nutrition services is evident in many states. As of 2005, 46 states, the District of Columbia, and Puerto Rico had enacted legislation regulating the practice of dietetics. Several states have passed legislation restricting the fraudulent use of the title "dietitian." Several other states have passed legislation to prohibit people from calling themselves nutritionists without a license that requires a background in dietetics.

The advantages of licensure are clear. Americans are accustomed to identifying *licensed* health professionals. States that regulate nutrition services by means of licensure or **certification** assure consumers, health professionals, and insurance companies that the people providing those services meet the specific professional standards established by the state's Department of Professional Regulation.

PRACTICE SETTINGS

The practice settings of community nutritionists include schools, worksites, universities, colleges, medical schools, voluntary and nonprofit health organizations, public health departments, home health care agencies, day care centers, residential facilities, fitness centers, sports clinics, hospital outpatient facilities, food companies, and homes (their own or those of their clients). Some community nutritionists work as consultants, providing nutrition expertise to government agencies, food companies, food service companies, or other groups who are planning community-based services or programs with a nutrition component.

Community nutritionists are also employed by world and regional health organizations. WHO's Division of Family Health, located at WHO headquarters in Geneva, Switzerland, includes an office of nutrition. Likewise, the North American regional WHO office in Washington, D.C., which is known officially as the Pan American Sanitary Bureau and coexists with the Pan American Health Organization (PAHO), has a strong nutrition mandate. The PAHO directs its efforts toward solving nutritional problems in Latin America and the Caribbean.[45] Another prominent organization in global community nutrition is the Food and Agriculture Organization (FAO) of the United Nations. The programs of the Food Policy and Nutrition Division of FAO are directed toward improving the nutritional status of at-risk populations and ensuring access to adequate supplies of safe, good-quality foods.[46]

For more about world hunger and international nutrition issues, see Chapter 13.

ROLES AND RESPONSIBILITIES

Community nutritionists play many roles: educator, counselor, advocate, coordinator, generator of ideas, facilitator, and supervisor. They interpret and incorporate new scientific information into their practice and provide nutrition information to individuals, specialized groups, and the general population. Their focus is normal nutrition, although they sometimes cover the principles of medical nutrition therapy and nutritional care in disease for certain groups (for example, HIV-positive children or people with diabetes).[47] In addition to serving the general public, community nutritionists refer clients to other health professionals when necessary and participate in professional activities. Community nutritionists who are registered dietitians are expected to have core competencies and community-based competencies such as the following:[48]

- Manage nutrition care for population groups across the life span.
- Participate in nutrition surveillance and monitoring of communities.
- Develop and implement community-based food and nutrition programs.
- Conduct outcome assessment and evaluation of community-based food and nutrition programs.
- Participate in community-based research.
- Participate in food and nutrition policy development and evaluation based on community needs and resources.
- Consult with organizations regarding food access for target populations.
- Develop and implement a health promotion/disease prevention intervention project.
- Participate in screening activities such as measuring hematocrit and cholesterol levels.

Community nutritionists are responsible for planning, evaluating, managing, and marketing nutrition services, programs, and interventions. Nutrition services range from individual counseling for weight management, blood cholesterol reduction, and eating disorders to consulting services provided for food companies and institutions such as residential centers and nursing homes. Nutrition programs may be national in focus, such as the WIC program, or local, such as "Healthy Start for Mom & Me," a Winnipeg prenatal nutrition program for low-income, high-risk pregnant women.[49] Community nutritionists who develop programs identify nutrition problems within the community, obtain screening data on target groups, locate information on community resources, develop education materials, disseminate nutrition information through the media, evaluate the effectiveness of programs and services, negotiate contracts

TABLE 1-6 Responsibilities of Community Nutritionists in Diverse Practice Settings

Position: Child Nutrition Specialist, State Department of Education **Responsibilities** 1. Interprets USDA's regulations, policies, and procedures related to the National School Lunch Program for users of the program 2. Trains program users in such areas as how to count meals and keep records accurately, how to determine that menus meet current nutrition requirements, and which foods can be ordered through the commodities program 3. Revises training manuals 4. Audits program user compliance with USDA's regulations and procedures, including assessing whether student eligibility for participation in the program was determined correctly, ensuring that proper accounting procedures have been followed, and creating an action plan when the program user has not complied with the USDA regulations and procedures	**Position:** Nutritionist, Special Supplemental Nutrition Program for Women, Infants, and Children (WIC) **Responsibilities** 1. Determines client's eligibility for program, using WIC program criteria (e.g., presence of anemia, underweight/overweight, prior pregnancies, inadequate diet) 2. Assesses the nutritional status of clients 3. Determines the adequacy of the client's diet 4. Provides one-on-one diet counseling 5. Conducts group sessions for clients on basic nutrition topics such as food sources of iron and calcium 6. Helps clients understand how to use the WIC-approved foods in their daily diets 7. Assists in developing educational materials as required 8. Assists in reviewing client records and monitoring health data posted to records as required
Position: Director of Health Promotion, First-Rate Spa and Health Resort **Responsibilities** 1. Develops, implements, and evaluates programs in the areas of nutrition, fitness, weight management, and risk reduction (e.g., blood cholesterol reduction, stress management, smoking cessation) 2. Assists the director in developing marketing strategies for programs 3. Prepares program budgets 4. Tracks program expenses 5. Teaches nutrition/fitness to groups of clients 6. Supervises dietitians and other staff involved in counseling clients and fitness assessments	**Position:** President/Owner, Millennium Nutrition Services **Responsibilities** 1. Sets business goals and objectives 2. Manages all aspects of company's programs and services, including developing programs and services, developing educational materials and teaching tools, and evaluating success of programs and services 3. Develops and evaluates a marketing plan 4. Identifies new business opportunities 5. Tracks income (including billings) and expenses for tax purposes 6. Maintains client records 7. Networks with colleagues in business and the community

for nutrition programs, train staff and community workers, and document program services.[50]

The job responsibilities of community nutritionists are similar across practice settings. The community nutritionists whose responsibilities are shown in Table 1-6 are all involved in assessing the nutritional status of individuals or identifying a nutritional problem within a community. They all have opportunities for teaching their clients about foods, diet, nutrition, and health and for addressing emerging issues in community nutrition (see the box "Emerging Issues in Community Nutrition" on page 23). Some are involved in the budget process and in developing marketing strategies; others are not. The community nutritionist in private practice does it all!

In today's market, community nutritionists are increasingly expected to manage projects, resources, and people. One survey of 350 community dietitians found that more

TABLE 1-7 Multiskilling in Community Dietetics

- Basic understanding of the epidemiology and surveillance of health conditions such as diabetes, including the origin and availability of data sets
- Knowledge of how to influence policy development
- Ability to do an evaluation that will drive program changes
- Knowledge of how to effect change by developing broad-based community partnerships
- Knowledge of how to write grant applications
- Familiarity with budget development, justification, and management skills
- Knowledge of how to work with different cultures and adapt approaches based on the specific needs of each culture
- Knowledge of how to promote health through social marketing
- Knowledge of how to facilitate health system change

Source: Public Health Nutrition Practice Group, *The Digest,* Fall 1998, p. 14.

than two-thirds had mid- to upper-level management responsibilities. Community dietitians in management positions reported spending more time planning, coordinating, and evaluating programs and less time interacting with clients than did those in lower-level positions.[51] A role delineation study conducted by the ADA found that RDs working in community settings also had advising and policy-setting roles.[52] The time allocated to these activities varied somewhat by practice level. Community dietitians reported having major responsibility for teaching students, other dietitians, and health professionals. Their roles overlapped significantly with those of RDs in clinical dietetics.

Community nutritionists are also expected to be multiskilled or cross-trained (see Table 1-7). Multiskilled practitioners perform more than one function, often in more than one discipline.[53] The multiskilled community nutritionist knows not only how to conduct a needs assessment and provide dietary guidance but also how to design and conduct a survey, use an Internet website for marketing health messages, and obtain funding to support a program's promotional plan. Survey design and analysis, marketing, Internet technology, and grant writing are new and important disciplines for modern community nutritionists. And in today's culturally diverse environment, bilingual community nutritionists are in demand. Being fluent in a language other than English helps the nutritionist gather information from at-risk or hard-to-reach populations and develop programs that meet their needs.

Entrepreneurship in Community Nutrition

Entrepreneurship is important in community nutrition. What is entrepreneurship? Who is an entrepreneur? How is entrepreneurship related to community nutrition? In the business world, entrepreneurship is defined as the act of starting a business or the process of creating new "values," be they goods, services, methods of production, technologies, or markets.[54] The essence of entrepreneurship is innovation. Consider the late Ray Kroc of McDonald's. He did not invent the hamburger, but he did develop an entirely new way of marketing and delivering it to his customers. In the process, he revolutionized the food service industry. **Entrepreneurship,** then, is the creation of something of value,

Entrepreneurship Creating something of value through the creation of organization.

be it a product or a service, through the creation of organization. In this context, *organization* means orchestration of the materials, people, and capital required to deliver a product or service. This definition encompasses the myriad actions of individuals—the entrepreneurs—who invent or develop some new product or service that is valued by the community or marketplace.[55]

Entrepreneurs and Intrapreneurs

An **entrepreneur** is an enterpriser, innovator, initiator, promoter, and coordinator. Entrepreneurs are change agents who seek, recognize, and act on opportunities. They ask, "What if?" and "Why not?" and translate their ideas into action. They tend to be creative, are able to see an old problem in a new light, and are willing to break new ground in delivering a product or service. When they spot an opportunity to fill a niche in the marketplace, they work to bring together the expertise, materials, labor, and capital necessary to meet the perceived need or want. Two entrepreneurs in community nutrition are Oklahoma dietitians Kellie Bryant, MS, RD, LD, and Mary S. Callison, MS, RD, LD, who developed a nutrition newsletter for Head Start programs. They observed that Head Start programs, particularly on Indian reservations and in rural areas, lacked practical information pieces on child health and on how to shop for, cook, store, and serve healthful foods. Their "Primarily Nutrition" newsletter is marketed to Head Start programs, which copy it for distribution to clients and their families.[56] The three registered dietitians—Gretchen Forsell, MPH, RD, LD, Jean Nalani Trobaugh, RD, and Kim Ziemer, MS, RD, CNSD—who launched *Dietetics Online* on the Web are also entrepreneurs.[57] They spotted a new technology trend and learned how to harness it for educating and informing dietitians and helping them network and become more efficient online.

Entrepreneurs share some common personality traits. They are achievers, setting high goals for themselves. They work hard, are good organizers, enjoy managing a project to completion, and accept responsibility for their ventures. They strive for excellence and are optimistic, believing that now is the best of times and anything is possible. Finally, entrepreneurs are reward oriented, seeking recognition and respect for their ventures and ideas. Recognition and respect are often more important than money to entrepreneurs.[58]

These qualities are typically applied to the self-starting, independent entrepreneur in business, but they also describe the **intrapreneur,** the corporate employee who is creative and innovative. Intrapreneurs are seldom solely responsible for the financial risk associated with a new venture, but they share the same entrepreneurial spirit as their more independent counterparts. Like entrepreneurs, intrapreneurs use innovation to exploit change as an opportunity for creating something of value. In other words, intrapreneurs seek to better the existing state of affairs within their organizations through creative problem solving.[59] A good example of intrapreneurship is the action of Dr. Cheryl Sonnenberg at the Economic Opportunity Board of Clark County in Las Vegas, Nevada. Dr. Sonnenberg had been struggling to find ways of delivering WIC services to clients in remote areas of Clark County. She had read of a Texas experiment in which a renovated Cadillac had been used to deliver WIC services in rural areas. When she discovered that the renovation costs far exceeded her budget, she began exploring other innovative delivery systems. Eventually, she arranged to buy a used Winnebago, which was renovated with the help of engineers in the transportation division. Her solution, launched in 1995, became known as "WIC on Wheels," or W.O.W. Her intrapreneurial initiative helped solve a problem and improve rural service delivery to 450 WIC clients.[60]

What do entrepreneurs do? (In this book, innovators in both the private [corporate] and public [government] sectors who embody the spirit and principles of entrepreneurship

Entrepreneur One who undertakes the risk of a business or enterprise.

Intrapreneur A risk taker whose job is located within a corporation, company, or other organization.

will be considered entrepreneurs.) One study of entrepreneurs identified at least 57 separate activities associated with launching a new venture—a clear indication of how complex entrepreneurial behavior can be. Entrepreneurs have wide-ranging competencies in areas such as planning, marketing, networking, budgeting, and team building, as shown in the margin. They turn their creative vision into deliberate decision-making and problem-solving actions to accomplish their goals. They are not just managers, although they typically "manage" themselves well. Their high self-esteem stems from a strong belief in their own personal worth, which strengthens their capacity for self-management. Successfully managing oneself means being in control (that is, having willpower), knowing one's personal strengths and weaknesses, and being willing to change one's own behavior and graciously make use of feedback and criticism.

Activities of Entrepreneurs and Intrapreneurs

- Identify an opportunity.
- Create the solution.
- Conduct market research.
- Establish business objectives.
- Set up an organizational structure.
- Determine personnel requirements.
- Prepare a financial plan.
- Locate financial resources.
- Prepare a production plan.
- Prepare a management plan.
- Prepare a marketing plan.
- Produce and test-market the product.
- Build an organization.

Community nutritionists who are RDs *and* entrepreneurs are expected to have competencies in addition to the core competencies expected of all dietitians. These business competencies include the following:[61]

- Perform organizational and strategic planning.
- Develop and implement a business or operating plan.
- Supervise procurement of resources.
- Manage the integration of financial, human, physical, and material resources.
- Supervise coordination of services.
- Supervise marketing functions.

What relevance does entrepreneurship have to community nutrition? The answer to this question will become increasingly clear as you read the remaining chapters of this book. Suffice it to say at this point that creativity and innovation—the essence of entrepreneurship—are as important to the discipline of community nutrition as to any other field. Consider the entrepreneurial activities listed in the margin above. Nearly all are relevant for the community nutritionist: recognizing an opportunity to deliver nutrition and health messages, developing an action plan for a target audience, building the team for delivering a nutrition program or service, developing a marketing plan, and evaluating the effectiveness of the nutrition program or service.

Community nutritionists who want to change people's eating habits must be able to see new ways of reaching desired target groups. The strategy that works well with Hmong young people in California will probably not work well with institutionalized elderly women living in Ohio. Community nutritionists must draw on theories and skills from the disciplines of sociology, educational psychology, medicine, communications, health education, technology, and business to develop programs for improving people's eating patterns. The twin stanchions of entrepreneurship—creativity and innovation—assist the community nutritionist in achieving the broad goal of improved health for all.

Emerging Issues in Community Nutrition

- Since the welfare law was passed in 1996, many advocacy groups have expressed concern that only certain legal immigrants and their families may receive Food Stamp and Social Security benefits. Should your state government restore benefits to all legal immigrants residing in the state?
- Back in 1977, the United States Department of Agriculture (USDA) restricted the sale of soft drinks and other snack items outside the school cafeteria. However, those regulations were overturned in the 1980s. Today, vending machines serve as a source of additional revenue for many schools and are typically stocked with candy, soda, chips, and other empty-calorie foods

that may be contributing to the increased number of overweight children in this country. In addition, many schools have opted to offer a la carte and fast-food options to increase revenue. If you were employed as a school food service director, what would you recommend to your school administrators and state legislators in order to foster a school nutrition environment that would support healthful eating?

- Health claims for dietary supplements such as chromium picolinate, pyruvate, selenium, and vitamin E are currently regulated under the Nutrition Labeling and Education Act of 1990. Some consumer groups believe that the Food and Drug Administration (FDA) should have more power in regulating health claims made on product labels for dietary supplements. In other words, they think the supplement industry should be required to prove that a product is safe, not that the FDA should be required to prove that a product is unsafe. If you were responding to a proposed rule in the *Federal Register,* would you favor giving the FDA more power in this area?
- Several recent studies have suggested that elevated plasma total homocysteine levels (above 12–15 micromoles per liter) are a strong predictor of mortality in patients with confirmed coronary heart disease (CHD). Homocysteine is a metabolite of methionine, an essential amino acid found mainly in meat and dairy products. High blood levels of homocysteine are associated with low availability of certain B vitamins such as folic acid, vitamin B_6, and vitamin B_{12}, although it has not been proved that B vitamin supplements reduce CHD risk. Given the current state of research, should the nutrition education materials used for clients in your organization's heart disease risk-reduction program be changed to reflect plasma homocysteine levels as a CHD risk factor?

By the Year 2020 . . .

- Another 50 to 80 million people will probably have been added to the U.S. population, which was 281 million in 2000.
- The population and labor force will continue to diversify, as immigration continues to account for a sizable part of the population growth. Certain states and cities, especially those on the East and West coasts, can be expected to receive a disproportionately large number of immigrants.
- The growing diversity of food choices is likely to echo the increasing diversity of the population. Between 1980 and 2020, the Hispanic population will have grown from 6.5 percent to 18 percent of the population; blacks will have grown from 11.6 percent to 12.9 percent of the population; and Asian or Pacific Islanders will have grown from 1.5 to 5 percent. For the same time period, whites will have declined from 79.9 percent to 62.5 percent of the population.
- The world's population will have grown by about 2.2 billion people. How to feed these additional people adequately will be a principal challenge facing the global system of agricultural production and trade.
- The number of people over 65 years of age will have grown from 35 million in 2000 to 54 million because of aging baby boomers and longer life expectancies.
- People aged 65–74 will have increased from 6 to 10 percent of the population, and those aged 75 and older will have increased from 6 to 7 percent.
- Total national spending on long-term health care will have risen to $207 billion—up from $115 billion in 1997; spending on home care will account for about one-third of spending on long-term care.

Sources: U.S. Department of Health and Human Services, Public Health Service, *Healthy People 2010: Understanding and Improving Health* (Washington, D.C.: U.S. Government Printing Office, 2000); U.S. Department of Health and Human Services, Hispanic Customer Service Home Page at www.dhhs.gov/about/heo/hisp.html; and U.S. Bureau of the Census, *Population Projections of the United States by Age, Sex, Race, and Hispanic Origin: 1995 to 2050,* Current Population Reports, P25-1130.

Social and Economic Trends for Community Nutrition

Recent social and economic trends have important consequences for community nutrition. For example, the demographic profile of many communities is changing rapidly, along with the client mix served by community nutritionists.

Leading Indicators of Change

A worldwide increase in the educational level of the workforce is anticipated. In 2000, 25 percent of the U.S. population were college graduates. By 2020, more than 55 percent are expected to have completed some college study.[62] Women have entered the global workforce in unprecedented numbers, a trend that has led to changes in traditional family norms and structures and thus may affect the format and delivery of nutrition programs and services. Consequently, a variety of alternative educational strategies will be needed to reach consumers whose education, training, income, language skills, time pressures, and economic potential will be highly diverse.[63]

See also the discussion in Chapter 7 of emerging policy issues, including complementary nutrition and health therapies, functional foods, genetics, and biotechnology.

AN AGING POPULATION

In North America, the aging of the population, coupled with a more ethnically diverse society, will challenge community nutritionists to develop new products and services. By 2010, 40 percent of our population—some 120 million people—will have a chronic disease or condition.[64] In the United States, for example, the fastest-growing segment of the nation's population consists of people over 65 years of age. Indeed, the most rapidly growing segment is the 80-plus age group, which, compared with younger groups, tends to be more frail, to have more chronic health problems, and to use more hospital health services such as inpatient and outpatient care.[65]

See Chapter 12 for more about nutrition issues and programs for older adults.

GENERATIONAL DIVERSITY

Many of the so-called baby boomers—those people born between 1946 and 1964—have reached their peak earning years and are expected to be a leading market force in redefining how to live as senior citizens.[66] At the same time, the younger generations are emerging with new values and attitudes about health, lifestyles, and society. The generations can differ in workplace values, lifestyle and social values, motivation, learning and communication styles, and technical competence (see Table 1-8).[67] Community dietitians will need to understand the characteristics of these distinct generations in order to develop skills, tools, and resources for communicating nutrition and health information most effectively to individuals throughout these groups.

INCREASING DEMANDS FOR NUTRITION AND HEALTH CARE SERVICES

As they move into their retirement years over the next two to three decades, baby boomers are expected to seek out information and opportunities to live healthier and longer lives. This represents a growing opportunity for community nutritionists to deliver sound nutrition and health promotion information to this emerging group of

See Chapter 9 for more about health care reform and related nutrition issues.

TABLE 1-8 Characteristics and Insights Regarding the Current American Generations

GENERATION AND BIRTH YEARS	CHARACTERISTICS AND INSIGHTS
Matures/Traditionalists Pre-1946	Respect authority; avoid challenging the system; place duty before pleasure; value honor, integrity, personal ties and relationships; give information on a "need to know" basis; not completely comfortable with technology-based delivery of information and services.
The Baby Boomers 1946–1964	Live to work; committed to climbing the ladder of success; optimistic; strive for convenience and personal fulfillment; first wave of dual-income, dual-career families; interested in interpersonal communication; gently question status quo; team- and process-oriented; want to see big picture of an organization; will crusade for a cause; desire to preserve their youth and be nostalgic about it; enjoy unprecedented influence on government, corporate, and organizational policies and consumer products because of their numbers; comfortable with technology.
Generation X (the Baby Bust) 1965–1980	Work to live, not live to work; first generation of latchkey kids; independent, resourceful, entrepreneurial, and focused on personal growth; desire versatility, challenging work, and substantial financial rewards; aggressively question status quo and authority, interested in removing outdated work models; believe in clear, consistent expectations.
Generation Y or Nexters (the Baby Boom Echo) 1981–1994	Live in the moment; earn to spend; rely on immediacy of technology; grew up with the Internet; comfortable in getting, using, and sharing information that is visual, fast-paced, and conceptual; prefer to be tech-savvy and multitasking; enjoy collaborative efforts; more culturally diverse than other generations; social-minded and altruistic; demand respect and often question everything; need clear and consistent expectations to ensure productivity.

Source: Adapted from J. Jarratt and J. B. Mahaffie, Key trends affecting the dietetics profession and the American Dietetic Association, *Journal of the American Dietetic Association* 102 (2002): 1825; C. Alexander, Understanding generational differences helps you manage a multi-age workforce, *The Digital* Edge, July 2001, www.digitaledge.org, and D. Brown, Ways dietitians of different generations can work together, *Journal of the American Dietetic Association* 103 (2003): 1461.

seniors. The aging population will probably place many demands on health care services, home care services, and food assistance programs—as will the millions of people who lack health insurance under existing insurance programs. An estimated 40 million persons in the United States, including almost 10 million children younger than 18 years of age, are uninsured.[68] Securing their access to nutrition and health services will be challenging.

INCREASING ETHNIC DIVERSITY

See Chapter 16 for more about gaining cultural competence in community nutrition practice.

Analysts predict that by the year 2010, the global workforce will be even more ethnically diverse, the result of massive relocations of people, including immigrants, refugees, retirees, temporary workers, and visitors, across national borders. As society becomes more ethnically diverse, more knowledge of health beliefs, culturals foods, and values is needed. An interesting consequence of the influx of immigrants into Canada and the United States is an anticipated change in consumer marketing strategies as a consequence of the family-oriented shopping behaviors of Asian and Hispanic consumers.

The cultural values and lifestyles of these ethnic groups tend to reinforce family decision making and collective buying behavior. Marketers will have to change their strategies to appeal to the decision style of the extended family. They will need to market their products and services to groups rather than individuals.[69]

CHALLENGES OF THE TWENTY-FIRST-CENTURY LIFESTYLE

The World Health Organization describes obesity as "an escalating epidemic" and one of the greatest neglected public health problems of our time. Many factors, including genetics, influence body weight, but excess calorie intake and physical inactivity are the leading causes of overweight and obesity and represent the best opportunities for prevention and treatment. Consider how the following lifestyle trends have either increased opportunities for poor nutrition (particularly excess calories) or decreased opportunities for physical activity:[70]

See Chapter 8 for a detailed discussion of the obesity epidemic and descriptions of current public health policies, as well as proposed policies and legislation to prevent obesity and overweight.

- Food portion sizes and obesity rates have grown in parallel. In the 1960s, an average fast-food meal of a hamburger, fries, and a 12-ounce cola provided 590 calories; today, many super-sized, "extra-value" fast-food meals deliver 1,500 calories or more.
- Vending machines selling soft drinks, high-fat snacks, and sweet snacks are common in schools and workplaces. Milk, juices, water, and healthful snacks are far less accessible than their unhealthful counterparts.
- Both adults and children spend more time in sedentary activities, such as watching television, sitting at the computer, playing video games, or commuting to and from work and school, and schools offer fewer physical education classes for children.
- Increasing numbers of families live in communities designed for car use, unsuitable (lack of green space for recreation) and often unsafe (lack of sidewalks, inadequate street lighting) for activities such as walking, biking, and running.

No doubt the causes of obesity are complex and many causes may contribute to the problem in a single person. Given this complexity, it is obvious that there is no panacea for successful weight maintenance. The top priority should be prevention, but where prevention has failed, the treatment of obesity represents a "crisis opportunity" for community dietitians to gain the public's attention by delivering effective health promotion and intervention programs.[71]

Watchwords for the Future

Several terms have surfaced repeatedly in this chapter: change, innovation, creativity, community, entrepreneurship. These watchwords herald the beginnings of a new century marked by unprecedented global social change. The world is growing smaller, and its peoples seem to be moving toward the birth of a single, global nation. "Citizen of the world," a phrase popular during the mid-twentieth century, takes on added meaning as we move forward in the twenty-first century. Where once a community was circumscribed by a distant ridge or the next valley, today the "information highway" links us via satellite and optic fibers to unseen faces on the other side of the earth. The growing connectedness of the human race promises to create new challenges for community nutritionists in their efforts to enhance the nutrition and health of all peoples.

Internet Resources

Check out these Internet addresses for resources relevant to community nutrition and public health.

Professional Organizations

American Dietetic Association	**www.eatright.org**
American Public Health Association	**www.apha.org**
Society for Nutrition Education	**www.sne.org**
Canadian Public Health Association	**www.cpha.ca**
Dietitians of Canada	**www.dietitians.ca**

Health Organizations

Food and Agriculture Organization of the United Nations	**www.fao.org**
Pan American Health Organization	**www.paho.org**
World Health Organization	**www.who.int**

Canadian Federal Government Agencies

Canadian Food Inspection Agency	**www.inspection.gc.ca**
Health Canada	**www.hc-sc.gc.ca**

U.S. Federal Government Agencies and Offices

Center for Food Safety and Applied Nutrition	**http://vm.cfsan.fda.gov/list.html**
Centers for Disease Control and Prevention	**www.cdc.gov**
Department of Health and Human Services	**www.hhs.gov**
Food and Drug Administration	**www.fda.gov**
Food and Nutrition Information Center	**www.nal.usda.gov/fnic**
Food Safety and Inspection Service	**www.fsis.usda.gov**
National Cancer Institute	**www.nci.nih.gov**
National Center for Chronic Disease Prevention and Health Promotion (CDC*)	**www.cdc.gov/nccdphp**
National Institutes of Health	**www.nih.gov**

Consumer Health Sites

CancerNet	**www.cancer.gov**
CHID[†] Online	**http://chid.nih.gov**
Healthfinder	**www.healthfinder.gov**
InteliHealth	**www.intelihealth.com**
National Health Information Center	**www.health.gov/nhic**
NIH[‡] Consumer Health Information	**http://health.nih.gov**
National Heart, Lung, and Blood Institute	**www.nhlbi.nih.gov**
Gateway to Nutrition Information from U.S. Government	**http://nutrition.gov**

Health Promotion Initiatives and Other Sites

Hispanic Customer Service Home Page	**www.haa.omhrc.gov/HAA2pg/customer.htm**
NCI's[§] "5 a Day" for Better Health Program	**www.5aday.gov**

Healthy People 2010	**www.healthypeople.gov**
President's Healthier U.S. Initiative	**www.healthierus.gov**
President's Council on Physical Fitness and Sports	**www.presidentschallenge.org**
National Center for Chronic Disease Prevention and Health Promotion	**www.cdc.gov/nccdphp/dnpa**
Obesity Education Initiative	**www.nhlbi.nih.gov/about/oei/index.htm**
The Weight-Control Information Network	**http://win.niddk.nih.gov**

*CDC = Centers for Disease Control and Prevention
†CHID = Combined Health Information Database
‡NIH = National Institutes of Health
§NCI = National Cancer Institute

Case *Study*

Ethics and You

By Alice Fornari, EdD, RD, and Alessandra Sarcona, MS, RD

Scenario

A dietetics student is placed in an urban community rotation as part of a field practicum that focuses on the nutritional needs of HIV clients. This city agency provides supportive services to the clients and also meals for clients who do not have assistance at home. This community practicum is a new exposure for the student in terms of the role of a nutritionist in an urban community setting. The many hats the nutritionist must wear in one day overwhelm her. The student had attended a suburban private high school and a small, rural college. She always thought of community nutrition as cooperative extension agencies reaching out to the local community. This type of community-based service to a specific at-risk population was certainly eye-opening. Of course she had studied HIV in her clinical nutrition courses and understood the disease process and nutrition implications. However, the diversity of the staff and clients was a new experience. In addition, she was taken aback by the professional peer group at the agency: many were gay.

As part of her practicum, there are biweekly reports to the faculty coordinator and presentations to the other students. During the first presentation the student makes disparaging remarks about the staff and clients, regarding sexual preference, past use of IV drugs by some clients, economic disadvantage, and the communal nature of HIV illness. Another student in the practicum class comments that her view seems provincial. The faculty coordinator encourages the student to be more open and to value the diversity of the placement and the professional challenges it entails.

The final evaluation completed by the site preceptor clearly indicates that the preceptor perceives a prejudice, on the part of this student, regarding HIV and the population affected. The student's self-evaluation indicates that she does not see her role as a nutritionist within this specialized practice setting.

Learning Outcomes

- Identify the three arenas of community nutrition practice.
- Recognize that a Code of Ethics by the American Dietetic Association (ADA) exists and is specific for the dietetics profession.
- Identify how a practice situation can violate or be supported by the ADA Code of Ethics.

Foundation: Acquisition of Knowledge and Skills

1. Outline the three arenas of community nutrition practice presented in this case. Include a description of each arena specific to the type of community agency presented.

2. Identify the determinants of health (see Table 1-1 in this chapter) that may contribute to HIV in the agency population presented in this case.
3. Access the ADA Code of Ethics by using the search function at www.eatright.org/ and the supporting narrative from the *Journal of the American Dietetic Association* 99 (1999): 109–13.

Step 1: *Identify Relevant Information and Uncertainties*

1. Why does the ADA Code of Ethics apply to the scenario? (All students enrolled in the practicum course are student members of ADA.)
2. Under what circumstances would the Code of Ethics not apply?
3. Which characteristics of the student make her vulnerable to bias and violation of the Code?
4. Specifically, which principle(s) in the Code of Ethics applies (or apply) to the issue raised?

Step 2: *Interpret Information*

1. How can the preceptor and faculty coordinator help the student recognize that she is violating the Code?
2. Brainstorm characteristics of individuals that can contribute to personal biases and affect decision making and actions.

Step 3: *Draw and Implement Conclusions*

1. To ensure that this is a learning experience for the student, she is asked to communicate her understanding of the ethical issue to her peers. Indicate one way she could do this as part of the practicum class.

Step 4: *Engage in Continuous Improvement*

1. From a student perspective, suggest one or two strategies that will help the student and her peers in the practicum class to reflect on the values and ethical principles supported by the ADA Code of Ethics.

Professional Focus

Community-Based Dietetics Professionals

There are many steps that dietetics professionals who are interested in working in a community setting can take to achieve and maintain a successful career. And for professionals already established in a nonclinical setting, there are key points to focus on when working to keep a high caliber of community dietitians.

Longtime community dietitian Karen Ensle, EdD, RD, is a family and consumer sciences educator and department head for Rutgers Cooperative Extension of Union County in New Jersey. An educator at the secondary and collegiate levels for 28 years, her chosen career path has led her to develop and teach programs in nutrition and family resource management. She is a project investigator for her county's Food Stamp Nutrition Education Program, where she supervises several community dietetics professionals and other staff members. The program reaches over 4,000 youth and adults every year. Ensle is also in charge of the Union County Senior Meals Program.

Ensle says a key component to succeeding in community dietetics is to follow the old adage "Know thyself." (For more on this topic, see the Professional Focus feature "Getting Where You Want to Go" in Chapter 2.) Professionals interested in working outside the clinical setting need to be flexible self-starters who are comfortable creating their own schedules and cooperating with many different types of people. On any given day, Ensle's job may take her from putting on a PowerPoint presentation for a local group of senior citizens to working with children in an inner-city school. "You have to be a very independent person with good people skills," says Ensle. "You've got to piece together ideas and look at the big picture—not micromanage."

Strong advocacy skills are also important. Ensle's Food Stamp program is funded by a $539,000 grant. Community dietetics professionals must be comfortable seeking and lobbying for funding on both the federal and state levels. Ensle says ADA's public policy workshop is so beneficial in this area that it should be required.

But for community dietetics professionals to make an impact, their numbers have to be strong. Ensle worries that the internship opportunities in community dietetics are not long enough to draw potential recruits into the field. "Three days [of an internship] in the community is not enough," says Ensle. "True, students don't always know what they want to do, but three days isn't enough to get their feet wet."

Echoing Ensle's concerns is Christine McCullum, PhD, RD. McCullum is currently an assistant professor with the Center for Health Promotion and Prevention Research at the University of Texas at Houston's School of Public Health. She recently worked on a strategic plan for the prevention and treatment of overweight children in Houston and Harris County, Texas, which was carried out under the auspices of St. Luke's Episcopal Health Charities.

"I think it's important to have the people who are overseeing the future of dietetics understand the importance of marketing the profession," says McCullum. "There's a tendency to market dietetics as a primarily clinically based occupation. In the future, we're going to have to do a better job of marketing a broader range of job opportunities, including those that are community based."

Like many dietetics professionals, when she began her training McCullum says she assumed she would enter the clinical arena. But after earning her master's degree in nutrition, she says she was drawn to the idea of policy work and prevention. Helping create the strategic plan for the prevention and treatment of overweight children was particularly enjoyable, because it enabled her to connect her academic research to a community building process. "It gives me a lot of energy," she says. "It keeps you in tune with the needs of people who live in underserved communities."

McCullum says that current community dietetics professionals can help seek out future colleagues by taking on an active role as a mentor in the dietetics community as well as the broader community. That may include everything from attending a high school career day to encouraging promising young dietetics professionals to consider doing an internship or taking a full-time position with a community-based organization.

She also adds that future dietetics professionals who may be interested in working with the community take sociology and political science courses or classes that teach communication and negotiation skills while they are still in school, along with their required biomedical and health science courses. They might also consider participating in a community service project. These experiences can benefit them later on, especially when working with a diverse group of people.

Because the country is becoming so ethnically diverse, being able to work with all kinds of populations is going to be key, says Karen Ensle, who adds that recruiting minorities to the profession is very important. Ensle recently worked with one intern of Sioux descent who created a food wheel that took into consideration the diet of her native people. The wheel was later presented at a USDA conference.

Although it is important to recruit new dietitians into the community setting, many dietetics professionals who started out in a clinical setting have successfully navigated the transition into a nonclinical one. One such professional is Tracy Fox, MPH, RD. Fox began her career in the United States Navy, doing outpatient work in naval hospitals for five years. As much as she enjoyed her clinical experience, she says her transition into the policy work that followed felt like a "natural progression." She simply found it compelling, so after her clinical experience she took an entirely different path, taking a job doing policy and analysis work for the Food and Nutrition Service (FNS) of the U.S. Department of Agriculture (USDA), where she helped develop and implement major school lunch regulations and served in other positions at FNS.

From there, she went to work for the Government Relations office of the American Dietetic Association in Washington, D.C., where she was responsible for regulatory affairs and helped define the association's position on a range of topics, including child nutrition, dietary supplements and functional foods, food safety, and food labeling.

Her experience led her to believe that undergraduates preparing to enter the field need to learn more about policy work. Agreeing with Ensle, Fox says it would be a terrific idea if all dietetics professionals, especially those just starting out, understood more about lobbying and whom they need to reach to influence change.

Fox says her diverse employment history helped prepare her to start her own consulting company that focuses on nutrition policy. Like many who choose a nonclinical setting, she found the flexible nature of the job appealing, along with the work itself. But that first year, Fox had a lot to learn about setting up her own business. She took a course specifically geared toward women starting their own companies and learned the nuts and bolts of certain important issues, such as how to market herself and how to bill clients.

During her first year of self-employment, Fox cast her potential client net far and wide by getting involved in her local community. She did everything from volunteering for her county's school health council to joining ADA task forces. Her past employment history also generated work. "My initial clients came from other jobs," says Fox. "Don't burn any bridges. You are marketing yourself anytime you talk to someone."

Currently, as president of Food, Nutrition & Policy Consultants, Fox utilizes her extensive experience in federal nutrition policy and the legislative and regulatory process. Her clients include federal, state and local agencies, such as the U.S. Department of Agriculture and the Centers for Disease Control and Prevention; nonprofit organizations, such as the Produce for Better Health Foundation and the Action for Healthy Kids Foundation; and public relations firms, where she provides advice and expertise on policy and nutrition initiatives.

Nancy Clark, MS, RD, an internationally known sports nutritionist and nutrition author, specializes in nutrition for exercise, health, and the nutritional management of eating disorders. Clark has a successful private practice in which she counsels casual exercisers and competitive athletes.

Along with marketing skills, she recommends getting a strong clinical background (i.e., working for two years in a clinical setting); the experience will be invaluable for an aspiring community and/or sports nutritionist who will be dealing with heart disease, cancer, pregnancy, diabetes, hypertension, and a myriad of health conditions. Although she had little interest in hospital dietetics, Clark saw the importance of developing a strong clinical background. Hence she worked for two years in a hospital environment. Next she worked with the New England Dairy and Food Council as a nutrition education consultant. Concurrent with her personal interests in hiking, biking, and other outdoor sports activities, she recognized a professional interest in counseling people active in sports. She next completed a master's degree in nutrition, with a special focus on exercise physiology.

In addition to individually counseling fitness exercisers and competitive athletes, Clark gives talks to local high school and college teams and sports clubs, presents workshops for health professionals, and writes sports nutrition articles. She also has created several teaching tools for nutrition professionals throughout the country and has written two popular books with information for the serious athlete or the active person who wants to eat optimally for health and energy.

For Ensle, McCullum, Fox, Clark, and others like them, working outside the traditional model of a clinically based dietetics professional has been rewarding, both personally and professionally. They hope that their career paths inspire others in the field to consider community dietetics. "The whole training process and mind set needs to shift," says Ensle. "We need to stimulate interest in this in young people. It's critical for the profession."

Sources: Adapted from Jennifer Mathieu, Community-Based Dietetics Professionals, *Journal of the American Dietetic Association* 103 (2003): 1126–27; and personal communication with Tracey Fox and Nancy Clark, www.nancyclarkrd.com.

Chapter *Two*

Assessing Community Resources

Learning *Objectives*

After you have read and studied this chapter, you will be able to:

- Describe seven steps in conducting a community needs assessment.
- Develop a statement that defines the nutritional problem within the community.
- Discuss the contribution of the target population to community needs assessment planning and priority setting.
- Define the terms *incidence* and *prevalence,* and explain how these concepts describe a population's and the community's health.
- Describe three types of data about the community that can be collected, and indicate where these data are found.

This chapter addresses such issues as needs assessment, evolving methods of assessing health status, collecting pertinent information for comprehensive nutrition assessments, and exploring current information technologies, which are Commission on Accreditation for Dietetics Education (CADE) *Foundation Knowledge and Skills* requirements for dietetics education.

Chapter *Outline*

Something To Think About...

Plans are everything.

– General Dwight D. Eisenhower

Introduction

Imagine that you have been asked to take a photograph of your city that will be used for the cover of a tourist brochure. In thinking about a photograph that best represents your city, you are immediately beset by choices. Should you photograph your city during a particular season and, if so, which one? Should you photograph your city's downtown area, showing its architectural and business diversity, or should you choose a popular city park? Should the photograph capture a historical landmark such as a statue or fountain, or should it focus on the *people* who live in your city and show a family picnicking at the fairgrounds or a baseball team at play? Your choice will probably be influenced by the expectations of the brochure sponsor, the time available to you for photographing various aspects of city life, and the budget for producing the brochure.

In many respects, conducting a community needs assessment is much like producing the "best" photograph of your city. It involves making choices to capture a picture of a nutritional problem or need within your community. Like photographing your city, it is a process influenced by the expectations of the people and organizations involved in the assessment, the time available for collecting and analyzing data, and the budget allocated for the assessment.

Consider the challenge of capturing a picture of hunger among Hispanics living in the United States. The Hispanic population is composed of about 40 million people, or nearly 13.7 percent of the U.S. population, making this population the nation's largest minority group.[1] Hunger in Hispanic communities is believed to derive from poverty, racism, and high unemployment,[2] factors that directly or indirectly limit access to nutritious foods. Community nutritionists who aim to improve the food intake of Hispanics begin by working with local community leaders, state and federal agencies, and other groups to determine why the Hispanic population experiences hunger. They ask many questions: How many Hispanics in our community experience hunger? Do Hispanics have a higher unemployment rate than other ethnic groups in the community? Are they more likely than other groups to have low-paying jobs? What is the mean income of Hispanic families? How many Hispanic families participate in food assistance programs? What are the barriers to their participation in these programs? What factors contribute to hunger in this population? How does the Hispanic population perceive this problem? Are existing community programs and services reaching Hispanics? If not, why not? The answers to these and other questions fill in some pieces of the puzzle of why some Hispanics experience hunger. The answers help community nutritionists gauge the extent of the problem in their community; identify resources available at the national, state, and local levels to alleviate it; determine where existing services and programs can be improved; and suggest areas where new programs and services are needed. These activities are part of the process of community needs assessment described in this chapter.

Community Needs Assessment

Community needs assessment is the process of evaluating the health and nutritional status of the community, determining what the community's health and nutritional needs are, and identifying places where those needs are not being met.[3] It involves systematically collecting, analyzing, and making available information about the health status and nutritional status of the community or some subgroup of it. The term **health status** refers to the condition of a population's or individual's health, including estimates of quality of life and of physical, and psychosocial functioning.[4] **Nutritional status** is defined as the condition of a population's or individual's health as affected by the intake and utilization of nutrients and nonnutrients.[5] The assessment is undertaken to find answers to basic questions: Who has a health or nutritional problem that is not being addressed? How did this problem develop? What programs and services exist to alleviate the problem? Why do existing programs and services fail to help the people who experience this problem? What can be done to improve the health and nutritional status of the population?

The assessment process is sometimes called "community analysis and diagnosis," "health education planning," or "mapping."[6] Its overall purpose is to provide a better understanding of how the community functions and how it addresses the public health and nutritional needs of its citizens. In some respects, the process is much like the clinical assessment of a patient's health, except that the community is the patient. This snapshot of the community identifies areas where it performs well (for example, local hospitals and clinics have good data on infant morbidity and mortality) and areas where it does not (although two food banks and several food assistance programs are available in the community, some families go without food).

What triggers a community needs assessment? Any number of factors may compel a city health department, state or federal agency, nonprofit organization, or other group to seek information about a community's health and nutritional status. There may be a need for new data on the community's health because existing data are several years old and may no longer accurately reflect a population's health and/or because data have never been collected on some segment of a population. Often a government agency at the state or federal level has a mandate to carry out such activities. Sometimes research findings are the impetus for taking action. For example, an article published in the *Journal of the American Dietetic Association* reported that the infants of teenage mothers who participated in the Higgins Nutrition Intervention Program weighed more and were less likely to have low birthweights than infants whose mothers did not participate in the program.[7] These study findings may prompt municipal health clinics to determine the number of low-birthweight infants born to adolescents and to study the feasibility of implementing the Higgins program in their communities. In other cases, a community leader or community action group may raise awareness about a health or nutritional issue and prompt action to undertake a community needs assessment. The availability of funds may also stimulate the collection of data on the community's health and nutritional problems. For whatever reason, a decision is made to gather information about a nutritional problem that is not being addressed adequately in the community.

Organizations approach the community needs assessment by first determining its purpose and then planning how it will proceed. The amount of time available to conduct the needs assessment, the staff members responsible for conducting it, and the scope of the assessment must all be specified. The scope deserves special mention. In some cases, the assessment is designed to identify the health and nutritional problems of a large population, such as all residents of the community, which might be the nation, state,

Community needs assessment An evaluation of the community in terms of its health and nutritional status, its needs, and the resources available to address those needs.

Health status The condition of a population's or individual's health, including estimates of quality of life and physical and psychosocial functioning.

Nutritional status The condition of a population's or individual's health as influenced by the intake and utilization of nutrients and nonnutrients.

province, or city. The assessment cuts across all income, educational, and geographical sectors, and it aims to identify the major causes of disease, disability, and death among the community's residents. The most recent Behavioral Risk Factor Surveillance System (BRFSS), for example, identified increases in overweight and diabetes as two nutrition-related problems in all demographic and geographical segments of the U.S. population.[8] Blacks had the highest rates of both obesity and diabetes among all races and ethnic groups, and people with less than a high school education had higher rates of both obesity and diabetes than those with a high school education. An assessment of New York State's population found that the conditions most urgently in need of attention were lung cancer among blacks, tuberculosis among blacks and Hispanics, HIV/AIDS, iron deficiency among blacks, prostate cancer among white males, and stroke. Other "second-priority" conditions included lung cancer among whites, anemia among blacks and whites, and diabetes mellitus, emphysema, and breast cancer among women.[9]

Large-scale assessments tend to be expensive, time-consuming efforts that may enlist the efforts of hundreds of people with expertise in public health, nutrition, epidemiology, statistics, management, and survey design and analysis. The team is pulled together from various departments within several agencies, all acting under the leadership of a single agency. Because they are costly and labor-intensive, such assessments may be undertaken only once in five or ten years.

More often than not, the community needs assessment is limited in scope, focusing on a particular subgroup of the community. The small-scale assessment is relatively inexpensive to conduct and can be undertaken by a small team of community nutritionists and other professionals from several organizations and agencies. For example, an assessment of the nutritional status of schoolchildren may involve experts from the municipal departments of public health and education, a local dietitian who directs the National School Lunch Program, faculty members of the hospital's community health department, and graduate students from the local university. In today's fiscal environment, where money for personnel, equipment, and other resources is scarce, the small-scale assessment is often the better—and sometimes the only—choice. It focuses on a high-risk group about which community nutritionists and other health professionals have some knowledge and concern.

Regardless of its scope, the purpose of the community needs assessment is to obtain information about the health and nutritional status of a particular group—namely, the **target population.** In the case of the large-scale assessment, the target population may be all residents of the nation. Because it is not practical or feasible to obtain information about every resident, large-scale assessments use statistical methods to select a sample of people whose age, sex, race, and other characteristics reflect the demographic profile of the entire population. The 1999–2002 National Health and Nutrition Examination Survey (NHANES), for instance, was conducted on a nationally representative sample of about 21,000 persons 2 months of age and older.[10] In a small-scale assessment, the target population may be low-birthweight infants born between December 1 and March 31 to mothers living in the state of Missouri, lactating women with type 1 diabetes living in Chicago, or adults with lactose intolerance who present to three city hospitals.

Basic Principles of Needs Assessment

Certain basic principles apply, no matter what the scope of the needs assessment. In this and the following chapter, we describe these principles and the process of conducting a community needs assessment. In this book, the community nutritionist is given primary responsibility for conducting it. She or he begins by defining the problem, setting the parameters of the assessment, and determining what types of data must be collected to paint a picture of the target population's nutritional problems. The steps are diagrammed in Figure 2-1 and described in the sections that follow.

Target population The population that is the focus of an assessment, study, or intervention.

Step 1	Define the nutritional problem.
Step 2	Set the parameters of the assessment.
Step 3	Collect data: about the community. about the background conditions. about individuals who represent the target population.
Step 4	Analyze and interpret the data.
Step 5	Share the findings of the assessment.
Step 6	Set priorities.
Step 7	Choose a plan of action.

FIGURE 2-1 Steps in Conducting a Community Needs Assessment

STEP 1: DEFINE THE NUTRITIONAL PROBLEM

In the course of living and working within the community and networking with colleagues, the community nutritionist probably has a notion about the target population's nutritional problems. She may not have a folder full of statistics at this point, but she can make a general statement about its main nutritional problem. What she must do now is develop a concise statement of the problem that concerns her. This statement is used to help plan the assessment and to motivate other agencies to join the assessment team. Thus the first step in the assessment process is to answer the question "What is the nutritional problem?" Consider the problems listed here:

▶ **Infants at nutritional risk.** A study of withdrawal rates from the Maryland Special Supplemental Nutrition Program for Women, Infants, and Children (WIC) found an overall rate of withdrawal from the program of 47 percent, with older infants (7 to 12 months of age) having higher withdrawal rates than 6-month-old infants.[11] Infants and children who are withdrawn from the WIC program are more likely to be at increased nutritional risk than those who participate fully in the program.[12] Withdrawal rates from WIC are believed to be high in this state, but there are no data that describe the problem or the factors that contribute to it among the state's WIC participants.

▶ **Obesity.** The prevalence of obesity increased more than 60 percent for men and women aged 18 to 70-some years between 1991 and 2002, according to data from the Behavioral Risk Factor Surveillance System, a telephone survey of health behaviors conducted among adults in 50 states.[13] Obesity contributes to increased morbidity, including increased risk of hypertension and diabetes mellitus. People who are successful at losing weight and maintaining the weight loss report using a combination of diet, exercise, and behavior modification to achieve their goals.[14] The prevalence of obesity in Johnson County is not known, nor is information available about the strategies used by overweight people in Johnson County to lose weight successfully.

See Chapter 8 for a detailed discussion of the obesity epidemic.

▶ **Low-birthweight infants.** National survey data reveal that the percentage of low-birthweight infants increased from 7.4 in 1996 to 7.6 in 1999, the highest level reported in more than two decades.[15] Low-birthweight infants have a higher risk of mortality and morbidity than normal-weight infants. The incidence of low-birthweight infants across all hospitals and clinics in this city is 8.2 percent, a figure higher than the national one. The reasons for this are not known.

▶ **Inadequate food intake of immigrants.** Results of the Ontario Health Survey indicate that only one-fifth of all immigrants consumed at least 75 percent of the minimum recommended number of servings of each food group in *Canada's Food Guide to Healthy Eating.*[16] Routine consumption of less than the minimum recommended number of servings of each food group may cause inadequate intakes of essential nutrients. No information is available on the food group intake of immigrants in this province.

▶ **Poor quality of diets of high school students.** A survey conducted by the Heart and Stroke Foundation of Manitoba found that one-third of students in 11 of the city's secondary schools eats mainly french fries for lunch and that only 1 in 20 students consumes a lunch containing foods from all food groups.[17] It is not known whether students in other secondary schools have similar eating patterns, and no information about the factors that influence the food choices of secondary school students in the province is available.

A problem that is inadequately defined is not likely to be solved.

– P. M. Kettner, R. M. Moroney, and L. L. Martin in *Designing and Managing Programs,* p. 36

For more about the study of epidemiology, refer to Chapter 4.

These statements of nutritional problems cover a range of nutritional issues, target populations, and age groups. Each statement indicates who is affected by the nutritional problem and how many people experience it. In most cases, information about the number of people who experience the problem is derived from published results of epidemiologic studies. Each statement indicates the impact of the problem on general health or nutritional status. All indicate areas where there are gaps in the community's knowledge of a nutritional problem. These statements serve as the starting point for undertaking an assessment of a nutritional problem within the community, whether the community consists of a city, county, state, province, or nation.

STEP 2: SET THE PARAMETERS OF THE ASSESSMENT

Before the community needs assessment is undertaken, certain parameters or elements must be determined. These parameters, which are described here, set the direction for the assessment. As you read the following material, consult Table 2-1, which describes the parameters for two assessments in the fictional city of Jeffers (population 612,000). One assessment (Case Study 1) focuses on the issue of women and coronary heart disease (CHD). The other (Case Study 2) focuses on issues surrounding the nutritional status of elderly persons living at home. These case studies, which form the basis for some of the discussion in this and the next chapter, illustrate two assessments that differ in scope and complexity.

Define "Community"

The scope of the "community" must be specified. The community might include the people who represent the target population and live within the city limits or within the greater metropolitan area bounded by certain suburbs. Sometimes the community is a geographical region, state, nation, province, or group of countries. In the case studies described in Table 2-1, the community is a typical municipal unit such as the city of Jeffers.

Determine the Purpose of the Needs Assessment

A needs assessment is undertaken to gather information about the social, political, economic, environmental, and personal factors that influence the nutritional problem and the population at risk. The community needs assessment may have one or more of the purposes listed here:[18]

- Identify groups within the community who are at risk nutritionally.
- Identify the community's or target population's most critical nutritional needs and set priorities among them.

TABLE 2-1 Parameters for Two Community Needs Assessments in the Fictional City of Jeffers, Population 612,000

PARAMETER	CASE STUDY 1: WOMEN AND HEART DISEASE	CASE STUDY 2: NUTRITIONAL STATUS OF THE ELDERLY
Lead Organization	State affiliate of the American Heart Association or, in Canada, provincial Heart & Stroke Foundation	City of Jeffers Health Department
Statement of Nutritional Problem	Coronary heart disease (CHD) is the leading cause of death among U.S. and Canadian women. Most women apparently do not realize that they are more likely to die from CHD than from cancer. There are no data on knowledge/awareness of CHD risk factors among women living in Jeffers or about existing CHD prevention programs and services in Jeffers.	The number of independent, noninstitutionalized elderly persons (75+ years) has increased nationally. In Jeffers, this population has increased 12% since the 1995 assessment. Several community-based social service agencies have perceived an increased demand for nutrition services by this population, but there are no data on the availability of such services or on the nutritional status of these persons.
Definition of Community	The metropolitan area of Jeffers, including the adjoining municipalities of Oakdale, Chambers, Kastle, and Morgan	City of Jeffers, bounded by the city limits as of July 31, 2005
Purpose of the Assessment	To obtain information about women and heart health to help determine whether a program or other intervention should be developed	To obtain information about the nutritional status of independent elderly persons $>$ 75 years of age and their needs for nutrition services
Target Population	Females over 18 years of age	Elderly persons $>$ 75 years of age living independently
Overall Goal of Assessment	Women's knowledge, attitudes, and practices related to CHD risk and existing programs and services designed to reduce CHD risk will be identified	The nutritional status of independent elderly ($>$ 75 years) and their use of nutritional services available through community-based agencies will be determined
Objectives of Assessment	On a sample of 250 women over the age of 18 years, within 3 months: • Assess women's knowledge and awareness of the leading causes of death and CHD risk factors. • Assess their practices to reduce CHD risk. • Identify existing services offered to women to help reduce their CHD risk. • Assess women's use of existing services designed to reduce risk. • Identify gaps in the delivery of such programs and services.	On a sample of 150 elderly persons aged 75 years and older, within 1 year: • Assess the nutritional status of independent elderly persons ($>$ 75 years). • Determine which community services are available for this population. • Identify the existing community services used by this population. • Identify gaps in program and service delivery.
Types of Data Needed: Community Conditions	• Mortality data for women (50+ years) • Morbidity data for women (50+ years) • Existing programs and services: Hospitals, medical clinics Fitness/sports centers Offered by health professionals in private practice • Educational materials available from: doctors' offices health professionals in private practice food/pharmaceutical companies bookstores, other	• Types of services offered by community organizations, including personal-care services, homemaker services, adult day care, home-delivered meals, hospice, and home health care services • Number of elderly persons who use these services • Number of elderly persons who participate in federal/state assistance programs, such as Food Stamps, Social Security, Medicare, Medicaid, Supplemental Security Income, Veteran's Benefits, assistance for housing and home heating, and home-delivered meals • Types of medical and social services offered by health professionals
Types of Data Needed: Background Conditions	• Advertising related to smoking • Health messages about CHD in magazines, newspapers, etc.	• Changes in eligibility for Medicare, Medicaid • Current funding of Older Americans Act
Types of Data Needed: Target Population	See Table 3-1 on page 65 in Chapter 3.	See Table 3-10 on page 81 in Chapter 3.

- Identify the factors that contribute to a nutritional problem within the community or target population.
- Determine whether existing resources and programs meet the community's or target population's nutritional needs.
- Provide baseline information for developing action plans to address the community's or target population's nutritional needs and for evaluating the program.
- Plan actions to improve the community's or target population's nutritional status, using methods that are feasible and focus on established health priorities.
- Tailor a program to a specific population.

Define the Target Population

The focus of the community needs assessment is the target population, whose health and nutritional status are affected by many community, environmental, and personal factors. The choice of the target population is influenced by the initial perception about the nutritional problem. Sometimes the needs assessment begins with one target population (for example, all women with infants under 1 year of age) and concludes with a more refined focus (for instance, teenagers with infants under 1 year of age). Usually, however, the target population remains constant over the course of the needs assessment (such as people with cancer, black men with hypertension, teenagers with eating disorders, persons with alcoholism, and people recovering from stroke).

Set Goals and Objectives for the Needs Assessment

This is an essential step, because the goals and objectives determine the types of data collected and how they will be used. **Goals** are broad statements that indicate what the assessment is expected to accomplish. **Objectives** are statements of outcomes and activities needed to reach a goal. Statements of objectives use a strong verb, such as *increase, reduce, begin,* or *identify,* that describes a measurable outcome. Each objective should state a single purpose.[19] The assessments described in Table 2-1 specify one overall goal and several objectives. Assessments may have 2, 3, or more goals and 10 to 15 objectives. The needs assessment cannot proceed without clearly defined goals and objectives.

Goals Broad statements of what the activity or program is expected to accomplish.

Objectives Statements of outcomes and activities needed to reach a goal.

Specify the Types of Data Needed

The types of data required in a needs assessment depend on the purpose, goals, and objectives of the assessment. Some data related to the target population may already exist in the literature or may be available from government sources. Other data must be collected directly from the target population. In general, data are used to identify a high-risk population, describe the nutritional problem within the target population, define areas where nutritional needs are not being met, identify duplication of services, or develop directories of services. Table 2-1 specifies the main types of data needed for the two assessments in the city of Jeffers. These lists are not meant to be comprehensive; rather, they represent the "first pass" in estimating the types of data needed about the target population and the nutritional problem it is perceived to have.

A health status snapshot of all 3,082 U.S. counties including causes of death, life expectancy, infectious diseases, teen mothers, and other indicators can be accessed from the *Community Health Status Indicators Reports (CHSI).* You can compare your county with similar "peer counties," with the nation, or with the *Healthy People 2010* national goals. A CD-ROM of CHSI reports is available from www.naccho.org/project2.cfm. You may also find useful data from the Internet Resources on page 56.

STEP 3: COLLECT DATA

When the parameters of the assessment have been set, the next step is to begin collecting data. In approaching the process of data collection, remember that the assessment's overall purpose is to paint a picture of the environment in which the target population lives to understand how the nutritional problem developed and what might be done to address it. Begin collecting data about the "big picture"—namely, the environment or community in which the target population lives and works. Then, when you have begun

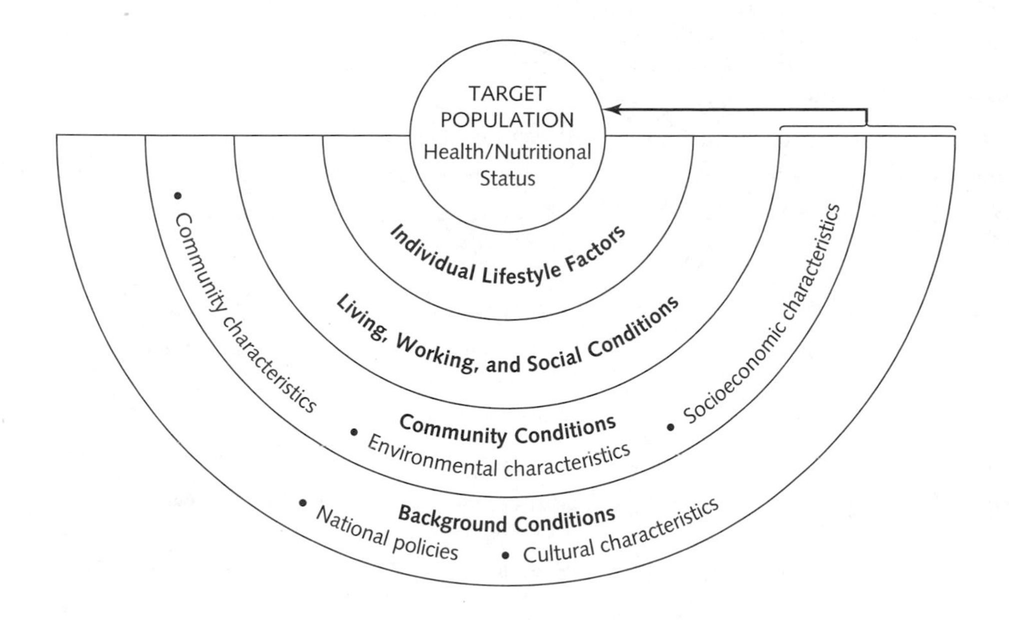

FIGURE 2-2 Types of Data to Collect about the Community

The focus of the community needs assessment is the target population, whose health and nutritional status are affected by many community and background conditions, as well as by individual characteristics such as lifestyle, living and working conditions, and social networks.

Source: Adapted from M. Whitehead, Tackling inequalities: A review of policy initiatives, in *Tackling Inequalities in Health: An Agenda for Action,* ed. M. Benzeval, K. Judge, and M. Whitehead (London: King's Fund, 1995), p. 23.

to get a sense of the big picture, collect data about individuals who make up the target population. Refer to Figure 2-2 as you read this section. The figure shows the general categories of data that might be collected over the course of the assessment and represents another way of thinking about the determinants of health shown in Table 1-1 on page 9. In this chapter, the discussion focuses on the outer layers of the figure—on collecting data about certain community and background conditions that affect the target population's health and nutritional status. Chapter 3 describes issues related to collecting data about the individual lifestyle factors and the living, working, and social conditions that influence the target population.

Collect Data about the Community

A good place to begin the community needs assessment is with the community itself. Both qualitative and quantitative data help describe the community and its values, health problems, and needs. **Qualitative data** such as opinions and insights may be derived from interviews with **key informants**—people who are knowledgeable about the community, its history, and its past efforts to address a nutritional or health problem—and with **stakeholders**—the people and organizations with vested interest in identifying and addressing the nutritional problem. Members of the target population itself can also provide information about the community. **Quantitative data** may be derived from a variety of databases, including registries of **vital statistics** (for instance, age at death and cause of death), published research studies, hospital records, and local health surveys.[20]

Qualitative data Data (such as opinions) that describe or explain, are considered subjective, and can be categorized or ranked but not quantified.

Key informants People who are "in the know" about the community and whose opinions and insights can help direct the needs assessment.

Stakeholders People who have a vested interest in identifying and addressing the nutritional problem.

Quantitative data Numerical data (such as serum ferritin concentration, birth rate, and income) that can be measured and are considered objective.

Vital statistics Figures pertaining to life events, such as births, deaths, and marriages.

This section describes the types of data that can be gathered about the community. These data are outlined in Table 2-2.

Community Characteristics Information is needed about how the community operates, how its population is distributed, and how healthy it is. Information about existing health services provides clues to the community's perception of its leading health and nutritional problems.

▸ **Community organizational power and structures.** Some qualitative data help the community nutritionist understand how the community operates politically and who

TYPE OF DATA	SOURCE OF DATA
Community Organizational Power and Structures	
• Organization of government (city, state, etc.)	Directory of municipal, state, etc., government
• Organization of health department (city, state, etc.)	Directory of municipal, state, etc., health department
• Local, state, and national organizations with a health mandate (e.g., American Heart Association, American Cancer Society, American Dietetic Association, American Public Health Association)	Yellow pages of the phone book, national or state directories
• Community groups and their leaders (e.g., United Way)	Yellow pages of the phone book, key informants
• Reporters and other people with the media	Local/national newspapers and magazines, television and radio stations
• Members of the Chamber of Commerce	Local Chamber of Commerce
Demographic Data and Trends	
• Total population by age, sex, race, marital status, etc.	Census Bureau, FedStats,* state data centers, data archives, libraries
• Distribution of population subgroups (e.g., percentage of population that is black, Hispanic, Asian, etc.; percentage that is foreign born)	Census Bureau, FedStats, state data centers, data archives
• Size and composition of households (e.g., number of family members in households, number of children in households, percentage of all households consisting of husband–wife families, percentage of single-parent families)	Census Bureau, FedStats, state data centers, data archives
Community Health	
• Mortality statistics (e.g., death rates according to age, sex, cause, location, etc.)	Census Bureau, NCHS†, FedStats, Public Health Service, state and municipal health departments
• Morbidity statistics (e.g., frequency of symptoms and disabilities, distribution of disease conditions)	Census Bureau, NCHS, FedStats, Public Health Service, state and municipal health departments
• Fertility and natality statistics (e.g., age and parity of mother, duration of pregnancy, percentage of mothers who get prenatal care, number of unmarried mothers, infant's birthweight, type of birth [i.e., single, twin], fertility rate, infant mortality)	Census Bureau, FedStats, state data centers, data archives
• Communicable diseases (e.g., incidence, distribution)	FedStats, CDC, published studies

*The Internet address for FedStats is www.fedstats.gov.
†The following abbreviations appear in this table: AOA = Administration on Aging; BRFSS = Behavioral Risk Factor Surveillance System; CDC = Centers for Disease Control and Prevention; NCHS = National Center for Health Statistics; NHANES = National Health and Nutrition Examination Survey; NNMRRP = National Nutrition Monitoring and Related Research Program; DRI = Dietary Reference Intake; YMCA/YWCA = Young Men's/Women's Christian Association.

TABLE 2-2 Types and Sources of Data about the Community

wields power within the system. The organization charts for government agencies and city hall provide information about how the community delivers health services and develops health policy. Knowing the key players in local health organizations and community, business, and media groups helps the community nutritionist identify the concerns of community leaders.

▸ **Demographic data and trends.** Demographic data help define the people who live in the community by sex, age, race, marital status, and living arrangements. Changes in the demographic profile of a community often serve as an early warning signal about potential gaps in services or undetected nutritional problems. For example, in an assessment of the nutritional and health status of persons aged 65 years and older in a rural community, the assessment team learned that the number of persons more than 85 years old who were

TYPE OF DATA	SOURCE OF DATA
Community Health—*continued*	
• Occupational diseases (e.g., incidence, distribution)	FedStats, National Institute for Occupational Safety and Health, published studies
• Leading causes of death	FedStats, NCHS, published studies
• Life expectancy	FedStats, NCHS, published studies
• Determinants and measures of health (e.g., vaccinations, disability days, cigarette smoking, use of selected substances [i.e., alcohol, marijuana, cocaine], hypertension, obesity, serum cholesterol concentrations, exposure to lead, occupational injuries, incidence of food-borne disease)	FedStats, CDC's *Vital and Health Statistics* series and the *Morbidity and Mortality Weekly Report*, NNMRRP surveys (e.g., NHANES, BRFSS)
• Food and nutrient intake (e.g., food group intake, nutrient and energy intakes, nutritional adequacy of diets compared with the DRI)	Agriculture Research Service, Elderly Nutrition Program (AOA)
• Use of health resources (e.g., frequency of patient contact with physicians, number of office visits to physicians and dentists)	State Department of Community Health Services
• Health care resources (e.g., persons employed in service, number of active physicians and other health personnel)	FedStats, Health Resources and Services Administration
• Inpatient care (e.g., days of care and average length of stay in hospitals, number and types of operations, number of nursing home residents)	Hospitals, nursing homes, etc.
• Facilities (e.g., short-stay and long-term hospitals, community hospital beds)	Yellow pages of phone book
Existing Community Services and Programs	
• Government-funded food assistance programs (e.g., number of referrals)	Related government agency
• Health and nutrition services and programs offered by hospitals, clinics, community health centers, sports/fitness centers, YMCA/YWCAs, the public health department, voluntary health organizations, schools, universities, colleges, civic groups	Hospitals, clinics, sports/fitness centers, directory of nutrition services (if available from state associations)
• Primary care services (e.g., location, accessibility)	Local hospitals, clinics, etc.
• Soup kitchens, food pantries	Yellow pages of phone book, municipal community services department
• Programs and services offered by nutritionists, dietitians, and other health professionals	Key informants, state associations

TABLE 2-2 Types and Sources of Data about the Community—*continued*

living alone at home had increased significantly since the 2001 assessment. This trend suggested a need to increase home care services for the 85-plus age group.

▸ **Community health.** A variety of health statistics are used to paint a picture of the community's health. Some health data describe the causes and rates of disease, disability, and death within the community; others focus on key life stages or events. Community health data help the community nutritionist describe the population's health and nutritional problems and identify persons who are malnourished. Some common health measures and related terms are listed in Table 4-1 on page 93.[21] For example, the infant mortality rate is an important measure of a nation's health and is used worldwide as an indicator of health status. The infant mortality rate in the United States has declined steadily over the past decades and was 7.0 per 1,000 live births in 2004. Even so, the U.S. infant mortality rate is higher than that of several industrialized countries (see Table 10-1 on page 311). The infant mortality rate for blacks was 13.8 per 1,000 in 2004, almost double that of the U.S. national average. American Indians and Alaska Natives also had infant mortality rates higher than the national average, signaling an urgent need to address this basic health issue.[22]

Information about population survey data, including data related to nutrition, can be obtained from sources in both government and the private sector, including

- Centers for Disease Control and Prevention (CDC) Wonder Database (http://wonder.cdc.gov)
- Combined Health Information Database (CHID) (http://chid.nih.gov)*

*CHID is a free bibliographic database of more than 100,000 entries, including teaching guides, audiotapes, videotapes, booklets, fact sheets, newsletters, journal articles, book chapters, and posters combining the resources of the National Institutes of Health, the Centers for Disease Control and Prevention, and other agencies of the Public Health Service.

▸ **Existing community services and programs.** Obtaining data on the community's existing health and nutrition services helps pinpoint gaps where services are needed. An inventory of the community's nutrition services and programs can be built by (1) identifying the nutrition services and programs available through government agencies, health organizations, and civic groups; (2) cataloging the educational services and materials offered by voluntary health organizations such as the American Red Cross, the National Council on Alcoholism, the American Heart Association, and the American Cancer Society in the United States and, in Canada, the Canadian Diabetes Association, the Canadian Arthritis and Rheumatism Society, and the Heart and Stroke Foundation of Canada, among others; and (3) identifying the programs and services delivered by local nutritionists, dietitians, and other health professionals. The United Way of America is a network of volunteers and local charities that also maintains directories of local community services and programs.[23] In addition, national information centers such as the U.S. National Health Information Clearinghouse can be contacted for general information about the availability of educational materials, programs, and referral services.

Environmental Characteristics Each target population lives and works within a particular physical environment. Certain aspects of this environment, described in Table 2-3, affect the target population's health and nutritional status. Access to medical clinics and ambulatory care services, which provide screening, diagnosis, counseling, follow-up, and therapy, influence the target population's health and nutritional status, as does access to nutritious foods.

Food availability is influenced by the community's geography and climate, which affect the length of the food-growing season; by the type of foods grown in commercial and family gardens; by the types of food storage systems needed to keep foods fresh; and by the location and types of grocery stores, convenience stores, and farmers' markets. Ready access to transportation, whether in the form of personal car, bus, or commuter train, improves the target population's access to medical services and supermarkets.

Socioeconomic Characteristics Certain economic and sociocultural data related to the community, as shown in Table 2-4 on page 46, are also useful. Information about the income of families and the number of families receiving public assistance provides a benchmark for comparing the target population's income with the community's mean or median family income. Information about the community's educational level, literacy rate, and major industries and occupations helps identify barriers to improving the health and nutritional status of the target population.

It is seldom necessary, expedient, or possible to collect and use all data available about the community. Consider a situation where the local health department is aware of the high prevalence of Type 2 diabetes in American Indians.[24] The department's health and wellness office wants to increase its initiatives to reduce Type 2 diabetes risk among American Indians, especially children. It perceives a need to develop a wellness program specifically for American Indian children, but it needs information about Type 2 diabetes in this population. The department's community nutritionist reviews the spectrum of community data that could be collected and decides to collect data about the structure of the tribal council, the distribution of Type 2 diabetes among American Indians by sex and age, the types of health care and medical services available to the population, its use of health care and medical services, the availability of diabetes education programs and the population's participation in these programs, food patterns of American Indians, mean family income and education level, and literacy. Data are not needed on housing, the water supply, recreation facilities, labor force characteristics, tangible wealth, or transportation systems.

TYPE OF DATA	SOURCE OF DATA
Food Systems	
• National, regional, and local food distribution networks; extent of emergency and supplemental feeding systems; food wholesale and retail systems; and amount of food grown locally	Census Bureau, data archives, crop reports by county or state agencies (can be accessed through FedStats*)
• Location, type, and number of grocery stores, supermarkets, convenience stores, and farmers' markets	Yellow pages of phone book
• Location, type, and number of restaurants (e.g., family style, fast food, etc.)	Yellow pages of phone book
• Location and number of health food stores	Yellow pages of phone book
Geography and Climate	Observation, state department of agriculture
Health Systems	
• Location, type, and number of hospitals, clinics, health maintenance organizations, long-term-care facilities	Yellow pages of phone book
• Types of ambulatory care	Annual reports of hospitals, clinics
Housing	
• Type of housing (e.g., percentage of year-round housing that is single dwelling units; housing characteristics, such as units in structure, year structure was built, number of rooms and bedrooms, plumbing, heating, and kitchen facilities)	Census Bureau, FedStats
• Condition of housing (e.g., percentage of standard housing with an exterior frame made of brick, wood, or concrete block)	Census Bureau, FedStats
Recreation	
• Location and number of fitness centers, sports facilities	Yellow pages of phone book
• Types of recreation available (swimming, golf, tennis, cross-country skiing, walking trails, etc.)	Observation, yellow pages of phone book
Transportation Systems	Municipal/state department of transportation or transit
Water Supply	Municipal/state department of water works and water quality
• Source of water, distance from residence, water quality	

*The Internet address for FedStats is www.fedstats.gov.

TABLE 2-3 Types and Sources of Data about the Environment

In this example, the focus of the assessment is fairly narrow, dealing with only one nutritional problem (Type 2 diabetes) experienced by a particular population (American Indian children and their caregivers living on a reservation). In contrast, consider the types of data required to evaluate the health and nutritional status of homeless people living in your community. Because homelessness cuts across all age and ethnic groups, a broad spectrum of data on homeless people and the community's resources for dealing with their health and nutritional problems is needed. In the next section, we describe how to locate information about the community and target population.

Sources of Data about the Community

It is said that the key to investing successfully in real estate is location, location, location. In community needs assessment, the key is to observe, observe, observe—and do a little legwork and networking. Consult Tables 2-2, 2-3, and 2-4 for information about sources of data related to the community.

TABLE 2-4 Types and Sources of Socioeconomic Data Related to the Community

TYPE OF DATA	SOURCE OF DATA
Sociocultural Data and Trends	
• Labor force characteristics (e.g., occupation, industry, class of workers, hours worked)	Census Bureau, Bureau of Labor Statistics
• Language spoken at home	FedStats*
• Education (e.g., median school years completed by individuals 25 years of age and older; individuals who completed high school)	FedStats, Census Bureau
• Literacy levels of school children and adults	FedStats, Census Bureau
Economic Data and Trends	
• Income of families and unrelated persons living in households	FedStats, Census Bureau
• Median incomes of families	FedStats, Census Bureau
• Percentage of families with incomes below the poverty level	FedStats, Census Bureau
• Number of families receiving Temporary Assistance for Needy Families	Administration for Children and Families (DHHS†)
• Number of participants in the WIC,‡ National School Lunch, and National School Breakfast programs	USDA,§ FedStats
• Number of individuals receiving food stamps	National Food Stamp Program
• Number of individuals receiving public assistance	Welfare office
• Unemployment statistics (e.g., percentage of households with one or more unemployed members)	FedStats, Census Bureau
• Tangible wealth (e.g., ownership of land and livestock, ownership of items such as personal computers and cellular phones)	Municipal, county, state records

*The Internet address for FedStats is www.fedstats.gov.
†DHHS = Department of Health and Human Services.
‡WIC = Special Supplemental Nutrition Program for Women, Infants, and Children.
§USDA = U.S. Department of Agriculture.

Where does one start to collect data about the community? There is no one right way to begin data collection, but observing the target population in its community setting—doing a walkabout—provides essential information. If your target population is the institutionalized elderly, then visit local nursing and residential homes, senior centers, and other facilities where the elderly live. If the population you are interested in is Muslim, visit local grocery stores where this group shops and find out about Halal food. If it is athletes, visit fitness centers, sports facilities, and schools with sports programs. Observe the target population in the community and ask questions: Where do these people shop for food? What kind of transportation is required for them to reach the supermarket or grocery store where they shop? What are the main occupations of this group? Do they live near a hospital or clinic? Which medical services do they use? Is a food bank, soup kitchen, or other emergency food assistance facility located near where they live? Do they grow some of their own food? What kinds of nutritional problems has this group experienced in the past? Take time to ask members of the target population how they perceive the nutritional problem. Indeed, you may be surprised to learn

Jeremy Horner/CORBIS

Visit grocery stores and supermarkets where your target population shops to learn about its food consumption and shopping practices. Your observation that few members of your target population drive cars, and most walk to the grocery store, is important information about their lifestyle and needs.

that some of them are not aware of a problem! The important thing is to listen to what the target population has to say about the problem, how it arose, and what might be done to address it.

Networking with colleagues also provides information about how the problem is perceived. Colleagues may know of newly published documents that provide recent health statistics about the target population, or they may help you locate unpublished data that are available in preliminary form. They may also be aware of similar needs assessments done in other cities, states, provinces, or regions.

Interviews with key informants are valuable. Key informants may be formal leaders of the community, such as the mayor, who have a broad knowledge of the community, or they may be informal leaders, such as the owner of a community center, whose opinions and connections provide valuable information about the community. Religious leaders, physicians, teachers, volunteers with nonprofit agencies, heads of social services, and members of the media are among those who can provide insights into how the community operates and how the target population perceives the nutritional problem. When arranging interviews with key informants, develop a short list of open-ended questions (that is, questions that require more than a simple "yes or no" answer); identify a few key informants whose opinions would be most useful to you; contact these informants for permission to interview them; conduct the interview in person, by mail, by e-mail, or by telephone; and summarize the results for other members of the assessment team.[25] It is always appropriate to thank the key informants for their time.

The next step is to turn to the Internet or the library to search the medical, nutrition, and public health databases for literature on the nutritional problem. The Internet Resources on page 56 address key data sources online. Also check the government documents section of public and university libraries for compilations of vital health statistics and reports on public health issues related to the community. Local historical records can be useful. Information available in newspaper archives, old maps, parish or county records, and other documents can provide a history of a local public health problem and the public attention given to it.

Many demographic and socioeconomic data can be located in publications of the U.S. Bureau of the Census, Bureau of Labor Statistics, Department of Agriculture, and Department of Health and Human Services. The decennial Census of the Population and Housing, for example, which is conducted by the Bureau of the Census, provides data on states, counties, local units of government, school districts, and congressional districts.[26] Census data typically describe age and sex distributions, births and deaths, labor force characteristics (for example, occupation, industry, hours worked), income, housing characteristics (for instance, year built, number of rooms, plumbing, heating, kitchen facilities), and other demographic variables. Many libraries are repositories for census data, and some databases are available on CD-ROM.

Additional legwork helps locate health statistics and related health reports from local, county, and state health departments; social welfare agencies; birth, death, marriage, and divorce registries; and courts. The annual reports of local hospitals, clinics, and health centers provide information on the types of health problems within the community, the existence of screening programs for detecting health and nutritional problems, and the resources available to deal with them.[27]

It is sometimes necessary to resort to secondary data sources such as data archives, which serve as repositories for thousands of surveys conducted over the last two decades. Data archive services have collected and cataloged the surveys of many communities. These services usually charge a fee for data tapes and supporting documentation. Examples of data archive services include the University of Michigan's Institute for Social Research, the largest university-based center for interdisciplinary research in the United States,[28] and the University of North Carolina's Institute for Research in Social Science, which is a source of nonproprietary opinion data, such as the Louis Harris public opinion data, national census data, and data from the World Fertility Surveys and the Demographic and Health Surveys.[29]

Collect Data about Background Conditions

Collect information about the broader environment in which the community is positioned. Many political, socioeconomic, cultural, and environmental factors at the national level operate in the background but have the potential to affect how the target population lives, the food choices it makes, and where it obtains medical services. The types and sources of background information are summarized in Table 2-5.

National policy, for example, affects eligibility for food assistance programs, minimum wage levels, and the distribution of commodity foods—all factors that may influence the target population's health and nutritional status. If the community consists of tribes of American Indians, the assessment team may review the health care policy of the Bureau of Indian Affairs, the U.S. federal agency charged with providing personal and public health services to American Indians and Alaska Natives. The assessment team may learn that the agency's policies have inadvertently created competition among tribes for money for health care services. This unexpected situation may result in less money being distributed to tribal communities to pay for expensive medical services,[30] an outcome that may affect the target population's health and nutritional status.

The broad culture also influences food intake and nutritional status. By **culture,** we mean the integrated pattern of human knowledge, beliefs, and behaviors that are learned and transmitted to succeeding generations.[31] Many of our food habits, attitudes, and practices arise from the traditions, customs, belief systems, technologies, values, and norms of the culture in which we live. For instance, in an assessment of the nutritional

Culture The knowledge, beliefs, customs, laws, morals, art, and any other habits and skills acquired by humans as members of society.

TABLE 2-5 Types and Sources of Information about Background Conditions

TYPE OF INFORMATION	SOURCE OF INFORMATION
National Policy	
General information	Articles published in journals, magazines, newspapers; commentary on TV, radio, Internet
• Agriculture	Department of Agriculture
• Economics	Department of Commerce
• Education	Department of Education
• Health	DHHS*
• Housing	Department of Housing and Urban Development
• Labor	Department of Labor
• Nutrition	DHHS (FDA, CDC, NIH, etc.)
• Social Security	DHHS
Cultural Conditions	
• Advertising	Television, radio, printed matter, Internet
• Health messages	Television, radio, printed matter, Internet, educational materials available from government, food companies, nonprofit groups, etc.
• Roles of women	Television, radio, printed matter, Internet
• Belief systems	Television, radio, printed matter, Internet, family, friends, other social contacts

*The following abbreviations appear in this table: CDC = Centers for Disease Control and Prevention; DHHS = Department of Health and Human Services; FDA = Food and Drug Administration; NIH = National Institutes of Health.

status of bulimic students, certain background conditions, such as advertising, society's emphasis on leanness, and cultural expectations about weight and body size, are likely to influence the students' food patterns and body image. This background information should be evaluated as part of the assessment.

Background information on the community's or region's health status is also important and can be obtained from international agencies such as the Food and Agriculture Organization (FAO) and the World Health Organization (WHO), which have regional offices that can furnish relevant population health data. Since 1954, for example, the Pan American Health Organization has published a series of quadrennial reports that document the health progress achieved by its members. Entitled *Health Conditions in the Americas,* these reports provide general information on the region's social and political climate, primary demographic characteristics, mortality data, and health conditions, focusing on women, children, and the elderly.[32]

Collect Data about the Target Population

When the community and background data have been gathered, the community nutritionist begins the process of collecting information about the target population. Although details of this process are described in Chapter 3, some general comments are needed here. When gathering information about the target population, the community nutritionist has two choices: (1) use existing data related to the target population or (2) collect data related to the target population because the needed data do not exist. Both options are described in this section.

See Chapter 6 for more about the activities of the National Nutrition Monitoring and Related Research Program.

Existing Data The most expedient and cost-effective way to obtain health statistics and behavioral information may be to use existing data. Data related to the target population can be gathered from large-scale population surveys, such as those conducted by the National Nutrition Monitoring and Related Research Program (NNMRRP) or from small surveys of special populations conducted in the immediate community or region. It is usually desirable to have both types of data on hand. NNMRRP survey data provide a national perspective on the target population and the factors that contribute to its nutritional needs in other regions of the country. In most cases, NNMRRP survey data can be obtained from the National Technical Information Service either in printed form or on tape. Some data tapes and publications are available from the sponsoring agency (for instance, the U.S. Department of Agriculture or the National Institute of Child Health and Human Development) or from the U.S. Government Printing Office. In addition, many health and nutrition data are published in periodicals such as the *Journal of the American Dietetic Association, Public Health Reports,* and the *New England Journal of Medicine.* Consult the Public Health Service's publications *Nutrition Monitoring in the United States: The Directory of Federal Monitoring Activities* and the *Third Report on Nutrition Monitoring in the United States,* Volumes 1 and 2, for a list of NNMRRP surveys and pertinent journal publications of survey findings and for a description of where NNMRRP survey data can be obtained.[33] Information about the availability of some nutrition monitoring and survey data is provided on the Centers for Disease Control and Prevention's website.[34]

The decision whether to use NNMRRP survey data will be influenced by the level of detail needed about the target population, the personnel and facilities available for sorting and analyzing the data, and budget constraints. The cost of purchasing data ranges from $100 to $1,500.

National survey data do not always reflect the nutritional status or food intake of the target population in a particular community. Consider this scenario: National data from the NHANES III Survey found that 27 percent of women aged 50 years and older met recommended intakes of calcium;[35] however, a 2004 survey of women living in your community found that only 19 percent of women aged 50 years and older met current calcium recommendations. In this situation, the national data do not reflect your community particularly well, but they may be a good benchmark against which to compare your target population. The question for a community nutritionist is why some women in your community have a calcium intake lower than the national average. Thus, small surveys carried out locally provide insights into the community's needs and values. They tend to be more relevant to defining the community's nutritional problems and needs than large, national surveys.

New Data In some situations, data related to the target population are not readily available, so one must choose a method for collecting them. The method selected for assessing nutritional needs must be simple, cost-effective, and able to be completed within a reasonable time frame. Chapter 3 describes some of the methods used to collect data about target populations, including nutrition surveys, health risk appraisals, screening tools, focus groups, interviews with key informants, and direct assessment techniques.

STEP 4: ANALYZE AND INTERPRET THE DATA

Analyze and examine the data collected about the community's nutritional problem. This requires first collating the data collected about the community itself and any background conditions that may have influenced the target population or the nutritional problem.

Then, data collected about the target population (described in Chapter 3) are merged with the community and background data to form a comprehensive report.

Data derived from the assessment are then used to diagnose the community. Four steps are involved in making the community diagnosis: (1) interpret the state of health of the target population within the community, (2) interpret the pattern of health care services and programs designed to reach the target population, (3) interpret the relationship between the target population's health status and health care in the community, and (4) summarize the evidence linking the target population's major nutritional problems to their environment. The summary describes the dimensions of the nutritional problem, including its severity, extent, and frequency; its distribution across the urban, rural, or regional setting and across age groups; its causes; and the mortality and morbidity associated with it. The summary should specify the major strengths of existing community resources and health care services as they relate to the target population, the areas where health problems seem to be concentrated, and the areas where health care delivery for the target population can be improved.[36] The summary may also indicate how the cost of treating the nutritional problem compares with the cost of preventing it and provide information about the social consequences of not intervening in the target population.[37]

The final step is to prepare an executive summary that captures the three or four key points that emerged from the assessment. The executive summary can be given to stakeholders and other interested parties who request information about the assessment outcomes. It can also be reformatted as a press release for the media.

STEP 5: SHARE THE FINDINGS OF THE ASSESSMENT

The results of the community needs assessment are often useful to agencies and organizations that were not directly involved in it. Sharing the assessment results with these other groups and stakeholders is cost-effective, prevents duplication of effort, and promotes cooperation among organizations and agencies. It also enlarges the sphere of awareness about the nutritional problem and increases the likelihood that more than one community group will choose to address the problem. However, releasing the results of an assessment to the community at large without seeking the support and approval of key stakeholders can create ill will. When in doubt, go back to the stakeholders and ask for permission to release sensitive material about the target population.

STEP 6: SET PRIORITIES

Setting priorities involves deciding who is to get what at whose expense.[38] When several nutritional problems are identified by the assessment—as often happens—the question asked by the community nutritionist is "Which health outcome is most important?" The term **health outcome** refers to the effect of an intervention on the health and well-being of an individual or population.[39] When a health outcome reflects a change, that change may have either a positive or a negative effect, or it may even cause no change in health status. It may be distinct, such as a drop in blood pressure or an increase in fiber intake, or it may be somewhat subjective, such as an increase in awareness of a health risk.[40] In other words, the community nutritionist is asking, "Given the several nutritional problems and needs of the target population, where should my organization direct its efforts to achieve the best health outcome?" In most cases, the best health outcome is an improvement in the nutritional status of the target population.

Health outcome The effect of an intervention on the health and well-being of an individual or population.

The challenge for the community nutritionist is deciding which of several nutritional problems or needs of the target population deserves immediate attention.

TABLE 2-6 Principles Involved in Setting Priorities

- Community priorities, preferences, and concerns should be given priority.
- Higher priority should be given to common problems than to rare ones.
- Higher priority should be given to serious problems than to less serious ones.
- The health problems of mothers and children that can easily be prevented should have a higher priority than those that are more difficult to prevent.
- Higher priority should be given to health problems whose frequencies are increasing over time than to those whose frequencies are declining or remaining static.

Source: Adapted from D. B. Jelliffe and E. F. P. Jelliffe, *Community Nutritional Assessment* (Oxford: Oxford University Press, 1990), p. 452.

Considering the fierce competition for scarce resources, how do the community nutritionist, the assessment team, and community leaders decide where to put their efforts and money? No one method exists for ranking problems or needs, although various scoring systems that rank risk factors by relative importance have been proposed. A few principles that provide guidance in identifying problems of the highest priority are listed in Table 2-6.

The priority-setting process begins with a review of the summary prepared in Step 4 (described previously). The findings of the community assessment can be compared with the *Healthy People 2010* focus areas of nutrition and overweight, physical activity and fitness, and maternal, infant, and child health, or other relevant *Healthy People 2010* objectives, to determine where improvements in health services should be made. The seriousness of each problem relative to other nutritional and health problems within the target population is considered. Members of the assessment team rank existing health and nutritional problems and make recommendations about where the community's resources should be directed. Perhaps only 55 percent of worksites with 50 or more employees offer nutrition or weight management classes, whereas the *Healthy People 2010* objective is 85 percent; or perhaps only 19 percent of teenage girls achieve their recommended intakes for calcium, whereas the *Healthy People 2010* objective is 75 percent. Key stakeholders or community leaders help determine which needs of the target population deserve immediate attention.

Healthy People 2010 lists some 467 national health objectives that cover 28 focus areas. These can be useful in forming a basis for program design, monitoring, and evaluation. The objectives can also help you convince funders, administrators, and other key personnel that your program is worthwhile.[41] *Healthy People in Healthy Communities* can be consulted for guidance on building community coalitions, creating a vision, measuring results, and forging partnerships for improving the health of the community.[42] Likewise, the *Healthy People 2010 Toolkit* offers guidance on helping the community translate the *Healthy People 2010* objectives into state-specific action plans.[43]

DATA 2010, the *Healthy People 2010* database (http://wonder.cdc.gov/DATA2010) An interactive database system with national data for tracking *Healthy People 2010* objectives. Users can construct tables that include baseline data, target, and any available updates for specific objectives or selected population groups.

In a perfect world, ample personnel, money, and other resources would be available to spend on each of the target population's problems. Setting priorities would not be necessary. In reality, however, there are never enough resources to address all public health problems, and the decisions about which problems receive attention are not always rational, right, or fair. The process of setting priorities is influenced by the community's political power base, federal and state public health priorities, public opinion, and the beliefs of key stakeholders. The final decisions about which areas to address generally reflect the community's ranking of the importance of public health problems and its assessment of the probable impact of its interventions.

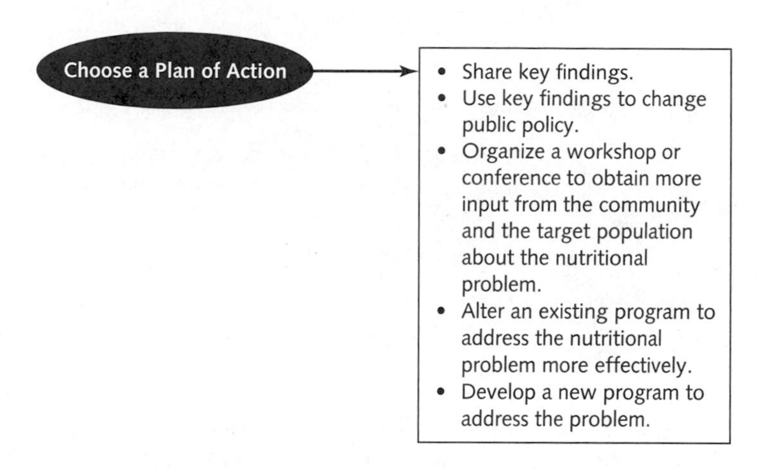

FIGURE 2-3 Choosing a Plan of Action

STEP 7: CHOOSE A PLAN OF ACTION

The community nutritionist is now ready to make a decision. He or she has on hand a definition of the nutritional problem, the results of the needs assessment, and a ranking of the nutritional problems and needs that most deserve attention. Now what? As shown in Figure 2-3, there are any number of options, but the most important thing is to do *something.* After all, the assessment required planning, team effort, and precious resources. Its findings are too valuable to ignore.

But what should be done? At the very least, the key findings of the assessment should be shared with community leaders and other people who are interested in the health and well-being of the target population. These people may use the findings within their own organizations to support interventions aimed at improving the target population's health and nutritional status. For example, the results of an assessment of the dietary changes made by pregnant teenagers can be shared with physicians, nurses, and other health care providers who need to know how and why teenagers change their food patterns during pregnancy.

Another action is to use the assessment's findings to advocate for a change in legislation or public policy that will ultimately improve the health potential of the target population. The term **advocacy** means building support for an idea, cause, or change. (Advocacy is discussed in Chapter 7.) Releasing the assessment's findings to the media is one way of increasing awareness of the problem and building support for policy changes that address the problem.

The community nutritionist and other team members may elect to organize a workshop or conference to obtain additional data on the problem or pull together community leaders and stakeholders to explore future actions. Or they might decide to alter an existing program by developing new educational materials, enlarging a marketing campaign, or changing the mechanism for delivering the program. They may develop a new program to address the nutritional problem of the target population, in which case they may write a grant proposal to apply for money to pilot test the new idea. (Grant writing is discussed in Chapter 20.) In reality, the community nutritionist and her or his team will probably take several actions simultaneously. Certainly, one or more actions should be taken to improve the target population's health and nutritional status through program planning and other activities. (The process of planning, marketing, and evaluating programs is described in Chapters 14 and 18.)

Advocacy Building support for an idea, cause, or change.

Programs in Action

The Nutrition Service Project

Television viewing can have negative health effects on children, according to the American Academy of Pediatrics.[1] Heavy viewing is correlated with low nutritional knowledge and incorrect perceptions about food commercials, as well as greater consumption of nonnutritious foods.[2,3,4] A positive relationship has been reported between television viewing and bad eating habits. A 1995 study determined that 91 percent of foods advertised during children's programming were high in fat, sugar, and/or salt—an issue of concern given that children are highly vulnerable to television messages.[5] Advertising, particularly television commercials, may play an important role in the formation of a child's view of which foods are healthful.[6] Research shows that childhood obesity, adolescent obesity, and poor body image are among the consequences of excess television watching.[7] A 1999 study found that adolescent girls who wanted to look like females on television were more likely to develop disordered eating.[8]

The Nutrition Service Project offers an example of a successful education program that utilizes college students to help adolescents and teens distinguish between sound and unhealthful nutrition information in the media and learn fundamentals of good nutrition.

Goals and Objectives

The primary goal of the Nutrition Service Project is to help children in underprivileged areas learn about health and nutrition. Its objectives include promoting volunteerism in the college environment, helping a community that had limited access to health promotion and education programs, and researching nutrition habits and media viewing among middle school and high school students.

Target Audience

The target audience was students attending grades 7 through 12 in a rural community in Pennsylvania.

Rationale for the Intervention

Health behaviors begin developing in childhood. By eighth grade, many students have made lifelong health choices.[9] In one study, students reporting lower activity also reported making fewer healthful food choices.[10] Early education is critical to helping children make sound decisions before their unhealthful behaviors begin to cluster.

Nutrition and food messages on television may influence the eating behaviors and attitudes of teens. A survey of 88 middle and high school students in rural Pennsylvania, conducted by the founder of the Nutrition Service Project, found an inverse relationship between television watching and fruit and vegetable consumption.

Methodology

The student-run Nutrition Service Project was founded in 1998. It provides college students with training by experts in nutrition education, child development, diversity, health promotion, and education as part of a year-long preparation. The students meet once a month and have guest lecturers speak about interventions, teaching methodologies, adolescent and child nutrition, and other relevant topics throughout the fall.

Students who have completed the training program spend their spring break teaching nutrition and media literacy to middle and high school children in a rural Pennsylvania community. One lesson, entitled "Hidden Messages," teaches

Entrepreneurship in Community Needs Assessment

Recall from Chapter 1 that entrepreneurship is the creation of something of value through the creation of organization.[44] In the case of the community needs assessment, that "something of value" is the snapshot of the nutritional and health problems of the community. Obtaining this valuable commodity, this snapshot, requires organization, vision, and new ideas—all aspects of the entrepreneurial process. Community nutritionists can apply the principles of entrepreneurship to community needs assessment by developing new strategies for collecting information about hard-to-reach populations such as new immigrants and the homeless; by forging new partnerships with food producers, retailers, distributors, and marketers to collect information about dietary patterns and beliefs at the local level; and by developing new methods of assessing nutritional needs and problems.

children to critically examine media sources of information, such as advertisements. The 2001 program, entitled "Navigating through Health Information in a Media World," helped students assess whether nutrition articles in newspapers or popular magazines were reliable and accurate.

The Nutrition Service Project consists of four components: a student director, a group of students in leadership roles for the project, a middle school or high school, and financial resources to cover training and program costs.

Results

The Nutrition Service Project has contributed approximately 2,500 volunteer hours and has reached 230 Pennsylvania youths. In 2001, 116 intervention students and 60 control students participated. A pre-survey assessed media viewing, perception of accuracy in reporting by different sources, and nutrition knowledge. A post-survey examined the ability to identify reliable media sources of health information and nutrition. According to survey data, the intervention group was significantly more likely to plan to increase fruit and vegetable intake and to make other positive lifestyle changes. Students wanted to learn more about nutrition; in the post-survey, they understood that diet is a factor in health. Preliminary data demonstrated that participants were not able to identify accurate nutrition information before the intervention. After the intervention, participants had greater understanding of how to view media critically and how to identify accurate articles about nutrition.

References

1. Committee on Public Education, American Academy of Pediatrics, Children, Adolescents, and Television (RE0043), *Pediatrics* (2001) 107: 423–6.
2. C. K. Atkin, Children's social learning from television advertising: Research evidence on observational modeling of product consumption, *Advances in Consumer Research* (1976) 3: 513–19.
3. C. K. Atkin, Effects of television advertising on children, in E. L. A. Dorr (ed.), Children's behavior responses to TV food advertisements, *Journal of Nutrition Education* (1980): 93–96.
4. N. Signorelli and M. Lears, Television and children's conceptions of nutrition: Unhealthy messages, *Health Communication* (1992) 4: 245–57.
5. H. L. Taras and M. Gage, Advertised foods on children's television, *Archives of Pediatric Adolescent Medicine* (1995) 149: 649–52.
6. Wolfkin, R., Introduction, in *Children's Television: An Analysis of Programming and Advertising,* ed. F. E. Barcus (New York: Praeger, 1977), pp. xx–xxvii.
7. American Academy of Pediatrics, 2001.
8. A. E. Field and coauthors, Relation of peer and media influences to the development of purging behaviors among preadolescent and adolescent girls, *Archives of Pediatric Adolescent Medicine* (1999) 153: 1184–9.
9. S. H. Kelder and coauthors, Community-wide youth nutrition education: Long-term outcomes of the Minnesota Heart Health Program, *Health Education Research* (1995) 10(2): 119–31.
10. L. A. Lytle and coauthors, How do children's eating patterns and food choices change over time? Results from a cohort study, *American Journal of Health Promotion* (2000) 14: 222–8.

Source: Community Nutritionary (White Plains, NY: Dannon Institute, Fall 2001). Used with permission. For more information about the *Awards for Excellence in Community Nutrition,* go to www.dannon-institute.org.

The COMPASS® tool kit developed by United Way of America is an example of entrepreneurship in community needs assessment. COMPASS® helps local United Ways strengthen communities by teaching volunteers how to forge partnerships across the community.[45] Another example is the President's Initiative on Race, a national strategy to eliminate racial and ethnic disparities in health by the year 2010—a goal that requires the concerted efforts of thousands of federal, state, and municipal staff, health professionals, and volunteers. Once again, partnerships among the Department of Health and Human Services, state and local governments, regional minority health and tribal organizations, and community groups are the key to achieving this goal.[46] Look at your community for examples of organizations or people who recognized an opportunity and took the initiative to improve the community's quality of life.

Internet Resources

The following Internet addresses provide data on vital statistics, income, unemployment, demographic characteristics, and other variables related to U.S. population groups.

FedStats **www.fedstats.gov**

More than 70 agencies provide data for this site. Search the site by keyword or alphabetically by topic. Key features at the site include

- Births
- Breastfeeding
- Chronic diseases
- Education
- Health
- Government
- Marriages
- Personal income
- Population
- Programs
- Unemployment
- Vital statistics

National Agricultural Statistics Service **www.usda.gov/nass**

This site offers a few databases on food consumption. Search under the "Food" keyword for the following topics:

- Changes in Food Consumption and Expenditures
- Expenditures of Food, Beverages, and Tobacco
- Food Away from Home
- Food Consumption
- Food Spending in American Households
- U.S. Food Expenditures

State and Local Governments **www.loc.gov/global/state/stategov.html**

Use this homepage to access information about your state and local governments. Some information on major cities is also available.

U.S. Census Bureau **www.census.gov**

This site provides national, state, and some city/county statistics. The site can be searched by word, place name, ZIP code, or map (geographical location). Browse the alphabetical listing of topics. Key features of the site include

- Census State Data Centers (Click on your state's name to access information about printouts, tapes, software, CD-ROM products and services, online data service, newsletters/technical journals, or maps. Link with your state's Census Data Center website.)
- Health statistics
- Household statistics (e.g., American Housing Survey data)
- Household economic statistics
    - Disability
    - Income
    - Labor force
    - Occupation
    - Poverty
    - Small-area income and poverty estimates
- Population topics (e.g., estimates and projections for the nation, states, and households and families)
- State profiles (e.g., census tables showing number of persons, number of families, households, races, and so on)

MEDLINE **www.nlm.nih.gov**

Search MEDLINE and other health databases from this website sponsored by the National Library of Medicine.

National Center for Health Statistics **www.cdc.gov/nchs**

This site offers a list of the latest publications and electronic products available from NCHS. Users can check out the new releases, fact sheets, and publications.

Kaiser Family Foundation **www.statehealthfacts.kff.org**

Offers individual state health profiles and comparisons of 50 states.

Getting Where You Want to Go

Imagine for a minute that circumstances require you to travel from Kansas City to Chicago. Spreading a map across your lap, you plot your course. You could take an interstate highway all the way, following I-35 north to Des Moines and then turning east toward Chicago on I-80. Or you might take I-70 to St. Louis and turn due north onto I-55, a course that would take you right into the Chicago Loop. Or you might decide to bypass the interstate highways altogether and stick to the so-called blue highways, those tiny threads on the map that snake along from town to town. Your decision about which route to take depends on many factors, including the purpose and urgency of your trip and how much time you can allocate for traveling.

In many respects, deciding what you do in life is much like plotting a journey by car to a distant city. Many choices confront you, and there is always the possibility that circumstances may compel or entice you to change your route along the way. Right now, you might be asking yourself the following questions: How can I get where I want to go? More important, how do I determine where I want to go in the first place? The answers to these questions are unique to each of you, because each of you is unique. As you read through this discussion, write down your thoughts to help clarify your vision.

Square 1: Know Yourself

The first step in determining where you want to go is to know yourself. Hold yourself up to the light (so to speak) so that you can see yourself from every angle. Evaluate both your strong points and the areas marked for improvement. (The "To Be Improved" areas are sometimes called weaknesses. Weaknesses are not personality defects or deficits. They are areas of personal development that you have not had the time or inclination to explore and strengthen.) Consider your personality, your view of the world, and what you want out of life. Do you like working with people? Do you enjoy tinkering with gadgets and gizmos? Do you value public service? Are you an optimist or a cynic? Would you describe yourself as impulsive, dependable, funny, unfocused, inquisitive, theatrical, or lazy? Write down the words that describe all aspects of your personality and character. There are no right or wrong answers.

Square 2: Define Your Dreams

Knowing who you are (and who you are not) will help you move to the next tier: defining your dreams. Your vision of your future lies in your dreams, for what you *imagine* yourself doing is what you are ultimately going to do. What do you see yourself doing? To help you define your dreams, answer the following questions:[1]

- What would you most like, ideally, to be?
- What would you most like, ideally, to do?
- What kinds of experiences help you feel complete?
- In what kinds of situations do you most want and tend to share yourself?

Let yourself dream freely and without constraints. Do not be concerned at this point about finances or family obligations. Give your dreams room to grow.

Square 3: Set Goals

Having dreams won't get you very far if you don't add some structure to them. As Henry David Thoreau stated so eloquently, "If you have built castles in the air, your work need not be lost; that is where they should be. Now put the foundations under them."[2] Setting specific goals for your future is one of the most challenging tasks you will undertake. There are many areas in which goal setting is desirable: economic, spiritual, social, physical, mental, emotional, educational, personal, and vocational.

For this exercise, set at least one goal for your personal life. The goal should be achievable but broad enough to accommodate your dreams. Joe D. Batten, author of the book *Tough-Minded Leadership,* wrote his personal goal as follows: "I will make the lives of others richer by the richness of my own."[3] Your personal goal might be similar or entirely different.

Another way to approach this exercise is to write your personal mission statement. A personal mission statement is much like a nation's constitution; it is a set of principles to live by. In his best-selling book *The 7 Habits of Highly Effective People,* Stephen R. Covey cites the personal mission statement developed by a friend. Here is a portion of it:

An Example of a Personal Mission Statement*

1. Succeed at home first.
2. Never compromise with honesty.
3. Be sincere yet decisive.
4. Develop one new proficiency a year.
5. Plan tomorrow's work today.
6. Maintain a positive attitude.
7. Keep a sense of humor.
8. Do not fear mistakes—fear only the absence of creative, constructive, and corrective responses to those mistakes.
9. Help subordinates achieve success.
10. Concentrate all abilities and efforts on the task at hand; do not worry about the next job or promotion.

*Adapted from S. R. Covey, *The 7 Habits of Highly Effective People* (New York: Simon & Schuster, 1989), p. 106.

Square 4: Develop an Action Plan

To paraphrase a Chinese proverb, if you don't know where you are going, then any road will take you there. To get where you want to go, you must develop an action plan. Action is the essence of achievement. Stephen Covey calls this "beginning with the end in mind." You must begin your journey with a clear picture of your destination. Use the following steps to develop an action plan for your personal life:

- Develop a picture in your mind's eye of what you want to do with your life. You may see yourself having a family and a career position with a major food company, or helping an isolated community in a developing country improve its standard of living, or starting your own business. The technique of mental imaging enables you to fine-tune your picture, so that when opportunities present themselves, you can determine whether they fit your action plan.
- Pretest your mental picture. If your mental picture shows you working with small animals as part of a research project, then find a way to test your decision before you commit yourself to this path. You may discover that you don't like working with rats or hamsters! Pretesting your decisions saves time and allows you to discard opportunities that are not useful or don't fit your action plan.
- Predetermine your alternatives. Have a backup plan to help you maximize your opportunities and forestall any crises. Explore your alternatives by talking to people who have pursued a similar dream.

Learn to Manage Yourself

Shirley Hufstedler, a lawyer who became the secretary of education, remarked, "When I was very young, the things I wanted to do were not permitted by social dictates. I wanted to do a lot of things that girls weren't supposed to do. So I had to figure out ways to do what I wanted to do and still show up in a pinafore for a piano recital, so as not to blow my cover. You could call it manipulation, but I see it as observation and picking one's way around obstacles. If you think of what you want and examine the possibilities, you can usually figure out a way to accomplish it."[4]

Getting where you want to go is nearly impossible if you don't learn to manage yourself—your goals, your time, your work. Aristotle observed that the hardest victory is the victory over self. Successful people have mastered themselves through discipline. For some people, *discipline* is a dirty word, but in truth, discipline means training. Any athlete will attest to the power of training, which builds, molds, and strengthens the body and mind for strong performance. Discipline is as important to life as it is to athletic competition. Without it, little can be accomplished. Acquiring discipline, the mastery of self, is a lifelong process for most of us, and there is no simple pattern by which it can be attained. The process involves developing a vision, setting goals, and following through on an action plan to reach those goals. The first step in acquiring discipline begins at square 1: know yourself.

References

1. J. D. Batten, *Tough-Minded Leadership* (New York: American Management Association, 1989), p. 177.
2. As cited in R. N. Bolles, *The Three Boxes of Life* (Berkeley, CA: Ten Speed Press, 1981), p. 34.
3. As cited in Batten, *Tough-Minded Leadership*, p. 179.
4. As cited in W. Bennis, *On Becoming a Leader* (Reading, MA: Addison-Wesley, 1989), pp. 53–54.

Chapter *Three*

Assessing the Target Population's Nutritional Status

Learning *Objectives*

After you have read and studied this chapter, you will be able to:

- Describe the types of data that might be collected about the target population specified in the community needs assessment.
- Describe a minimum of eight methods for obtaining data about the target population.
- Discuss the issues of validity and reliability as they apply to data collection.
- Discuss cultural issues that are considered when choosing a method for obtaining data about the target population.

This chapter addresses such issues as needs assessment, evolving methods of assessing health status and screening individuals for nutritional risks, collection of pertinent information for comprehensive nutrition assessments, and current information technologies, which are Commission on Accreditation for Dietetics Education (CADE) *Foundation Knowledge and Skills* requirements for dietetics education.

Chapter *Outline*

Something To Think About...

I think it could be plausibly argued that changes of diet are more important than changes of dynasty or even of religion.

– GEORGE ORWELL

Introduction

The purpose of the community needs assessment is to obtain answers to basic questions: What is the nutritional problem experienced by the target population? How does the target population perceive the problem? Which factors contribute to it? Where does this group live, work, seek medical care, and buy their groceries? Why do existing services fail to help this group? How can their health and nutritional status be improved?

Answers to some of these questions were found during the community phase of the assessment (described in Chapter 2). Consider a target population consisting of teenagers aged 13 to 19 years who live in Carlson County. The community assessment might have determined that the mortality rate for this group is similar to the national average, their level of alcohol and substance abuse exceeds the state average, they access the health care system through pediatric practice and family practice settings, and most live in households earning $22,500 to $35,000 annually. National data on energy and nutrient intakes for this age group are available from the most recent National Health and Nutrition Examination Survey (NHANES). Even so, questions about the health and nutritional status of Carlson County teenagers remain. How many teenagers living in Carlson County are overweight? How many have eating disorders? What are their attitudes about seeking medical attention for these conditions? Are their usual dietary patterns nutritionally adequate? Do Carlson County teenagers take vitamin, mineral, herbal, or other dietary supplements routinely?

When key questions about the target population's health and nutritional status remain unanswered, the community nutritionist must identify those data elements that are still needed and choose one or more methods for obtaining them. This chapter describes a plan for collecting data, the types of data that might be gathered about the target population, methods used commonly to obtain such data, and several issues to consider when choosing a method of data collection.

A Plan for Collecting Data

The plan for collecting data about a target population helps determine which questions to ask. Some aspects of the plan, such as personnel and budget management, are activities that fall within the realm of project management. (See Chapter 19 for this discussion.) Other aspects, such as sample size, data management, quality control, and statistical analysis, are beyond the scope of this book. For our purposes, the following planning activities are most important and should be completed before data collection begins:

Step 1: Review the purpose, goals, and objectives of the needs assessment.
Step 2: Develop a set of questions related to the target population's nutritional problem, how it developed, and/or the factors that influence it.
Step 3: Choose a method for obtaining answers to these questions.

Recall that at this point in the assessment process, the community nutritionist has already obtained information about the community in which the target population lives and the background issues that influence its nutritional and health status. Decide which questions about the target population are most important, which methods can be used to obtain answers to these questions, and whether the answers are readily quantifiable.

Types of Data to Collect about the Target Population

One key activity in community needs assessment is asking questions. In Chapter 2, we asked questions about the community and environmental factors that affect the health and nutritional status of the target population (the outer layers of influence shown in Figure 2-2 on page 41). In this chapter, we pose questions about the lifestyle choices, dietary patterns, working conditions, and social networks that affect the health and nutritional status of the target population (the inner layers shown in Figure 2-2). The sections that follow offer a brief review of how these factors influence the target population's health and nutritional status and of which questions might be asked in these areas.

INDIVIDUAL LIFESTYLE FACTORS

Most community needs assessments are concerned with some aspect of the target population's lifestyle and diet. They ask questions about how the target population's food intake, nutrient intake, and dietary patterns influence their nutritional status.

Lifestyle

Lifestyle factors include such areas as how physical activity level, leisure activities, stress management techniques, smoking status, and drugs and alcohol usage influence health and nutritional status. For example, people who are physically active have lower blood cholesterol concentrations and a lower risk of coronary heart disease (CHD) than sedentary people.[1] Women who become pregnant unexpectedly are less likely to breastfeed their infants than women whose pregnancies are planned.[2] People who are convinced that eating fruits and vegetables protects against cancer are more likely to consume these foods than people who see no positive consequences to eating them.[3] People who smoke have higher body levels of oxidized products, including oxidized ascorbic acid, than nonsmokers.[4] The effects of these and other lifestyle choices on health and nutritional status are complex.

When conducting an assessment, the community nutritionist can ask broad questions about lifestyle: What effect does a particular lifestyle choice have on the target population's nutritional status? Why does the target population behave this way? How can this behavior be changed? Alternatively, she or he can ask specific questions: Do people who begin an exercise program also make other positive lifestyle choices? Do people who routinely have their blood pressure checked also consume low-sodium diets? Do people who consume excessive amounts of alcohol also smoke and eat high-fat and salty foods? The point of these questions is to determine the extent to which a target population's nutritional status is affected by its lifestyle.

Diet

For the community nutritionist, diet is a key individual factor to be analyzed, because nutritional status is affected directly by nutrient intake and utilization and indirectly by the food supply and other factors. The relationships among these factors are shown in Figure 3-1. Along the top of the conceptual model are various stages (national food supply, food intake, and so forth) in which the effects of food and nutrient intake on nutritional status or other health outcomes can be measured. This model is not comprehensive; some

FIGURE 3-1 A Conceptual Model of the Relationships of Food to Health

This conceptual model shows the relationships among food choices, food and nutrient intake, nutritional and health status, and the factors that influence them. It suggests the major areas where measurements of the effects of food and nutrient intake on nutritional and health status can be made. The model is not inclusive, and many factors that affect health status are not shown here.

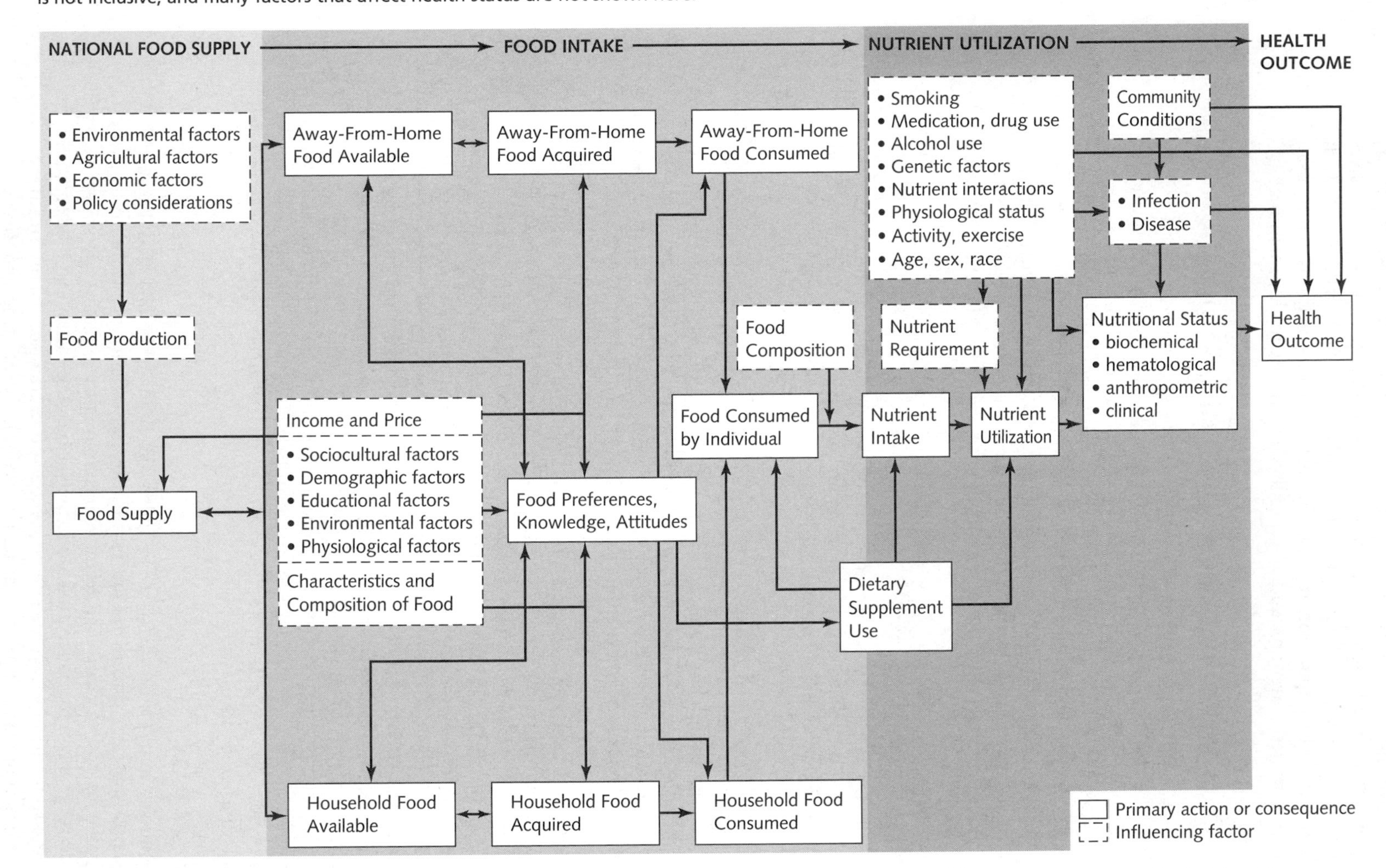

Source: Adapted from Federation of American Societies for Experimental Biology, Life Sciences Research Office, prepared for the Interagency Board for Nutrition Monitoring and Related Research, *Third Report on Nutrition Monitoring in the United States,* Vol. 1 (Washington, D.C.: U.S. Government Printing Office, 1995), p. 4.

factors and interrelationships that influence nutritional and health status are not shown.[5] Even so, the figure serves as a good overview of the areas where the community nutritionist is likely to ask questions about the target population's food-related behavior and nutritional status.

The community nutritionist may ask questions about the food supply and its effects on the target population's nutritional status. The food supply determines which foods are available to the target population. It is a product of the geographical area, climate, soil conditions, labor, and capital available for building the agricultural base. When assessing the food intakes of low-income families, the community nutritionist may ask questions about how foods are distributed throughout the region or how government regulates the food supply. In Zimbabwe, Africa, for example, an assessment found that low-income families lacked good-quality, affordable foods because the government's agricultural policy allocated so much land to crops grown for export. This policy reduced the amount of land available for families to grow their own food, forcing them to buy expensive foods in commercial markets.[6] In this example, the question was: What effect does government agriculture policy have on food availability in this region?

Food intake is influenced by a constellation of biological, psychosocial, cultural, and lifestyle factors, as well as by our personal food preferences, **cognitions, attitudes,** and health beliefs and practices. This stage presents many opportunities for asking questions about the target population. Consider an assessment designed to determine the nutrient intakes of pregnant teenagers and to identify what they perceive as their motivations for making dietary changes during pregnancy.[7] The questions that might be asked about this target population include these: Are the diets of pregnant teenagers similar to those of nonpregnant teenagers? What dietary changes do pregnant teenagers make? Do they experience food cravings and aversions and, if so, to which foods? Do they use dietary supplements? Why do pregnant teenagers change their dietary habits? What are their motivations for increasing or decreasing their food and nutrient intakes? Do they follow the health advice given by their health care providers? If not, why not? What are their health concerns during pregnancy? These questions reflect the many factors shown in Figure 3-1 that influence food choices and food intake.

Activity level, smoking status, dietary supplement use, drug–nutrient interactions, and physiological status (for example, growth or pregnancy) affect nutrient utilization. A community nutritionist involved in an assessment of the dietary practices of male cigarette smokers may have many questions about nutrient utilization in this target population:[8] How does cigarette smoking affect serum retinol, β-carotene, and α-tocopherol concentrations in this group? How does alcohol intake or the use of vitamin supplements affect these variables? Does the number of cigarettes smoked daily affect nutrient utilization? Other questions that might arise in this assessment are linked with health outcome, the final stage shown in Figure 3-1. The questions asked here might focus on nutritional status or some measure of health status: What is the nutritional status of male cigarette smokers? Do male cigarette smokers who take vitamin supplements have a lower incidence of cancer than those who don't take supplements? Do male cigarette smokers with high dietary intakes of vitamins A and E have lower mortality rates than those with low intakes of these vitamins? Whatever the assessment's goals and objectives, many questions can be asked about the important health outcomes for the target population.

Cognitions The knowledge and awareness that people have of their environment and the judgments they make related to it.

Attitude An individual's positive or negative evaluation of performing a behavior or engaging in an activity.

LIVING, WORKING, AND SOCIAL CONDITIONS

The manner in which the target population lives and works affects their health and nutritional status. Education, occupation, and income all have powerful effects on health. Individuals who have few job skills or who are poor and uneducated tend to have more health problems than those with job training and education.[9] Low socioeconomic status is linked with high prevalence rates of chronic conditions, high stress levels, reduced access to medical care for the diagnosis and treatment of diseases, and poor outcomes following treatment.[10] Among children, poverty—even more than family structure and race—has the strongest association with health. Single mothers, black children, and those living below the poverty level are more likely to be in poor health than those living in two-parent families.[11] Low literacy is also a predictor of poor health.[12]

What questions might be asked about socioeconomic status during an assessment? A good example comes from an assessment of the food and nutrition situation of low-income Latino children living in Hartford, Connecticut.[13] The assessment team asked questions about the children's family situation, including the education and employment status of their caretaker, the number of people in the household, and the family's access to a car. A question was posed about the family's use of various household appliances, such as a dishwasher, refrigerator, stove, toaster, television, and microwave. Other questions focused on whether caretakers used emergency food assistance or received food stamps. These questions were designed to obtain information about how these low-income Hispanic children and their caregivers live and how the family's living conditions affected the food intake of these children.

Primary **social groups** such as families, friends, and work groups also influence health and nutritional status. The family, not surprisingly, exerts the most influence over an individual's health; it is the first social group to which an individual belongs, and it is usually the group to which he or she belongs for the longest period of time.[14] The family is a paramount source of values for its members, and its values, attitudes, and traditions can have lasting effects on their food choices and health. For example, Nigerian women feed their newborn infants water from household water pots—a potential source of bacteria that cause infant diarrhea. This practice is passed down from grandmothers and other respected older women to new mothers.[15] Here, the community nutritionist might ask questions about how these practices developed and why women do not perceive the benefits of exclusive breastfeeding to prevent infant diarrhea. Questions involving social networks should be designed to help the nutritionist understand why particular rituals and customs are important and how they influence health and nutritional status.

Case Study 1: Women and Coronary Heart Disease

In this and a later section, we return to the case studies originally described in Chapter 2. We consider the first one here as an example of how the questions asked about the target population are tied to the assessment's objectives. We begin with this case study because it is fairly simple.

Recall that Case Study 1 (summarized in Table 2-1 on page 39) described an assessment aimed at obtaining information about women and coronary heart disease (CHD). The community consisted of the city of Jeffers and four adjoining municipalities. The target population was women over 18 years of age. In the assessment's community phase, data were collected about community morbidity and mortality associated with CHD; types of educational materials available from nonprofit organizations, doctors' offices, and other sources; and existing programs and services designed to reduce CHD risk. The community nutritionist now reviews the assessment's objectives and, for each objective, develops a list of questions about this population's knowledge, attitudes, and practices related to CHD risk. These questions are shown in Table 3-1.

Social group A group of people who are interdependent and share a set of norms, beliefs, values, or behaviors.

TABLE 3-1 Case Study 1: Women and Coronary Heart Disease (CHD)

OBJECTIVE (Refer to Table 2-1)	QUESTION(S) ASKED*	TYPES OF DATA (Refer to Figure 2-2)	METHOD OF OBTAINING THE ANSWER
• Assess women's knowledge and awareness of the leading causes of death and CHD risk factors.	1. Do women know that CHD is the leading cause of death among women in the United States?	Knowledge/awareness (cognitions)	Survey
	2. Can women identify four major risk factors for CHD?	Knowledge/awareness (cognitions)	Survey
	3. Where do women obtain information about health? About CHD?	Health practices	Survey
• Assess women's practices related to reducing CHD risk.	1. Are women eating a heart-healthy diet?	Health practices	24-hour recall
	2. Do women exercise regularly?	Health practices	Survey
	3. Do women smoke? If yes, how many packs/day?	Health practices	Survey
• Assess women's use of existing services designed to reduce CHD risk.	1. Which community services related to CHD do women use?	Community conditions	Survey
	2. What aspects of these services are most important to women?	Attitudes	Survey

*The population in this case study is women over 18 years old living in the fictional city of Jeffers and four adjoining municipalities.

There are two things we should consider when we examine this table. First, not all of the data that might be collected about the target population are shown. Some data, particularly demographic data such as age, education level, and income, are collected as a matter of course. These data allow the community nutritionist to compare the findings of this assessment with those of other assessments or studies. Decisions about which demographic data to collect are made with the help of a statistician. Second, the questions posed in column 2 of the table are asked about the target population as a whole—in this case, all women over 18 years of age who live in the fictional city of Jeffers and four adjoining municipalities. The answers, however, are obtained from individuals—called the **sample**—who represent the target population. Once again, the advice of a statistician is required to ensure that the individuals who are sampled represent the target population.

In this case study, the community nutritionist has developed a list of questions, each one tied to an objective. She then considers the types of data that might be collected to answer the question (knowledge and awareness, for example, and health practices) and chooses a method for obtaining them. Her choices include a survey and a 24-hour recall. What is the purpose for placing this information in a table? The main purpose is to help organize data collection. She sees, for example, that a survey can be used to obtain answers to seven of the questions. In other words, a single survey instrument or questionnaire can be designed to answer all seven questions. The 24-hour recall method should be a separate tool.

Sample A group of individuals whose beliefs, biological characteristics, or other features represent those of a larger population.

This approach is part of the planning process, and it helps streamline data collection. Ways of obtaining answers to questions about the target population are described in the next section.

Methods of Obtaining Data about the Target Population

A variety of methods exist for collecting data related to the target population. The methods range from simple surveys and screening tools to interviews with key informants.

SURVEY

A **survey** is a systematic study of a cross section of individuals who represent the target population. It is a relatively inexpensive way to collect information from a large group of people. Surveys can be used to collect qualitative or quantitative data in formal, structured interviews; by phone, mail, or online; or from individuals or groups. Some survey instruments (such as questionnaires) are self-administered, whereas a trained interviewer administers others.[16]

Designing a questionnaire and conducting a survey involve more than heading out with a clipboard and a list of questions to interview people as they come into your clinic, office, or community center. Survey design and analysis is a discipline in itself, and the process of conducting a survey usually requires a team of experts with knowledge of survey research, statistics, epidemiology, public health, and nutrition. Although a detailed discussion of survey methodology is beyond the scope of this book, a few comments are in order about the issues to consider when designing a **nutrition survey** or adapting an existing survey for your own purposes.

"Planning a survey consists of making a series of scientific and practical decisions."[17] The first step is to determine the purpose of the survey. Most nutrition surveys are carried out to assess the food consumption of households or individuals, evaluate eating patterns, estimate the adequacy of the food supply, assess the nutritional quality of the food supply, measure the nutrient intake of a certain population group, study the relationship of diet and nutritional status to health, or determine the effectiveness of an education program.[18] A nutrition survey does not have to be complex and gargantuan to be meaningful, but it must have a well-defined purpose.

Readability can be assessed by formulas such as the SMOG formula (see Appendix C) and the Fog Index or by computer programs designed for this purpose.

Next, decisions must be made about who will design the survey, who will conduct it, and how it will be carried out. The survey instrument must be designed and pretested, and the sample must be chosen. The personnel responsible for conducting the survey and analyzing data must be trained. Numerous other decisions must be made about the feasibility of the survey; the quality of data obtained by the survey; the costs of carrying it out, the readability of the survey, literacy issues, and the manner in which data will be analyzed and used. At every step in the planning process, there are practical constraints related to time and money.[19]

Surveys are important tools in assessing the health and nutritional status of individuals, but they must be designed and carried out carefully to provide valid and reliable information. Consult Table 3-2 for a list of questions to ask when designing a survey for your community needs assessment.

Survey A systematic study of a cross section of individuals who represent the target population.

Nutrition survey An instrument designed to collect data on the nutritional status and dietary intake of a population group.

Health Risk Appraisal

The health risk appraisal (HRA) is a type of survey instrument used to characterize a population's general health status. We mention it here in a separate category because it is widely used in worksites, government agencies, universities, and other organizations as a health education or screening tool. The HRA is a kind of "health hazard chart" that asks questions about the lifestyle factors that influence disease risk,[20] and it has been used successfully to improve health behaviors.[21]

TABLE 3-2 Questions to Ask When Designing a Survey

- **Is the survey valid and reliable?** Will it measure what it is intended to measure, and, assuming that nothing changed in the interim, will it produce the same estimate of this measurement on separate occasions?
- **Are norms available?** That is, are reference data or population standards available against which the data from your target group can be compared?
- **Is the survey suitable for the target population?** A survey designed to obtain health and nutrition data on free-living elderly people may not be appropriate for the institutionalized elderly.
- **Are the survey questions easy to read and understand?** Survey questions must be geared to the target population and its level of literacy, reading comprehension, and fluency in the primary language. Having a readable survey is especially important if it is to be self-administered.
- **Is the format of the questionnaire clear?** If the questionnaire is not laid out carefully, respondents may become confused and inadvertently skip questions or sections.
- **Are the responses clear?** A variety of scales and responses may be used in designing a survey. Some questions may require filling in blanks or providing simple yes/no or true/false answers. Others may ask respondents to rank-order their responses from "seldom/never use" to "use often/always." The trick when selecting such scales is to choose one that allows you to discriminate between responses but doesn't provide so many categories that respondents are overwhelmed.
- **Is the survey comprehensive but brief?** Often the length of the survey must be limited to ensure that respondents complete it within a reasonable time frame. With long questionnaires, respondents are likely to answer questions hurriedly and mark the same answer to most questions.
- **Does the survey ask "socially loaded" questions?** Each survey question should be evaluated for how it is likely to be interpreted. Questions that imply certain value judgments or socially desirable responses should be rewritten. This is especially important when dealing with respondents from cultures other than your own.

Source: Adapted from L. Fallowfield, *The Quality of Life* (London: Souvenir, 1990), pp. 40–45.

The HRA instrument consists of three components: a questionnaire, certain calculations that predict risk of disease, and an educational message or report to the participant.[22] A typical HRA asks questions about age, sex, height, weight, marital status, size of body frame, exercise habits, consumption of certain foods (for example, fruits and vegetables) and ingredients (such as sodium), intake of alcohol, job satisfaction, hours of sleep, smoking habits, and medical checkups or hospitalizations. An HRA has been developed and tested for people aged 55 years and older.[23] A portion of the Healthier People Network's questionnaire is shown in Figure 3-2 on page 68.

HRAs are used to alert individuals to any risky health behaviors that they engage in and to inform them how such behaviors might be modified, usually through a lifestyle modification program.[24] For example, a health risk questionnaire was used at a trucker trade show to assess the health status of truck drivers. Truckers who stopped at the trade show enjoyed free food, music, raffles, and other events. At one booth, truckers received free blood pressure measurements and health education materials and were asked to complete a survey of their health risk factors, health status, and driving patterns. An analysis of the survey results showed that truck drivers tend to smoke cigarettes, to be sedentary and overweight, and to be unaware of it if they have high blood pressure.[25]

SCREENING

Screening is an important preventive health activity designed to reverse, retard, or halt the progress of a disease by detecting it as soon as possible. Screening occurs in both clinical practice and community settings, and it entails procedures that are safe, simple, and cheap.

FIGURE 3-2 A Portion of a Health Risk Appraisal Form

Source: Used with permission of The Healthier People Network, Inc.

The HEALTHIER PEOPLE NETWORK, Inc.

. . . linking science, technology, & education to serve the public interest . . .

IDENTIFICATION NUMBER

The health risk appraisal is an educational tool, showing you choices you can make to keep good health and avoid the most common causes of death (for a person of your age and sex). This health risk appraisal is **not** a substitute for a check-up or physical exam that you get from a doctor or nurse; however, it does provide some ideas for lowering your risk of getting sick or injured in the future. It is NOT designed for people who already have HEART DISEASE, CANCER, KIDNEY DISEASE, OR OTHER SERIOUS CONDITIONS; if you have any of these problems, please ask your health care provider to interpret the report for you.

DIRECTIONS:
To get the most accurate results, **answer as many questions as you can**. If you do not know the answer leave it blank.

The following questions must be completed or the computer program cannot process your questionnaire:

1. SEX 2. AGE 3. HEIGHT 4. WEIGHT 15. CIGARETTE SMOKING

Please write your answers in the boxes provided. (Examples: ☒ or 98)

1.	**SEX**	1 ☐ Male 2 ☐ Female
2.	**AGE**	☐ Years
3.	**HEIGHT** (Without shoes) (No fractions)	☐ Feet ☐ Inches
4.	**WEIGHT** (Without shoes) (No fractions)	☐ Pounds
5.	Body frame size	1 ☐ Small 2 ☐ Medium 3 ☐ Large
6.	Have you ever been told that you have diabetes (or sugar diabetes)?	1 ☐ Yes 2 ☐ No
7.	Are you now taking medicine for high blood pressure?	1 ☐ Yes 2 ☐ No
8.	What is your blood pressure now?	
★38.	Do you eat some food every day that is high in fiber, such as whole grain bread, cereal, fresh fruits or vegetables?	1 ☐ Yes 2 ☐ No
★39.	Do you eat foods every day that are high in cholesterol or fat, such as fatty meat, cheese, fried foods, or eggs?	1 ☐ Yes 2 ☐ No
★40.	In general, how satisfied are you with your life?	1 ☐ Mostly satisfied 2 ☐ Partly satisfied 3 ☐ Not satisfied

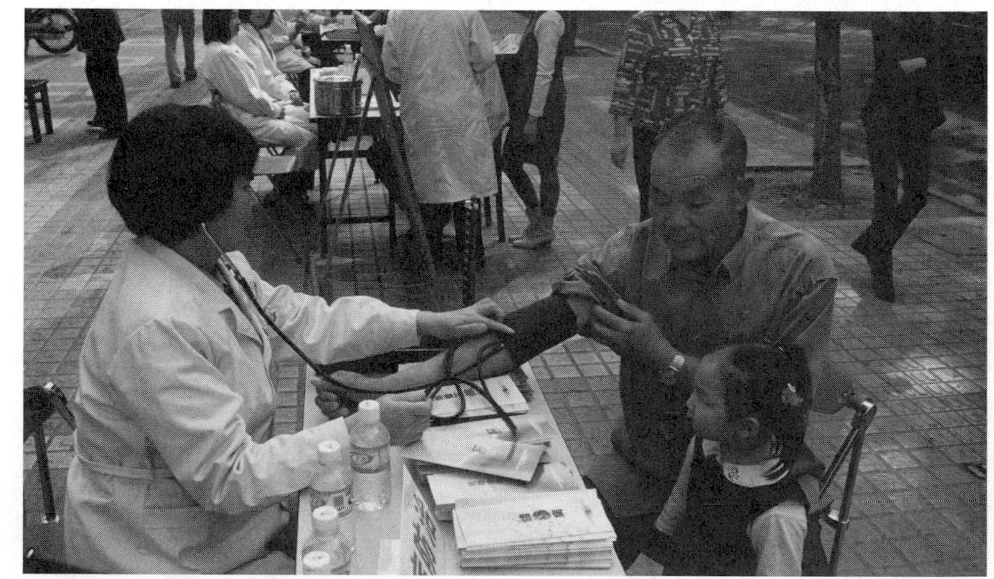

David Butow/CORBIS SABA

Screening is one method of identifying people with high blood pressure. People whose screening value indicates high blood pressure are referred for medical diagnosis and treatment, which may include nutrition counseling.

TABLE 3-3 Common Screening Procedures in Clinical Practice and Community Settings

SCREENING PROCEDURE	TARGET POPULATION
• **Clinical practice**	
Taking a medical history	All ages, both sexes
Height and weight	All ages, both sexes
Phenylketonuria (PKU)	Newborn infants
Posture (for detection of scoliosis)	Children over 3 years of age
Vision	Children over 3 years of age
Hearing	Children over 3 years of age
Tuberculin test	Children over 1 year of age
• **Community settings**	
Health risk appraisal	Primarily adults over 18 years of age
Blood pressure	Adults over 18 years of age
Blood cholesterol level	Adults over 18 years of age

Sources: Adapted from J. M. Last, *Public Health and Human Ecology,* 2nd ed. (New York, NY: McGraw-Hill Professional, 1998); and *The Merck Manual,* ed. M. Beers and R. Berkow, 17th ed. (Whitehouse Station, NJ: Merck Research Laboratories, 1999).

Table 3-3 lists some common screening procedures. In community screening programs, people from the community are invited to have an assessment made of a health risk or behavior. Their screening value is then compared with a predetermined cutpoint or risk level. For example, during Heart Month, a popular shopping mall sponsors a health fair that offers screening for high blood cholesterol concentrations. People who come through the screening booth give a finger-stick blood sample, which is analyzed on site by a machine designed for this purpose. Their blood cholesterol concentrations are classified as high, borderline-high, or desirable, on the basis of the classification scheme developed by the National Cholesterol Education Program. Individuals whose screening value suggests an elevated risk are referred for medical diagnosis and treatment. Screening programs are not meant to substitute for a health care visit or routine medical monitoring for people already receiving treatment, but they do have educational value and serve to identify high-risk persons.[26]

Screening programs abound. One example is the Nutrition Screening Initiative (NSI), a project designed to promote nutrition screening and improved nutritional care in the United States, especially among older adults. The NSI uses an educational tool—the one-page DETERMINE Checklist—to identify persons at increased risk of poor nutritional status. The acronym DETERMINE is used to help remind respondents of the warning signs for nutritional risk: Disease, Eating poorly, Tooth loss/mouth pain, Economic hardship, Reduced social contact, Multiple medicines, Involuntary weight loss/gain, Needs assistance in self-care, and Elder years above age 80. The checklist consists of a set of basic questions that are written at the fourth- to sixth-grade reading level and address the categories of nutritional risk specified by the NSI. The checklist can be self-administered or completed by anyone who interacts with older family members or friends. It helps increase awareness of the factors that influence nutritional health.[27]

More information about the Nutrition Screening Initiative is provided in Chapter 12. (See Figures 12-4 and 12-5.)

FOCUS GROUPS

One method of obtaining information about the target population is to hold focus group interviews. **Focus groups** usually consist of 5 to 12 people who meet in sessions lasting about 1 to 3 hours. The group members are brought together to talk about their concerns, experiences, beliefs, or problems. Focus groups are used to obtain advice and insights about new products and services, research data and information about key variables used

Focus group An informal group of about 5 to 12 people who are asked to share their concerns, experiences, beliefs, opinions, or problems.

FIGURE 3-3 Key Focus Group Questions

Adolescents participating in a focus group designed to assess their perceptions of factors influencing their food choices were first asked to complete a worksheet on which they recorded what they ate over the previous 24-hour period and their reasons for choosing the foods that they ate. ("Why did you choose those foods and not others?") Afterward, there was a focus group discussion regarding factors influencing their food choices, while the interviewer recorded these factors on a flip chart. The focus group questions are listed here.

Source: D. Neumark-Sztainer, M. Story, C. Perry, and M. A. Casey, Factors influencing food choices of adolescents: Findings from focus-group discussions with adolescents, *Journal of the American Dietetic Association* 99 (1999): 930.

- Is one of the factors you mentioned as influencing your food choices particularly strong? Does this change from one time to another? In what situations does the reason you eat sometimes change?
- I hear you talking a lot about fast foods. Why do you like to east fast foods?
- People are different. Our ethnicity, religion, or family traditions may influence what we eat. Do these things influence what you eat? If so, how?
- Here is a list of some national guidelines for how we should eat: Increase fruit and vegetable intake. Increase dairy food intake. Decrease fat intake. Why don't kids eat like this? What makes it hard to eat like this? What makes it easier to eat like this?
- Would you serve fruits and vegetables instead of or in addition to potato chips at a party? Why or why not?
- Think about any changes you've made in what you eat over the last few years. What triggered that change? What brought it about?
- Our goal is to help young people eat better foods. What advice do you have for us?

in quantitative studies, and opinions about products or creative concepts such as advertising campaigns or program logos.[28]

A trained moderator who is skilled at putting people at ease and promoting group interaction leads focus group sessions. Listening is the most important skill used during focus group interviews. Like a good teacher, the focus group leader must be able to listen on several levels, concentrating on what participants say—and do not say—*and* on the progress being made during the interview. A good session leader explores a topic without making participants feel guilty or defensive, avoids asking leading questions, doesn't interrupt participants when they are talking, pays attention to nonverbal cues such as body language, and keeps participants focused on the session's topic. Asking open-ended questions—for example, "How do you make decisions about buying milk, cheese, and other dairy products?"—allows participants to take any direction they want and to reconstruct particular experiences.[29] Key focus group questions used by researchers to assess adolescents' perceptions of factors influencing their food choices are listed in Figure 3-3. The information obtained from focus group interviews is then used to provide direction for the needs assessment or to change a marketing strategy, product, or existing program.[30]

Focus group interviews provide qualitative information that helps community nutritionists understand how an existing nutritional problem developed and whether the target population perceives a problem. They are less expensive to conduct than face-to-face interviews, and they help community nutritionists obtain information of a sensitive nature or information that might otherwise be difficult or costly to get.[31] The Johns Hopkins Hospital staff, for example, used focus groups to help them understand why some clients were not complying with cardiovascular health promotion programs to control hypercholesterolemia. Participants were asked questions about health, high blood cholesterol, current diet, food preferences, grocery shopping, and the types of education programs and materials they usually used. The staff learned that clients preferred to be taught by professionals in small groups and liked to hear from individuals who had successfully lowered their blood cholesterol. The focus group interviews helped the assessment team understand why a certain behavior developed; they also suggested ways to address the problem.[32]

INTERVIEWS WITH KEY INFORMANTS

Key informants—the people "in the know" about the community—can also provide information about the target population. Key informants may have worked with the target population in the community or in some other part of the world, or they may have conducted research related to the population's health status, beliefs, or practices and may thus be familiar with its attitudes and opinions about the nutritional problem. Informant interviews can be used to complete a cultural assessment of the target population, a process described later in this chapter. Interviews with key informants may also reveal whether the target population perceives a nutritional problem and which actions for addressing the problem are culturally appropriate. For example, if the target population is obese persons, then interviews with physicians in family practice, internal medicine, and endocrinology will provide information about how this group is managed medically, what advice it is given about weight management, and whether it follows advice received from physicians.[33] If the target population is teenagers with a high risk of HIV/AIDS, then interviews with teenagers who work in peer education programs will provide valuable information about how these high-risk teenagers perceive risky behavior, what they know about preventing and transmitting HIV/AIDS, how susceptible they are to peer pressure, and what their overall health beliefs are.[34]

DIRECT ASSESSMENT OF NUTRITIONAL STATUS: AN OVERVIEW OF METHODS

Another method of determining nutritional needs is to conduct a direct assessment of individuals. Direct assessment methods use dietary, laboratory, anthropometric, and clinical measurements of individuals to identify those with malnutrition or a nutritional deficiency state. These methods may be used alone or in combination.

In her comments at the Second International Conference on Dietary Assessment Methods, Elizabeth Helsing, of the World Health Organization Regional Office for Europe, remarked that "we still have a lot to learn from one another in the area of dietary assessment about the slow and arduous process of collecting vast amounts of data, the battle to mold them into some shape, and finally the wrestling of meaning from them."[35] Her comments allude to the challenge of collecting meaningful information about what people eat and why they make the food choices they do.

Dietary methods are used to determine an individual's or population's usual dietary intake and to identify potential dietary inadequacies. Dietary inadequacies represent stage 1 of the nutrient depletion scheme shown in Table 3-4. The primary

TABLE 3-4 General Scheme for the Development of a Nutritional Deficiency

STAGE	DEPLETION STAGE	METHOD(S) USED TO IDENTIFY
1	Dietary inadequacy	Dietary
2	Decreased level in reserve tissue store	Biochemical
3	Decreased level in body fluids	Biochemical
4	Decreased functional level in tissues	Anthropometric/biochemical
5	Decreased activity in nutrient-dependent enzyme	Biochemical
6	Functional change	Behavioral/psychological
7	Clinical symptoms	Clinical
8	Anatomical sign	Clinical

Source: Reprinted with permission from D. E. Sahn, R. Lockwood, N.S. Scrimshaw, *Methods for the Evaluation of the Impact of Food and Nutrition Programmes.* Tokyo: United Nations University, 1984.

TABLE 3-5 Issues to Consider When Selecting a Method to Assess Dietary Intake

Program Objectives
- Degree of accuracy needed
- Type of intake data needed
  - Food intake
  - Nutrient intake
  - Dietary pattern
  - Food pattern

Study Population
- Sample size needed
- Ability of respondents
  - Age
  - Literacy level
  - Language skill
- Willingness to cooperate
- Time constraints

Financial Issues
- Cost of analysis software
- Cost of training staff to use software
- Cost of intake forms (e.g., printed versus computer-generated)
- Cost of analyzing records (e.g., data entry by humans versus computer-scanned data)
- Number of interviewers and coders to process data

Implementation Requirements
- Time required for respondents to complete intake form
- Ease of completion of intake form
- Need for support materials (e.g., instruction forms, food records, scales for weighing food, measuring cups)
- Training and skill level of interviewer

Analysis Requirements
- Quality of nutrient database
- Quality of analysis software
- Training and skill level of food coder
- Level of quality control

Source: Adapted from J. M. Karkeck, Improving the use of dietary survey methodology, *Journal of the American Dietetic Association* 87 (1987): 869–71.

methods of measuring the food consumption of individuals include the 24-hour recall method, the food diary or daily food record, the diet history interview, and the food frequency questionnaire. At the household level, food records and inventory methods are used to estimate food consumption; population dietary intakes are estimated using food balance sheets and market databases (discussed in Chapter 4).[36] The method chosen depends on many factors, some of which are shown in Table 3-5. All have strong and weak points—and advantages and disadvantages. Table 3-6 provides a summary of dietary assessment tools. The trick is to choose the most valid method, given the financial resources and personnel available to collect and analyze the dietary intake information.[37]

METHODS	STRENGTHS	LIMITATIONS
Client assessment questionnaire/historical data form: A preliminary nutritional assessment form usually divided into sections for administrative data, medical history, medication data, psychosocial history, and food patterns	Provides clues to strengths and potential barriers	May seem invasive, may not be culturally sensitive
Food diary/daily food record: A written record of the food and beverages consumed by an individual over a period of time, usually 3 to 7 days	Does not depend on memory Provides accurate intake data Provides information about food habits	Requires literacy Requires a motivated client Recording process may influence food intake Requires ability to measure/judge portion sizes Time-consuming
Twenty-four-hour recall: A dietary assessment method in which an individual is requested to recall all food and beverages consumed in a 24-hour period	Quick Easy to administer No burden for respondent Does not influence usual diet Literacy not required	Relies on memory May not represent usual diet Requires ability to judge portion sizes Under/overreporting occurs
Food frequency: A method of analyzing a diet based on how often foods are consumed (i.e., servings per day/week/month/year)	Furnishes overall picture of diet Not affected by season	Requires ability to judge portion sizes No meal pattern data
Usual diet: Clients are led through a series of questions to describe the foods typically consumed in a day	May be more of a typical representation than a 24-hour recall	Not useful if diet pattern varies considerably
Diet history interview: A conversational assessment method in which clients are asked to review their normal day's eating pattern	Provides clarification of issues	Relies on memory Requires interview training

Source: K. Bauer and C. Sokolik, *Basic Nutrition Counseling Skill Development* (Belmont, CA: Wadsworth/Thomson Learning, 2002), p. 96.

TABLE 3-6 Summary of Methods, Strengths, and Limitations of Selected Dietary Assessment Tools

Diet History Method

Bertha Burke, with the Department of Child Hygiene of the Harvard School of Public Health, reported one of the earliest descriptions of diet analysis. In studies of pregnant women and their infants and children, conducted during the 1930s, Burke and her colleagues developed a diet history questionnaire to assess usual dietary intake. Their studies showed statistically significant relationships between a mother's diet during pregnancy and the condition of her infant at birth. In addition, there was a correlation between the dietary ratings of the children's diets and objective measures of nutritional status, such as hemoglobin concentration.[38]

The diet history method has the advantage of being easy to administer, although it is time-consuming and requires a trained interviewer. For these reasons, it is not practical for large population studies. A true validation of the diet history method is probably not possible, although the method allows for reasonable confidence in classifying respondents according to some preset dietary criteria (for example, the Dietary Reference Intakes).[39] When undertaken repeatedly at different times, this method is fairly reliable. It is perhaps best used to provide qualitative, not quantitative, data.

See Figure 6-3 on page 175 for a description of the *multiple-pass* 24-hour dietary recall, which is designed to limit the extent of under-reporting of food intake in national surveys.

Twenty-four-Hour Recall Method

The 24-hour recall method is one of the most widely used diet assessment methods. It is easy to administer in person or by phone and lends itself to large population studies, mainly because it requires little time from either the respondent or the interviewer.

Its appropriateness for assessing the intake of individuals—what is called validity, or the ability of the instrument to measure what it is intended to measure—has repeatedly been questioned, however.[40] There are indications that 24-hour recall data are subject to recall bias (that is, respondents cannot accurately recall the foods they ate during the previous 24-hour period and thus either overestimate or underestimate their dietary intakes). Gender differences in recalling dietary intakes have also been reported. Thus, a single 24-hour recall does not provide an accurate estimate of an individual's usual dietary intake.

The **validity** of the 24-hour recall can be improved by administering repeated recalls. Seven or eight 24-hour recalls of an individual's dietary intake over a period of two or three weeks are more likely than a single recall to provide a reasonable estimate of that person's usual intake. Even so, 24-hour recall data are best suited to describing the intakes of populations, not individuals.[41]

Diet Record Method

Diet records or food records have been considered the "gold standard" of diet assessment methods. Completed over a period of three, four, or seven days, or even as long as a year, they have the advantage of providing detailed information about food products, including brand names and methods of food preparation, and they eliminate the uncertainty that goes with trying to recall the foods eaten. The amount of food consumed can be estimated or calculated by weighing. If the respondents are properly trained, diet records give a reasonably accurate picture of usual dietary intake. However, the diet record method replaces errors in recall with errors in recording, and the possibility always exists that the act of recording food intake changes the actual foods chosen for recording (for example, choosing meals that are easy to record or not snacking in order to ease the task of recording). Accurate food records can be obtained if the respondents are highly motivated, literate, and well trained.

Food Frequency Method

One of the first large-scale uses of the food frequency questionnaire was in the Nurses' Health Study, a study of a cohort of more than 95,000 female registered nurses being followed for the occurrence of coronary heart disease and cancer. A semiquantitative food frequency questionnaire—sometimes called "Willett's FFQ" after the study's principal investigator, Dr. Walter C. Willett—was developed to categorize individuals by their intake of selected nutrients (for instance, vitamin A, vitamin C, and animal fat). The validity and precision of this instrument were evaluated using four sets of seven-day food records and a one-year diet record.[42]

Other food frequency questionnaires besides Willett's are available. Most have been tested and used with large groups of people participating in epidemiologic studies, where the use of more labor-intensive instruments such as food records is not practical. A few food frequency questionnaires have been used with varying success among minority populations in the United States.[43]

The food frequency questionnaire offers several advantages: It is self-administered, requires only about 15 to 30 minutes to complete, and is analyzed at a reasonable cost. Thus it has been used in a variety of population studies where no other instrument could have been administered practically. It suffers, however, from the same limitations as any other recall method, in that the accurate reporting of intake depends on memory. In addition, the food list must necessarily be short and thus may fail to include some foods commonly consumed by the population being surveyed. Controversy over the appropriate uses of the food frequency questionnaire continues.[44]

Validity The accuracy of the diet assessment instrument. Validity reflects the ability of a diet assessment instrument to measure what it is intended to measure; that is, a valid instrument accurately measures an individual's usual or customary dietary intake over a period of time.

TABLE 3-7 Questions to Explore Cultural Food Behavior

Questions to aid in understanding of food habits and to assist in completing a dietary assessment:

Traditional Foods

- What traditional foods do you eat?
- What is your favorite cultural or traditional food?
- How often do you eat traditional foods?
- If you do not eat them, why not?

Foods and Health

- Are there foods you won't eat? Why?
- Can what you eat help cure your sickness? Or make it worse?
- Are there foods you eat to keep healthy? To make you strong?
- Are there foods you avoid to prevent sickness?
- Do you balance eating some foods with other foods?

New Foods

- What new foods have you tried since coming to this community?
- Do you eat them regularly?
- Which foods do you dislike, and why?

Food Acquisition

- Where do you get most of your family's food? (Examples: neighborhood supermarket, convenience store, open market)
- How do you get to the market? Does anyone go with you? Do they speak English?

Amount and Quality of Food

- How do you describe the quality of the food you buy?
- Do you have enough food to feed your family each day?
- How do you divide up the food among family members if you do not have enough?
- Are you able to get the types of foods and beverages needed by everyone in your family?

Food Preparation

- Do you have enough time to prepare the kinds of foods your family enjoys?
- Do you have the equipment you need for cooking and preparing the kinds of foods your family likes to eat?
- Do you have any trouble preparing food?

Family Interaction Around Food

- Do the children in your family like the foods enjoyed by the adults?
- Do your school-age children like the school meals?
- Do you have recipes that your family enjoys?

Source: Adapted from D. E. Graves and C. W. Suitor, *Celebrating Diversity: Approaching Families Through Their Food* (Arlington, VA: National Center for Education in Maternal and Child Health, 1998).

Other Diet Assessment Methods

One recent innovation in the diet assessment arena is the use of photography to record dietary intake. Photography reportedly has provided more valid and precise results than food records.[45] Another method of estimating food intake is the picture-sort approach, in which participants sort into categories various cards on which pictures or drawings of foods appear. The method was developed for use with a diverse population of older adults with low education or literacy levels. It can be self- or interviewer-administered. When tested among participants in the Cardiovascular Health Study, the method was an easy and quick way to obtain data on food patterns.[46] Table 3-7 shows questions for exploring cultural food behavior when completing a dietary assessment. Computers, not surprisingly, are also used to conduct personalized interviews and collect data about dietary intake.[47]

Laboratory Methods

Laboratory methods can be used to identify individuals at risk of a nutrient deficiency (stages 2 to 5 in Table 3-4), because tissue stores of nutrients gradually become depleted over time. The depletion may result in alterations in the level or activity of some nutrient-dependent enzymes or in the levels of metabolic products. Thus laboratory methods are used to detect subclinical deficiencies. Static biochemical tests measure a nutrient in biological tissues or fluids or the urinary excretion rate of the nutrient. Examples of static tests include the platelet concentration of α-tocopherol and urinary 3-hydroxyproline excretion. Functional biochemical tests measure the biological importance of a nutrient and the consequences of the nutritional deficiency. Functional biochemical tests include taste acuity, a measure of zinc status; dark adaptation, a measure of vitamin A status; and capillary fragility, a measure of vitamin C deficiency. Functional tests are generally too invasive and expensive to employ in most field surveys of nutrition status.[48]

Anthropometric Methods

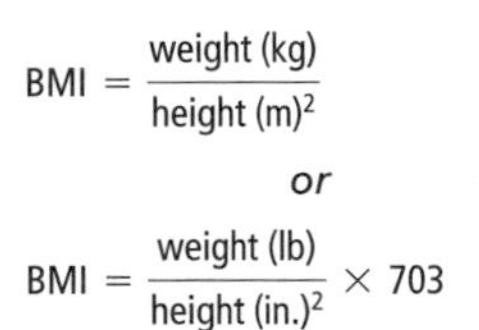

BMI can also be determined from the BMI table on the inside back cover of this text.

Measurements of the body's physical dimensions and composition are used to detect moderate and severe degrees of malnutrition and chronic imbalances in energy and protein intakes. The most common growth indices are measurements of stature (height or length), weight, and circumference of the head. Measurements of skinfolds, mid-upper-arm circumference, and waist circumference are used to derive equations that predict muscle and fat mass. The body mass index is a commonly used indirect measure of overweight and obesity in adults.[49] Anthropometric measurements are useful in large-scale community assessment programs, because they involve simple, safe, noninvasive procedures; require inexpensive and portable equipment; and produce accurate and precise data when obtained by trained personnel. Such measurements are used to estimate an individual's long-term nutritional history. They do not provide data on short-term nutritional status, nor can they provide information about specific nutritional deficiencies.[50]

Clinical Methods

Clinical assessment of health status consists of a medical history and a physical examination to detect physical signs and symptoms associated with malnutrition. The medical history includes a description of the individual and his or her living situation (for example, married or single, number of children, and nature of employment). It typically obtains information about existing clinical conditions, previous bouts of illness, presence of congenital conditions, smoking status, existence of food allergies and intolerances, use of medications, and usual levels of physical activity. In the physical examination, the clinician evaluates the major organic systems: skin, muscular and skeletal, cardiovascular, gastrointestinal, and nervous. The hair, face, eyes, lips, tongue, teeth and gums, and nails are also examined for signs associated with malnutrition.[51]

Issues in Data Collection

The choice of method for obtaining information about the target populations is influenced by practical, scientific, and cultural considerations. These issues are described in this section.

Practical Issues

The choice of assessment method is influenced by practical issues such as the number of staff available to collect and analyze data, the cost of administering the test, and the amount of time needed to identify and interview or sample members of the target population. For example, it may be impractical to use food diaries to collect data about the food patterns of low-income, immigrant women and their children who live in a city's public housing. This particular population may read English poorly, be uncomfortable with record keeping, and lack transportation to the clinic for training on how to keep food records. An interviewer-administered 24-hour dietary recall is a better choice of diet assessment method with this group.

The assessment method chosen should be simple to administer, take only a few minutes to complete, be inexpensive, and be safe. Blood pressure measurements are ideal screening tests, for example, because they are cheap, quick, need little advance preparation and setup, and are not uncomfortable or unsafe for participants.

Scientific Issues

Scientific issues such as the validity and reliability of the assessment method and the nature of dietary variation also influence the choice of assessment method. Key scientific issues to consider when choosing an assessment method are described here.

SENSITIVITY VERSUS SPECIFICITY

Two issues to consider when choosing an assessment method are sensitivity and specificity. **Sensitivity** is the proportion of subjects with the disease or condition who have a positive test result. A sensitive test rarely misses people with the disease or condition, and it is often used in screening situations in which the purpose of the test is to detect a disease or condition in people who appear to be asymptomatic.[52] For example, a screening test that uses phlebotomy to detect high blood cholesterol concentrations is more sensitive—that is, it is more likely to identify an individual's true blood cholesterol concentration and to yield a positive result in the presence of a high blood cholesterol concentration—than a finger-prick cholesterol test.

Specificity is the proportion of subjects without the disease or condition who have a negative test. Specific tests are used to confirm a diagnosis, and they are important tests to administer when not properly identifying a disease or condition might harm the patient or subject. The oral glucose tolerance test is a highly specific test for diagnosing diabetes mellitus.

Ideally, it is desirable to have an assessment method that is both highly sensitive and highly specific, but this is seldom possible in practice. The trick is to choose a method that correctly classifies subjects into a particular group and misclassifies few of them.

VALIDITY AND RELIABILITY

Two questions arise when evaluating assessment methods: One concerns the instrument's validity and the other its reliability, or reproducibility. An assessment instrument's validity is its ability to measure what it is intended to measure. Another word for validity is *accuracy*. In the case of a diet assessment tool, a valid instrument accurately measures an individual's usual or customary dietary intake over a period of time—one day, three days, seven days, one month, one year. An instrument's validity can be affected by many factors:[53]

- Characteristics of the respondent—literacy level, education level, conscientiousness in completing the instrument, ability to follow instructions

Sensitivity The proportion of individuals in the sample with the disease or condition who have a positive test for it.

Specificity The proportion of individuals in the sample without the disease or condition who have a negative test for it.

TABLE 3-8 Examples of Survey Questions Designed to Obtain Information about Fiber and Fat Intake

SURVEY QUESTION	RESPONSE
• Not counting juices, how many fruits do you usually eat per day or week?	________ fruits per ________ day, week
• How often do you use fat or oil in cooking?	________ times per ________ day, week, month
• How many servings, on average, do you eat of mayonnaise or salad dressing (serving size = 2 tablespoons)?	________ never or < 1 per month ________ 1–3 per month ________ 1–4 per week ________ 5–7 per week ________ 2 or more per day

Source: The first two questions were taken from the National Cancer Institute's Health Habits and History questionnaire (Bethesda, MD: National Cancer Institute, 1987), p. 4. The third question was adapted from the University of Minnesota Healthy Worker Project questionnaire (Minneapolis: University of Minnesota, 1987), p. 3.

- Questionnaire design—difficulty of instructions, ease of recording intake, number and types of foods listed, portion sizes given (if any)
- Adequacy of reference data—obtaining a sufficient number of days of intake data to estimate the "true" nutrient intake; completeness of the nutrient database used to calculate nutrient intake
- Accuracy of data input and management—quality control of keypunching or coding of food items

Consider the validity of a survey instrument used during a telephone interview or mailed to respondents. One threat to the survey's validity is ambiguous language in the survey questions.[54] Writing a survey question that is easy for respondents to understand *and* that yields the health or nutrition information you want is no simple task. Examine the questions in Table 3-8. The first question is fairly straightforward. Most consumers know what fruits are, and they can estimate their fruit intake with a reasonable degree of accuracy. The second question is more difficult to answer. Respondents must know what a fat is and must recognize butter, stick margarine, tub margarine, lard, fatback, bacon fat, and vegetable oils and shortening as fats used in cooking. They must then estimate how frequently they use one or more of these fats in cooking, a complex calculation for most consumers. The third question likewise presents a problem. Respondents must decide what the survey question means by mayonnaise and salad dressing. Is the question asking about all such products, be they low-fat, full-fat, or low-cholesterol, or is it asking only about regular, full-fat mayonnaise and salad dressings? This third question thus illustrates that some survey instruments used to assess dietary intake are not valid because only full-fat foods are indicated on the survey form, and respondents cannot indicate that they usually consume reduced-fat or low-fat foods. These survey instruments tend to overestimate fat intake.

An individual's true usual diet cannot be known with certainty. It is not possible to follow respondents around all day and night and surreptitiously record every morsel they consume. And the very act of recording food intake can have a subtle influence, because the respondents may make choices they might not have made otherwise. (Although people's food and beverage intake can be monitored with precision on a metabolic ward, their dietary intake cannot be considered "usual" under these circumstances.) However, doing everything possible to create and/or use an instrument high in validity ensures that respondents can be placed with great accuracy along a distribution of intake, from low to high consumption.

TABLE 3-9 Factors to Consider during a Cultural Assessment

Religion

- **Belief system**—Do this population's religious beliefs affect their food choices, use of alcohol, or other lifestyle decisions?
- **Food rituals**—Are particular foods consumed only during religious events/periods? Are certain foods taboo?

Etiquette and Social Customs

- **Typical greeting**—What is a proper form of address? Is a handshake appropriate? Are shoes worn in the home?
- **Social customs**—Should certain social exchanges occur before the interview or other assessment is undertaken? Are refreshments offered?
- **Direct and indirect communications**—Should a senior household member be expected to answer a question before a junior member does? Are some questions considered taboo by this ethnic population?

Nonverbal Communications

- **Eye contact**—Is eye contact considered polite or rude?
- **Tone of voice**—Does a soft voice have a particular meaning in this culture? A hard voice?
- **Facial expressions**—What do smiles and nods mean?
- **Gestures**—Are certain hand gestures considered rude and offensive? Is it acceptable to cross your legs?
- **Personal space**—Is the realm of personal space wider or narrower than in the North American culture?
- **Touch**—Is touch appropriate? If so, when, where, and by whom?

Source: Adapted from M. C. Narayan, Cultural assessment in home healthcare, *Home Healthcare Nurse* 15 (1997): 665.

The second concern is the **reliability** of an assessment instrument—that is, its ability to produce the same estimate of dietary intake, for example, on two separate occasions, assuming that the diet did not change in the interim. (Other words for reliability are *precision*, *repeatability*, and *reproducibility.*) This issue is different from validity. It is possible for an instrument to give reproducible results that are also incorrect! An instrument's reliability can be affected by the respondents' ability to estimate their dietary intake reliably, by real dietary changes that occurred between the two assessment periods, and by inaccuracies in the coding of dietary data.

Cultural Issues

A cultural assessment of the target population is needed before data collection begins. The cultural assessment is undertaken to identify appropriate and inappropriate behaviors within the target population's culture. This is especially important if the assessment involves one-on-one interviews with members of the target population conducted in their home. Issues to consider during the cultural assessment are listed in Table 3-9. The manner in which strangers greet each other, the types of questions that are appropriate to ask, body language during interviews, and other customs differ among cultures.[55] In Japan, for example, a visitor who crosses her legs and shows the bottom of her shoe is being disrespectful. On the Canadian prairie, it is impolite for a guest to wear street shoes in someone's home. The people of North Africa value hospitality and time-honored rituals of eating, including hand washing, clapping, and eating with one's fingers. Visitors

Reliability The repeatability or precision of an assessment instrument.

to the home of a North African are prepared to eat, whether they are hungry or not, for to refuse food would be insulting.[56]

For more about cultural assessment, see Chapter 16.

Survey questions must also be culturally appropriate. For instance, a questionnaire designed to obtain information about the health attitudes and beliefs of Canadian aboriginal adults of the Mi'kmaq First Nation might have included a question that asked, "Do you worry about your diabetes?" If most aboriginal people responded no to this question, a researcher who was unfamiliar with the belief system and culture of the Mi'kmaq people might have concluded that aboriginal people were apathetic about their health. In fact, the question is culturally inappropriate, for in the traditional Mi'kmaq culture, the word *worry* has no meaning. A native person who gave a negative response to the question would probably be saying, "I pay attention to my diabetes, but it is a part of my life. I take it day by day." This meaning is far different from what an uninformed researcher might have concluded.[57] Cultural context is important when survey questionnaires are being designed. The Programs in Action on page 82 shows how cultural diversity dictates that programs be targeted toward particular populations and tailored to their needs.

Case Study 2: Nutritional Status of Independent Elderly Persons

Case Study 2, introduced in Chapter 2, focused on assessing the nutritional status of an elderly population in the fictional city of Jeffers. The target population was people older than 75 years of age who live at home. (Review Table 2-1 on page 39.) The community phase of the assessment obtained information about the types of existing community services aimed at addressing the needs of this group and how many members of the target population used these services. As with the first case study, the community nutritionist is now positioned to collect data about the target population. He first reviews the purpose, goals, and objectives of the assessment and then develops a set of questions aimed at measuring the nutritional status of this target population. These questions are listed in Table 3-10.

This needs assessment is more complex, requires a larger data set, and involves more people than the first case study. The nutritional status assessment methods in this case study include measurements of laboratory, clinical, anthropometric, and dietary outcomes, plus measures of functional status and other risk factors for poor nutritional status. Data about the population's use of community services are also needed. Remember, the assessment's purpose is to (1) obtain data about the nutritional status of these independent elderly persons and (2) link these outcomes with the services described in the community phase of the assessment to determine where services can be improved.

As in the first case study, each group of questions is tied to a specific objective, and the material in the table helps the community nutritionist plan the data collection. How does the community nutritionist make a decision about the assessment method, given the important factors described in this chapter? Several strategies can be used in decision making. First, he might turn to the medical or nutrition literature for information about accepted, standard assessment tests, such as serum retinol for vitamin A status and serum ferritin for iron deficiency. Or he might conduct a search of MEDLINE or other databases for articles that describe studies of the method's validity, reliability, sensitivity, and variation. (Consult the Internet Resources on page 85.) Next, he might speak with colleagues who are working with similar populations to learn about their approach to the assessment of the target group.

Nutritional status indicator A quantitative measure used as a guide to screen, diagnose, and evaluate interventions in individuals.

He might consider a standard measure of health or nutritional status. These measures, called **nutritional status indicators,** are quantitative measures that serve as guides "to

TABLE 3-10 Case Study 2: Nutritional Status of Independent Elderly Persons

OBJECTIVE (Refer to Table 2-1)	QUESTION(S) ASKED*	TYPES OF DATA	METHOD OF OBTAINING THE ANSWER
• Assess the nutritional status of independent elderly persons (> 75 years).	1. Has this population experienced a significant weight loss over time?	Anthropometric data (height, weight)	Scale
	2. Is their weight low or high for their height?	Anthropometric data (height, weight)	Scale
	3. Have they experienced a significant change in skinfold measurement?	Anthropometric data (skinfolds)	Calipers
	4. Has this population experienced a significant reduction in serum albumin?	Laboratory data	Phlebotomy
	5. Does this population have nutrition-related disorders (e.g., osteoporosis, arthritis)?	Clinical data	Medical chart
	6. Does this population have an inappropriate food intake?	Dietary data	24-hour recall
	7. Does this population use any dietary supplements?	Dietary data	24-hour recall and survey
	8. Does this population use nutrition support?†	Dietary data	Survey
	9. What is the functional status‡ of this population?	Clinical data	Survey
	10. What kinds of medications does this population take?	Clinical data	Survey
• Determine which community services are used by this population.	1. How many members of this population use: social services? mental health services? nutrition support services? home health care? home-delivered meals? federal/state assistance programs? housing and home heating assistance programs? other?	Community conditions	Survey
	2. How many members use services provided by dietitians and other health professionals?	Community conditions	Survey

*The population about which questions are asked consists of elderly persons over 75 years old who live at home in the fictional city of Jeffers.
†Nutrition support refers to enteral and total parenteral nutritional support.
‡Functional status refers to the individual's ability to bathe, dress, groom, use the toilet, eat, walk or move about, travel (outside the home), prepare food, and shop for food or other necessities.

screen, diagnose, and evaluate interventions in individuals."[58] Nutrition status indicators are often used to estimate the magnitude of a nutrition problem, its distribution within the population, its cause, and the effects of programs and policies designed to alleviate the problem. Researchers, program planners, health professionals, and policy makers use them for analyzing problems in health and nutrition.

Because there is no single "best" indicator, several may be used in the nutritional needs assessment. For example, major indicators of poor nutritional status in older U.S. adults include a weight loss of 5 percent or more of body weight in 1 month, being underweight or overweight, and having a serum albumin below 3.5 grams per deciliter, an inappropriate food intake, a midarm muscle circumference below the 10th percentile, and folate deficiency, among others.[59] Several of these nutritional status indicators were used in this case study. Nutritional status indicators exist for vitamin, mineral, and protein status and for energy intake. Efforts arc under way to develop accurate and reliable indicators of hunger

Programs in Action

The Project L.E.A.N. Nutrition Campaign

The nutrition education of preschoolers requires the involvement of the young child's primary role models: parents, caregivers, and teachers. Socioeconomic status and cultural traditions strongly influence eating habits and therefore affect health, nutrition status, and disease rates. Indeed, it has been said that socioeconomic status exerts a greater influence on disease rates than any other single factor.[1] Socioeconomic and cultural diversity in this country dictates that health education programs be targeted toward particular populations and tailored to their needs.[2]

Assessment of parents' or caregivers' level of knowledge and needs can begin shortly after a child enrolls in preschool. Parents and caregivers who may not have received nutrition education in the past, who are of low socioeconomic status, or for whom English is a second language may require special attention. A food frequency questionnaire, to be filled out by the parents or caregivers when the child enrolls in preschool, can help educators tailor nutrition education to the parents' or caregivers' needs and level of knowledge. The food list on the questionnaire must include foods typically consumed in the home[3] and may need to be translated into the parents' or caregivers' native language. It may also be necessary to provide verbal instructions in a common language. Bilingual parents or community volunteers can be a valuable source of information on the foods and eating habits of the target population. This can help parents and caregivers describe the typical diet of the family.

Family members and caregivers need both knowledge and skills in order to change their eating and cooking behaviors.[4] A step-by-step demonstration of a new healthful behavior—for example, low-fat cooking—can help them develop new food skills. Parents, other family members, and caregivers should be given the opportunity to practice and successfully master new skills, which builds self-confidence and self-efficacy, a feeling that can be critical in changing and maintaining a behavior. Teachers also strongly influence the preschooler's eating behaviors.

Providing teachers with materials helps ensure the accuracy and consistency of nutrition information provided to parents, caregivers, and young children. The attitudes and behaviors of preschool caregivers during meals and the information that they teach children in the classroom can influence the development of a child's lifelong eating habits.[5] Optimal caregiver mealtime behaviors include eating the same foods as the children do, encouraging children to try new foods, encouraging conversation during the meal, and using mealtime as an opportunity to provide nutrition education.

Preschool children themselves can be taught healthful eating habits in the classroom. The preschool years are a time when children become aware of health-related behaviors. They do not yet understand the concept of health, but they will imitate health-related behaviors they see at home and school.[6] Children can understand the concept of wellness. A positive emphasis works best—for example, discussion of foods that are good to put in our bodies. Children of this age learn better by participating in activities than they do by sitting and listening.

The Project L.E.A.N. (Low-fat Eating for America Now) Nutrition Campaign of Orange County Head Start, Inc. (OCHS) was directed toward a target audience of over 3,500 low-income 3- and 4-year-old children and their families, who were enrolled at the 40 OCHS centers throughout Orange County in California. Seventy-six percent of the target audience were members of the Hispanic community.

Goals and Objectives

The Project L.E.A.N. Nutrition Campaign aimed to improve the quality of life and reduce the incidence of nutrition-related chronic diseases among Head Start children and families through low-fat eating and regular physical activity.

Methodology

A preschool-level Project L.E.A.N. nutrition lesson plan (in English and Spanish) and food activity for the Head Start

and overall nutritional status. Table 3-11 lists the core nutritional status indicators recommended for the assessment of populations that are difficult to survey, such as homeless people, migrants, and institutionalized persons.[60]

Finally, the community nutritionist considers any factors particular to the target population that should influence his decision. Consider questions 6, 7, and 8 under the first objective in Table 3-10, which address the target population's nutritional intake. The community nutritionist reviewed the literature related to dietary intake methods and spoke with colleagues who routinely dealt with this age group in their practices. He decided on an interviewer-administered 24-hour recall. His main concern was that some classroom were developed by OCHS nutrition specialists. The nutrition lesson emphasized low-fat food choices and physical activity. Teachers showed examples of low-fat foods, using food models, picture flash cards, flannel boards, books, and Food Guide Pyramid models and posters. Food activities included filling plastic straw "blood vessels" with butter (to show how blood vessels can clog as a consequence of a high-fat diet) and helping make low-fat tortilla chips. An obstacle course was developed to encourage physical activity; children felt their own heartbeat before and after playing on the obstacle course.

Additionally, nutrition specialists at the OCHS centers conducted Project L.E.A.N. workshops for parents. At a Low-Fat Fiesta education program, staff encouraged label reading, offered tastings of a low-fat casserole, and supplied healthful recipes. The staff at a Get Fit Breakfast supplied parent participants with a low-fat breakfast, discussed the importance of exercise, and encouraged parents to participate in a warm-up stretch and morning walk.

Results

The nutrition lesson/food activity was conducted in at least 34 OCHS classrooms and reached over 700 children. Close to two-thirds of OCHS teachers who used the classroom nutrition lesson reported that children were able to name two to four low-fat and high-fat foods after the lesson. All students understood the importance of physical activity. The nutrition lesson and activities have been incorporated into the OCHS nutrition education curriculum for upcoming years.

A total of 207 OCHS parents and families participated in the nutrition workshops. Over 90 percent of the parents stated that they gained new knowledge and found the workshop they attended to be worthwhile and easy to follow. The parents reported plans to use the nutrition information from the workshop to cook and eat more low-fat foods and indicated that they intended to read labels and increase exercise. Several parents started walking groups.

Lessons Learned

Students modeled their behavior to match that of their teachers. For example, at the tortilla chip tasting, teachers remarked that they preferred the baked chips, so students said the same. This is why it is so important for teachers to be positive about the foods they eat with their students.

Factors such as limited income, lack of transportation, child care needs, work schedule, culture, and language must be considered when offering services to the parents of a culturally diverse, low-income preschool population. Collaboration with other professionals can help consolidate nutrition education and other topics into fewer classes that are easier for parents to attend.

References

1. R. I. Pasick, Socioeconomic and cultural factors in the development and use of theory, in *Health Behavior and Health Education,* ed. K. Glanz, F. M. Lewis, and B. K. Rimer (San Francisco: Jossey-Bass, 1997), pp. 423–40.
2. Ibid.
3. K. Bettin and J. V. Anderson, Designing a client-administered food frequency questionnaire, *JADA* 96 (1996) 505–8.
4. E. W. Maibach and D. Cotton, Moving people to behavior change, in *Designing Health Messages,* ed. E. Maibach and R. L. Parrott (Thousand Oaks, CA: Sage, 1995), pp. 41–64.
5. M. Nahikian-Nelms, Influential factors of caregiver behavior at mealtime: A study of 24 child-care programs, *JADA* 97 (1997) 505–9.
6. E. W. Austin, Reaching young audiences, in *Designing Health Messages,* ed. E. Maibach and R. L. Parrott (Thousand Oaks, CA: Sage, 1995), pp. 114–51.

Source: Community Nutritionary (White Plains, NY: Dannon Institute, Spring 1999). Used with permission. For more information about the *Awards for Excellence in Community Nutrition,* go to www.dannon-institute.org.

individuals in the sample might not remember to record all the food and beverages they consumed during the day (as is required for completing food diaries) or might not be able to estimate their food intake on a food frequency questionnaire. After considering all these factors, he decided that the interviewer-administered 24-hour recall was the dietary method most likely to yield good estimates of usual dietary intake in this target population in this setting. The community nutritionist also elected to develop a diet questionnaire to be administered by the trained interviewer at the same time as the 24-hour recall. This questionnaire would ask specific questions about the individual's intake of dietary supplements (for example, vitamins, minerals, phytochemicals, and

TABLE 3-11 Indicators for the Assessment of Nutritional Status among Migrant Workers, the Homeless, and Other Difficult-to-Sample Populations

NUTRITION CONCERN	INDICATOR
Obesity	Weight for length Body mass index (BMI)
Hypercholesterolemia	Serum total cholesterol
Hypertension	Brachial artery pressure
Iron status	Complete blood count
Food insecurity	Coping strategies • Food security survey module Food consumption: • Patterns over time • Types of foods • Frequency of consumption • Variety of foods Self-reports of: • Food sufficiency • Constraints to obtaining food Sources of food
Drug–nutrient interactions	Assessment of specific nutrient(s) affected by drugs and alcohol
Protein–energy malnutrition	Height Weight for height Weight change Skinfolds Body circumference Edema
Folate status as a marker for quality of diets limited in variety and quantity of foods	Serum and red blood cell folate concentrations
Vitamin A status	Serum retinol

Source: Adapted from S. A. Anderson, Core indicators of nutritional state for difficult-to-sample populations. © *Journal of Nutrition:* Vol. 120, p. 1585, American Institute of Nutrition.

herbal products) and use of nutrition support. Thus decisions about methods used are arrived at by searching the scientific literature, talking with colleagues, and considering any challenges posed by the target population (such as their not speaking English).

Putting It All Together

This phase of the community needs assessment focuses on obtaining data about the target population. It is designed to find answers to questions about how extensive the nutritional problem is and how it developed. The data collected during this phase are coded, entered into the computer, checked for errors, and analyzed using accepted statistical methods. Working carefully during the data collection process is important and ensures that the assessment's findings will be valid.[61]

Once the data are analyzed, a new issue arises: the choice of reference data against which the assessment's outcomes can be compared. For example, if the nutritional problem being evaluated is the nutritional status of children whose mothers participate in the WIC program, and the nutritional status indicators are height, weight, and hemoglobin, the community nutritionist might choose reference growth data for children published by the U.S. National Center for Health Statistics (NCHS) to evaluate the children's growth

Internet Resources

To locate published studies about the target population or diet assessment methods:

Current Bibliographies in Medicine	**www.nlm.nih.gov/pubs/resources.html**
Dietary Assessment Calibration/Validation Register	**www-dacv.ims.nci.nih.gov**
Dietary Survey Questionnaire Validity References	**www.nutritionquest.com/validation.htm**
Find Health Literature References	**http://health.nih.gov**
Food Composition Resource List for Professionals	**www.nal.usda.gov/fnic/foodcomp**
MEDLINE	**www.nlm.nih.gov**
Community Nutrition Mapping Project *CNMap*	**www.barc.usda.gov/bhnrc/cnrg/cnmapfr.htm**
National Nutrition Summit Information Resources	**www.nlm.nih.gov/pubs/cbm/nutritionsummit.html**

and development. The NCHS curves for evaluating the physical growth of children are based on a large, nationally representative sample of U.S. children. For evaluating the iron status of children, the community nutritionist might use reference data for hemoglobin from the most recent NHANES survey for children of the same age and sex.[62] The values obtained for the target population during the community needs assessment can be compared against the reference data. In Case Study 2 on page 81 (see Table 3-10), reference data for evaluating skinfold measurements among the independent elderly sample could be compared with skinfold standards derived from NHANES.[63]

Nutrient intake data are usually compared with the Dietary Reference Intakes (DRIs). Reference data for comparison purposes can also be obtained from countries where national surveys of dietary intakes have been carried out. In situations in which no reference data for a country exist, the FAO and/or WHO requirements for nutritional intake can be used.[64]

Statements drawn from the data collected about the target population and from comparisons with reference data are then organized and added to the final report described in Chapter 2. In this manner, the final report contains information about the target population and about the community in which it lives and works.

The Community Nutrition Mapping Project is an online resource for checking a state's nutritional health. It includes information on nutrient intakes, physical activity and body weight, healthful eating patterns, and food security. Access the CNMap at www.ba.ars.usda.gov/cnrg/services/cnmapfr.html.

Professional Focus

Lighten Up—Be Willing to Make Mistakes and Risk Failure

If you knew that you had only a few months to live, would you live differently than you do right now? Would you find time to take dancing lessons, learn inline skating, snorkel off the Great Barrier Reef, study the stock market, get a pilot's license, or try your hand at papier-mâché? Would you take more chances and worry less about your image?

Most of us would probably answer that last question in the affirmative. We would choose to live differently if we knew that only a few grains of sand were left in the hourglass. Nadine Stair, at the age of 85, said the same thing: "If I had my life to live over again, I'd try to make more mistakes next time. I would relax. I would limber up. I would be sillier than I have been this trip. I know of a very few things I would take seriously. I would take more trips. I would climb more mountains, swim more rivers and watch more sunsets. I would do more walking and looking. I would eat more ice cream and less beans. . . . If I had it to do over again, I would go places, do things and travel lighter than I have. . . . I'd pick more daisies."[1]

Notice that the first thing Nadine Stair said she would do differently "next time" was to try to make more mistakes. Most people work hard to *avoid* making mistakes, and few of us are willing to undertake a venture so risky that mistakes are almost guaranteed and the probability of failure looms large. In our culture, these activities are to be avoided at all costs.

Virtually every successful entrepreneur, adventurer, and risk taker has made mistakes and has failed at some point in his or her struggle to reach a personal or professional goal. In an essay in *Science,* Harold T. Shapiro, president of Princeton University, remarked that "the world too often calls it failure if we do not immediately reach our goals; true failure lies, rather, in giving up on our goals."[2] What would our world be like if the following individuals had given up on their goals?

- In 1842, at a time when most young British women of position were concerned mainly with parties and pending marriages, Florence Nightingale felt a call to perform some lifework. Although she sensed that her destiny "lay among the miserable of the world," the precise nature of this vocation eluded her for many years. Not until she was in her early thirties did she begin to pursue a career in nursing despite the persistent objections of her mother and sister, a cultural bias against nursing care, and the resistance of the traditional medical establishment. Over a lifetime of hard work, her determination and vision radically altered the practice and professionalism of nursing. Nightingale was among the first to document and describe hospital conditions, and she became an expert on sanitation. She reformed the health administration of the British army in response to the brutal mortality of the Crimean War and thereby influenced medical practice for years to come. Her reports on proper hospital construction, the training of nurses, and patient care led to the establishment of sanitary commissions and eventually to the public health service.[3]
- Thomas Alva Edison, born in 1847, has been described as "one of the outstanding geniuses in the history of technology."[4] He received very little formal schooling, having been expelled by a schoolmaster as "addled," and was taught history, science, and philosophy primarily by his mother. His fascination with the wireless telegraph as a young boy led to a lifelong enjoyment of experimentation and research. At his death in 1931, he held 1,093 patents on such devices as the incandescent electric lamp, the phonograph, the carbon telephone transmitter, and the motion picture projector.

  At one point in his career, Edison struggled to develop a storage battery. "I don't think Nature would be so unkind as to withhold the secret of a *good* storage battery if a real earnest hunt for it is made," he said. "I'm going to hunt."[5] This was no mean feat. He knew what was required of a good battery—it must last for years, should not lose capacity when recharged, and needed to be nearly indestructible. He began by testing one chemical after another in a series of experiments that spanned a decade. When 10,000 experiments failed to give the desired results, Edison remarked, "I have not failed. I've just found 10,000 ways that won't work."[6] He eventually succeeded.
- When he was a 20-year-old student at Yale University, Fred Smith wrote a term paper that analyzed freight services existing at the time. He concluded that there was a market for a company that moved "high-priority, time-sensitive" goods such as medicines and electronic components. He believed the existing system was cumbersome and failed to respond quickly to consumers' needs. In his paper, he proposed an overnight delivery service based on a "hub-and-spokes" air freight system. His professor was unimpressed with Smith's proposal, citing a restrictive regulatory environment and competition from airlines as major barriers to implementing such a service. The paper earned a grade of C.

  Smith did not give up on his idea, although he could not do anything about it for several years. Eventually, at the age of 29, he founded a company, Federal Express, designed to deliver packages "absolutely, positively overnight." In March 1973, his first planes flew over the eastern United States, carrying a total of six packages. One month later the volume had increased to 186 packages. In its first years, the company nearly folded from a lack of capital, concerns about Smith's leadership, and a formal charge of fraud against him. The company—with Smith at its helm—survived this difficult period. By 1983, its earnings were more than $1 billion. Today, FedEx delivers 3.2 million packages to 220 countries every day and has altered American business practices substantially.[7]

The Secret of Success

Risk takers make mistakes and sometimes fail. If they all share one feature, however, it is their willingness and determination to persevere, sometimes against great odds. In our professional lives, it is impossible to avoid risk and the chance of failure. The trick is to learn how to minimize risk and capitalize on your mistakes. Here are a few points to keep in mind when you are next faced with undertaking a risky venture:

1. Do your homework. There are risks and there are calculated risks. The difference between the two is substantial. To prepare for a calculated risk, talk with people who have undertaken similar ventures. Find out about the unexpected problems they experienced and how they handled them.
2. Write down your options and the potential outcome of each. This activity will help you focus on the option that may stand the best chance for success. Then write down the worst possible thing that could happen if you proceeded with that option. Is it something you can live with? If not, how can the option be changed to protect you or your employees?
3. Learn from your mistakes or failures. We all make them, but we don't all learn from them. In the business world,

bankruptcy is often viewed as the ultimate failure. One entrepreneur commented that "if you hadn't been bankrupt at least once, you hadn't really learned much about business."[8] Although it is certainly painful, business failure can be an opportunity to learn new lessons both personally and professionally. The successful entrepreneur and risk taker has the ability to learn from her experiences and regain control of her destiny.

4. Be committed to your goal. A high level of commitment to the work at hand is one element that distinguishes the successful entrepreneur from the also-rans.[9]

Words to Work By

In his book of wildlife portraits, the artist Robert Bateman remarked, "A great master teacher once said, 'In order to learn how to draw you have to make two thousand mistakes. Get busy and start making them.'"[10] These are apt words to keep in mind as you begin traveling your career path.

References

1. As cited in R. N. Bolles, *The Three Boxes of Life—And How to Get Out of Them* (Berkeley, CA: Ten Speed Press, 1981), p. 377.
2. H. T. Shapiro, The willingness to risk failure, *Science* 250 (1990): 609.
3. C. Woodham-Smith, *Florence Nightingale* (New York: McGraw-Hill, 1951). The quotation in this paragraph was taken from p. 31.
4. Edison, in *The New Encyclopaedia Britannica: Macropaedia,* Vol. 17, 15th ed. (Chicago: Encyclopaedia Britannica, 1985), pp. 1049–51.
5. W. A. Simonds, *Edison—His Life, His Work, His Genius* (London: Kimble & Bradford, 1935), p. 278.
6. As cited by Shapiro, Willingness to risk failure.
7. W. Davis, *The Innovators* (New York: American Management Association, 1987), pp. 361–65; and www .fedex.com/us/about. Accessed April 28, 2005.
8. As cited in B. J. Bird, *Entrepreneurial Behavior* (Glenview, IL: Scott, Foresman, 1989), pp. 354–55.
9. Bird, *Entrepreneurial Behavior,* pp. 366–67.
10. R. Bateman, *The Art of Robert Bateman* (Ontario: Penguin Books Canada, 1981), p. 19.

Chapter *Four*

Principles of Epidemiology

Learning *Objectives*

After you have read and studied this chapter, you will be able to:

- Define epidemiology.
- Describe various vital statistics used by epidemiologists to monitor a population's health status.
- Explain prevalence rates and how they differ from incidence rates.
- Describe the strengths and weaknesses of various types of epidemiologic studies.
- Explain why the day-to-day variation in an individual's nutrient intake can have important implications for nutritional epidemiologic studies.
- Discuss the advantages and disadvantages of various dietary assessment methods.

This chapter addresses such issues as research methodologies, the scientific method, interpreting current research, collecting pertinent information for nutrition assessments, and current information technologies, which are Commission on Accreditation for Dietetics Education (CADE) *Foundation Knowledge and Skills* requirements for dietetics education.

Chapter *Outline*

Something To Think About...

The argument that we should wait for certainty [in making public health recommendations] is an argument for never taking action—we shall never be certain.

– M. G. Marmot

Introduction

In 1855 Dr. John Snow, of London, wrote an account of his discovery of the link between contaminated water and a local outbreak of cholera, one of the greatest scourges of modern times. Snow observed that people who had drunk water from a pump on Broad Street in central London were attacked by the disease. He formulated a hypothesis about the mode of cholera transmission, suggesting that the intestinal discharges of people sick with cholera were seeping into the Thames River—the source of Broad Street drinking water—and were turning it into a cesspool. To test his hypothesis, he tracked down cholera cases and interviewed family members about their source of drinking water. He identified common events that linked cholera patients and established the ways in which they differed from healthy individuals in the same neighborhood. Snow observed that the rate of cholera was significantly lower in communities consuming water drawn from the Thames River upstream, which he therefore presumed was uncontaminated. His recommendation for stopping the cholera epidemic was simple: Remove the handle from the Broad Street pump so that it could not be used. Medical publishers were so skeptical of Snow's theory that he was forced to publish the particulars of his discovery at his own expense.[1]

Snow described the mode of transmission of this dreaded infection nearly 30 years before the cholera bacterium, *Vibrio cholerae*, was discovered. Thanks in part to Snow's careful observations, London took a major step toward controlling cholera epidemics and improving the public's health by installing new water supply systems. This story is an example of how the classic epidemiologic method was applied to determining the cause and course of a disease outbreak. In fact, the word **epidemiology** is derived from *epidemic*, which, translated from the Greek, means "upon the people." As this derivation indicates, the epidemiologic method was initially used to investigate, control, and prevent epidemics of infectious diseases such as cholera, plague, smallpox, typhoid, tuberculosis, and poliomyelitis. Although investigating infectious disease outbreaks is still part of epidemiology, today it is also applied to the study of injuries, especially those occurring in the home and at work; chronic diseases such as cancer, coronary heart disease, and arthritis; and social problems such as teenage pregnancies, homicide, alcoholism, and cocaine abuse.

The Practice of Epidemiology

The discipline of epidemiology has expanded from its origin as the study of epidemics to include the control and prevention of all types of health problems. It is similar to clinical medicine and laboratory science in its concern with understanding the processes of health and disease in humans, but it differs from these disciplines in its focus on the

Epidemiology From the Greek word meaning "upon the people"; the study of epidemics.

health problems of populations rather than of individual patients. A panel of international experts offers the following definition of epidemiology:[2]

> Epidemiology is the study of the distribution and determinants of health-related states and events in specified populations and the application of this study to the control of health problems.

This definition requires some explanation. The term *distribution* refers to the relationship between the health problem or disease and the population in which it exists. The distribution includes the persons affected and the place and time of the occurrence. It also encompasses such population parameters as age, sex, race, occupation, income and educational levels, exposure to certain agents, and other social and environmental features. Distribution is concerned with population trends or patterns of disease or exposure to a specific agent among groups of people.

The term *determinants* refers to the causes and factors that affect the risk of disease. Infectious diseases have a single, necessary cause. Shigellosis, for example, is an infection of the bowel caused by *Shigella* organisms. Rubella, or German measles, is a highly contagious disease caused by a virus spread by airborne droplets or close contact. Many conditions, however, have multiple determinants, as in the contribution of gender, race, dietary intake, and hormonal status to osteoporosis. Determinants of disease are typically divided into two groups: (1) host factors, such as age, sex, race, genetic makeup, nutrition status, and physiologic state, which determine an individual's susceptibility to disease; and (2) environmental factors, such as living conditions, occupation, geographical location, and lifestyle, which determine the host's exposure to a specific agent.

Investigating Causes of Diseases

Whereas the clinician is concerned with an individual patient and with the host and environmental factors that affect that patient's health status, the epidemiologist is concerned with groups of individuals or populations. The epidemiologist measures or counts those elements that are common to individuals, so that the magnitude and effects of individual variation within a population can be accounted for in studying a disease process. Differences in rates of disease in populations are then used to formulate hypotheses about the cause of a health problem or to assess exposure to a specific agent. Working closely with clinicians, laboratory scientists, biostatisticians, and other health professionals, the epidemiologist works to identify the causes of disease and to propose strategies for controlling or preventing health problems.

For example, in the 1990s, epidemiologic studies established that women could reduce their risk of bearing a child with a neural tube birth defect such as spina bifida by increasing their intake of folic acid.[3] Consequently, the Food and Drug Administration now mandates that grain products be fortified with folic acid to improve intakes in the United States. Epidemiologic studies have also associated low intakes of fruits and vegetables with increased risks of some types of cancer. Although the mechanisms by which these foods may protect against cancer are not completely known, the epidemiologic evidence provides a reasonable basis for action, and the National Cancer Institute recommends that people increase their consumption of fruits and vegetables in an effort to prevent cancer.

EXAMINING A COMMUNITY'S HEALTH STATUS

Chapter 2 provides a detailed discussion on the assessment of a community's health status.

In addition to enhancing our understanding of health and disease and searching for the causes of disease, epidemiology has several other uses. The epidemiologic method can be used to describe a community's particular health problems and to determine whether a

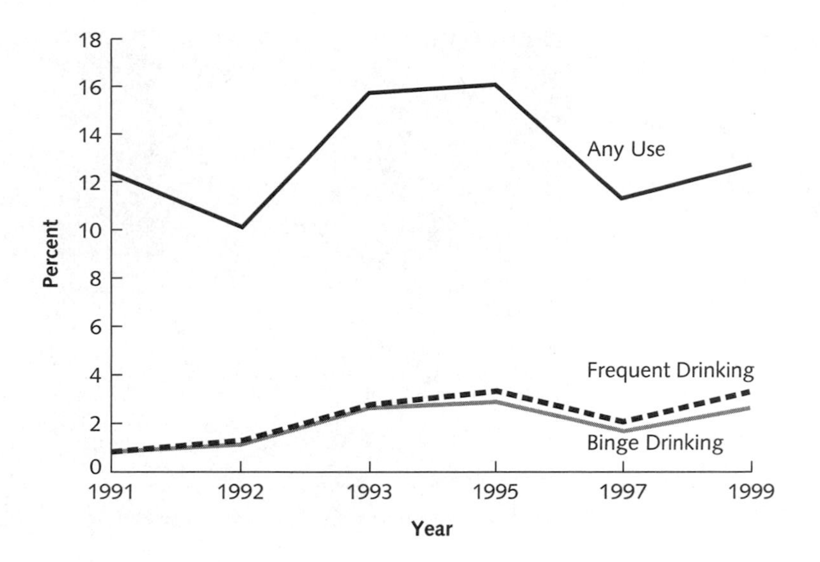

FIGURE 4-1 Percentage of Pregnant Women, Aged 18 to 44 Years, Who Reported Alcohol Use, United States, 1991–1999
Prenatal exposure to alcohol is one of the leading preventable causes of mental retardation in the United States. One of the *Healthy People 2010* national health objectives is to increase to 94 percent the percentage of pregnant women abstaining from alcohol use.
Source: CDC.

community's overall health is improving or getting worse. Diagnosing a community's health problems is important to health agencies. The public health department, for example, may need to know whether the number of low-birthweight infants born within the community decreased, increased, or remained the same compared with the previous year. The agency may want to compare the community's current rate of births of underweight infants with the rate from the previous decade or with the state or national average. This information can be used to evaluate the agency's current programs and services and thereby determine whether they are meeting the health needs of pregnant women in the community. Potential health problems may also be spotted through this ongoing examination of a community's health status. For example, in the early 1980s an unusual illness was observed mostly among gay men living in San Francisco, Los Angeles, and New York. The finding of an unusually high number of cases of Kaposi's sarcoma, a rare cancer affecting primarily middle-aged men of Jewish or Italian ancestry, led to the discovery of the acquired immunodeficiency syndrome (AIDS).

SURVEILLANCE AND RELATED ACTIVITIES

Over the past decade, epidemiologic data have also been used to develop surveillance methods for identifying women at high risk for giving birth to a child with **fetal alcohol syndrome (FAS)** and to design and implement prevention activities.[4] In order to characterize trends in alcohol use among women of childbearing age, the Centers for Disease Control and Prevention (CDC) analyzed representative survey data from the Behavioral Risk Factor Surveillance System (BRFSS) from 1991 to 1999. As illustrated in Figure 4-1, the rate of any alcohol use (for example, at least one drink) during pregnancy has declined since 1995. However, rates of binge drinking (for example, five or more drinks on any one occasion) and frequent drinking (for example, seven or more drinks per week) during pregnancy have not declined and remain higher than the *Healthy People 2010* objectives.[5] Additionally, the rates of binge drinking and frequent drinking have not declined among nonpregnant women of childbearing age. As a result of these findings, and to ensure more uniform dissemination of prenatal alcohol prevention messages, CDC, in collaboration with the Association of Schools of Public Health, is conducting targeted media campaigns to increase public awareness of the adverse effects of alcohol use during pregnancy among diverse geographical, racial, and ethnic populations and among younger women.

Fetal alcohol syndrome (FAS) A disorder characterized by growth retardation, facial abnormalities, and central nervous system dysfunction, and caused by a woman's use of alcohol during pregnancy; FAS is potentially 100 percent preventable.

Targeted media campaigns can help increase public awareness of the adverse effects of alcohol use during pregnancy. Because levels of binge and frequent drinking among nonpregnant women have not declined, all women of childbearing age should be warned about the adverse effects of alcohol use in order to avert early prenatal exposure before women become aware of pregnancy. Additional information about the CDC's activities to prevent alcohol-exposed pregnancies is available at www.cdc.gov/ncbddd/fas.

Courtesy of the ARC.

Examples of surveillance systems relevant to the community nutritionist include the Pediatric Nutrition Surveillance System (PedNSS), which monitors growth indicators and anemia status of children, and the Pregnancy Nutrition Surveillance System (PNSS), which monitors pre-pregnancy weight, pregnancy weight gain, behavioral risk factors, anemia status, birth outcome, and infant feeding practices. These CDC-managed systems monitor low-income women, infants, and children participating in government health and nutrition assistance programs.

Based on the **vital statistics** (such as age at death and cause of death) recorded on death certificates, the epidemiologic method can also be used to calculate an individual's risk of dying before a certain age. Table 4-1 provides indices commonly used in determining key types of vital statistics. This type of risk assessment is the foundation of the actuarial tables developed by life insurance companies. A related use occurs in clinical decision making when, for example, the effectiveness of a particular drug or surgical intervention in treating a certain condition is evaluated, as well as whether its use is associated with side effects or other health risks. Sometimes the epidemiologic method is used to identify the characteristics of a disease and determine whether some syndromes are related to one another or represent distinct conditions. A good example of this use was the development of a working case definition that outlined the clinical, behavioral, and physiological manifestations of chronic fatigue syndrome.[6]

Basic Epidemiologic Concepts

To put it simply, "the basic operation of the epidemiologist is to count cases and measure the population in which they arise" in order to calculate rates of occurrence of a health problem and compare the rates in different groups of people.[7] The primary reason for investigating and analyzing these health problems is to work toward controlling and preventing them, typically through the formulation of specific health policies. This section presents key epidemiologic concepts that illustrate how data about disease processes are obtained and analyzed.

Rates and Risks

Vital statistics Figures pertaining to life events, such as births, deaths, and marriages.

Case A particular instance of a disease or outcome of interest.

A middle-aged woman is admitted to the emergency room, complaining of chest pain, nausea, and dizziness. The attending physician orders a diagnostic workup, which confirms that the woman has had a heart attack. For the emergency room physician, this woman is a patient requiring immediate critical care. For the epidemiologist studying the factors that contribute to coronary heart disease, this woman is a **case**—a single individual with a confirmed diagnosis of myocardial infarction. The physician and the epidemiologist share a common concern about this woman—her risk status. Does she possess certain

TABLE 4-1 Vital Statistics: Equations for Commonly Used Population Data

1. $\text{Crude birth rate}^* = \dfrac{\text{Number of live births during year}}{\text{Average (midyear) population}} \times 1000$

2. $\text{Crude death rate}^* = \dfrac{\text{Number of deaths during year}}{\text{Average (midyear) population}} \times 1000$

3. $\text{Age-specific death rate} = \dfrac{\text{Number of deaths to people in a particular age group}}{\text{Average (midyear) population in specified age group}} \times 1000$

4. $\text{Cause-specific death rate} = \dfrac{\text{Number of deaths due to a particular cause during year}}{\text{Average (midyear) population}} \times 1000$

5. $\text{Infant mortality rate} = \dfrac{\text{Number of deaths to infants} < 1 \text{ yr during year}}{\text{Number of live births in same year}} \times 1000$

6. $\text{Neonatal mortality rate} = \dfrac{\text{Number of deaths to infants} < 28 \text{ days during year}}{\text{Number of live births in same year}} \times 1000$

7. $\text{Fetal death rate} = \dfrac{\text{Number of fetal deaths (} > 20 \text{ weeks gestation) during year}}{\text{Number of live births and fetal deaths in same year}} \times 1000$

8. $\text{Maternal mortality rate} = \dfrac{\text{Number of pregnancy-related deaths during year}}{\text{Number of live births in same year}} \times 100{,}000$

*In general, *crude rates* apply to an entire population and are not useful for comparisons, because population characteristics may differ greatly, particularly with respect to age.

Source: Adapted from CDC; available at www.cdc.gov/nchs/datawh/nchsdefs/rates.htm.

characteristics that placed her at high risk for heart attacks? If so, can these characteristics be modified to reduce her chance of having another? If the woman does not appear to belong in a high-risk group, why did she have a heart attack at this particular time?

In epidemiology, **risk** refers to the likelihood that people who are without a disease, but are exposed to certain **risk factors,** will acquire the disease at some point in their lives.[8] These risk factors may be inherited. Others are found in the physical environment in the form of infectious agents, toxins, or drugs. Some risk factors are derived from the social environment—that is, the family, community, or culture. Others are behavioral, such as not wearing seat belts or smoking.

The *relative risk (RR)* is a comparison of the risk of some health-related event, such as disease or death, in two groups. The two groups might be differentiated by gender, age, or exposure to a suspected risk factor (for example, tobacco or high dietary intake of saturated fat). The risk for disease or death will be greater in the exposed group if an exposure is harmful (as in the case of a diet high in saturated fat) or will be smaller if an exposure is protective (as in the case of a diet high in soluble fiber).

If the relative risk is greater than 1, people exposed to the factor have an increased risk of the outcome under investigation.[9] If the relative risk is less than 1, people exposed to the factor have a decreased risk of the outcome. For example, in a Spanish study of bladder cancer, subjects with high intake of saturated fat had a relative risk of 2.25, meaning they had more than double the risk of developing bladder cancer that those with low intake of saturated fat had. In an Italian study of colorectal cancer, subjects with high intakes of β-carotene had a relative risk of 0.38, which means that they had about one-third the risk of developing colorectal cancer compared to those with low β-carotene intakes. Relative risks can be used to compare the strengths of different associations. The relative risk of lung cancer in people with low fruit and vegetable intake compared to those with high intake is about 2.0. The relative risk of lung cancer in smokers compared to nonsmokers is at least 10.0. Clearly, the association of cancer with smoking is stronger than its association with low intake of fruits and vegetables.

Risk The probability or likelihood of an event occurring—in this case, the probability that people will acquire a disease.

Risk factors Clinically important signs associated with an increased likelihood of acquiring a disease.

A general formula for relative risk is as follows:

$$\text{Relative risk}^* = \frac{\text{Risk of disease or death for exposed persons}}{\text{Risk of disease or death for unexposed persons}}$$

*If RR = 1.0, the risk in the exposed group and that in the unexposed group are the same.

If RR > 1.0, the exposed group is at greater risk of dying than the unexposed group.

If RR < 1.0, the exposed group has smaller risk than the unexposed group—possibly because of a protective effect from the "exposure."

TABLE 4-2 Characteristics of Incidence Rates and Prevalence Rates

RATE	NUMERATOR	DENOMINATOR	TIME	HOW MEASURED
Incidence	New cases occurring during the follow-up period in a group initially free of the disease	All susceptible individuals present at the beginning of the follow-up period (often called the population at risk)	Duration of the follow-up period	Cohort study
Prevalence	All cases counted in a single survey or examination of a group	All individuals examined, including cases and noncases	Single point in time	Prevalence or cross-sectional study

Source: R. H. Fletcher, S. W. Fletcher, and E. H. Wagner, *Clinical Epidemiology—The Essentials,* 3rd ed.

Some of us are exposed to certain risk factors more than other individuals are.[10] For example, workers in photographic laboratories, dry-cleaning establishments, and some industrial processing plants are exposed daily to chloroform, a highly volatile liquid that has been shown to cause tumors in mice and rats. Workers in office buildings and schools are typically not exposed to chloroform. These different rates of exposure to a risk factor (in this case, chloroform) allow for basic comparisons of disease rates among individuals. One expression of how frequently a disease occurs in a population is **incidence,** the fraction or proportion of a group initially free of a disease or condition that develops the disease or condition over a period of time. Incidence is measured by a two-step process:

1. Identify a group of susceptible people who are initially free of the disease or condition.
2. Examine them periodically over a period of time to discover and count the new cases of the disease that develop during that interval.

Another common measure of frequency of occurrence of an event is **prevalence,** or the fraction or proportion of a group possessing a disease or condition at a specific time. The prevalence rate is measured by a single examination or survey of a group. The characteristics of these two measures are summarized in Table 4-2. Incidence and prevalence rates describe the frequency with which particular events occur. By calculating and comparing rates, epidemiologists can determine the strength of the association between risk factors and the health problem being studied.

Incidence The number of *new* cases of a disease during a specific time period in a defined population.

Prevalence The number of existing cases of a disease or other condition in a given population. *Point prevalence* is the amount of a particular disease present in a population at a particular point in time—usually the time a survey was done. *Period prevalence* is the amount of a particular disease in a population over a period of time. The *prevalence rate* is the proportion (usually the percentage) of persons in a population who have a particular disease or attribute at a particular point in time.

The Epidemiologic Method

In its study of disease processes, epidemiology uses a variety of tools: clinical, microbiological, pathological, demographic, sociological, and statistical. None of these is exclusive to epidemiology, but the manner in which they are used uniquely defines the epidemiologic method. The following example uses the investigation of the "diet–heart" hypothesis to illustrate the rigorous, scientific approach of the epidemiologist:

1. **Observing.** Many years of investigation had shown that atherosclerosis could be induced in laboratory animals, particularly rabbits and monkeys, by feeding them a diet rich in fats and cholesterol. Physicians working in India, Africa, and Latin America also observed that coronary heart disease (CHD) was fairly rare in human populations whose diet tended to be high in vegetables and grains.[11]

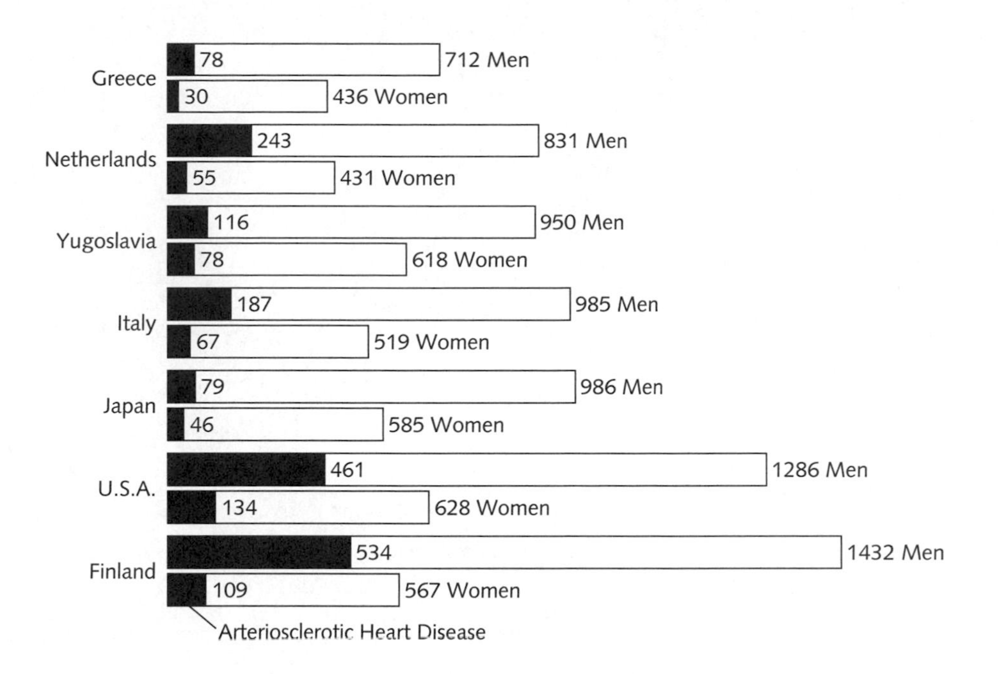

FIGURE 4-2 Deaths per 100,000 in 1965 from CHD and from All Causes for Ages 35 to 64 Years (age-standardized)

Source: A. Keys, Coronary heart disease in seven countries, *Circulation* 41 and 42 (Supplement 1) (1970): 1–4. Reprinted by permission of Lippincott, Williams & Wilkins.

2. **Counting cases or events.** Vital statistics obtained from the World Health Organization showed marked differences in deaths from CHD among countries, with Finland having one of the highest and Greece and Japan among the lowest CHD death rates, as shown graphically in Figure 4-2. The habitual diets of the peoples in these countries differed as well.

3. **Relating cases or events to the population at risk.** Public health officials in the United States became increasingly concerned about the number of deaths attributable to CHD, which was (and remains) the leading cause of death in the U.S. population. Identifying the risk factors for CHD became a public health priority.

4. **Making comparisons.** One of the first large population studies to examine the relationship between blood cholesterol levels and risk of CHD was the Seven Countries Study.[12] This project was undertaken in the late 1950s to examine the effects of differences in lifestyle, culture, diet, and general health habits on risk factors such as serum cholesterol levels, morbidity, and mortality from CHD. Sixteen population groups, involving about 11,000 men aged 40 to 59 years, were studied in seven countries that had reliable census data and well-organized medical systems: Greece, the Netherlands, Yugoslavia, Italy, Japan, the United States, and Finland. The **cohort** of adult men was followed for 10 years. One of the main study findings, shown in Figure 4-3, was that the population risk of CHD death, after adjusting for age, rose in a stepwise manner with increasing total blood cholesterol level. In addition, an analysis of dietary intakes of the study population showed that the serum cholesterol level increased progressively as the level of saturated fat in the diet rose.

Figure 4-4 shows that Finland, with the highest reported intake of saturated fat (expressed as a percentage of total calories), had the highest mean population serum cholesterol level (about 260 mg/dL). (*Note:* The r value in the lower right-hand corner is the correlation coefficient. The correlation coefficient is a number between -1 and $+1$ that is used to quantify the strength of an association between two variables—in this case, dietary saturated fat and serum cholesterol level. The stronger the relationship of the two variables, the closer r is to 1; the weaker the relationship, the closer r is to 0. An r value of 0.89 means that although dietary saturated fat intake is strongly correlated with serum cholesterol level, the relationship between the two variables isn't perfect

Cohort A well-defined group of people who are studied over a period of time to determine their incidence of disease, injury, or death.

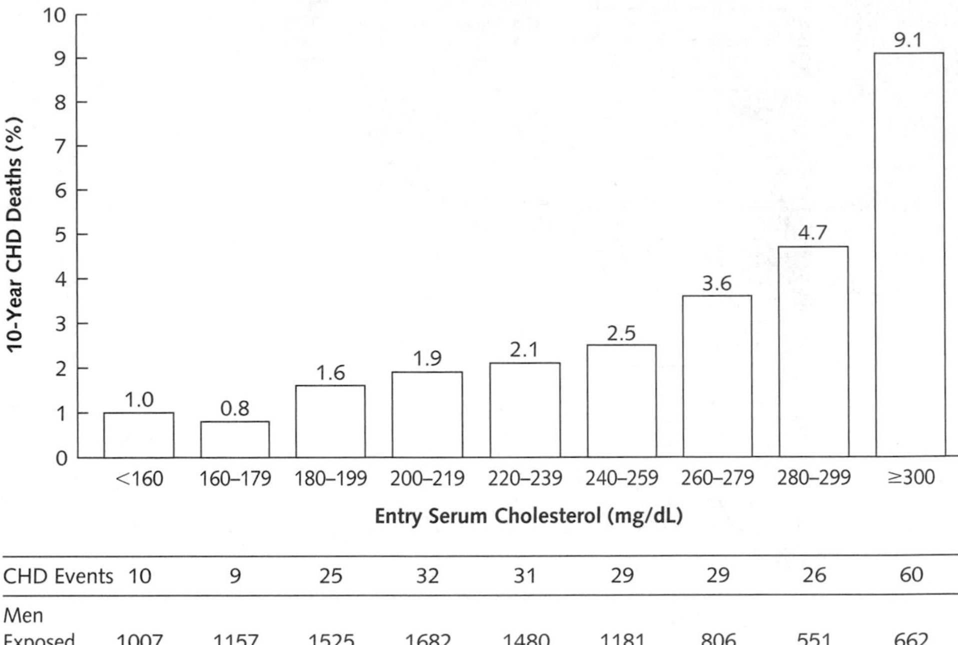

	<160	160–179	180–199	200–219	220–239	240–259	260–279	280–299	≥300
CHD Events	10	9	25	32	31	29	29	26	60
Men Exposed	1007	1157	1525	1682	1480	1181	806	551	662

FIGURE 4-3 Total Blood Cholesterol and CHD Deaths in Men

(*Note:* A recent analysis of 30-year follow-up data revealed that longevity in men aged 50 years or more was not related to serum cholesterol in any of the cohorts, indicating the importance of factors other than those related to cholesterol. —Ancel Keys, personal communication)

Source: Reprinted by permission of the publishers from *Seven Countries: A Multivariate Analysis of Death and Coronary Heart Disease* by Ancel Keys. Cambridge, MA: Harvard University Press. Copyright © 1980 by the President and Fellows of Harvard College.

because of measurement errors [for example, in calculating the dietary intake of saturated fat or determining the serum cholesterol level] or other factors [for example, genetic determinants].[13])

5. **Developing the hypothesis.** The results of animal studies and large-scale population studies such as the Seven Countries Study were used to formulate the hypothesis that a diet high in saturated fat increases blood cholesterol levels and contributes to the development of CHD.

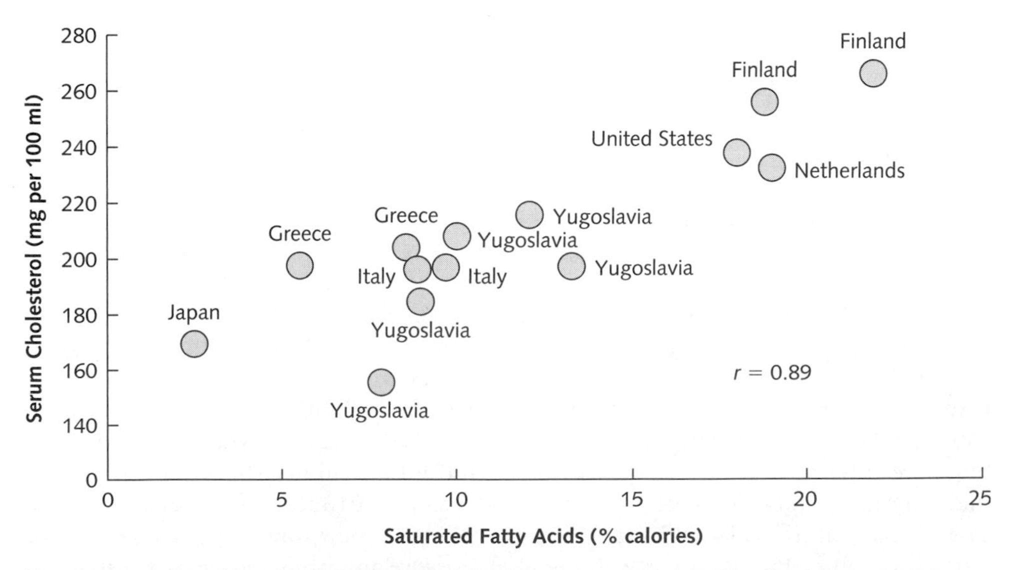

The seven countries cohort was drawn from the following seven regions and cities: Finland = east, west; Greece = Crete, Corfu; Italy = Crevalcore, Montegiorgio; Japan = Tanushimaru; Netherlands = Zutphe; U.S.A = U.S. railroad; Yugoslavia = Belgrade, Dalmatia, Slavonia, Velika Krsna, Zrenjanin.

FIGURE 4-4 Median Serum Cholesterol Values of the Seven Countries Cohort Plotted versus the Percentage of Total Calories from Saturated Fat

Source: Adapted from A. Keys, Coronary heart disease in seven countries, *Circulation* 41 and 42 (Suppl 1) (1970): 1–170. Reprinted by permission of Lippincott, Williams & Wilkins.

6. **Testing the hypothesis.** Many studies were undertaken to test this hypothesis. In one migration study, groups of Japanese living in Japan were compared with groups of Japanese who were originally from the same regions in Japan but had emigrated to Hawaii or California.[14] The study design used standardized protocols for measuring blood cholesterol and the number of existing and new cases of CHD. It showed that the intake of saturated fat as a percentage of total calories in the three populations varied widely, with Japanese living in Japan having a saturated fat intake of 7 percent of total calories and Japanese living in California having an intake of 26 percent. (The value for Japanese living in Hawaii was 23 percent.) Japanese living in Japan had the lowest blood cholesterol level, followed by those living in Hawaii; Japanese living in California had the highest blood cholesterol level. These differences in blood cholesterol paralleled the rates of CHD, suggesting that changes in diet after migration were partly responsible for the differences in blood cholesterol levels and the incidence of CHD.

Other population studies, such as the Western Electric Study[15] and the Puerto Rico Heart Health Program,[16] helped confirm the relationship of diet to blood cholesterol levels and CHD risk.

7. **Drawing scientific inferences.** The results of these and other large-scale population studies provided convincing evidence that the rates of CHD differ among populations and that these differences are due partly to environmental (diet) factors. However, the data have not been entirely consistent. The famous Ireland–Boston brothers study provided only weak support for the diet–heart hypothesis. This study did not find significant differences in CHD mortality between Irish brothers who were born and remained in Ireland and those who, although born in Ireland, had lived in Boston for at least 10 years. It did, however, underscore the increased risk of death from CHD among study subjects consuming a diet high in fat. Furthermore, it found that the consumption of a high-fiber, vegetable-rich diet decreased the risk of CHD death.[17]

It is unrealistic to expect total agreement among the results of epidemiologic or clinical studies. Many genetic, environmental, social, and experimental factors affect a study's outcome. On balance, however, many studies conducted over the last four decades have strengthened the link among diet, blood cholesterol, and CHD.

8. **Conducting experimental studies.** Studies with rabbits, pigs, and nonhuman primates showed that these species are susceptible to diet-induced hypercholesterolemia. In addition, the blood cholesterol levels of some laboratory animals were affected by the types of fatty acids in the diet. In primates, for example, blood cholesterol levels rose when a diet rich in cholesterol and saturated fatty acids was consumed; a diet rich in polyunsaturated fatty acids lowered them.[18] Animal studies provided compelling support for a role of dietary fat and cholesterol in the atherogenic process.

9. **Intervening and evaluating.** One of the most ambitious interventions undertaken to modify CHD risk factors was the Multiple Risk Factor Intervention Trial (MRFIT), commonly called Mister Fit. This six-year clinical trial was directed toward the primary prevention of CHD among middle-aged men determined to be at high risk for CHD because they had hypertension or high blood cholesterol or smoked cigarettes. The 12,866 eligible men were randomly assigned to a special-intervention group or a usual-care group. Men in the special-intervention group received extensive nutrition counseling to help them make dietary changes to reduce their fat and cholesterol intake. Participants in the special-intervention group who smoked received counseling to help them stop smoking.[19] Although the mean reduction in blood cholesterol in the intervention group was less than the predicted response, MRFIT showed that serum cholesterol reductions and dietary changes could be sustained over a period of several years.[20]

TABLE 4-3 Leading Risk Factors for Heart Disease

RISK FACTOR	HOW TO MINIMIZE THE RISK
High LDL cholesterol; Low HDL cholesterol	Limit intake of cholesterol, saturated fat, and *trans* fat. Increase your intake of soluble fiber, soy foods, and omega-3 fats. Maintain a physically active lifestyle.
High blood pressure	Control high blood pressure with medication and a heart-healthy diet. Maintain a healthy weight.
Cigarette smoking	Stop smoking. Nicotine constricts blood vessels and forces your heart to work harder. Carbon monoxide reduces oxygen in blood and damages the lining of blood vessels.
Diabetes	Maintain proper weight. Eat high-fiber foods. Limit saturated fat and sugar. Get regular exercise.
Physical inactivity	Get at least 60 minutes of moderate-paced physical activity on most days of the week.
Obesity	Maintain a healthy weight and exercise. Being only 10 percent overweight increases heart disease risk.
"Atherogenic" diet	Keep saturated fat to under 10 percent of daily calories. Substitute olive and canola oils for saturated fat. Increase fiber intake by eating cereal grains, legumes, fruits, and vegetables.
Stress	Get regular exercise. Avoid excessive caffeine and alcohol. Practice relaxation techniques.
RISK FACTORS YOU CAN'T CHANGE	
Age	Men over age 45 and women over age 55 are at increased risk.
Gender	Men are at higher risk. Estrogen may protect women before menopause.
Genetics	Increased risk if you have a father or brother under age 55 or a mother or sister under 65 who had heart disease.

Source: M. Boyle and S. L. Anderson, *Personal Nutrition,* 5th ed. (Belmont, CA: Wadsworth/Thomson Learning, 2004) p. 127.

The *Dietary Guidelines for Americans 2005* contains additional recommendations for fats. The full document is available at www.healthierus.gov/dietaryguidelines.

The mounting experimental (animal) and epidemiologic evidence supporting the diet–heart hypothesis led to the formulation of dietary advice designed to reduce CHD risk in the general population. Voluntary and nonprofit health agencies, such as the American Heart Association[21] and the American Medical Association,[22] published statements for health professionals describing the known risk factors for CHD and strategies for reducing CHD risk (see Table 4-3). These were followed by the report of the National Cholesterol Education Program on the detection, evaluation, and treatment of high blood cholesterol in adults.[23] The science supporting the diet–heart link was eventually incorporated into the Dietary Guidelines for Americans, which most recently advised the public to "Limit intake of fats and oils high in saturated and/or *trans* fatty acids, and choose products low in such fats and oils" as a means of reducing the risk of chronic disease and improving health.[24]

Hypothesis Testing

Another precept of the epidemiologic method is hypothesis testing. Its importance to the experimental process cannot be overstated. In planning an experimental trial, the investigator identifies a cause–effect comparison to be tested as the research hypothesis. For example, a study might be designed to determine whether a vaccine for hepatitis A is effective among children or whether a protein in cow's milk is responsible for triggering

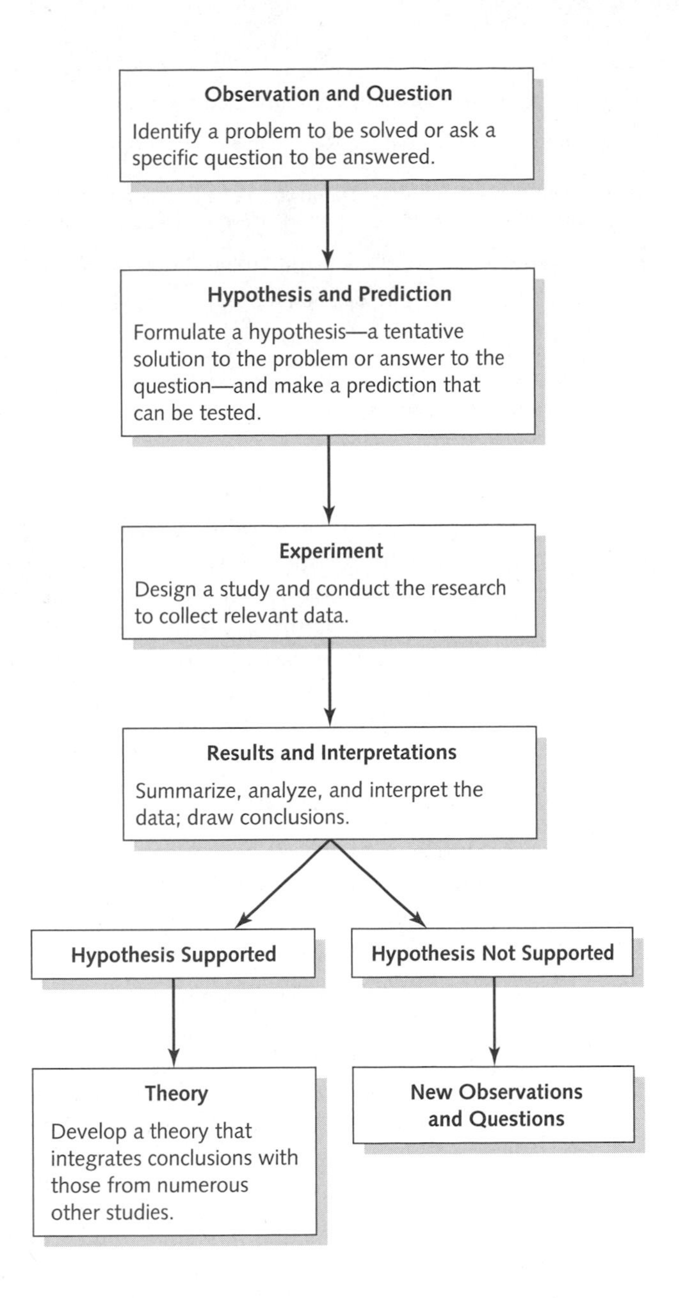

FIGURE 4-5 The Scientific Method: Hypothesis Testing

Source: E. Whitney and S. Rolfes, *Understanding Nutrition*, 10th ed. (Belmont, CA: Wadsworth/Thomson Learning, 2004) p. 12.

Type 1 diabetes mellitus. In a community-based nutrition study, one hypothesis might be stated thus: "There will be a significant decrease in the calories from fat and an increase in the grams of fiber consumed by employees in the eight intervention worksites receiving nutrition programming compared with employees in the eight control worksites."[25]

Once the specific research question (hypothesis) has been formulated, the investigators design a study to obtain information that will enable them to make inferences about the original hypothesis, as illustrated in Figure 4-5. In some epidemiologic studies, however, not all hypotheses are specified at the beginning. Rather, one or more hypotheses are generated retrospectively, after the research data have been collected and analyzed. The temptation to do this is compelling. A typical epidemiologic study produces reams of data about individuals (age, sex, race, educational level, occupation), specific agents (diet, vitamins, tobacco, alcohol, drugs, pesticides), and outcomes (death, heart attack,

TABLE 4-4 Possible Explanations for Research Observations

Bias	The observation is incorrect because a systematic error was introduced by: • Selection bias—the method by which patients or study subjects were selected for observation • Measurement bias—the method by which the observation or measurement was made • Confounding bias—the presence of another variable that accounts for the observation
Chance	The observation is incorrect because of error arising from random variation.
Truth	The observation is correct. This explanation should be accepted only after the others have been excluded.

Source: R. H. Fletcher, S. W. Fletcher, and E. W. Wagner, *Clinical Epidemiology—The Essentials,* 3rd ed. Copyright © 1996 Lippincott, Williams & Wilkins. Reprinted by permission.

colon cancer, impairment of renal function). When the data are downloaded into a computer for statistical analysis, the investigator may consider rummaging around in the data, searching for statistical associations among various groups that may suggest a cause–effect relationship. This activity, sometimes called "data dredging," is worth avoiding. The statement of a clear, precise hypothesis (or hypotheses) at the beginning of the study ensures that the appropriate data are collected to answer the research question(s) and avoids the pitfall of drawing spurious conclusions from the data set.[26]

Explaining Research Observations

An important aspect of the epidemiologic method—and, indeed, of the scientific method in general—is determining whether the data are valid. That is, do the data represent the true state of affairs or are they distorted in some fashion? Research data can have three possible explanations, as shown in Table 4-4. Consider the following research project:

> A study was designed to determine whether a worksite-sponsored exercise program resulted in a lower percentage of body fat and lower blood lipids among overweight adults. Employees in two companies were invited to enroll in the three-month program. Those who agreed to participate underwent several clinical measurements at the beginning and end of the study: height, weight, skinfolds, and total blood cholesterol. The 156 employees who participated in the exercise program showed a reduction in percentage of body fat and total blood cholesterol, compared with 210 adults who did not participate.

One possible explanation for these results is that the data are incorrect because they are biased; in other words, a systematic error was made in measuring one or more outcome variables, or there were systematic differences in the populations studied. Although there are probably dozens of possible biases, most fall into one of the three categories listed in Table 4-4. More than one bias may operate at one time. In the study just described, *selection bias* may have occurred because the study participants were self-selected. In other words, employees who where interested in improving their fitness level or losing weight were more inclined to join the program than those who were less interested in improvements in fitness or weight loss. *Measurement bias* may have occurred if the technician responsible for obtaining skinfold measurements was an enthusiastic supporter of vigorous exercise and took more pains to make careful measurements in the exercising adults than in the sedentary adults. Finally, *confounding bias* may have existed because the study focused on exercise as the sole determinant of changes in total blood cholesterol and body fat. It is possible that employees entering the exercise program made significant, but unconscious, alterations in their eating patterns that contributed to the changes in blood lipids and body composition. Thus the presence of (unmeasured or unidentified) **confounding factors** may have influenced the study outcome.

Another possible explanation for the study results is that they are due simply to chance and do not represent the true state of affairs; that is, the observations made on these employees arose from random variation within the sample. Many factors could have contributed to the differences between the groups, including the manner in which the clinical or laboratory measurements were made and differences in the makeup of the workforce (for example, a manufacturing plant versus a company that develops computer software programs). Random variation cannot be totally eliminated and usually occurs in tandem with some form of bias. The influence of these two sources of error can be reduced by careful study design and statistical analysis.

The final possible explanation for the study results is that they represent the truth: Participating in an exercise program resulted in reductions in body fat and total blood cholesterol among employees in these two companies. To say that the data are valid, then, means that they are neither biased nor incorrect due to chance and that they represent the true state of affairs regarding the physiologic effects of exercise.

Types of Epidemiologic Studies

When a nutritional problem is suspected, an investigation is undertaken "to find out something." The hypothesis to be tested is defined, the population to be studied is selected, measurements are taken or cases are counted, and the data are analyzed to determine whether the established facts support the hypothesis. The type of investigation undertaken depends to a large extent on the research question being asked and the type of data needed to answer the question. The major types of nutritional epidemiological studies are described below.[27]

Ecological or Correlational Studies

Ecological or correlational studies compare the frequency of events (or disease rates) in different populations with the per capita consumption of certain dietary factors (for example, saturated fat, total fat, and β-carotene). The dietary data collected in this type of study are usually *disappearance data*—that is, the national figures for food produced for human consumption minus the food that is exported, fed to animals, wasted, or otherwise not available for human consumption.

An example of an ecological study is an investigation of the correlation between fish consumption and breast cancer incidence and mortality rates in humans.[28] In this study, incidence and mortality rates were derived from the cancer registries of various countries. Food consumption for each country was estimated from food availability data averaged over three years. Although there were some exceptions, as Figure 4-6 shows, the risk for breast cancer tended to be high in countries where the relative proportion of calories derived from fish was low and the consumption of animal fat was high, such as the United States, Canada, and Switzerland. In countries where fish was consumed frequently and the animal fat intake was low, such as Japan, the breast cancer death rate was low.

Do these results mean that diets high in animal fat *cause* breast cancer, whereas diets high in fish protect against cancer? No, the data from an ecological study cannot be used to draw conclusions about the role of foods or nutrients in the development of cancer (or other diseases). Why not? One reason is that the dietary data obtained in such a study are based on population food disappearance data and are therefore not particularly specific. Ecological studies can be used, however, to generate hypotheses about the relationship of dietary components to the disease process, and these hypotheses can then be tested with a more rigorous study design.

Confounding factor (confounder) A "hidden" factor or characteristic that is distributed differently in the study and control groups and may cause an association that the researchers attribute to other factors. Common confounding factors, such as age, gender, race, ethnicity, and dietary or lifestyle factors, can make it difficult to distinguish between a response to treatment and the effect of some other factor. Confounders may be due to bias or to chance.

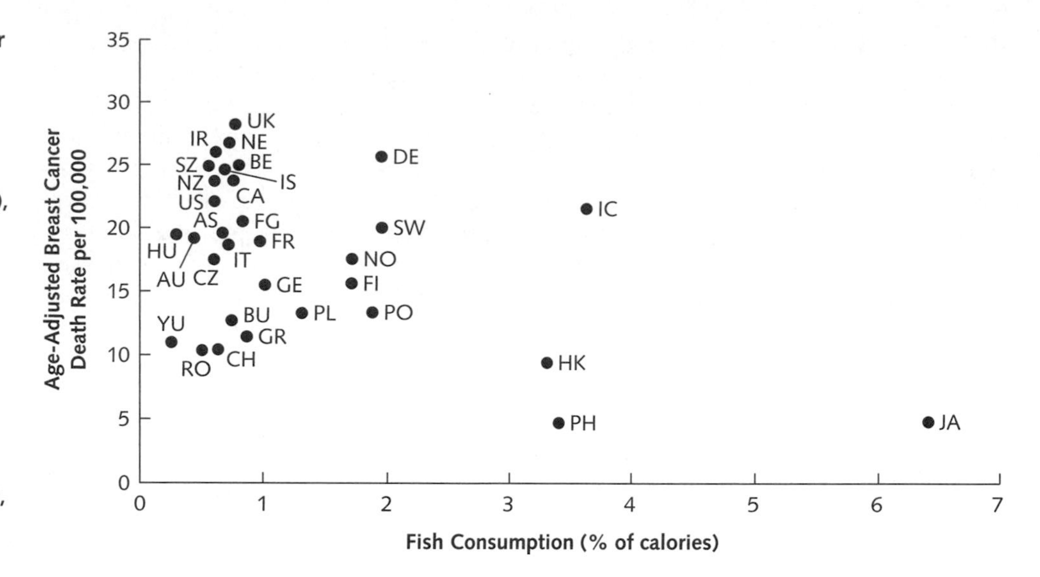

FIGURE 4-6 Breast Cancer Death Rate (per 100,000) Plotted versus Fish Consumption (percentage of caloric intake)

Points include Australia (AS), Austria (AU), Belgium (BE), Bulgaria (BU), Canada (CA), Chile (CH), Czechoslovakia (CZ), Denmark (DE), Federal Republic of Germany (FG), Finland (FI), France (FR), German Democratic Republic (GE), Greece (GR), Hong Kong (HK), Hungary (HU), Iceland (IC), Ireland (IR), Israel (IS), Italy (IT), Japan (JA), Netherlands (NE), Norway (NO), New Zealand (NZ), Philippines (PH), Poland (PL), Portugal (PO), Romania (R), Sweden (SW), Switzerland (SZ), United Kingdom (UK), United States (US), and Yugoslavia (YU).

Source: From "Fish Consumption and Breast Cancer Risk: An Ecological Study" by L. Kaizer, N. F. Boyd, V. Kriukov, and D. Tritchler, 1989, *Nutrition and Cancer* 12, p. 63. Copyright 1989 by Lawrence Erlbaum Associates, Inc. Reprinted with permission.

Cross-sectional or Prevalence Studies

Cross-sectional or prevalence studies examine the relationships among dietary intake, diseases, and other variables as they exist in a population at a particular time. This type of study is much like a camera snapshot. It gives a picture of what is happening within a particular population at a specific time.

An example of a cross-sectional study is the investigation of nutrient intakes and growth of predominantly breastfed infants living in Capulhauac, Mexico, a farming community of about 5,500 Otomi Indians. A representative sample of four- to-six-month-old infants, together with their mothers, was enrolled in the study. The infants' weight, length, and intake of human breast milk were measured, and information about their health and morbidity (for example, diarrhea or respiratory illness) was recorded. The study found that despite these infants' having milk intakes similar to or greater than those of infants from more economically privileged populations (for example, infants living in Houston, Texas), the nutrient intake from breast milk among the Otomi infants was not sufficient for normal growth. Growth faltering was evident by six months of age in this population.[29]

Cohort Studies

Whereas cross-sectional studies are like a snapshot, cohort studies are like moving pictures of events occurring within populations. In these studies, a group of people free of the disease or condition of interest is identified and examined. This group is called the *cohort.* The members of the cohort are followed for months or even years, during which time they are examined periodically to determine which individuals develop the characteristics of interest and which do not. Cohort studies may look back in time to reconstruct exposures and health outcomes; such studies are called *retrospective cohort studies.* Those that follow a group into the future are called *prospective cohort studies.* Consult Table 4-5 for a list of the advantages and disadvantages of this type of study.

Cohort study A type of observational analytic study that can be retrospective or prospective in nature. Enrollment in the study is based on exposure characteristics (for example, past exposure to risk factors) or on membership in a group. Disease, death, and/or other health-related outcomes are then determined and compared.

An example of a **cohort study** (in this case, a prospective study) is the Framingham Heart Study, which was begun in 1949 in Framingham, Massachusetts. This study was undertaken to identify the factors associated with increased risk of coronary heart disease. A representative sample of 5,209 men and women, 30 to 59 years of age, was selected from

TABLE 4-5 Advantages and Disadvantages of Cohort and Case–Control Studies

COHORT STUDIES	CASE–CONTROL STUDIES
Advantages	**Advantages**
May provide complete data on cases, stages	Excellent way to study rare diseases and diseases with long latency periods
Allow study of more than one effect of exposure	Relatively quick
Can calculate and compare rates in exposed and unexposed	Relatively inexpensive
Choice of factors available for study	Require relatively few study subjects
Quality control of data	Can often use existing records
In prospective studies, exposure is assessed prior to diagnosis of disease	Can study many possible causes of disease
Disadvantages	**Disadvantages**
Need to study large numbers	Rely on recall or existing records about past exposures
May take many years	Difficult or impossible to validate data
Circumstances may change during study	Control of extraneous factors incomplete
Expensive	Difficult to select suitable comparison group
Control of extraneous factors may be incomplete	Cannot calculate rates
Rarely possible to study mechanism of disease	Cannot study mechanisms of disease
May provide incomplete data from subject nonresponse and loss to follow-up	Disease may affect the exposure being studied

Source: Adapted from C. W. Tyler, Jr., and J. M. Last, Epidemiology, in *Maxcy-Rosenau-Last Public Health & Preventive Medicine,* 14th ed., ed. J. M. Last and R. B. Wallace (Norwalk, CT: Appleton & Lange, 1998).

the 10,000 or so residents of that age group living in the Framingham community. Of these, 5,127 were free of CHD when they were first examined. The cohort underwent complete physical examinations every 2 years for more than 30 years. The study has shown that the risk of developing CHD is associated with cigarette smoking, serum cholesterol, glucose intolerance, and blood pressure.[30] An analysis of the cohort's high-density lipoprotein cholesterol (HDL-C) levels taken over a period of 12 years revealed that HDL-C is a consistent predictor of CHD risk in both men and women. Individuals with a high HDL-C level (> 60 mg/dL) had a low risk of CHD, regardless of their total cholesterol level.[31] Individuals with a low total blood cholesterol level (< 200 mg/dL) had a high risk of CHD if their HDL-C levels were also low (< 40 mg/dL). The risk of CHD increases with higher total cholesterol levels, but even more strongly with low HDL levels. The ratio of total cholesterol to HDL cholesterol (total cholesterol divided by HDL cholesterol) provided the strongest prediction of relative risk. Thus, in the words of the investigators, "a low total cholesterol level per se does not necessarily indicate a low risk of developing CHD."

Case–Control Studies

In a **case–control study,** a group of persons or cases with the disease or condition of interest are compared with a group of persons without the disease or condition. For example, in a study investigating the effects of high blood levels of homocysteine on risk of CHD, researchers found higher blood levels of homocysteine in people with CHD (cases) than in healthy individuals (controls).[32] Case–control studies are also useful when a rare condition is being studied (see Table 4-5).

Case–control study A type of observational analytic study; enrollment in the study is based on presence (case) or absence (control) of disease. Characteristics, such as previous exposure to a factor, are then compared between cases and controls.

One of the most thorough case–control studies of diet and cancer was a Canadian study published in 1978.[33] In this study, the diets of 400 Canadian women with breast cancer were compared with 400 women from similar neighborhoods who did not have breast cancer (the latter were called "neighborhood controls"). The investigators used three types of diet assessment techniques: the 24-hour recall, a 4-day dietary record, and

A Sampling of Recent Diet-Related Studies from the National Institutes of Health*

Name/Type of Study/Support/Year Begun	Purpose/Objectives
Diabetes Prevention Program: Clinical trial involving 4,000 people at risk of Type 2 diabetes (NIDDK), 1997 (http://diabetes.niddk.nih.gov/dm/pubs/preventionprogram/).	Prevent or delay the development of Type 2 diabetes in people who are at high risk of the disease because of impaired glucose tolerance, by means of intensive lifestyle changes and drug interventions; follow subjects for 3 to 6 years to observe the potential development of diabetes.
Nurses' Health Study I (1976): Cohort study of 121,700 married female registered nurses aged 30 to 55 years; *Nurses' Health Study II* (1989): Cohort study of 116,686 female registered nurses aged 25 to 42 years (NHLBI, NCI) (www.nurseshealthstudy.org).	Investigate diet and lifestyle risk factors for major chronic diseases in women; investigate the potential long-term consequences of the use of oral contraceptives
Health Professionals Follow-Up Study: Cohort study of 51,529 male health professionals aged 40 to 75 years (NHLBI, NCI), 1986 (www.hsph.harvard.edu/hpfs/).	Evaluate a series of hypotheses about men's health, relating nutritional factors to the incidence of serious illnesses, such as cancer, heart disease, and other vascular diseases.
Child and Adolescent Trial for Cardiovascular Health (CATCH): randomized school-based clinical trial involving 5,106 third-grade boys and girls enrolled in 96 ethnically and racially diverse elementary schools in four locations (California, Louisiana, Minnesota, and Texas) (NHLBI), 1994 to 1996 (www.nhlbi.nih.gov).	School-based study to determine whether multicomponent health promotion efforts targeting both children's behaviors and the school environment, including classroom curricula, school food service modifications, physical education changes, and family reinforcement, would reduce cardiovascular disease risk factors later in life.
Pathways Project: Randomized school-based clinical trial involving third-, fourth-, and fifth-grade students (21 schools in intervention group and 20 schools in control group) (NHLBI), 1997 (www.cscc.unc.edu/path).	School-based study for prevention of obesity in Native-American children using four intervention components: physical activity, family involvement, food service, and classroom curriculum.
Honolulu Heart Program: Cohort study of 8,006 American men of Japanese ancestry born between 1900 and 1919 and living in Hawaii—having either migrated from Japan or been born in Hawaii (NHLBI), 1965–1996 (www.nhlbi.nih.gov/resources/deca/descriptions/honolulu.htm).	A 30-year prospective study to investigate relationships among disease frequencies, pathologic fndings, and disease predictors in cohort and compare to other populations (for example, men living in Japan); assess dietary factors (for example, intake of fish, milk and calcium, coffee and caffeine, cholesterol, alcohol, and antioxidants) and incidence of CHD.

a diet history questionnaire. The investigators found that only the intake of total calories differed between cases and controls when the diet was analyzed by the 24-hour recall method. There were no differences in the intakes of total fat and saturated fat between cases and controls as a function of type of diet assessment method. This study did not support the hypothesis that fat composition of the diet is associated with the incidence of breast cancer. The investigators concluded that all three diet assessment methods may have been imperfect measures of dietary intake of fat.

Controlled Trials

The randomized controlled trial or clinical trial conducted as a double-blind experiment is the most rigorous evaluation of a dietary hypothesis. The primary drawback of the controlled trial is its expense. The MRFIT study cited earlier is an example of a randomized controlled trial.

A Sampling of Recent Diet-Related Studies from the National Institutes of Health* —*continued*

Name/Type of Study/Support/Year Begun	Purpose/Objectives
Women's Health Initiative (WHI): Clinical trial involving 93,726 postmenopausal women aged 50 to 79 years (NHLBI, NIAMS, NCI, NIA), 1991 (www.nhlbi.nih.gov/whi).	A 15-year study to examine the extent to which diet, hormone replacement therapy, and calcium and vitamin D might prevent heart disease, breast and colorectal cancer, and osteoporosis in postmenopausal women; identify new risk factors for these conditions, compare risk factors and presence of disease at start of study with new occurrences of disease over extended period of follow-up. Includes randomized control trial of (HRT),† dietary modification (DM), and calcium/vitamin D supplementation (CaD); observational study (OS); and a study of community approaches to developing healthful behaviors (Community Prevention Study—CPS).
Antioxidants, Zinc, and the Age-Related Eye Disease Study (AREDS): Clinical trial involving 4,757 participants aged 55 to 80 years (NEI), 1997 (www.medicalophthalmics.com/brochures/areds_fullstudylink.pdf).	Assess clinical course, prognosis, and risk factors for age-related macular degeneration (AMD) and cataracts; subjects receive high-dose antioxidants or zinc with follow-up for 7 years.
Beat Osteoporosis: Nourish and Exercise Skeletons (BONES) Project: Randomized after school intervention trial involving 1,500 elementary school children aged 6 to 9 years from 84 after school programs in 33 diverse communities (NICHD), 1999 (www.clinicaltrials.gov/ct/show/NCT00065247).	Implement and evaluate an after school program with exercise, education, and diet components designed to improve bone quality and muscle strength in children. After school programs were randomized to either the BONES Project, to the BONES Project plus a parent/caregiver component, or to a no-intervention control group.

*NIDDK = National Institute for Diabetes and Digestive and Kidney Diseases; NHLBI = National Heart, Lung, and Blood Institute; NIA = National Institute on Aging; NEI = National Eye Institute; NIAMS = National Institute of Arthritis, Musculoskeletal and Skin Diseases; NCI = National Cancer Institute; NICHD = National Institute of Child Health and Human Development.

†The estrogen plus progestin (HRT) trial was stopped early in July 2002 after an average follow-up of 5.2 years on the recommendation of the Data and Safety Monitoring Board.

Sources: Adapted from J. Pennington and coauthors, Update: Diet-related trials and observational studies supported by the National Institutes of Health, *Nutrition Today* 35 (2000): 158–160; and information from http://clinicaltrials.gov.

Nutritional Epidemiology

The epidemiologic method lends itself to the study of the relationship of diet to health and disease. Historically, one of the first applications of epidemiology to nutrition science was James Lind's controlled trial investigating the curative effects of citrus fruits among sailors with scurvy.[34] Like Snow's proposed intervention to control cholera epidemics, Lind's suggestion that sea vessels carry a supply of limes, oranges, and other citrus fruits to prevent scurvy came 150 years before researchers proved in the laboratory that scurvy results from a dietary deficiency.[35] Similar investigations into the health effects of vitamin C and the other 40-odd essential nutrients are being conducted today.

Epidemiology has other applications in the nutrition arena. It can be used to monitor and describe the food consumption, nutrient intake, and nutrition status of populations or specific subgroups of a population. The information obtained from large population

FIGURE 4-7 Risk Factors and Chronic Diseases

Source: Adapted from E. Whitney and S. Rolfes, *Understanding Nutrition*, 10th ed. (Belmont, CA: Wadsworth/Thomson Learning, 2004) p. 619.

	Diet Risk Factors						Other Risk Factors					
Chronic Diseases	Diet high in fat, saturated fat, and/or *trans* fat	Excessive alcohol intake	Low complex carbohydrate/ fiber intake	Low vitamin and/or mineral intake	High sugar intake	High intake of salty or pickled foods	Genetics	Age	Excess weight	Sedentary lifestyle	Smoking and tobacco use	Stress
Cancers	✔	✔	✔	✔		✔	✔	✔	✔	✔	✔	
Hypertension	✔	✔		✔		✔	✔	✔	✔	✔	✔	✔
Diabetes (type 2)	✔		✔				✔	✔	✔	✔		
Osteoporosis		✔		✔			✔	✔		✔	✔	
Atherosclerosis	✔	✔	✔	✔			✔	✔	✔	✔	✔	✔
Obesity	✔	✔	✔		✔		✔		✔	✔		
Stroke	✔	✔	✔				✔	✔	✔	✔	✔	✔
Diverticulosis	✔		✔	✔				✔		✔		
Dental and oral disease				✔	✔		✔				✔	

This chart shows that the same risk factor can affect many chronic diseases. Notice, for example, how many diseases have been linked to a sedentary lifestyle. The chart also shows that a particular disease, such as atherosclerosis, may have several risk factors.

surveys is then used to develop specific programs and services for groups whose nutrition status appears to be compromised. Furthermore, the epidemiologic method can be used to evaluate nutrition interventions. This usually involves monitoring the nutrition and health status of a high-risk group of individuals for a period of several months or even years.

See Chapter 6 for more about national nutrition monitoring and related activities, including the National Health and Nutrition Examination Survey (NHANES).

Nutritional epidemiology is a fairly new area of study. Whereas its focus was once the deficiency diseases such as scurvy, beriberi, and rickets, today it is primarily concerned with the major, chronic diseases of the so-called Western world. Chronic diseases, including cardiovascular diseases, diabetes, obesity, cancer, and respiratory diseases, are now the major cause of death and disability worldwide—accounting for 59 percent of the 57 million deaths annually worldwide.[36] Unlike nutritional deficiency states, chronic diseases tend to have many, sometimes interrelated, causes, as illustrated in Figure 4-7.

A relatively few risk factors—high cholesterol, high blood pressure, obesity, physical inactivity, insufficient consumption of fruits and vegetables, smoking, and alcohol use—play a key role in the development of chronic diseases.[37] Epidemiologic evidence suggests that a change in dietary habits and physical activity can substantially influence several of these risk factors.[38]

CDC's state-based Behavioral Risk Factor Surveillance System (BRFSS) can be a source of information on behaviors that increase the risk for chronic diseases, such as CHD.[39] As described in Chapter 6, this system gathers information from adults in all 50 states on knowledge, attitudes, and behaviors related to key health issues, such as tobacco use, dietary patterns, levels of leisure-time physical activity, and use of preventive services. Similarly, CDC's Youth Risk Behavior Surveillance System (YRBSS) provides key data, nationally and by state, about the prevalence of health risk behaviors among young people—including tobacco use, lack of physical activity, and poor nutrition. Using BRFSS and YRBSS data, states can monitor changes in health risk behaviors over time and can better target health promotion efforts to populations most at risk.

The Nature of Dietary Variation

One challenge to the study of the relationship of diet to disease is the complexity of our diets. The foods we consume each day are complex mixtures of chemicals, some of which are known to be important to human health and some of which have not even been identified or measured. The chemicals found in or on foods include essential nutrients, structural compounds (for example, cellulose), additives, microbes, pesticides, inorganic compounds such as heavy metals, natural toxins (nicotine, for example), and other natural compounds such as phytochemicals. The sheer diversity of the chemicals found in foodstuffs creates problems for investigators studying the relationship of diet to disease processes. When assessing the relationship of vitamin A intake to the development of lung cancer, for example, an investigator must calculate the population's intake of compounds with vitamin A activity—that is, preformed vitamin A (retinol), retinal, retinoic acid, and the carotenoids. And not only must the intake of these compounds from foods be considered, but also the intake from vitamin supplements and other sources if they exist. In addition, for most epidemiologic investigations, it is the long-term dietary intake of foodstuffs that is important. In the case of a disease such as lung cancer, which may take 10 to 20 years to develop, the lifelong intake of vitamin A must be estimated.

Another difficulty is that people don't eat the same foods every day. Work and school schedules, illnesses, holidays, the seasons of the year, weekends, personal preferences, availability of foods, social and cultural norms, and numerous other factors influence our daily food choices. As a result, our nutrient intake varies from day to day. The magnitude of the variation differs with the nutrient.[40] Scientists with the Human Nutrition Information Services of the U.S. Department of Agriculture conducted a study to determine the number of days of food intake records needed to estimate the "true" average nutrient intake of a small number of adults.[41] The 29 men and women participating in the study completed detailed food records for 365 consecutive days. The range of days, and the average number of days, of food records required to estimate the "true" average intake for individuals are shown, for several nutrients, in Table 4-6. Note the gender

See Chapter 15 for more about factors that influence our daily food choices.

TABLE 4-6 Range and Average Number of Days Required to Estimate the True Average Intake* for an Individual

	RANGE AND AVERAGE NUMBER OF DAYS REQUIRED					
	Males (n = 13)			Females (n = 16)		
COMPONENT	Minimum	Average	Maximum	Minimum	Average	Maximum
Food energy	14	27	84	14	35	60
Protein	23	36	72	23	48	70
Fat	34	57	131	32	71	114
Carbohydrate	10	37	177	16	41	77
Iron	18	68	130	28	66	142
Calcium	30	74	140	35	88	168
Sodium	27	58	140	36	73	116
Vitamin A	115	390	1,724	152	474	1,372
Vitamin C	90	249	900	83	222	328
Niacin	27	53	89	48	78	126

*The "true" average intake was defined as the 365-day average for individuals.

Source: P. P. Basiotis and coauthors, Number of days of food intake records required to estimate individual and group nutrient intakes with defined confidence. © *J. Nutr.:* (Vol. 117, p. 1640), American Institute of Nutrition.

TABLE 4-7 Number of Days Required to Estimate the True Average Intake* for Groups of Individuals

COMPONENT	ESTIMATED NUMBER OF DAYS REQUIRED FOR EACH GROUP	
	Males (n = 13)	Females (n = 16)
Food energy	3	3
Protein	4	4
Fat	6	6
Carbohydrate	5	4
Iron	7	6
Calcium	10	7
Sodium	6	6
Vitamin A	39	44
Vitamin C	33	19
Niacin	5	6

*The "true" average intake was defined as the 365-day average for groups of individuals.

Source: P. P. Basiotis and coauthors, Number of days of food intake records required to estimate individual and group nutrient intakes with defined confidence. © *J. Nutr.:* (Vol. 117, p. 1641), American Institute of Nutrition.

differences in the average values. For example, more food records were needed to estimate the true average intake of calories by women than by men. Note, too, that there were differences between nutrients. Compare the average values for calories with those for vitamins A and C, sodium, and fat. Finally, note the wide ranges in food records necessary to estimate true intakes. For a reliable estimate of vitamin A intake, 115 days of food records were required for some men, whereas 1,724 days (nearly 5 years) were required for others! As these data demonstrate, a relatively large number of days of food intake records are needed to achieve a certain level of statistical significance for an individual. If the individual data are combined into groups, however, fewer days of food intake records are required, as shown in Table 4-7.

The day-to-day variation in an individual's nutrient intake (called within-person variation) has important implications for nutritional epidemiologic studies. If only one day's intake is determined, then the true long-term nutrient intake may be misrepresented, and, for example, an individual whose vitamin C intake over a period of several days is in fact adequate may be classified as having a low vitamin C intake. The effects of within-person variation on dietary intake must be considered when designing and evaluating studies of the relationship of diet to disease.

FOOD CONSUMPTION AT THE NATIONAL LEVEL

The primary method of assessing the available food supply at the national level is based on food balance sheets. **Food balance sheets** do not measure the food actually ingested by a population. Rather, they measure the food *available* for consumption from imports and domestic food production, less the food "lost" through exports, waste, or spoilage, on a per capita basis. The per capita figures are obtained from the population estimate for the country.

Food balance sheets National accounts of the annual production of food, changes in stocks, imports and exports, and distribution of food over various uses within the country.

Food balance sheets tend to be affected by errors that arise in calculating production, waste, and consumption. Hence, they are not used to describe the nutritional inadequacies of countries. They can be used to formulate agricultural policies concerned with food production and consumption.[42]

FOOD CONSUMPTION AT THE HOUSEHOLD LEVEL

Methods of assessing **household food consumption** consider the per capita food consumption of the household, taking into account the age and sex of persons in the household (or institution), the number of meals eaten at home or away from home, income, shopping practices, and other factors. In most cases, no record is made of food obtained outside the household food supply or of food wasted, spoiled, or fed to pets.

Household food consumption The total amount of food available for consumption in the household, generally excluding food eaten away from home unless taken from home.

FOOD CONSUMPTION BY INDIVIDUALS

A variety of methods are available for estimating dietary intake: diet history, 24-hour dietary recall, food record or diary, and food frequency questionnaire. None of the methods is perfect.[43] All have strong points, weak points, advantages, and disadvantages, as shown in Table 3-6 on page 73.

Dietary recalls are appropriate for assessing the intakes of groups of people, but a single 24-hour recall may not give an adequate picture of a specific individual's usual intake.[44] Food records are often considered the best method of assessing dietary intake, but they are time consuming, and the results may not be accurate if subjects modify their eating habits during the time of the study. Diet histories can provide detailed information, but they require subjects to make judgments about their usual food habits. Food frequency questionnaires provide less detailed information, but they are well suited for use with large groups. These questionnaires should include all important population-specific food sources of the nutrients under investigation.

See Chapter 3 for a review of the most common dietary assessment methods, including dietary recalls, food records, diet histories, food frequency questionnaires, and the use of photography to record dietary intake.

Epidemiology and the Community Nutritionist

The science of epidemiology may seem far removed from the job responsibilities of the community nutritionist, but in fact, it is absolutely essential to the delivery of effective nutrition programs and services. Recall from Chapter 1 that the key roles of the community nutritionist include *identifying* nutritional problems within the community and *interpreting* the scientific literature—especially experimental, clinical, and epidemiologic nutrition research findings—for the public and other health professionals. The community nutritionist must be able to critically evaluate the scientific literature before formulating new nutrition policies or offering advice about eating patterns.

How is this accomplished, considering the number of research findings published and the complexity of the diet–disease relationship? As the criteria in Table 4-8 indicate, it helps to consider certain elements in judging the strength of epidemiologic associations. Interpreting epidemiologic data basically involves two steps: (1) Evaluate the criterion for a causal association carefully, and (2) assess the causal association critically for the presence of bias and the contribution of chance. Competence in this area is achieved to some degree by experience and dogged determination. This chapter's Professional Focus feature describes some of the journals and newsletters that will be important to you as a community nutritionist and explains how you can critically analyze a study's results.

TABLE 4-8 Criteria for Evaluating the Strength of the Association between Variables or Outcomes

Chronological Relationship	Exposure to the causative factor must occur before the onset of the disease.
Strength of Association	The association is strong if all those with a health problem have been exposed to the agent believed to be associated with this problem and only a few in the comparison group have been so exposed.
Intensity or Duration of Exposure	The association is likely to be causal if those with the most intense, or longest, exposure to the agent have the greatest frequency or severity of illness, while those with less exposure are not as ill. This is also referred to as a dose–response relationship.
Specificity of Association	Does the removal of the agent or risk factor lead to a reduction in risk of the disease? The likelihood of a causal association is increased if an agent, or risk factor, can be isolated from others and shown to produce changes in the frequency of occurrence or severity of the disease.
Consistency of Findings	Have similar results been shown in other studies? An association is consistent if it is confirmed by different investigators, in different populations, or using different methods of study.
Plausibility	Is the association consistent with other knowledge? Evidence from experimental studies and evidence from other forms of observation are among the kinds of evidence to be considered.

Source: Adapted from C. W. Tyler, Jr., and J. M. Last, Epidemiology, in *Maxcy-Rosenau-Last Public Health & Preventive Medicine,* 14th ed., ed. J. M. Last and R. B. Wallace (Norwalk, CT: Appleton & Lange, 1998).

Internet Resources

Arbor Nutrition Guide *A directory of food and nutrition resources.*	**www.arborcom.com**
Centers for Disease Control and Prevention *Morbidity and Mortality Weekly Report, Surveillance Summaries.*	**www.cdc.gov/mmwr**
Directory of Nutrition Resources *Resources from the Food and Nutrition Information Center of the National Agricultural Library.*	**www.nal.usda.gov/fnic**
National Agricultural Library *Includes online publications, nutrient databases, software, and information centers.*	**www.nalusda.gov**
National Institutes of Health *Links to the various institutes, centers, and divisions of the NIH.*	**www.nih.gov**
Nutrition Organizations *A list compiled by the American Society for Clinical Nutrition.*	**www.ascn.org/relate.htm**
Tufts University Health Sciences Library, Nutrition Research Guide *Nutrition links include guides, databases, electronic journals, government resources, listservs, newsgroups, and more.*	**www.library.tufts.edu/hsl/subjectGuides/nutrition.html**

Case *Study*

Epidemiology of Obesity

By Alice Fornari, EdD, RD, and Alessandra Sarcona, MS, RD

Scenario

You have been hired as a nutritionist by a county public health agency in southern California. This agency has never had a nutritionist. Health professionals at the agency would like to have the nutritionist implement new programs. The population served by this agency is 10 percent Caucasian, 15 percent African American, and 75 percent Mexican American. Your first task is to review the health programs already in place at the agency. Second, you research nutrition education programs instituted at other local community agencies. Your next step is prioritizing the nutritional needs of your target population. In your review of health records at the agency, you note a high incidence of overweight and obesity, especially among the Mexican-American population. You also note that the incidence is very high in preschool-aged children. You calculate that 30 percent of these children have a BMI that exceeds the 85th percentile of U.S. standards for body mass index (BMI). Another critical issue is the occurrence of baby bottle tooth decay (BBTD) among this age group. The majority of your population are low-income (100 percent) and speak limited English, with Spanish being their primary/native language (80 percent).

Learning Outcomes

- Utilize data from epidemiologic research to identify nutrient needs and concerns of a target population.
- Identify and prioritize the needs of a target population.
- Construct goals with intervention strategies for a target population.
- Demonstrate the process of program development.

Foundation: Acquisition of Knowledge and Skills

1. Define epidemiology.
2. What are the U.S. standards for defining overweight and obesity in the pediatric population? Compare the American Academy of Pediatrics (AAP) definition of *at risk of overweight* and *overweight* in the pediatric population (see www.aap.org/obesity) to the definition of childhood obesity given by the Institute of Medicine (IOM). Go to www.iom.edu and view: *Preventing Childhood Obesity: Health in the Balance,* 2005, or www.iom.edu/object.file/master/22/606/0.pdf to see: *Childhood Obesity in the U.S.: Facts and Figures.*

Step 1: *Identify Relevant Information and Uncertainties*

1. Go to www.cdc.gov/nchs/data/nhanes/databriefs/adultweight.pdf and outline the trends of overweight and obesity from the National Health and Nutrition Examination Surveys: NHANES II, NHANES III and NHANES 1999–2000.
2. Go to www.nhlbi.nih.gov/meetings/workshops/hispanic.htm and scroll down to *Obesity and Physical Activity*. What is the prevalence of obesity among Mexican Americans?
3. What are some of the determinants of obesity in the Hispanic population that may be relevant to your target population? Use the websites above and other resources relevant to your target population.
4. What are some of the uncertainties not presented in the data or the case?

Step 2: *Interpret Information*

1. Communicate, in a memo to your agency supervisor, that you wish to begin a program on overweight prevention for preschool-aged Mexican-American children. To support your program, include a brief review of recent epidemiologic studies that reveal the national epidemic of obese children and the health risks that may afflict these children, with a special emphasis on data for your target population.

Step 3: *Draw and Implement Conclusions*

1. Identify the most critical nutritional needs of your target population, and set priorities among them based on the data reviewed and information presented in the case.

Step 4: *Engage in Continuous Improvement*

1. Outline three major goals for your target population, including intervention strategies that coincide with these goals. What may be some limitations in carrying out your intervention strategies?

Professional Focus

The Well-Read Community Nutritionist

Here is a sobering statistic: More than 20,000 biomedical journals are published each month. If each journal published just 20 research articles, the conscientious community nutritionist would need to browse through 400,000 articles per month, or more than 13,000 articles per day, to stay abreast of current scientific findings! And this figure doesn't include editorials, review articles, and letters to the editor—all valuable reading. How can you, as a busy community nutritionist, handle this volume of information, keep yourself informed, and maintain your sanity? We don't have all the answers, but we offer the following suggestions to help you cope with the onslaught of medical, health, and nutrition information. Let's begin with a few good reasons for reading the literature.

Ten Good Arguments for Reading Journals

Consider the following ten good reasons for reading journals regularly:[1]

1. To impress others
2. To keep abreast of professional news
3. To understand pathophysiology
4. To find out how a seasoned health practitioner handles a particular problem
5. To find out whether to use a new or an existing diagnostic test, survey instrument, or educational tool with your patients or clients
6. To learn the clinical features and course of a disorder
7. To determine etiology or causation
8. To distinguish useful from useless or even harmful therapy
9. To sort out claims concerning the need for and the use, quality, and cost-effectiveness of clinical and other health care
10. To be titillated by the letters to the editor

Regular reading of the literature, especially in your area of specialization, is a must. There is no other way to learn about the latest scientific findings, the merits of a particular intervention or assessment instrument, or current legislation and its potential impact on your programs and clients. In short, to be an effective community nutritionist, you must constantly increase and update your knowledge base through regular perusal of journals, which brings us to our next question.

Which Journals Should You Read?

There is no hard and fast rule about the "best" journals to read, because so much depends on the type of work you are doing and the needs of your clients. Some journals will appear on your "must read regularly" list; others can be spot-checked every month or two. Although nutrition journals will take priority, other specialty journals, particularly in the disciplines of epidemiology, health education, and medicine, are important. Presented here is a list of journals, newsletters, and other publications that will help you stay abreast of current developments that may be useful to you in delivering community programs.

Nutrition Journals

American Journal of Clinical Nutrition
Canadian Journal of Dietetic Practice and Research
Human Nutrition: Applied Nutrition
Journal of the American College of Nutrition
Journal of the American Dietetic Association
Journal of Hunger and Environmental Nutrition
Journal of Nutrition
Journal of Nutrition Education and Behavior
Journal of Nutrition for the Elderly
Journal of Obesity Research
Nutrition Research
Nutrition Reviews
Nutrition Today
Topics in Clinical Nutrition

Nutrition Newsletters

Nutrition Week
Dairy Council Digest
FDA Consumer
Harvard Health Letter
Nutrition Action Health Letter
Nutrition and the M.D.
Tufts University Health & Nutrition Letter
University of California at Berkeley Wellness Letter

Specialty Journals

American Journal of Epidemiology
American Journal of Health Promotion
American Journal of Public Health
Annals of Internal Medicine
Health Education
Health Education Quarterly
Food, Nutrition, and Agriculture
Journal of the American Medical Association
Journal of Human Lactation
Journal of Medicine and Science in Sports and Exercise
Journal of the National Cancer Institute
Lancet

New England Journal of Medicine
Pediatrics
Preventive Medicine
Public Health Reports
Science

Other Publications

American Council on Science and Health Reports
Bulletin of the World Health Organization
Science News
Read consumer magazines, newspapers, and journals in other disciplines. Read anything you can get your hands on!

How to Get the Most Out of a Journal

There is no best way to "read" a journal. Although you will want to develop your own reading style, consider the following points. A glance at the table of contents will direct you to pertinent research articles and briefs for in-depth reading. In journals that you subscribe to personally, highlight special-interest articles in the table of contents with a colored marking pen. (This simple act may make it easier for you to remember, for example, where you spotted that intriguing article on monounsaturated fat and Type 2 diabetes mellitus.)

After scanning the table of contents, check the professional updates and news features to keep informed about key players, committees, conferences, and events in your discipline. Consult review articles for extensive coverage of current issues. The journal's editorials will expose you to the controversies surrounding a study's findings or the implications of the findings for practitioners. Regular reading of the letters to the editor will help you appreciate the flaws that plague some study designs and will expose you to the questions raised by scientists and practitioners in interpreting study results.

In choosing articles for in-depth reading, be selective and discriminating. You have only so much time. Select those articles that appear to be directly relevant to your needs, but allow time for other articles of interest. Refrain from agonizing because you don't have time to read as much as you think you should. Be organized and disciplined in your reading, but accept the fact that no one can read everything.

How to Tease Apart an Article

Most research articles have the same basic format with the following specific sections to help you assimilate the material:[2]

- **Abstract or summary.** Provides an overview of the study, highlights the results, and indicates the study's significance. It should contain a precise statement of the study's goal or purpose.
- **Introduction.** Presents background information such as the history of the problem or relevant clinical features. It reviews the work of other scientists in the area and describes the rationale for the study.
- **Methods.** Describes the study design, selection of subjects, methods of measurement (for example, the diet assessment instrument used), specific hypotheses to be tested, and analytical techniques (for instance, the method used to measure blood cholesterol or the method of statistical analysis).
- **Results.** Details the study's outcomes. The results are typically presented in tables, graphs, charts, and figures that help summarize the study's findings.
- **Discussion.** Provides an analysis of the meaning of the findings and compares the study's findings with those of other researchers. The discussion includes a critique of the work: What were the limitations of the study design? What problems occurred that may have affected the study outcome? What were the study's strengths?
- **Conclusions/implications.** Some, but not all, articles include a short section that summarizes the findings or considers how the study results can be applied to practice. This section may also comment on directions for future research.
- **References/bibliography.** Cites the relevant work of other scientists that the present investigators considered in conducting their study or interpreting the results.

Reading an article involves more than simply scanning the abstract and flipping to the last paragraph of the discussion section for the author's summary statements. If you have decided that the article is important to you, take time to read the methods sections carefully, for the substance of the work is outlined there. Any new information presented in the article is only as good as the method by which it was obtained. Was the hypothesis clearly stated? Was the study design clearly described? Were the methods used appropriate for testing the hypothesis? How were the data collected and analyzed? Once you learn to review the methods section critically, you will find that in some cases you do not need to read further. The study was so poorly designed or seriously flawed that the results lack validity.

One other important precept remains: Learn to form your own opinions about the study findings presented in the articles you read. Do not automatically assume that the findings are valid merely because the study was published by a leading researcher or research team. Reading the letters to the editors and talking about the study results with your colleagues are good ways to help you assess the validity of a study.

What Else Should You Read?

What else should you read to be an informed, effective community nutritionist? Everything! Well, everything you can get your hands on: newsletters, books, consumer magazines, food labels, newspapers, advertising, menus, junk mail, Internet websites.

Are we serious? Absolutely. We said at the outset (in Chapter 1) that one of your roles as a community nutritionist is to improve the nutrition and health of individuals living in communities. To do this well, you must be able to draw on many diverse elements within your community and culture to shape a program that meets a nutritional or health need. Let's say that you have been asked to design, implement, and evaluate a program to reduce the prevalence of obesity among schoolchildren in your community. To develop a nutrition and fitness program that appeals to children, you must be able to speak their language and get inside *their* culture. Which reading material will help you find the right approach, the right action figures, the right tone? Everything from the *Journal of the American Dietetic Association* to children's books, advertising inserts in your local newspaper, the newspaper comics, scripts of popular TV programs, fast-food menus, T-shirt logos, a *Newsweek* article on latchkey children, a government publication on quick snack ideas—the list is endless. You never know when something you read in a totally unrelated area will be just the thing you need to help convey a nutrition message to your clients.

References

1. D. L. Sackett, How to read clinical journals. I. Why to read them and how to start reading them critically, *CMA Journal* 124 (1981): 555–58.
2. S. H. Gehlbach, *Interpreting the Medical Literature—A Clinician's Guide* (New York: Macmillan, 1982), pp. 1–15.

Chapter *Five*

Food Insecurity and the Food Assistance Programs

Learning *Objectives*

After you have read and studied this chapter, you will be able to:

- Communicate the current status of food security in the United States.
- Understand the complexity of domestic food insecurity.
- Explain the significance and relevance of food security to dietetics professionals.
- Describe current food security and hunger policy initiatives.
- Describe the purpose, status, and current issues related to the U.S. food assistance programs.
- Describe actions that individuals might take to eliminate food insecurity.

This chapter addresses such issues as the health behaviors and educational needs of diverse populations; economics and nutrition; the availability of food and nutrition programs in the community; food availability and access for the individual, family, and community; local, state, and national food security policy; food and nutrition laws, regulations, and policies; the influence of socioeconomic, cultural, and psychological factors on food and nutrition behavior; and current information technologies, which are Commission on Accreditation for Dietetics Education (CADE) *Foundation Knowledge and Skills* requirements for dietetics education.

Chapter *Outline*

Something To Think About...

Everyone has the right to a standard of living adequate for the health and well-being of himself [herself] and his [her] family, including food, clothing, housing, and medical care and necessary social services, and the right to security in the event of unemployment, sickness, disability, widowhood, old age, or other lack of livelihood in circumstances beyond his [her] control.

– THE UNITED NATIONS GENERAL ASSEMBLY'S UNIVERSAL DECLARATION OF HUMAN RIGHTS, ADOPTED MORE THAN 50 YEARS AGO

Introduction

Although problems of overnutrition—obesity, heart disease, diabetes, cancer, and others—plague society, not everyone shares these problems. People in developing nations, as well as people in the less privileged parts of developed nations, suffer the problems of undernutrition. Characterized by chronic debilitating hunger and malnutrition, undernutrition has been a problem throughout history, and despite numerous development and assistance programs, hunger and malnutrition are not disappearing. They can be found among people of both genders and of all ages and ethnic backgrounds. Even so, these problems hit some groups harder than others. Severe forms of undernutrition are not characteristic of the United States, but the problem of households not being able to consistently access food persists.

Everyone has known the uncomfortable feeling of hunger pains, which pass with the eating of the next meal. But many people know hunger more intimately, because often a meal does not follow to quiet the signal. For them, hunger is a constant companion, bringing ceaseless discomfort and weakness—the continuous lack of food and nutrients. People who live with chronic hunger may have too little food to eat or may not receive an adequate intake of the essential nutrients from the foods available to them; either way, malnutrition ensues.[1] The conceptual model found in Figure 5-1 shows the large number of interrelated factors associated with food insecurity and its outcomes.

Today this phenomenon is discussed in terms of **food security or food insecurity with or without hunger.**[2] Food security—access by all people at all times to enough food for an active, healthy life—is one of several conditions necessary for a population to be healthy and well nourished.[3] The concept of food security includes five components:[4]

- Quantity: Is there access to a sufficient quantity of food?
- Quality: Is food nutritionally adequate?
- Suitability: Is food culturally acceptable and the capacity for storage and preparation appropriate?
- Psychological: Do the type and quantity of food alleviate anxiety, lack of choice, and feelings of deprivation?
- Social: Are the methods of acquiring food socially acceptable?

A writer in Boston described food insecurity in more personal terms:[5]

> I've had no income and I've paid no rent for many months. My landlord let me stay. He felt sorry for me because I had no money. The Friday before Christmas he gave me $10. For days I have had nothing but water. I knew I needed food; I tried to go out but I was too weak to walk to the store. I felt as if I were dying. I saw the mailman and told him I thought I was starving.

Food security Access by all people at all times to sufficient food for an active and healthy life. Food security includes at a minimum the ready availability of nutritionally adequate and safe foods and the ability to acquire them in socially acceptable ways (without resorting to emergency food sources, scavenging, stealing, or other coping strategies to meet basic food needs).

Food insecurity Limited or uncertain ability to acquire or consume an adequate quality or sufficient quantity of food in socially acceptable ways (for example, not knowing where one's next meal is coming from constitutes food insecurity).

Food insecurity with hunger The uneasy or painful sensation caused by a recurrent or involuntary lack of food, which can lead to malnutrition over time.

- *Moderate:* hunger among adults but not among children.
- *Severe:* hunger among children and more severe hunger among adults.

FIGURE 5-1 Conceptual Model of Factors Associated with Food Insecurity and Its Outcomes

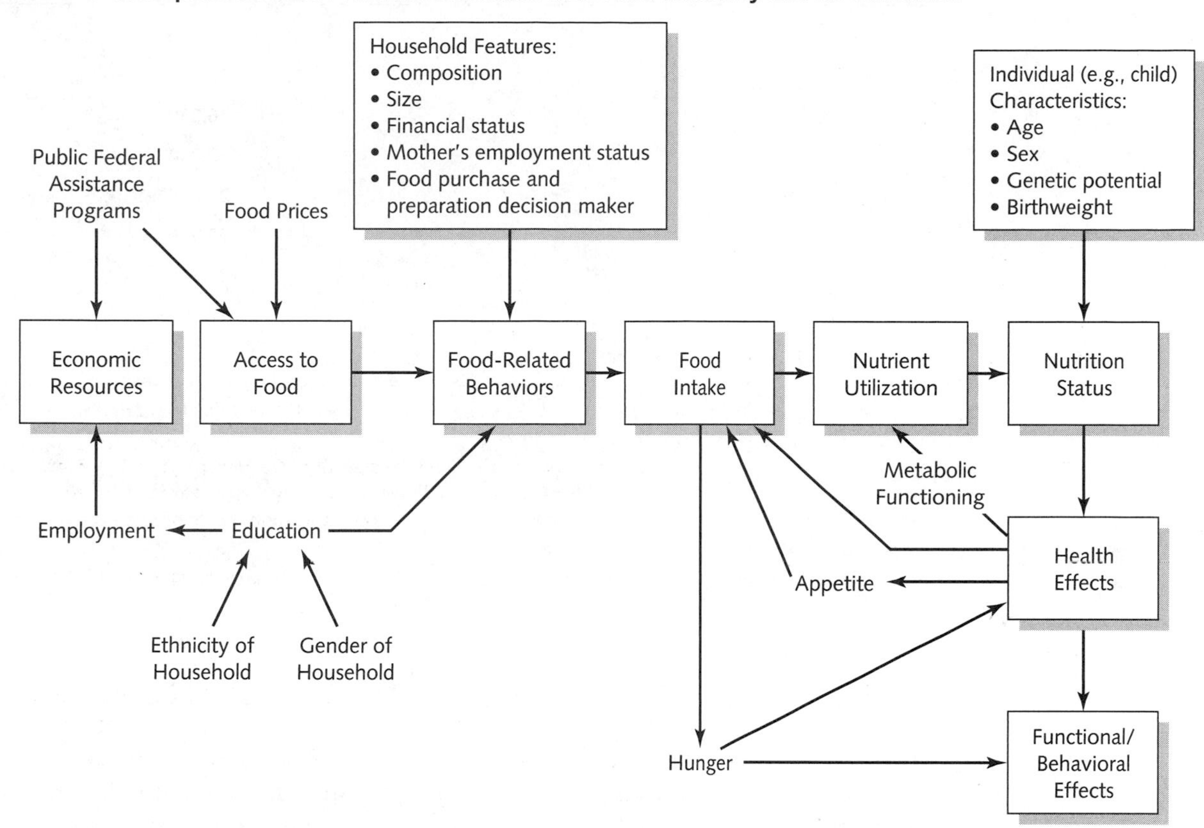

Source: Adapted from C. A. Wehler, R. I. Scott, and J. J. Anderson, The Community Childhood Identification Hunger Project: A model of domestic hunger—demonstration project in Seattle, Washington. Reprinted with permission. *Journal of Nutrition Education* 24 (1992): 325. Society for Nutrition Education.

He brought me food and then he made some phone calls and that's when they began delivering these lunches. But I had already lost so much weight that five meals a week are not enough to keep me going.

I just pray to God I can survive. I keep praying I have the will to save some of my food so I can divide it up and make it last. It's hard to save because I am so hungry that I want to eat it right away. On Friday, I held over two peas from the lunch. I ate one pea on Saturday morning. Then I got into bed with the taste of food in my mouth and I waited as long as I could. Later on in the day I ate the other pea.

Today I saved the container that the mashed potatoes were in and tonight, before bed, I'll lick the sides of the container. When there are bones I keep them. I know this is going to be hard for you to believe and I am almost ashamed to tell you, but these days I boil the bones until they're soft and then I eat them. Today there were no bones.

Food insecurity without hunger The condition prevailing in households that experience uncertain access to sufficient food, are concerned about inadequate resources to buy enough food, and cannot afford to eat balanced meals.

This chapter examines the extent of hunger and food insecurity in the United States; Chapter 13 examines the incidence of hunger around the world. Both chapters offer suggestions for personal involvement with the issues presented. As you read, challenge yourself with the following questions: What problems would you attack first in solving the problem of hunger and food insecurity? Should we solve our own hunger problems before tackling problems related to international hunger? Although these issues are complex and often overwhelming from an individual's standpoint, the virtual elimination of hunger, as well as the reduction of poverty, is within our nation's reach.[6]

Healthy People 2010

The *Healthy People 2010* initiative set a goal of increasing the rate of food security among U.S. households from 88 percent to 94 percent and, in so doing, reducing hunger by the end of the decade.

TABLE 5-1 Annual Poverty Guidelines, 2005

HOUSEHOLD SIZE	POVERTY GUIDELINE* (100% Poverty)†
1	$ 9,570
2	12,830
3	16,090
4	19,350
5	22,610
6	25,870
7	29,130
8	32,390
For each additional family member, add $3,260.	

*The poverty guideline for Alaska starts at $11,950 and rises by increments of $4,080, and that for Hawaii starts at $11,010 and rises by increments of $3,750.

†This table shows income levels equal to the poverty line (100 percent of the poverty line). Programs sometimes set program income eligibility at some point above the poverty line. For example, if a program sets income eligibility at 130 percent of the poverty line then the cutoff for a family of two living in the 48 contiguous states is $12,830 × 130% = $16,679.

Source: 2005 HHS Poverty Guidelines, *Federal Register,* Vol. 70, No. 33, February 18, 2005, pp. 8373–8375.

Counting the Hungry in the United States

The United States is the world's biggest food exporter and one of the wealthiest nations in the world. Yet food insecurity and hunger persist, especially among the impoverished. In 2003, 12.5 percent of people in the United States (almost 36 million people) lived in poverty.[7] The poverty line was developed in 1963–1964 from the food budgets conceived by the U.S. Department of Agriculture for economically stressed families and considered the proportion of their income spent on food.[8] This work became the basis for the current poverty thresholds—one of two slightly different versions of the federal poverty line—which are adjusted annually to reflect changes in the consumer price index for urban consumers (CPI-U). The *poverty thresholds,* which are the dollar amounts below which a family would be viewed as living in poverty, are used for calculating all official poverty population statistics (for example, estimating the number of children and families in poverty each year). The food budget used in the poverty threshold calculation reflects a diet that is just barely adequate—one designed for short-term use when funds are extremely low.[9] Therefore, people with incomes below the poverty threshold undoubtedly have a difficult time buying nutritionally adequate foods even for a short-term, emergency diet.

The *poverty thresholds* are available on the Census Bureau website at www.census.gov. The *poverty guidelines* are sometimes loosely referred to as the "federal poverty level" or "poverty line." Updates to the poverty guidelines can be found at http://aspe.hhs.gov/poverty/index.shtml.

The *poverty guidelines* are the other version of the federal poverty measure. The poverty guidelines are a simplified version of the poverty thresholds and are used for administrative purposes (for example, determining eligibility for the Food Stamp Program). The Department of Health and Human Services issues the poverty guidelines each year based on the previous year's poverty thresholds. The 2005 poverty guidelines listed in Table 5-1 are approximately equal to the 2004 poverty thresholds. If you wanted to assess a person's income by comparing it with the "poverty line," you could use either the thresholds or the guidelines.

Despite their inadequacies, the official poverty guidelines define eligibility for many federal assistance programs; however, government aid programs do not have to use the official poverty measure as their eligibility criterion.[10] "Needy" individuals with incomes a certain amount above the poverty line are automatically ineligible for programs such as food stamps or free and reduced-price school meals. Such criteria mean that some

TABLE 5-2 Measuring Food Security: Questions from the Food Security Survey Module Used in the Current Population Survey and Other Surveys/Studies[a,b]

1. We worried whether our food would run out before we got money to buy more.
2. The food that we bought just didn't last, and we didn't have money to get more.
3. We couldn't afford to eat balanced meals.
4. We relied on only a few kinds of low-cost food to feed our children because we were running out of money to buy food.
5. We couldn't feed our children a balanced meal, because we couldn't afford that.
6. Our children were not eating enough because we just couldn't afford enough food.
7. In the last 12 months, did you or other adults in your household ever cut the size of your meals or skip meals because there wasn't enough money for food?
8. How often did this happen?[c]
9. In the last 12 months, did you ever eat less than you felt you should because there wasn't enough money for food?
10. In the last 12 months, were you ever hungry but didn't eat because you couldn't afford enough food?
11. In the last 12 months, did you lose weight because you didn't have enough money for food?
12. In the last 12 months, did you or other adults in your household ever not eat for a whole day because there wasn't enough money for food?
13. How often did this happen?[c]
14. In the last 12 months, did you ever cut the size of your children's meals because there wasn't enough money for food?
15. In the last 12 months, were the children ever hungry but you just couldn't afford more food?
16. In the last 12 months, did your children ever skip a meal because there wasn't enough money for food?
17. How often did this happen?[c]
18. In the last 12 months, did your children ever not eat for a whole day because there wasn't enough money for food?

[a]The questionnaire for households with no children has 10 items (1–3 and 7–13).
[b]To score the questionnaire for a household with children, a "yes" response to any 0 to 2 items is considered food secure; a "yes" response to any 3 to 7 items is food insecure without hunger; 8 to 12 items, food insecure with moderate hunger; 13 to 18 items, food insecure with severe hunger. For households with no children, a "yes" response to any 0 to 2 items is considered food secure; 3 to 5 items, food insecure without hunger; 6 to 8 items, food insecure with moderate hunger; and 9 to 10 items, food insecure with severe hunger.
[c]Counted as a "yes" if it occurred in 3 or more months during the previous year.

Source: J. S. Hampl and R. Hall, Dietetic approaches to U.S. hunger and food insecurity, *Journal of the American Dietetic Association* 102 (2002): 921. For further information on measuring household food security status, see G. Bickel and coauthors, *Guide to Measuring Household Food Security, Revised 2000* (Alexandria, VA: U.S. Department of Agriculture, Food and Nutrition Service, 2000).

households are unable to participate in particular programs, which may prevent them from escaping poverty.

Since 1995, the USDA has monitored food security through an annual survey of 50,000 to 60,000 households, conducted as a supplement to the U.S. Census Bureau's nationally representative Current Population Survey.[11] Households without children are asked a series of 10 questions; households with children are asked 18 questions (see Table 5-2). The questions address such issues as

- Fear and anxiety related to the insufficiency of the food budget to meet basic needs;
- Experiencing food shortages without having the money to purchase more;
- Perceived quality and quantity of food eaten by household members;
- Atypical food usage (substituting fewer or cheaper foods); and
- Episodes of reduced food intake, hunger, or weight loss by household members.[12]

These questions reflect the different stages that households go through as food insecurity worsens—from worrying about running out of food to children missing meals for

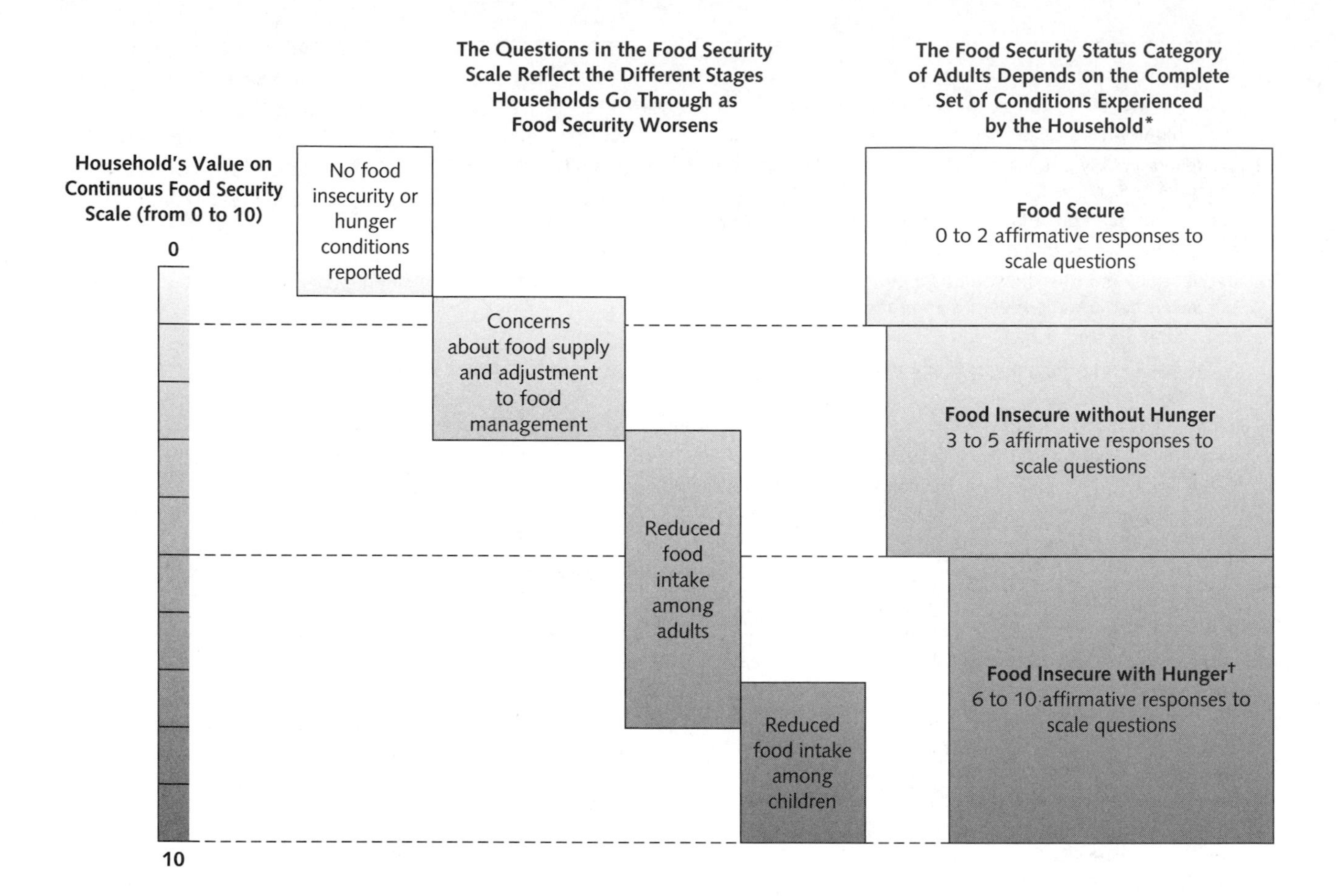

FIGURE 5-2 Continuous and Categorical Measures of Household Food Security

*To categorize a household with children, a "yes" response to any 0 to 2 items is considered food secure; a "yes" response to any 3 to 7 items is food insecure without hunger; 8 to 18 items, food insecure with hunger (8 to 12 items, food insecure with moderate hunger; and 13 to 18 items, food insecure with severe hunger).

†For households with no children, this category is further divided: A "yes" response to 6 to 8 items, food insecure with moderate hunger; and a "yes" response to 9 to 10 items, food insecure with severe hunger.

Source: J. F. Guthrie and M. Nord, Federal activities to monitor food security, *Journal of the American Dietetic Association* 102 (2002): 906.

a whole day. A scale measuring the food security status of each household is calculated on the basis of the household's answers to the questions. This food security scale locates each household along a continuum extending from fully "food secure" at one end to "food insecure with hunger" at the other end, as shown in Figure 5-2. Households are categorized in one of three ways:[13]

1. *Food secure:* Households with no or minimal evidence of food insecurity.
2. *Food insecure without hunger:* Households experiencing uncertain access to sufficient food, concerned about inadequate resources to buy enough food, and who couldn't afford to eat balanced meals.
3. *Food insecure with hunger:* Households in which the adults have decreased the quantity as well as the quality of food they consume (because of lack of money) to the point where they show clear evidence of a repeated pattern of hunger. This category includes households who have indicated that because of constrained resources, their children were not eating enough and that they had, at times, been forced to cut the size of their children's meals in order to make ends meet.

If there are children in the household, the full set of 18 questions is used to identify food insecure households and a subset of the questions is used to identify those with hunger among children (see questions 4–6 and 14–18 in Table 5-2). Households with hunger among children are those in which children's food intake had been reduced even further, and children were hungry because of lack of financial resources.[14] In this case, adults' food intakes are typically also severely reduced.

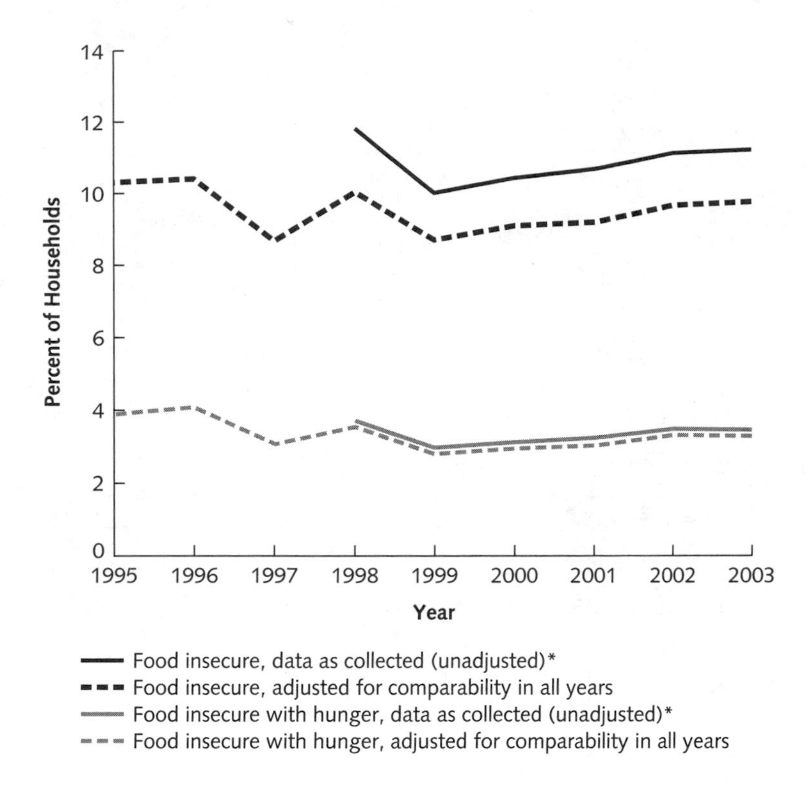

FIGURE 5-3 Trends in U.S. Household Food Insecurity and Hunger, 1995–2003

Food insecurity in the United States has decreased since its first measurement in 1995.

*Data collected in 1995–1997 are not directly comparable with data collected in 1998–2003.

Source: M. Nord and coauthors, *Household Food Security in the United States, 2003* (Alexandria, VA: Economic Research Service, Food and Rural Economics Division, 2004), Report No. 42 (FANRR-42); also available at www.ers.usda.gov/Briefing/FoodSecurity/trends/.

Who Are the Hungry in the United States?

Over the past two decades, numerous studies on food security have been conducted throughout the United States, supporting the conclusion that food insecurity continues to be a serious and persistent domestic problem, which may lead to physical, social, and mental health problems, including overweight and obesity.[15] Over 12 million households, representing over 36 million people, experienced hunger or the risk of hunger in 2003 because of lack of resources.[16] Figure 5-3 summarizes the trends in food insecurity and hunger in the United States from 1995 to 2003 and shows that food insecurity has decreased over the years, since its first measurement in 1995. Poverty and hunger affect certain socioeconomic, geographical, and demographic groups more than others. As illustrated in Figure 5-4, the millions who experience food insecurity and hunger today in the United States include the following.

THE POOR

The most compelling single reason for this hunger is **poverty.**[17] In 2003, as noted in Figure 5-4, those living below the poverty threshold experienced food insecurity and hunger at over 3 times the national average. However, with an income of 185 percent of the poverty threshold and over, food insecurity and hunger plummeted to 4.9 percent, less than half the national average.[18]

Poverty The state of having too little money to meet minimum needs for food, clothing, and shelter. As of 2005, the U.S. Department of Health and Human Services defined a poverty-level income as $19,350 annually for a family of four.

THE WORKING POOR

A job that pays the minimum wage does not lift a family above the federal poverty threshold, and many such jobs fail to provide fringe benefits to help meet rising health care costs. States differ in the minimum wage they allow, but the federal minimum wage is $5.15 per hour.[19] Some states that require hourly rates greater than the federal

FIGURE 5-4 Prevalence of Food Insecurity and Hunger in U.S. Households, 2003

Source: Prepared by ERS (Economic Research Service) based on data from the December 2003 Current Population Survey Food Security Supplement. [M. Nord and coauthors, *Household Food Security in the United States, 2003* (Alexandria, VA: Economic Research Service, Food and Rural Economics Division, 2004), Report No. 42 (FANRR-42).]

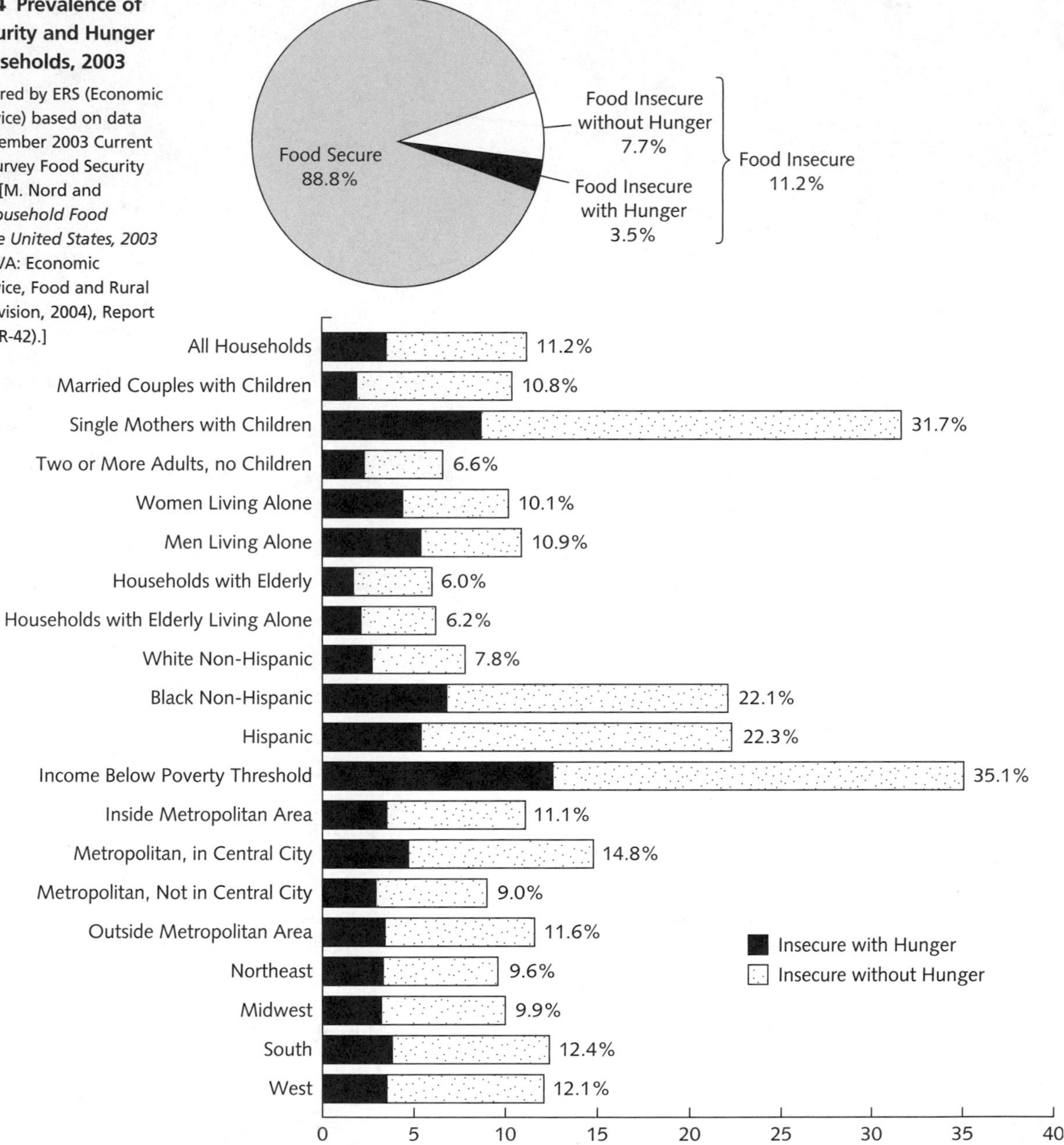

minimum wage recognize that the minimum wage, even at 40 hours per week, is insufficient for a family's survival needs.[20] However, some 10 million or more workers earn the federal minimum wage. One-third of those who earn between $5.15 and $6.15 an hour—working mostly in the retail and service trades—are the sole income producers for their families; 75 percent are 21 or older; and 60 percent of these workers are women.[21] Not surprisingly, 39 percent of emergency food recipient households (those served in soup kitchens, food pantries, and shelters) have at least one adult working.[22] Consider the hypothetical budget scenario shown in the box.

FOOD BUDGET EXERCISE*

Imagine that you are a single parent with three children. You take a job paying $9 an hour (more than minimum wage), without benefits, in the fast-food industry. You work 40 hours per week for 52 weeks. Is your income above or below the poverty threshold (refer to Table 5-1)? After you subtract the monthly expenses shown here for rent, utilities, transportation, and child care, how much money remains in your budget to feed yourself for one day?

1. Total number of household members ___4___
2. Monthly income $ ___1,560___ /month

Monthly expenses:
3. Rent (or mortgage) $ ___725___
4. Heat/electricity $ ___150___
5. Transportation $ ___65___
6. Phone $ ___25___
7. Child care $ ___475___
8. TOTAL (add lines 3 through 8) $ ___1,440___

To calculate money for food for yourself for one day:

9.	Monthly income	(line 2) $ ______
10.	Total expenses	(minus line 8) $ ______
11.	Money left over for food	(equals) $ ______
12.	Divided by number of people in household	(divide by line 1) $ ______
13.	Food money for one day	(divide by 30 days) $ ______

Money available for food for yourself, per day (line 13) $ ______

Calculations do not include emergencies, sick or vacation days, medical care, alimony, leisure, education, or other expenses, and it is assumed that the Earned Income Tax Credit would be approximately the same as the federal income tax, thus canceling out that cost.

*This exercise is meant to give you an idea of the difficulties of living on a low-wage budget.

Source: Hunger: A Picture of Washington, January 2002; available at www.childrensalliance.org. Based on the Northwest Harvest website, www.northwestharvest.org/minwage.htm.

THE YOUNG

Over 13 million children were hungry or at risk of hunger in 2003; in fact, 16.7 percent of households with children experienced food insecurity and hunger—more than twice the rate of households without children (8.2 percent).[23] Not all persons living in food insecure or hungry households experience food insecurity and hunger, however. In fact, children are typically protected by the adults in the households. As previously discussed, using a subset of the questions summarized in Table 5-2 (questions 4–6 and 14–18), the prevalence of households classified as "food insecure with hunger among children" can be measured.[24] In 2003, less than 1 percent of children in the nation (0.6 percent, or 420,000 children) lived in food insecure households with hunger among children.[25] Research shows that children living in food insufficient households have poorer health, even after controlling for confounding factors such as poverty. A child's growth, cognitive development, academic achievement, and physical and emotional health are negatively affected by living in a family that does not have enough food to eat.[26] Additional forms of deprivation may accompany poverty, including poor housing and lack of adequate medical care.[27]

LOW-INCOME WOMEN

Many women live in poverty, struggling to provide child care while working for the minimum wage. Food insecurity rates are higher than average in households with children that are headed by a single woman.[28]

ETHNIC MINORITIES

Although the majority of the poor in America in 2003 were white, the median income of black and Hispanic households was lower than that of white households. The national data reveal marked disparity of hardship among racial and ethnic groups. In 2003, the poverty rate was 10.5 percent for white households (8.2 percent for white, non-Hispanic), 24.4 percent for black households, and 22.5 percent for Hispanic households.[29] Overall, non-Hispanic black and Mexican-American children are more likely than non-Hispanic white children to be poor, food insufficient, and in poor health.[30] Table 5-3 summarizes national poverty and hunger trends for the past three decades, with a focus on racial and ethnic disparities.

THE ELDERLY

Social Security and other programs have pulled many older people out of poverty, but large numbers of older people who cannot work and have no savings or families to turn to are facing rising bills for housing, utilities, food, and health care. In 2003, 10.2 percent of all Americans age 65 and over were poor, that is, they lived below the poverty threshold.[31] Although a lower proportion of households with older adults experience hunger and food insecurity than other groups, not all households with seniors are food secure.[32] Low-income older adults often have health-related problems that necessitate special diets and costly drugs as part of their treatment, forcing many to choose between paying for rent, medication, and/or food. Food insecurity can prevent the elderly from adhering to their recommended therapies, often with dire health consequences.[33]

INNER-CITY AND RURAL DWELLERS

The prevalence of food insecurity in households in inner-city and rural areas substantially exceeded that in suburbs and other metropolitan areas in 2003 (see Figure 5-4).[34] A lack of adequate transportation and limited access to quality supermarkets can be problems for dwellers in low-income inner-city communities, as well as for dwellers in remote rural areas. Nutrition and consumer education may be one of the keys to improving household food security status. A recent study showed that mothers from rural, low-income households using food and financial skills (including managing bills, making a budget, stretching groceries, and preparing meals) tended to have food secure households, compared to mothers who used fewer of these types of skills.[35] Having a garden also appears to be positively related to household food security status of some rural families.[36]

CERTAIN SOUTHERN AND WESTERN STATES

Percentages of food insecurity vary from state to state and, for 2001–2003, ranged from 6.2 percent in Massachusetts to 15.5 percent in Arkansas. Hunger also varied and ranged from 1.8 perent in Delaware to 5.2 percent in Oklahoma.[37] States in the Northeast and Midwest had lower levels of food insecurity, whereas those in the South and West generally had higher-than-average rates (see Figure 5-5).[38]

MANY FARMERS

Changes in the domestic economy have caused problems for producers as well as for consumers.[39] U.S. farmers today lack significant control over the products they produce, the prices they must pay for supplies, and the prices they receive for their commodities. While the costs for seed, fertilizer, equipment, and loans have steadily risen, crop prices

TABLE 5-3 National Poverty and Hunger Trends, 1970–2003

	1970	1980	1990	1993	1996	1999	2000	2003
TOTAL POPULATION (Millions)	**205.1**	**227.8**	**249.4**	**257.8**	**265.3**	**272.7**	**281.4**	**290.8**
Food Insecurity Prevalence Estimates								
All U.S. households—food insecure (%)	—	—	—	—	10.4	10.1	10.5	11.2
Without hunger	—	—	—	—	6.3	7.1	7.3	7.7
With hunger	—	—	—	—	4.1	3.0	3.2	3.5
Adult members (total)—food insecure (%)	—	—	—	—	9.6	9.5	10.1	10.8
Without hunger	—	—	—	—	6.0	7.0	7.3	7.7
With hunger	—	—	—	—	3.6	2.5	2.8	3.1
Child members (total)—food insecure (%)	—	—	—	—	—	16.9	18.0	18.2
Without hunger	—	—	—	—	—	16.2	17.2	17.6
With hunger	—	—	—	—	—	0.7	0.8	0.6
Percent of Federal Budget Spent on Food Assistance	**0.5**	**2.4**	**1.9**	**2.5**	**2.43**	**1.94**	**1.83**	**2.0**
Total Infant Mortality Rate (per 1,000 live births)	**20.0**	**12.6**	**9.1**	**8.4**	**7.3**	**7.1**	**6.9**	**7.0***
White	17.8	11.0	7.7	6.8	6.1	5.8	5.7	5.8
African American	32.6	21.4	17.0	16.5	14.7	14.6	14.0	14.4
Hispanic	—	—	7.8	—	5.9	5.8	5.6	—
American Indian	—	—	—	—	—	8.1	7.4	—
Asian or Pacific Islander	—	—	—	—	—	3.9	4.1	—
Total Poverty Rate (%)	**12.6**	**13.0**	**13.5**	**15.1**	**13.7**	**11.8**	**11.3**	**12.5**
Northeast	—	—	10.2	13.3	12.7	10.9	10.3	11.3
Midwest	—	—	11.9	13.4	10.7	9.8	9.5	10.7
South	—	—	15.9	17.1	15.1	13.1	12.5	14.1
West	—	—	11.6	15.6	15.4	12.6	11.9	12.6
White	9.9	10.2	10.7	12.2	11.2	9.8	9.4	10.6
Non-Hispanic	—	—	—	—	—	7.7	7.5	8.2
African American	33.5	32.5	31.9	33.1	28.4	23.6	22.1	24.3
Hispanic	—	25.7	28.1	30.6	29.4	22.8	21.2	22.5
American Indian/Alaska Native	—	—	—	—	—	25.9	25.9	20.0
Asian and Pacific Islander	—	—	—	—	—	10.7	10.8	11.8
Elderly (65 years and older)	24.6	15.7	12.2	12.2	11.5	9.7	10.2	10.2
Female-headed households	38.1	36.7	33.4	38.7	32.6	27.8	24.7	28.0
Total Child Poverty Rate (%) (18 years and under)	**15.1**	**18.3**	**20.6**	**22.7**	**20.5**	**16.9**	**16.2**	**17.6**
White	—	13.9	15.9	17.8	16.3	13.5	13.0	14.3
African American	—	42.3	44.8	46.1	39.9	33.1	30.9	34.1
Hispanic	—	33.2	38.4	40.9	40.3	30.3	28.0	29.7
Asian and Pacific Islander	—	—	17.6	18.2	19.5	11.8	14.5	12.7
Unemployment Rate (%)	**4.9**	**7.1**	**5.6**	**6.9**	**5.4**	**4.2**	**4.0**	**6.0**
White	4.5	6.3	4.8	6.1	4.7	3.7	3.5	5.2
African American	—	14.3	11.4	13.0	10.5	8.0	7.6	10.8
Hispanic	—	10.1	8.2	10.8	8.9	6.4	5.7	7.7

*Infant mortality data are from 2002.

Source: Bread for the World Institute, *Hunger Report 2005: Strengthening Rural Communities* (Washington, D.C.: Bread for the World Institute, 2005). Used by permission.

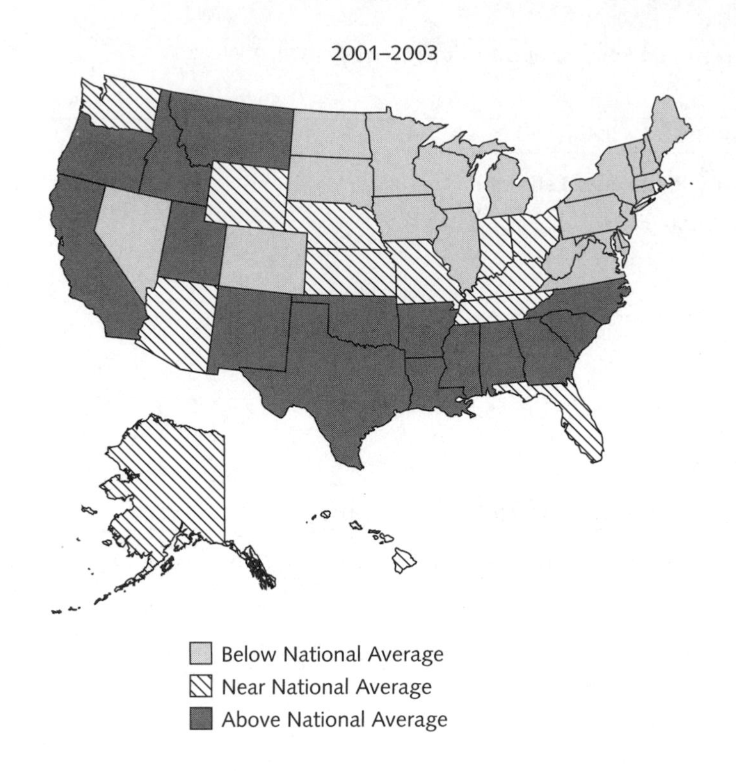

FIGURE 5-5 Prevalence of Food Insecurity and Hunger by State, Average 2001–2003

Source: M. Nord and coauthors, *Household Food Security in the United States, 2003* (Alexandria, VA: Economic Research Service, Food and Rural Economics Division, 2004), Report No. 42 (FANRR-42); also available at www.ers.usda.gov/Briefing/FoodSecurity/trends.

have declined. Today, thousands of U.S. farmers are hungry, frustrated, and desperate about their debt. The number of hungry farm families is not known, but agencies that provide aid to the rural poor say the demand for food assistance is increasing.

THE HOMELESS

The hungry and food insecure are often faced with making choices between food and other necessities, including housing.[40] As many as 2 million people—700,000 people on any given night—experienced homelessness during 1999.[41] Figure 5-6 shows the average composition of the survey cities' homeless population.[42] Who are the homeless? Forty percent are families with children; 23 percent are mentally ill; 17 percent are employed; and 10 percent are veterans. Factors contributing to the problem, in order of predominance, according to the 2003 survey, include lack of affordable housing, mental illness and lack of needed services, low-paying jobs, substance abuse and the lack of needed services,

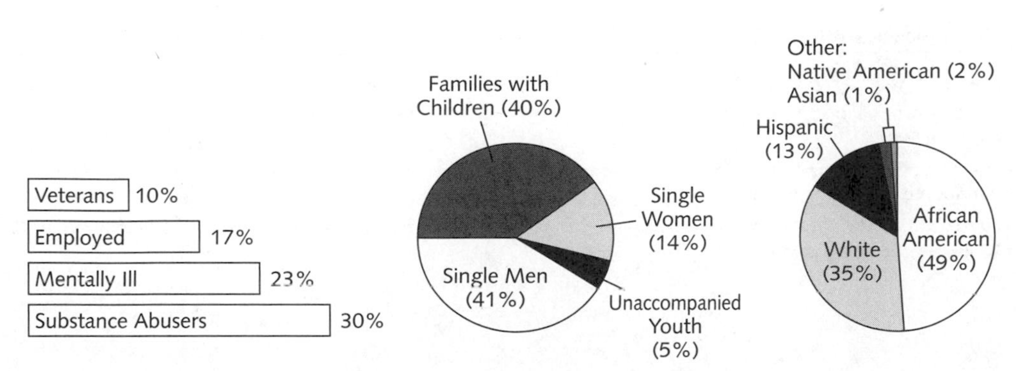

FIGURE 5-6 Demographics of the Homeless Population in 25 Survey Cities

Source: Adapted from the U.S. Conference of Mayors–Sodexho Hunger and Homelessness Survey, *A Status Report on Hunger and Homelessness in America's Cities: 2003.*

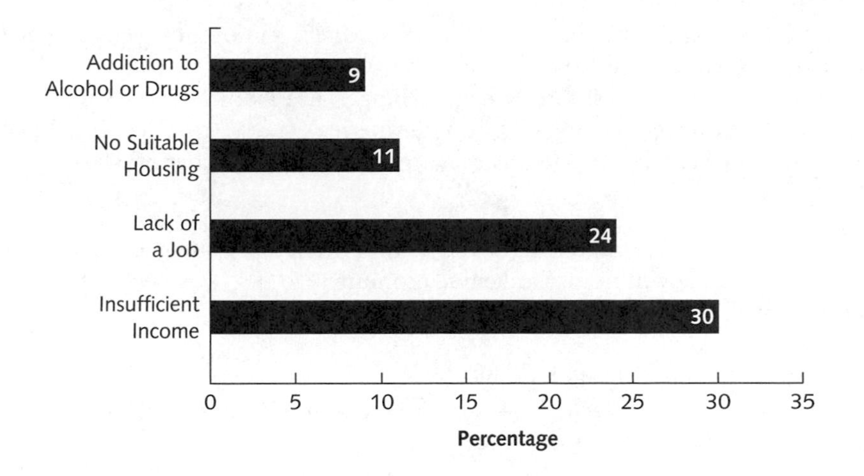

FIGURE 5-7 Factors Considered the Most Important Contributors to Homelessness*

*Percentages shown here reflect the percentage of some 4,207 homeless clients interviewed in the National Survey of Homeless Assistance Providers and Clients who identified factors contributing to their homelessness.

Source: V. Peterson, *Homeless: Struggling to Survive* (Farmington Hills, MI: Gale Group, 2000), p. 13.

unemployment, domestic violence, poverty, and prison release. Figure 5-7 shows the factors considered most important to overcome in order to reduce the survey cities' homeless population. Considering the 37 percent increase in the housing wage—that is, the hourly wage that must be earned for a 40-hour work week in order to afford a two-bedroom housing unit—since 1999 (see Figure 5-8),[43] it is not surprising that lacking affordable housing and insufficient income are two of the factors cited most often as contributing to homelessness.

Lack of food, inadequate diet, poor nutrition status, and nutrition-related health problems (stunted growth, failure to thrive, low-birthweight babies, infant mortality, anemia, and compromised immune systems) are common among homeless persons.[44] Increasing numbers of people living with the HIV virus are homeless because of the high costs of health care or lack of supportive housing.[45] Tuberculosis is spreading at alarming rates among the homeless because of the close sleeping arrangements in shelters and on the streets.

Causes of Hunger in the United States

Solving food insecurity and hunger in America is paramount. "It is the position of The American Dietetic Association that systematic and sustained action is needed to bring an end to domestic food insecurity and hunger and to achieve food and nutrition security for all in the United States. The Association believes that immediate and long-range interventions are needed, including adequate funding for and increased utilization of food and nutrition assistance programs, the inclusion of food and nutrition education in all programs providing food and nutrition assistance, and innovative programs to promote and support the economic self-sufficiency of individuals and families, to end domestic hunger and food insecurity."[46] Understanding the root causes of food insecurity and hunger, however, is necessary before a solution can be devised.

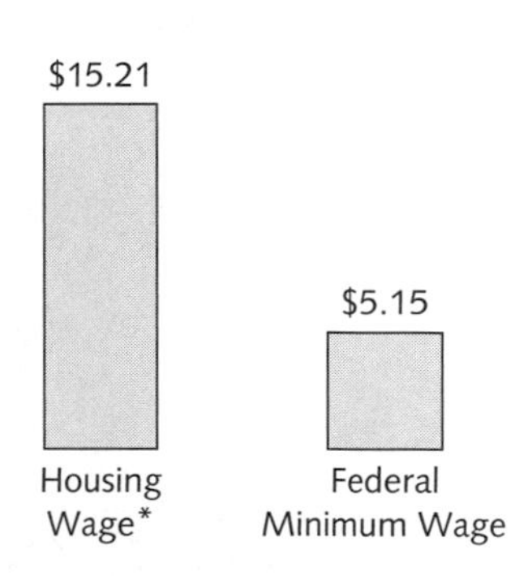

FIGURE 5-8 The Gap Between Affordable Housing and Low-Income Wages

*Hourly wage, at 40 hours per week, needed to afford U.S. median fair market rent for a two-bedroom unit.

Source: Out of Reach, 2003, National Low Income Housing Coalition, Washington, D.C., 2003.

Because poverty is the major cause of domestic food insecurity and hunger, improving poverty in the United States is vital. Although a one-to-one relationship between poverty and hunger does not exist,[47] 33 percent of all food insecure households and 43 percent of all hungry households have incomes below the poverty threshold; in fact, 45 percent of all food insecure households and 54 percent of all hungry households have incomes below 130 percent of the poverty threshold.[48] Events that stress a household's budget, such as losing a job, gaining a household member, or losing food stamp benefits, often precipitate food insecurity.[49] Households must then make tough choices that compromise their ability to buy food, including having to choose between food and other necessities, such as utilities and heating fuel (45 percent), rent or mortgage (36 percent), or medicine or medical care (30 percent).[50]

The U.S. Conference of Mayors–Sodexho Hunger and Homelessness Survey 2003, a 25-city survey on hunger and homelessness in America's cities,[51] revealed that the causes of hunger are many and interrelated. The factors most frequently identified by the survey cities included

- Poverty or lack of income, unemployment and other employment-related problems, low-paying jobs, downturn or weakening economy;
- Housing, medical or health, and child care costs;
- Homelessness;
- Substance abuse, mental health problems;
- Reduced public benefits, lack of information about food stamp benefits;
- Limited life skills, lack of nutrition education;
- Transportation expenses or lack of transportation; and
- Increase in senior population.

Considering the reports on food insecurity and hunger cited previously in this chapter, it appears that

- Food insecurity is a chronic cause of hunger in the United States.
- The nutritional adequacy and balance of diets of food insecure and hungry people are inadequate.
- Food insecurity can lead to physical, social, and mental health problems, including overweight and obesity (see the discussion on page 129 about the paradox of hunger and obesity). Likewise, health problems, including those of old age, chronic disease, alcoholism, or substance abuse, may precipitate an inability to purchase and prepare food, leading to food insecurity and hunger.
- Low-paying jobs result in incomes inadequate to meet the costs of housing, utilities, health care, and other fixed expenses; these items compete with and may take precedence over food, leading to food insecurity and hunger, which may be problematic among some ethnic minorities because they are overrepresented in the low-income population.
- Individuals stave off hunger by using a variety of coping skills. Poor management of limited family resources may contribute to food insecurity and hunger. A lack of education and employment skills make it difficult for the impoverished to recover.
- Families and individuals rely on emergency food assistance facilities both in emergencies and as a steady source of food over long periods of time. However, insufficient community food resources are available to the hungry. Likewise, insufficient community transportation systems are linked to hunger and food insecurity.
- Individuals self-select to participate or not to participate in food assistance programs. The reluctance of some people, including the elderly, to accept what they perceive as "welfare" or "charity" may delay their receiving food stamps and other public/private assistance benefits, which may lead to hunger. Likewise, individuals and households may forgo seeking benefits because of intimidation, ineligibility, complicated paperwork, and other reasons.
- Private charity cannot solve the food insecurity problem. Voluntary activities may be limited in expertise, time, and resources.

Concern for food insecurity and hunger back in the 1930s and 1960s resulted in the creation of food assistance programs. Let us look first at how programs were developed to handle the problems of hunger and poverty in those times and then at how those programs are working now.

The Paradox of Hunger and Obesity in America[1]

Hunger and food insecurity have been called America's "hidden crisis." At the same time, and apparently paradoxically, obesity has been declared an epidemic. Both obesity and food insecurity are serious public health problems, sometimes coexisting in the same families and the same individuals. Their coexistence sounds contradictory, but those with insufficient resources to purchase adequate food can still be overweight, for reasons that researchers now are beginning to understand.

The need to maximize caloric intake. Without adequate resources for food, families must make decisions to stretch their food money as far as possible and maximize the number of calories they can buy so that their members do not suffer from frequent hunger. Low-income families therefore may consume lower-cost foods with relatively higher levels of calories per dollar to stave off hunger when they lack the money or other resources (such as food stamp benefits) to purchase a more healthful balance of more nutritious foods.

The tradeoff between food quantity and quality. Research on coping strategies among food insecure households shows that, along the continuum of typical coping strategies, food *quality* is generally affected before the *quantity* of intake. Households reduce food spending by changing the quality or variety of food consumed before they reduce the quantity of food eaten.[2] As a result, although families may get enough food to avoid feeling hungry, they also may be poorly nourished because they cannot afford a consistently adequate diet that promotes health and averts obesity. In the short term, the stomach registers that it is full, not whether a meal was nutritious.

Overeating when food is available. In addition, obesity can be an adaptive response to periods when people are unable to get enough to eat. Research indicates that chronic ups and downs in food availability can cause people to eat more, when food is available, than they normally would.[3] When money or food stamps are not available for food purchases during part of the month, for example, people may overeat during the days when food is available. Over time, this cycle can result in weight gain.[4] Research among food insecure families also shows that low-income mothers first sacrifice their own nutrition by restricting their food intake during periods of food insufficiency in order to protect their children from hunger.[5] This practice may result in eating more than is desirable when food is available, thereby contributing to obesity among poor women.

Sources: 1. This discussion is adapted from *The Paradox of Hunger and Obesity in America*, developed by Center on Hunger and Poverty (www.centeronhunger.org) and Food Research and Action Center (www.frac.org). 2. K. L. Radimer and coauthors, Understanding hunger and developing indicators to assess it in women and children, *Journal of Nutrition Education* 24 (1992): 36S–45S. 3. J. Polivy, Psychological consequences of food restriction, *Journal of the American Dietetic Association* 96 (1996): 589–592. 4. M. S. Townsend and coauthors, Food insecurity is positively related to overweight in women, *Journal of Nutrition* 131 (2001): 1738–1745. 5. K. L. Radimer and coauthors, 1992.

Historical Background of Food Assistance Programs

During the Great Depression of the 1930s, concern about the plight of farmers who were losing their farms and the economic problems facing families in the United States, in general, led Congress to enact legislation giving the federal government the authority to buy and distribute excess food commodities. A few years later, Congress initiated an experimental Food Stamp Program to enable low-income people to buy food. Then, in 1946, it passed the National School Lunch Act in response to testimony from the surgeon general that "70 percent of the boys who had poor nutrition 10 to 12 years ago were rejected by the draft."

Despite these programs, in the 1960s large numbers of people were still going hungry in the United States, and some of them suffered seriously from malnutrition as a result.

As evidence accumulated during the 1960s and 1970s that hunger was prevalent in the United States, poverty and hunger became national priorities. Old programs were revised and new programs were developed in an attempt to prevent malnutrition in those people found to be at greatest risk. The Food Stamp Program was expanded to serve more people. School lunch and breakfast programs were enlarged to support children nutritionally while they learned. Feeding programs were started to reach senior citizens. To provide food and nutrition education during the years when nutrition has the most crucial impact on growth, development, and future health, a supplemental food and nutrition program (WIC) was established for low-income pregnant and breastfeeding women, infants, and children who were nutritionally at risk.

As a result of these efforts, hunger diminished as a serious problem for the United States. Several studies, including comparative observations made ten years apart, documented the difference the food assistance programs had made.[52] In a baseline study in the late 1960s, a Field Foundation report stated,

> Wherever we went and wherever we looked we saw children in significant numbers who were hungry and sick, children for whom hunger is a daily fact of life, and sickness in many forms, an inevitability. The children we saw were . . . hungry, weak, apathetic . . . visibly and predictably losing their health, their energy, their spirits . . . suffering from hunger and disease, and . . . dying from them.

Ten years later, in 1977, the same group reported as follows:

> Our first and overwhelming impression is that there are far fewer grossly malnourished people in this country today than there were ten years ago. . . . This change does not appear to be due to an overall improvement in living standards or to a decrease in joblessness in those areas . . . . But in the area of food there is a difference. The Food Stamp Program, school lunch and breakfast programs, and the Women-Infant-Children programs have made the difference.

By 1973, the poverty rate in the United States had reached an all-time low of 11.1 percent. However, in the 1980s, federal spending for antipoverty programs was reduced in an attempt to reduce the national debt. Likewise, the Personal Responsibility and Work Opportunity Reconciliation Act of 1996 reduced federal spending on programs such as the Food Stamp Program that support needy families and children.[53] Poverty in the United States dropped from 15.1 percent in 1993 to 11.3 percent in 2000 and gradually increased to 12.5 percent in 2003. The national trends in poverty are shown in Figure 5-9.

Today, despite the strong economy of the 1990s, hunger affects all segments of the population without regard to age, marital status, previous employment or successes, family ties, or efforts to change the situation. Increasingly, national surveys show that hunger isn't just a problem for those who are homeless or unemployed. People who lack access to a variety of resources—not just food—are most at risk of hunger. Other causes of hunger cited in the surveys, in order of frequency, include unemployment, high housing costs, food stamp cuts, poverty or lack of income, economic downturn or weakening of the economy, utility costs, welfare reform, escalating health care costs, mental health problems, and the fact that available resources are failing to reach many groups.[54]

Welfare Reform: Issues in Moving from Welfare to Work

Welfare in the United States changed in the mid-1990s when the challenge to reform the welfare system resulted in the Personal Responsibility and Work Opportunity Reconciliation Act of 1996 (PRWORA).[55] Table 5-4 summarizes key aspects of PRWORA.

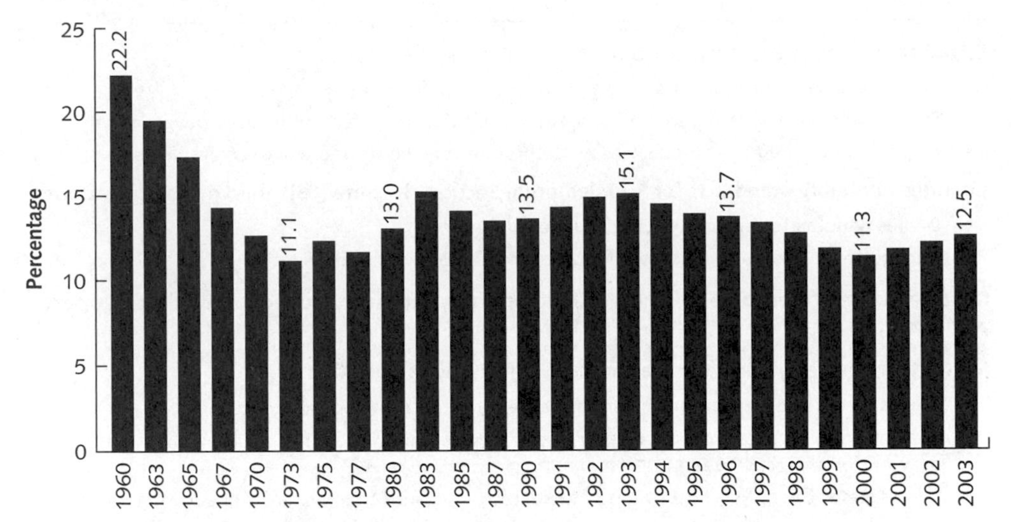

FIGURE 5-9 Percentage of Total U.S. Population in Poverty

Source: U.S. Census Bureau, *Income, Poverty, and Health Insurance Coverage in the United States, 2003.* Adapted from http://publicagenda.org.

The challenge of the welfare reform legislation was to change the welfare system from an income-support-based system to a work-based system with a five-year time limit on benefits. This welfare reform law was intended to encourage self-sufficiency. It sought to promote personal responsibility by promoting work, reducing nonmarital births, and strengthening and supporting marriage. It resulted in a decline in the numbers of those on welfare, a dramatic increase in employment of low-income mothers, increases in earnings by females heading low-income households, and a decline in child poverty.[56] Figure 5-10 shows what former welfare recipients were doing five years after enactment of the law.[57]

As a result of the welfare reform law, the Temporary Assistance for Needy Families (TANF) Program replaced the former Aid to Families with Dependent Children (AFDC) Program. Under TANF, states determine the eligibility of needy families and the benefits and services those families will receive. The welfare reform law allows states greater flexibility in creating opportunities for job training and economic security for households with low incomes.[58] Single women with children are a major target group for job placement and training. Two critical issues for states to consider in their efforts to enhance the long-term successful placement of these women in jobs are transitional child care assistance and the maintenance of health care benefits.[59]

The Welfare Indicators Act of 1994 requires that the Department of Health and Human Services prepare annual reports to Congress on indicators and predictors of welfare dependence.[60] The 2003 report provided welfare dependence indicators through 2000, reflecting changes that have taken place since PRWORA, focusing on benefits under TANF, the Food Stamp Program, and the Supplemental Security Income (SSI) program. It proposed that a family be considered to be dependent on welfare if greater than 50 percent of its annual income is from TANF/AFDC, food stamps, and/or SSI and if this welfare income is not associated with work activities. This construct is difficult to measure because of data collection constraints and limited availability of data.[61] Highlights of the *Indicators of Welfare Dependence: Annual Report to Congress, 2003* follow.

- In 2000, 3.0 percent of the total population was dependent on welfare, down from 5.2 percent in 1996 (5.4 million fewer were dependent on welfare).
- Preliminary data from 2001 suggest that the dependence rate will remain at 3.0 percent.

TABLE 5-4 Key Provisions of the Personal Responsibility and Work Opportunity Reconciliation Act (PRWORA)

*A *block grant* is a grant from the federal government to states or local communities for broad purposes (for example, community services or social services) as authorized by legislation.

Establishes Temporary Assistance for Needy Families (TANF) that:

- Replaces former entitlement programs with federal block grants*
- Shifts authority and responsibility for welfare programs from federal to state government
- Emphasizes moving from welfare to work through time limits and work requirements

Changes eligibility standards for Supplemental Security Income (SSI) child disability benefits:

- Denies benefits to certain formerly eligible children
- Changes eligibility rules for new applicants and eligibility redetermination

Requires states to enforce a strong child support program for collection of child support payments.

Restricts aliens' eligibility for welfare and other public benefits:

- Denies illegal aliens most public benefits, except emergency medical services
- Restricts most legal aliens from receiving food stamps and SSI benefits until they become citizens or work for at least 10 years
- Allows states the option of providing federal cash assistance to legal aliens already in the country
- Restricts most new legal aliens from receiving federal cash assistance for 5 years
- Allows states the option of using state funds to provide cash assistance to nonqualifying aliens

Provides resources for foster care data systems and national child welfare study.

Establishes a block grant to states to provide child care for working parents.

Alters eligibility criteria and benefits for child nutrition programs:

- Modifies reimbursement rates
- Makes families (including aliens) that are eligible for free public education also eligible for school meal benefits

Tightens national standards for food stamps and commodity distribution:

- Institutes an across-the-board reduction in benefits
- Caps the standard deduction at fiscal year 1995 level
- Limits receipt of benefits to 3 months in every 3 years for childless, able-bodied adults aged 18–50 [years] unless working or in training

Source: Economic Research Service, United States Department of Agriculture. Food and Nutrition Assistance Programs: Welfare Reform and Food Assistance Briefing Room (http://ers.usda.gov/Briefing/FoodNutritionAssistance/gallery/keyprovisions.htm).

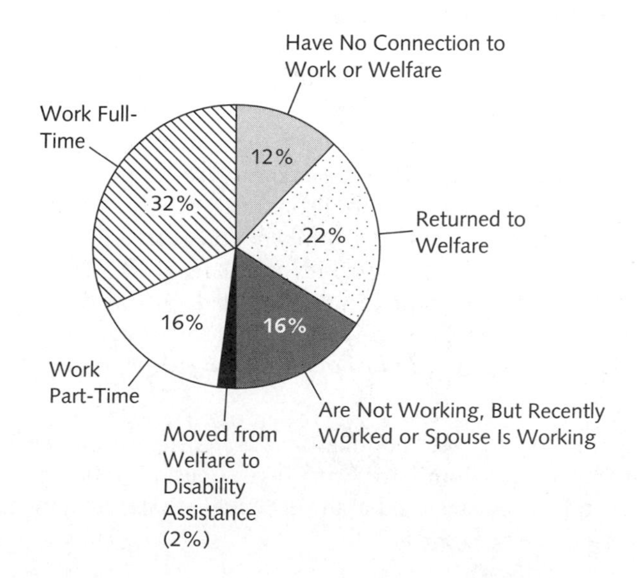

FIGURE 5-10 Welfare Reform Five Years Later: How Are Former Recipients Doing?

Source: Urban Institute, 2001.

- The drop in welfare dependence parallels the drop in TANF and food stamp program caseloads, which fell from 4.6 to 1.9 percent and from 9.5 to 5.7 percent, respectively, between 1996 and 2001.
- In 2000, 59 percent of TANF, 56 percent of food stamp, and 37 percent of SSI recipients lived in families with at least one family member in the labor force.
- While welfare dependence fell, the poverty rate also fell from 13.7 percent in 1996 to 11.3 percent in 2000.

When PRWORA was enacted, economic conditions were very strong in the United States, so it is likely that both the economy and welfare reform influenced the decline in food stamp expenditures.[62] As summarized in the position paper of the American Dietetic Association on domestic food and nutrition security, "distinguishing between the influence of the strong economy and program changes on food assistance program participation is difficult. If a strong economy was primarily responsible for the declines of the late 1990s, a less strong economy should cause participation to rebound. Conversely, if program changes were responsible for the decline, decreased participation should continue with maintenance of the current policy."[63] The economic prosperity of the 1990s ended in fiscal year 2000, with unemployment at about 4 percent and about 17.1 million people receiving food stamp benefits. The 2001 recession was accompanied by an increase in both unemployment and the food stamp caseload. From the end of 2001 to the middle of 2003, the unemployment rate rose slightly, with the food stamp caseload continuing to increase. Since the middle of 2003, the unemployment rate has had a downward trend, with the numbers of those receiving food stamps continuing to increase.[64]

PRWORA could negatively impact food access by families. Consider the U.S. household food security status, in light of the enactment of PRWORA and recent related research.[65] Examining 1995 to 2003 (Figure 5-3) reveals a downward trend in food insecurity from 1995 to 2000 and then an increase in food insecurity in 2001 and 2002. The 2003 levels were unchanged from 2002. Recent studies point to the negative effects of welfare reform on food security, including negative effects on young children.

Food pantry usage can also be examined. Since 2000, 2.4 percent of U.S. households obtained food from pantries one or more times during the year, and the use of pantries consistently increased among households to 3.0 percent in 2002. In fact, use among food insecure households increased from 16.7 percent to 19.3 percent during that same period. With time, the impact of welfare reform on the well-being of Americans will become more clear.

Federal Domestic Nutrition Assistance Programs Today

The U.S. Department of Agriculture (USDA) has implemented an array of programs targeted at different populations with different nutritional needs to provide needy individuals with access to a more nutritious diet, to improve the eating habits of the nation's children, and to help farmers in the United States by providing an outlet for the distribution of food purchased under farmer assistance authorities.[66] These programs, coupled with programs administered by other agencies, including the Departments of Health and Human Services and Homeland Security, are often referred to as a food or nutrition "safety net." Households with limited resources employ a variety of methods to help meet their food needs, including participating in one or more federal food assistance programs and/or obtaining food from emergency food providers in their community; in fact, about one in five Americans participated in a USDA food assistance program in 2003.[67]

Ultimately, the goal of food assistance programs is to improve the food security, nutritional status, and health of Americans. In 2003, 56.0 percent of all "food insecure"

Food pantry Usually attached to existing nonprofit agencies, a food pantry distributes bags or boxes of groceries to people experiencing food emergencies. Distributed foods are prepared and consumed elsewhere. Pantries often require referrals or proof of need.

Bettmann/CORBIS

Bettmann/CORBIS

Feeding the hungry in the United States: then and now.

households participated in the Food Stamp Program, free or reduced-price school meals, or the Special Supplemental Nutrition Program for Women, Infants, and Children (WIC), and 50.3 percent of "food insecure with hunger" households participated. Ultimately, participation in food assistance programs appears to help protect the health of children from low-income households, particularly girls.[68] Current U.S. federal food assistance and nutrition education programs and initiatives are summarized in Table 5-5.

Food assistance programs are typically administered by federal agencies, and benefits are often delivered through state and local agencies such as welfare offices, schools, and public health clinics. The individual states determine most details regarding distribution of food benefits and eligibility of participants. In 2003, expenditures for the USDA Food and Nutrition Service's food assistance programs were $41.6 billion—up from $33 billion

(Text discussion continues on page 140.)

TABLE 5-5 Federal Food and Nutrition Assistance and Nutrition Education-Related Programs and Initiatives

PROGRAM, YEAR STARTED, ADMINISTRATION,* AND PURPOSE(S)	TYPE OF ASSISTANCE PROVIDED	ELIGIBILITY REQUIREMENTS AND WEBSITE ADDRESS
Food Stamp and Related Education Program		
Food Stamp Program; 1961; USDA, Food and Nutrition Service (FNS) *Purpose(s):* Improve the diets of low-income households by increasing access to food/food purchasing ability.	Direct payments in the form of electronic benefits transfer (most states) or coupons redeemable at most retail food stores.	Household eligibility and allotments are based on household size, income, assets, housing costs, work requirements, and other factors. Online prescreening tool available on program website; www.fns.usda.gov/fsp.
Food Stamp Nutrition Education (in some states called Family Nutrition Program); 1986; USDA, FNS *Purpose(s):* Improve the likelihood that those eligible for food stamps will make healthful choices within a limited budget and choose active lifestyles consistent with the Dietary Guidelines for Americans.	Optional program, not offered in every state. States are reimbursed 50 percent by USDA of the allowable administrative costs deemed necessary to operate the program. (Cooperative Extension System typically is the state sponsoring agency.)	Persons eligible to receive food stamps; www.fns.usda.gov/fsp/nutrition_education/default.htm.
Food Distribution Programs		
Commodity Supplemental Food Program (CSFP); 1969; USDA, FNS *Purpose(s):* Improve the health and nutritional status of low-income pregnant and breastfeeding women, other new mothers up to one year postpartum, infants, children up to age 6, and older adults at least 60 years of age by supplementing their diets with nutritious USDA commodity foods.	Provides food and administrative funds to states.	Eligible persons must meet age and income requirements and must be determined to be at nutritional risk by a competent health professional at the local agency. (Eligible participants cannot participate in USDA's Special Supplemental Nutrition Program for Women, Infants, and Children (WIC) at the same time that they participate in CSFP); www.fns.usda.gov/fdd/programs/csfp.
Food Distribution Program on Indian Reservations (FDPIR); 1976; USDA, FNS *Purpose(s):* Improve, through provision of commodity foods and nutrition education, the dietary quality of low-income households, including the elderly, living on Indian reservations, and of Native American families residing in designated areas near reservations.	Provides food and administrative funds to Indian tribal organizations and states.	Eligible households have at least one person who is a member of a federally recognized tribe. Must meet income and resource criteria and be low-income American Indian or non-Indian households that reside on a reservation or in approved areas near a reservation or in Oklahoma. May not participate in the FDPIR and Food Stamp Program in the same month; www.fns.usda.gov/fdd/programs/fdpir.
The Emergency Food Assistance Program (TEFAP); 1981; USDA, FNS *Purpose(s):* Supplement the diets of low-income needy persons, including elderly people, by providing them with emergency food and nutrition assistance.	Provides commodity foods to state distributing agencies, which are typically food banks, that then distribute foods to the public through soup kitchens and food pantries.	Needy individuals, usually including those who have low incomes, are unemployed, or receive welfare benefits; www.fns.usda.gov/fdd/programs/tefap.
Nutrition Services Incentive Program (NSIP); 1978; DHHS, Administration on Aging (AOA) (with financial support from USDA) *Purpose(s):* Provide incentives to states and tribes for the effective delivery of nutritious meals to older adults.	Cash and/or commodities to agencies for meals served.	People 60 years of age or over and their spouses (or a younger age in tribes that define "older" adults differently). Disabled people under age 60 who live in elderly housing facilities where congregate meals are served; disabled persons who reside at home and accompany elderly participants to meals; and volunteers who assist in the meal service may also receive meals through NSIP; www.fns.usda.gov/fdd/programs/nsip or www.aoa.gov/eldfam/Nutrition/Nutrition_services_incentive.asp.

Continued

PROGRAM, YEAR STARTED, ADMINISTRATION* AND PURPOSE(S)	TYPE OF ASSISTANCE PROVIDED	ELIGIBILITY REQUIREMENTS AND WEBSITE ADDRESS
Food Distribution Programs—*continued*		
Schools/Child Nutrition Commodity Programs; 1961; USDA, FNS *Purpose(s):* Support American agricultural producers by providing cash reimbursements for meals served in schools and to provide nutritious, USDA-purchased food for the NSLP, CACFP, and SFSP.	Food commodities to eligible programs.	NSLP-participating schools or institutions participating in CACFP or SFSP are eligible to receive USDA-donated commodities; www.fns.usda.gov/fdd/programs/schcnp.
Food Distribution Disaster Assistance (Food Assistance in Disaster Situations); 1977; USDA, FNS. [Disaster Feeding Situations are administered by DHS's Federal Emergency Management Agency (FEMA).] *Purpose(s):* Supply food to disaster relief organizations such as the Red Cross and the Salvation Army for mass feeding or household distribution in a disaster situation, such as a storm, earthquake, civil disturbance, or flood.	Provides commodity foods for shelters and other mass feeding sites, distributes commodity food packages directly to households in need, and issues emergency food stamp benefits.	The USDA authorizes states to release commodity food stocks to disaster relief agencies for shelters and mass feeding sites; if the president declares a disaster, states can also, with USDA approval, distribute commodity foods directly to households (typically when normal commercial food supply channels have been disrupted, damaged or destroyed or can't function for some other reason); those who might not ordinarily qualify for food stamps may be eligible under the disaster food stamp program if they have had disaster damage to their homes or expenses related to protecting their homes, if they have lost income as a result of the disaster, or if they have no access to bank accounts or other resources; those already participating in the regular Food Stamp Program may also be eligible for certain benefits under the disaster food program; each household's circumstances must be reviewed by the certification staff to determine eligibility; www.fns.usda.gov/fdd/programs/fd-disasters.
Child Nutrition and Related Programs		
National School Lunch Program (NSLP); 1946; USDA, FNS *Purpose(s):* Assist states in providing nutritious free or reduced-price lunches to eligible children.	Schools receive cash subsidies and USDA commodities for each meal served. (Meals must meet federal requirements.)	Public or nonprofit private schools of high school grade or under and public or nonprofit private residential child care institutions may participate. Eligibility standards for children: • All students attending schools where the program is provided may participate. • Children from households with incomes at or below 130 percent of poverty guidelines are eligible for free meals. • Children from households with incomes between 130 and 185 percent of the poverty guidelines are eligible for reduced-price meals. • Children from households with incomes over 185 percent of the poverty guidelines pay full price, a price set by the school (food service operations must be nonprofit, however; www.fns.usda.gov/cnd/Lunch/default.htm.

PROGRAM, YEAR STARTED, ADMINISTRATION* AND PURPOSE(S)	TYPE OF ASSISTANCE PROVIDED	ELIGIBILITY REQUIREMENTS AND WEBSITE ADDRESS
Child Nutrition and Related Programs—*continued*		
After-School Snack in the NSLP; 1998; USDA, FNS *Purpose(s):* Assist school-based after-school educational or enrichment programs in providing healthful snacks to children through age 18; an expansion of the NSLP.	Available through NSLP. Schools receive cash subsidies for each snack served. Snacks must contain at least two different components of the following four: a serving of fluid milk; a serving of meat or meat alternative; a serving of vegetable(s) or fruit(s) or full-strength vegetable or fruit juice; a serving of whole grain or enriched bread or cereal.	After-school snacks are provided to children on the same income eligibility basis as NSLP; however, programs that operate in areas where at least 50 percent of students are eligible for free or reduced-price meals serve all snacks free; www.fns.usda.gov/cnd/Afterschool/default.htm.
School Breakfast Program; 1966; USDA, FNS *Purpose(s):* Assist states in providing nutritious breakfasts to children. Free and reduced-price meals must be offered to eligible children.	Schools and institutions receive cash subsidies for each meal served. (Meals must meet federal requirements.)	Public and nonprofit private schools and residential child care institutions may participate. Operates in the same manner as NSLP; www.fns.usda.gov/cnd/Breakfast/default.htm.
Special Milk Program; 1955; USDA FNS *Purpose(s):* Encourage fluid milk consumption by children.	Schools receive reimbursement for milk served to children eligible for free milk and cash subsidies for each half-pint of milk sold. Pasteurized fluid types of unflavored or flavored whole milk, low-fat milk, skim milk, and cultured buttermilk that meet state and local standards may be served. All milk should contain vitamins A and D at levels specified by the FDA.	Schools, child care institutions, and eligible camps that do not participate in other federal meal service programs may participate; however, an institution may participate to provide milk to children in half-day pre-kindergarten and kindergarten programs where children do not have access to the school meal programs. Milk programs must be offered on a nonprofit basis. Any child from a family that meets income guidelines for NSLP free meals is eligible for free milk. www.fns.usda.gov/cnd/Milk/default.htm.
Summer Food Service Program (SFSP); 1968; USDA, FNS *Purpose(s):* Ensure that children in lower-income areas continue to receive nutritious meals during long school vacations, when they do not have access to school lunch or breakfast. All meals are served free to eligible children.	Approved sponsors receive reimbursement for serving meals that meet federal nutritional guidelines; payments are received through state agencies, based on the number of meals served and documented costs of running the program.	May be sponsored by organizations capable of managing a food service program: public or private nonprofit schools, units of local, municipal, county, tribal, or state government, private nonprofit organizations (including eligible emergency shelters), public or private nonprofit camps, and public or private nonprofit universities or colleges. Types of sites served include areas/programs with a majority of children eligible for free and reduced-price school meals, residential or day camps, migrant worker communities, and National Youth Sports Programs. All children 18 years of age or younger who come to an approved open site or to an eligible enrolled site may receive meals. At camps, only children who are eligible for free and reduced-price school meals may receive SFSP meals. People over age 18 who are enrolled in school programs for persons with disabilities may also receive meals. www.fns.usda.gov/cnd/Summer/default.htm.

Continued

PROGRAM, YEAR STARTED, ADMINISTRATION* AND PURPOSE(S)	TYPE OF ASSISTANCE PROVIDED	ELIGIBILITY REQUIREMENTS AND WEBSITE ADDRESS
Child Nutrition and Related Programs—*continued*		
Child and Adult Care Food Program (CACFP); 1968; USDA, FNS *Purpose(s):* Improve the quality and affordability of day care for low-income families by providing nutritious meals and snacks to children, to provide meals and snacks to adults who receive care in nonresidential adult day care centers, and to provide meals to children residing in homeless shelters, and snacks and suppers to youths participating in eligible after-school care programs.	Cash reimbursement for meals served that meet federal nutritional guidelines and reimbursement of associated administrative costs. (CACFP meal patterns vary by children's age and type of meal served.) Agricultural commodities or cash in lieu of commodities is also available.	Eligible institutions include public or private nonprofit child care centers, outside-school-hours care centers, Head Start programs, some day care homes, community-based programs that offer enrichment activities for at-risk children and teenagers after the regular school day ends, public or private nonprofit emergency shelters that provide residential and food services to homeless families, and public or private nonprofit adult day care facilities that provide structured, comprehensive services to nonresidential adults who are functionally impaired or aged 60 and older. Participant eligibility standards for free and reduced-price meals are the same as the NSLP; www.fns.usda.gov/cnd/Care/CACFP/cacfphome.htm.
Team Nutrition; 1995; USDA, FNS *Purpose(s):* Improve the health of children by having school meals reflect federal dietary guidelines.	Schools receive technical training and assistance to help school food service staff prepare healthful meals and provide nutrition education to help children understand the link between eating/physical activity and health.	Eligible schools include those who participate in school meal programs; other partners found in federal, state, and local programs, agencies, and organizations; www.fns.usda.gov/tn.
Fresh Fruit and Vegetable Program (FFVP); 2002; USDA, FNS *Purpose(s):* Promote fresh fruit and vegetable consumption among the nation's schoolchildren. After being piloted during the 2002–2003 school year, the program gained permanent status in 2004 with passage of the Child Nutrition Reauthorization Bill.	Provide fresh and dried fruits and fresh vegetables free to children in 107 elementary and secondary schools—100 schools in 4 states (25 each in Indiana, Iowa, Michigan, and Ohio) and 7 schools in the Zuni Indian Tribal Organization in New Mexico. Program expanded in 2004 to include Pennsylvania, North Carolina, Washington, and two Indian reservations in Arizona and South Dakota.	Go to www.ers.usda.gov/Briefing/ChildNutrition/fruitandvegetablepilot.htm or www.fns.usda.gov/cnd.
Programs for Women and Young Children†		
Special Supplemental Nutrition Program for Women, Infants, and Children (WIC); 1972; USDA, FNS *Purpose(s):* Safeguard the health of low-income women, infants, and children up to age 5 who are at nutritional risk.	Provides nutritious foods to supplement diets, nutrition education and counseling, and screening/referrals to other health, welfare, and social services.	Pregnant, breastfeeding, and post-partum women, infants up to 1 year of age, and children up to 5 years of age are eligible if they are individually determined by a qualified health professional to be in need of the special supplemental foods provided by the program because they are nutritionally at risk (having a medical-based or dietary-based condition), and if they meet an income standard (gross income at or below 185 percent of the poverty guidelines); www.fns.usda.gov/wic.
WIC Farmers' Market Nutrition Program; 1992; USDA, FNS *Purpose(s):* Provide fresh, unprepared, locally grown fruits and vegetables to WIC recipients, and to expand the awareness, use of, and sales at farmers' markets.	FMNP coupons to purchase a variety of fresh, nutritious, unprepared, locally grown fruits, vegetables, and herbs (each state agency develops a list of the fresh fruits, vegetables and herbs eligible for purchase).	Eligiblity is the same as WIC, but infants must be over 4 months of age (not operated in all states, territories, or tribal organizations); www.fns.usda.gov/wic/FMNP/FMNPfaqs.htm.

PROGRAM, YEAR STARTED, ADMINISTRATION* AND PURPOSE(S)	TYPE OF ASSISTANCE PROVIDED	ELIGIBILITY REQUIREMENTS AND WEBSITE ADDRESS
Programs for Women and Young Children†—*continued*		
Head Start/Early Head Start (EHS); 1965/1994; DHHS, Administration for Children and Families *Purpose(s):* Increase the school readiness of young children in low-income households. Also, to promote healthy prenatal outcomes, enhance the development of infants and toddlers, and foster healthy family functioning.	Comprehensive, child-focused child development programs (including home-based programs) serve children from birth to age 5, pregnant women, and their families. Health, education, nutrition, and social services are provided and are responsive and appropriate to each child's and each family's heritage and experience; services encompass all aspects of a child's development and learning.	Eligible participants reside in households with incomes below the official poverty guidelines; www.acf.dhhs.gov/programs/hsb.
Programs for Older Adults†		
Elderly Nutrition Program (ENP); 1965; DHHS, AOA *Purpose(s):* Improve the dietary intakes and nutritional status of participating older adults and to offer them opportunities to form new friendships and create informal support networks.	Congregate and home-delivered meals and other nutrition services. Nutrition screening, assessment, education, and counseling to identify these older adults' general and special nutritional needs are provided in a variety of settings, such as senior centers, schools, and individual homes. Meals served must provide at least one-third of recommended intakes established by the Food and Nutrition Board. Also provides an important link to other needed supportive in-home and community-based services (homemaker or home health aide services, transportation, fitness programs, and home repair and home modification programs).	Adults 60 years of age and older may participate. No means test for participation; however, services are targeted to older people with the greatest economic or social need, with special attention given to low-income minorities. Others who may receive services include a spouse of any age, disabled persons under age 60 who reside in housing facilities occupied primarily by the elderly where congregate meals are served, disabled persons who reside at home and accompany older persons to meals, and nutrition service volunteers; www.aoa.gov/press/fact/alpha/fact_elderly_nutrition.asp or www.aoa.gov/press/fact/pdf/fs_nutrition.pdf.
Senior Farmers' Market Nutrition Program; 2001; USDA, FNS *Purpose(s):* Provide resources in the form of fresh, nutritious, unprepared, locally grown fruits, vegetables, and herbs from farmers' markets, roadside stands, and community-supported agriculture programs to low-income seniors, to increase the domestic consumption of agricultural commodities by expanding or aiding in the expansion of these domestic agriculture programs, and to develop or aid in the development of new and additional farmers' markets, roadside stands, and community-supported agriculture programs.	During the harvest season, coupons are provided that can be exchanged for eligible foods at farmers' markets, roadside stands, and community-supported agriculture programs. (Fresh, nutritious, unprocessed fruits, vegetables, and fresh-cut herbs can be purchased; items not eligible for purchase include dried fruits and vegetables, potted fruit or vegetable plants, potted or dried herbs, wild rice, nuts of any kind, honey, maple syrup, cider, and molasses.)	Older adults at least 60 years of age with incomes (generally) no more than 185 percent of the poverty guidelines, www.fns.usda.gov/wic/SeniorFMNP/SFMNPmenu.htm.

Continued

PROGRAM, YEAR STARTED, ADMINISTRATION* AND PURPOSE(S)	TYPE OF ASSISTANCE PROVIDED	ELIGIBILITY REQUIREMENTS AND WEBSITE ADDRESS
Other Nutrition Education and Food/Nutrition-Related Programs		
The Expanded Food and Nutrition Education Program (EFNEP); 1968; USDA, Cooperative State Research, Education, and Extension Service *Purpose(s):* Assist adults and youth with limited resources in acquiring the knowledge, skills, attitudes, and changed behavior necessary for nutritionally sound diets and to contribute to their personal development and the improvement of the total family diet and nutritional well-being.	Adults participate in a series of 10–12 or more lessons, often over several months, delivered by paraprofessionals and volunteers (trained by county extension home economists), many of whom are indigenous to the target population, using a hands-on, learn-by-doing approach to imparting practical skills. Youth programs take various forms, including nutrition education at schools as an enrichment of the curriculum, in after-school care programs, and through 4-H EFNEP clubs, day camps, residential camps, community centers, neighborhood groups, and home gardening workshops. In addition to lessons on nutrition, food preparation, budgeting, and food safety, youth topics may also include fitness, avoidance of substance abuse, and other health-related topics.	Recruitment typically through referrals from neighborhood contacts and community agencies (such as the Food Stamp Program and WIC); www.csrees.usda.gov/nea/food/efnep/efnep.html.

*Programs are typically administered by federal agencies but often deliver benefits through state and local agencies such as welfare offices, schools, and public health clinics.

†The Food Stamp Program and some food distribution and nutrition education-related programs/initiatives also serve women, infants, children, and older adults.

in 1992 and $1.1 billion in 1969, the first year of the agency's operation (see Figure 5-11). Five programs—the Food Stamp Program, the National School Lunch Program, the Special Supplemental Nutrition Program for Women, Infants, and Children (WIC), the School Breakfast Program, and the Child and Adult Care Food Program—together accounted for 94 percent of all federal expenditures for food assistance in 2003.[69]

Other food assistance programs and programs providing a nutrition component—the Nutrition Services Incentive Program, the Elderly Nutrition Program, Head Start/Early Head Start—are administered by the Department of Health and Human Services (DHHS). In addition, disaster feeding is administered by the Department of Homeland Security (DHS). In addition to the information in Table 5-5, many food assistance programs are discussed in the sections that follow.

FOOD STAMP AND RELATED PROGRAMS/INITIATIVES

Nutrition surveys in the United States have demonstrated consistently that the lower a family's income, the less adequate its nutrition status. Similarly, food insecurity negatively impacts the diet and health of individuals.[70] Improving food access and purchasing ability, as well as providing education to foster improved food choices within a limited budget, is one avenue to improving the diet and health of Americans.

The Food Stamp Program and Food Stamp Nutrition Education

The Food Stamp Program is currently authorized by the Food Stamp Act of 1977 (PL 95-113).

The Food Stamp Program (FSP) dates back to the food assistance programs of the Great Depression—a time when farmers were burdened with surplus crops they could not sell, while thousands stood in breadlines, waiting for something to eat.[71] To help both farmers and consumers, the government began distributing the surplus farm foods to hungry citizens. Milo Perkins, the first administrator of the program that preceded the current

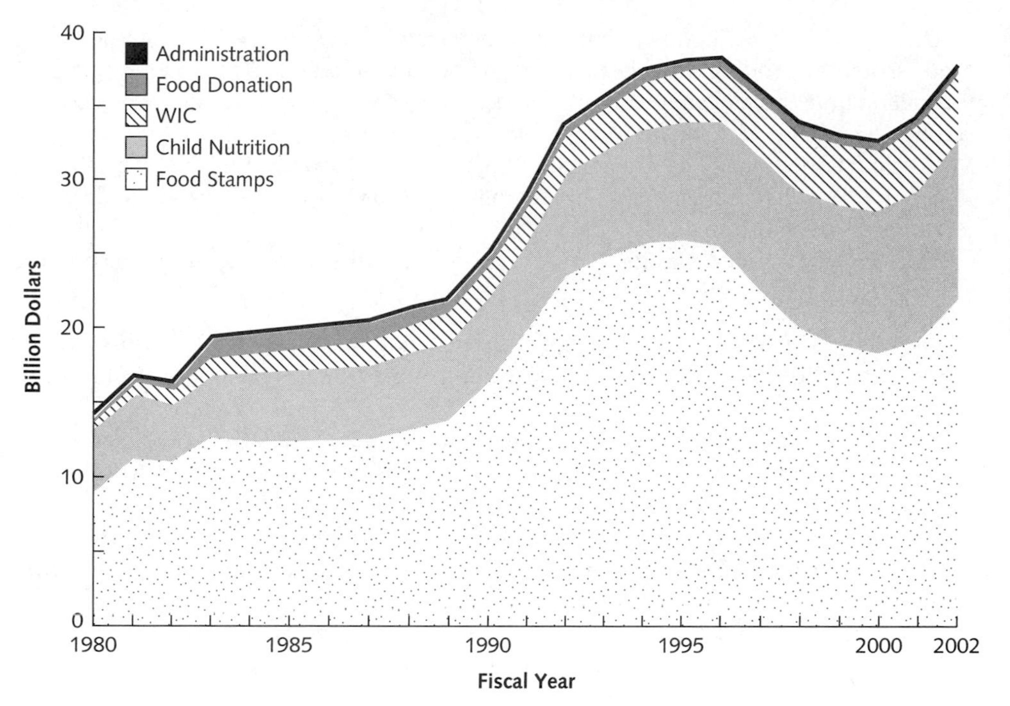

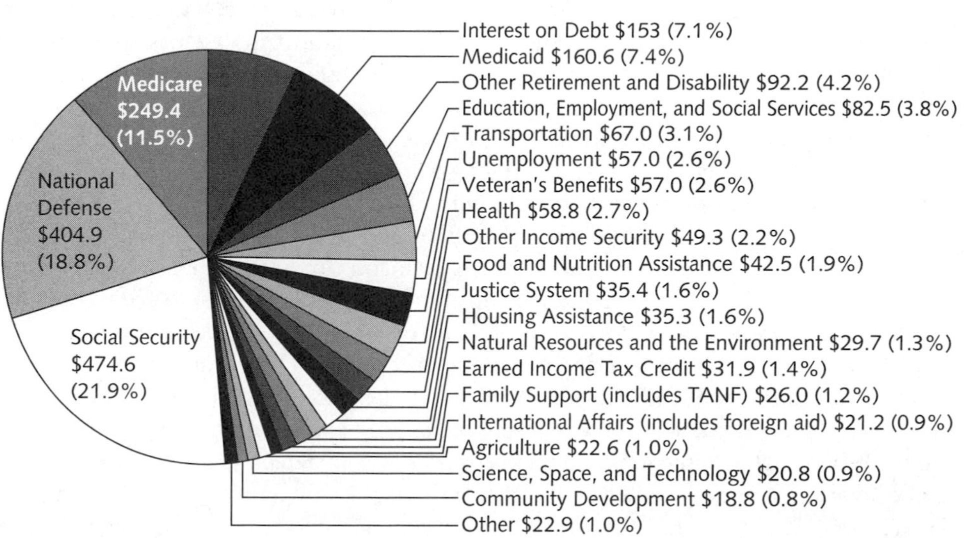

FIGURE 5-11 Food Program Costs, 1980–2002

Sources: Economic Research Service, USDA (chart); Fiscal 2005 Budget of the U.S. Government, as illustrated at www.publicagenda.org (pie chart). Copyright © 2004 Public Agenda. Reprinted by permission.

program by two decades, said, "We got a picture of a gorge, with farm surpluses on one cliff and under-nourished city folks with outstretched hands on the other. We set out to find a practical way to build a bridge across that chasm."[72] Today, the Food Stamp Program enables recipients to buy approved food items at authorized food stores, with the goal of improving the diets of low-income households by increasing access to food and food-purchasing ability. The program operates in the 50 states, the District of Columbia, Guam, and the Virgin Islands.

Use the Food Stamp Program (FSP) Map Machine to illustrate FSP participation and benefit levels for a county, a state, or the nation. Go to http://maps.ers.usda.gov/fsp/.

Work requirements: At the time of application and once every 12 months, all able-bodied household members between 18 and 60 years of age and 16- and 17-year-old heads of households who are not in school must register to work. Many adult recipients must participate in employment and training programs. Generally, able-bodied adults between 18 and 50 who do not have any dependent children can get food stamps only for 3 months in a 36-month period if they do not work or participate in a workfare or employment and training program other than job search.

Net income is all the household's income that counts in figuring food stamps minus the deductions for which the household is eligible. Most households may have up to $2,000 in countable resources (cash, bank accounts, stocks/bonds, and so forth). House and property lot values are not included in this assets limit, and vehicles are in a special category. Households may have $3,000 if at least one person is age 60 years or older or is disabled.

During 2003, Food Stamp Program spending totaled $23.7 billion, 57 percent of USDA food assistance spending. Average participation per month was 21.3 million people (from 9.2 million households), about 1 in 13 Americans. However, this participation is below the record 27.5 million participants in 1994. Monthly benefits averaged $83.91 per person and $194.92 per household.[73] The Food Stamp Program (FSP) Map Machine[74] is an invaluable, interactive, Web-based mapping utility that illustrates program participation and benefit levels down to the county level.

Household characteristics for 2002 give insight into the demographic and economic circumstances of households receiving food stamps.[75]

- Over half (51 percent) of all participants were children (18 or younger), and 9 percent were age 60 or older. Working-age women accounted for 28 percent of the caseload, and working-age men accounted for 12 percent.
- Twenty-one percent of households received TANF benefits, and 30 percent received SSI. Social Security benefits were received by 24 percent, and 11 percent had no cash income of any kind.
- Over one-quarter (28 percent) of households had earnings from employment.
- Only 12 percent had incomes above the poverty line; in fact, 36 percent had incomes at or below half the poverty line.
- The typical household had a monthly income of $633 and received $173 in food stamps.
- The average household possessed only $134 in countable resources (including value of checking and saving accounts and other items).
- Average household size was 2.3 persons, with households with children averaging 3.3 and those with older adult members averaging 1.3.

The FSP is an **entitlement program,** which means that anyone who meets eligibility standards is entitled to receive benefits. Eligibility and allotments are based on income, household size, assets, housing costs, work requirements, and other factors (see Table 5-5). The program's website outlines eligibility standards, including current income guidelines.[76] In 2002, benefits were restored to adult legal immigrants—previously dropped from the program in 1996 under welfare reform rules—who have been in the United States for at least five years, and to all children of legal immigrants.

In the past, households received, from state welfare or human services agencies, monthly allotments of coupons that were redeemable for food at authorized grocery stores. However, since October 2002, benefits in most states have been issued in the form of electronic benefits on a debit card—known as the EBT card (Electronic Benefits Transfer card). The EBT card acts in much the same way as a bank card to transfer funds from a food stamp recipient's account to a food retailer's account. With an EBT card, food stamp recipients purchase groceries without the use of paper coupons. EBT keeps an electronic record of all transactions and makes fraud easier to detect. The amount of benefits that a household receives (called an allotment) varies according to its size and income (see Figure 5-12). Recipients may use the benefits to purchase food and seeds at stores authorized to accept them. They cannot buy food that will be eaten in the store, ready-to-eat hot foods, vitamins or medicines, pet foods, tobacco, cleaning items, alcohol, or nonfood items (except seeds and garden plants) with food stamp benefits.

Entitlement program A government program that provides cash, commodities, or services to all qualifying low-income individuals or households.

Although the program potentially increases a household's ability to purchase nourishing foods, food stamps may be used to buy most available human foods. Therefore, the effect of the food stamp purchases on nutrient intakes of participants will vary depending on the nutritional composition of the foods they select. An optional program for states is Food Stamp Nutrition Education, which is intended to improve the likelihood that the program participants will make healthful choices within a limited budget and choose active lifestyles consistent with the Dietary Guidelines for Americans.

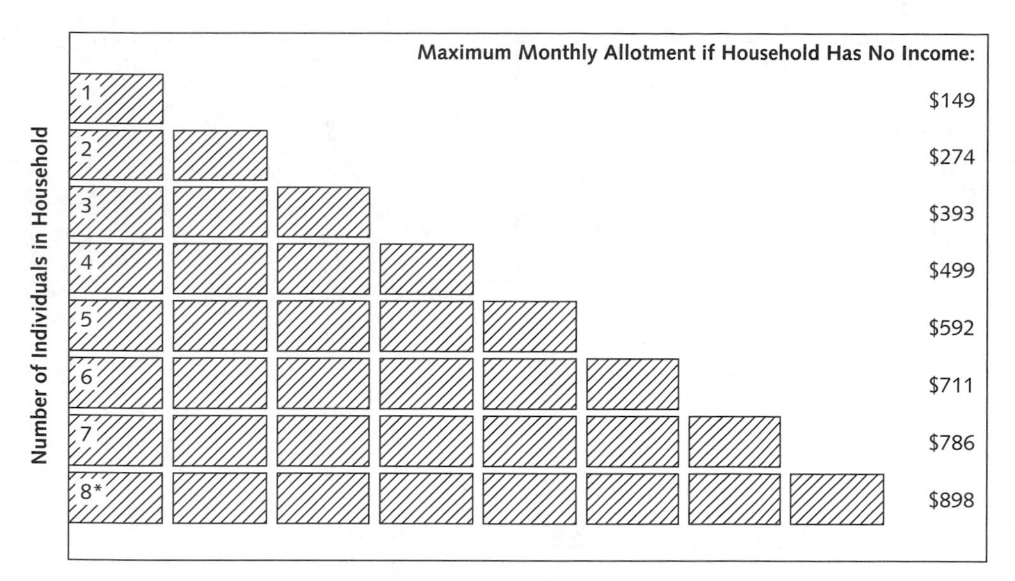

FIGURE 5-12 Food Stamp Allotments (2005)

Subtract 30 percent of net monthly income from the maximum food stamp allotment to determine the allotment. For example, a 4-person household with $556 in available income would receive $332 in food stamps. 30% of income = $167. Maximum monthly allotment for a family of 4 = $499 − $167 = $332 food stamp allotment for a full month.

*For each additional person, add $112.

Source: United States Department of Agriculture.

The *Thrifty Food Plan* (TFP) serves as a food guide for a nutritious diet at a minimal cost, is used as the basis for food stamp allotments, and is maintained by the USDA's Center for Nutrition Policy and Promotion. A sampling of TFP menus and recipes demonstrating nutritious meals and snacks on a minimal-cost budget are available at www.usda.gov/cnpp/Pubs/Cookbook/thriftym.pdf.

Two problems with the FSP are that benefit allotments are insufficient to meet needs and that environmental factors, including access to supermarkets and to high-quality, healthful foods, impact the dietary choices of participants.[77] Many households receiving food stamps still need emergency food by the end of the month because their benefits rarely last the entire month. Two studies give insight into this problem.[78] In a study of 1,922 households, monthly food expenditures of FSP participants, including cash, food stamps, and WIC benefits, averaged just under 80 percent of the value of the food guide for low-income families known as the *Thrifty Food Plan*. Results from the national surveys have shown that only 12 percent of people purchasing food valued at 100 percent of the Thrifty Food Plan were eating nutritionally adequate diets, indicating that the FSP households spending only 80 percent of the value of the Thrifty Food Plan may be at nutritional risk.

Whereas food stamp participation tends to mirror economic conditions (see Figure 5-13), a second major problem is that many households who are eligible for the FSP and are in need do not participate. In 2001, an estimated 32.9 million people were living at income levels at or below the poverty line. These people were all potentially eligible to receive food stamps, yet only some 17.3 million people participated in the program.[79] In 2003, 30.8 percent of all food insecure households participated in the Food Stamp Program.[80] Potential reasons for nonparticipation include embarrassment about receiving assistance, complex rules and requirements, confusing paperwork, caseworker hostility, and lack of public information about eligibility requirements.

The USDA is trying to improve participation rates by those eligible through a number of outreach efforts targeted at low-participation groups, such as persons who are elderly, working poor, non-English-speaking, homeless, or living in rural areas. Outreach efforts include training community workers and volunteers to refer families to food stamp offices, community education and mass media campaigns, and individualized client assistance.[81]

To improve the FSP's ability to meet the needs of low-income households, the following steps are recommended:[82]

- Improve and expand outreach about the FSP.
- Lower administrative barriers to participation in the FSP.

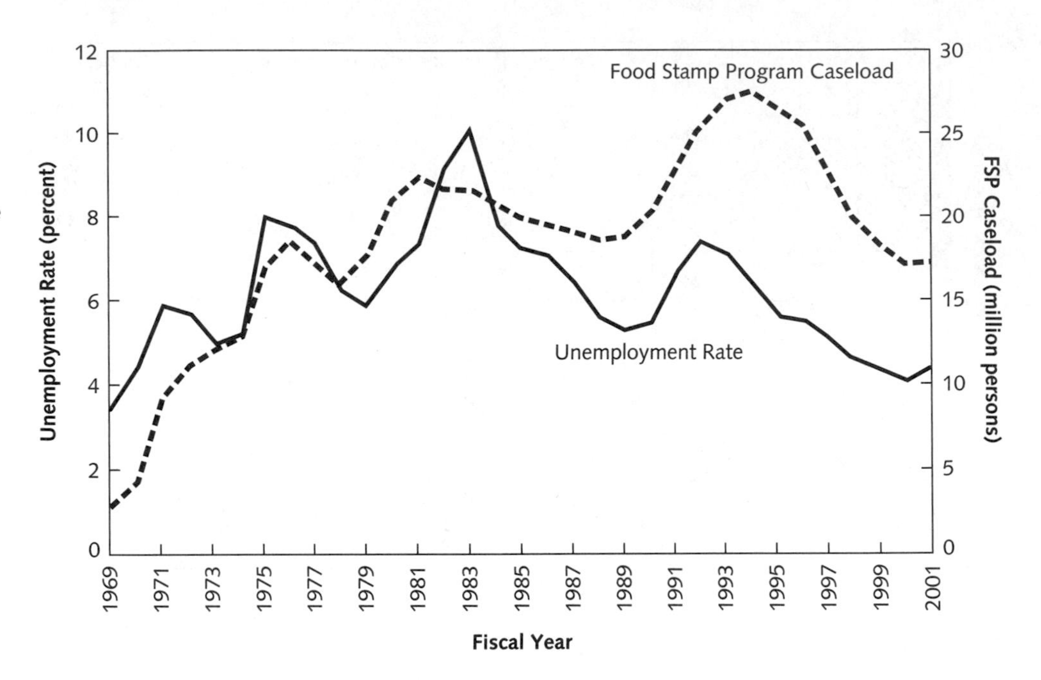

FIGURE 5-13
Unemployment Rate and Food Stamp Program Enrollment, 1969–2001

Food Stamp Program caseloads tend to mirror the unemployment rate.

Source: Bureau of Labor Statistics, Current Population Survey. As summarized by the USDA; available at www.ers.usda.gov/Briefing/GeneralEconomy/gallery.htm.

- Increase allotments so that families can afford to eat a nutritionally adequate diet throughout each month.
- Provide nutrition education materials to households that receive food stamps.

Nutrition Assistance Programs in Puerto Rico, American Samoa, and the Commonwealth of the Northern Marianas Islands

Instead of the Food Stamp Program, Puerto Rico, American Samoa, and the Commonwealth of the Northern Marianas Islands receive block grant funds that allow these United States territories to operate food assistance programs designed specifically for their low-income residents.* Overall, instead of food stamps and commodities, cash or food vouchers are provided. The Food Stamp Program in Puerto Rico was replaced in 1982 by the Nutrition Assistance Program. The same program was instituted in the Northern Marianas in 1982 and in American Samoa in 1994. These programs were essentially unaffected by PRWORA.[83]

FOOD DISTRIBUTION PROGRAMS

Food distribution programs are another aspect of the U.S. national effort to improve access to food. Food distribution programs are intended to strengthen the nutrition safety net through commodity distribution and other nutrition assistance to low-income households, emergency feeding programs, Indian reservations, and older adults.[84]

Commodity Supplemental Food Program

The CSFP was authorized by the Agriculture and Consumer Protection Act of 1973.

The Commodity Supplemental Food Program (CSFP) is a direct food distribution program. It is intended to improve the health and nutritional status of low-income pregnant and breastfeeding women, other new mothers up to one year postpartum, infants, children

*A *block grant* is a grant from the federal government to states or local communities for broad purposes (for example, maternal or child health or social services) as authorized by legislation. Recipients have great flexibility in distributing such funds as long as the basic purposes are fulfilled.

up to age six, and elderly people at least 60 years of age by supplementing their diets with nutritious USDA commodity foods. Eligibility is reviewed in Table 5-5. Recipients may not participate in both WIC and CSFP. The commodities available vary from one year to another and are subject to market conditions. Food packages are designed to suit the nutritional needs of participants and may include canned fruit juice, canned fruits and vegetables, ready-to-eat cereals, nonfat dry milk, evaporated milk, dry beans, peanut butter, canned meat, poultry or tuna, dehydrated potatoes, pasta, rice, cheese, butter, honey, and infant cereal and formula. Distribution sites make packages available on a monthly basis to participants, including more than 389,000 older adults and more than 66,000 women, infants, and children, with expenditures of $118 million in 2003.[85]

Food Distribution Program on Indian Reservations

This program is intended to improve the dietary quality, through provision of commodity foods and nutrition education, of low-income households, including the older adults, living on Indian reservations and of Native American families residing in designated areas near reservations. Currently, over 240 tribes are served through 98 Indian tribal organizations and 5 state agencies. The USDA commodities include canned meats and fish products; vegetables, fruits, and juices; dried beans; peanuts or peanut butter; milk, butter, and cheese; pasta, flour, or grains; adult cereals; corn syrup or honey; and vegetable oil and shortening. On most reservations, participants can select fresh produce instead of canned fruits and vegetables. Nutrition counseling and classes, cooking classes, and tips for using and storing the commodities are among the nutrition education activites that can be provided as part of the program.

Participants may choose from month to month whether they will participate in the Food Stamp Program or the food distribution program. Many prefer the Food Distribution Program on Indian Reservations if they do not have easy access to grocery stores. Monthly participation in 2003 averaged more than 108,000 people, and expenditures were $69 million. In 2004, additional funding was appropriated to the program ($4 million) for a special purchase of bison meat.[86] This may be a positive step in offering culturally appropriate foods to diverse groups.

Emergency Food Assistance Program

The Emergency Food Assistance Program (TEFAP) provides commodity foods to state distributing agencies, typically **food banks,** which then distribute foods to the public through **soup kitchens** and food pantries. This is not disaster relief. TEFAP reduces the level of government-held surplus commodities and supplements the diets of low-income needy persons, including elderly people. Eligibility is summarized in Table 5-5. The types of foods vary with agricultural market conditions. Foods typically available include canned and dried fruits, fruit juice, canned vegetables, meat, poultry, fish, rice, grits, cereal, peanut butter, nonfat dried milk, and pasta products. In 2003, 522 million pounds of food were distributed, with a total food cost of $396 million. Total cost of the program was $465 million in 2003.[87]

Food banks Nonprofit community organizations that collect surplus commodities from the government and edible but often unmarketable foods from private industry for use by nonprofit charities, institutions, and feeding programs at nominal cost.

Soup kitchen Small feeding operations attached to existing organizations, such as churches, civic groups, or nonprofit agencies, that serve prepared meals that are consumed on-site. Soup kitchens generally do not require clients to prove need or show identification.

Nutrition Services Incentive Program

The Nutrition Services Incentive Program (NSIP) is administered by the Department of Health and Human Services (as of 2002) and receives financial support from the USDA. This program provides incentives for effective delivery of nutritious meals to older adults. Cash and/or commodities are provided to local agencies for meals served. Eligibility is summarized in Table 5-5. USDA costs are limited to the value of the commodities distributed, which was $3 million in 2003. In 2001, the total cost of the program was $152 million, and 253 million meals were served.[88]

Food Distribution Disaster Assistance

This program provides food to state relief agencies and organizations (for example, the Red Cross and the Salvation Army) in times of emergency such as civil disturbances, hurricanes, earthquakes, floods, and winter storms. FNS may provide commodity foods for distribution to shelters and mass feeding sites, distribute commodity food packages directly to persons in need, or approve issuance of emergency food stamps. The program is administered by the Federal Emergency Management Agency (FEMA) in the Department of Homeland Security. Expenditures for this program totaled $0.4 million in 2003, with most disaster relief provided through the Food Stamp Program.[89]

CHILD NUTRITION AND RELATED PROGRAMS

More about the Child Nutrition Programs appears in Chapter 11.

Many federal programs address the special nutritional needs of children. The following sections describe the diversity of such programs. Table 5-5 hightlights the programs below, as well as Team Nutrition and the Fresh Fruit and Vegetable Program.

National School Lunch Program, School Breakfast Program

The NSLP was authorized in 1946 by the National School Lunch Act (PL 79-396). The School Breakfast Program was authorized by the Child Nutrition Act of 1966 (PL 89-642).

The National School Lunch Program (NSLP) and the School Breakfast Program assist schools in providing nutritious lunches and breakfasts, respectively, to children. Participating schools get cash subsidies on the basis of the number of meals served and also receive food commodities through the NSLP. Program schools must serve meals that meet specified nutritional guidelines and must offer free or reduced-price meals to eligible students. These programs enable students in households with incomes at or below 130 percent of the poverty guidelines to receive meals at no cost and allow students from households with incomes between 130 and 185 percent of the poverty guidelines to receive a reduced-price meal. Children whose families participate in the Food Stamp Program are automatically eligible for free school meals. Eligibility is summarized in Table 5-5. In 2003, more than 28 million children participated in the NSLP each schoolday (58 percent of all children attending a participating school or institution); 49 percent of these children received their meal free, and over 9 percent received meals at a reduced price. Total meals served numbered almost 4.8 billion each month. Total cost of the NSLP, including the after-school snacks served (see the next section) was $7,186 million ($6,337 million in cash payments and $849 million in commodity costs) in 2003. Approximately 8.4 million children participated in the School Breakfast Program each schoolday (22 percent of all children attending eligible institutions). Over 1.4 billion meals were served in 2003, with 74 percent of the children served receiving free and 9 percent receiving reduced-price breakfasts, at a total cost of over $1.6 billion in 2003.[90]

After-School Snack Program

School-based after-school programs can provide healthful snacks to children through age 18 via this expansion of the NSLP. Eligibility is summarized in Table 5-5. Because this program is available through the NSLP, the expenditures for the 140 million snacks served monthly in 2003 are reflected in the previous section.[91]

Special Milk Program

The SMP was incorporated into the Child Nutrition Act of 1966 (PL 89-642).

The Special Milk Program (SMP) encourages fluid milk consumption by children by providing cash reimbursement for each half-pint of milk served to children in schools and child care institutions that are not participating in the NSLP (see Table 5-5). Nearly 108 million half-pints of milk costing $14 million were served in 2003.[92]

Summer Food Service Program for Children

The Summer Food Service Program (SFSP) ensures that children in lower-income areas can continue to receive nutritious meals during long school vacations, when they do not have access to school lunch or breakfast. All meals are served free to eligible children. The program operates in areas where half or more of the children are from households with incomes at or below 185 percent of the poverty guidelines or where half or more of the participants in a program are from households at that same income level. These are called "open" sites. "Migrant" sites primarily serve children of migrant workers, and "camp" sites offer food service as part of a residential or day camp. Finally, "NYSP" sites operate through local colleges and universities offering the National Youth Sports Program. In the community, sponsors of the program include local schools or colleges, government units (for example, parks and recreation departments), summer camps, community action agencies, and other nonprofit organizations. The SFSP offers meals that meet the same nutritional standards as those provided by the NSLP at no cost to all children, up to age 18, who attend the program site (see Table 5-5). Almost 2.1 million children participated daily in the SFSP in 2003, for a total of 117 million meals served, with expenditures of $256 million.[93]

The SFSP was authorized in 1975 as an amendment to the National School Lunch Act (PL 94-105).

Child and Adult Care Food Program

The Child and Adult Care Food Program (CACFP) is designed to help public and private nonresidential child and adult day care programs provide nutritious meals for children up to age 12, the elderly, and certain people with disabilities. Sponsors may also receive USDA commodity foods or cash in lieu of commodities. Eligibility guidelines are summarized in Table 5-5. The average daily attendance was over 2.9 million in 2003, with a total of almost 1.8 billion meals served in 2003 at child care centers (over 1.0 billion meals), day care homes (694 million meals), and adult care centers (49 million meals). Of the meals served, over 83 percent were free or reduced-price. Expenditures for this program totaled over $1.9 billion in 2003.[94]

The CACFP was permanently authorized in 1978 (PL 94-105). The 1987 amendments to the Older Americans Act authorized the Child Care Food Program to change its name and to expand its service to include the elderly and persons with disabilities.

PROGRAMS FOR WOMEN AND YOUNG CHILDREN

Many federal programs address the special nutritional needs of women and their young children. The following sections review some of these programs. Table 5-5 gives the highlights of the programs below, as well as of Head Start and Early Head Start.

Special Supplemental Nutrition Program for Women, Infants, and Children (WIC)

WIC provides supplemental foods to infants, children up to age five, and pregnant, breastfeeding, and nonbreastfeeding postpartum women who qualify financially and are at nutritional risk. Financial eligibility is determined by income (between 100 percent and 185 percent of the poverty guidelines or below) or by participation in the Food Stamp Program or Medicaid. Nutritional risks, determined by a health professional, may include one of three types: medically based risks (anemia, underweight, maternal age, history of high-risk pregnancies), diet-based risks (inadequate dietary pattern), or conditions that make the applicant predisposed to medically based or diet-based risks, such as alcoholism or drug addiction. Homelessness and migrancy are also considered nutritional risks for purposes of WIC.

The WIC program serves both a remedial and a preventive role. Its services include

- Food packages or vouchers for supplemental food to provide specific nutrients (protein, calcium, iron, and vitamins A and C) known to be lacking in the diets of the target population
- Nutrition education
- Referrals to health care services

More about WIC appears in Chapter 10.

Authorization for WIC: Congress created a pilot WIC project in 1972 (PL 92-433) and authorized WIC as a national program as part of the National School Lunch and Child Nutrition Act Amendments of 1975 (PL 94-105).

Unlike most of the other food assistance programs, WIC is not an entitlement program. Over 7.6 million women, infants, and children received monthly WIC benefits in 2003, with women accounting for 24 percent, infants younger than 1 year accounting for 26 percent, and children 1–4 years accounting for 50 percent of participants. Because of caps on allocated federal funds, WIC is currently unable to reach all eligible persons. The average monthly WIC food cost per person in 2003 was $35.28, for a total food cost of over $3.2 billion. Total expenditures for the program (food benefits, nutrition services and administrative funds, WIC Farmers' Market Nutrition Program, and other items) was over $4.5 billion in 2003.[95]

WIC Farmers' Market Nutrition Program

Congress created a pilot FMNP through the Hunger Prevention Act of 1988 (PL 100-435) and awarded grants to ten states (Connecticut, Iowa, Maryland, Massachusetts, Michigan, New York, Pennsylvania, Texas, Vermont, and Washington). The FMNP was authorized in 1992 (PL 102-314) as a national program through an amendment to the Child Nutrition Act of 1966.

The WIC Farmers' Market Nutrition Program (FMNP) was created to accomplish two goals:

- Provide, from farmers' markets, fresh, nutritious unprepared fruits and vegetables to low-income, at-risk women, infants, and children.
- Expand the awareness and use of farmers' markets and increase sales at such markets.

In 2003, over 2.3 million WIC participants received benefits from this program. In 2004, 44 state agencies offered the program; they included Washington, D.C., Guam, Puerto Rico, 5 Indian tribal organizations, and 36 states. Coupons (not less than $10 but not more than $20 per year per participant) are issued to eligible recipients, who can use their coupons at authorized farmers' markets.[96]

PROGRAMS FOR OLDER ADULTS

Like people at other stages of the life span, older adults have unique needs. This section and Table 5-5 outline some of the programs designed to meet those needs.

Elderly Nutrition Program

The Congregate Meals Program and the Home-Delivered Meals Program were authorized by the Older Americans Act of 1965 (PL 89-73).

The Elderly Nutrition Program (ENP) (Title III) is intended to improve older people's nutritional status and enable them to avoid medical problems, continue living in communities of their own choice, and stay out of institutions. Its specific goals are to provide[97]

- Low-cost, nutritious meals
- Opportunities for social interaction
- Nutrition education and shopping assistance
- Counseling and referral to other social services
- Transportation services

More about the Elderly Nutrition Program appears in Chapter 12.

One of the Title III efforts is the Congregate Meals Program. Administrators try to select sites for congregate meals that will attract as many of the eligible elderly as possible. Through the Home-Delivered Meals Program, meals are delivered to those who are homebound either permanently or temporarily. The home-delivery program ensures nutrition, but its recipients miss out on the social benefits of the congregate meal sites; every effort is made to persuade them to come to the shared meals if they can. The DHHS's Administration on Aging administers these programs.

All persons 60 years and older (and spouses of any age) are eligible to receive meals from these programs, regardless of their income level. Priority is given to those who are economically and socially needy. In 2001, total funding for the Title III congregate and home-delivered meal programs was almost $530 million (almost $152 million for home-delivered meal programs and over $378 million for congregate programs).[98]

Senior Farmers' Market Nutrition Program (SFMNP)

Under SFMNP, low-income seniors are provided with coupons that may be used to purchase fresh, unprepared, locally grown fruits, vegetables, and herbs at farmers' markets, roadside stands, and community-supported agriculture programs. In 2002, products were available from over 11,000 farmers at 1,600 farmers' markets, as well as 1,500 roadside stands and more than 200 community-supported agriculture programs. In some areas, participants are provided with transportation to and from the markets by partnering with senior centers. Others have arranged to have local farmers transport their produce directly to senior housing locations. Table 5-5 lists eligibility guidelines. In 2003, $16.7 million was available to operate SFMNP in 35 states, 3 Indian tribal organizations, Puerto Rico, and the District of Columbia.[99]

Filling in the Gaps to Strengthen the Food Resource Safety Net

Eight-year-old Jack dreams of becoming a doctor; Helen, a 69-year-old grandmother, dreams of seeing her first great-grandson turn 5; and Meg, a 26-year-old single mother, dreams of getting a college degree. These very different people have one thing in common. None of them ate dinner last night.

—America's Second Harvest Annual Report

Despite all of the federal food assistance programs, emergency shelters and community food programs are straining to meet the rising requests for food. The public demand for emergency food assistance has increased in every region of the United States since 1980. The country has become a "soup kitchen society" to an extent unmatched since the breadlines of the Great Depression. The demand for emergency food assistance has increased steadily over the past two decades (see Figure 5-14). One survey of 27 major cities indicated that in 2001, requests for emergency food assistance increased by an average of 23 percent.[100] A subsequent (2003) study of 25 cities estimated that requests for emergency food assistance had increased by about 17 percent, with 88 percent of the cities reporting an increase. Much of the increased public demand for emergency food assistance is coming from the "working poor" and families with children, who report having to choose between food and other necessities, such as rent, utilities, or medicine.

As we noted when we discussed welfare reform, use of food pantries has consistently increased among households to 3.0 percent in 2002, with use among food insecure households increasing to 19.3 percent.[101] Usage by households at all income levels has increased since the beginning of the twenty-first century.

The Rising Tide of Food Assistance Need

To help fill the gaps in the federal programs, concerned citizens are working through community programs and churches to provide meals to the hungry. **Second Harvest,** the nation's largest supplier of surplus food, distributes over 1.8 billion pounds of food to food banks and other agencies for direct distribution around the nation annually (see Figure 5-15 on page 151). An estimated 23.3 million people relied on food banks, soup kitchens, and other agencies for emergency food in 2001—a 9 percent increase over the number of people served in 1997. Approximately 40 percent of these households have one employed person, 39 percent are children, 11 percent are elderly, 64 percent have incomes at or below the poverty guidelines, and 10 percent are homeless.[102]

Second Harvest A national network hunger-relief organization to which the majority of food banks belong.

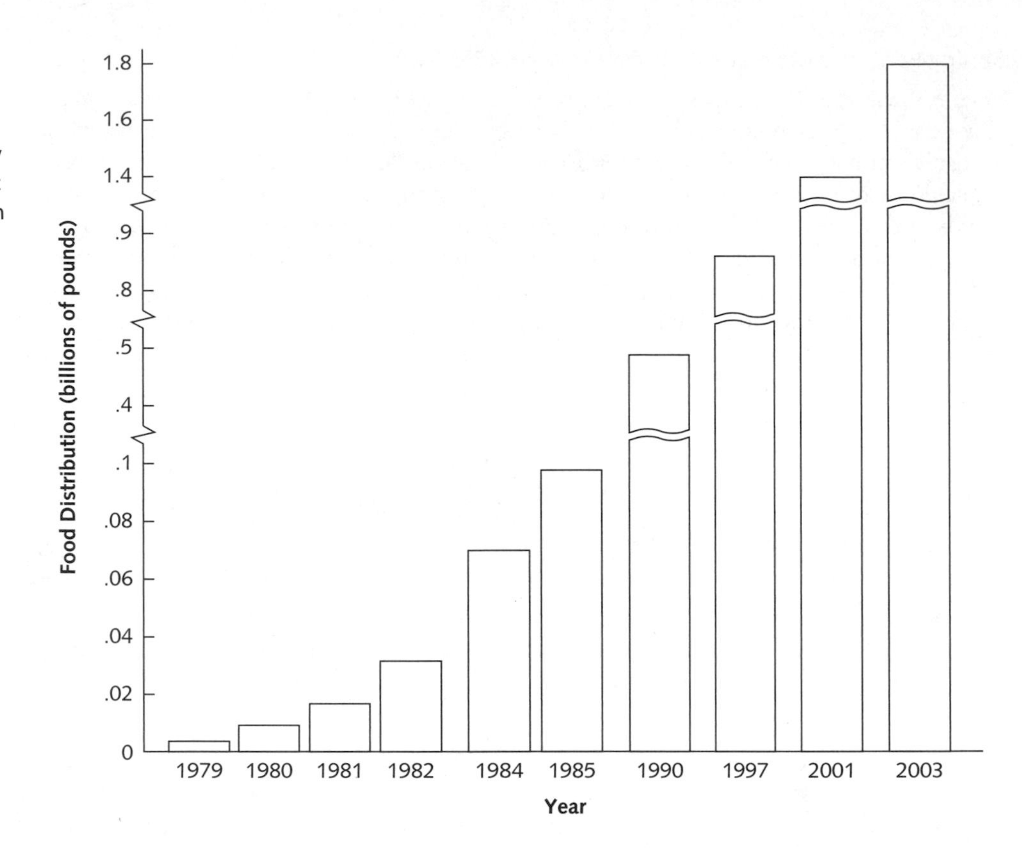

FIGURE 5-14 Demand for Emergency Food Assistance

Donated foods distributed by the America's Second Harvest network rose from 2.5 million pounds in 1979 to 100 million pounds in 1985 and to well over 860 million pounds in 1997. By 2003, Second Harvest was collecting and distributing 1.8 billion pounds of food nationwide, reflecting the efforts of private organizations to cope with the public's demand for emergency food assistance.

Emergency food services:

- Soup kitchens
- Church charities
- Surplus food giveaways
- Food banks
- Food pantries
- Prepared and perishable food programs

Source: Second Harvest; www.secondharvest.org.

However, even the dramatic increases in the number of food banks, food pantries, soup kitchens, **prepared and perishable food programs (PPFPs),** and other emergency food assistance programs cannot keep pace with the growing number of hungry people seeking food assistance.[103] It is estimated that on average, 14 percent of the requests for emergency food assistance go unmet.[104] Some facilities have had to decrease the number of bags of food provided and/or the number of times people can receive food. Moreover, each day's supply of meals lasts only for that day, leaving the problem of poverty unsolved. Furthermore, one out of every five needy people is not even receiving meals. These people must scavenge garbage, steal food or money to buy food, or continue to starve.

Prepared and perishable food programs (PPFPs) Nonprofit programs that help to feed people in need by linking sources of unused, unserved cooked and fresh food—such as caterers, restaurants, hotel kitchens, and cafeterias—with social service agencies that serve meals to people who would otherwise go hungry.

COMMUNITY FOOD SECURITY: ENHANCING LOCAL FOOD ACCESS

In an effort to reduce hunger, the USDA partnered with states, local municipalities, nonprofit groups, and the public sector to create a Community Food Security Initiative. The Professional Focus section of this chapter focuses on **community food security,** a relatively new concept with its roots in such disciplines as community nutrition, public health, nutrition education, sustainable agriculture, antihunger advocacy, and community development.[105] Community-based initiatives—such as farmers' markets and community gardens established on vacant city lots—can increase the availability of affordable, high-quality foods and serve to reconnect local farmers with urban consumers.[106] This chapter's Programs in Action feature describes a successful community garden venture in Wisconsin. The Community Food Security Initiative focuses on specific goals, including:[107]

Community food security The development and enhancement of sustainable, community-based strategies to ensure that all persons in a community have access to culturally acceptable, nutritionally adequate food through local nonemergency sources at all times.

- Building and enhancing local infrastructures to reduce hunger and food insecurity in communities

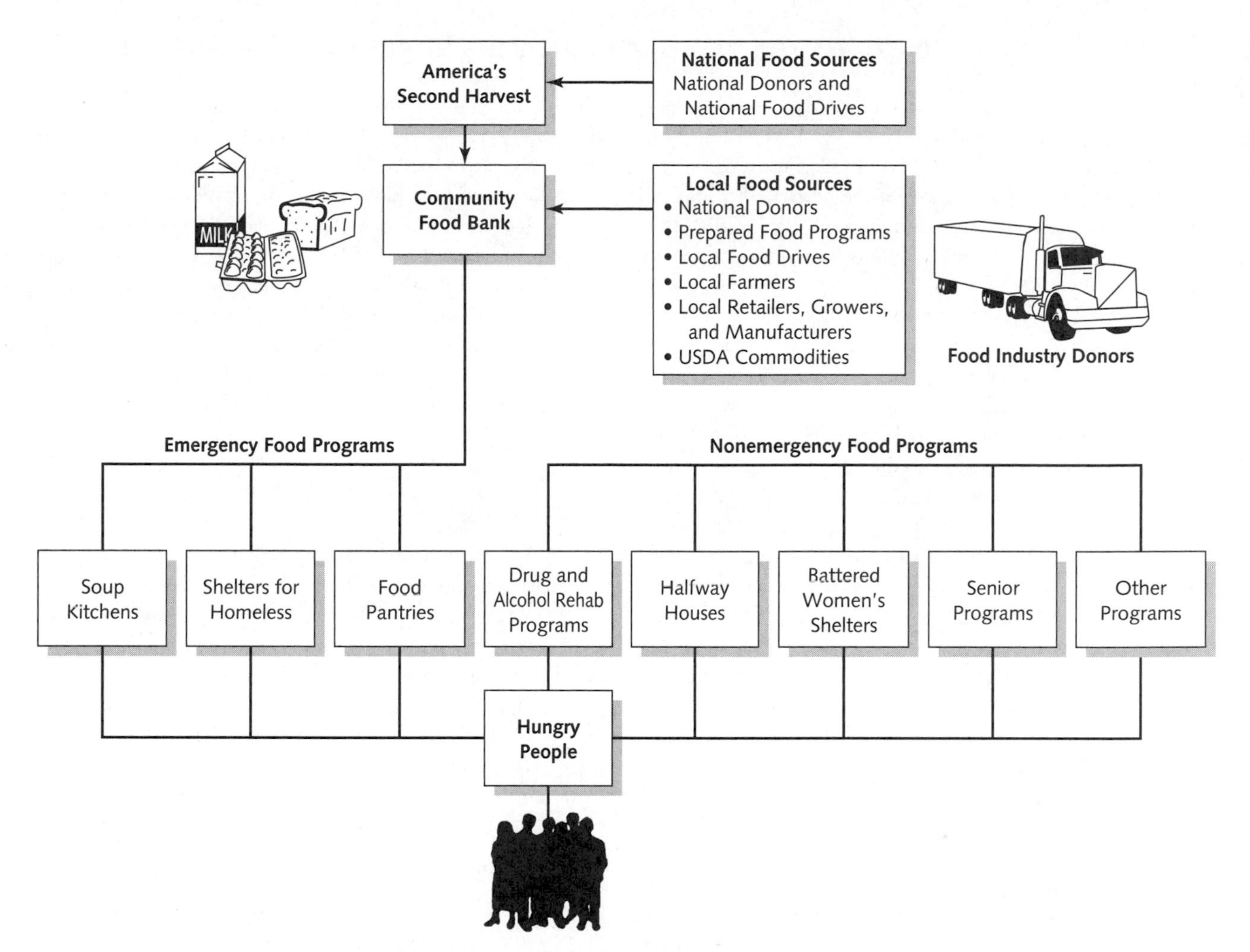

- Industry donors contact the food bank when they have products to contribute: goods that are mislabeled, in damaged packages, underweight, close to expiration, or in oversupply.
- The food bank picks up large-scale donations, cases to trailerloads, from the plant, distribution center, public warehouse, or retailer.
- At the food bank, nonprofit, charitable groups select products to stock their emergency food pantries, soup kitchens, shelters, child care centers, senior citizen programs, rehabilitation centers, and summer camps.
- Donated food reaches people in need: children and the elderly, the unemployed and low-wage earners, the homeless and frail homebound, and the disabled and ill.

FIGURE 5-15 What Is a Community Food Bank? With extra groceries, hot meals, or day care snacks, hundreds of thousands of individuals benefit from a Food Bank's partnerships with industry and the community.

- Increasing economic and job security for low-income people by helping people locate living-wage jobs and achieve self-sufficiency
- Strengthening the federal food and nutrition assistance safety net
- Bolstering supplemental food provided by nonprofit groups by assisting or developing local food recovery, gleaning, and donation efforts
- Improving community food production and marketing by aiding community projects that grow, process, and distribute food locally
- Boosting education and raising awareness about nutrition, food safety, and food security among community residents

Programs in Action

Overcoming Barriers to Increasing Fruit and Vegetable Consumption

Even though messages abound telling us to eat a minimum of five servings of fruits and vegetables each day, the average American eats only three and a half servings.[1] People give many reasons for this, but most of the barriers they note are self-imposed.[2] Low-income families have cited difficulty of preparation, expense, and perishability as barriers to consuming more fruits and vegetables.[3]

Barriers to eating fruits and vegetables can be tackled in several ways. Programs such as WIC and the National School Lunch Program provide food vouchers for free or reduced-price foods, including produce, to members of lower-income families, thereby alleviating the barriers related to price and accessibility. Point-of-purchase education, such as the supermarket-based *5 a Day* program, positively influences eating behaviors and can increase fruit and vegetable consumption.[4] Building self-efficacy—in this case, confidence in one's abilities to make healthful choices—is associated with increased intake of fruits and vegetables.[5] Demonstrating techniques for preparing fruits and vegetables and providing opportunities for hands-on practice may enhance self-efficacy and increase consumption. Encouraging family members to participate in intervention programs has also been shown to increase fruit and vegetable consumption.[6]

Community gardens can break down barriers to eating fruits and vegetables. They offer numerous other benefits to participants as well. "Community gardens are common ground for growing plants that feed, heal and give aesthetic pleasure. They are civic spaces where people work and recreate to nourish themselves, their families and friends. Most gardeners take satisfaction in having filled some part of their diet with food they have grown themselves."[7] The Kane Street Community Garden brought together educators, volunteers, and families to increase fruit and vegetable consumption among a low-income population in Wisconsin.

Inadequate fruit and vegetable consumption was common among low-income parents participating in the WIC program in La Crosse County, Wisconsin. Barriers to consumption of fresh fruits and vegetables included the high cost of produce in a limited budget, lack of knowledge of preparation methods, unfamiliarity with numerous fresh fruits and vegetables, and family members' disliking fruits and vegetables. Program initiatives were established to overcome these barriers.

Goals and Objectives

The Kane Street Community Garden was established to increase the consumption of fresh fruits and vegetables among low-income residents of La Crosse County. Program objectives were to establish a community garden on a vacant site, begin free distribution of community garden produce, and provide opportunities for learning to prepare fresh fruits and vegetables. The time frame for planting the garden, harvesting, hosting cooking classes, and evaluating the program ran from June to October 1998.

Methodology

In the winter of 1997–1998, the Hunger Task Force of La Crosse County established a subcommittee to oversee the planting of a community garden in a low-income neighborhood. The City Planning Department donated a parcel of land

- Improving research, monitoring, and evaluation efforts to help communities assess and strengthen food security

Examples of strategies and activities included under the label of community food security that fight hunger and strengthen local food systems are

- *Farmers' Markets.* Boost the income of small local farmers and increase consumers' access to fresh produce.
- *Community-Supported Agriculture Programs.* Partnerships between local farmers and consumers who buy shares in the farm in exchange for weekly supplies of fresh produce. Provide small-scale farmers with economic stability, while ensuring consumer participants high-quality produce, usually at below-retail prices.
- *Farm-to-School Initiatives.* Help local farmers sell fresh fruits and vegetables directly to school meals programs.
- *Food Stamp Outreach Programs.* Help increase the number of eligible households that participate in the Food Stamp Program.

for the garden. Volunteers prepared the garden for planting as part of a local community volunteer day. AmeriCorps volunteers were recruited to help the garden committee oversee garden maintenance and harvesting. Produce was harvested and distributed to low-income families free of charge two nights a week throughout the harvest season; any remaining produce was donated to local food pantries and free meal sites. Cooperative extension staff members at the garden provided preparation tips, recipes, and fact sheets on harvest nights. Produce samples were offered to encourage families to take home unfamiliar fruits and vegetables. Local media aided in publicizing the garden and soliciting donations of money and materials. The garden was publicized through WIC sites, the Salvation Army, the local food pantry, community centers, and senior meal sites.

Results

The 1998 growing season ended with 5,006 pounds of produce grown and distributed. One hundred twenty-five community residents volunteered to work or help in the garden, and 95 low-income families helped harvest. Six community organizations distributed surplus produce to needy families. Of the low-income families that responded to a survey, 71 percent stated that their fruit and vegetable consumption had increased.

Lessons Learned

As with many new projects, securing sufficient funding was and continues to be one of the biggest challenges. The subcommittee decided to expand the garden and sell a portion of the additional produce to raise additional funding for on-going support. It also decided to hire a part-time coordinator to help protect the volunteers from burn-out.

References

1. 5-A-Day for Better Health: A Baseline Study of America's Fruit & Vegetable Consumption, 1995.
2. L. Harnack et al., Association of cancer prevention–related nutrition knowledge, beliefs, and attitudes to cancer prevention dietary behavior, *Journal of the American Dietetic Association* (1997), 97: 957–65.
3. K. Treiman et al., Attitudes and behaviors related to fruits and vegetables among low-income women in the WIC Program, *Journal of Nutrition Education* (1996), 28: 149–56.
4. S. B. Foerster et al. California's "5 A Day—for Better Health!" campaign: An innovative population-based effort to effect large-scale dietary change, *American Journal of Preventative Medicine* (1995), 11: 124–31.
5. S. Havas et al., Factors associated with fruit and vegetable consumption among women participating in WIC, *Journal of the American Dietetic Association* (1998), 98: 1141–48.
6. G. Sorenson et al., Increasing fruit and vegetable consumption through worksites and families in the Treatwell 5-A-Day Study, *American Journal of Public Health* (1999), 89: 54–60.
7. City of Madison Advisory Committee on Community Gardens, *Growing a Stronger Community with Community Gardens: An Action Plan for Madison* (Madison, WI: City of Madison, 1999).

Source: Community Nutritionary (White Plains, NY: Dannon Institute, Spring 2000). Used with permission. For more information about the *Awards for Excellence in Community Nutrition,* go to www.dannon-institute.org.

- *Community Gardens.* Connect local consumers to locally grown food; the food may be made available to members of the community or to local food banks and food pantries; improve access to fresh produce.
- *Food Recovery and Gleaning Programs.* The terms **food recovery** and **gleaning** refer to programs that collect, for delivery to hungry people, excess wholesome food that would otherwise be thrown away. Approximately 96 billion pounds of food—or over one-quarter of the 356 billion pounds of food produced in this country for human consumption—is lost at the retail and food service levels.[108] The USDA is working with the National Restaurant Association to produce a food recovery handbook for its members, enabling schools to donate excess food from the School Lunch Program, encouraging airlines to donate unserved meals, working with the Department of Transportation to develop a comprehensive way to transport recovered foods, facilitating the donation of excess food from the Department of Defense, and providing technical assistance to community-based groups who seek to help.

Food recovery Such activities as salvaging perishable produce from grocery stores; rescuing surplus prepared food from restaurants and caterers; and collecting nonperishable food from manufacturers, supermarkets, or people's homes. The items recovered are donated to hungry people.

Gleaning The harvesting of excess food from farms, orchards, and packing houses to feed the hungry.

- *Food-Buying Cooperatives.* Help families save money by pooling resources (money, labor, purchasing, and distribution) to buy food in bulk quantities at reduced cost.
- *Directory of Supply and Demand for Community Food Surplus.* Creating a community directory can link farmers, retailers, and other sources of food surplus with local food pantries, soup kitchens, or other sites in need of emergency food.[109]

Beyond Public Assistance: What Can Individuals Do?

Solutions to the hunger problem depend on the willingness of people to take action and work together. Can we realistically hope to end hunger in the United States in the near future? Dr. J. Larry Brown, director of the Center on Hunger at Brandeis University provides our response:

> The chief answer to those who question whether we can eliminate hunger lies in the fact that we virtually did so in our recent past. The programs created by the nation in the late 1960s and early 1970s worked. The evidence indicates that hunger significantly declined in the face of a national commitment expressed through the vehicles of school meals, food stamps, WIC, and elderly feeding programs. By fully utilizing these existing programs, we could again end hunger.
>
> The larger question—the truly complicated one—is how to eliminate the cause of hunger: poverty. The United States pays a high price for poverty, and hunger is only one part of it. From a public health perspective it is the height of folly to permit such a significant risk factor for illness and premature mortality to persist. From a moral perspective the prevalence of poverty in one of the world's wealthiest nations is yet another matter.
>
> Our nation has the ability to end hunger in a matter of months once we determine to do so. The larger issue is whether we will address poverty and, in doing so, not only eliminate hunger but prevent much untimely disease and death as well.[110]

Regardless of the type and level of involvement a person chooses, each person can make a difference. The government programs described in this chapter need people's support in a number of ways. Individual people can

- Assist in these programs as volunteers. Community nutritionists can educate providers about safe food-handling practices and healthful diets, identify the most effective means of providing the needed nutrients from limited resources, and teach participants how to shop for the most economical nutritious foods.[111]
- Help develop means of informing low-income people about food-related federal and local services and programs for which they are eligible—from the Food Stamp Program, Medicaid, and WIC, to rent and utility assistance programs, to job-training programs that will help provide a living wage.[112]
- Help increase the accessibility of existing programs and services to those who need them.
- Document the hunger-related needs that exist in their own communities.
- As a nutrition professional, monitor the household food security of the clients you serve.
- Support local food production such as farmers' markets, roadside stands, community gardens, and community-supported agriculture programs.[113]
- Join with others in the community who have similar interests. Use the "hunger-free" criteria listed in Table 5-6 as guidelines for implementing comprehensive food assistance networks. The criteria are meant to provide a useful means of evaluating local antihunger networks.
- Conduct or participate in research to document the effectiveness of food assistance programs.[114]
- Represent nutrition issues at community health planning meetings.
- Document the impact of welfare reform on food security in local communities.

TABLE 5-6 Fourteen Ways to Reduce Hunger in Communities

1. Establish a community-based emergency food delivery network.
2. Assess community hunger problems and evaluate community services. Create strategies for responding to unmet needs.
3. Establish a group of individuals, including low-income participants, to develop and implement policies and programs to combat hunger and the threat of hunger; monitor responsiveness of existing services; and address underlying causes of hunger.
4. Participate in federally assisted nutrition programs that are easily accessible to targeted populations.
5. Integrate public and private resources, including local businesses, to relieve hunger.
6. Establish an education program that addresses the food needs of the community and the need for increased local citizen participation in activities to alleviate hunger.
7. Provide information and referral services for accessing both public and private programs and services.
8. Support programs to provide transportation and assistance in food shopping, where needed.
9. Identify high-risk populations and target services to meet their needs.
10. Provide adequate transportation and distribution of food from all resources.
11. Coordinate food services with parks and recreation programs and other community-based outlets to which residents of the area have easy access.
12. Improve public transportation to human service agencies and food resources.
13. Establish nutrition education programs for low-income citizens to enhance their food purchasing and preparation skills and make them aware of the connections between diet and health.
14. Establish a program for collecting and distributing nutritious foods—either agricultural commodities in farmers' fields or prepared foods that would have been wasted.

Source: House Select Committee on Hunger, legislation introduced by Tony P. Hall, excerpted in *Seeds,* Sprouts edition, January 1992, p. 3, with permission, © SEEDS Magazine, P.O. Box 6170, Waco, TX 76706.

- Learn more about the problem of food insecurity and become familiar with organizations that advocate for sustainable solutions to the problems associated with food insecurity.
- Follow food security legislation; call and write legislators about food insecurity issues; lobby to draw political attention to employment and wages.

Besides individual actions, all persons who are concerned about the problems of poverty and undernutrition in the United States can exercise their right to affect the political process. Anyone can decide what she or he thinks local, state, and national governments should do to help and can communicate these ideas to elected officials, urging them to support needed legislative changes. Individuals who volunteer their efforts and express their convictions to improve food assistance programs can also make a difference. Consider the following words, spoken over a hundred years ago:

I am only one,

But still I am one.

I cannot do everything,

But still I can do something;

And because I cannot do everything

I will not refuse to do the something that I can do.[115]

Internet Resources

See Table 5-5 for additional resources.

American Gardening Association	**www.communitygarden.org**
Bread for the World Institute	**www.bread.org**
Center on Hunger and Poverty	**www.centeronhunger.org**
The Community Food Security Coalition	**www.foodsecurity.org**
Community Supported Agriculture (CSA)	**www.nal.usda.gov/afsic/csa**
Congressional Hunger Center	**www.hungercenter.org**
CSREES Hunger and Food Security Information, including Community Food Projects	**www.csrees.usda.gov/index.html**
Empty Bowls Project	**www.emptybowls.net**
Food and Nutrition Service (USDA)	**www.fns.usda.gov/fncs**
Food Assistance and Nutrition Research Program (Briefing Room)	**www.ers.usda.gov/Briefing/FoodNutritionAssistance/FANRP**
Food Research and Action Center	**www.frac.org**
Food Security in the United States (Briefing Room)	**www.ers.usda.gov/briefing/foodsecurity**
Heifer International	**www.heifer.org**
HungerWeb	**http://nutrition.tufts.edu/academic/hungerweb/index.html**
Mazon	**www.mazon.org**
National Center for Children in Poverty	**www.nccp.org**
Recipes and Tips for Healthy Thrifty Meals	**www.usda.gov/cnpp/Pubs/Cookbook/thriftym.pdf**
Rural Income, Poverty, and Welfare (Briefing Room)	**www.ers.usda.gov/Briefing/IncomePovertyWelfare**
Samaritan's Purse	**www.samaritanspurse.org/home.asp**
Second Harvest	**www.secondharvest.org**
SERVEnet	**www.servenet.org**
Share Our Strength	**www.strength.org**
USDA Food and Nutrition Information Center's Hunger Resources	**www.nal.usda.gov/fnic**
USDA Food Recovery and Gleaning (Guide)	**www.usda.gov/news/pubs/gleaning/content.htm**

Hunger in an At-Risk Population

By Alice Fornari, EdD, RD, and Alessandra Sarcona, MS, RD

Scenario

As an Americorp volunteer with an educational background in foods and nutrition, you are asked by the social service agency sponsoring you to help a current client budget for food within the allowance that she receives to feed herself and her children. This single African-American mother with six children is renting a home in an urban environment; she has no access to a car and depends on public transportation. She currently is working 40 hours a week at a fast-food restaurant for minimum wage. Because the property she is renting is being sold, she has called the social service emergency housing program. No apartments are currently available, so she and her six children must occupy a motel room subsidized by the county social service agency. There is a small refrigerator and no cooking facilities. She received a $340 monthly food stamps allocation when she was renting a home. Since moving to the motel, she has depended on local restaurants and take-out food as her main supply of food for the family.

Learning Outcomes

- Identify food assistance programs and federal policies that are applicable to the family described in the case, including Internet sites providing this information; identify Internet sources of reliable statistics on hunger.
- Create a conceptual model (see the example on page 117) describing this victim of hunger, considering household features (including nutritional needs of children), cultural food-related behaviors, access, and economics.
- Identify variables that affect the food intake of the family, and characterize each of these variables as modifiable or not modifiable.
- Develop a plan for the family to incorporate adequate nutrition in their daily diet within the constraints of their socioeconomic status.

Foundation: Acquisition of Knowledge and Skills

1. Describe four food assistance programs and identify criteria for eligibility.
2. Access Internet sites supporting assistance for families that are hungry and homeless (refer to page 156).

Step 1: *Identify Relevant Information and Uncertainties*

1. Create a conceptual model (see the example on page 117) showing the interrelated factors associated with food insecurity and its outcomes for the family described in this case.
2. Identify information that the nutritionist does not have readily available and information that needs to be uncovered by the professional to help this family move forward.

Step 2: *Interpret Information*

1. Propose possible solutions for the family, supporting access to food and adequate nutrition; include federally funded food assistance programs that they are eligible for on the basis of their socioeconomic status. If applicable, include community outreach programs supported by local agencies.
2. Prioritize issues and variables impacting solutions.

Step 3: *Draw and Implement Conclusions*

1. Describe the limitations of the proposed solution(s).
2. Design a memo to social service staff communicating possible solutions and suggestions applicable to the family; design a comparable communication for the head of the household in language that is "consumer-friendly."

Step 4: *Engage in Continuous Improvement*

1. Explain how conditions might change for the family in such a way as to impact the current proposed solutions (for example, if this mother decides to relocate to a suburban community and has limited access to public transportation, or she decides to move in with a family member).
2. Establish a plan for monitoring the implementation of the solution(s) over time.

Professional Focus

Moving toward Community Food Security

Over the past decade, various entities within the United States, Canada, and Europe have applied a food systems approach to build community food security (CFS).[1] Using a food systems approach to build CFS requires understanding how communities interact with resources in their social and physical environments over extended periods of time. It also draws on strategies that address broad systemic issues affecting food availability, affordability, accessibility, and quality.[2] This Professional Focus discusses a three-stage continuum of evidence-based strategies and activities that applies a food systems approach to build community food security.

Community Food Security

CFS is an evolving concept that emphasizes long-term, systemic, and broad-based approaches to addressing food insecurity.[3] CFS can be defined as "A situation in which all community residents obtain a safe, culturally acceptable, nutritionally adequate diet through a sustainable food system that maximizes self-reliance and social justice."[4] A combination of practical activities and policy development is required to achieve CFS.[5]

Food Systems, Sustainability, and Sustainable Community Food Systems

A *food system* is a set of interrelated functions that includes food production, processing and distribution; food access and utilization by individuals, communities and populations; and food recycling, composting and disposal.[6] Food systems operate and interact at multiple levels, including community, municipal, regional, national, and global. Sustainability is defined as society's ability to shape its economic and social systems to maintain both natural resources and human life.[7] A sustainable community food system improves the health of the community, environment, and individuals over time, involving a collaborative effort in a particular setting to build locally based, self-reliant food systems and economies.[8]

Evidence-Based Strategies to Build Community Food Security

Examples of evidence-based strategies and activities that dietetics professionals can use to build CFS are arranged on a continuum related to the time frame of the expected outcome (short to long term) (see Figure 5-16). These strategies and activities fall into three progressive stages: initial food systems change, food systems in transition, and food systems redesign for sustainability.

In Stage 1, participants create small but significant changes to existing food systems. In Stage 2, food systems change is progressing, and efforts are directed toward facilitating and stabilizing that change. In Stage 3, efforts are made to institutionalize food systems change through advocacy and development of public policy. Data collection, monitoring, and evaluation are conducted at all stages. A detailed discussion of the strategies and activities outlined in the three-stage CFS continuum follows and is summarized in Figure 5-16.

Stage 1: Initial Food Systems Change

Stage 1 of the CFS continuum focuses on strategies and activities that create small but significant changes in existing food systems. An example of a strategy that dietetics professionals can use to facilitate initial food systems change is client counseling to maximize access to existing programs providing food and nutrition assistance, social services, and job training.[9] Furthermore, dietetics professionals can collect data on the nutritional adequacy of foods served in emergency food programs, because there is evidence that people who rely regularly on such sources may have inadequate nutritional intake.[10] There is also some evidence that people who live in low-income neighborhoods may not have easy access to food retail outlets that sell a variety of affordable and healthful foods.[11] Therefore, it is valuable to determine whether pricing and food quality inequities exist in food stores located in low-income neighborhoods.[12] Overall, this type of research and documentation is a useful early step to ensure that all community residents have access to nutritionally adequate and affordable foods.

Dietetics professionals can also facilitate initial change in food systems by educating consumers and institutions about the benefits of purchasing locally produced, seasonally available, and organically grown food. When consumers purchase foods that have been produced locally, a greater proportion of the profits remain with local farmers, providing them with a livable income while supporting local economies.[13] Purchasing locally produced foods protects the environment by reducing the use of fossil fuel and packaging materials.[14] The benefits of organic farming systems extend to farmers, consumers, and the environment. The results of a Washington State study showed that organic apple production provided similar yields, better-tasting fruit, and higher profitability and that it was more environmentally sound and energy efficient than producing apples by conventional practices.[15]

Stage 2: Food Systems in Transition

Stage 2 of the CFS continuum focuses on strategies and activities that support food systems in transition toward initiatives that have not traditionally been utilized by the current

FIGURE 5-16 Evidence-Based Strategies and Activities Associated with a Three-Stage Community Food Security Continuum

Stage 1: Initial Food Systems Change	Stage 2: Food Systems in Transition	Stage 3: Food Systems Redesign for Sustainability*
Strategies and Activities	**Strategies and Activities**	**Strategies and Activities**
• Counsel clients to maximize access to existing programs providing food and nutrition assistance, social services, and job training. • Document the nutritional value of emergency foods. • Identify food quality and price inequities in low-income neighborhoods. • Educate consumers and institutions about the benefits of locally produced, seasonally available, and organically produced foods.	• Connect emergency food programs with local urban agriculture projects. • Create multisector partnerships and networks. • Facilitate participatory decision making and policy development by serving on food policy councils and organizing community-mapping processes and multi-stakeholder workshops.	• Advocate for minimum wage increase and more affordable housing. • Advocate for food labeling standards about product history (e.g., place of origin, organic certified, Fair Trade certified†). • Mobilize governments and communities to institutionalize: 1. land use policies that facilitate large-scale urban agriculture; 2. market promotion and subsidies; 3. tax incentives and financing mechanisms to attract local food businesses to low-income neighborhoods.
Time Frame: Short-term	**Time Frame:** Medium-term	**Time Frame:** Long-term

Evaluation
Data collection, monitoring, and evaluation are conducted at all stages of the CFS continuum.‡

*Sustainability is defined as society's ability to shape its economic and social systems to maintain both natural resources and human life.
†Fair Trade is an innovative, market-based approach to sustainable development. Fair Trade helps family farmers in developing countries to gain direct access to international markets, as well as to develop the business capacity to compete in the global marketplace. In the United States, TransFair USA places the "Fair Trade Certified" label on coffee, tea, cocoa, bananas, and other fruits. For more information, see www.transfairusa.org.
‡This three-stage CFS continuum was adapted from a framework that was originally developed by R. J. MacRae and coauthors, Policies, programs, and regulations to support transition to sustainable agriculture in Canada, *American Journal of Alternative Agriculture* 5 (1990): 76–92.

Source: C. McCullum, E. Desjardins, V. I. Kraak, P. Ladipo, and H. Costello, Evidence-based strategies to build community food security, *Journal of the American Dietetic Association* 105 (2005): 279. Reprinted by permission of American Dietetic Association.

food systems. In this stage, the social infrastructure needed to connect various food system processes is established or strengthened through capacity building and multi-sector partnerships and networks. Stage 2 involves connecting private food distribution activities (e.g., food banks) with public spaces (e.g., community gardens and community-supported agriculture [CSA] farms), and promoting economic renewal projects and job creation through farmers' markets and small-scale food businesses. Several types of transition strategies and activities are described below.

Connecting Emergency Food Programs with Urban Agriculture Projects

Urban agriculture involves producing food closer to where most consumers live; it is an increasingly important strategy for achieving food security in the twenty-first century, because the world is becoming more urbanized.[16] Urban agriculture offers many potential benefits, such as reducing energy costs and pollution from food transportation and storage, absorbing greenhouse gas emissions, offering a viable use for urban waste as compost, and creating employment and economic development opportunities.[17]

An example of a successful effort to link urban agriculture projects with emergency food programs is the Michigan Food Bank Project, which administers 18 community gardens in the Lansing area. All garden participants receive supplies and training, which enable them to grow and preserve their own fresh vegetables. A second initiative of this project organizes volunteers to harvest surplus fruits and vegetables from local farms and distribute them to residents of low-income housing projects.[18]

Another successful example involves linking emergency food programs with CSA farms, an innovative strategy designed to connect local farmers with local consumers, develop a regional food supply and strengthen a local economy, maintain a sense of community, encourage land stewardship, and honor the knowledge and experience of local food producers.[19] CSA members pay a fee or volunteer their time in exchange for a share of the CSA farm's produce each week during the harvest season.[20]

Creating Multi-Sector Partnerships and Networks

Dietetics professionals can support food systems in transition by creating or joining multi-sector partnerships and networks that result in mutually beneficial programs and projects. For example, partnerships and networks are created by providing nutrition education at farmers' markets and conducting research on barriers to establishing, accessing, and participating in farmers' markets within low-income communities.[21] Farmers' markets improve consumers' access to fresh produce through reduced prices, while stimulating the vitality and sustainability of the local economy.[22]

Research in Michigan revealed that the maximum positive impact on fruit and vegetable consumption was achieved among WIC Farmers' Market Nutrition Program (FMNP) participants when nutrition education accompanied the coupons that were distributed as incentives to improve affordability.[23] Evaluation of consumer participation in the Seniors Farmers' Market Nutrition Program (SFMNP) in South Carolina suggested that participants receiving vouchers reported an intention to eat more fruits and vegetables year-round.[24]

Dietetics professionals can also create multi-sector partnerships that involve urban agriculture projects (e.g., CSA farms and community gardens) and farm-to-school programs. For example, one urban agricultural partnership in Colorado connected CSA farms with the WIC program to promote both fruit and vegetable consumption and physical activity for WIC participants.[25] Urban agricultural partnership projects such as community gardening exemplify an integrated approach to health promotion by increasing community networks, expanding green space, lowering crime rates in urban neighborhoods, and providing employment opportunities.[26] Farm-to-school partnerships provide local markets for farmers and integrate education about local food and farming issues with local foods served in school cafeterias. These partnerships may also lead to arranging special events with local farm organizations, creating nutrition curricula around school gardens, and providing opportunities for field trips to local farms. Farm-to-school programs have been shown to promote greater fruit and vegetable consumption.[27]

Facilitating Participatory Decision-Making Processes and Policy Development

Community residents must participate in decision-making processes and policy development in order to increase their access to resources.[28] Participatory decision-making and policy development can promote social cohesion and reduce inequities by building connections between local food production and consumption.[29] Participatory CFS strategies and activities such as *food policy councils* (FPCs), *community-mapping processes*, and *multi-stakeholder workshops* offer a planning framework and tools to involve local residents in defining and analyzing their community's issues and in mobilizing community action around a range of food system problems.[30] Each of these strategies and activities is described in more detail below.

A food policy council (FPC) is an officially sanctioned body representing various segments of a state, city, or local food system. It is composed of diverse stakeholders representing a wide range of interests related to agriculture, food, nutrition, and health. The goal of an FPC is to foster a comprehensive and systematic examination of agriculture, food, nutrition and health policies.[31]

The Toronto FPC has been instrumental in placing CFS and food policy development on the municipal agenda. Among its notable accomplishments, it has worked with the Economic Development Committee, Board of Health, and Parks and Recreation to develop strategies for featuring farmers' markets at various civic centers, and has chaired the School Garden and Compost Committee at the Toronto Board of Education, which entailed conducting 25 gardening workshops and developing a manual for school garden and compost projects.[32]

Dietetics professionals can also facilitate participatory decision-making processes and policy development through organizing *community-mapping processes* and *multi-stakeholder workshops*. A community-mapping process involves analyzing the community environment, examining the causes and consequences of food insecurity, and implementing strategies for improving local CFS.[33] Diverse food system stakeholders—including urban planners, food producers and retailers, volunteers in food access projects, food insecure individuals, and other concerned citizens—convene to engage in a process that examines how a local community food system can meet household and community needs by identifying available local food resources, food prices, transportation options, and employment opportunities. For example, the Portland–Multnomah County FPC has partnered with the regional government to create a geographical information system (GIS) map of grocery stores, farmers' markets, emergency food locations, and community gardens in the county.[34]

The purpose of a multi-stakeholder workshop is to provide a common vision and a platform for building consensus among diverse participants who may have divergent or competing interests.[35] One evaluation in upstate New York suggests that in their efforts to build CFS, practitioners may benefit from skills in facilitation, negotiation, and conflict resolution in order to transform conflict into greater capacity, equity, and justice.[36] See the Professional Focus feature in Chapter 18 for more about developing negotiation skills.

Stage 3: Food Systems Redesign for Sustainability

Stage 3 of the CFS continuum provides examples of strategies and activities in which citizens and government institutions play a larger role in building CFS. This stage involves advocacy and public policies that integrate different policy fields (such as education, labor, economic development, agriculture, food, social welfare, and health) in order to increase a community's food self-reliance and achieve nutritional goals (see Figure 5-16).[37] Integrated policies should ensure that all community members have the capacity to buy healthful foods rather than relying regularly on charitable food sources. It is also important that the proportion of the locally based food supply increase over time for the entire population, a goal that may be achieved through land use policies, market promotion and subsidies, and tax incentives and financing mechanisms.

Norway is an example of a country that has used integrated policy instruments to redesign its food system. Norwegians aspired to increase their domestic food self-reliance from 39 to 52 percent of total calories and to achieve macronutrient intakes appropriate for a healthful diet using policy tools such as production and consumer subsidies, market promotion, consumer education, food labeling, and penalties for unhealthful foods.[38] By 1988, Norway had reached 50 percent food self-reliance and had increased whole grain consumption and the quality of locally produced grains and potatoes. Greater improvements were limited by the lack of human and financial resources.[39]

In summary, dietetics professionals can use a three-stage continuum of evidence-based strategies and activities that applies a food systems approach to build community food security. Stage 1 creates small but significant changes in existing food systems through such strategies as identifying any inequities in food quality and pricing in low-income neighborhoods and educating consumers about the need and the possibilities for alternative food systems. Stage 2 stabilizes and augments change for food systems in transition by developing social infrastructure through multi-sector partnerships and networks and fostering participatory decision making and initial policy development. Based on these changes, Stage 3 involves advocacy and integrated policy instruments to redesign food systems for sustainability. Data collection, monitoring, and evaluation are key components of all stages of the CFS continuum.

References

1. M. Winne, H. Joseph, and A. Fisher, eds., *Community Food Security: A Guide to Concept, Design and Implementation* (Venice, CA: Community Food Security Coalition, 2000); M. Hora and J. Tick, *From Farm to Table: Making the Connection in the Mid-Atlantic Food System* (Washington, DC: Capital Area Food Bank, 2001); B. Cohen, *Community Food Security Assessment Toolkit* (United States Department of Agriculture, Economic Research Service; Washington, D.C., E-FAN-02-013), July 2002; C. McCullum, D. Pelletier, D. Barr, and J. Wilkins, Use of a participatory planning process as a way to build community food security, *Journal of the American Dietetic Association* 102 (2002): 962–967; M. W. Hamm and A. C. Bellows, Community food security and nutrition educators, *Journal of Nutrition Education and Behavior* 35 (2003): 37–43; Toronto Food Policy Council, *Reducing Urban Hunger in Ontario: Policy Responses to Support the Transition from Food Charity to Local Food Security,* Toronto Food Policy Council Discussion Paper Series, Discussion Paper 1, Toronto Food Policy Council, Toronto, 1994; Ontario Public Health Association Food Security Workgroup Position Paper, *A Systemic Approach to Community Food Security: A Role for Public Health* (Ontario, Canada: Ontario Public Health Association, 2002); and R. M. Pederson, *Urban Food and Nutrition Security: Participatory Approaches for Community Nutrition* (Copenhagen, Denmark: World Health Organization Regional Office for Europe), 2001.
2. Ibid. K. H. Brown and A. Carter, *Urban Agriculture and Community Food Security in the United States: Farming in the City Center to the Urban Fringe* (Venice, CA: Community Food Security Coalition, 2002), pp. 1–30;

and R. J. MacRae and coauthors, Policies, programs, and regulations to support transition to sustainable agriculture in Canada, *American Journal of Alternative Agriculture* 5 (1990): 76–92.

3. M. Winne, 2000.
4. M. Hamm, 2003.
5. Ibid.
6. K. Dahlberg, *What Are Food Systems? Local and Regional Food Systems: A Key to Healthy Cities.* Paper presented at: International Healthy Cities Conference, December 1993, San Francisco, California.
7. Position of the American Dietetic Association: Addressing world hunger, malnutrition, and food insecurity, *Journal of the American Dietetic Association* 103 (2003): 1046–1057.
8. J. Wilkins, Seasonal and local diets: Consumers' role in achieving a sustainable food system, *Research in Rural Sociology and Development* (1995) 6: 150–152; G. Feenstra, Local food systems and sustainable communities, *American Journal of Alternative Agriculture* 12 (1997): 26–28; and J. Peters, Community food systems: Working toward a sustainable future, *Journal of the American Dietetic Association* 97 (1997): 955–956.
9. Toronto Food Policy Council, 1994; and D. H. Holben and W. Myles, Food insecurity in the United States. Its effect on our patients, *American Family Physician,* March 1, 2004.
10. M. Hamm, 2003; V. S. Tarasuk and G. H. Beaton, Women's dietary intakes in the context of household food insecurity, *Journal of Nutrition* 129 (1999): 672–679; and U. O. Akobundu and coauthors, Vitamins A and C, calcium, fruit, and dairy products are limited in food pantries, *Journal of the American Dietetic Association* 104 (2004): 811–813.
11. K. Morland and coauthors, Neighborhood characteristics associated with the location of food stores and food service places, *American Journal of Preventive Medicine* 22 (2002): 23–29.
12. Toronto Food Policy Council, 1994; and S. Crixell and B. J. Friedman, Food insecurity in the barrio: Availability of affordable and nutritious foods in local grocery stores, *Journal of the American Dietetic Association* 103 (2003): A-45.
13. B. Storper, Moving toward healthful sustainable diets, *Nutrition Today* 38 (2003): 57–59.
14. J. Walsh and N. de Beaufort, Dietitians and local farmers: A unique alliance to change the way people think about food, *Hunger and Environmental Nutrition Dietetic Practice Group Newsletter,* Spring 2003:1, 3; Position of the American Dietetic Association: Dietetics professionals can implement practices to conserve natural resources and protect the environment, *Journal of the American Dietetic Association* 101 (2001): 1221–1227.
15. J. P. Reganold and coauthors, Sustainability of three apple production systems, *Nature* 410 (2001): 926–930.
16. Brown, 2002; and United Nations Development Programme, *World Urbanization Prospects: The 1999 Revision* (New York: United Nations Development Programme, 1999).
17. S. C. Chaplowe, Sustainable prospects in urban agriculture, in J. P. Madden and S. G. Chaplowe, eds. *For All Generations: Making World Agriculture More Sustainable* (Glendale, CA: OM Publishing, 1997), pp. 70–100; L. Horrigan, R. S. Lawrence, and P. Walker, How sustainable agriculture can address the environmental and human health harms of industrial agriculture, *Environmental Health Perspective* 110 (2002): 445–456.
18. Brown, 2002.
19. T. Murphy, R. E. Valenzuela, and I. Tsagarakis, Partnering community supported agriculture and emergency food programs, *Journal of the American Dietetic Association* 102 (2002): A44; and University of Massachusetts Extension, *Community Supported Agriculture and How Does it Work?* (Amherst, MA: University of Massachusetts Extension, 1999).
20. University of Massachusetts Extension, 1999.
21. Peters, 1997; J. C. V. Klotz and J. Steiner, *Improving and Facilitating a Farmers Market in a Low Income Urban Neighborhood: A Washington, DC Case Study* (Washington, D.C.: U.S. Department of Agriculture, Agricultural Marketing Service, 2001); and E. J. Conrey and coauthors, Integrated program enhancements increased utilization of a farmers' market nutrition program, *Journal of Nutrition* 133 (2003): 1841–1844.
22. H. Festing, *Farmers' Markets: An American Success Story* (Bath, UK: Ecologic Books, 1999); D. Hilchey, T. Lyson, and G. Gillespie, *Farmers' Markets and Rural Economic Development* (Ithaca, NY: Farming Alternatives Program, Cornell University; 1995).
23. J. V. Anderson and coauthors, 5 A Day fruit and vegetable intervention improves consumption in a low income population, *Journal of the American Dietetic Association,* 101 (2001): 195–202.
24. M. E. Kunkel, B. Luccia, and A. C. Moore, Evaluation of the South Carolina Seniors Farmers' Market Nutrition Education Program, *Journal of the American Dietetic Association* 103 (2003): 880–883.
25. S. H. Swartz and coauthors, Urban gardening yields benefits for low-income families, *Journal of the American Dietetic Association,* 103 (2003): A48.
26. J. Twiss and coauthors, Community gardens: Lessons learned from California Healthy Cities and Communities, *American Journal of Public Health* 93(9) (2003): 1435–1438; D. Armstrong, A survey of community gardens in upstate

New York: Implications for health promotion and community development, *Health and Place* 6 (2000): 319–327.

27. K. Sanger and L. Zenz, *Farm-to-Cafeteria Connections: Marketing Opportunities for Small Farmers in Washington State* (Olympia, WA: Washington State Department of Agriculture, 2004).
28. Pederson, 2001; M. A. Altieri, Agroecology: The science of natural resource management for poor farmers in marginal environments, *Agriculture, Ecosystems & Environment* 93 (2002): 1–24; World Health Organization, *Urban and Peri-Urban Food and Nutrition Action Plan, 2001* (Copenhagen, Denmark: World Health Organization Regional Office for Europe, 2001).
29. McCullum, 2002; Pederson, 2001; World Health Organization, 2001; Common Ground, *Mapping Food Matters: A Resource on Place-Based Food System Mapping* (Victoria, British Columbia, Common Ground and Victoria Chapter of Oxfam Canada, 2001).
30. Ibid.
31. N. D. Hamilton, Putting a face on our food: How state and local food policies can promote the new agriculture, *Drake Journal of Agricultural Law* 7 (2002): 407–443.
32. City of Toronto, *Public Health: Food Policy Council*. Available at www.toronto.ca/health/tfpc_index.htm. Accessed May 7, 2004.
33. Webster, 2000; World Health Organization, 2001.
34. Portland–Multnomah Food Policy Council, *Food Policy Recommendations. Portland-Multnomah Food Policy Council* (Multnomah County, City of Portland, Office of Sustainable Development, Portland, Oregon, 2003).
35. Drake University's Agricultural Law Center, Des Moines, Iowa, *State and Local Food Policy Council Project Partners*. Available at http://www.statefoodpolicy.org/partnerships.htm. Accessed February 12, 2004.
36. C. McCullum and coauthors, Mechanisms of power within a community-based food security planning process, *Health Education Behavior* 31(2) (2004): 206–222.
37. Toronto Food Policy Council, 1994.
38. K. Ringen, Norwegian food and nutritional policy, *American Journal of Public Health* 67 (1977): 550–551.
39. N. Milio, *An Analysis of the Implementation of Norwegian Nutrition Policy, 1981–87,* Report Prepared for the World Health Organization 1990 Conference on Food and Nutrition Policy, University of North Carolina, Chapel Hill, 1988.

Source: This Professional Focus is adapted from C. McCullum, E. Desjardins, V. I. Kraak, P. Ladipo, and H. Costello, Evidence-Based Strategies to Build Community Food Security, *Journal of the American Dietetic Association* 105 (2005): 278–283.

Chapter *Six*

A National Nutrition Agenda for the Public's Health

Learning *Objectives*

After you have read and studied this chapter, you will be able to:

- Describe the relationship of nutrition research and nutrition monitoring to U.S. national nutrition policy.
- Describe five key components of the National Nutrition Monitoring and Related Research Program.
- Discuss the Dietary Reference Intakes and explain how they are used to plan and assess diets.
- Describe appropriate uses of current dietary guidance systems.

This chapter addresses such issues as public policy development, the role of food in the promotion of a healthful lifestyle, current information technologies, evolving methods of assessing health status, diets for health promotion and disease prevention activities, and the scientific method, which are Commission on Accreditation for Dietetics Education (CADE) *Foundation Knowledge and Skills* requirements for dietetics education.

Chapter *Outline*

Something To Think About...

Action is the proper fruit of knowledge.

– Thomas Fuller, MD

Introduction

Time was when scientists investigating the role that diet plays in health zeroed in on the consequences of getting too little of one nutrient or another. Until the end of World War II in 1945, in fact, nutrition researchers concentrated on eliminating deficiency diseases such as goiter and pellagra. Since then, however, there has been a paradigm shift: The dietary recommendations now focus on preventing chronic disease. Although an abundant food supply and the practice of enriching and fortifying foods with essential nutrients have virtually eliminated deficiency diseases in North America, diseases related to dietary excess and imbalance are widespread. Four of the leading causes of death—heart disease, cancer, stroke, and diabetes—have been linked to diet. Nearly 40 percent of deaths can be attributed to smoking, physical inactivity, poor diet, or alcohol misuse, not to mention the contribution of these factors to hospitalizations, time lost on the job, and poor quality of life.[1] Dietary excesses and imbalances contribute to other ills as well, including high blood pressure, dental disease, osteoporosis, and obesity. Overweight and obesity alone account for many thousands of preventable deaths annually and cost $117 billion in 2000 [$61 billion in direct (health care) costs and $56 billion in indirect costs (lost wages, for example)].[2]

How do we know the prevalence of obesity and how do we know that dietary imbalances exist in the United States? Likewise, what population subgroups are at risk of malnutrition? And having determined who is malnourished, what guidelines can community nutritionists and other health professionals use to address the nutritional needs of malnourished groups? The answers to these questions are found in the policy arena, for nutrition policy dictates the strategies used to determine who is malnourished and formulates the appropriate dietary guidance to improve nutritional intake.

National Nutrition Policy

By **national nutrition policy,** we mean a set of nationwide guidelines that specify how the nutritional needs of the American people will be met and how the issues of hunger, malnutrition, food safety, food labeling, food fortification, sustainable agricultural practices, and nutrition research will be addressed.[3] Does the United States have a national nutrition policy? The answer is both yes and no. It is no in the sense that no one federal body or agency has as its sole mandate to establish, implement, and evaluate national nutrition policy. This deficiency in national policy making and planning was recognized more than 25 years ago, when Senator George McGovern, the chairman of the Senate Select Committee on Nutrition and Human Needs, called for the formation of such a body:

> We need a Federal Nutrition Office. The White House Conference on Food, Nutrition and Health recommended such an Office more than five years ago. . . . We cannot continue to operate on the assumption that the increasingly complex threads affecting nutrition policy will automatically weave themselves together into a coherent plan.[4]

National nutrition policy A set of nationwide guidelines that specify how the nutritional needs of the population will be met.

The function of such an entity would be to coordinate and direct federal nutrition policy. This federal nutrition office would follow through on the commitments made by federal agencies in implementing a national nutrition plan, help develop surveillance systems to monitor the population's overall health and nutritional status, and guarantee that any secondary nutritional implications of major policy decisions would be recognized and published in a "nutrition impact statement."[5]

More than three decades later, there is no Federal Nutrition Office, and nutrition policy in the United States is still fragmented. The problem with formalizing federal policy decisions in the nutrition arena lies in determining which agency should be the "power center" responsible for final decisions. This task is both complex and politically sensitive, because nutrition policy cuts across several policy areas, including agriculture, exports, imports, commerce, foreign relations, public health, and even national defense. No one federal agency can claim exclusive jurisdiction over nutrition issues. It is interesting that the comments made by McGovern in 1975 are still relevant in today's health care and policy environment:

> Nutrition is . . . [the] neglected [component] of income maintenance programs which themselves are woefully inadequate. This narrow conception virtually denies the nutrition dimension in comprehensive health care, or even that nutrition is a health issue. This parochial view ignores disturbing questions about misleading food advertising and other issues totally unrelated to income inequality. It fails to grapple with the reality that even wealthy Americans are often nutritionally illiterate, and that arteriosclerosis and other diseases associated with the aging process affect more than the poor. These and other issues germane to the health and well-being of the American people go far beyond the perils of poverty, and require a much broader Federal conception of the nation's nutritional policy requirements.[6]

And yet, even though the United States utilizes a more decentralized system with no single Federal Nutrition Office, the country can still be said to have a national nutrition policy. Palumbo remarked that "we can assume that no matter what was intended by government action, what is accomplished *is* policy."[7] Thus, national nutrition policy in the United States manifests itself in food assistance programs, national nutrition and health objectives as found in the *Healthy People 2010* initiative, regulations to safeguard the food supply and ensure the proper handling of food products, dietary guidance systems, monitoring and surveillance programs, food labeling legislation, and other activities in the nutrition arena.

The activities that form the basis of the nation's agenda to improve the public's health are outlined in Figure 6-1.[8] Research results and data obtained from nutrition monitoring provide information that helps in decision making—and hence, policy making—within the two main federal agencies that deal with food and nutrition issues, the U.S. Department of Agriculture (USDA) and the Department of Health and Human Services (DHHS). Some policy decisions affect the types of data collected during nutrition surveys and research related to human nutrient needs. Some aspects of nutrition policy, such as food assistance programs, are discussed in Chapter 5 and later chapters; this chapter focuses on three elements of U.S. nutrition policy: national nutrition monitoring and surveillance activities, nutrient intake standards, and dietary guidance systems.

National Nutrition Monitoring

Nutrition screening A system that identifies specific individuals for nutrition or public health intervention, often at the community level.

Most nations monitor the health and nutritional status of their populations as a means of deciding how to allocate scarce resources, enhance the quality of life, and improve productivity. National nutrition policies are typically guided by the outcomes of food and

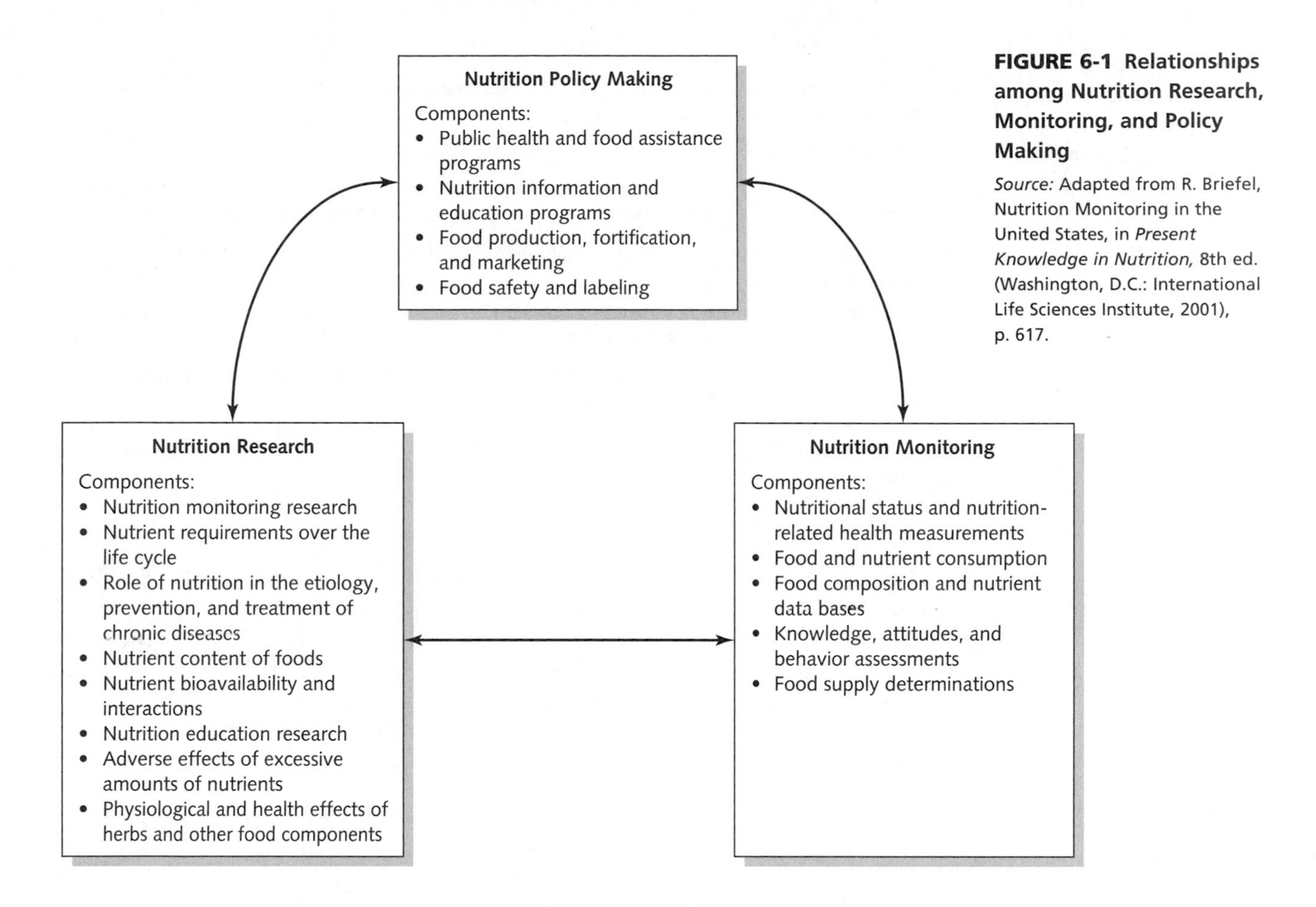

FIGURE 6-1 Relationships among Nutrition Research, Monitoring, and Policy Making

Source: Adapted from R. Briefel, Nutrition Monitoring in the United States, in *Present Knowledge in Nutrition,* 8th ed. (Washington, D.C.: International Life Sciences Institute, 2001), p. 617.

health surveys, which are designed to obtain data on the distribution of foodstuffs, the extent to which people consume food of sufficient quality and quantity, the effects of infectious and chronic diseases, and the ways in which these factors are related to human health. Such information can be derived by any of several methods, including **nutrition screening, nutrition assessment, nutrition monitoring,** and **nutrition surveillance.**[9] These methods are sometimes treated together under the rubric "nutrition monitoring." Such activities provide, at regular intervals, information about nutrition in populations and the factors that influence food consumption and nutritional status. The objectives of any national nutrition monitoring system, as outlined by the United Nations Expert Committee Report, are shown in Table 6-1.[10] The U.S. Congress has defined nutrition monitoring and related research as "the set of activities necessary to provide timely information about the role and status of factors that bear on the contribution that nutrition makes to the health of the people of the United States."[11]

Nutrition assessment The measurement of indicators of dietary status and nutrition-related health status to identify the possible occurrence, nature, and extent of impaired nutritional status (ranging from deficiency to toxicity).

Nutrition monitoring The assessment of dietary or nutritional status at intermittent times with the aim of detecting changes in the dietary or nutritional status of a population.

Nutrition surveillance The continuous assessment of nutritional status for the purpose of detecting changes in trends or distributions so that corrective measures can be taken.

BACKGROUND ON NUTRITION MONITORING IN THE UNITED STATES

The U.S. government has been involved in tracking certain elements of the food supply and food consumption for more than nine decades, beginning with the USDA's Food Supply Series undertaken in 1909. In the 1930s, the first USDA Household Food Consumption Survey was conducted (in 1965 this survey became known as the Nationwide Food Consumption Survey). In the late 1960s, concerns about the "shocking" nutritional status

TABLE 6-1 Objectives of a National Nutrition Monitoring and Surveillance System

- To describe the health and nutrition status of a population, with particular reference to defined subgroups who may be at risk.
- To monitor changes in health and nutrition status over time.
- To provide information that will contribute to the analysis of causes and associated factors and permit selection of preventive measures, which may or may not be nutritional (for example, smoking).
- To provide information on the interrelationship of health and nutrition variables within population subgroups.
- To estimate the prevalence of diseases, risk factors, and health conditions and of changes over time, which will assist in the formulation of policy.
- To monitor nutrition programs and evaluate their effectiveness in order to determine met and unmet needs related to target conditions under study.

Source: G. B. Mason and coauthors, *Nutritional Surveillance* (Geneva: World Health Organization, 1984). Used with permission.

of Mississippi schoolchildren and widespread chronic hunger and malnutrition led to the nation's first comprehensive nutrition survey, the Ten-State Nutrition Survey, conducted between 1968 and 1970 in California, Kentucky, Louisiana, Massachusetts, Michigan, New York, South Carolina, Texas, Washington, and West Virginia.[12] In the 1970s other surveys, such as the National Health and Nutrition Examination Surveys (NHANES I and II) and the Pediatric Nutrition Surveillance System, were added to the roster of methods used to obtain information about the nutritional status of the population.

More recently, in 1990, the U.S. Congress passed legislation (PL 101-445) that established the **National Nutrition Monitoring and Related Research Program (NNMRRP).**[13] The legislation specified that the USDA and the DHHS would jointly implement and coordinate the activities of the NNMRRP to obtain, through surveys, surveillance, and other monitoring activities, data about the dietary, nutritional, and nutrition-related health status of the U.S. population; the relationship between diet and health; and the factors that influence nutritional and dietary status. The NNMRRP takes a multidisciplinary approach to monitoring the nutritional and health status of Americans in general and of high-risk groups (such as low-income families, pregnant women, and minorities) in particular.[14] Today, the NNMRRP includes more than 50 surveillance activities that monitor and evaluate the health and nutritional status of the U.S. population.[15]

THE NATIONAL NUTRITION MONITORING AND RELATED RESEARCH PROGRAM

The NNMRRP includes all data collection and analysis activities of the federal government related to (1) measuring the health and nutritional status, food consumption, dietary knowledge, and attitudes about diet and health of the U.S. population and (2) measuring food consumption and the quality of the food supply.[16] Overall, the NNMRRP has the following goals:[17]

- Provide the scientific foundation for the maintenance and improvement of the nutritional status of the U.S. population and the nutritional quality and healthfulness of the national food supply.
- Collect, analyze, and disseminate timely data on the nutritional and dietary status of the U.S. population, the nutritional quality of the food supply, food consumption patterns, and consumer knowledge and attitudes concerning nutrition.

National Nutrition Monitoring and Related Research Program (NNMRRP) The set of activities that provides regular information about the contribution that diet and nutritional status make to the health of the U.S. population and about the factors affecting diet and nutritional status.

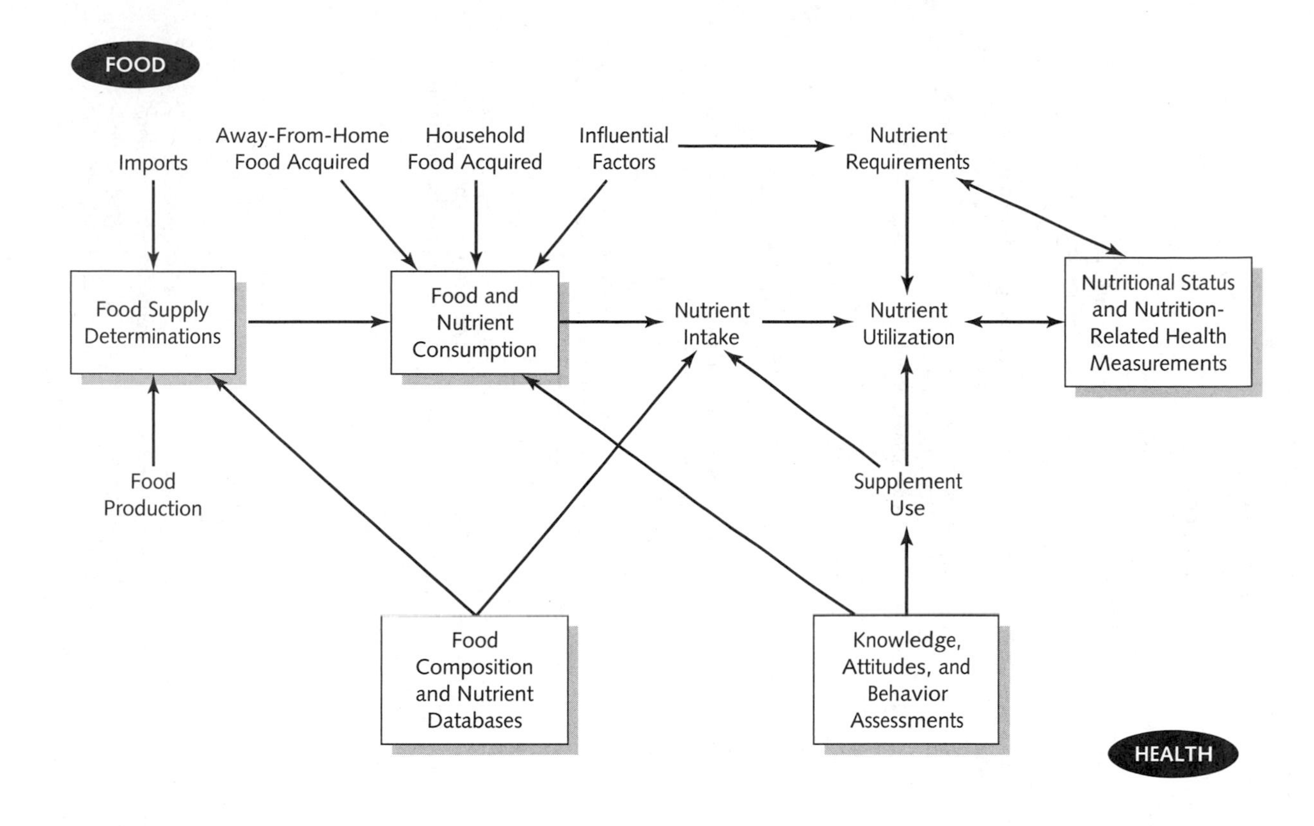

FIGURE 6-2 A Conceptual Model of the Relationships of Food to Health, Showing the Major Components of the National Nutrition Monitoring and Related Research Program

Source: Federation of American Societies for Experimental Biology, Life Sciences Research Office, prepared for the Interagency Board for Nutrition Monitoring and Related Research, *Third Report on Nutrition Monitoring in the United States,* Vol. 1 (Washington, D.C.: U.S. Government Printing Office, 1995), p. 6.

- Identify high-risk groups and geographical areas, as well as nutrition-related problems and trends, to facilitate prompt implementation of nutrition intervention activities.
- Establish national baseline data and develop and improve uniform standards, methods, criteria, policies, and procedures for nutrition monitoring.
- Provide data for evaluating the implications of changes in agricultural policy related to food production, processing, and distribution that may affect the nutritional quality and healthfulness of the U.S. food supply.

The NNMRRP surveys can be grouped into the five areas as shown in Figure 6-2: nutritional status and nutrition-related health measurements; food and nutrient consumption; knowledge, attitudes, and behavior assessments; food composition and nutrient databases; and food supply determinations. The next sections, which describe the major surveys in each area, are based on the directory of federal and state nutrition monitoring activities compiled by the Interagency Board for Nutrition Monitoring and Related Research.[18] Consult Figure 6-2 to see how the various surveys are used to obtain information about the relationship of food to health. The figure indicates that an individual's health and nutritional status is influenced by his or her food intake, which includes the food prepared and eaten at home and away from home. Food intake is influenced by the individual's knowledge about the relationship between diet and health, use of supplements, nutrient requirements, and attitudes about food and dietary and health practices. The composition and types of food available in the food supply also influence food choices. The boxes in the figure represent the five major component areas of the NNMRRP.

Refer to Table 6-2 for a description of the major surveys described in the section that follows. Other major NNMRRP surveys not mentioned in this chapter are described in the Directory of Federal and State Nutrition Monitoring Activities. Table 6-2 and the

(*Text discussion continues on page 173.*)

TABLE 6-2 Sources of Data from the Five Component Areas of the NNMRRP Considered in the Third Report on Nutrition Monitoring*

COMPONENT AREA AND SURVEY OR STUDY	SPONSORING AGENCY (Department)	DATE	POPULATION	DATA COLLECTED
Nutritional Status and Nutrition-Related Health Measurements				
Pediatric Nutrition Surveillance System (PedNSS) www.cdc.gov/pednss/	NCCDPHP, CDC (HHS)	1973, Continuous	Low-income, high-risk children, birth–17 years of age, with emphasis on birth–5 years of age	Demographic information; anthropometry (height and weight), birthweight, and hematology (hemoglobin, hematocrit), breastfeeding
Pregnancy Nutrition Surveillance System (PNSS) www.cdc.gov/PNSS	NCCDPHP, CDC (HHS)	1973, Continuous	Convenience population of low-income, high-risk pregnant women	Demographic information; pregravid-weight status, maternal weight gain during pregnancy, anemia (hemoglobin, hematocrit), pregnancy behavioral risk factors (smoking and drinking), birthweight, breastfeeding, and formula-feeding data
Annual National Health and Nutrition Examination Survey (NHANES) www.cdc.gov/nchs/nhanes.htm	NCHS, CDC (HHS)	1999–2004, Annual	Civilian, noninstitutionalized population 2 months of age and older. Oversampling of adolescents, African Americans, Mexican Americans, and adults ≥ 60 years	Survey elements resembling those of NHANES III and the National Health Interview Survey
Third National Health and Nutrition Examination Survey (NHANES III) www.cdc.gov/nchs/nhanes.htm	NCHS, CDC (HHS)	1988–94	Civilian, noninstitutionalized population 2 months of age and older. Oversampling of non-Hispanic blacks and Mexican Americans, children < 6 years of age, and adults aged ≥ 60 years	Dietary intake (one 24-hour recall and food frequency), socioeconomic and demographic information, biochemical analyses of blood and urine, physical examination, body measurements, blood pressure measurements, bone densitometry, dietary and health behaviors, and health conditions. Two additional 24-hour recalls for participants 50 years of age and older
Hispanic Health and Nutrition Examination Survey (HHANES) www.cdc.gov/nchs/nhanes.htm	NCHS (HHS)	1982–84	Civilian, noninstitutionalized Mexican Americans in five southwestern states, Cuban Americans in Dade County, FL, and Puerto Ricans in New York, New Jersey, and Connecticut; 6 months–74 years of age	Dietary intake (one 24-hour recall), food frequency, socioeconomic and demographic information, dietary and health behaviors, biochemical analyses of blood and urine, physical examination, body measurements, and health conditions
Food and Nutrient Consumption				
5 a Day for Better Health Baseline Survey http://dccps.nci.nih.gov/5ad_3_origins.html	NCI (HHS)	1991	Adults 18 years of age and older in the United States	Demographic information; fruit and vegetable intake; knowledge, attitudes, and behaviors regarding fruit and vegetable intake
Continuing Survey of Food Intakes by Individuals (CSFII) www.barc.usda.gov/bhnrc/foodsurvey	ARS, HNIS[†] (USDA)	1994–96 1989–91	Individuals in households in the 48 conterminous states. The survey was composed of two separate samples: households with incomes at any level (basic sample) and households with incomes ≤ 130% of the poverty thresholds (low-income sample)	One-day and 3-day food intakes by individuals of all ages, times of eating occasions, and sources of food eaten away from home. Data collected over 3 consecutive days by use of a 1-day recall and a 2-day record. Data available for 28 food components

COMPONENT AREA AND SURVEY OR STUDY	SPONSORING AGENCY (Department)	DATE	POPULATION	DATA COLLECTED
Food and Nutrient Consumption—*continued*				
Total Diet Study (TDS) http://vm.cfsan.fda.gov/~comm/tds-toc.html	FDA (HHS)	1961, Annual	Representative diets of specific age and gender groups	Chemical analysis of nutrients and contaminants in the U.S. food supply. Food composition data are merged with food consumption data to estimate daily intake of nutrients and contaminants
Nationwide Food Consumption Survey (NFCS) www.barc.usda.gov/bhnrc/foodsurvey	HNIS (USDA)	1987–88	Households in the 48 conterminous states and individuals residing in those households. The survey was composed of two samples: a basic sample of all households and a low-income sample of households with incomes ≤ 130% of the poverty threshold. This survey has been discontinued.	For households: quantity (pounds), money value (dollars), and nutritive value of food used. For individuals: 1-day and 3-day food intakes by individuals of all ages, times of eating occasions, and sources of food eaten away from home. Data collected over 3 consecutive days using a 1-day recall and a 2-day record. Data available for 28 food components
Vitamin and Mineral Supplement Use Survey (no website)	FDA (HHS)	1980	Civilian, noninstitutionalized adults 16 years of age and older in the United States	Prevalence of use, sociodemographic characteristics of the users, intakes of 24 nutrients (12 vitamins and 12 minerals) and other miscellaneous substances, and supplement use behaviors of the users by telephone interview
Knowledge, Attitudes, and Behavior Assessments				
Behavioral Risk Factor Surveillance System (BRFSS) www.cdc.gov/brfss/	NCCDPHP, CDC (HHS)	1984, Continuous	Adults 18 years of age and older residing in households with telephones in participating states	Demographic information; height, weight, smoking, alcohol use, weight control practices, diabetes, preventable health problems, mammography, pregnancy, cholesterol-screening practices, awareness, treatment, and modified food frequencies for dietary fat, fruit, and vegetable consumption by telephone interview
Weight Loss Practices Survey (WLPS) (no website)	FDA and NHLBI (HHS)	1991	Individuals currently trying to lose weight, 18 years of age and older	Demographic information; body mass index; diet history and other health behaviors; self-perception of overweight by telephone interview
Youth Risk Behavior Survey (YRBS) www.cdc.gov/HealthyYouth/yrbs/index.htm	NCCDPHP, CDC (HHS)	1990, Annual	Youths attending school in grades 9–12 in the 50 states, District of Columbia, Puerto Rico, and the Virgin Islands	Demographic information; smoking, alcohol use, weight control practices, exercise, and eating practices information
Diet and Health Knowledge Survey (DHKS) www.barc.usda.gov/bhnrc/foodsurvey	HNIS (USDA)	1989–91 1994–96	Main meal planner and preparer in households in the 48 conterminous states who participated in CSFII 1989–91 and 1994–1996	Self-perceptions of relative intake levels, awareness of diet–health relationships, use of food labels, perceived importance of following dietary guidance for specific nutrients and food components, beliefs about food safety, and knowledge about food sources of nutrients. These variables can be linked to data on individuals' food and nutrient intakes from CSFII 1989–91

Continued

COMPONENT AREA AND SURVEY OR STUDY	SPONSORING AGENCY (Department)	DATE	POPULATION	DATA COLLECTED
Food Composition and Nutrient Databases				
Nutrient Data Laboratory www.nal.usda.gov/fnic/foodcomp/				
• National Nutrient Data Bank (NNDB)	ARS (USDA)	NA[‡]	NA[‡]	Nutrient content of foods. Data from the National Nutrient Data Bank are used in the USDA Survey Nutrient Database for analysis of national dietary intake surveys and are also made available in published tables of food composition and as computerized databases
• USDA Nutrient Database for Standard Reference	ARS (USDA)	NA[§]	NA[§]	A computer file for *Agriculture Handbook No. 8* (USDA, 1992) produced from the National Nutrient Data Bank and the main source of the data for the USDA Survey Nutrient Database. This database includes data on food energy, 28 food components, and 18 amino acids for about 6,839 food items
• USDA Survey Nutrient Database	ARS (USDA)	NA	NA	The database is used for analysis of nationwide dietary intake surveys. It is updated continuously and includes data on food energy and 28 food components for more than 7,100 food items
Food Label and Package Survey (FLAPS) www.cfsan.fda.gov/~dms/lab-flap.html	FDA (HHS)	1977–2001 Biennially	NA	Use of nutrition labeling; declaration of selected nutrients and ingredients; nutrition claims; label statements and descriptors; nutrient analysis of a representative sample of packaged foods with nutrition labels
Food-Supply Determinations				
U.S. Food Supply Series http://209.48.219.54/default.htm	ARS, CNPP, and ERS[#] (USDA)	1909, Annual	U.S. total population	Quantities of foods available for consumption on a per capita basis; quantities of food energy, nutrients, and food components provided by these foods (calculated)

*Within each component area, entries are listed in reverse chronological order. NCCDPHP, National Center for Chronic Disease Prevention and Health Promotion; CDC, Centers for Disease Control and Prevention; HHS, Department of Health and Human Services; NCHS, National Center for Health Statistics; USDA, U.S. Department of Agriculture; NCI, National Cancer Institute; HNIS, Human Nutrition Information Service; FDA, Food and Drug Administration; ARS, Agricultural Research Service; NHLBI, National Heart, Lung, and Blood Institute; CNPP, Center for Nutrition Policy and Promotion; ERS, Economic Research Service; NA, not available.

[†]Legislation passed on February 20, 1994, transferred the functions and staff of the USDA's Human Nutrition Information Service (HNIS) to the existing Agricultural Research Service (ARS) of that department.

[‡]The work leading to the establishment of the NNDB was initiated in 1892 and has been maintained by the USDA since that time. The sponsoring agency is currently the ARS.

[§]The USDA Nutrient Data Base for Standard Reference was initiated in 1980 and has been maintained by the USDA since then. The sponsoring agency is currently ARS.

[#]On December 1, 1994, the U.S. Food Supply Series work conducted by the ARS was transferred to the Center for Nutrition Policy and Promotion (CNPP).

Source: Federation of American Societies for Experimental Biology, Life Sciences Research Office, prepared for the Interagency Board for Nutrition Monitoring and Related Research, *Third Report on Nutrition Monitoring in the United States,* Vol. 1 (Washington, D.C.: U.S. Government Printing Office, 1995), pp. 7–15.

Internet Resources at the end of this chapter provide addresses for accessing online survey data and other documents.

The Directory of Federal and State Nutrition Monitoring and Related Research Activities is a comprehensive summary of federal and state nutrition monitoring activities and a resource for finding nutrition monitoring data sources, as well as published research using these data. The summary is available at www.cdc.gov/nchs/pressroom/00facts/nutrit.htm and provides extensive links to other federal websites.

Nutritional Status and Nutrition-Related Health Measurements

The surveys that form the basis of the health and nutritional status component of the NNMRRP target a variety of specific population groups, including noninstitutionalized civilians over the age of 55 years, children aged 2 to 6 years, women of reproductive age, and individuals residing in nursing homes. The surveys collect data on diverse issues, such as family structures, community services, risk factors associated with cancer, aspects of family planning and fertility, and the causes of low birthweight among infants. The NHANES series and the Pediatric Nutrition Surveillance System (PedNSS) are included in this component. The PedNSS is a surveillance program that monitors the nutritional status of low-income children from birth to 17 years of age at high risk for nutrition-related problems, through the collection of measurements such as height, current weight, weight at birth, hemoglobin, and hematocrit.

The following surveys are among the most important of the health and nutritional status measurements. The NHANES series uses a sample that is representative of the civilian, noninstitutionalized population, and it has a good **response rate.** Consequently, it has dramatically influenced several aspects of health, including development of the 2000 CDC Growth Charts, awareness of dietary and lifestyle strategies associated with high cholesterol, and evidence related to high lead levels among Americans, evidence largely responsible for the current use of lead-free gasoline.[19]

National Health and Nutrition Examination Survey (NHANES I) Conducted in 1971–1974, the NHANES I was designed to collect and disseminate data that could be obtained best or only by direct physical examination, laboratory and clinical tests, and related measurements. The target population was civilian noninstitutionalized persons aged 1 through 74 years. The measures included dietary intake (one 24-hour recall), body composition, hematologic tests, urine tests, X-rays of the hand and wrist, dental examinations, and other measurements.

NHANES II Conducted in 1976–1980, this program targeted civilian noninstitutionalized persons aged 6 months through 74 years. It collected the same types of data as the NHANES I.

Hispanic Health and Nutrition Examination Survey (HHANES) This survey, conducted in 1982–1984, was designed to collect and disseminate data obtained from physical examinations, diagnostic tests, anthropometric measurements, laboratory analyses, and personal interviews of Mexican Americans, Puerto Ricans, and Cubans. Dietary intake was assessed by one 24-hour recall and a food frequency questionnaire. The target population consisted of "eligible" Hispanics aged 6 months through 74 years. "Eligible" Hispanics were limited to Mexican Americans living in five southwestern states; Cubans living in Dade County, Florida; and Puerto Ricans living in New York, New Jersey, and Connecticut.

NHANES III The NHANES III was conducted in 1988–1994 on a nationwide sample of about 34,000 persons aged 2 months and over. The survey was divided into two 3-year surveys (phase 1 and phase 2) so that national estimates could be produced for each 3-year period and for the entire 6-year period.[20] Its target population was civilian noninstitutionalized persons aged 2 months and over. Dietary intake was assessed by one 24-hour recall and a food frequency questionnaire. The examination components vary by age. Infants aged 2 to 11 months underwent a physician's exam, body measurements, and an

Response rate The value obtained by multiplying the participation rates for each survey component.

assessment of tympanic impedance; a dietary interview was conducted with the child's caregiver or parent. Between the ages of 1 and 19 years, additional assessments were added, including a dental exam, vision test, cognitive test, allergy skin test, and spirometry (a measurement of lung capacity). Participants over 20 years of age underwent these assessments, plus an oral glucose tolerance test, bone density test, fitness test, electrocardiogram, and other tests.[21] NHANES III data were used to develop new, nationally representative equations to predict stature for non-Hispanic white, non-Hispanic black, and Mexican-American adults aged 60 years and older.[22]

Compared with other surveys, the response rate for the NHANES III was good. In NHANES III, 100 percent of households were screened, and 86 percent of those screened were later interviewed. The survey's overall response rate was 73 percent. NHANES III also investigated the effects of the environment on health. Data were gathered to measure the levels of pesticide exposure, the presence of certain trace elements in the blood, and the amounts of carbon monoxide present in the blood.[23] For example, NHANES III found that nearly 9 out of 10 nonsmoking Americans were exposed to smoke either at home or on the job.

Current NHANES

Beginning in 1999, the NHANES program took a new direction as a continuous survey that can be linked to related government surveys of the U.S. population—specifically, the National Health Interview Survey (NHIS) and the USDA Continuing Survey of Food Intakes by Individuals (CSFII). NHANES is linked to NHIS with regard to questionnaire content of the household interview, for selected topics.

Early in 2000, the USDA and the DHHS announced a decision to integrate NHANES and CSFII into a single, more cost-effective survey—the National Food and Nutrition Survey (NFNS).[24] The merger is viewed as the most efficient use of the limited government funding available for nutrition monitoring purposes. As of January 2002, the dietary portion of the new integrated survey—"What We Eat in America"—is administered as part of the NHANES. The National Center for Health Statistics (NCHS) of DHHS is responsible for sample design and survey operations. The Agricultural Research Service (ARS) of the USDA has responsibility for processing the dietary data derived from 2-day food recalls.

For updates on the integrated survey process and answers to frequently asked questions, visit www.cdc.gov/nchs/nhanes.htm.

The integrated nutrition survey collects data about diet and health annually rather than periodically as it was done in the past.[25] The sample for the survey is representative of the civilian, noninstitutionalized population of the United States. The survey interviews and examines about 5,000 people in a 12-month period. There are two parts to the NHANES survey: the home interview and the health examination. During the in-home interview, participants are asked questions about their health status, disease history, and diet. Dietary 24-hour recalls are completed in person or by phone using a research-based, multiple-pass approach. Figure 6-3 reviews this five-step approach used in NHANES. Community nutritionists can use aspects of this approach to improve the quality of dietary information gathered in practice.

The health examination is performed in a mobile exam center (MEC). Table 6-3 is a summary of the health exam tests part of NHANES. As in NHANES III, the examinations that a participant will have depend on that participant's age and gender. As in the previous NHANES, there is oversampling of adolescents (12–19 years), older persons (60 years and over), low-income persons (less than 130 percent of the poverty threshold), pregnant women, African Americans, and Mexican Americans. The oversampling allows greater precision for estimates of food and nutrient intakes for these groups. The sample design allows for limited estimates to be produced annually and for more detailed estimates to be determined on 3-year samples (for example, 1999–2001).

Courtesy of WESTAT

The annual NHANES provides estimates of the health of Americans by examining a representative sample of people. To accomplish this, survey teams travel to approximately 15 sites per year across the United States in specially equipped mobile examination centers (MEC). Each MEC consists of four large, interconnected trailer units. Each survey team consists of one physician, one dentist, two dietary interviewers, three medical technologists, five health technicians, one phlebotomist, two interviewers, and one computer data manager. To take a virtual tour of the MEC, go to www.cdc.gov/nchs/about/major/nhanes/mectour.htm.

Food and Nutrient Consumption

These surveys collect data that are used to describe food consumption behavior and to evaluate the nutritional content of diets in terms of their implications for food policies, marketing, food safety, food assistance, and nutrition education. The Vitamin and Mineral Supplement Use Survey, for example, assesses quantitatively the nutrient intake from vitamin and mineral supplements and the characteristics of supplement users. The 5 a Day for Better Health Baseline Survey, conducted in 1991, obtained data on the fruit and vegetable intake and knowledge, attitudes, and behaviors regarding fruit and vegetable intake of U.S. adults. The following are the other major surveys of this NNMRRP component.

- Computerized
- Interviewer-administered, in person or by phone
- Used in NHANES
- 5-step approach designed to enhance complete and accurate food recall and reduce respondent burden
- Uses research-based strategies
  - Respondent-driven, allowing initial recall to be self-defined (as opposed to starting with "breakfast," for example)
  - Association with a 24-hour day's events
  - Probes for frequently forgotten foods
- Companion Food Model Booklet for portion size estimation
- Utilizes Food and Nutrient Database for Dietary Studies
- 5-steps of the multiple-pass approach
  - *Quick list*. Collect a list of foods and beverages consumed the previous day.
  - *Forgotten foods*. Probe for foods forgotten during the quick list.
  - *Time and occasion*. Collect time and eating occasion for each food.
  - *Detail cycle*. For each food, collect detailed description, amount, and additions. Review 24-hour day.
  - *Final probe*. Final probe for anything else consumed.

FIGURE 6-3 Overview of the Automated Multiple-Pass Method (AMPM) for 24-Hour Dietary Recalls

Source: USDA Automated-Pass Method; available at www.barc.usda.gov/bhnrc/foodsurvey/ampm_features.html.

TABLE 6-3 A Sampling of NHANES Health Exam Tests and Measurements

HEALTH MEASUREMENTS	LABORATORY MEASUREMENTS
• Physician's exam	• Anemia
• Blood pressure	• Cholesterol
• Body fat	• Glucose measures
• Bone density	• Markers of immunization status
• Dentist's oral health exam	• Kidney function tests
• Vision test	• Lead
• Hearing test	• Cadmium
• Fitness test	• Mercury
• Height, weight, and other body measures	• Liver function tests
• Balance	• Nutrition status
• Leg circulation and sensation	• Exposure to environmental chemicals
• Skin conditions (hand dermatitis and psoriasis)	• Infectious diseases

Source: Health Exam Tests. See complete list of NHANES tests by participant age and gender available at www.cdc.gov/nchs/about/major/nhanes/testcomp.htm.

Continuing Survey of Food Intakes by Individuals (CSFII), 1985 and 1986 The CSFII was designed to provide timely data on U.S. diets in general, the diets of high-risk population subgroups, and changes in dietary patterns over time. The target population was persons of selected sex and age in private households with incomes at any level (basic survey) and incomes at or below 130 percent of the poverty threshold (low-income survey). The 1985 survey focused on women and men aged 19 to 50 years and children aged 1 to 5 years. The 1986 survey focused on women and children and did not include men. Dietary measures included food intakes from multiple 24-hour recalls collected by interview.

Continuing Survey of Food Intakes by Individuals (CSFII), 1989–1991, 1994–1996, and 1998 Popularly known as the "What We Eat in America Survey," these surveys resemble the CSFII series conducted in 1985 and 1986. The target population is men, women, and children of all ages. The dietary component includes the kinds and amounts of food ingested both at home and away from home by household members for 3 consecutive days as measured by a 24-hour recall (obtained by personal interview) and a 2-day food diary.

Total Diet Study (TDS) This survey, conducted annually by the Food and Drug Administration (FDA), is designed to assess the levels of various nutritional components and organic and elemental contaminants of the U.S. food supply. The Selected Minerals in Food Survey, a component of the TDS, estimates the levels of 11 essential minerals in representative diets. The target population is eight age–sex groups: infants, young children, male and female teenagers, male and female adults, and male and female older persons. The design includes collecting 234 foods from retail markets in urban areas, preparing them for consumption, and analyzing them for nutritional elements and contaminants four times a year.

Knowledge, Attitudes, and Behavior Assessments

Surveys in this component of the NNMRRP gather data on weight loss practices; the general public's knowledge about the relationship of diet to health problems such as hypertension, coronary heart disease, and cancer; awareness among the general public, physicians, nurses, and dietitians of the risk factors of high blood cholesterol and coronary heart disease; and knowledge and attitudes about cancer prevention and lifestyle risk factors. For example, the Diet and Health Knowledge Survey (DHKS) was used by the USDA as a follow-up to the CSFII to measure consumers' awareness of diet–health relationships and dietary guidance, knowledge of food sources of nutrients, use of food labels, and beliefs about food safety. The Behavioral Risk Factor Surveillance System (BRFSS) is a unique system active in all 50 states. It is the main source of information on risk behaviors among adult populations. The BRFSS collects data related to health status, access to health care, tobacco and alcohol use, injury control (for example, use of seat belts), use of prevention services such as immunization and breast cancer screening, HIV and AIDS, weight control practices, treatment for high blood cholesterol, and frequency of intake of dietary fat, fruits, and vegetables. The BRFSS surveys adults by telephone interview.[26]

The Weight Loss Practices Survey (WLPS) also uses telephone interviews to obtain information about dieting history, health behaviors, and self-perception of overweight from adults trying to lose weight. The high-risk behaviors of youths, such as smoking, alcohol use, eating practices, and weight control practices, are assessed in the Youth Risk Behavior Survey.

To find data about risk factors and health behaviors of U.S. adults for 1 or all 50 states, the District of Columbia, or Puerto Rico, go to www.cdc.gov/brfss. You can choose the way you want to see the behavior data (e.g., How many people eat the recommended five fruits and vegetables a day?) or risk factor data (e.g., How many people lack health insurance?) by selecting among the following options.

- Grouped by age, gender, race, income, or education
- Comparisons of different states
- Comparisons of different years

Food Composition and Nutrient Databases

Information about the nutrient content of foods is provided by four different activities. The Food Label and Package Survey (FLAPS), undertaken biennially by the FDA in the DHHS, is designed to monitor the labeling practices of U.S. food manufacturers; it analyzes about 300 foods to check the accuracy of nutrient values on food labels. Other USDA activities are part of the Nutrient Data Laboratory and include the following:

National Nutrient Data Bank This continuous activity of the Agricultural Research Service of the USDA compiles and disseminates data on the nutrient composition of foods. Sources of nutrient data include government-funded university research, scientific publications, food processors and trade groups, and the results of food analyses by the Nutrient Composition Laboratory.

Nutrient Database for Standard Reference This database was initiated in 1980 and is sponsored by the Agricultural Research Service. It consists of computerized data of the nutrient composition of foods and is published as *Agriculture Handbook 8*.

Survey Nutrient Database This database is updated continuously and is used to analyze nationwide dietary intake surveys. It includes data on food energy and 28 food components for more than 7,100 food items.

Food disappearance data The amount of food remaining, from the total available food supply, after subtracting nonfood uses such as exports and industrial uses. These data represent the food that "disappears" into the marketing system and is available for human consumption.

Food Supply Determinations

Food available for consumption by the U.S. civilian population is determined by the USDA's Center for Nutrition Policy and Promotion through its Food Supply Series surveys. These food supply data or **food disappearance data** have been available annually since 1909.[27] The nutrient content of the available food supply is determined using food composition data and then used to estimate the nutrient content of the food supply on a per capita basis.

NHANES Data Have Been Used to:

- Get lead removed from gasoline
- Update the pediatric growth charts
- Establish baseline estimates for blood cholesterol levels in the United States

USES OF NATIONAL NUTRITION MONITORING DATA

The primary purpose of national nutrition monitoring activities is to obtain the information needed to ensure a population's adequate nutrition. The collected data are used in health planning, program management and evaluation, and timely warning and

TABLE 6-4 Uses of Data from the National Nutrition Monitoring and Related Research Program

Assessment of Dietary Intake

- Provide detailed benchmark data on food and nutrient intakes of the population
- Monitor the nutritional quality of diets
- Determine the nature of populations at risk of having diets low or high in certain nutrients
- Identify socioeconomic and attitudinal factors associated with diets

Monitoring and Surveillance

- Identify high-risk groups and geographical areas with nutrition-related problems to facilitate implementation of public health intervention programs and food assistance programs
- Evaluate changes in agricultural policy that may affect the nutritional quality and healthfulness of the U.S. food supply
- Assess progress toward achieving the nutrition and health objectives in *Healthy People 2010*
- Evaluate the effectiveness of nutritional initiatives for military feeding systems
- Recommend guidelines for the prevention, detection, and management of nutrition and health conditions
- Monitor food production and marketing

Regulatory

- Develop food labeling policies
- Document the need for food fortification policies and monitor such policies
- Establish food safety guidelines

Food Programs and Guidance

- Develop food guides and dietary guidance materials that target nutritional problems in the U.S. population
- Identify educational strategies to increase the knowledge of nutrition and to improve the eating habits of Americans
- Identify factors affecting participation in some food programs, and estimate the effect of participation on food expenditures and diet quality
- Identify populations that might benefit from intervention programs
- Identify changes in food and nutrient consumption that would reduce health risks
- Develop food guides and plans that reflect food consumption practices and meet nutritional and cost criteria
- Determine the amounts of foods that are suitable to offer in food distribution programs

Scientific Research

- Establish nutrient requirements (e.g., Dietary Reference Intakes)
- Study diet–health relationships and the significance of knowledge and attitudes toward dietary and health behavior
- Conduct national nutrition monitoring research
- Conduct food composition analysis

Historical Trends

- Correlate food consumption and dietary status with incidence of disease over time
- Follow food consumption through the life cycle
- Predict changes in food consumption and dietary status as they may be influenced by economic, technological, and other developments
- Track use and understanding of food labels and their effect on dietary intakes

Source: Adapted from Food Surveys Research Group, Beltsville Human Nutrition Research Center, Agricultural Research Service, U.S. Department of Agriculture website at www.barc.usda.gov/barc/bhnrc/foodsurvey/.

intervention efforts to prevent acute food shortages.[28] Data related to the population's nutritional status and dietary practices, obtained through national nutrition monitoring activities, are then used to direct research activities and make a variety of policy decisions involving food assistance programs, nutrition labeling, and education. For instance, the BRFSS allows for comparisons among states and between individual states and the nation. BRFSS data are also used to help states set priorities among health issues, develop strategic plans, monitor the effectiveness of public health interventions, measure the achievement of program goals, and create reports, fact sheets, press releases, and other publications to help educate the public, health professionals, and policy makers about disease prevention and health promotion. National policy makers use BRFSS data to monitor the nation's progress toward the *Healthy People 2010* objectives.[29] Specific uses of NNMRRP surveys are listed in Table 6-4. Congress, in particular, needs the data from nutrition monitoring activities to formulate nutrition and health policies and programs, assess the consequences of such policies, oversee the efficacy of federal food and nutrition assistance programs, and evaluate the extent to which federal programs result in a consistent and coordinated effort (a significant activity, considering that no Federal Nutrition Office exists at the present time).[30]

Nutrient Intake Standards

Merely collecting data on a population's nutrient intake and eating habits is not enough. Such data are meaningless on their own; to be valuable, they must be compared with some national standard related to nutrient needs. In the United States, the Food and Nutrition Board was established in 1940 to study issues of national importance pertaining to the safety and adequacy of the nation's food supply, to establish principles and guidelines for adequate nutrition, and to render authoritative judgment on the relationships among food intake, nutrition, and health, at the request of various agencies. The Food and Nutrition Board (FNB), a unit of the Institute of Medicine (IOM), is part of the National Academies, a private, nonprofit corporation created by an Act of Congress, with a charter signed in 1863 by President Abraham Lincoln. The IOM, chartered in 1970, acts as an adviser to the federal government on issues of medical care, research, and education.[31]

Dietary Reference Intakes (DRIs) for all age groups are listed on the inside front cover. The DRI reports can be accessed via www.nap.edu.

The major focus of the FNB is to evaluate emerging knowledge of nutrient requirements and relationships between diet and the reduction of risk of common chronic diseases and to relate this knowledge to strategies for promoting health and preventing disease in the United States and internationally; and to assess aspects of food science and technology that affect the nutritional quality and safety of food and thereby influence health maintenance and disease prevention. The inside front cover of this text summarizes the Dietary Reference Intakes (DRIs), which can be used in assessing and planning diets. The next section reviews the DRIs.

Health Canada is the federal department responsible for helping the people of Canada maintain and improve their health. For information related to food and physical activity, visit www.hc-sc.gc.ca.

DIETARY REFERENCE INTAKES (DRIs)

The **DRIs** are nutrient goals to be achieved over time.[32] They can be used to set standards for food assistance programs and for licensing group facilities such as day care centers and nursing homes, to design nutrition education programs, and to develop new food products.[33]

The DRIs consist of reference values developed by Health Canada and the Food and Nutrition Board of the Institute of Medicine to be used in planning and assessing the diets of individuals and groups. The DRIs include the **Estimated Average Requirement, Recommended Dietary Allowance, Adequate Intake, Estimated Energy Requirement, Acceptable Macronutrient Distribution Range,** and **Tolerable Upper Intake Level.** These terms are defined in the miniglossary on page 181.[34]

Uses of the DRI Values for Assessing and Planning Diets

Diets of Individuals
Assessment: Use EAR, EER, RDA, AI, UL, AMDR
Planning: Use RDA, AI, UL, AMDR

Diets of Groups
Assessment or Planning: Use EAR, EER, AI, UL, AMDR

Measuring Health Risks Among Adults: CDC's Unique Surveillance System

In the early 1980s, CDC worked with states to develop the Behavioral Risk Factor Surveillance System (BRFSS). Now active in all 50 states, 3 territories, and the District of Columbia, the BRFSS is the primary source of information on health-related behaviors of Americans. States use standard procedures to collect data through a series of monthly telephone interviews with adults. Questions are related to chronic diseases, injuries, and infectious diseases that can be prevented. A strong focus has been on the following behaviors, which are linked with heart disease, stroke, cancer, and diabetes—the nation's leading killers:

- Not getting enough physical activity
- Eating a high-fat, low-fiber diet
- Using tobacco and alcohol
- Not getting medical care that is known to save lives (for example, mammograms, Pap smears, colorectal cancer screening, and flu shots)

The surveys have given us a wealth of knowledge about these and other harmful behaviors—how common they are, whether they are increasing over time, and which people might be most at risk. Such information is essential to public health agencies at the national, state, and local levels.

State and local health departments rely heavily on data from the BRFSS to

- Determine priority health issues and identify populations at highest risk for morbidity
- Develop strategic plans and target prevention programs
- Monitor the effectiveness of interventions and progress in meeting prevention goals
- Educate the public, the health community, and policy makers about disease prevention
- Support community policies that promote health and prevent disease

BRFSS data also help public health professionals monitor progress in meeting the nation's health objectives outlined in *Healthy People 2010.* BRFSS information is used by researchers, voluntary and professional organizations, and managed care organizations to target prevention efforts.

- The BRFSS data can be analyzed according to age, sex, education, income, race, ethnicity, and other variables. This enables states to find groups at highest risk for health problems and make better use of scarce resources to prevent these problems.
- The BRFSS is designed to examine trends over time. For example, state-based data from the BRFSS have revealed a national epidemic of obesity.
- States can readily address urgent and emerging health issues. Questions may be added for a wide range of important health issues, such as diabetes, arthritis, tobacco use, folic acid consumption, health care coverage, and terrorism. You can learn about and view questionnaires at www.cdc.gov/brfss/questionnaires/index.htm. You can view historical BRFSS questions at http://apps.nccd.cdc.gov/BRFSSQuest.

The BRFSS is flexible in that it allows states to add timely questions specific to their needs. Yet standard core questions enable health professionals to make comparisons between states and reach national conclusions. BRFSS data have highlighted wide state-to-state differences in key health issues. In 2000, for example, the percentage of adults who smoked ranged from a low of 13 percent in Utah to a high of 30 percent in Kentucky.

Source: Centers for Disease Control and Prevention; available at www.cdc.gov/brfss.

Miniglossary of DRI Terms

Dietary Reference Intakes (DRIs) A set of reference values for energy and nutrients that can be used for planning and assessing diets for healthy people.

Estimated Average Requirement (EAR) The amount of a nutrient that is estimated to meet the requirement for the nutrient in half of the people of a specific age and gender. The EAR is used in setting the RDA.

Recommended Dietary Allowance (RDA) The average daily amount of a nutrient that is sufficient to meet the nutrient needs of 97 to 98 percent of healthy individuals of a specific age and gender.

Adequate Intake (AI) The average amount of a nutrient that is assumed to be adequate for individuals when there is not sufficient scientific research to calculate an RDA. The AI exceeds the EAR and possibly the RDA.

Estimated Energy Requirement (EER) The average calorie intake that is predicted to maintain energy balance in a healthy adult of a defined age, gender, weight, height, and level of physical activity, consistent with good health.

Acceptable Macronutrient Distribution Range (AMDR) A range of intakes for a particular energy source (carbohydrates, fat, protein) that is associated with a reduced risk of chronic disease while providing adequate intakes of essential nutrients.

Tolerable Upper Intake Level (UL) The maximum amount of a nutrient that is unlikely to pose any risk of adverse health effects to most healthy people. The UL is not intended to be a recommended level of intake.

Both the Institute of Medicine and Health Canada's Office of Nutrition Policy and Promotion have information about the DRIs available. The DRIs are a good example of policy making in action. Compared with the former policies related to nutrient intake recommendations, the DRIs represent a major shift in thinking about nutrient requirements for humans—a shift from prevention of nutrient deficiencies to prevention of chronic disease.[35] They also herald new thinking about the role of dietary supplements in achieving good health. Overall, the DRIs represent a new approach to dietary guidance and have the potential to influence the dietary messages provided to clients and communities, affect food fortification policy, and stimulate the development of new food products by industry.

DIETARY RECOMMENDATIONS OF OTHER COUNTRIES AND GROUPS

Various nations and international groups have published different sets of standards similar to the DRIs. Among the most widely used recommendations are those of the Food and Agriculture Organization (FAO) and the World Health Organization (WHO). The FAO/WHO recommendations are considered sufficient for the maintenance of health in nearly all people. They differ from the DRIs because they are based on slightly different judgment factors and serve different purposes. The FAO/WHO recommendations, for example, assume a protein quality lower than that commonly consumed in the United States. They also take into consideration that people worldwide are generally smaller and more physically active than the U.S. population. For the most part, the various recommendations fall within the same general range.

Nutrition Recommendations from WHO

- Energy: Sufficient to support normal growth, physical activity, and body weight (BMI = 20–22).
- Total fat: 15–30% of total energy.
  - Saturated fatty acids: 0–10% of total energy
  - Polyunsaturated fatty acids: 3–7% of total energy
  - Dietary cholesterol: 0–300 milligrams per day
- Total carbohydrate: 55–75% of total energy
  - Complex carbohydrates: 55–75% of total energy
  - Dietary fiber: 27–40 grams per day
  - Refined sugars: 0–10% of total energy
- Protein: 10–15% of total energy
- Salt: Upper limit of 6 grams per day

Nutrition Survey Results: How Well Do We Eat?

To track the nation's progress in reaching the *Healthy People 2010* healthy weight objectives, visit the interactive *Healthy People 2010* database at http://wonder.cdc.gov/DATA2010.

Food Components as Current Public Health Issues

- Food energy
- Total fat
- Saturated fatty acids
- *Trans* fatty acids
- Cholesterol
- Alcohol
- Calcium
- Iron
- Sodium

Potential Public Health Issues Requiring Further Study

- Dietary fiber
- Sugars
- Monounsaturated fatty acids
- Polyunsaturated fatty acids
- Vitamin A
- Antioxidant vitamins
- Folate
- Vitamin B_6
- Vitamin B_{12}
- Fluoride
- Iodine

What do the results of national surveys tell us about the nutritional status and dietary patterns of Americans? How well do we eat? The answers to these questions are mixed. Although we are well nourished, we are also generally overfat, underexercised, and beset to some extent with nutrient deficiencies. Chapter 8 reviews the obesity epidemic in the United States, but what else do we know about the health and nutrition of people in the United States, according to NHANES data?

Caution must be exercised in interpreting the NHANES data. On the one hand, when average nutrient intakes are examined, severe deficiencies in individuals can be missed. On the other hand, findings based on a single day's intake—as the NHANES findings were—can overestimate the extent of undernutrition.[36] What do survey results indicate about U.S. dietary patterns? Table 6-5 summarizes the intake of ten key nutrients for public health according to 1999–2000 NHANES data.

Another way to look at the 1999–2000 data is to use the USDA Healthy Eating Index (HEI), a summary measure of overall diet quality compared to the Dietary Guidelines for Americans and the food guidance system (food guide pyramid for 1999–2000) The HEI has ten components: the degree to which the diet meets the recommended intake of the five major food groups of the food guidance system (grains, vegetables, fruits, milk, and meat); total fat, saturated fat, total cholesterol, and sodium intakes; and variety of the diet.[37] According to the 1999–2000 HEI report,[38] American diets have not changed since 1996, at which point they had showed an improvement over 1989. One hundred is a perfect HEI score, with scores over 80 implying a "good" diet, 51 to 80 implying a diet in need of improvement, and 50 and less implying a "poor" diet. Overall, most people need dietary improvement—10 percent of Americans had a good diet, 16 percent had a poor diet, and 74 percent needed improvement. Fruit and milk product intakes were especially in need of improvement. Figure 6-4 illustrates the HEI component mean scores for 1999–2000, and Table 6-6 summarizes the overall mean scores by selected characteristics. Lower-quality diets were especially of concern for 15–18-year-old males, non-Hispanic blacks, low-income groups, and those with a high school diploma or less education. The HEI findings provide insight into how Americans need to improve their diets.

Information is still accumulating from ongoing analyses of the data yielded by the most recent surveys, but a considerable gap remains between current nutrition recommendations and consumers' practices. Undoubtedly, trends such as eating away from home and increased portion sizes in grocery and restaurant items negatively affect American eating habits. In the 1960s, an average fast-food meal of a hamburger, fries, and a 12-ounce cola provided 590 calories; today, many super-sized, extra-value fast-food meals deliver 1,500 calories or more.[39]

Dietary Guidance Systems

One approach to improving the public's knowledge about healthful eating involves the dissemination of dietary guidance in the form of dietary guidelines and food group plans. In his report to the American Society for Clinical Nutrition on "The Evidence Relating Six Dietary Factors to the Nation's Health," former assistant secretary for health and surgeon general Julius Richmond commented, "Individuals have the right to make informed choices and the government has the responsibility to provide the best data for making good dietary decisions."[40] Dietary guidance systems are methods by which the federal government helps the American people make prudent dietary decisions.

TABLE 6-5 United States, 1999–2000: Dietary Intake of Ten Key Nutrients

This table summarizes the mean intake of ten key nutrients by people living in the United States by gender.

NUTRIENT	MALES AND FEMALES	MALES	FEMALES
Energy (Calories)	2146	2475	1833
Calories from Protein (%)	14.7	14.9	14.6
Calories from Carbohydrate (%)	51.9	50.9	52.8
Calories from Fat (%)	32.7	32.7	32.6
Calories from Saturated Fat (%)	11.2	11.2	11.1
Cholesterol (mg)	265	307	225
Calcium (mg)	863	966	765
Folate (mcg)	361	405	319
Iron (mg)	15.2	17.2	13.4
Zinc (mg)	11.4	13.3	9.7
Sodium (mg)	3375	3877	2896

Source: J. D. Wright, C. Wang, J. Kennedy-Stephenson, and R. B. Ervin, Dietary intake of ten key nutrients for public health, United States: 1999–2000. Advance data from vital and health statistics; No. 334 (Hyattsville, Maryland: National Center for Health Statistics, 2003).

DIETARY GUIDELINES

Dietary guidelines are typically broad plans that focus on goal statements related to overall nutrient intake and daily eating patterns. The first attempt to formulate national dietary guidelines was undertaken in 1977, with the publication of the "Dietary Goals for the United States," a report of the U.S. Senate Select Committee on Nutrition and Human Needs.[41] In the 1980s and 1990s, so many dietary guidelines were published that consumers were overwhelmed by the dietary advice, not knowing which group to listen to.[42] In this section, we describe the major dietary guidelines published by government agencies and nongovernment organizations.

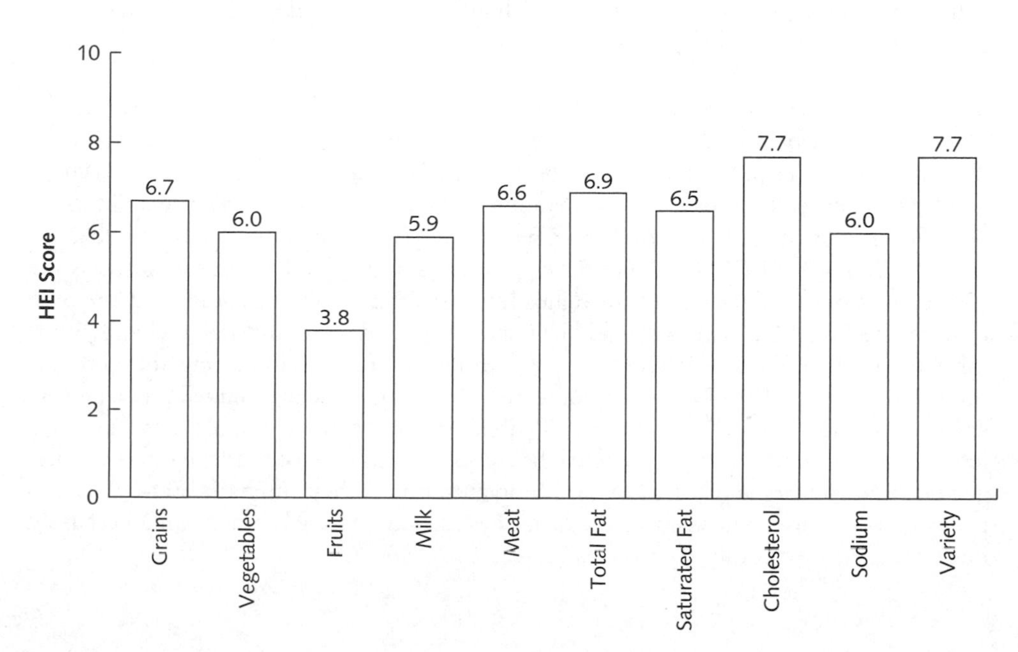

FIGURE 6-4 United States, 1999–2000: Healthy Eating Index Component Mean Scores

In 1999–2000, the overall score was 63.8/100 for Americans. Healthy Eating Index (HEI) component scores range from 0 to 10, with greater scores indicating intakes closer to the recommended range or amount.

Source: P. P. Basiotis and coauthors, *The Healthy Eating Index: 1999–2000*, U.S. Department of Agriculture, Center for Nutrition Policy and Promotion (2002). CNPP-12.

TABLE 6-6 Healthy Eating Index Overall Scores by Selected Characteristics, 1999–2000

This table summarizes overall scores by gender/age, race/ethnicity, education level, and income level.

	Overall Score	
	Male	Female
Gender/Age	63.2	64.5
15 to 18 years	59.9	61.7
19 to 50 years	61.3	63.2
over 50 years	65.2	66.6

Race/Ethnicity	Score
Non-Hispanic White	64.2
Non-Hispanic Black	61.1
Mexican American	64.5
Other Race (Asian, Pacific Islander, American Indian, Alaska Native)	63.4
Other Hispanic	64.2

Education (age 25 and over only)	Score
No High School Diploma	61.1
High School Diploma	63.0
More than a High School Diploma	65.3

Income	Score
< 100% of Poverty Level	61.7
100 to 184% of Poverty Level	62.6
> 184% of Poverty Level	65.0

Source: P. P. Basiotis, A. Carlson, S. A. Gerrior, W. Y. Juan, and M. Lino, *The Healthy Eating Index: 1999–2000;* U.S. Department of Agriculture, Center for Nutrition Policy and Promotion. CNPP-12.

Government Guidelines

Although the federal government's decision to promote a set of dietary goals or guidelines to improve the public's health would seem to be a straightforward matter of setting national nutrition policy, there has been much discussion about whether the government *should* establish such guidelines for the entire population. Concerns have focused on whether scientists could achieve a consensus on research outcomes and the relationship of diet to disease processes. Other questions also arose: What obligation do nutrition scientists and educators have to the general public to explain study outcomes and the differences that sometimes arise in interpreting study results? Who decides when research results are firm enough to warrant incorporating them into dietary guidance systems for the public? Is it appropriate to make general dietary recommendations for the *entire* population, when only a portion may be at risk for a specific disease condition? In what ways will the availability of national dietary guidelines—or the lack of them—affect the delivery of health care and educational efforts in the public health arena?[43]

www.nutrition.gov is a government resource that provides easy access to all online government information on nutrition and dietary guidance.

Dietary Guidelines for Americans By law (PL 101-445), the Dietary Guidelines for Americans must be revised every five years, beginning in 1995.[44] A short history of the development of this policy tool can be found at the Internet address shown in this chapter's list of Internet Resources. The sixth edition of the *Dietary Guidelines for Americans* was published jointly by the USDA and the DHHS in 2005 (earlier editions had been published in 1980, 1985, 1990, 1995, and 2000).[45] The principles of healthful eating promoted by the 2005 edition of the Dietary Guidelines, illustrated in Table 6-7, are science-based advice to promote health and to reduce the risk for major chronic diseases through diet and physical activity; these principles grew out of a variety of reports indicating the need for improved eating patterns in the U.S. population.[46] Thus the Dietary Guidelines include 41 key recommendations—23 for the general public and 18 for special populations—grouped into 9 general topics and were designed to be consistent with the recommendations of other government agencies and voluntary health organizations. They emphasize variety, calorie control, moderation, and safety of the diet. Physical activity is also emphasized because it increases energy expenditure and helps in weight control.[47]

(*Text discussion continues on page 188.*)

TABLE 6-7 The Dietary Guidelines, 2005

The 2005 Dietary Guidelines for Americans include 41 key recommendations, 23 for the general public and 18 for special populations, in 9 general topics.

ADEQUATE NUTRIENTS WITHIN CALORIE NEEDS

Key Recommendations

- Consume a variety of nutrient-dense foods and beverages within and among the basic food groups while choosing foods that limit the intake of saturated and *trans* fats, cholesterol, added sugars, salt, and alcohol.
- Meet recommended intakes within energy needs by adopting a balanced eating pattern, such as the USDA's MyPyramid food guidance system (www.mypyramid.gov) or the DASH Eating Plan (www.nhlbi.nih.gov).

Key Recommendations for Specific Population Groups

- *People over age 50*. Consume vitamin B_{12} in its crystalline form (e.g., fortified foods or supplements).
- *Women of childbearing age who may become pregnant*. Eat food high in heme–iron and/or consume iron-rich plant foods or iron-fortified food with an enhancer of iron absorption, such as vitamin C-rich foods.
- *Women of childbearing age who may become pregnant and those in the first trimester of pregnancy*. Consume adequate synthetic folic acid daily (from fortified food or supplements) in addition to food forms of folate from a varied diet.
- *Older adults, people with dark skin, and people exposed to insufficient ultraviolet band radiation (i.e., sunlight)*. Consume extra vitamin D from vitamin D-fortified foods and/or supplements.

WEIGHT MANAGEMENT

Key Recommendations

- To maintain body weight in a healthy range, balance calories from foods and beverages with calories expended.
- To prevent gradual weight gain over time, make small decreases in food and beverage calories and increase physical activity.

Key Recommendations for Specific Population Groups

- *Those who need to lose weight*. Aim for a slow, steady weight loss by decreasing calorie intake while maintaining an adequate nutrient intake and increasing physical activity.
- *Overweight children*. Reduce the rate of body weight gain while allowing growth and development. Consult a health care provider before placing a child on a weight-reduction diet.
- *Pregnant women*. Ensure appropriate weight gain as specified by a health care provider.
- *Breastfeeding women*. Moderate weight reduction is safe and does not compromise weight gain of the nursing infant.
- *Overweight adults and overweight children with chronic diseases and/or on medication*. Consult a health care provider about weight loss strategies prior to starting a weight-reduction program to ensure appropriate management of other health conditions.

PHYSICAL ACTIVITY

Key Recommendations

- Engage in regular physical activity and reduce sedentary activities to promote health, psychological well-being, and a healthy body weight.

  To reduce the risk of chronic disease in adulthood: Engage in at least 30 minutes of moderate-intensity physical activity, above usual activity, at work or home on most days of the week.

Key Recommendations for Specific Population Groups

- *Children and adolescents*. Engage in at least 60 minutes of physical activity on most, preferably all, days of the week.
- *Pregnant women*. In the absence of medical or obstetric complications, incorporate 30 minutes or more of moderate-intensity physical activity on most, if not all, days of the week. Avoid activities with a high risk of falling or abdominal trauma.

Continued

PHYSICAL ACTIVITY—*continued*

Key Recommendations

- For most people, greater health benefits can be obtained by engaging in physical activity of more vigorous intensity or longer duration.
- To help manage body weight and prevent gradual, unhealthful body weight gain in adulthood: Engage in approximately 60 minutes of moderate to vigorous activity on most days of the week, while not exceeding caloric intake requirements.
- To sustain weight loss in adulthood: Participate in at least 60 to 90 minutes of daily moderate-intensity physical activity while not exceeding caloric intake requirements. Some people may need to consult with a health care provider before participating in this level of activity.
- Achieve physical fitness by including cardiovascular conditioning, stretching exercises for flexibility, and resistance exercises or calisthenics for muscle strength and endurance.

Key Recommendations for Specific Population Groups

- *Breastfeeding women*. Be aware that neither acute nor regular exercise adversely affects the mother's ability to successfully breastfeed.
- *Older adults*. Participate in regular physical activity to reduce functional declines associated with aging and to achieve the other benefits of physical activity identified for all adults.

FOOD GROUPS TO ENCOURAGE

Key Recommendations

- Consume a sufficient amount of fruits and vegetables while staying within energy needs. Two cups of fruit and $2\frac{1}{2}$ cups of vegetables per day are recommended for a reference 2,000-calorie intake, with higher or lower amounts depending on the calorie level.
- Choose a variety of fruits and vegetables each day. In particular, select from all five vegetable subgroups (dark green, orange, legumes, starchy vegetables, and other vegetables) several times a week.
- Consume 3 or more ounce-equivalents of whole-grain products per day, with the rest of the recommended grains coming from enriched or whole-grain products. In general, at least half the grains should come from whole grains.
- Consume 3 cups per day of fat-free or low-fat milk or equivalent milk products.

Key Recommendations for Specific Population Groups

- *Children and adolescents*. Consume whole-grain products often; at least half the grains should be whole grains. Children 2 to 8 years old should consume 2 cups per day of fat-free or low-fat milk or equivalent milk products. Children 9 years of age and older should consume 3 cups per day of fat-free or low-fat milk or equivalent milk products.

FATS

Key Recommendations

- Consume less than 10 percent of calories from saturated fatty acids and less than 300 mg/day of cholesterol, and keep *trans* fatty acid consumption as low as possible.
- Keep total fat intake between 20 and 35 percent of calories, with most fats coming from sources of polyunsaturated and monounsaturated fatty acids, such as fish, nuts, and vegetable oils.
- When selecting and preparing meat, poultry, dry beans, and milk or milk products, make choices that are lean, low-fat, or fat-free.
- Limit intake of fats and oils high in saturated and/or *trans* fatty acids, and choose products low in such fats and oils.

Key Recommendations for Specific Population Groups

- *Children and adolescents*. Keep total fat intake between 30 to 35 percent of calories for children 2 to 3 years of age and between 25 to 35 percent of calories for children and adolescents 4 to 18 years of age, with most fats coming from sources of polyunsaturated and monounsaturated fatty acids, such as fish, nuts, and vegetable oils.

CARBOHYDRATES

Key Recommendations

- Choose fiber-rich fruits, vegetables, and whole grains often.
- Choose and prepare foods and beverages with little added sugars or caloric sweeteners, such as amounts suggested by the USDA Food Guide and the DASH Eating Plan.
- Reduce the incidence of dental caries by practicing good oral hygiene and consuming sugar- and starch-containing foods and beverages less frequently.

Key Recommendations for Specific Population Groups

- *Older adults*. Since constipation may affect up to 20 percent of people over 65 years of age, older adults should choose to consume foods rich in dietary fiber.
- *Children*. Carbohydrate intakes of children need special considerations with regard to obtaining sufficient amounts of fiber, avoiding excessive amounts of calories from added sugars, and preventing dental caries.

SODIUM AND POTASSIUM

Key Recommendations

- Consume less than 2,300 mg (approximately 1 tsp of salt) of sodium per day.
- Choose and prepare foods with little salt. At the same time, consume potassium-rich foods, such as fruits and vegetables.

Key Recommendations for Specific Population Groups

- *Individuals with hypertension, blacks, and middle-aged and older adults*. Aim to consume no more than 1,500 mg of sodium per day, and meet the potassium recommendation (4,700 mg/day) with food.

ALCOHOLIC BEVERAGES

Key Recommendations

- Those who choose to drink alcoholic beverages should do so sensibly and in moderation—defined as the consumption of up to one drink per day for women and up to two drinks per day for men.
- Alcoholic beverages should not be consumed by some individuals, including those who cannot restrict their alcohol intake, women of childbearing age who may become pregnant, pregnant and lactating women, children and adolescents, individuals taking medications that can interact with alcohol, and those with specific medical conditions.
- Alcoholic beverages should be avoided by individuals engaging in activities that require attention, skill, or coordination, such as driving or operating machinery.

Key Recommendations for Specific Population Groups

- *Adults*. Consuming more than one drink per day for women and two drinks per day for men increases the risk for motor vehicle accidents, other injuries, high blood pressure, stroke, violence, some types of cancer, and suicide.
- *Note:* Children and adolescents, women of childbearing age who may become pregnant, pregnant and lactating women, individuals who cannot restrict alcohol intake, individuals taking medications that can interact with alcohol, and individuals with specific medical conditions should not drink at all.

FOOD SAFETY

Key Recommendations

- To avoid microbial foodborne illness:
  - Clean hands, food contact surfaces, and fruits and vegetables. Meat and poultry should not be washed or rinsed.
  - Separate raw, cooked, and ready-to-eat foods while shopping, preparing, or storing foods.
  - Cook foods to a safe temperature to kill microorganisms.
  - Chill (refrigerate) perishable food promptly and defrost foods properly.
  - Avoid raw (unpasteurized) milk or any products made from unpasteurized milk, raw or partially cooked eggs or foods containing raw eggs, raw or undercooked meat and poultry, unpasteurized juices, and raw sprouts.

Key Recommendations for Specific Population Groups

- *Infants and young children, pregnant women, older adults, and those who are immunocompromised.* Do not eat or drink raw (unpasteurized) milk or any products made from unpasteurized milk, raw or partially cooked eggs or foods containing raw eggs, raw or undercooked meat and poultry, raw or undercooked fish or shellfish, unpasteurized juices, and raw sprouts.
- *Pregnant women, older adults, and those who are immunocompromised.* Only eat certain deli meats and frankfurters that have been reheated to steaming hot.

Source: United States Department of Agriculture and Department of Health and Human Services, *2005 Dietary Guidelines for Americans*; available at www.healthierus.gov/dietaryguidelines.

Other Government Dietary Guidelines A variety of U.S. government agencies, including the Surgeon General's office, the National Cancer Institute, and the National Heart, Lung, and Blood Institute, have issued dietary guidelines over the past two decades. Some of these recommendations are quite general, whereas others contain very specific recommended intakes of certain dietary components, such as saturated fat, polyunsaturated fat, and fiber.

Nongovernment Dietary Recommendations

Dietary recommendations are also issued by a variety of nonprofit health organizations, such as the American Heart Association and the American Cancer Society.[48] These groups are motivated to provide dietary guidance for the public because of the growing scientific evidence linking certain dietary patterns to increased risk for heart disease and some types of cancer. These recommendations, which are listed in Table 6-8, represent attempts to create broad, noncontroversial recommendations for dietary patterns in the United States.

FOOD INTAKE PATTERNS/FOOD GROUP PLANS

Dietary guidance in the form of food intake patterns or **food group plans** have been provided to the U.S. population since the turn of the century. The early forms of dietary recommendations tended to focus on the consumption of adequate amounts of the foods needed to provide nutrients and energy intake for good health.[49] The USDA, for example, published its first dietary guidance plan in 1916. Between 1916 and the 1940s, a variety of food group plans, featuring anywhere from 5 to 12 food groups, were published by federal agencies and voluntary health organizations. In 1943, the Bureau of Home Economics published a food guide that promoted seven food groups as part of the USDA's National Wartime Nutrition Program. This guide served as the basis for nearly all nutrition education programs for more than a decade. Then, in 1955, the Department of Nutrition at the Harvard School of Public Health published a recommendation that these seven food groups be collapsed into four.[50] This format was adopted in 1956 by the USDA and promoted as the Four Food Group plan. The USDA added a fifth group (fats, sweets, and alcohol) to the Basic Four plan in 1979 to draw attention to foods targeted for moderation, but it did not offer specific limitations for the fifth group.

The most recent USDA food intake pattern is the MyPyramid food guidance system which was released in 2005 and can be found inside the back cover.

Since the publication of the Basic Four plan, however, the emphasis of nutritional guidance has expanded from just meeting nutrient needs to also eating a diet low in saturated fat and cholesterol and generous in fruits, vegetables, and whole grains. With the publication of the *Surgeon General's Report on Nutrition and Health* in 1988, the overconsumption of certain dietary constituents became a major concern. For this reason, the Food Guide Pyramid was introduced to reinforce our understanding of the nutrient composition of foods, human nutrient needs, and the relationship of diet to health. The Food Guide Pyramid conveyed five of the essential components of a healthful daily diet: adequacy, balance, moderation, energy control, and variety.[51]

Implementing the Recommendations: From Guidelines to Groceries

The challenge today is to help consumers put the wide assortment of dietary recommendations into practice.[52] This requires translating the recommendations into food-specific guides that consumers can implement in their homes, at the grocery store, and in restaurants.[53] For example, consider this dietary guideline: Limit intake of fats and oils high in saturated and/or *trans* fatty acids. This broad guideline must first be translated into food-specific behaviors for consumers, who must acquire the knowledge and skill to change their eating patterns. Table 6-9 lists three food-specific behaviors that can be derived from the guideline and the knowledge and skill set required for each behavior.

Food group plan A diet-planning tool that sorts foods of similar origin and nutrient content into groups and then specifies that the individual eat a certain number of foods from each group.

TABLE 6-8 Guidelines from Nonprofit Health Organizations

THE AMERICAN HEART ASSOCIATION'S DIETARY GUIDELINES FOR HEALTHY AMERICANS

- Eat a variety of fruits and vegetables. Choose five or more servings per day.
- Eat a variety of grain products, including whole grains. Choose six or more servings per day.
- Include fat-free and low-fat milk products, fish, legumes (beans), skinless poultry, and lean meats.
- Choose fats and oils with 2 grams or less saturated fat per tablespoon, such as liquid and tub margarines, canola oil and olive oil.
- Balance the number of calories you eat with the number you use each day. (To find that number, multiply the number of pounds you weigh now by 15 calories. This represents the average number of calories used in one day if you're moderately active. If you get very little exercise, multiply your weight by 13 instead of 15. Less active people burn fewer calories.)
- Maintain a level of physical activity that keeps you fit and matches the number of calories you eat. Walk or do other activities for at least 30 minutes on most days. To lose weight, do enough activity to use up more calories than you eat every day.
- Limit your intake of foods high in calories or low in nutrition, including foods like soft drinks and candy that have a lot of sugars.
- Limit foods high in saturated fat, *trans* fat and/or cholesterol, such as full-fat milk products, fatty meats, tropical oils, partially hydrogenated vegetable oils and egg yolks. Instead choose foods low in saturated fat, *trans* fat and cholesterol from the first four points above.
- Eat less than 6 grams of salt (sodium chloride) per day (2,400 milligrams of sodium).
- Have no more than one alcoholic drink per day if you're a woman and no more than two if you're a man. "One drink" means it has no more than 1/2 ounce of pure alcohol. Examples of one drink are 12 oz of beer, 4 oz of wine, $1\frac{1}{2}$ oz of 80-proof spirits or 1 oz of 100-proof spirits.

AMERICAN CANCER SOCIETY'S NUTRITION AND PHYSICAL ACTIVITY GUIDELINES FOR CANCER PREVENTION

Recommendations for Individual Choices

1. Eat a variety of healthful foods, with an emphasis on plant sources.
   - Eat five or more servings of a variety of vegetables and fruits each day.
   - Choose whole grains in preference to processed (refined) grains and sugars.
   - Limit consumption of red meats, especially those high in fat and processed meats.
   - Choose foods that help maintain a healthful weight.
2. Adopt a physically active lifestyle.
   - Adults: engage in at least moderate activity for 30 minutes or more on 5 or more days of the week; 45 minutes or more of moderate to vigorous activity on 5 or more days per week may further enhance reductions in the risk of breast and colon cancer.
   - Children and adolescents: engage in at least 60 minutes per day of moderate to vigorous physical activity at least 5 days per week.
3. Maintain a healthful weight throughout life.
   - Balance caloric intake with physical activity.
   - Lose weight if currently overweight or obese.
4. If you drink alcoholic beverages, limit consumption.

Recommendations for Community Action

Public, private, and community organizations should work to create social and physical environments that support the adoption and maintenance of healthful nutrition and physical activity behaviors.

- Increase access to healthful foods in schools, worksites, and communities.
- Provide safe, enjoyable and accessible environments for physical activity in schools, and for transportation and recreation in communities.

Sources: American Heart Association and American Cancer Society.

Consumers who follow the Dietary Guidelines for Americans eat plenty of fruits, vegetables, and whole grains and balance their food intake with physical activity.

Ariel Skelley/CORBIS

Successfully choosing such foods means having some knowledge of food composition, recognizing major food sources of fat, knowing which cooking techniques to use, knowing how to adapt recipes, and being able to read a restaurant menu. The process of breaking down a dietary guideline into specific behaviors can be carried a step further. The guideline can be translated into an actual eating pattern, as shown in Table 6-10. The eating pattern

TABLE 6-9 Translating a Dietary Guideline into Food-Specific Behaviors

DIETARY GUIDELINE	FOOD-SPECIFIC BEHAVIOR	KNOWLEDGE AND/OR SKILLS REQUIRED TO SUPPORT THE BEHAVIOR
Limit intake of fats and oils high in saturated and/or *trans* fatty acids.	• Choose low-fat foods more often than high-fat foods.	• Know major sources of fat, saturated fat, *trans* fat, and cholesterol. • Be able to read food product labels. • Know low-fat cooking techniques. • Know how to adapt recipes. • Know which foods to substitute for higher-fat foods. • Know how to read restaurant menus.
	• Choose lean meats, fish, chicken, and turkey.	• Know lean cuts of meat. • Know low-fat cooking methods. • Remove skin from chicken and turkey before cooking.
	• Choose low-fat dairy products.	• Know which dairy products are low in fat. • Be able to read food product labels.

TABLE 6-10 Eating Pattern for a 1,900-kcal Diet with about 30 Percent of Calories from Fat*

FOODS		ENERGY (kcal)	FAT (g)
Breakfast			
Grapefruit, raw, white	1/2 medium	38	0.1
Sugar	1 tsp	16	0
Whole-wheat cereal, hot, cooked	1 cup	160	0.9
Milk (1% low-fat)	1.5 cup	153	3.6
Coffee	1 cup	2	0.1
Lunch			
Hotdog, regular, with bun	1	242	14.5
Baked beans with tomato sauce and pork	1/2 cup	119	1.2
Coleslaw made with:			
Cabbage, shredded	1/2 cup	8	0.1
Carrot, grated	1 small	2	0.1
Mayonnaise, low-fat	2 tbsp	98	9.8
Diet soft drink	1, 12 oz can	4	0
Snack			
Whole-wheat crackers	8	76	3.4
Swiss cheese	1 oz	108	7.9
Milk (1% low-fat)	1 cup	102	2.4
Dinner			
Roast beef, lean only	4 oz	201	8.4
Corn (frozen), boiled	1/2 cup	66	0.6
Salad:			
Spinach, raw	1 cup	7	0.1
Tomato, red, raw	1/2 medium	11	0.1
Italian dressing, fat-free	1 tbsp	7	0.1
Rice, cooked	1/2 cup	103	0.2
Dinner rolls, whole wheat	2	149	2.6
Margarine, soft spread	1 tsp	25	2.9
Peach melba:			
Peach halves, fresh	1 large peach	61	0.4
Raspberries, fresh	1/4 cup	16	0.2
Ice cream, vanilla, light	1/2 cup	125	3.7
Iced tea, unsweetened	16 oz	5	0
TOTAL		**1,922**	**63.4**

*The eating pattern was adapted from U.S. Department of Agriculture, Human Nutrition Information Service, *Preparing Foods and Planning Menus Using the Dietary Guidelines*, Home and Garden Bulletin No. 232-8 (Washington, D.C.: U.S. Government Printing Office). The energy and fat values were taken from U.S. Department of Agriculture, Agricultural Research Service, *USDA Nutrient Database for Standard Reference*, Release 17, 2004, available on the Nutrient Data Laboratory Home Page at www.nal.usda.gov/fnic/foodcomp.

TABLE 6-11 Connection Between Policy Tools and Policy Decisions

DECISION	POLICY TOOL	SOURCE OF POLICY TOOL
Instructor will be a registered dietitian (RD).	Credentials	Commission on Dietetic Registration
Program should include a discussion of healthful eating patterns.	Dietary Guidelines for Americans	U.S. Department of Agriculture; U.S. Department of Health and Human Services
Program segment on osteoporosis should include recommendations for obtaining calcium from foods, including fortified foods, and dietary supplements.	Dietary Reference Intakes (DRIs)	Food and Nutrition Board of the National Academy of Sciences
Program should include a segment on reading food product labels.	Labeling regulations	Food and Drug Administration
All overheads must show the company logo.	Company policy	FirstRate Spa and Health Resort
Program segment on dietary fats should include information about omega-3 fatty acids and alpha-linolenic acid.	Dietary Reference Intakes (DRIs)	Food and Nutrition Board of the National Academy of Sciences

should reflect the basic principles of the dietary guidelines and should focus on variety—eating a selection of foods from all of the food groups; proportionality—eating appropriate amounts of foods to meet nutritional needs; moderation—enjoying all foods but avoiding eating patterns that are associated with chronic disease; and usability—being flexible and practical enough to accommodate individual food preferences and meet nutrient needs.

POLICY MAKING IN ACTION

Students often have difficulty visualizing the connection between policy making and their work as community nutritionists.[54] Let's describe an example of this connection, drawing on the job responsibilities of the director of health promotion for the FirstRate Spa and Health Resort (described in an earlier chapter). The director decides to update a risk-reduction program that will help spa clients make more healthful eating choices. The new program will address eating strategies to reduce the risk of coronary heart disease, stroke, cancer, and osteoporosis.

In organizing the content of the program, the director realizes that clients need to be given information about healthful eating patterns; recommendations related to fat, calcium, and antioxidant vitamin intake; and instructions on how to read food product labels, among other topics. Table 6-11 shows six decisions the director must make in developing the program's content, the policy "tool" to be used for each decision, and the source of each tool. The first decision is to choose a program instructor who is a registered dietitian. This decision enhances the image of the program and ensures that accurate and timely information will be offered to clients.

Deciding to use the Dietary Guidelines for Americans is a fairly obvious choice, because it is a widely accepted policy tool of the federal government. The company's requirement that all visual aids show the logo of the FirstRate Spa and Health Resort is also a matter of policy. Every organization, agency, and institution has its own policies that affect the practice of community nutritionists.

The decision about how much material to present to clients on the essential fatty acid, alpha-linolenic acid, and other omega-3 fatty acids—hot topics for consumers today—is more complex. A DRI for the intake of omega-3 fatty acids has been recently published but is not currently used in the spa's education materials. Should the DRI be used, in this situation, to justify the additional expense for developing educational materials on this topic? Ah, this is precisely where we see policy making in action, for the director is making a policy decision when choosing to include the DRI for omega-3 fatty acids in program materials. Many decisions made by community nutritionists are opportunities for making and implementing policy.

POLICY MAKING DOES NOT STAND STILL

This is an exciting time in community nutrition. New legislation related to Medicare reform, as discussed in Chapter 9, third-party reimbursement for medical nutrition therapy, food and supplement labeling requirements, new science-based dietary guidelines, and market forces in the health care field and the food industry promise many opportunities for community nutritionists to serve as liaisons between policy makers and the general public. The one guarantee is that nutrition policy will continue to change.

The task of formulating a national nutrition agenda is daunting. Government must make a commitment to develop and promote a coordinated plan to improve the nation's nutritional and health status, scientists must reach a consensus on the interpretation of scientific findings and appropriate dietary advice for all Americans, and sufficient financial resources must be allocated to implementing the policy. Consider the policy-making machinery that has geared up to ensure that all Americans are properly fed. As an outgrowth of the 1996 World Food Summit in Rome, the United States developed a blueprint, titled "U.S. Action Plan for Food Security," to strengthen the commitment of the U.S. government and civil society to reducing hunger and malnutrition both at home and abroad by the year 2015.[55] (Check the Internet addresses below for the U.S. and Canadian action plans.) The actions of the U.S., Canadian, and other governments regarding the issues of hunger and malnutrition are good examples of the importance of policy making in community nutrition.

Internet Resources

Canada

Canada's Food Guide to Healthy Eating	**www.hc-sc.gc.ca**
The National Plan of Action for Nutrition	**www.hc-sc.gc.ca**

United States

The Interactive Healthy Eating Index	**www.usda.gov/cnpp**
History of the Dietary Guidelines for Americans	**www.nal.usda.gov/fnic/Dietary/12dietapp1.htm**
The Dietary Guidelines for Americans 2005	**www.healthierus.gov/dietaryguidelines**

Nutrition Surveys

National Health and Nutrition Examination Survey: NHANES Data Briefs	**www.cdc.gov/nchs/about/major/nhanes/Databriefs.htm**

National Health and Nutrition Examination Survey: Survey Results and Products	**www.cdc.gov/nchs/about/major/nhanes/survey_results_and_products.htm**
U.S. Department of Agriculture: Agricultural Research Service	**www.barc.usda.gov/bhnrc/foodsurvey/home.htm**
U.S. Action Plan on Food Security	
Framework for the U.S. Action Plan on Food Security	**www.fas.usda.gov/icd/summit/framewor.html**
Discussion Paper on Domestic Food Security	**www.fas.usda.gov/icd/summit/discussi.html**
Government-Sponsored Websites for Public Information Regarding Lifestyle Habits	
Healthier US	**www.healthierus.gov**
Diet	**www.nutrition.gov**
Body Weight, Cholesterol, Blood Pressure, and Others	**www.nhlbi.nih.gov/subsites**
Smoking Cessation	**www.cdc.gov/tobacco/sgr/sgr_2000/index.htm**

Case *Study*

From Guidelines to Groceries

By Alice Fornari, EdD, RD, and Alessandra Sarcona, MS, RD

Scenario

As a nutrition consultant and public relations chair of your local district dietetic association, you are approached by the local health department to prepare a series of educational announcements for the adult clinic waiting room. It is up to you to determine whether the announcements should be communicated as a written document or as video clips.

The focus of the announcements is to acquaint consumers with current public health issues as determined by national data bases and surveys. In addition, you are asked to translate the *Dietary Guidelines for Americans* into behaviors that consumers could adopt to implement the recommendations.

To further assess your audience, you decide it is important to meet clients who would be hearing or reading the announcements. A visit to the Department of Health indicates to you that communication with clients is achieved by written publications, video monitors, and a visible bulletin board. You also notice that most of the clients are African-American women of various ages and their children.

You take time during the visit to approach some of the clients waiting for services. These brief conversations reveal that there is no preliminary knowledge of the Dietary Guidelines and definitively no application of these guidelines to personal food choices. Here are some comments by clients:

"I use sweetened iced tea as our beverage. All the kids like it."
"Milk is not healthy; it gives everyone stomachaches."
"Fresh fruits and vegetables are too expensive. I use canned ones."

A sample recall from a middle-aged woman reveals the following:

Breakfast: Donut or egg on a biscuit, and sweetened coffee
Lunch: Fast-food hamburgers with french fries and a sweetened soft drink
Dinner: Fried chicken, potatoes or rice, peas or corn
Snacks: Chips, ice cream, snack cakes, iced tea

These conversations reveal that your task is much more complex than you anticipated. You must begin by identifying a goal for your educational announcement. As a first attempt, you decide the video monitor will be the most effective tool for communication.

Learning Outcomes

- Identify and access, via the Web, USDA/Economic Research Service (ERS) food consumption survey data and the *Dietary Guidelines for Americans,* 2005.
- Use survey data to identify at-risk groups that could be targeted for intervention by a Department of Health.
- Prepare an educational announcement translating data into positive food-specific behaviors for a target population.

Foundation: Acquisition of Knowledge and Skills

1. Outline the *Dietary Guidelines for Americans:* www.healthierus.gov/dietaryguidelines and www.health.gov/dietaryguidelines
2. What do Americans think about dietary guidance issues, and how does the American diet measure up to the *Dietary Guidelines*? These questions are addressed at the following sites: www.ific.org/research/newconvres.cfm and www.ers.usda.gov/publications/aib750

   Access *descriptive text* on USDA surveys from the Web:
   www.ers.usda.gov/briefing/consumption
   www.ers.usda.gov/briefing/consumption/Individual.htm

Step 1: *Identify Relevant Information and Uncertainties*

1. After your visit to the health clinic, and considering the *Dietary Guidelines for Americans,* create a list of priority nutrition issues for the population who would be accessing the educational announcements.
2. Explain why the clinic population may have difficulties following the *Dietary Guidelines for Americans.* That is, judging on the basis of your observations and interactions with the clinic clientele, identify barriers to following the guidelines.

Step 2: *Interpret Information*

1. Connect the nutrition issues identified for the target population to the survey data reported on the Web by USDA.
2. Identify knowledge and/or skills that this target population will need to support positive behavior changes.

Step 3: *Draw and Implement Conclusions*

1. Refer to the chapter discussion of translating a dietary guideline into food-specific behaviors. Identify a dietary guideline, not sampled in the text, that appears to be one that this clinic population is having difficulty following. Link the selected dietary guideline to positive food-specific behaviors; in addition, identify the knowledge and/or skills necessary to support positive behavior changes in the target population.
2. Outline an educational announcement to be viewed in the clinic, through the waiting room video monitor.

Step 4: *Engage in Continuous Improvement*

1. If you had the opportunity to continue working on publications for the clinic, how could you monitor the impact of your efforts?
2. For use on the clinic bulletin board, develop a shopping guide to be used as a resource for a hands-on tour of a local supermarket. The tour will help clients visualize healthful food and beverage purchases and strategies for incorporating the new *Dietary Guidelines for Americans.*

Professional Focus

Evaluating Research and Information on Nutrition and Health

Nutrition and health policy is based on sound research. However, at the same time that science has shown that to some extent we really are what we eat, many consumers are more confused than ever about how to translate the steady stream of new findings about nutrition into healthful eating.[1] Each additional nugget of nutrition news that comes along raises new concerns: Is caffeine bad for me? Does feverfew prevent headache? Should I take vitamin supplements? Will diet pills work? Can a sports drink improve my performance? Do pesticides pose a hazard? As a community nutritionist, you will find that many consumers turn to you as the resident expert on these questions and more.

Some manufacturers of food and nutrition-related products, as well as many members of the media, compound the confusion by offering a myriad of unreliable products and misleading dietary advice targeted to health-conscious consumers.

Unfortunately, many consumers fall prey to this barrage of misinformation. Americans spend more than $30 billion annually on medical and nutritional health fraud and quackery, up from $1 billion to $2 billion in the early 1960s.[2] At the same time, the sale of weight-loss programs and products—not all of them sound—has become a $33 billion industry. Media attention to "hot" foods and nutritional supplements generally causes spending on those items to soar. This Professional Focus provides a sieve with which you can help consumers separate valid from bogus nutrition information.

Money down the drain is just one of the problems stemming from misleading dietary information. Some fraudulent claims about nutrition are harmless and make for a good laugh, but others can have tragic consequences. False claims about nutritional products have been known to bring about malnutrition, birth defects, mental retardation, and even death in extreme cases. Negative effects can happen in two ways. One is that the product in question causes direct harm. Even a seemingly innocuous substance such as vitamin A, for instance, can cause severe liver damage over time if taken in large enough doses. The other problem is that spurious nutritional remedies build false hope and may keep a consumer from obtaining sound, scientifically tested medical treatment. A person who relies on a so-called anticancer diet as a cure, for example, might forgo possible lifesaving interventions such as surgery or chemotherapy.

Part of the confusion stems from the way the media interpret the findings of scientific research. A good case in point is the controversy over whether a high-fiber diet protects against colon cancer—a disease that affects some 130,000 Americans each year. This issue dates back to the early 1970s, when scientists observed that colon cancer was extremely uncommon in areas of the world where the diet consisted largely of unrefined foods and little meat. The researchers theorized that dietary fiber may be protective against colon cancer by binding bile and speeding the passage of wastes and potentially harmful compounds through the colon. Since then, other studies have suggested that those who eat a high-fiber diet have a lower risk of colon and rectal cancers.[3]

However, in 1998, a flurry of headlines threatened to pull the pedestal out from under the popular fiber theory, asking, "Fiber: Is It Still the Right Choice?" A Harvard-based study published in the *New England Journal of Medicine* suggested that fiber did nothing to prevent cancer.[4] The 16-year trial of almost 90,000 nurses—called the Nurses' Health Study—found that the nurses who ate low-fiber diets (less than 10 grams daily) were no more likely to develop colon cancer than those who ate higher levels of fiber (about 25 grams daily). As a result, the researchers concluded that the study provided no support for the theory that fiber could reduce the risk of colon cancer.

This surprising news reinforces the need for research studies to be duplicated. All studies have their limitations, and a number of questions can be raised regarding the conclusions of the Nurses' Health Study. For example, the study relied on participants to recall their eating habits accurately. Is this type of self-reported dietary information reliable? Back in 1980, the nurses were asked for information about their intakes of "dark" breads. However, food labels at that time did not list the fiber content of breads, and some wheat breads on the shelf had amounts of fiber similar to those found in white bread. Did the nurses mistakenly consider "dark" bread the same as 100 percent whole-wheat bread?

Another question to ask is "What are the optimal levels of fiber intake for colon cancer protection?" Some experts believe that it may take more than 25 grams of fiber a day to show cancer-protective effects, which might explain the lack of effect noted in the Nurses' study.

This fiber story illustrates how news reports based on only one study can leave the public with the impression that scientists can't make up their minds. It seems as though one week scientists are saying that fiber is good, and the next week the word is that fiber doesn't do any good at all.

Contrary to what some headlines imply, reputable scientists do not base their dietary recommendations for the public on the results of one or two studies. Scientists are still conducting research to determine whether fiber does in fact help to prevent colon cancer, and if so, what types of fiber and in what amount. Scientists design their research to test theories, such as the notion that eating a high-fiber diet is associated with lower risk of cancer. Other factors, however, often confound the matter at hand. The study of fiber and colon cancer is complex because many other factors, such as inactivity, obesity, saturated fat intake, and low calcium or folate intake, are linked with the development of colon cancer.

You can critique the nutrition news you read by asking a series of questions (see Table 6-12). Consider the following points as a checklist for separating the bogus news stories from those worth your attention.

- *Where is the study published?*

The study being described in the news story should be published in a peer-reviewed journal—a journal that uses experts in the field to review research results. These reviewers serve to point out any flaws in the research design and can challenge the researcher's conclusions before the study is published.

- *How recent is the study?*

The science of nutrition continues to develop. New studies employ state-of-the-art methods and technology and benefit from the scrutiny of experts in the particular field of study.

TABLE 6-12 Questions to Ask about a Research Report

- Was the research done by a credible institution? A qualified researcher?
- Is this a preliminary study? Have other studies reached the same conclusions?
- Was the study done with animals or humans?
- Was the research population large enough? Was the study long enough?
- Who paid for the study? Might that affect the findings? Is the science valid despite the funding source?
- Was the report reviewed by peers?
- Does the report avoid absolutes, such as "proves" or "causes"?
- Does the report reflect appropriate context (for example, how the research fits into a broader picture of scientific evidence and consumer lifestyles)?
- Do the results apply to a certain group of people? Do they apply to someone your age, gender, and health condition?
- What do follow-up reports from qualified nutrition experts say?

Source: *Journal of the American Dietetic Association* (102) 2002: 264.

- *What research methods were used to obtain the data?*

Are the results from an epidemiologic study or an intervention study? Epidemiologic studies examine populations to determine food patterns and health status over time. These population studies are useful in uncovering correlations between two factors (for example, whether a high calcium intake early in life reduces the incidence of bone fractures later in life). However, they are not considered as conclusive as intervention studies. A correlation between two factors may *suggest* a cause-and-effect relationship between the factors but does not prove it.

Intervention studies examine the effects of a specific treatment or intervention on a particular group of subjects and compare the results to a similar group of people not receiving the treatment. An example is a cholesterol-monitoring study in which half the subjects follow dietary advice to lower their blood cholesterol and half do not. Ideally, intervention studies should be randomized and controlled—that is, subjects are assigned to either an experimental group or a control group by means of a random selection process. Each subject has an equal chance of being assigned to either group. The experimental group receives the "treatment" being tested; the control group receives a placebo, or neutral substance. If possible, neither the researcher nor the participants should know which group the subjects have been assigned to until the end of the experiment. A randomized, controlled study helps to ensure that the study's conclusions are a result of the treatment and minimizes the chances that the results are due to a placebo effect or to bias on the part of the researcher.

- *What was the size of the study?*

In order to achieve validity—accuracy in results—studies must generally include a sufficiently large number of people, such as 50 or more in intervention studies. This reduces the chances that the results are simply a coincidence and justifies generalizing the conclusions of the study to a wider audience.

- *Who were the subjects?*

Look for similarities between the subjects in the study and yourself. The more you have in common with the participants (age, diet pattern, gender, etc.), the more pertinent the study results may be for you.

- *Does a consensus of published studies support the results reported in the news?*

Even if an experiment is carefully designed and carried out perfectly, its findings cannot be considered definitive until they have been confirmed by other research. Testing and retesting reduce the possibility that the outcome was simply the result of chance or an error or oversight on the part of the experimenter. Every study should be viewed as preliminary until it becomes just one addition to a significant body of evidence pointing in the same direction.

When making dietary recommendations for the public, experts pool the results of different types of studies, such as analyses of food patterns of groups of people and carefully controlled studies on people in hospitals or clinics. Before drawing any conclusions, they then consider the evidence from all of the research. The bottom line is that if you read a report in the newspaper or watch one on television that advises making a dramatic change in your diet or lifestyle on the basis of the results of one study, don't take it to heart. The findings may make for a good story, but they're not worth taking too seriously.

Consumers sometimes ask why the government doesn't prevent the media from disseminating misleading nutrition information, but the government lacks the power to do so. The First Amendment guarantees freedom of the press, which means that people may express whatever views they like in the media, whether these opinions are sound, unsound, or even dangerous. By law, writers cannot be punished for publishing misinformation unless it can be proved in court that the information has caused a reader bodily harm.

Fortunately, most professional health groups maintain committees to combat the spread of health and nutrition misinformation. In addition, many professional organizations have banded together to form the National Council Against

Health Fraud (NCAHF), which has branches in many states. The NCAHF monitors radio, television, and other advertising and investigates complaints. The Internet websites listed at the end of this feature can serve as sources for your own inquiries about the authenticity of information in nutrition and health-related areas. Remember, too, that although the Internet is an excellent source of an incredible amount of health information, it has also become one of the fastest-growing outlets for health fraud.

As the Internet continues to grow and becomes integrated into community nutrition practice, community nutritionists will increasingly be expected to evaluate information obtained from the Internet. Perhaps your clients will have questions about information they found on the World Wide Web, or your organization may ask you to choose 5–10 sites to link to its own Web page, or you may wish to include websites on a handout for program participants. For whatever reason, community nutritionists must have skills for evaluating Internet information.

Information is rampant on the Internet. In a sense, the Internet *is* information, and the information is continually being revised and created. Internet information exists in many forms (for instance, facts, statistics, stories, and opinions). This information is created for many purposes (for example, to entertain, to inform, to persuade, to sell, and to influence), and it varies in quality from very good to worthless or even dangerous. One method for determining whether the information found on the Internet is reliable and of good quality is the CARS Checklist.[5] The acronym CARS stands for credibility, accuracy, reasonableness, and support.

- *Credibility.* Check the credentials of the author (if there is one!) or sponsoring organization. Is the author or organization respected and well known as a source of sound, scientific information? There being no posted author and the presence of misspelled words or bad grammar can be taken as evidence of a lack of credibility. A credible sponsor will use a professional approach to designing the website.
- *Accuracy.* Check to ensure that the information is current, factual, and comprehensive. If important facts, consequences, or other information is missing, the website may not be presenting a complete story. There being no date on the document, the use of sweeping generalizations, and the presence of outdated information are evidence of a lack of accuracy. Watch for testimonials masquerading as scientific evidence. This is a common method for promoting questionable products on the Internet.
- *Reasonableness.* Evaluate the information for fairness, balance, and consistency. Does the author present a fair, balanced argument supporting the ideas presented? Are the arguments offered rational? Has the author maintained objectivity in discussing the topic? Does he or she have an obvious—or hidden—conflict of interest? Evidence of a lack of reasonableness includes gross generalizations ("Foods not grown organically are all toxic and shouldn't be eaten") and outlandish claims ("Kombucha tea will cure cancer and diabetes").
- *Support.* Check to see whether supporting documentation is cited for scientific statements. Are there references to legitimate scientific journals and publications? Is it clear where the information came from? An Internet document that fails to indicate the sources of its information is suspect.

It's not always easy to separate the nutrition wheat from the chaff, given that many misleading claims are supposedly backed by scientific-sounding statements, but you can offer your clients some tips to help them tell whether a product is bogus. The following red flags can help you spot a quack:

- *The promoter claims that the medical establishment is against him or her and that the government won't accept this new "alternative" treatment.*

If the government or medical community cannot accept a treatment, it is because the treatment has not been proved to work. Reputable professionals do not suppress knowledge about fighting disease. On the contrary, they welcome new remedies for illness, provided that the treatments have been carefully tested.

- *The promoter uses testimonials and anecdotes from satisfied customers to support claims.*

Valid nutrition information comes from careful experimental research, not from random tales. A few persons' reports that the product in question "works every time" are never acceptable as sound scientific evidence.

- *The promoter uses a computer-scored questionnaire for diagnosing "nutrient deficiencies."*

Those programs are designed to show that just about everyone has a deficiency that can be reversed with the supplements the promoter just happens to be selling, regardless of the consumer's symptoms or health.

- *The promoter claims that the product will make weight loss easy.*

Unfortunately, there is no simple way to lose weight. In other words, if a claim sounds too good to be true, it probably is.

- *The promoter promises that the product is made with a "secret formula" available only from this one company.*

Legitimate health professionals share their knowledge of proven treatments so that others can benefit from it.

- *The treatment is offered only in the back pages of magazines, over the phone, or by mail-order solicited by ads in the form of news stories or 30-minute commercials (known as infomercials) in talk-show format.*

Results of studies on credible treatments are reported first in medical journals and are administered through a physician or other health professional. If information about a treatment appears only elsewhere, it probably cannot withstand scientific scrutiny.

Internet Addresses

Center for Food Safety and Applied Nutrition
http://vm.cfsan.fda.gov/list.html
National Council Against Health Fraud
www.ncahf.org
Quackwatch
www.quackwatch.org

References

1. This discussion was adapted from M. A. Boyle and S. L. Anderson, *Personal Nutrition,* 5th ed. (Belmont, CA: Wadsworth, 2004), pp. 21–27.
2. Position of the American Dietetic Association, Food and nutrition misinformation, *Journal of the American Dietetic Association* 102 (2002): 260–266.
3. World Cancer Research Fund and American Institute for Cancer Research, *Food, Nutrition, and the Prevention of Cancer: A Global Perspective* (Washington, D.C.: American Institute for Cancer Research, 1997).
4. S. Fuchs and coauthors, Dietary fiber and the risk of colorectal cancer and adenoma in women, *New England Journal of Medicine* 340 (1999): 169–176.
5. R. Harris, Evaluating Internet research sources online, available at www.virtualsalt.com/evalu8it.htm; see also *Bibliography on Evaluating Web Information* at www.lib.vt.edu/help/instruct/evaluate/evalbiblio.html.

Chapter *Seven*

The Art and Science of Policy Making

Learning *Objectives*

After you have read and studied this chapter, you will be able to:

- Describe the policy-making process.
- Explain how laws and regulations are developed.
- Describe the federal budget process.
- Identify a minimum of four emerging policy issues in the food and nutrition arena.
- Prepare a letter addressed to your congressperson.
- Summarize the importance of policy making to nutritionists working in the community.
- Identify three ways in which the community nutritionist can influence policy making.

This chapter addresses such issues as public policy development, food and nutrition laws and regulations, dietary supplements, and current information technologies, which are Commission on Accreditation for Dietetics Education (CADE) *Foundation Knowledge and Skills* requirements for dietetics education.

Chapter *Outline*

Something To Think About...

Each of the great social achievements of recent decades has come about not because of government proclamations but because people organized, made demands and made it good politics to respond. It is the political will of the people that makes and sustains the political will of governments.

– James P. Grant, former executive director, UNICEF

Introduction

In bold letters the newspaper headline proclaimed, "Little Ones Doomed: Child Malnutrition a 'Silent Emergency.' "[1] According to Stephen Lewis, deputy executive director of UNICEF, who was quoted in the accompanying article, "The silent emergency of malnutrition is so shocking and simultaneously unnecessary that it must be brought to public attention." His comments coincided with the release of one of UNICEF's *State of the World's Children* reports, which indicated that more than 200 million children in developing countries under the age of 5 years are malnourished. Two weeks after the newspaper article appeared, *Time* magazine featured a two-page column, donated jointly by *Time* and Canon, that called attention to UNICEF's message about the silent emergency of malnutrition.[2]

What relevance do these documents have for community nutritionists? What do they have to do with policy, the topic of this chapter? The answer to both questions is "A great deal." The newspaper article and magazine column are examples of how an organization (UNICEF) used a particular strategy (a press release to the media and partnerships with two companies, *Time* magazine and Canon) to help convince people that a serious problem (global childhood malnutrition) exists. Addressing problems is the core activity of the policy-making process. Whether the issue is regulating stem cell research, controlling air and water pollution, or providing quality health care for all citizens, policy making is an ongoing process that affects our lives daily. As a community nutritionist, you will find that both local and national policy issues affect the way you work, how you deliver nutrition services, and the dietary messages you give to clients in your community. Consider how you would respond to the following issues in the nutrition policy arena:

- In order to create a more healthful nutrition environment in elementary and high schools, should nutrition policies be developed to control when and what foods are sold in vending machines, concession stands, and à la carte by school food service programs?[3]
- Are dietary guidelines developed for adults appropriate for children?
- Should U.S. manufacturers of baby formula be allowed to sell their products in developing countries where the use of such products may undercut breastfeeding practices?
- Should television stations be required to run public service announcements that feature healthful food messages for children to balance current food-related advertising, which tends to promote high-fat, high-sugar foods?
- Should mandatory labeling of foods be required of manufacturers who include genetically modified foods or food ingredients in their products so that consumers can know the source of the foods they choose? Or should the labeling of genetically modified foods be required only when a specific health concern—such as an allergy to specific proteins—needs to be communicated to consumers?

- Because overweight and obesity are viewed as public health problems, should we tax foods based on their nutrient value per calorie (e.g., full-fat ice creams and sodas might be taxed heavily, whereas fruits and vegetables might be tax-free)?
- Should lower health insurance premiums be made available to persons who maintain a healthy weight and participate in moderate exercise for 30 minutes at least five times a week?
- Should the costs incurred by people losing weight in order to achieve a healthy weight be tax deductible?
- Should penalties be imposed on makers of both exercise equipment and supplements that claim their products melt away pounds effortlessly?
- Is the recommendation to consume moderate amounts of alcohol to help prevent coronary heart disease appropriate, considering that excessive alcohol intake is associated with increased risk of some types of cancer?[4]

There are no simple answers to these difficult policy questions. This is not surprising, for public policy is complex and ever-changing. The purpose of public policy is to fashion strategies for solving public problems. In the nutrition arena, the strategies for solving problems typically include food assistance programs, dietary recommendations, and reimbursement mechanisms for nutrition services. This chapter describes the policy-making process, examines emerging policy issues, and discusses the policy-making activities of community nutritionists.

The Process of Policy Making

If dietetics is your profession, politics is your business.
– American Dietetic Association

You may not think of yourself as a political animal. You may think of politics as being confined to Senate hearings, city council meetings, and elections. But if you have ever lobbied a professor to allow you to take an exam at a later date or signed a petition calling for increased funding for a local food bank, then you have walked onto the political stage. You have tried to get something you want by presenting compelling reasons why an existing policy should be changed.

Recall that policy was defined in Chapter 1 as the course of action chosen by public authorities to address a given problem. A **problem** is a "substantial discrepancy between what is and what should be."[5] When public authorities state that a problem exists, they are recognizing the gap between current reality and the desired state of affairs. Policies, then, are guides to a range of activities designed to address a problem.[6]

Policy making is the process by which authorities decide which actions to take to address a problem or set of problems, and it can be viewed as a cycle, as shown in Figure 7-1.[7] This diagram has the advantage of simplifying the policy-making process but also makes it a little too simple and neat. In reality, the various stages often overlap as a policy is fine-tuned. Sometimes the stages occur out of sequence, as when agenda setting leads directly to evaluation. For instance, when the issue of hunger became a part of the national policy agenda in the 1960s, as discussed in Chapter 5, existing nutrition and welfare programs were evaluated to discover why they were not reaching hungry children, and this approach effectively by-passed the policy design and implementation stages. Nevertheless, viewing the policy-making process as a cycle enables us to see how policies evolve over time.

Problem A significant gap between current reality (the way things are) and the desired state of affairs (the way things should be).

Policy making The process by which authorities decide which actions to take to address a problem or set of problems.

The discussion that follows focuses on policy making at the national level, because the laws that arise from federal policy may affect some aspects of community nutrition practice. As you study this section, think of ways in which the policy cycle can be applied to lower levels of government, such as your state or municipal government, and to institutions, such

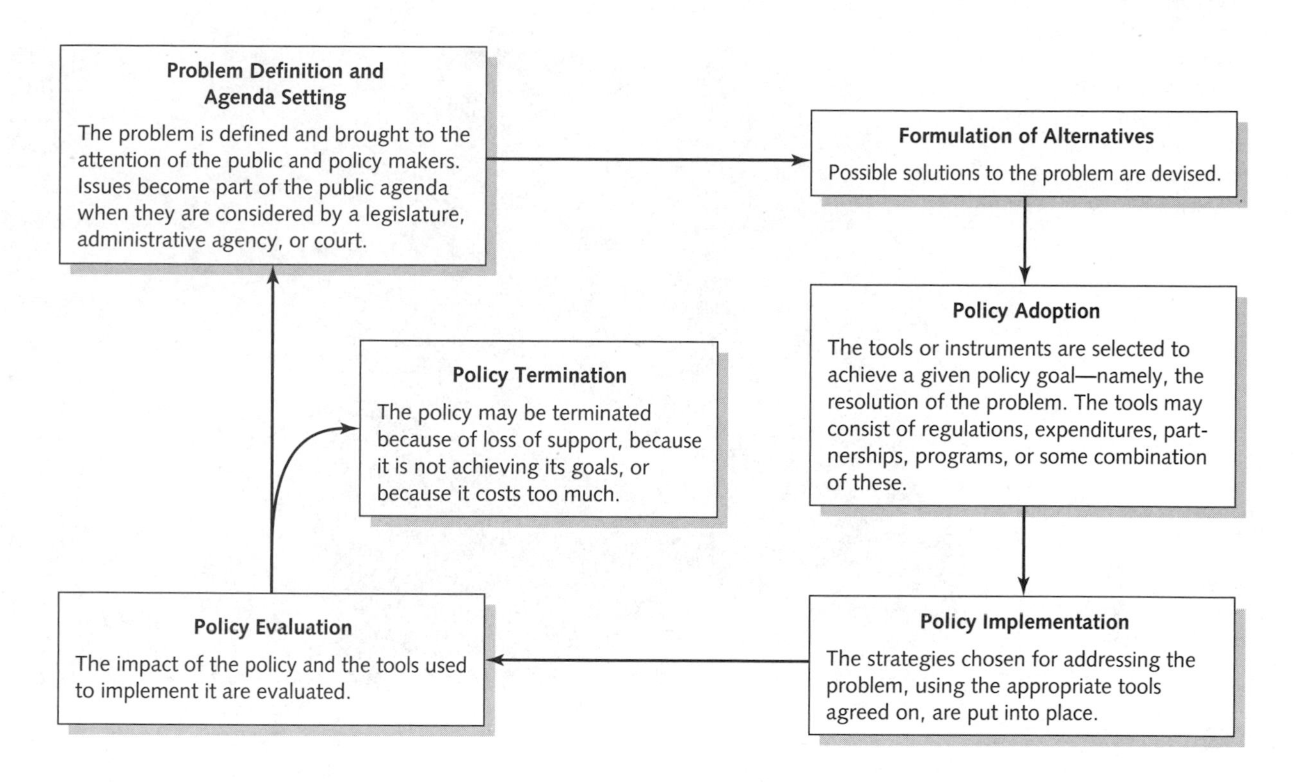

FIGURE 7-1 The Policy Cycle

Source: Adapted from W. Lyons, J. M. Scheb II, and L. E. Richardson, Jr., *American Government: Politics and Political Culture,* p. 468. Copyright © 1995 West Publishing Company. Used by permission of Wadsworth Publishing Co.

as your college or university or place of employment. As a student, for example, your life is affected by your school's policies on course requirements for graduation, residency, use of campus libraries, and many other activities. Consider how the policy cycle described here reflects the process by which your school formulated its policies.

1. **Problem definition and agenda setting.** The first step in the policy process is to convince other people that a public problem exists. For example, the fact that approximately 45 million people in the United States do not have medical insurance could be seen as either a public problem or a private (individual) problem.[8] Before a problem can be addressed, then, a majority of people must be convinced that it is a public issue. A clear statement of the problem is derived by asking questions: Who is experiencing the problem? Why did this problem develop? How severe is the problem? What actions have been taken in the past to address the problem? What action can be taken now to solve the problem? What resources exist to help alleviate the problem?[9] The manner in which the problem is defined will probably determine whether it succeeds in capturing the public's attention and whether action is taken to address it.

Once a problem is defined and gains attention, it is placed on the **policy agenda.** This agenda is not a written document or book but a set of controversial issues that exist within society. Agenda setting is a process in which people concerned about an issue work to bring the issue to the attention of the general public and policy makers. Getting policy makers to place a problem on the official agenda can be difficult. An issue may be so sensitive that policy makers or public attitudes work to keep it from reaching the agenda-setting stage. Consider, for example, the question of whether gay couples should be allowed to marry legally or adopt children. In the absence of widespread public support, some gay couples have taken this issue to the courts as a means of accessing the policy agenda. In other situations, an issue may be perceived as a problem only by a small number of people who lack the political clout required to get the issue onto the agenda. Sometimes a major catastrophe, such as an earthquake, assassination, riot, or unusual human event, triggers a

Policy agenda The set of problems to which policy makers give their attention.

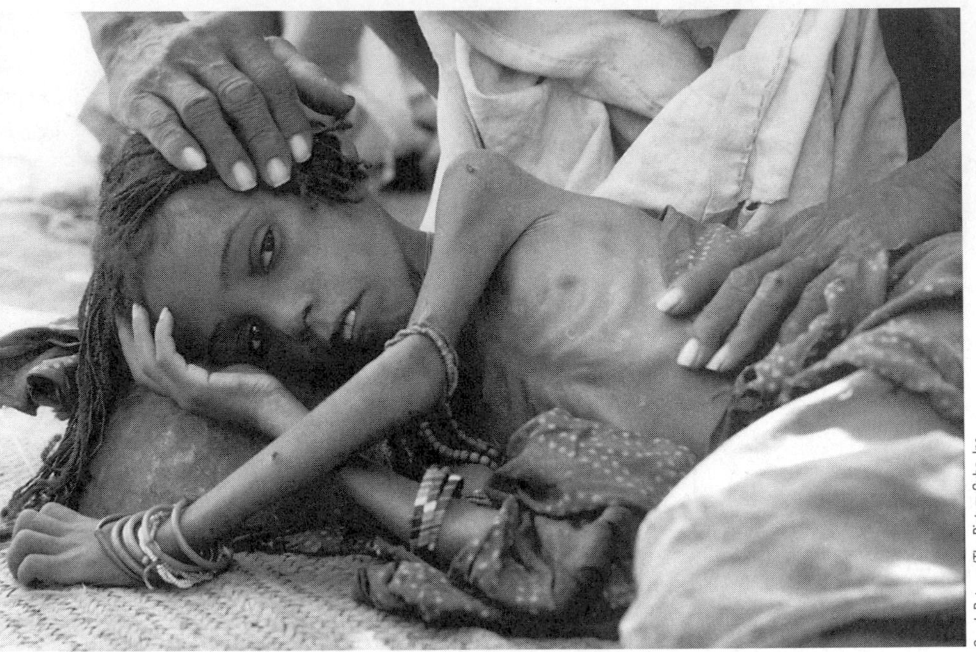

A photograph such as this one can be an effective method of bringing an urgent issue to the public's attention. Convincing people that a problem exists is the first step in the policy process.

Sarah Putnam/The Picture Cube, Inc.

public outcry and pushes an issue onto the policy agenda. When CBS aired its television report "Hunger in America" in May 1968, the problem of malnutrition and the failure of major government feeding programs for families became vividly real. Millions saw the televised images of starving and dying American children. The public reaction was swift and angry, pushing Congress to form the Senate Select Committee on Nutrition and Human Needs to investigate hunger and malnutrition among America's poor.[10]

Public opinion in this country is everything.
– Abraham Lincoln

How are issues placed on the policy agenda? The first step is to build widespread public interest for the issue that deserves government attention. One of the most effective ways to build public interest and support for an issue is to work through the media—radio, television, newspapers, and the Internet. Because the media can both create and reflect public issues, they are one of the most powerful tools for setting the public policy agenda. For example, the publication in 1962 of Rachel Carson's book *Silent Spring* opened the public's eyes to environmental dangers and launched the modern environmental movement.[11] UNICEF's success at getting newspapers and magazines to carry the message about childhood malnutrition is another example of using the media to increase public awareness of an issue and seek support for efforts to address it. UNICEF also uses the Internet to keep the issue in the public eye. Its website provides information about its annual *State of the World's Children* reports.

UNICEF's Internet address is www.unicef.org.

However, it is not enough merely to bring the issue to the attention of policy makers and the public through the media. The issue must get onto the **institutional agenda** defined by each legislative body of the government (for example, Congress, state legislatures, city councils). This is accomplished by winning support for the issue among what some political scientists have called the "iron triangles," because they often exert enormous control over the policy-making process. An iron triangle is made up of three powerful participants in the policy-making process: interest groups, congressional committees or subcommittees, and administrative agencies. These groups are not formal, recognized units of government; rather, they consist of anyone interested in policy issues and outcomes, such as government administrators, members of Congress and their staffs,

Institutional agenda The issues that are the subject of public policy.

bureau chiefs, interest groups, professionals (for example, dietitians, physicians, bankers, real estate agents), university faculty members, governors, and members of state and local governments, coalitions, and networks.

2. **Formulation of alternatives.** Possible solutions to the problem are devised in this, the most creative phase of the policy-making process. How can the WIC program be modified to meet the needs of working women? How can school food programs be improved to feed more children and teenagers? Should families experiencing food insecurity be given greater access to food banks and food programs? Or should they receive job training to help them increase their income?

Discussion of possible solutions to public problems—what is sometimes called "policy formulation"—often begins at the grassroots level. Interest groups, coalitions, and networks of experts and people interested in the issue craft a set of possible solutions and bring them forward to policy makers, who continue the discussion of solutions in legislative assemblies, government agencies, other institutions, congressional hearings, town hall meetings, and even focus groups. These forums give the general public an opportunity to express its opinions about possible solutions, potential costs and benefits of various alternatives, and the "best" course of action. A key consideration is whether the best proposed solution—in other words, the action that will become policy—is reasonable. During World War II, for example, American and British military officers were stymied in their efforts to stop German submarine attacks on Allied ships. In exasperation, they turned to an operations researcher for a solution. He thought for a moment and then responded, "That's easy—all you have to do is boil the ocean." The military officers replied, "But how do we do that?" The operations researcher answered, "I don't know. I only make policy. It's your job to implement it."[12] In this example, the suggested policy was not realistic or achievable, and the solution proposed by the operations researcher reflected no consideration of how the policy would be implemented or even whether it *could* be implemented. The act of making policy was uncoupled from the process of implementing it—an unworkable situation in real life. Policies designed to address food- and nutrition-related problems are nearly always workable, although they change as circumstances, information, and priorities change.

In the United States, policy is formulated by the legislative, executive, and judicial branches of the government at the national, state, and local levels. Three examples of national policies formulated by Congress are the Federal Food, Drug, and Cosmetic Act, the legislation that regulates the U.S. food supply;[13] the Nutrition Labeling and Education Act (NLEA), the legislation that specifies national uniform food labels and mandatory nutrition labeling information on nearly all foods marketed to U.S. consumers;[14] and the Dietary Supplement Health Education Act (DSHEA), the legislation that severely restricts the FDA's authority over dietary and herbal supplements. The DSHEA legislation is discussed later in this chapter.

3. **Policy adoption.** In this step, the tools or instruments for dealing with the problem are chosen. Examples of policy "tools" include regulations, cash grants, loans, tax breaks, certification, fines, price controls, quotas, public promotion, public investment, and government-sponsored programs.[15] These tools are wielded by federal, state, and municipal departments and agencies that are responsible for implementing policy.

At the federal level, two departments are important for our purposes: the Department of Health and Human Services (DHHS) and the U.S. Department of Agriculture (USDA). The mission of the DHHS is to promote, protect, and advance the nation's physical and mental health. Its organizational chart is shown in Figure 7-2. The DHHS includes more than 300 programs, covering a broad spectrum of activities from conducting medical science research and preventing outbreaks of infectious disease to ensuring food and drug

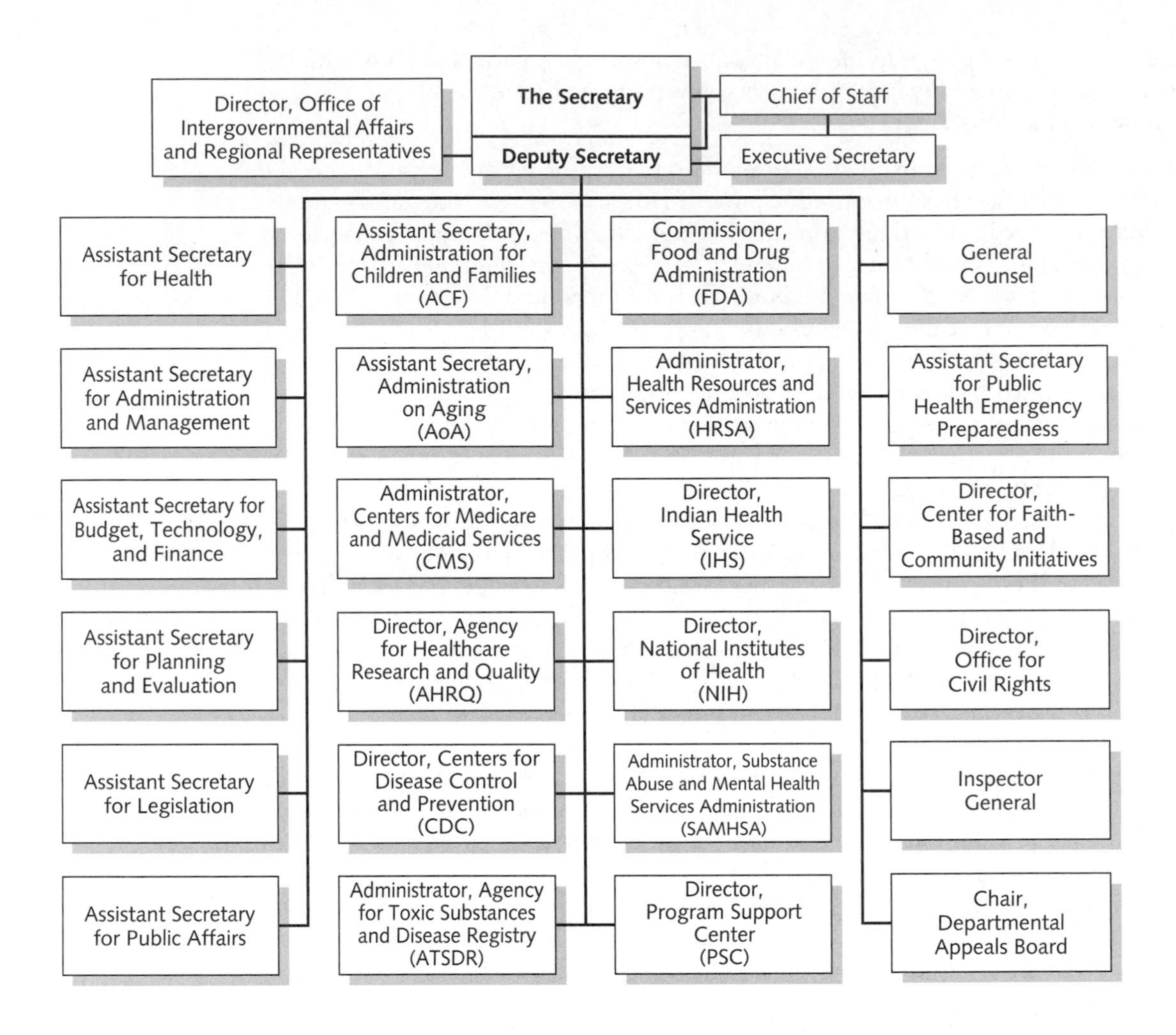

FIGURE 7-2 Organization of the Department of Health and Human Services

Source: The U.S. Department of Health and Human Services; available at www.hhs.gov/about/orgchart.html.

safety and providing financial assistance for low-income families and older Americans. The DHHS works closely with state and local governments, and many of its services are provided by state or county agencies. The Public Health Service operating division of the DHHS includes the National Institutes of Health, which houses 27 separate health institutes and centers such as the National Library of Medicine and supports about 35,000 research projects worldwide; the Food and Drug Administration, which works to ensure the safety of the food supply and cosmetics, and the safety and efficacy of pharmaceuticals; and the Centers for Disease Control and Prevention, which collects national health data and works to prevent and control disease. The Human Resources operating division includes the Centers for Medicare and Medicaid Services (CMS), which administers the Medicare and Medicaid programs; the Administration for Children and Families, which administers some 60 programs for needy children and families, including the Head Start program; and the Administration on Aging, which provides services, including some 240 million meals each year, to the elderly.[16]

The USDA is also concerned with some important aspects of public health and policy making. Its overall mission is to provide leadership on food, agriculture, natural resources, and related issues. The USDA seeks to enhance the quality of life for all Americans by working to ensure a safe, affordable, nutritious, and accessible food supply; reducing food insecurity in America; and supporting the production of agriculture. Its organization is illustrated

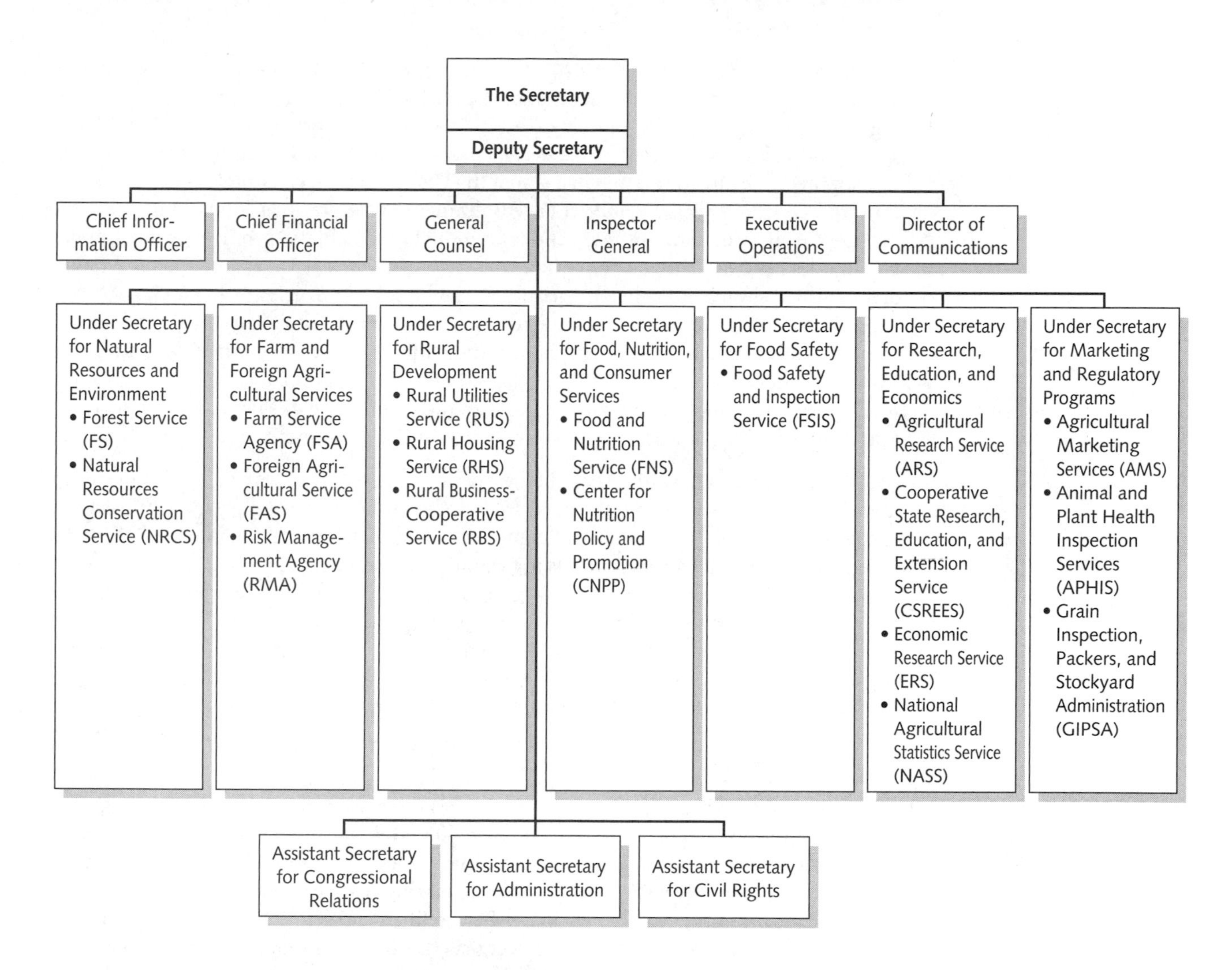

FIGURE 7-3 Organization of the U.S. Department of Agriculture

Source: The U.S. Department of Agriculture; available at www.usda.gov.

in Figure 7-3. The USDA's Food Safety mission strives to make sure that the nation's meat and poultry supply is safe for consumption, wholesome, and packaged and labeled properly. The agency responsible for carrying out this mission is the Food Safety and Inspection Service.

The mission of the USDA's Food, Nutrition, and Consumer Services is to ensure access to nutritious, wholesome food and healthful diets for all Americans and to provide dietary guidance to help them make healthful food choices. Two agencies within this division are important to community nutritionists. The Food and Nutrition Service administers nutrition assistance programs such as the Food Stamp Program and the National School Lunch Program (described along with other food assistance programs in Section Two of this book). The Center for Nutrition Policy and Promotion coordinates nutrition policy within the USDA and provides national leadership in educating consumers about nutrition. This center works with the Department of Health and Human Services to review, revise, and disseminate the *Dietary Guidelines for Americans* (described in Chapter 6).

The mission of the USDA's Research, Education, and Economics division is to develop innovative technologies that improve food production and food safety. This area includes three agencies of interest to community nutritionists: The Agricultural Research Service, which works to solve broad agricultural problems and ensure an adequate supply of food for all consumers; the Cooperative State Research, Education, and Extension Service (CSREES), which strives to develop national priorities in research, extension services, and higher education; and the Economic Research Service, which produces economic and social science data to help Congress make decisions about the practice of agriculture and rural development. The main research division of the USDA is the Agricultural Research Service. It oversees research related to nutrient needs throughout the life cycle, food trends, the composition of the diet, nutrient interactions, and the bioavailability of nutrients. It compiles data on the nutrient composition of foods through its Nutrient Databank; the food values are published electronically in the Agriculture Handbook No. 8 series, *Composition of Foods.*

Policy adoption also occurs at the state level. States are the basic unit for the delivery of public health services and the implementation of health policies. They determine the form and function of local health agencies, select and appoint local health personnel, identify local health problems, and guarantee a minimum level of essential health services. State and federal public health services have a similar organization.

4. **Policy implementation.** After the best solution to the problem has been agreed on and the tools for dealing with the problem have been chosen, the policy is modified to fit the needs, resources, and wants of the implementing agencies and the intended clientele. Implementation is the process of putting a policy into action. In Tacoma, Washington, for example, the Tahoma Food System, a nonprofit organization, implemented a policy to promote a sustainable food system. Its policy supported a community coalition called Bridging Urban Gardens Society (BUGS), which helps create "greening" projects such as organic gardens on unused, littered vacant lots.[17] The implementors of public policy in the United States number in the millions and include employees of federal, state, and local governments who work with private organizations, interest groups, and other parties to carry out government policy.

5. **Policy evaluation.** As soon as public policies move into the agenda-setting stage, the evaluation process begins. The purpose of policy evaluation is to determine whether a program is achieving its stated goals and reaching its intended audience, what the program is actually accomplishing, and who is benefiting from it.

Consider, for example, whether the welfare reform legislation known as the Personal Responsibility and Work Opportunity Reconciliation Act of 1996 has been a success. This legislation created a new program—Temporary Assistance to Needy Families (TANF)—to replace the old welfare "safety net" system and dramatically changed the requirements for receiving assistance. To measure TANF's success, the Children's Defense Fund and other nonprofit social justice organizations have conducted a number of post–welfare-reform surveys.[18] Survey results show that although welfare caseloads have been significantly reduced, the number of people with neither jobs nor welfare assistance has risen by as much as 27 percent. Additionally, surveys of former welfare recipients indicate that many working parents had been unable to pay rent, buy food, or afford medical care and had had their telephone and electric service disconnected. As a result, hunger advocacy groups now seek legislation guaranteeing a living wage, universal health care, and affordable housing, along with the assistance that enables people to move successfully from welfare to work (e.g., help with child care, transportation, and education).

From almost the moment of their conception, public policies undergo both formal and informal evaluations by citizens, legislators, administrative agencies, the news media, academicians, research firms, auditors, and interest groups. Ideally, public policies should be evaluated *after* they have been implemented, using the best available research methods and according to a systematic plan. In the real world, policy evaluation seldom works this way. Instead, it is often undertaken without a preconceived plan and before the strategy chosen to solve a problem has been fully implemented.

6. **Policy termination.** A policy or program may be terminated for any of several reasons: the public need was met, the nature of the problem changed, government no longer had a mandate in the area, the policy lost political support, private agencies relieved the need, a political system or subgovernment ceased to function, or the policy was too costly. Determining when a policy should be terminated is somewhat subjective. At what point do you decide that the public's need has been met? What measures do you use to conclude that the problem was solved? Typically, policy termination represents a process of adjustment in which the policy makers shift their focus to other policy concerns. Whereas some policy systems go out of existence, others survive and expand, bringing the policy cycle full circle to a redefinition of the public problem.[19]

The People Who Make Policy

The stages of the policy cycle outlined in Figure 7-1 do not correspond directly to the agencies and institutions involved in making government policy. We tend to assume that policy is formulated by legislatures and implemented by administrative agencies, such as the Public Health Service. In fact, administrative agencies sometimes formulate policy and legislatures become involved in policy implementation.

This point leads us to ask, "Who makes policy?" The authorities who "make policy" may be executives, administrators, or committees of an organization or company; elected officials; officers and employees of municipal, state, or federal agencies; members of Congress and state legislatures; and even **street-level bureaucrats:** welfare workers, public health nurses, police officers, schoolteachers, sanitation workers, housing authority managers, judges, and many other people working in government agencies. In the course of carrying out their jobs, these street-level bureaucrats daily make policy decisions by interpreting government laws and regulations for citizens. For example, as a community nutritionist, you will find yourself making policy decisions when you tailor a program to meet a particular client's needs or when you recommend a calcium supplement for a client on the basis of your interpretation of the Dietary Reference Intakes (DRIs).

Legitimizing Policy

Once it has been decided that a policy should be put into effect, a choice must be made about *how* it will be implemented. This is not a trivial decision. Consider, for example, the decision by the FDA to allow food product labels to carry health claims such as those listed in Table 7-1. Some consumers, scientists, and food companies objected to this policy, believing that some health claims on food product labels might distort research findings or be extravagant in presenting evidence of a link between food components and health. Others believed that the policy would help the public make healthful food choices. Thus, a policy may be perceived as benefiting some citizens and working to the detriment of others. Because it is impossible to achieve universal agreement on a policy and its effects, careful attention must be given to the process by which policy decisions are made. This is the point at which legitimizing policy is important.

Legitimacy is "the belief on the part of citizens that the current government represents a proper form of government and a willingness on the part of those citizens to accept the decrees of that government as legal and authoritative."[20] In this sense, legitimacy is mainly in the mind, for it depends on a majority of the population accepting that the government has the right to govern. In the case of FDA's health claim policy, the appearance of health claims on food labels indicates that consumers and food companies accept the FDA's *authority* to allow this action. In contrast, the failure of health care reform in President Clinton's administration was a signal that the American people did not consider the federal government a legitimate provider of comprehensive health care.[21]

Street-level bureaucrats Individuals within government who have direct contact with citizens.

TABLE 7-1 A Sampling of Health Claims Currently Authorized by the U.S. Food and Drug Administration Under the Nutrition Labeling and Education Act of 1990

DIET-DISEASE RELATIONSHIP	MODEL CLAIM
Dietary saturated fat and cholesterol and risk of coronary heart disease	While many factors affect heart disease, diets low in saturated fat and cholesterol may reduce the risk of this disease.
Fruits and vegetables and cancer	Low-fat diets rich in fruits and vegetables may reduce the risk of some types of cancer, a disease associated with many factors.
Folate and neural tube birth defects	Healthful diets with adequate folate may reduce a woman's risk of having a child with a brain or spinal cord birth defect.
Foods that contain fiber from whole oat products and coronary heart disease	Diets low in saturated fat and cholesterol that include soluble fiber from whole oats may reduce the risk of heart disease.
Whole-grain foods and risk of heart disease and certain cancers	Diets rich in whole-grain foods and other plant foods and low in total fat, saturated fat, and cholesterol may reduce the risk of heart disease and some cancers.

Source: U.S. Food and Drug Administration, Center for Food Safety and Applied Nutrition, *A Food Labeling Guide,* Appendix C, available at www.cfsan.fda.gov.

Government, then, must somehow legitimize each policy choice. Several mechanisms exist for legitimizing policies: the legislative process; the regulatory process; the court system; and various procedures for direct democracy, such as referenda, which put sensitive issues directly before the people. In the next section, we will explore the legislative and regulatory processes in greater detail, because most policies in the areas of health, food, and nutrition arise through these mechanisms.*

The Legislative and Regulatory Process

Governments can use any number of instruments to influence the lives of their citizens: taxes and tax incentives, services such as defense and education, price supports for commodities, unemployment benefits, and laws, to name only a few. Laws are a unique tool of government. In the United States, we traditionally associate lawmaking with Congress, the primary legislative body. It is Congress that sets policy and supplies the basic legislation that governs our lives.

Laws and Regulations

The laws passed by Congress tend to be vague. A law defines the broad scope of the policy intended by Congress. For example, the Special Supplemental Nutrition Program for Women, Infants, and Children (WIC) was authorized by Public Law 92-433 and approved on September 26, 1972. This law authorized a two-year, $20 million pilot program for each of the fiscal years 1973 and 1974. It gave the secretary of agriculture the

*The use of the court system for legitimizing policy in the nutrition arena is not discussed in this chapter, although it is important in formulating food and nutrition policy. Regulations *are* challenged through the court system. When the FDA issued a final rule establishing nutrition labeling regulations in 1973, a portion of the regulations dealing with special dietary foods was challenged in the courts by the National Nutritional Foods Association (see *National Nutritional Foods Association v. FDA* and *National Nutritional Foods Association v. Kennedy,* as cited in the *Federal Register* 1990 [July 19]: 29476–77).

authority to make cash grants to state health departments or comparable agencies to provide supplemental foods to pregnant and lactating women, infants, and children up to four years of age who were considered at "nutritional risk" by competent professionals. (The WIC program presently covers children up to five years of age.) As written, this law, like most others, was too vague to implement. It did not define or specify which professionals would determine the eligibility of clients. It did not define the concept of "nutritional risk" and other eligibility requirements, the method by which clients would obtain food products, or other aspects of the proposed program. Sorting out these details was left to the USDA.

Thus, once a law is passed, it is up to administrative bodies such as the USDA to interpret the law and provide the detailed regulations or rules that put the policy into effect. These regulations are sometimes called "secondary legislation." The total volume of this activity is enormous, as is apparent in the size of the *Federal Register,* a weekly publication that contains all regulations and proposed regulations, and the *Code of Federal Regulations (CFR),* a compendium of all regulations currently in force. When the WIC program was started, the details of the regulations were not published in the *Federal Register* until July 11, 1973, nearly 9 months after Congress passed the law. Over time, new laws and amendments to the existing law were enacted to increase the amount of money allocated for the WIC program, specify the means by which the program should be implemented, and authorize the continuation of the program for additional budget years.[22] (A detailed discussion of the WIC program appears in Chapter 10.)

How an Idea Becomes Law

All levels of government pass laws. (At the local level, laws are sometimes called ordinances or bylaws.) The process by which an idea becomes law is complicated. It may take many months, or even years, for an idea or issue to work its way onto the policy agenda. Then, once it reaches the legislative body empowered to act on it, the formal rules and procedures of that body can delay decision making on a proposed bill or scuttle it altogether, especially at the end of the legislative sessions. Lewis Carroll, writing in *Alice in Wonderland,* could have been speaking of the U.S. Congress when he wrote, "I don't think they play at all fairly, and they quarrel so dreadfully one can't hear oneself speak—and they don't seem to have any rules in particular: at least, if there are, nobody attends to them—and you've no idea how confusing it is."[23]

The general process by which laws are made is outlined in Figure 7-4, which shows the path a bill would take on its way through Congress. The process is much the same for bills introduced into state legislatures. The process begins when a concerned citizen, group of citizens, or organization brings an issue to the attention of a legislative representative at the local, state, or national level. Typically, the issue is presented to private attorneys or the staff of the legislative counsel, who draft the bill in the proper language and style. A bill is introduced by sending it to the clerk's desk, where it is numbered and printed. It must have a member as its sponsor. Simple bills are designated as either "H.R." or "S.," depending on the house of origin. For instance, the Medicare Medical Nutrition Therapy Act of 2005 was designated "H.R. 1582," indicating that the bill was introduced into the House of Representatives. When the bill is introduced, the bill's title is entered in the *Congressional Journal* and printed in the *Congressional Record.*[24]

For information on the status of legislation or the legislative process, visit http://thomas.loc.gov or www.house.gov/house/Tying_it_all.html.

As bills work their way through the House and Senate, they are considered by several committees and subcommittees, which may hold public hearings and seek the testimony of interested persons or experts before deciding whether to move the bill forward. The bill is revised during a **markup session.** If the bill approved by the Senate is identical to

Markup session A congressional committee session during which a bill is put into its final form before being reported out of committee.

FIGURE 7-4 How a Bill Becomes a Law

When a bill is introduced in the Senate (or House), it is assigned to a committee. The committee usually refers the bill to a subcommittee for hearings, revision, and approval. The subcommittee sends the bill back to the full committee, which may write revisions or amendments to the bill. The full committee (or Rules Committee in the House) decides whether to send it to the floor of its chamber for approval. The leaders of the chamber then schedule the bill for debate and vote. The bill is debated, amendments may be offered and voted on, and a final vote is taken.

Note: If the two chambers pass different versions of the bill, then a conference committee, composed of members of each chamber, will work out a compromise bill. The bill is returned to each chamber for a vote on the revised bill. The president signs or vetoes the bill. If signed, the bill becomes a law; if the president vetoes the bill, it cannot become law unless it is passed by a two-thirds vote of both chambers.

Source: Adapted from W. Lyons, J. M. Scheb II, and L. E. Richardson, Jr., *American Government: Politics and Political Culture*, p. 360. Copyright © 1995 West Publishing Company. Used by permission of Wadsworth Publishing Co.

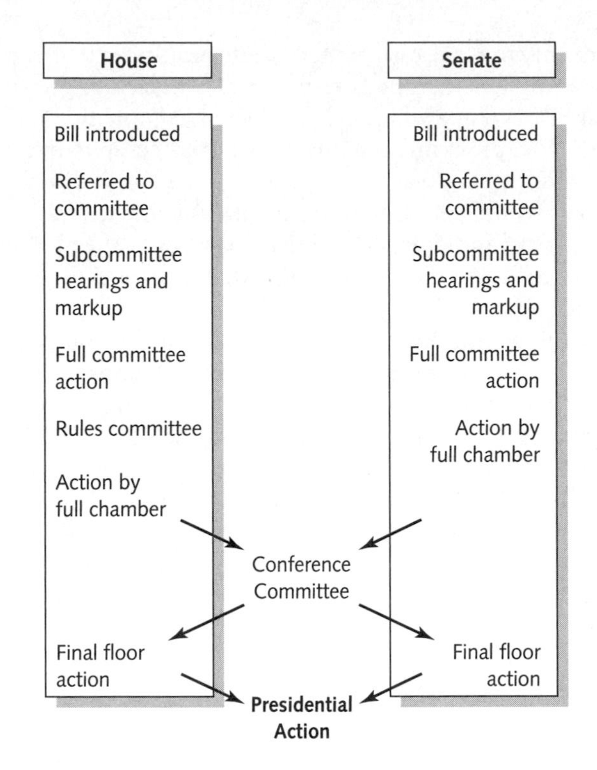

A list of standing and select congressional committees can be accessed from www.gpoaccess.gov/legislative.html.

the one passed by the House, it is sent to the president to be signed. If the two versions differ, a joint House–Senate conference committee is formed to modify the bill by mutual agreement. Once both houses agree on the compromise bill, it is sent to the president, who may sign it into law, allow it to become law without his or her signature, or veto it. When Congress overrides the president's veto by a two-thirds majority vote in both houses, the bill becomes law without the president's signature.[25] (In state legislatures, of course, the bill is sent to the governor.) When the bill is signed into law by the president, it becomes an *act* and is given the designation "PL," which stands for Public Law, and a number: the first two or three digits indicate the number of the congressional session in which the law was enacted, and the remaining digits represent the number of the bill. Recall that the bill authorizing the WIC supplemental feeding program became Public Law 92-433 (that is, bill number 433, enacted by the 92nd Congress). Table 7-2 lists the typical opportunities for input in the legislative process.

Before a law enacted by Congress goes into effect, it is reviewed by the appropriate federal agency, which is responsible for issuing guidelines or regulations that detail how the law will be implemented and what penalties may be imposed if the law is violated. These regulations, such as the final rule shown in Figure 7-5 on page 214, are published in the *Federal Register.* Because federal law requires that the public have the opportunity to comment on an agency's proposed guidelines, the agency first issues "proposed regulations." During the comment period that follows the publication of proposed regulations, the general public, experts, companies, and interested organizations submit their written comments and, in some cases, present their views at public hearings. From 30 to 60 days are allowed for comment, depending on the type and complexity of the regulations. At the end of the comment period, the agency reviews all comments, both positive and negative, before issuing its final regulations, which are incorporated into the *CFR.* The *CFR* and the *Federal Register* are available at most local libraries and county courthouses;

TABLE 7-2 Opportunities for Input in the Legislative Process

Bill Introduction/Sponsorship

State or federal legislators can be encouraged to introduce a bill to address a specific issue or to cosponsor a bill introduced by another legislator. Obtaining a large number of cosponsors on a bill is one strategy for gaining attention and credibility for an issue.

Subcommittee

The most important time for constituent involvement is the subcommittee stage. Legislators are not yet committed to specific bills or legislative language. Grassroots advocates can communicate their positions on the issue and suggest specific provisions or language. Action by constituents of subcommittee members can be very effective at this point.

Committee

Grassroots advocacy at the committee stage is also important. Communications may focus on supporting or opposing specific language developed by the subcommittee; encouraging legislators to sponsor amendments; and asking committee members to vote for or against the bill. Again, action by constituents of committee members can be effective.

Floor

Constituent communication with all senators and representatives is important when it comes to the floor vote. Grassroots efforts at this stage focus on encouraging a legislator to vote either for or against the bill, to sponsor a floor amendment, or to vote for or against a floor amendment offered by another legislator.

Conference

Opportunities are more limited at the conference stage. The conference committee works out the differences between similar bills passed by the House and Senate. Communication at this point may influence whether the House or the Senate provision is accepted in the compromise bill.

Floor

Once a conference committee has worked out differences, passage of a bill is normally routine and is not affected by further constituent communication.

Source: Data from The legislative process, in *Influencing Public Policy at the Grassroots, A Guide for Hospital and Health System Leaders,* American Hospital Association: 1999; and E. Winterfeldt, Influencing public policy, *Topics in Clinical Nutrition* 16 (2001); 10 (Aspen Publishers, 2001).

recent rules and regulations can be found on the Internet. (See the list of government Internet addresses at the end of this chapter.)

The Federal Budget Process

Laws and regulations will have no effect unless there are funds to enforce them. Congress must enact bills to fund the programs and services mandated by federal legislation. The federal budget process has been described as "fractured, contentious, and chaotic," mainly because it forces the president and Congress to negotiate and agree on the problems that deserve top priority.[26] Budgets are designed to count and record income and expenditures, to demonstrate the government's intention regarding the funding (and, more important, the priority) of programs, and to control and shape the activities of government agencies. In its simplest form, the budget process has two stages: The president proposes a budget, and then Congress reacts to the president's proposal. The actual budget process is complex and cumbersome, with the final budget reflecting the distribution of power among competing concerns and groups within the current political system.[27]

FIGURE 7-5 A Portion of a Final Rule Published by the Food and Drug Administration in the *Federal Register*

On July 9, 2003, the FDA issued a regulation requiring manufacturers to list *trans* fatty acids, or *trans* fat, on the Nutrition Facts panel of foods and some dietary supplements. This requirement gives consumers more information with which to make food choices that could lower their consumption of *trans* fat as part of a heart-healthy diet. Scientific reports have confirmed the relationship between *trans* fat and an increased risk of coronary heart disease. For more information, go to www.fda.gov/oc/initiatives/transfat/backgrounder.html.

DEPARTMENT OF HEALTH AND HUMAN SERVICES

Food and Drug Administration

21 CFR Part 101

[Docket No. 94P-0036]

RIN 0910-AB66

Food Labeling: *Trans* Fatty Acids in Nutrition Labeling, Nutrient Content Claims, and Health Claims

AGENCY: Food and Drug Administration, HHS.

ACTION: Final rule.

SUMMARY: The Food and Drug Administration (FDA) is amending its regulations on nutrition labeling to require that *trans* fatty acids be declared in the nutrition label of conventional foods and dietary supplements on a separate line immediately under the line for the declaration of saturated fatty acids. This action responds, in part, to a citizen petition from the Center for Science in the Public Interest (CSPI). This rule is intended to provide information to assist consumers in maintaining healthy dietary practices. Those sections of the proposed rule pertaining to the definition of nutrient content claims for the "free" level of *trans* fatty acids and to limits on the amounts of *trans* fatty acids wherever saturated fatty acid limits are placed on nutrient content claims, health claims, and disclosure and disqualifying levels are being withdrawn. Further, the agency is withdrawing the proposed requirement to include a footnote stating: "Intake of *trans* fat should be as low as possible." Issues related to the possible use of a footnote statement in conjunction with the *trans* fat label declaration or in the context of certain nutrient content and health claims that contain messages about cholesterol-raising fats in the diet are now the subject of an advance notice of proposed rulemaking (ANPRM) which is published elsewhere in this issue of the *Federal Register*.

DATES: This rule is effective January 1, 2006.

FOR FURTHER INFORMATION CONTACT: Julie Schrimpf, Center for Food Safety and Applied Nutrition (HFS-832), Food and Drug Administration, 5100 Paint Branch Pkwy.,

THE LANGUAGE OF THE BUDGET

The budget is the president's financial plan for the federal government. It indicates how government funds have been raised and spent, and it proposes financial policy choices for the coming fiscal year and sometimes beyond. Financial policy, or fiscal policy, consists of the government's plan for taxation and spending on the economy in general.[28] The budget describes the government's **receipts or revenue,** the **budget authority,** and **budget outlays.**

The budget allocates funds to cover two types of spending. *Mandatory spending* is required by law for **entitlements**—that is, programs that require the payment of benefits to any person who meets the eligibility requirements established by law.[29] For these programs, Congress provides whatever money is required from year to year to maintain benefits for eligible people. Many of these programs are indexed to the cost of living or similar measures, resulting in steady increases in benefits with inflation. Entitlements such as Social Security, Medicare, food stamps, agricultural subsidies, and veterans' benefits—so-called "uncontrollable expenditures"—command a major portion (about two-thirds) of the federal budget. The remainder of the federal budget consists of *discretionary spending*—that is, the budget choices that can be made in such areas as defense, energy assistance, nutrition assistance (for example, the WIC program and nutrition programs for seniors) and education after the mandatory allocations have been made.[30]

Receipts or revenue Amounts that the government expects to raise through taxes and fees.

Budget authority Amounts that government agencies are allowed to spend in implementing their programs.

Budget outlays Amounts actually paid out by government agencies.

Entitlements Programs that require the payment of benefits to all eligible people as established by law.

PRINCIPLES OF FEDERAL BUDGETING

The federal fiscal year begins on October 1 and runs through September 30 of the following year. The fiscal year is named for the year in which it ends; thus fiscal year 2005, or FY05, begins October 1, 2004 and ends September 30, 2005. (States and municipalities

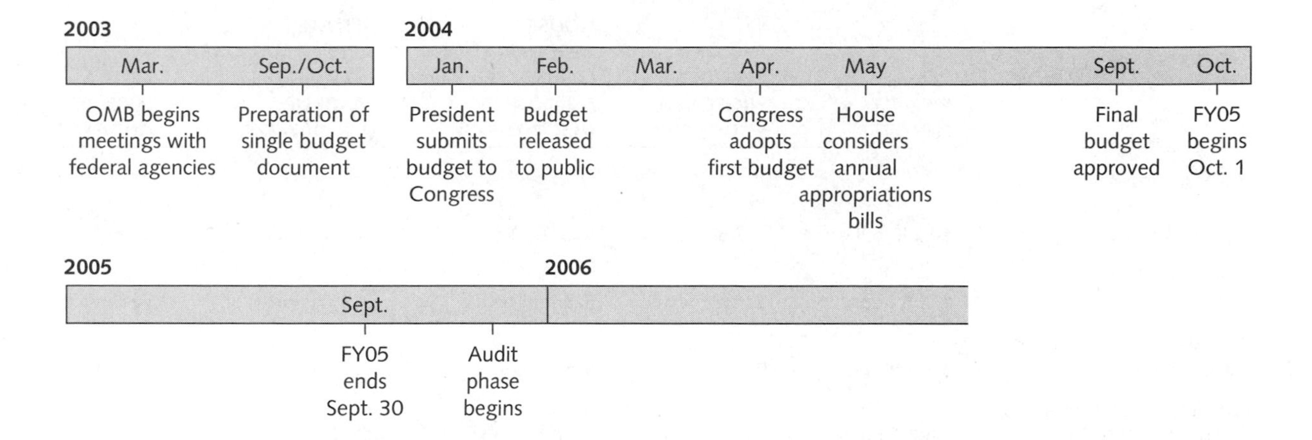

FIGURE 7-6 Federal Budget Cycle for 2005 (FY05)

differ in terms of fiscal years; some follow a calendar year [January–December], some start on July 1, and others have two-year fiscal cycles.[31]) The fiscal year is the year in which money allocated in the budget is actually spent, but important steps must be taken both before and after the fiscal year that can affect the government's or an agency's programming. These steps form the budget cycle and include budget formulation, approval, implementation (sometimes called execution), and audit.

Figure 7-6 shows the federal budget cycle for FY05. The first step—budget formulation—begins 15 to 18 months before the start of the fiscal year. (For FY05, the budget process began in March 2003). During this phase the Office of Management and Budget (OMB), the central budget office at the federal level, works with federal agencies to outline their funding projections for new and ongoing programs. After this consultation process, a single budget document is prepared and released, usually in September. This document is revised by the president in November and December and becomes the basis of his budget message—the most important statement of the administration's priorities and concerns—submitted to Congress in late January or early February.

As shown in Table 7-3, Congress can approve, disapprove, or modify the president's budget proposal, adding or eliminating programs or altering methods of raising revenue. After the president's budget is submitted to Congress, it is reported to committees and subcommittees that must make decisions about budget authority, taxes, appropriations, and the reconciliation of the budget. In this process, Congress passes revenue bills that specify how funds to support the government's activities will be raised. In the House, the Ways and Means Committee has jurisdiction over revenue bills, a responsibility that makes it one of the most powerful committees in Congress. In terms of spending, congressional committees must pass bills to authorize government programs. An **authorization** defines the scope of a program and sets a ceiling on how much money can be spent on it. Before money can be released to a program, however, an **appropriation** bill must be passed. The appropriation for a program may cover a single year, several years, or an indefinite period of time.

All revenue and appropriations bills passed by the House are forwarded to the Senate for consideration. Differences between the two houses are worked out in *conference committee,* and ultimately a *reconciliation bill* is passed. The end result of the authorization and appropriations work is that Congress adopts its version of the budget in a *budget resolution.* The first budget resolution is usually passed by May 15, and the second, after all spending bills have been passed, by September 15. If Congress is unable to pass a budget by the beginning of the fiscal year, it may adopt *continuing resolutions,* which authorize expenditures at the same level as in the previous fiscal year, until a budget agreement can be reached.

Authorization A budget authorization provides agencies and departments with the legal ability to operate.

Appropriation A budget appropriation is the authority to spend money.

	FY 2002 FUNDING	PRESIDENT'S PROPOSED BUDGET	HOUSE APPROPRIATIONS COMMITTEE MARKUP (H.R. 107-623)	SENATE APPROPRIATIONS COMMITTEE MARKUP (S.R. 107-223)	ACTUAL 2003 BUDGET
Special Supplemental Program for Women, Infants, and Children (WIC)	$4.348 billion	$4.751 billion	$4.776 billion	$4.751 billion	$4.661 billion
WIC Farmers' Market Program	$25 million	$25 million	$25 million	$25 million	$25 million
Senior Farmers' Market Nutrition Program (FMNP)*	$10 million	0	$15 million	$20 million	$15 million
Commodity Supplemental Food Program (CSFP)	$102.8 million	$95 million	$120 million	$107 million	$104 million
The Emergency Food Assistance Program (TEFAP)	$150 million	$150 million	$200 million	$195 million	$190 million
Food Stamp Program	$22.992 billion	$24.4 billion	$26.3 billion	$26.3 billion	$25.487 billion
School Lunch Program	$5.759 billion	$6.420 billion	$6.074 billion	$6.074 billion	$6.2 billion
School Breakfast Program	$1.580 billion	$1.661 billion	$1.661 billion	$1.661 billion	$1.6 billion
Child and Adult Care Food Program (CACFP)	$1.878 billion	$1.904 billion	$1.904 billion	$1.904 billion	$1.7 billion

*The budget appropriations process represents another opportunity for your input into the legislative process. Note here that the Bush administration eliminated the Senior Farmers' Market Nutrition Program for the 2003 budget year. This prompted concerned advocacy organizations and individuals to contact members of the House and Senate Appropriations Committees to voice their support for future funding for the Farmers' Market Nutrition Program.

TABLE 7-3 The President's 2003 Budget, the Senate and House Agriculture Appropriations for Selected Nutrition Assistance Programs, and the Actual 2003 Budget

Fiscal year 2005 began October 1, 2004, marking the implementation stage of the budget cycle when government agencies execute the agreed-on policies and programs. At the end of FY05, the audit phase begins, during which time the agencies' operations are examined and verified. In recent years, the audit phase has come to include performance auditing, or determining whether the agency's goals and objectives were met and whether the agency made the best use of its resources.

The Political Process

The complexities of the legislative and policy-making process present many challenges, and years may be required to reach a critical mass in public support for a policy change. As a case in point, consider that until 1906 no *federal* legislation regulated the nation's food supply and protected consumers against food adulteration, mislabeling, and false advertising. Even then, more than 25 years of persistent pressure from consumers and the agricultural community had been required to achieve this legislative milestone.[32] Check the boxed insert on page 218 for a description of the imperfect process leading to the passage of the Food and Drugs Act.

A more recent example of the legislative process is the campaign by the American Dietetic Association (ADA) supporting Medical Nutrition Therapy (MNT). MNT is a service provided by a registered dietitian or nutrition professional that includes counseling, nutrition support, and nutrition assessment and screening to improve people's health and quality of life. The ADA launched its MNT campaign in 1992 with the publication of

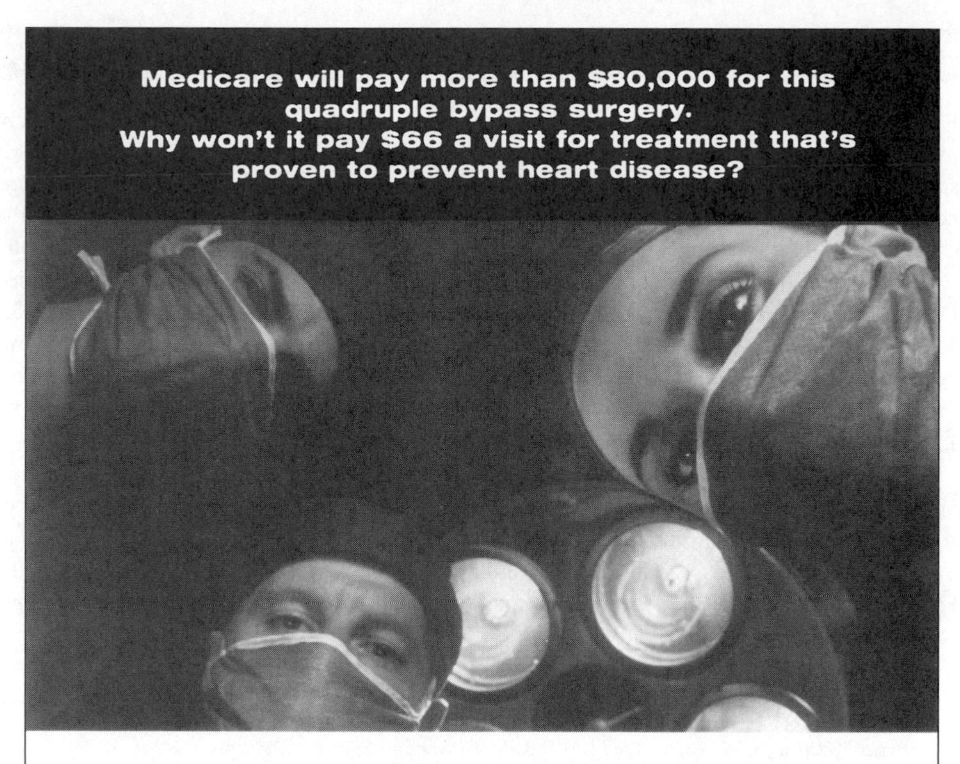

FIGURE 7-7 An Advertisement for Medical Nutrition Therapy

Source: Property of American Dietetic Association. Used with permission.

a position paper on health care services[33] and a report on the importance of including reimbursable nutrition services as part of any health care reform legislation.[34] It organized a grassroots campaign among its members to raise the visibility of the dietetics profession on Capitol Hill and to lobby Congress to support MNT,[35] and it formed coalitions with other health-oriented organizations to develop a uniform position on health care reform.[36] ADA members wrote letters to legislators, met with Washington lobbyists and key members of Congress,[37] and testified at a hearing held by the Health Subcommittee of the House Ways and Means Committee.[38] Advertisements, such as the one shown in Figure 7-7, helped educate consumers and policy makers about the importance and potential impact of MNT. The ADA's efforts resulted in the introduction of the Medical Nutrition

The Legislative Process in Real Life

Between 1880 and 1906, nearly 200 bills designed to protect consumers against food adulteration, misbranding, and false advertising were introduced into Congress without success. One such bill was submitted to the House and passed in 1904; the Senate began debating and eventually passed a similar bill, with amendments, in 1906. During the debates on the bill, Dr. Harvey W. Wiley, who was the chief chemist of the Department of Agriculture, testified about adulterated foods, bringing examples to show before Congress. At his urging, women who were concerned about food safety also lobbied Congress to pass the bill. When the bill went to the conference committee to iron out differences between the two houses, tensions were high. At one point, President Theodore Roosevelt felt compelled to lean on Congress and express his support for the pure foods bill. Finally, the bill passed both the House and Senate on June 27, 1906, and was signed into law by President Roosevelt on June 30, 1906. The Food and Drugs Act became law effective January 1, 1907.

Unfortunately, the law was defective. In the beginning, Congress failed to pass appropriation bills to provide the funds to enforce the law. Congress also failed to authorize the development of standards of food composition and quality, an omission that made it difficult for authorities to prove in court that a food was an imitation and not the genuine food product. Thus, because the Food and Drugs Act lacked teeth, food adulteration remained a threat to public health. Attempts to strengthen the law failed until 1938, when Congress passed the Federal Food, Drug, and Cosmetic Act, which was signed into law by President Franklin D. Roosevelt in 1938. It became effective one year later and is the primary legislation by which our food supply is regulated today.

Source: Adapted from H. W. Schultz, *Food Law Handbook* (Westport, CT: Avi, 1981), pp. 3–21.

Therapy Act (H.R. 2247 and S. 1964) during the 104th Congress. Although the House version had 91 cosponsors and the Senate version had 4 cosponsors, both bills expired when Congress adjourned.[39] Even so, the bills' support encouraged the ADA to continue its efforts. On April 17, 1997, the Medicare Medical Nutrition Therapy Act of 1997 (H.R. 1375 and S. 597) was introduced in the 105th Congress,[40] and ADA launched a new effort, the Majority by March Campaign, to secure a majority of congressional members as cosponsors of the legislation by March 1998.[41] During the following two years, the ADA continued to work for passage of legislation that, in the words of the Honorable John E. Ensign, who introduced the bill in the House, "will help to save Medicare, and most importantly, to save lives."[42]

After eight years of effort, victory was finally achieved on December 21, 2000, when President Clinton signed legislation (Public Law 106-554) that included the provision for creating a new Medicare MNT benefit for patients with diabetes or kidney disease (predialysis). The MNT benefit was part of a package of Medicare provisions (H.R. 5661) that was incorporated into a massive omnibus bill (H.R. 4577) that Congress approved December 15. ADA president Jane White acknowledged the tangible importance of the MNT victory: "As we consider the significance of this victory, we can say without reservation that our work to bring recognition to dietetic professionals working in all settings is producing results. The ability of dietetic professionals to serve in the food and health care systems is growing in meaningful ways." With Clinton's signature, the implementation process for the MNT benefit began. The final rules developed by the Centers for Medicare and Medicaid Services (CMS) of DHHS to govern the application of the benefit were published in the *Federal Register* in November 2001 and took effect on January 1, 2002. This process determined the payment levels, provider qualifications, and other coverage details. The ADA is now continuing its efforts to advocate that Congress expand

Medicare MNT Legislation

S. 604—Medicare Medical Nutrition Therapy Act of 2005—Introduced March 11, 2005
Sponsor: Senator Larry Craig
H.R. 1582—Medicare Medical Nutrition Therapy Act of 2005—Introduced April 12, 2005
Sponsor: Representative Fred Upton

The Medicare Medical Nutrition Therapy Act of 2005 is a bill seeking to amend Title XVIII of the Social Security Act to authorize expansion of Medicare coverage of medical nutrition therapy services. The bill will allow the Secretary of Health and Human Services, through the Director of the Centers for Medicare and Medicaid Services, not Congress, to determine future expansions of MNT coverage when scientific evidence shows it would be cost-effective in treating conditions and diseases such as hypertension, dyslipidemia, obesity, certain cancers, HIV/AIDS, or other diseases in order to help prevent the onset or progression of more serious conditions.

Stay up-to-date on MNT legislation: Go directly to http://thomas.loc.gov.

---

the MNT benefit to cover cardiovascular disease, cancer, osteoporosis, HIV-AIDS, and other conditions.[43]

Current Legislation and Emerging Policy Issues

In this chapter, we have seen that policies, and the laws and regulations derived from them, are the means by which public problems are addressed. Like the discipline of community nutrition itself, policies are dynamic. In the food and nutrition arena, existing policies are constantly being challenged by market forces, scientific knowledge, and consumer practices and attitudes. Current legislation and emerging issues have the potential to affect the delivery of food and nutrition programs and the way in which community nutritionists work.

The American Dietetic Association is currently addressing six public policy issues: medical nutrition therapy (MNT), aging, child nutrition, nutrition research, nutrition monitoring, and obesity.[44]

On the Pulse is a weekly online newsletter informing ADA members of developments affecting food, nutrition, and health. Topics reflect the ADA's legislative and regulatory priorities in Washington and the states, as well as reimbursement, science, and practice-related matters. You can subscribe as an ADA member at www.eatright.org.

- *Medical Nutrition Therapy.* Ensure access to medical nutrition therapy for a range of conditions in Medicare. Support H.R. 1225 and S. 632, which establish a new cardiovascular benefit in traditional Medicare.
- *Aging.* Applied nutrition and aging research is needed to determine the nutritional needs and optimal diets for nursing homes, home care, and the elderly. Improved care and broadened access to nutrition services can be achieved by expanding and funding federal and state nutrition services in home and community-based programs. Nutrition services such as risk reduction, home care, nutrition education, health promotion, wellness, and caregiver training are needed in the wide variety of settings in which older adults live, dine, and receive health care.[45] These include community health agencies, congregate meal sites, adult day care, assisted living, rehabilitation, and nursing home facilities. Recruitment and retention in allied health professions is needed to meet future projected health care demands for serving the elderly.
- *Child Nutrition.* Improve school nutrition environments to help children make sound choices for healthful eating and physical activity. The House-passed bill H.R. 3873 places trained nutrition professionals in decision-making roles that affect what foods and beverages schools serve; requires local districts to establish wellness policies that address the nutrition environment; and supports and funds nutrition education.

- *Nutrition Research.* The ADA supports doubling food and agriculture research over the next five years. Human nutrition issues are highly complex, and greater knowledge is needed to understand diet, behavior, and external influences affecting them.
- *National Nutrition Monitoring.* Support the National Health and Nutrition Tracking Act. Build health literacy to help people in all stages of life make better individual health decisions and to guide public policy initiatives. Programs that gather and analyze information such as the national nutrition monitoring system, including NHANES need support and funding.
- *Obesity.* Seek changes to obesity legislation to reflect the important and unique role that dietetics professionals play in preventing and treating this national epidemic. Designate obesity as a disease so that obesity prevention and treatment may be covered by insurance. Support the Improved Nutrition and Physical Activity Act (IMPACT) of 2003 (H.R. 716). This legislation addresses many obesity-related areas, including the training of health professionals in the use of evidence-based interventions to diagnose, treat, and prevent obesity.[46]

See Chapter 8 for a detailed discussion of the obesity epidemic and descriptions of current public health policies, as well as proposed policies and legislation to prevent obesity and overweight.

STATE LICENSURE LAWS

Go to www.eatright.org/Public/GovernmentAffairs/98_12914.cfm for a map giving information about requirements to enter the profession of dietetics in each of the states.

Licensure is a state regulatory action that establishes and enforces minimum competency standards for individuals working in regulated professions such as dietetics. Licensure is designed to help the general public identify individuals qualified by training, experience, and testing to provide nutrition information and medical nutrition therapy. Forty-six states currently have statutory provisions regarding professional regulation of dietitians and/or nutritionists. Enacting licensure laws in those states that still do not have such a law remains a high priority of the American Dietetic Association in the area of state affairs. State legislatures are charged with protecting the health and safety of the public, so every state regulates occupations that have an impact on the public's health and safety.[47]

According to the American Dietetic Association, licensing of dietitians and nutritionists ensures that individuals disseminating nutrition advice have the appropriate education and experience.[48] Because medical nutrition therapy is used in the treatment of various diseases, individuals seeking nutritional advice deserve assurance that the individual treating them has the requisite education and experience. Licensure laws protect the public from unqualified individuals who would portray themselves as nutrition experts.

BIOTERRORISM AND FOOD SAFETY

As part of a heightened awareness of bioterrorism threats, the FDA has created a special website that links to information about bioterrorism, as well as to other information sources. See www.fda.gov/oc/opacom/hottopics/bioterrorism.html.

Within hours of the September 11, 2001, attacks on the United States, the nation's food and water supplies were identified as likely targets of terrorists. The term *food safety* refers to foods free of foodborne pathogens, as well as a food supply free of contamination via bioterrorism.[49] As part of its efforts to address food safety more effectively, the USDA has called for better coordination of its numerous federal food safety programs. Twelve agencies are involved in regulating the food supply. Table 7-4 lists them and identifies each agency's responsibilities in ensuring the safety of the nation's food. The Safe Food Act of 2004 called for the formation of a single, independent agency—the Food Safety Administration—that would be responsible for food safety, inspection, and labeling.[50] The Safe Food Act would consolidate the USDA's Food Safety and Inspection Service and the FDA's Center for Food Safety and Applied Nutrition, along with the Commerce Department's National Marine Fisheries Service, the Environmental Protection Agency's Office of Pesticide Programs, and possibly other offices designated by the president.

TABLE 7-4 Federal Agencies Involved in Food Safety

AGENCY	FOOD SAFETY RESPONSIBILITIES
FDA Center for Food Safety and Applied Nutrition (CFSAN)	Ensures safety of domestic and imported food products (except meat, poultry, and processed egg products), animal drugs, and animal feed
CDC	Conducts surveillance for foodborne diseases; develops new methods to enhance surveillance and detection of outbreaks; assists local, state and national efforts to identify, characterize, and control foodborne hazards
USDA's Food Safety and Inspection Service (FSIS)	Ensures that meat, poultry, and some eggs and egg products are safe, wholesome, and correctly marked, labeled, and packaged
USDA's Animal and Plant Health Inspection Service	Ensures the health and care of animals and plants to protect them against pathogens or diseases that pose a risk for humans
USDA's Grain Inspection, Packers and Stockyards Administration	Reports to FDA any grain, rice, or food products that are considered objectionable for consumption
USDA's Agricultural Marketing Service	Establishes quality standards for grading dairy, fruit, vegetable, livestock, meat, poultry, and egg products
USDA's Agricultural Research Service	Conducts food safety research
National Marine Fisheries Service	Conducts voluntary safety and inspection programs for seafood products meant for human consumption
Environmental Protection Agency	Regulates all pesticide products and sets maximum allowed residue levels for pesticides on food and animal feed
Federal Trade Commission	Prevents representations about food that are meant to deceive consumers
U.S. Customs Service	Assists FDA and FSIS in carrying out their regulatory roles in food safety
Bureau of Alcohol, Tobacco, and Firearms	Administers and enforces laws covering the production, use, and distribution of alcoholic beverages

Source: General Accounting Office report GAO-02-47T, Food safety and security: Fundamental changes needed to ensure safe food.

BIOTECHNOLOGY

See Chapter 13 (page 428) for more about the debate over genetically engineered foods and crops.

The introduction of biotechnology-derived foods and crops has created challenges for scientists, regulatory agencies, and consumers worldwide. Early applications of food biotechnology include the production of vinegar, alcoholic beverages, sourdough, and cheese. Today, biotechnology has many applications in the dairy, baking, meat, enzyme, and fermentation industries.[51] In the last decade, plant biotechnology has been applied to producing plants that are resistant to viruses, insects, fungi, and herbicides. From a regulatory standpoint, the development and testing of genetically modified plants are monitored by the FDA, the USDA, the Environmental Protection Agency (EPA), and most state governments.[52] The FDA has approved as safe several genetically engineered crops, including herbicide-resistant soybeans, potatoes that resist a damaging beetle, and virus-resistant squash.[53] Future biotechnology goals are to improve the nutritional quality of plants, increase harvest yield, and produce special oils, carbohydrates, and proteins.[54] However, new advances in this area will continue to challenge existing regulations, and public concern about the safety and environmental impact of genetically engineered foods remains. Some consumers and advocacy groups urge mandatory labeling that discloses the use of genetic engineering. Others advocate more stringent testing of these products before marketing. Still others want a ban on all genetically engineered foods.[55]

COMPLEMENTARY AND ALTERNATIVE MEDICINE

Complementary and alternative medicine (CAM) is commonplace in many parts of the world, where it is accepted as appropriate therapy. In North America, CAM has emerged more recently as a potential adjunct approach to traditional Western medicine, mainly owing to its adoption by consumers who have embraced it as an alternative to the invasive treatments typical of Western medical practice today. The National Institutes of Health defines CAM as "those treatments and health care practices not taught widely in medical schools, not generally available in hospitals, and not usually reimbursed by medical insurance companies."[56] CAM practices include acupuncture, homeopathy, herbal therapy, manual healing methods such as reflexology and chiropractic, methods of controlling the mind and body such as meditation and biofeedback, pharmacological and biological treatments such as chelation therapy, and dietary therapies such as macrobiotics and nutritional supplements. The main objection to CAM is that few controlled, clinical studies of its safety and efficacy have been conducted.[57] The growth in CAM practices is likely to challenge existing policies related to health care delivery and the practice of dietetics.

Appendix B provides more information about complementary nutrition and health therapies.

In October 1998, Congress established the National Center for Complementary and Alternative Medicine (NCCAM), a division of the National Institutes of Health. The center is devoted to conducting and supporting basic and applied research and training, and it disseminates information on complementary and alternative medicine to practitioners and the public. Some of the herbs that are currently undergoing research investigations in the United States are garlic, St. John's wort, *Ginkgo biloba,* saw palmetto, echinacea, hawthorn, and cranberry.

FUNCTIONAL FOODS AND NUTRACEUTICALS IN THE MAINSTREAM

Functional foods—foods that may provide health benefits beyond basic nutrition—are increasingly evident on grocery store shelves. The Reuters news agency recently reported that health-conscious baby boomers see links between diet and health and are increasingly buying functional foods or nutraceuticals.[58] The term *nutraceutical* is used to describe food products created by new technologies and scientific developments. A proposed definition for *nutraceutical* is "any substance that may be considered a food or part of a food and provides medical or health benefits, including the prevention and treatment of disease."[59] Under this definition, nutraceuticals would include nutrients, dietary supplements, herbal products, genetically engineered "designer" foods, and some processed foods. According to Datamonitor, its research suggests that simple vitamin supplements will decline in significance in the future as consumers turn increasingly to products promising cures and prevention of specific conditions. The wide range of foods and ingredients classified as nutraceuticals has stimulated some controversy over whether they are really "healthful" or "healing" foods.

THE GROWING DIETARY AND HERBAL SUPPLEMENT MARKETS

The dietary and herbal supplement markets continue to grow despite minimal government oversight and a profusion of questions regarding the quality and reliability of the products available in today's marketplace. In 1994, Congress passed the Dietary Supplement and Health Education Act (DSHEA), which severely restricted the FDA's authority over virtually any product labeled "supplement" so long as the product made no claim to affect a disease.[60]

DSHEA allowed herbal medicines to be marketed without prior approval from FDA. What DSHEA did allow manufacturers to state on a label is a description of how the product affects a structure or function of the body, such as the claim that the herbal product can "support," "promote," or "maintain" health. DSHEA states that a product cannot claim that it affects disease, and a manufacturer cannot state on the label that the herbal product will "prevent," "treat," "diagnose," "mitigate," or "cure" disease. A disclaimer must always be included on the label stating that "This product has not been evaluated by the Food and Drug Administration. This product is not intended to diagnose, treat, cure, or prevent any disease." Therefore, herbal products are not obliged to meet any standards of effectiveness or safety that have been established for other medicines, which require extensive laboratory and clinical trials before approval. Today a supplement is presumed safe until the FDA receives well-documented reports of adverse reactions.

Consumers and public interest groups have been lobbying for the regulation of herbal supplements. DSHEA did give the FDA the power to require that supplement makers follow "good manufacturing practices." This would seem to specify standards for sanitation, but not necessarily for efficacy or purity. The FDA has not yet mandated these practices, but it is considering doing so.

The Office of Nutritional Products, Labeling, and Dietary Supplements hosts a dietary supplement website: www.cfsan.fda.gov/~dms/supplmnt.html. The site was developed to help consumers sort out the increasing amounts of information about dietary supplements, by providing tips for searching the Web and evaluating research.

THE HUMAN GENOME AND THE POTENTIAL OF GENETIC SCREENING

The *Human Genome Project* is an international effort to locate and identify the sequencing of all of the human genes—together known as the human **genome**—a blueprint that consists of about 100,000 genes that encode the templates for more than 30,000 proteins.[61] **Genetic disorders** can be passed on to family members who inherit a genetic variation or abnormality. When inherited, these variations can cause alterations in nutrient absorption, digestion, metabolism, and other individual differences. A small number of rare disorders (such as sickle-cell disease and cystic fibrosis) are caused by a mistake in a single gene. But most disorders involving genetic factors—such as heart disease and most cancers—arise from a combination of small variations in genes, often in concert with environmental influences.

Recent advances in genetics research are likely to alter disease management. By 2010, for example, it may be possible to complete a series of genetic tests to determine your personal disease susceptibilities and receive nutritional counseling tailored to your genetic characteristics.[62] Nutrition and health professionals must update their knowledge of genetics in order to meet the coming challenges of counseling clients affected by genetic disorders. One of the ADA's recent partnerships is with the Human Genome Education Model II Project, a program designed to increase the knowledge and skill levels of dietitians, occupational therapists, psychologists, social workers, and other professionals in providing genetic counseling.[63] These activities enhance the ADA's efforts to educate the public about how food choices prevent and delay the progression of chronic disease.

Policies in the food and nutrition arena will continue to evolve as our knowledge of foods and their relationship to health expands and the issues of public concern change. The broad scope of food and nutrition policy provides ample opportunity for you to become involved in the policy process.

The *Human Genome Project (HGP)* is an international research effort to determine the DNA sequence of the entire human genome. Contributors to the HGP include the National Institutes of Health (NIH), the U.S. Department of Energy (DOE), numerous universities throughout the United States, and international partners in the United Kingdom, France, Germany, Japan, and China.

For more information on genetics, visit the Human Genome Research Institute at www.genome.gov; see also www.cdc.gov/genomics.

Genome A term that combines the words *gene* and *chromosome*; the genetic material, in the chromosomes of the cell, that contains the complete set of instructions (DNA) for making an organism.

Genetic disorder A disease caused in whole or in part by a variation or mutation of a gene.

The Community Nutritionist in Action

Whether the issue is food safety legislation, health care reform, licensure of registered dietitians, or funding of the School Lunch Program, there are many ways in which you, as a community nutritionist, can influence the policy-making process. Whether

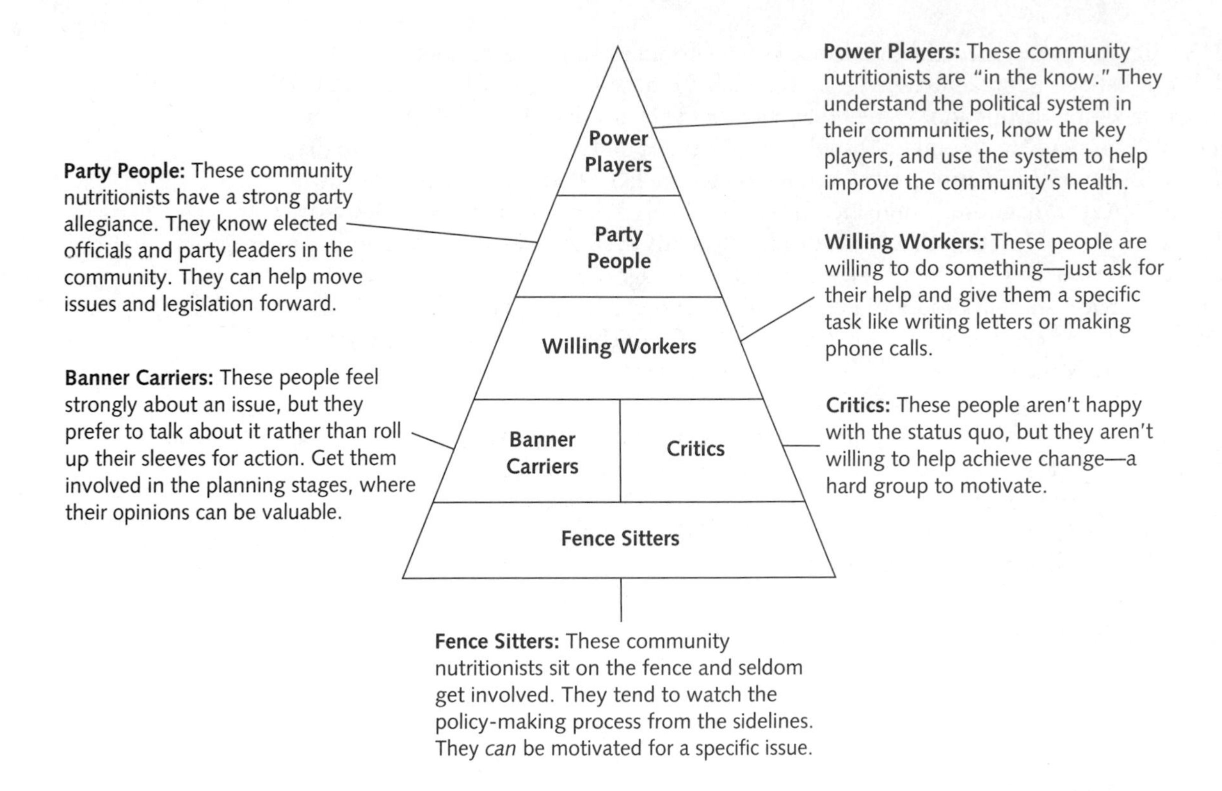

FIGURE 7-8 Grassroots Pyramid

Source: Adapted from the American Dietetic Association's Grassroots Targeting (1997).

you are new to this effort or not, consider which category of involvement shown in the grassroots pyramid in Figure 7-8 best describes your current level of involvement.[64] Are you a fence sitter who doesn't think about the big issues in community nutrition? Or are you a "banner carrier" who has strong opinions about issues but would rather talk about them than take action? What kind of player will you be in the future? No matter where you fall on the grassroots pyramid, there are opportunities in community nutrition for you to move up one level of involvement—or even two! Some strategies for influencing and becoming involved in the political process are described next.

Make Your Opinion Known

Expressing your opinion about an issue is one way of influencing the political process. When the issue is important to you, present your ideas and opinions at a public meeting or write a letter to the editor of a newspaper, magazine, or scientific journal. For example, an article in the Toronto *Globe and Mail* prompted a registered dietitian to submit a letter to the editor decrying the article's scare tactics about canola oil and its role in the Canadian diet. Most of the letter's contents, along with comments from other concerned professionals, were published the following week and helped educate the newspaper's readers about the safety of canola oil. The letter was written on business stationery, mentioned the date of the original article, and was short.

Use these principles of letter writing when responding to proposed rules and regulations. When writing the FDA, for example, refer to the proposed rule's publication in the *Federal Register,* outline your concerns, and keep your comments as short and direct as possible. Use personal, institutional, or company letterhead, as appropriate.

Become Directly Involved

When you become directly involved in policy making, *you* become a political actor and run for political office, sponsor a referendum, or initiate a campaign to bring an issue to the attention of the general public or policy makers. For example, you might seek an elected position within your state or the national dietetic association as one means of influencing the practice of dietetics, or you might participate on a local advisory board that is in a position to influence the political process in your community. You might organize the collection of signatures for a petition to be sent to your state legislature or city council, or you might work to elect a candidate to political office. Perhaps you help raise campaign funds for a local politician. Getting involved directly in the political process requires time and energy, but it is rewarding.

Join an Interest Group

Joining an interest group is another way of becoming involved in policy making. **Interest groups** are pressure groups that try to influence public policy in ways that are favorable to their members. They consist of people who work together in an organized manner to advance a shared political interest.[65] Interest groups may exert pressure by persuading government agencies and elected officials to reach a particular decision, by supporting or opposing certain political candidates or incumbents, through litigation, or by trying to shape public opinion. Although interest groups are sometimes accused of being bureaucratic and power-hungry and of not always representing their constituents fairly, they do contribute to the political process. Among other things, they encourage political participation and strengthen the link between the public and the government. In addition, they provide government officials with valuable technical and policy information that may not be readily obtained elsewhere.[66]

There are different types of interest groups. Business, for example, is well represented on Capitol Hill by lobbyists from food companies, such as Kellogg, Procter & Gamble, and the Coca-Cola Company, and from trade associations, such as the American Meat Institute, the Sugar Association, and the National Soft Drink Association. Professional groups, such as the American Dietetic Association and the American Medical Association, also have certain policy concerns. "Public interest" groups are a special category, in that they work to achieve goals that do not directly benefit their membership but serve to inform, educate, and influence the legislative process.[67] The Sierra Club, Common Cause, Bread for the World, and the Food Research and Action Center are well-known public interest groups. Still other interest groups may represent particular segments of the population, such as blacks (for instance, the National Association for the Advancement of Colored People) and women (such as the National Organization for Women). Activities to influence the political process even occur within the government itself. The National Governors' Association and the Hall of the States, for example, have been active in trying to set and direct the policy agenda.

WORK TO INFLUENCE THE POLITICAL PROCESS

Interest groups and their memberships use a number of tactics to influence the political process. Litigation is increasingly being used to change policy through the court system. Filing a class action suit in court on behalf of all persons who might benefit from a court action is one example. Another tactic is the public relations campaign, in which interest groups try to influence public opinion favorably. Three other common tactics are political action committees, lobbying, and building coalitions.

Interest group A body of people acting in an organized manner to advance shared political interests.

Political Action Committees (PACs)

A political action committee, or PAC, is the political arm of an interest group. It has the legal authority to raise funds from its members or employees to support candidates or political parties. The purpose of a PAC is to help elect candidates whose views are favorably aligned with the group's mission or goals. PACs also work to keep the lines of communication open between policy makers and the interest group's membership.[68] The PAC of the American Dietetic Association strives to influence the policy-making process and its outcome. ADA member contributions support a special fund that makes contributions to the election or reelection campaigns of selected candidates. ADAPAC is run by a five-person board of directors appointed by the ADA president. The ADAPAC Board selects candidates for contributions on the basis of whether they are in a position to assist the ADA on priority issues, what committee assignments or leadership positions they hold, and whether they are supportive of the ADA's positions.

Lobbying

Lobbying is often the method of choice when trying to influence the political system. "It is probably the oldest weapon, and certainly one of the most criticized."[69] Lobbying has acquired a negative connotation, giving rise to images of backdoor, professional power brokers who bend the political process through large campaign contributions. Although this image may be appropriate in some situations, it does not apply to all lobbyists. Remember, **lobbying** means talking to public officials and legislators to persuade them to consider the information you provide on an issue you believe is important.[70] Lobbyists' experience, knowledge of the legal system, and political skills make them an important part of the political process. They provide technical information to policy makers, help draft laws, testify before committees, and help speed (or slow) the passage of bills.

When this tactic is used, three issues are important: deciding how to lobby, knowing whom to lobby, and determining when to lobby. One of the first decisions is whether to lobby directly or to hire a registered lobbyist. In some cases, you or your organization may not need a registered lobbyist to achieve your goal; in other situations, you may not be successful without the knowledge of the political machine and its players that a registered lobbyist offers.

Knowing whom to lobby and when is also critical. To lobby successfully, identify the politicians or elected officials who are in a position to act on your concern by studying the formal structure of the government and its agencies, reading the newsletters of interest groups, and talking with policy makers who share your concerns. Sometimes, reporters who cover certain events can help you identify the proper authority figures. Once you know whom to lobby, choose the right time. You will accomplish little if you lobby a legislator when a bill is up for the final reading instead of when it is being considered in subcommittee. Use a triggering event to bring an issue to a politician's attention. If possible, turn the triggering event into an opportunity for placing the issue on the policy agenda. Above all, remember that you will have to lobby for as long as it takes, even if that means years. Establishing rapport and building recognition for your concern among elected officials take time.

Having determined where and when to lobby, you must consider how to do it most effectively. Four points are helpful when trying to influence public officials:

1. Show that you are concerned about the official's image. You want to make it easy for the politician to give you what you want, and *at the same time,* you want to make the politician look good. You must not appear to be applying force directly. Instead, provide compelling reasons why the politician should support your proposal.

Lobbying Providing information to elected officials.

2. Accept the constraints under which elected officials work. Politicians must deal daily with many pressing issues and diverse groups, including their constituents and staff, lobbyists, party politics, committee leadership, fund raising, campaigning, and the media. All of these place demands on the official's time. Effective lobbyists recognize the cross pressures that politicians face and, when possible, develop strategies for reducing those pressures.
3. Consider reaching elected officials indirectly. Politicians can be influenced through their staff, campaign workers, former colleagues, financial contributors, business associates, and friends. Approaching someone among these groups may enhance your ability to refine your proposal and avoid creating a problem for the politician.
4. Provide information for the official through letters, phone calls, and meetings. Ralph Nader, the consumer advocate, once remarked, "Talking frequently to legislators is the best way to persuade them of your position; the importance of this simple method cannot be overstated."[71]

Building Coalitions

An organization is sometimes too small or isolated to influence the political system effectively. It can better achieve policy changes by working with other organizations toward a common goal. Depending on the scope or depth of the cooperative effort, the joint venture may be a formal coalition or a more informal network or alliance. Formal coalitions tend to arise in geographical areas where problems affect many people across different communities. Coalitions may bring together social service organizations, church groups, professional associations, neighborhood groups, and businesses to develop a long-term, joint commitment to solving problems. The challenge is getting such diverse groups to agree on which problems deserve immediate attention.

Networks tend to arise when different organizations across the country share a variety of problems. A network may support a permanent staff in one location and have a common training and information system, but individual members of the network may pursue different problems. In general, network members share a common philosophy about how to mobilize people for action. Alliances, by comparison, tend to bring together organizations that are dispersed geographically to address one specific problem. The level of participation in alliance activities tends to wax and wane according to the urgency of the issue.

Coalitions, networks, and alliances are usually formed to increase the pressure on the political system. By joining forces, organizations can launch more effective public information and media campaigns to mobilize public support for an issue and bring it to the attention of policy makers. The ADA, for example, has more than 200 partners in government, industry, and health care. In Manitoba, Canada, an Alliance for the Prevention of Chronic Disease was formed to improve the health of Manitobans, prevent chronic disease, and reduce total spending on health care. The alliance consists of six nonprofit health charities that focus on five major chronic diseases: cancer, diabetes, and heart, kidney, and lung diseases. The first organization of its kind in Canada, the alliance plans to work with regional health authorities to help shift the emphasis from the traditional medical model to disease prevention and health promotion.[72]

TAKE POLITICAL ACTION

Community nutritionists *are* lobbyists at the local, state, and national levels. Here are a few strategies to keep in mind when trying to influence the political process.

Write Effective Letters

A personal letter to an elected official from a constituent can be a powerful instrument for change. Public opinion is important to politicians, and they take note of the number of letters received on a particular issue. Consider the following points when writing to your elected official:[73]

- Get the elected official's name right! Misspelling his or her name detracts from your credibility.
- Limit your letter to one page. Letters should be typed or handwritten legibly.
- Write about a single issue.
- Refer to the legislation by bill number and name. Note the names of sponsors, and refer to hearings that have been held.
- Explain how a legislative issue will affect your work, your organization, or your community.
- Use logical rather than emotional arguments in support of your position. Let the facts speak for themselves.
- Ask direct questions and request a reply.
- Be cooperative. Offer to provide further information. Do not seek confrontation, but don't hesitate to ask for your legislator's position on the issue.
- Follow up. Congratulate your legislator for a positive action or express concern again if the legislator acted contrary to your view on the issue.
- Whenever possible, use an example from your local community to draw attention to the issue.
- Write as an individual rather than as a member of the ADA (although you should identify yourself as a registered dietitian).

Many state dietetic associations prepare sample letters to send to elected officials. The letter in Figure 7-9, for instance, was prepared by the New Jersey Dietetic Association as a sample letter to Senator Jon Corzine seeking his support for expanded medical nutrition therapy legislation. An identical letter could be sent to a member of the House of Representatives, except that the salutation should be changed to read "Dear Congressman [or "Congresswoman"] ____________." In either case, use the sample letter as your guide, but change it to make it more personal. Elected officials do not enjoy getting dozens of form letters in the mail!

Identifying Members of Congress
www.congress.org
Find your senators: www.senate.gov/senators/senator_by_state.cfm
Find your House representative(s): www.house.gov/writerep

Making Contact by Phone
For the Senate: www.senate.gov/contacting/index.cfm
For the House: clerk.house.gov

Making Contact by Mail
For the Senate: www.senate.gov/contacting/index.cfm
For the House—write to any member of the House at this address:
The Honorable (full name)
U.S. House of Representatives
Washington, DC 20515

Making Contact by E-mail
For the Senate: www.senate.gov/contacting/index.cfm
For the House: www.house.gov/writerep/wyrfaqs.shtml

Make Effective Telephone Calls

Getting through to a legislator or other elected official can be difficult. When the opportunity presents itself, remember the following points when phoning an elected official:

- Write down the points you wish to make, your arguments supporting them, and the action you want the legislator to take.
- Don't expect to speak directly with the legislator. Contact the staff person responsible for the issue you want to address.
- Request a written response so that you will have a record of the legislator's position on the subject.

Use E-mail Effectively

E-mail is an acceptable way to communicate with your elected officials.[74] Be sure to use the same rules as for letter writing, including formality in salutations and content. Try to limit your message to two or three brief paragraphs. Be specific about the issue or bill number and about what you are asking the member of Congress to do. Make sure you include your name and street address on all correspondence to confirm that you are a constituent of the member of Congress.

The Honorable Jon Corzine
U.S. Senate
Washington, DC 20510

Dear Senator Corzine:

Medicare patients need your help! The number one killer of Americans is cardiovascular disease. Many of these diseases can be managed through proper diet. A bill (S.960) is pending before Congress to expand Medical Nutrition Therapy or MNT to cardiovascular diseases for Medicare recipients. Passage of this legislation would mean that Medicare patients could get help from nutrition professionals in managing their disease through proper diet.

After nearly a ten year fight, Congress passed a bill in December 1999 to provide Medical Nutrition Therapy for Medicare recipients with renal and diabetes diseases. Those benefits became available to Medicare patients in January 2002. Now is the time to expand this vital coverage to cardiovascular diseases. Already over 120 Representatives and 15 Senators have co-sponsored the legislation. And yet, while thousands of people die every year as a result of high blood pressure, high cholesterol, and other disorders caused (at least in part) from a poor diet, the cardiovascular MNT bills languish in subcommittee with no hearings scheduled.

They need for you to co-sponsor S. 960 and work with the sponsor of the bill, Senator Bingaman to move these bills through Congress. The 2,300 members of the New Jersey Dietetic Association strongly urge you to add your name to those supporting this legislation. Please use your influence with the Chairman of the Senate Finance Committee to report the bill.

When the prestigious Institute of Medicine reviewed the MNT for cardiovascular disease this was their conclusion:

The clinical literature contains evidence that nutrition therapy reduces mortality and morbidity through reduced complications of diabetes and reduced incidence of heart failure and cardiovascular disease. Given data limitations, it is difficult to reliably estimate the budgetary implications of such averted costs for the Medicare program. However, economic benefits to the Medicare program and to its beneficiaries are likely to be significant. Given reasonable assumptions regarding treatment efficacy and service use, initial estimates indicate that averted costs due to reduced incidence of coronary heart disease could range from $52 million to $167 million for patients with hypertension, $132 million to $330 million for patients with diabetes, or $54 million to $164 million for patients with dyslipidemia.

Save money and save lives – support Cardiovascular MNT.

Sincerely,

Jesse B. Struble

Jesse B. Struble, R.D.
196 Briarcliff Lane
Belmar, N.J. 07719

FIGURE 7-9 Sample Letter to a Senator Asking for Support of the Medicare Medical Nutrition Therapy Amendment Bill

Source: Used with the permission of the New Jersey Dietetic Association.

Work with the Media

Reporters and journalists with radio and television stations and newspapers can help build support for your position on an issue. Get to know the key media representatives in your community. When they call about an issue, be prepared to answer their questions about the issue and its impact on the community. When appropriate, prepare a press release to alert them to activities related to an issue. Press releases should be short (no longer than two pages, double-spaced) and concise, and they should provide one or two quotes by a key spokesperson on the issue and give a contact's name, address, and phone number. The Professional Focus feature on page 233 offers tips for working with the media.

Political Realities

Never doubt that a small group of thoughtful, committed citizens can change the world; indeed, it's the only thing that ever has.
– Margaret Mead

Reread the quotation at the beginning of this chapter. What does it mean? In the political realm, it means that an issue isn't important on Capitol Hill until it is important back home. It means that constituents can have more influence over elected officials than party officials have. It means that a local example of a problem carries more weight with policy makers than a national statistic. It means that elected officials measure the importance of an issue by the number of messages they receive about it—which explains why your letters and political activities count.

Getting involved in the policy-making process is one way to strengthen your connections with other people and with your community. It can be chaotic at times, but knowing that your effort as an individual improved the health or well-being of people in your community can also provide great personal satisfaction. You *can* make a difference in your community by understanding the policy-making process, taking time to express your opinion, and being persistent and patient.

Internet Resources

Canada

Agriculture and Agri-Food Canada Online	**www.agr.gc.ca/index_e.phtml**
Government of Canada	**http://canada.gc.ca**

United States

Information Locators

FedWorld	**www.fedworld.gov**
GPO* Access	**www.access.gpo.gov**
Thomas Legislative Information	**http://thomas.loc.gov**

Federal Agencies

Office of Management and Budget	**www.whitehouse.gov/omb**
Department of Health and Human Services	**www.hhs.gov**
Administration for Children and Families	**www.acf.dhhs.gov**
Administration on Aging	**www.aoa.dhhs.gov**
Centers for Disease Control and Prevention	**www.cdc.gov**
Food and Drug Administration	**www.fda.gov**
Centers for Medicare and Medicaid Services	**www.cms.hhs.gov**
Health Resources and Services Administration	**www.hrsa.gov**
Indian Health Service	**www.ihs.gov**
National Institutes of Health	**www.nih.gov**
U.S. Department of Agriculture	**www.usda.gov**
Agricultural Research Service	**www.ars.usda.gov**
Center for Nutrition Policy and Promotion	**www.usda.gov/cnpp**
Cooperative State Research, Education, and Extension Service	**www.csrees.usda.gov**
Economic Research Service	**www.ers.usda.gov**
Food and Nutrition Service	**www.fns.usda.gov**
Food Safety and Inspection Service	**www.fsis.usda.gov**

The Hill Sites	
U.S. House of Representatives	**www.house.gov**
U.S. Senate	**www.senate.gov**
White House	**www.whitehouse.gov**
Political Party Sites	
Democratic Party Headquarters	**www.democrats.org**
Republican National Committee	**www.rnc.org**
Government Publications Online	
Code of Federal Regulations	**www.gpoaccess.gov/cfr/index.html**
Federal Register	**www.gpoaccess.gov/fr/index.html**
State and Local Government	
Library of Congress State and Local Government Information	**www.loc.gov/global/state/stategov.html**
U.S. State and Local Government Gateway	**http://firstgov.gov/Government/State_Local.shtml**

*GPO = Government Printing Office.

Case *Study*

Food Safety as a Food Policy Issue

By Alice Fornari, EdD, RD, and Alessandra Sarcona, MS, RD

Scenario

There has been a series of articles in the local newspaper asserting that food pantries have empty shelves and soup kitchens do not have enough meals to feed the number of patrons who need their services. The president of the dietetics club at a nearby college decides to ask the professor in charge of the foods laboratory what is done with the leftover prepared and unprepared food from all the food courses (basic foods, cultural foods, and experimental foods). Unfortunately, the response indicates all leftover food is discarded daily and that leftover nonperishable food is discarded at the end of the semester.

This is astonishing to the student, who is very moved by the articles that she has read recently regarding hunger among local residents. She decides she must mobilize her fellow nutrition students to promote a new policy for the foods courses that would result in distributing leftover food, perishable and nonperishable, to the local food pantries or soup kitchens.

This seems to the student to be a simple task to accomplish. She proposes that all recipes be doubled to ensure that there are perishable leftovers to be distributed. All leftovers can be wrapped and refrigerated. Leftover nonperishables will be collected in a box or crate for distribution to either pantries or soup kitchens. Deliveries would be three times per week, and students would be required to assume this responsibiity on a rotating schedule. It would be a service learning activity required in the program. The club would make the delivery schedule and distribute the schedule in classes and electronically.

Our idealistic student is very diligent and prepares a written proposal, which she submits to the head of the Foods and Nutrition Department. This individual is impressed with the proposal and arranges a meeting with the student to congratulate her on her increasing awareness of the connection between dietetics and hunger—but also to raise the following questions with the student:

- What are the food safety issues associated with distribution of prepared and nonperishable food items?
- Do the food pantries and soup kitchens have an existing written policy on accepting food?

- Does the Department of Health have a policy on accepting prepared food?
- Does the government have a policy or guidelines on recycling food? Which government agency addresses this issue? Is it a federal or a state body?
- Is there a written policy in the departmental foods laboratory manual on this issue? If not, are you prepared to write one?
- Where will funds come from to double recipes? Are there ways in which the dietetics club could fund the increase in ingredient purchases?

The student is overwhelmed that a "simple" idea could generate so many questions. She needs to begin the process of answering the questions in preparation for a follow-up meeting with the entire faculty.

Learning Outcomes

- Identify how a community nutritionist can influence the policy-making process.
- Communicate food safety issues specific to food distribution.
- Link food safety issues to policy.
- Develop a written policy, specific to the department food courses, that addresses the food safety issues relevant to food distribution of prepared and nonperishable food items.

Foundation: Acquisition of Knowledge and Skills

1. Access the ADA position paper on domestic hunger and food insecurity and other professional literature on food safety: www.eatright.org, www.eatright.org/Member/PolicyInitiatives/index_21047.cfm, www.homefoodsafety.org, and www.eatright.org/Member/PolicyInitiatives/index_21043.cfm.
2. Access government documents related to food safety, such as www.fightbac.org/main.cfm, www.fstea.org, www.fsis.usda.gov, and www.foodsafety.gov/~fsg/ednet.html.
3. Identify why food safety is a critical issue to food distribution in a community setting.
4. Outline the basic steps involved in the process of policy making—this task can vary at different levels of an organization.
5. On the basis of your reading of this chapter, identify components of the policy-making cycle.

Step 1: *Identify Relevant Information and Uncertainties*

1. Consult experts and/or explore literature, on- and off-line, to create a list of food safety issues that may arise from food distribution in a community setting.
2. Explain how commodity-based food distribution—of both raw and prepared food—can violate food safety parameters.

Step 2: *Interpret Information*

1. With the goal of distributing food to hungry individuals, and to ensure that food safety is an integral component of any food distribution plan, include a description of food safety measures that are required. If applicable, prioritize the measures.

Step 3: *Draw and Implement Conclusions*

1. Prepare a written policy for food distribution.

Step 4: *Engage in Continuous Improvement*

1. Identify any limitations of the new policy; consider budget implications to the department; and determine how the dietetics club could be involved.
2. Devise an implementation plan or procedure for the new policy to make sure that all faculty teaching food courses are aware of the food distribution guidelines.
3. Identify three indicators to assess whether the food being distributed is appropriate and usable to the community.

Building Media Skills

In America the President reigns for four years, and journalism governs for ever and ever.

– Oscar Wilde

Louis Pasteur said, "Chance favors the prepared mind." When working with the media, always be prepared to provide credible nutrition information. Besides keeping current with the various scientific and trade journals and the popular press, you'll need to consider radio and television news and talk shows. (The Professional Focus feature in Chapter 4 offers tips for becoming a media monitor.) When you provide accurate information, everyone benefits—the media by providing valuable information to the audience, the public by receiving accurate information, and you as a community nutritionist by enhancing your image and visibility as a professional.

The American Dietetic Association's latest survey of consumers' attitudes about nutrition tells us that although dietetics professionals are among the most valued resources of nutrition information, consumers get the majority of their nutrition and health information from various media outlets.[1] This Professional Focus offers tips on developing media skills. The more that nutrition professionals succeed in developing a media presence, the better the access that consumers have to positive, science-based messages. As you work with the media, consider the following tips:[2]

- Be sure that the information you supply is accurate—check and recheck all names, dates, facts, and figures. This helps establish your credibility with the media.
- Become familiar with the format and types of coverage of the various media—television and radio news and talk shows, newspapers, newsletters and local publications, trade publications and magazines, and the Internet. Adapt your messages to the format of the media you choose. Identify how a particular audience will benefit from the information you provide. Know what the media cover in your area, who covers it, when they cover it, why they cover it, and how they cover it. This will help you to pitch your ideas to the appropriate media market.[3]
- Present scientific information in a concise, understandable manner so that your audience can readily use and remember it. Avoid technical language. Be able to communicate a *single overriding communications objective* (SOCO).[4] For example, in a news story on osteoporosis prevention, the SOCO might be for the audience to know that healthy bones require calcium. Next, develop three key points that will support the SOCO and provide the foundation for the news story. These points represent the key messages that you want your audience to retain. In this example, the key points might be (1) Diet can make a difference in lowering your risk of osteoporosis, (2) Weight-bearing exercise can enhance the benefits of a calcium-rich diet, and (3) Consider calcium supplements if dietary changes cannot or will not be made.
- Be consumer-oriented. Keep your audience in mind as you prepare your media information. Be both an authority on nutrition information and (when appropriate) an entertainer. Get the audience interested and involved with your message. Consider using visuals to support your messages.

General Guidelines for Working with the Media

- Nurture good press relations with the media contact people (reporters, editors, program directors, and producers) in your area. Keep a list of the names, e-mail addresses, phone and fax numbers, and deadlines of the media in your community. They are the "gatekeepers" to your target audiences. Occasionally, you may send people on your contact list an FYI (for your information) piece about a research report or recently published article. This alerts them to newsworthy topics or events that they may find useful either now or at a later date. Make yourself available to them for follow-up information or assistance.
- When working with television or radio, consider your appearance—dress with professional style. Practice your presentation as much as possible beforehand. You'll want your facial expressions, body language, and voice to show animation and enthusiasm for your topic.

There are several things you can do to encourage media to cover nutrition issues. Here are some examples:[5]

- Send out a news release with a newsworthy local story.
- Write a letter "pitching" a story to a TV station, newspaper or local magazine, including background information.
- Write a letter to the editor.
- Listen to call-in radio shows that cover current news or health issues, and call in to voice your opinion on nutrition-related topics.

Issuing a News Release

A well-written news release can generate significant publicity at relatively little cost by convincing a reporter or producer to cover a story.[6] Opportunities to issue a news release include tying into any major medical research announcement that involves a nutrition component; an RD visiting a legislator or a legislator visiting an RD's workplace; RDs receiving recognition from employers or patient groups; or a particular human interest angle emerging from an RD–patient relationship. Keep in mind the following tips when writing a news release:

- Write a gripping headline that conveys, "This is NEWS!"
- Writers learn to capture their readers' attention with a news "hook" or "spin" in the first paragraph, followed by answers to the five Ws: who, what, when, where, and why.
- Write the news release in the third person; keep it brief and to the point—usually around two pages in length.
- Make the opening statement strong to get the attention of the journalist.
- Include important quotations in the second or third paragraph to tie the story quickly to individuals who live in your area.
- List a contact name and telephone number.
- Include extra information in an accompanying fact sheet or brief backgrounder. Provide easily understood summaries of scientific information.

Writing a Letter to the Editor or an Op-ed Piece

A letter to the editor is most likely to be used when it is timed to respond quickly to events such as an article that appeared in the newspaper, remarks made by an elected official at a public event, or activities in the capital.[7] A letter to the editor is written to a newspaper editor expressing an opinion, issuing a call to action, or citing an issue or situation that needs attention or change (see Figure 7-10). Here are some guidelines for writing to newspaper editorial pages:

- Look at similar pieces in the newspaper or magazine before writing to see what topics are covered, how similar pieces are written, and how long the average letter that is printed seems to be.
- Contact the newspaper or magazine to see whether it has specific guidelines for letters to the editor.
- Develop a strong news slant or a local angle, with examples of real individuals who are affected.
- Include your name and affiliation if you are writing on behalf of your state or student dietetic association; speaking for a group carries more weight.

Pitching Your Ideas On-line

As a community nutritionist, you may have an opportunity to help develop your organization's website, contribute to existing

FIGURE 7-10 A Sample Letter to the Editor

Source: American Dietetic Association Advocacy Guide: Food and nutrition matters: Effective nutrition and health policy begins with you (Chicago, IL: American Dietetic Assocaition, 2004), p. 38.

xx/xx/xxxx

Dear (Editor's name):

Everyone agrees that the future of the Medicare Trust Fund is in jeopardy, while very few agree on how to fix it. While the Bipartisan Commission on the Future of Medicare and leaders in Congress investigate ways to fix the Medicare system, it makes sense to focus on solutions that save money over the long run and also improve the quality of care.

Medical nutrition therapy (MNT) is one solution. An important study from The Lewin Group, an independent health policy research firm, projects that savings to Medicare would be greater than cost after just three years. The study, conducted for The American Dietetic Association, projected savings to grow steadily in following years. It just makes common sense to adopt this solution. If people who need medical nutrition therapy can get access to it regularly without having to worry about how they will pay for it, they can avoid more complicated, more expensive health problems.

Medical nutrition therapy is the service a registered dietitian (RD) provides in many medical cases. It begins with assessing a patient's overall nutrition status, followed by prescribing a personalized course of treatment. An RD may consider a range of factors including medications, food/drug interactions, physical activity, other complex therapies such as chemotherapy, and the patient's ability to feed himself or herself. MNT is a medically necessary and cost-effective way of treating and controlling life-threatening diseases and medical conditions, including diabetes, heart disease, cancer, AIDS, kidney disease, and severe burns. It saves money because it reduces the need for medicines, it reduces hospital admissions, it reduces the length of stay in the hospital, and it reduces many painful and dangerous complications.

It's time to look at all possible solutions that can save Medicare for future generations, while also protecting the quality of health care to all Americans. Medical nutrition therapy is a solution that saves.

(name, RD)

websites, or you may want to design and post your own homepage on the World Wide Web. Whatever the case, there are basic questions to ask and issues to consider when contributing to or designing a website. Answer as many of the following questions as possible before designing a website. The more time you spend thinking about what you want—and don't want—the easier the design process will be. Also go to www.usability.gov for more guidelines for writing for the Web and designing useful websites.

- *Who Will Develop the Website?* If you are helping design a website for your organization, the chances are good that a webmaster has been hired to develop, test, update, and maintain the site. In this case, your role may be one of choosing the site's content. If it is your own website, you must decide whether to develop it yourself or hire a web page designer to do it for you.
- *What Is the Purpose of the Website?* Your site may be designed to sell a product or service, present information to a certain audience, or provide a collection of links to other websites. Perhaps it will do all three of these things. Specify the purpose of your site right from the beginning to help control costs and design time.
- *Who Is the Intended Audience?* Your site may be designed to reach clients, customers, people who already know something about the subject matter, or people who are unfamiliar with the topic. Specifying the intended audience helps determine how much background information must be provided and the terminology that must be explained. When answering this question, consider the typical user of your website and the type of problem he or she is trying to solve. In other words, think about why this user accessed your site in the first place.
- *How Long Should the Website Pages Be?* One frustrating aspect of web browsing is scrolling up and down long pages. A rule of thumb is to keep page lengths to one window. This translates to about $1\frac{1}{2}$ screenfuls of text. And remember, short pages are easier to maintain than long ones.

Other Issues to Consider

The number one consideration in website design is ease of use. If a website is cluttered, disorganized, or takes several minutes to download, users may seek information elsewhere—and you will have lost an opportunity to get your message across. To prevent clutter, use graphics wisely. Use only those that are essential to the website's purpose. To help keep the site organized, use document and chapter headings and put a title heading on each page.

The second consideration is quality. A website should be as presentable as an educational brochure, journal article, or textbook. Check spelling. Write well. Test every link. Update the site's pages often and date the pages.

Finally, pay attention to "netiquette." Do not publish registered trademarks or copyrighted material without permission. Do not publish a link to someone else's website without permission. Take time to respond to the people who send you queries about the information on your website. Provide good customer service to ensure that users keep coming back.

A Final Word About Content

Whatever the format, keep these guidelines in mind when preparing your media information:[8]

- Make your message easy to understand and recall.
- Focus on the positive.
- Be certain your information is based on sound scientific research.
- Be practical: Zoom in on specific nutrition facts that can be easily applied.
- Be sure that your presentation exhibits cultural sensitivity and that your content is relevant.
- Tailor the information to your audience.

References

1. *Nutrition and You: Trends 2002* (Chicago, IL: American Dietetic Association, 2002), available at www.eatright.org; and K. T. Ayoob, Commentary on dietitians and the media: Consumers want us and we need to be there, *Topics in Clinical Nutrition* 16 (2001): 1–7.
2. The list of tips is adapted from M. Hermann and G. A. Levey, Media presentations, in *Communicating as Professionals,* ed. R. Chernoff (Chicago: American Dietetic Association, 1994), pp. 35–42; and M. L. Chin and J. Horbiak, Pursuing the potential of the media, in *The Competitive Edge: Advanced Marketing for Dietetics Professionals,* 2nd ed. (Chicago: American Dietetic Association, 1995), pp. 95–102.
3. B. McManamon and N. Pazder, Pitching Your Ideas to the Media, *Journal of the American Dietetic Association* 100 (2000): 1451–53.
4. M. Hermann and G. Levey, 1994.
5. American Dietetic Association, *Working with the Media: A Handbook for Members of the American Dietetic Association.* Available at www.eatright.org.
6. The discussion about issuing a news release is from American Dietetic Association, *Working with the Media.*
7. The discussion about writing a letter to the editor is adapted from American Dietetic Association, *Working with the Media.*
8. U.S. Department of Health and Human Services, *Building Media Skills for Better Nutrition* (Atlanta, GA: Centers for Disease Control and Prevention, 1994), pp. 6–29.

Chapter *Eight*

Addressing the Obesity Epidemic: An Issue for Public Health Policy

Deanna M. Hoelscher, PhD, RD, LD, CNS, and Christine McCullum-Gómez, PhD, RD, LD

Learning *Objectives*

After you have read and studied this chapter, you will be able to:

- Define the terms *obesity* and *overweight* as they apply to adults.
- Define the terms *overweight* and *at risk for overweight* as they apply to children.
- Describe the epidemiology of obesity and overweight among adults and children.
- Explain how to assess and survey obesity and overweight in the population.
- List and discuss determinants of obesity and overweight.
- Discuss various interventions and intervention strategies for the prevention and treatment of obesity and overweight among adults and children.
- Describe potential public health strategies to prevent obesity, including examples of current and proposed policies and legislation.

This chapter addresses such issues as the influence of socioeconomic and psychological factors on food and nutrition behavior, assessment of nutritional health risks, public policy, food and nutrition laws and regulations, and current information technologies, which are Commission on Accreditation for Dietetics Education (CADE) *Foundation Knowledge and Skills* requirements for dietetics education.

Chapter *Outline*

Something To Think About...

Here is a great irony of 21st-century global public health: While many hundreds of millions of people lack adequate food as a result of economic inequities, political corruption, or warfare, many hundreds of millions more are overweight to the point of increased risk for diet-related chronic diseases. Obesity is a worldwide phenomenon, affecting children as well as adults and forcing all but the poorest countries to divert scarce resources away from food security to take care of people with preventable heart disease and diabetes.

– Marion Nestle, *Science*, **February 7, 2003**

Introduction

During the past 15 years, obesity has emerged as a significant public health problem because of its increasing prevalence, as well as the morbidity and mortality attributed to it, both in the United States and in other countries.[1] This increase has been noted in both adults and children.[2] Although the development of obesity is a clinical problem at the individual level, the increasing morbidity and mortality attributed to the metabolic abnormalities associated with excess body weight, as well as the increased health care costs, ultimately affect all of the population, making it a significant public health issue.[3]

Why are overweight and obesity so prevalent now? No one underlying cause has been determined, but because it takes extended periods of time to change the genetic makeup of an organism or population, it is not likely that the increased rates of obesity are due only to heritable factors.[4] More likely, the rapid increases in obesity over the past three decades are due primarily to societal and environmental factors.[5] Factors in the environment that may contribute to the obesity epidemic include increased caloric intake due to technological innovations in food production and transportation; other technological changes; increased portion sizes; increased consumption of foods and meals away from homes; urban sprawl and other changes in the built environment; and poverty. Because many factors contribute to obesity, the solutions must also be multifaceted and will probably involve both societal and environmental changes. Since obesity is linked with chronic diseases, such as Type 2 diabetes, cardiovascular disease, and some forms of cancer, it can be projected that the increases in rates of obesity will lead to increased death and disability.[6] Currently, the epidemic of obesity is among the most important public health challenges in the United States.

Body mass index (BMI) An index of a person's weight in relation to height that correlates with total body fat content. BMI = (weight in kg)/[(height in m)^2], where kg = kilogram, and m = meter.

Conversions:

1 kilogram = 2.2 pounds
1 inch = 2.54 centimeters
100 centimeters = 1 meter

Body mass index can be estimated using pounds and inches with the following equation:

[Weight in pounds/(height in inches × height in inches)] × 703

Defining Obesity and Overweight

The terms *obesity* and *overweight* both refer to the accumulation of excess adipose tissue; obesity is the more severe form.[7] Although many measures of excess body fat exist, the most common criterion for screening and monitoring at the population level is the **body mass index (BMI),** with a BMI of 25.0 to 29.9 defined as *overweight* and a BMI of 30.0 or more defined as *obese* for adults. Obesity can be further divided into categories, such as *morbidly obese* (see Table 8-1). A BMI of 30 translates into about 30 extra pounds for an adult. For example, a woman 5 feet 6 inches tall who weighs 186 pounds would be considered obese.

TABLE 8-1 Classification of Overweight and Obesity by Body Mass Index (BMI) in Adults

CLASSIFICATION	OBESITY CLASS	BMI (kg/m^2)
Underweight		<18.5
Normal		18.5–24.9
Overweight		25–29.9
Obese	I	30.0–34.9
	II	35–39.9
Extreme obesity	III	≥40

Source: National Heart, Lung, and Blood Institute Expert Panel on the Identification, Evaluation, and Treatment of Overweight and Obesity in Adults, Executive summary of the clinical guidelines on the identification, evaluation, and treatment of overweight and obesity in adults. *Journal of the American Dietetic Association* 98 (1998): 1178–1191.

Central obesity Characterized by an "apple-shaped" body with large fat stores around the abdomen; a strong risk factor for Type 2 diabetes, heart disease, and other problems. Central fat is also known as *android fat* to distinguish it from fat deposition on thighs and hips (*gynoid, or peripheral, fat*).

Waist circumference A measure used to assess abdominal (visceral) fat. Substantially increased risk of obesity-related health problems is associated with waist circumference measures as follow:

Men: > 40 inches
Women: > 35 inches

Weight Classifications in Children

Overweight: BMI-for-age ≥ 95th percentile

At Risk for Overweight: BMI-for-age = 85th to < 95th percentile

Unlike BMI values for adults, BMI values for children and adolescents are gender- and age-specific and can be calculated using growth charts for children and adolescents from 2 to 20 years old. The CDC Growth Charts are available at www.cdc.gov/growthcharts/.

Other measures of obesity exist, but most are expensive to use and not practical for large population-based studies. Although BMI has been shown to correlate highly with body fat, the measure may be inaccurate for persons with large muscle mass (such as weight lifters) or for those with low muscle mass, such as the elderly. Body fat distribution is also important, because **central obesity** (fat around the abdomen) is more highly associated with metabolic disturbances and health problems. Excessive fat is indicated by **waist circumference,** with measures of greater than 40 inches for men and greater than 35 inches for women considered abnormal.[8]

In children, obesity is defined using different terms, because body fat deposition changes during different developmental phases and varies by gender. The term analogous to adult obesity is **overweight,** which is defined as a BMI at or above the CDC growth chart criterion of 95th percentile based on gender and age standards (see Table 8-2).[9] **At risk of overweight** is similar, in children, to the designation *overweight* in adults; it reflects a BMI above the CDC growth chart criterion of 85th percentile, but lower than the 95th percentile based on gender and age standards. For a fourth-grade child (9 years of age), overweight is approximately 10 pounds over ideal weight, whereas for an eighth-grade child (13 years of age), overweight is approximately 20 pounds over ideal weight.*[10]

Epidemiology of Obesity and Overweight

In the United States, data for obesity are generally obtained from two national surveys: (1) the *National Health and Nutrition Examination Survey (NHANES)*, and (2) the *Behavioral Risk Factor Surveillance System (BRFSS)*. NHANES is conducted by the National Center for Health Statistics and involves a nationally representative sample.[11] For NHANES data collection, traveling measurement trailers obtain subjective or self-report information about dietary and physical activity behaviors, as well as objective measures on body size. Data collection used to be conducted periodically, but since 1999 data have been collected annually. NHANES staff members measure and weigh participants, so this is considered an *objective* measurement. Other biological and behavioral data are collected for NHANES, including blood measures for cholesterol, glucose, and B vitamins; dietary recalls for intake data; physical activity information; other anthropometric data, such as waist circumference; and demographic data such as gender, age, and household size (see Chapter 6).

*Criteria for child overweight and obesity have also been set by the International Obesity Task Force (IOTF). For the IOTF criteria, excess weight in children and adolescents may be classified as obese and overweight. These values differ slightly from the values on the CDC growth charts, because they are based on reference standards from several countries, including the United States, whereas the CDC National Center for Health Statistics (NCHS) growth charts are based on U.S. data only.

TABLE 8-2 Classification of Overweight in Children and Adolescents by Body Mass Index (BMI)

CLASSIFICATION	BMI (kg/m²)
Underweight	BMI-for-age < 5th percentile
At Risk of Overweight	BMI-for-age 85th percentile to < 95th percentile
Overweight	BMI-for-age ≥ 95th percentile

Source: CDC, 2004.

Data from NHANES show that the prevalence of overweight and obesity in the United States has increased substantially in the last two decades.[12] By 2001–2002, the age-adjusted prevalence of adult obesity was 30.6 percent compared with 22.9 percent in 1988–1994 (NHANES III). The percentage of overweight adults also increased during this time; 65.7 percent of U.S. adults were overweight in 2001–2002. In 1999–2002, the prevalence of overweight was 68.8 percent and 61.6 percent for men and women over age 20, respectively, while the prevalence of obesity was 27.6 percent and 33.2 percent for men and women. In women, the prevalence of overweight and obesity is higher in non-Hispanic black and Mexican-American populations compared to non-Hispanic white populations; in men, non-Hispanic black men have a lower prevalence of overweight than non-Hispanic white and Mexican-American populations, although there are no significant differences in the prevalence of obesity among racial/ethnic groups in males. Not only has the prevalence of overweight and obesity increased, but the prevalence of people who are at extreme obesity levels (BMI ≥40) has also increased; current data suggest that the prevalence of extreme (or morbid) obesity is 4.9 percent, with the highest rates among non-Hispanic black women (13.5 percent).[13]

The increasing prevalence of overweight has been noted in children as well. Child overweight has increased dramatically in the United States, from 5 to 6 percent in the 1970s and 1980s to 16 percent in 1999–2002 among children aged 6 to 19, and 10.3 percent among children aged 2 to 5 years old (see Figure 8-1).[14] These increases are greater in certain ethnic/racial groups (Hispanics and African Americans), as well as in certain regions of the country (the South).[15] As in adults, excess body weight in children is associated with increased morbidity and mortality and psychological disorders, as well as with increased health care costs.[16]

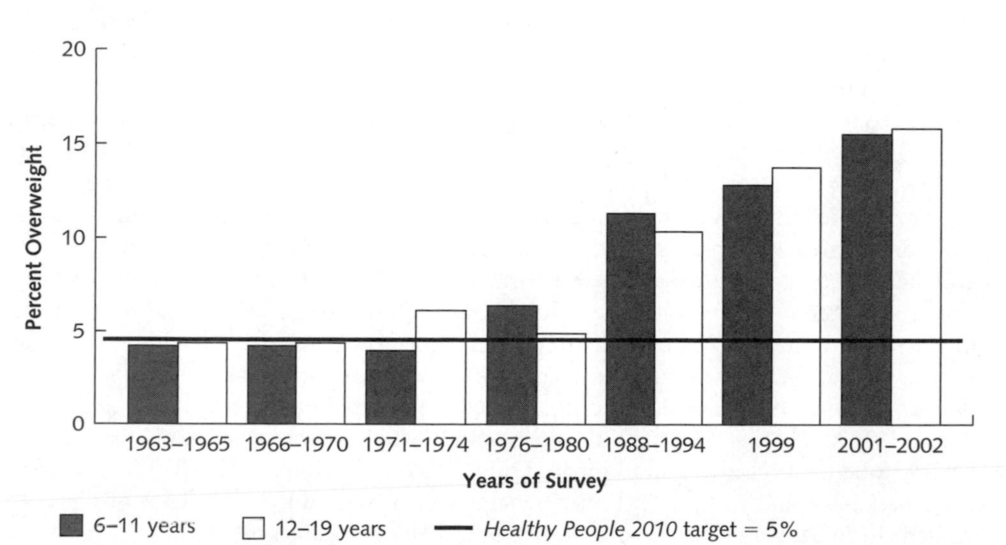

FIGURE 8-1 Trends in Overweight Among Children and Adolescents

Source: National Center for Health Statistics, *Health U.S., 2003*; and A. A. Hedley and coauthors, Prevalence of overweight and obesity among U.S. children, adolescents, and adults, 1999–2002, *Journal of the American Medical Association* 291 (2004): 2840–50.

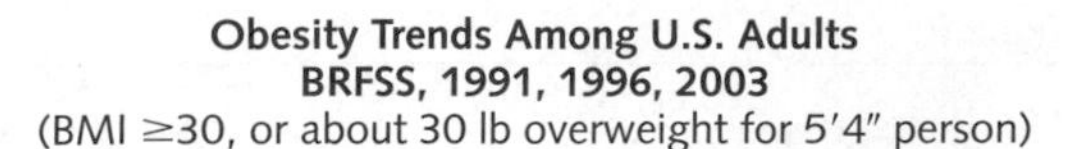

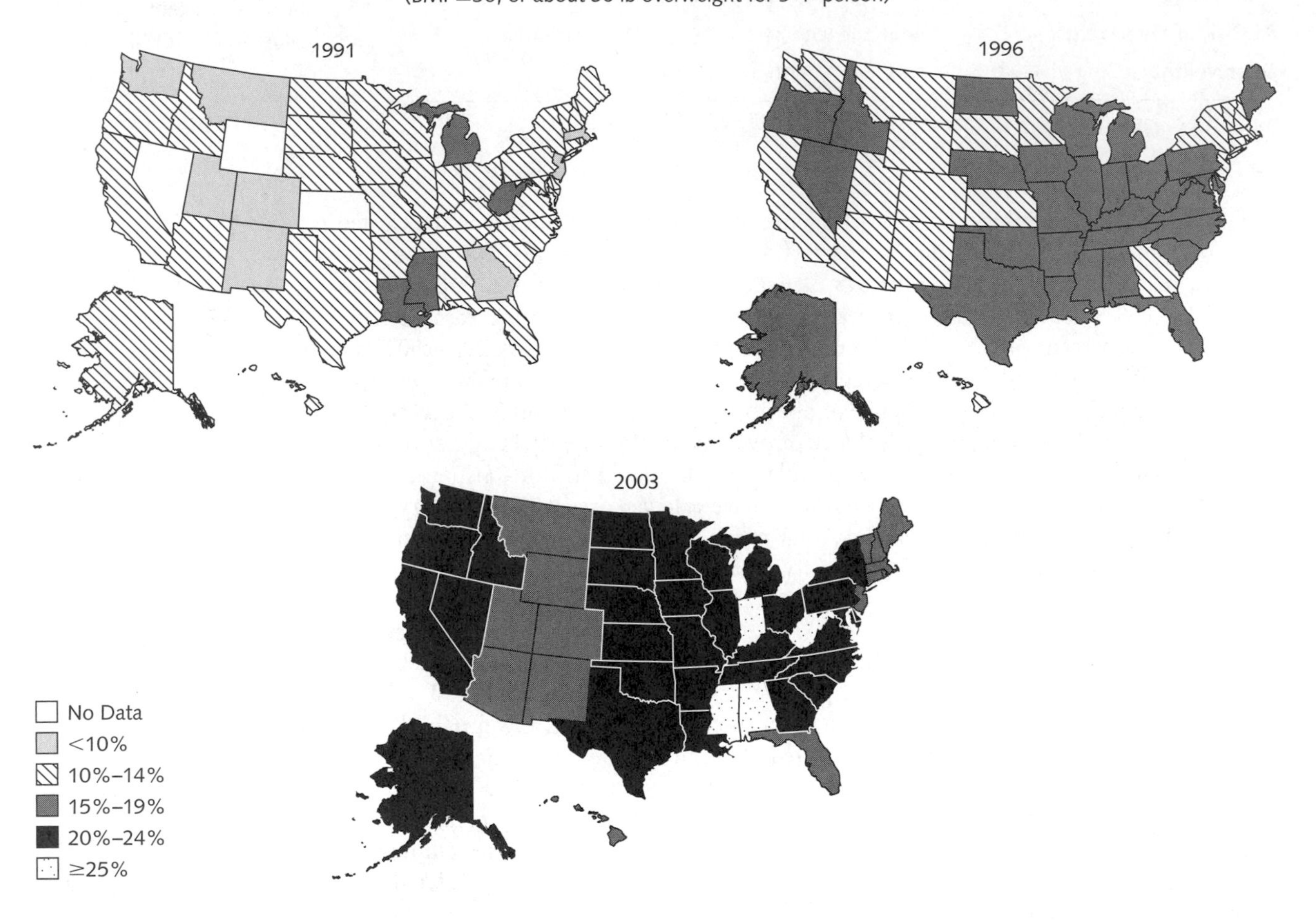

FIGURE 8-2 Prevalence of Obesity by State, 1991, 1996, and 2003

Source: CDC, Obesity Trends Among Adults, 1985–2003; and A. H. Mokhad and coauthors, *Journal of the American Medical Association* 289 (2003): 1.

For the BRFSS, data are collected from individual states through a random-digit dial telephone interview with adults aged 18 and older. Thus, data on height and weight collected for the BRFSS constitute a *subjective* measurement, in which participants report their own measurements without objective confirmation. Weight is often underreported in self-reports, so it can be assumed that the actual prevalence of overweight and obesity obtained from BRFSS data is higher than the reported values. Because the data collected for the BRFSS are representative at the state level, trends in overweight and obesity among adults can be obtained for each of the states. As Figure 8-2 shows, the increases in obesity among U.S. adults have been significant since 1991. The most recent data (from 2003) showed several states (Mississippi, Alabama, West Virginia, Indiana) with obesity prevalence rates above 25 percent.[17] The prevalence of adult obesity in the states in 2002 ranged from a high of 27.6 percent in West Virginia to a low of 16.5 percent in Colorado.[18] In addition to state prevalence of obesity, several large metropolitan areas are also assessed during BRFSS data collection; these regions are called Selected Metropolitan/Micropolitan Area Risk Trends (SMART) areas.[19] Of these, the areas with the highest prevalence of obesity include Charleston, West Virginia, and Toledo, Ohio. Data collection for the BRFSS includes information on diet and physical activity behaviors, as well as the prevalence of obesity-related chronic diseases, such as diabetes (see page 180 in Chapter 6).

Although relatively good estimates of child overweight are available at the national level through NHANES, there are few monitoring systems at the state level. The *Youth Risk Behavior Surveillance System* (YRBSS) does provide the prevalence of youth BMI by state, but the data are self-reported and limited to high school students.[20] Of the 31 state health departments that reported data for high school students' overweight levels in the year 2003, 22 reported rates above 10 percent.[21] The highest overweight levels for high school students were in Mississippi (15.7 percent) and Tennessee (15.2 percent). The lowest overweight levels for high school students were in Utah (7 percent), Wyoming (7.2 percent), and Idaho (7.4 percent). The median overweight level for high school students across the states was 11.1 percent. Another 14.5 percent were at risk of becoming overweight.[22]

Several states are conducting more extensive statewide sampling among different age groups to determine the prevalence of child overweight by grade or age for the state. For example, Texas conducted a statewide survey of children in grades 4, 8, and 11.[23] This survey, the School Physical Activity and Nutrition (SPAN) project, selected a random sample of children from Texas public schools and obtained measured height and weight to calculate BMI, as well as self-reported information on diet and physical activity behaviors. Children in Texas had a higher prevalence of overweight compared to national data, with the highest prevalence rates among fourth-grade children (22 percent), Hispanic/Latino boys in eighth grade (32.6 percent), and fourth-grade African-American girls (30.8 percent).

The state of Arkansas recently conducted a statewide surveillance of BMI among children at all grades through school nurses.[24] Overall, 19 percent of school-age children in Arkansas were overweight, and 18 percent were at risk for overweight. Hispanic children had a higher prevalence of overweight (20 percent) than African-American and Caucasian children (17 percent for both). The highest prevalence of overweight by grade was among fourth- through eighth-grade children (23 percent). In general, the prevalence of overweight among the children increased from pre-kindergarten through fourth grade, reached a plateau through eighth grade, and then decreased through twelfth grade.

The CDC and states do not survey information about the general population of preschool-age children. Instead, the health trends of children younger than age 5 are often derived from CDC's Pediatric Nutrition Surveillance System (PedNSS), which monitors the nutritional status of low-income children in federally funded maternal and child health programs, such as the Special Supplemental Nutrition Program for Women, Infants, and Children (WIC). Using CDC PedNSS data from 1989, 1994, and 2000, the number of states ($n = 30$) that reported a child overweight prevalence of more than 10 percent increased from 11 in 1989 to 28 in 2000. Trend analyses showed significant increases in preschool child overweight in 30 states and decreases in underweight in 26 states, although no geographical trend was apparent.[25] Although it might be argued that participation in the WIC program contributed to the increase in overweight, analyses of NHANES III data documented insignificant differences in weight status between WIC participants and nonparticipants.[26] Furthermore, using data from CDC's 2004 PedNSS, researchers at the *Trust for America's Health* reported that of the 38 states reporting data, 33 states and Washington, D.C., had an overweight prevalence of 10 percent or more among low-income, young children (aged 2 to 5 years old). Six states reported a preschool child overweight prevalence of over 15 percent, with an average of 14.3 percent overweight across states.[27]

Medical and Social Costs of Obesity

The Surgeon General's Report (2001) estimated the total economic burden of obesity to be $117 billion in 2000.[28] Annual U.S. medical expenditures attributable to obesity have been estimated to be $75 billion in 2003 dollars, with approximately half of these expenditures being financed by government and, ultimately, the taxpayers through Medicare

TABLE 8-3 Problems Associated with Obesity in Adults and Children

CHILDREN	ADULTS
Accidents	Abdominal hernias
Complications with surgical procedures	Accidents
Decreased quality of life	Certain cancers: colon, rectum, prostate, breast, uterus, cervical, ovarian
Depression	Complications during pregnancy
Difficulties with pubertal development	Complications with surgical procedures
Gallbladder and liver disease	Decreased longevity
High blood cholesterol levels	Decreased quality of life
Hormonal imbalances	Depression
Hypertension	Fertility problems
Injury to weight-bearing joints	Gallbladder and liver disease
Kidney abnormalities	Gout
Metabolic syndrome	Heart disease
Poor self-esteem	High blood cholesterol levels
Respiratory problems	Hormonal imbalances
Sleep disturbances	Hypertension
Type 2 diabetes	Injury to weight-bearing joints
	Kidney abnormalities
	Metabolic syndrome
	Osteoarthritis
	Poor self-esteem
	Respiratory problems
	Sleep disturbances
	Type 2 diabetes
	Varicose veins

Source: Adapted from A. Must and coauthors, The disease burden associated with overweight and obesity, *Journal of the American Medical Association* 282 (1999): 1523–1529; A. Must and R. S. Strauss, Risks and consequences of childhood and adolescent obesity, *International Journal of Obesity & Related Metabolic Disorders,* 23 (1999): S2–S11; and M. Boyle and S. L. Anderson, *Personal Nutrition,* 5th ed. (Belmont, CA: Wadsworth/Thomson Publishing, 2004), p. 261.

and Medicaid.[29] State expenditures for obesity vary, ranging from $87 million in Wyoming to $1.7 billion in California.[30] Compared to people of normal weight, obese people spent approximately 37 percent more on health care costs in 2001.[31] Of this increase, 38 percent was related to spending for diabetes, 22 percent was related to spending for hyperlipidemia, and 41 percent was related to spending for heart disease. These higher costs, along with the increased prevalence of obesity, account for 27 percent of the growth in health expenditures from 1987 to 2001. Costs are also associated with obesity and obesity-related diseases in children and adolescents. For children, obesity-associated hospital costs per year, based on 2001 constant dollars, were $35 million during 1979–1981 and increased to $127 million during 1997–1999.[32]

Obesity is costly to society because it is associated with chronic diseases, including cardiovascular disease, Type 2 diabetes, hypertension, stroke, dyslipidemia, osteoarthritis, selected cancers, gallbladder disease, sleep-breathing disorders, musculoskeletal disorders, and all-cause mortality (see Table 8-3).[33] Obesity is associated with increased blood lipid levels, increased blood glucose levels, and hypertension; together, these risk factors can be characterized as **metabolic syndrome.**[34] The prevalence of metabolic syndrome (age-adjusted) in the United States was 23.7 percent based on results from NHANES III, and this prevalence is higher among Mexican Americans.[35] Recent data show that metabolic syndrome can also be found among youth, especially those who are overweight.[36]

Metabolic syndrome
A syndrome associated with development of Type 2 diabetes and cardiovascular disease and is defined by abnormal values of three of the following indicators: waist circumference, serum triglycerides, HDL-cholesterol, blood glucose, and blood pressure:

- Central obesity as measured by waist circumference: Men $>$ 40 inches; Women $>$ 35 inches
- Fasting blood triglycerides greater than or equal to 150 mg/dL
- Blood HDL cholesterol: Men $<$ 40 mg/dL; Women $<$ 50 mg/dL
- Blood pressure greater than or equal to 130/85 mm Hg
- Fasting blood glucose greater than or equal to 110 mg/dL

Obesity and overweight can result in social costs as well, for both adults and children. Overall quality of life is often worse with increasing obesity, and obese people experience prejudice and discrimination.[37] Children who are overweight are often seen as lazy and unmotivated and can have lower self-esteem and depression.[38] Recent research has found that children who are overweight tend to have fewer friends and social networks; in addition, students who were overweight were more likely to be teased about their weight, and the teasing behavior was associated with depressive symptoms and low self-esteem.[39]

Determinants of Obesity

Determinants of obesity can be related to either dietary intake or physical activity or to both, and they can be *genetic, psychosocial, behavioral,* or *environmental.* In addition, determinants of obesity can vary, depending on other factors, such as age of the person. Although in the past, much emphasis was placed on individual risk factors for obesity, recent reviews and recommendations have focused on the contribution of environmental factors to the development of obesity.[40] In fact, researchers refer to the environment in which we live as an **obesogenic** (obesity-promoting) **environment** or as a **toxic environment** because of the decrease in opportunities for physical activity as well as the increasing supply of highly palatable, energy-dense, low-nutrient foods.[41] Some researchers argue that increasingly "obesogenic" environments are probably the main driving force behind the obesity epidemic.[42]

Excess weight accumulation occurs with an imbalance in energy, caused by either a surplus of energy intake (calories from food) or a lack of energy expenditure (physical activity).[43] Energy needs depend on the individual's **basal metabolic rate (BMR),** which is influenced by fat-free body mass (muscle and skeletal tissue), age, genetics, temperature, hormones, growth, and other factors. BMR is the largest influence on energy expenditure (60–65 percent), followed by volitional physical activity (25–35 percent) and the energy needed for digestion of food (the thermic effect of food, 5–10 percent). Theoretically, a small energy imbalance each day can result in a large weight gain over time. For example, consuming just one 12-oz sweetened soda per day in excess of a person's energy requirements would mean taking in 150 extra calories per day. *Provided that energy expenditures remain constant, these extra calories would result in a 16-pound weight gain over a year!*

GENETIC RISK FACTORS

Although some variation in body size has been shown to be related to genetic factors, no one gene has been associated with an obesity phenotype.[44] Instead, predisposition to obesity may be caused by a complex interaction among at least 250 obesity-associated genes.[45] With familial associations, it is sometimes difficult to disentangle the effects of the environment from the genetic influences. However, the variable response to diet that is found among humans is probably due to differences in genetic factors, especially for genes related to lipid and lipoprotein metabolism.[46] It is likely that several different genes contribute to the development of obesity and that the expression of these genes is strongly associated with or influenced by behavioral risk factors, such as diet and physical activity.* [47]

PSYCHOSOCIAL RISK FACTORS

Research has begun to highlight the association between certain psychosocial risk factors (such as depression or stress) and the risk of obesity.[48] A recent study showed that childhood depression was associated with an increased BMI in adulthood.[49] This association

*Genes that are likely to be associated with obesity include those for leptin, uncoupling protein, apolipoprotein B and E, resistin, and others.

Obesogenic Obesity-promoting.

Toxic environment Modern eating and exercise environment that contributes to obesity; includes the wide availability of food and technological innovations that contribute to increases in sedentary activity and decreases in physical activity.

Basal metabolic rate The rate at which the body expends energy to support its basal metabolism.

persisted after controlling for other socioeconomic factors. Other researchers found that "depressed adolescents are at increased risk for the development and persistence of obesity during adolescence. These researchers concluded that ". . . the shared biological and social determinants linking depressed mood and obesity may inform the prevention and treatment of both disorders."[50] It has been postulated that in adults, overconsumption of high-fat, high-carbohydrate comfort foods, such as macaroni and cheese or baked goods, may be used as a way to reduce chronic stress.[51] Individuals may eat these foods in an attempt to counter the body's response to stress and the resulting anxiety. These mechanisms may help explain how psychosocial factors (such as stress and depression) are contributing to obesity in society.[52]

BEHAVIORAL RISK FACTORS

There are various behavioral risk factors to take into consideration as determinants of obesity. These range from caloric intake to lethargy from a sedentary lifestyle.

Caloric Intake

Information on per capita energy consumption has shown that energy intake has increased during the past several decades.[53] Data from the annual loss-adjusted, per capita analysis of the U.S food supply in 2000 indicated an increase in food consumption of approximately 300 calories, or 12 percent, over the 1985 levels.[54] Most of the increase in calories is attributable to increases in consumption of grain products (largely refined grains), added fats, and added sugars. Other national food consumption data show similar trends, with adult women eating 335 more calories in 2000 than in 1971, and adult men consuming 168 more calories over the same time period. Most of the increase in calories noted in this study is from carbohydrates.[55]

Types of Food Consumed

A recent analysis of NHANES data (NHANES III, NHANES 1999–2000) found that the number-one contributor to energy intake was soft drinks, which supplied 7.1 percent of energy intake in 1999–2000.[56] Similarly, using data from the Nationwide Food Consumption Survey and the Continuing Survey of Food Intake by Individuals, investigators reported that between 1977 and 2001, total daily energy intake from soft drinks rose on average from 2.8 percent to 7 percent, representing nearly a tripling of calories, while energy intake from fruit drinks per person grew from 1.1 percent to 2.2 percent. Milk supplied 5 percent of energy for all age groups, down from 8 percent over the 24 years. Servings of sweetened beverages also increased for every age group, while servings of milk decreased for all.[57] An increased consumption of sweetened drinks, snacks, snack food items (such as potato chips), and foods consumed away from home by children and young adults has been shown to be associated with obesity and weight gain.[58]

Physical Activity

Data on trends in energy expenditure, or physical activity, are not as clear, largely because of lack of adequate surveillance. Although the dietary reference intakes (DRI) recommend increased physical activity, data suggest that neither adults nor children achieve suggested levels of physical activity, which may have a negative effect on weight status.[59] For example, a recent study found that failing to meet the guideline suggesting 60 minutes per day of moderate to vigorous physical activity was associated with overweight status for both adolescent girls and boys aged 11 to 15 years.[60] An analysis of data from the National Human Activity Pattern Survey revealed that leisure-time physical activity contributed only 5 percent of the population's total energy expenditure. Not counting sleep, the largest contributor to energy expenditure was "driving a car," followed by "office work" and "watching TV."[61] Women, the majority of racial/ethnic minority populations in the United States, and older adults have the greatest prevalence of leisure-time physical inactivity.[62]

Use of Television, Video Games, and Computers

Although both adults and children spend much time in sedentary activities, more data are available for children and adolescents. A recent nationally representative media study found that youth aged 2 to 18 spend an average of 5 hours and 29 minutes per day using various types of media.[63] In a recent longitudinal study, television viewing between ages 5 and 15 years remained a significant predictor of adult BMI, even after adjustment for childhood socioeconomic status.[64] In another study, however, the direct relationship between hours of television viewing and overweight disappeared after controlling for ethnicity and socioeconomic status.[65] Thus, although research supports the link between television viewing and obesity, it is likely that the relationship is complex and may be modified by other factors, such as the effects of the media on food choices.

ENVIRONMENTAL RISK FACTORS

Technological Innovations in Food Production and Transportation

Some researchers have hypothesized that an increase in calorie intake, not a decline in physical activity, is the major factor behind increased obesity in the United States, indicating that innovations in food production and transportation that have reduced the real cost of prepared foods are responsible for a major part of the obesity epidemic.[66] In particular, these researchers indict the mass production and preparation of ready-to-eat meals, which have replaced home food preparation. Central production of ready-to-eat foods can lower average food cost and eventually lead to reduction of the retail prices of these foods for the consumer.

Other Technological Changes

Using multiple datasets, a recent analysis concluded that about 40 percent of the increase in weight over the last few decades may be due to expansion of the food supply, partly through agricultural innovation, which leads to a decline in food prices.[67] Technological change may lead to an expanding food supply, putting downward pressure on the price of food, which in turn stimulates consumption at the consumer level. If agricultural technology is determined to be a major factor driving the trend toward increases in obesity, then economic incentives, rather than nutrition education alone, may be important.[68]

Portion Sizes

Food portion sizes have increased over time in the United States. This trend toward larger marketplace portions has occurred in parallel with rising rates of obesity.[69] Portion sizes vary by food source, the largest portions being consumed at fast-food establishments.[70] Larger portion sizes encourage people to eat more, although consumers may not always understand this intuitive connection.[71] As one nutrition professor explained, "*Many people seem to view a soft drink as a soft drink, no matter how big it is. When I explain that a 64 ounce soft drink container could provide as much as 800 kcal, audiences gasp.*"[72]

Eating Away from Home and Consumption of "Fast Foods"

An increase in the per capita number of restaurants may also play a role in the rising rates of obesity. A recent economic analysis of adult obesity concluded that the per capita number of restaurants has positive and significant effects on BMI and the probability of being obese.[73] The study's authors also noted that growth in the per capita number of restaurants, especially fast-food restaurants, may be largely a response to the increasing scarcity and value of household or nonmarket time.

A 2004 study reported that prevalence of fast-food restaurants in the United States was associated with obesity on a statewide basis.[74] Frequency of fast-food use may also be associated with increased risk of obesity, through its effects on promoting positive energy balance and weight gain.[75] For example, in 3,031 young (aged 18 to 30 years in 1985–86) black and white adults, researchers reported that after adjustment for lifestyle factors, baseline fast-food frequency was directly associated with changes in body weight in both black and white persons.[76] A healthful lifestyle score was strongly inversely associated with fast-food intake in white but not black persons. A recent geographical analysis found that predominantly black neighborhoods have 2.4 fast-food restaurants per square mile, compared to 1.5 such restaurants in predominantly white neighborhoods.[77] These researchers concluded that "the link between fast-food restaurants and black and low-income neighborhoods may contribute to the understanding of environmental causes of the obesity epidemic in these populations."[78]

Fast food is ubiquitous in society, even in public schools and hospitals.[79] Fast-food consumption in children has increased from 2 percent of total calories in 1977–78 to 10 percent of calories in 1994–96.[80] Fast foods and other unhealthful foods such as soda, candy, savory snacks, and baked goods are among the most frequently advertised products on television, and children are often the target audience for these advertisements.[81] Young children are especially vulnerable to commercial promotion because they lack the skills to understand the difference between information and advertising.[82] A national household survey reported that when socioeconomic and demographic variables are controlled, increased fast-food consumption by children was associated with boys, increased age, increased household incomes, non-Hispanic black race/ethnicity, and residing in the southern United States.[83] Children who ate fast food consumed more total energy and poorer diet quality (more added sugars, more sugar-sweetened beverages, less milk, and fewer fruits and nonstarchy vegetables) on days with fast food than on days without it.

Maternal Employment

Using matched mother–child data from the National Longitudinal Survey of Youth, investigators reported that a 10-hour increase in average hours that the mother works per week increases the overall probability that a child is overweight by 0.5 to 1 percentage point.[84] Children of more highly educated mothers are significantly more likely to be overweight if their mothers work more hours per week. For mothers in the highest socioeconomic status category, a 10-hour increase in average hours worked per week since a child's birth increases by 1.3 percentage points the likelihood that the child will be overweight. However, when controlling for race/ethnicity (black, Hispanic and white), the authors found that maternal employment significantly predicts higher weight only for white children. Further research is needed to investigate the association between maternal employment and childhood overweight, and factors that affect this relationship.

Mixed land use A layout of land that allows multiple types of use together, such as commercial use and residential use. This is in contrast to single-use zoning, which allows land use for only one specific purpose.

Walkability A neighborhood characteristic defined by residential density, mixed land use, and street connectivity.

Urban Sprawl and the Built Environment

Urban sprawl has been defined as "an overall pattern of development across a metropolitan area where large percentages of the population live in lower-density residential areas."[85] Various investigators have found an association between obesity and urban sprawl as well as other aspects of the built environment, such as **mixed land use** patterns and **"walkability,"** which is defined by residential density, mixed land use, and street connectivity.[86] The **"built environment"** encompasses a variety of community design elements such as street layout, zoning, transportation options, stairs, public and green spaces (walking paths, parks/recreation areas, playgrounds, community gardens), and business areas.[87]

The effect of the built environment on overnutrition in industrialized countries is only now beginning to be recognized.[88] One group of researchers found that after controlling for individual factors such as gender, age, race/ethnicity, and education, urban sprawl was associated with leisure-time walking, obesity, and hypertension, but not with overall physical activity, diabetes, or coronary heart disease.[89] Other researchers have reported that residents of high-walkability neighborhoods had a lower prevalence of obesity (adjusted for individual demographics) than residents of low-walkability neighborhoods.[90] Likewise, each additional hour spent in a car per day has been associated with a 6 percent increase in the likelihood of obesity.[91]

Poverty

For more about the paradox of poverty and obesity, see Chapter 5.

The rates of obesity in the United States follow a socioeconomic gradient, such that the burden of disease falls disproportionately on people with limited resources, racial/ethnic minorities, and the poor.[92] Individuals who have not gone to college, women with low incomes, single mothers, and men living in rural areas are significantly more likely to be overweight.[93] And although the prevalence of overweight and obesity in the United States has continued to increase in both sexes, at all ages and educational levels, and in all races, the highest prevalence of overweight and obesity occurs among low-income groups, women, and ethnic/racial minority groups.[94]

Some experts argue that the association between obesity and poverty may be mediated, in part, by the low cost of energy-dense foods—that is, foods composed of refined grains, added sugars, or fats, which may in turn promote overconsumption, and that consumption of these foods may be reinforced by the high palatability of sugar and fat.[95] Foods that are energy-dense (high in calories per serving) and highly palatable are associated with diminished feelings of fullness and overconsumption of fats and sweets, which lead to higher energy intakes.[96] On the other hand, foods that are high in fiber, with a high water content, promote a feeling of satiation, which leads to a reduced energy intake.[97] The energy density of foods is related to water content, so foods high in fiber and water content, such as raw fruits and vegetables, have a lower energy density than foods such as potato chips, chocolate, and doughnuts.[98] In fact, eating more fruit has been shown to be associated with a lower BMI.[99]

Various studies suggest that low-income households, particularly those located in rural areas and poor central cities, have less access to reasonably priced, high-quality food than other households.[100] For example, investigators have reported a lower prevalence of independently owned grocery stores in low-wealth and predominantly black neighborhoods, as well as a greater proportion of households without access to private transportation in these neighborhoods.[101] Another study found that after controlling for confounding variables, easy access to supermarket shopping was associated with increased household use of fruits in a nationally representative sample of participants who were enrolled in the Food Stamp program.[102] Furthermore, distance from home to food store was inversely associated with fruit use by households. Other researchers have reported that availability of more healthful food products in grocery stores was found to be associated with increased consumption of more healthful food products by individuals living near these stores.[103] In addition, another study found that the presence of a supermarket within a census tract was associated with a 32 percent increase in fruit and vegetable intake, compared with neighborhoods without supermarkets.[104]

Built environment The built environment encompasses a variety of community design elements such as street layout, zoning, transportation options, stairs, public and green spaces, and business areas.

Obesity Prevention and Treatment Interventions

In public health applications, interventions that address body weight are often preventive, rather than treatment-oriented. The goal of an *obesity prevention* program is to maintain a stable weight and not increase body size over time, in contrast to an *obesity treatment* program, in which the primary goal is to lose weight over time. Current recommendations for obesity treatment range from lifestyle therapy, which includes dietary therapy (low-calorie diets, very-low-calorie diets, vegetarian diets, and other regimens designed to lower energy intake), increases in physical activity, and behavioral therapy (for instance, use of behavioral strategies such as goal setting) to clinical therapies such as pharmacotherapy and weight-loss surgery. Generally, pharmacotherapy is not recommended unless patients are obese (BMI $\geq$30) with no related risk factors or have a BMI *greater than* 27 with related risk factors, such as hypertension or Type 2 diabetes. Weight-loss surgery, which includes gastric bypass procedures, is recommended for patients with severe obesity: BMI of 40 or greater, or BMI between 35 and 40 with related risk factors such as hypertension or Type 2 diabetes.[105] Public health approaches target lifestyle therapies, which are less invasive and more appropriate for population-level interventions. Other public health intervention approaches can include screening for high-risk patients, such as those with extreme obesity, and subsequent referral for clinical treatment.

For an effective intervention in the public health setting, the program has to have significant effectiveness in clinical settings, broad reach in the population of interest, and consistent implementation in real-life settings.[106] Thus, most obesity prevention programs are derived from initial clinical work and are oriented around a site where people congregate, such as churches, worksites, and community settings and schools. Because effectiveness and reach can vary depending on age, interventions can also be divided into adult-based programs and programs for children and adolescents.

The recidivism rate for obesity is high, and a significant number of patients regain weight within 3–5 years after the end of the intervention.[107] In addition, there are few community interventions that have produced significant decreases in body weight or BMI, either in adults or in children. The National Weight Control Registry (NWCR) is a study that examines the habits of people who have lost 30 or more pounds and kept it off for at least a year.[108] As part of their weight-loss plan, most registry participants had made significant and consistent changes in dietary intake and physical activity, were engaged in regular exercise (predominantly walking), and ate breakfast regularly.[109]

Adult Interventions

Most adult-based obesity interventions have revolved around clinical approaches to obesity treatment, as described above, and are not practical for population-based prevention approaches. Public health interventions tend to focus on lifestyle approaches, such as dietary changes and physical activity. In general, dietary interventions produce modest weight loss, and caloric content seems to be more important than macronutrient composition.[110] Physical activity can also be useful in weight loss, but the most effective interventions include both dietary alterations and increases in physical activity. Most community-based studies in adults have not been effective, and few studies have evaluated environmental interventions.[111] This trend is changing, though, largely thanks to increased research funding in this area and to initiatives such as the Robert Wood Johnson Foundation Active Living Research program.[112]

Many of the public health interventions for adults have been implemented through worksites.[113] In general, worksite health promotion programs have shown modest effects on weight in the short term, but these effects are generally limited.[114] Through a program on *Overweight and Obesity Control at Worksites*, the NIH is encouraging the development of new studies to test interventions that emphasize environmental approaches or a combination of environmental and individual approaches at worksites to promote weight control in adults. Environmental strategies include programs, policies, or organizational practices to influence health behaviors by changing the physical environment, such as increasing the availability of healthful food choices and facilities for physical activity, or by creating a supportive social climate.[115]

Child and Adolescent Interventions

In general, most public health interventions for overweight prevention in children and adolescents have been conducted through the schools.[116] Public health interventions that have been shown to be most effective among children tend to include a component on decreasing television viewing.[117] School-based programs that increase physical activity also appear to have a significant effect on reducing body size, especially when physical activity is for an hour or more per day.[118] For example, expanding physical education programs in elementary schools has been shown to be an effective intervention for combating obesity in elementary schools.[119] In the United Kingdom, a targeted, school-based education program produced a modest reduction in the number of carbonated drinks consumed, which was associated with a reduction in the number of overweight and obese children.[120] Programs conducted in middle schools and high schools tend to be more successful than those conducted in elementary schools, especially if the person implementing the program is not a teacher.[121] Only a small number of environmental interventions targeted toward children and adolescents have been reported, and few programs have been conducted in community settings outside the school.

Several child and adolescent interventions involve a family component. For small children, the parents are often "gatekeepers" for diet and physical activity, inasmuch as they control the types of foods in the house, the access to and availability of those foods, opportunities for activity or inactivity, and meal patterns. In addition, parents often serve as role models for their children. Although family is important for preschool and elementary school children, adolescents are often more interested in peer relationships and influences, so interventions for adolescents should include a peer component as well.[122]

Public Health Policy Options for Addressing the Global Obesity Epidemic

Obesity is a significant public health issue that will affect the entire population through increased health care costs, as well as increased morbidity and mortality. As Dr. William Dietz of the Centers for Disease Control and Prevention acknowledged in his statement before the Subcommittee on Public Health of the Committee on Health, Education, Labor, and Pensions for the U.S. Senate on May 21, 2002,

> Given the size of the population that we are trying to reach, we obviously cannot rely solely upon individual interventions that target one person at a time. Instead, the prevention of obesity will require coordinated policy and environmental changes that affect large populations simultaneously.[123]

Efforts to control obesity at the public policy level in the United States need improvement. A 2005 evaluation conducted by the University of Baltimore's Obesity Initiative assessed the efforts of individual states to control obesity in adults and children.[124] States were evaluated on legislation that targeted obesity initiatives, specifically (1) mandates on nutrition standards, (2) vending machine usage, (3) BMI measurements in schools, (4) recess and physical education, (5) obesity programs and education, (6) obesity research, (7) obesity treatment in health insurance, and (8) legislature-established obesity commissions. No state received an A, 11 states received a B, 23 received a C, 11 received a D, and 5 received an F rating for no action. The report card for policies targeted at child obesity included the first five initiatives listed above, because those are the mandates that specifically focus on children. For state efforts to control child obesity, one state (California) received an A, 15 states received a B, 21 received a C, 7 received a D, and 6 received an F. Clearly, much work remains to be done in state efforts to control obesity, both at the general population level and specifically for children.

It is possible to achieve significant health and medical savings in the population by investing in obesity prevention and reduction. But for this to occur, federal and state governments must recognize the impact that political decisions have had on the obesity epidemic and make a serious commitment to addressing this issue. The various policy options for addressing the obesity epidemic can be broken down into six different categories: obesity surveillance and monitoring efforts; awareness building, education, and research; regulating environments; private enforcement; pricing policies, such as subsidies and taxing; and societal-level solutions.

Many of the policy initiatives address children rather than adults, for a number of reasons. First, because obesity is so intractable, it is preferable and more cost-effective to focus on prevention. The rates of overweight are increasing in children as well as in adults, so prevention of obesity at a young age can potentially attenuate the development of chronic disease, which comes about over a long period of time. Food habits are still fairly malleable for children, so it makes sense to target policies and programs at this age group. Finally, it is easy to reach children, because the majority of children attend schools, and school programs often include physical activity and nutrition elements that can be targeted.[125]

Obesity Surveillance and Monitoring Efforts

Healthy People 2010 outlines a set of health objectives for the nation to achieve over the first decade of the twenty-first century.[126] As we noted in Chapter 1, the *Healthy People 2010* objectives for the nation have two overarching goals: (1) to increase quality and years of healthy life and (2) to eliminate health disparities. *Healthy People 2010* also identifies a wide range of public health priorities and specific measurable objectives. Three specific objectives from *Healthy People 2010*—Objectives 19.1, 19.2, and 19.3—are related to overweight and obesity in adults and children (see Table 8-4).

According to a report released by the nonprofit organization Trust for America's Health, 41 states and the District of Columbia have adult obesity levels that exceed 20 percent.[127] Furthermore, no state is currently on track to meet the goal of reducing the proportion of adults who are obese to the *Healthy People* goal of 15 percent or lower by 2010. Current YRBSS data also suggest that every state will fail to meet the national goal of reducing the proportion of children and adolescents who are overweight or obese to 5 percent or lower by 2010.[128] In its 2003 policy statement, the American Academy of Pediatrics (AAP) recommended BMI testing for children and adolescents to diagnose overweight and obesity.[129] The Institute of Medicine (IOM) has also recommended that health care professionals routinely track BMI in children and youth and offer appropriate counseling and guidance to children and their families.[130]

HEALTHY PEOPLE (HP) *2010* OBJECTIVE NUMBER	HP 2010 OBJECTIVE	BASELINE (Year)	CURRENT DATA	SOURCE OF DATA	HP 2010 TARGET
19–1	Increase the proportion of adults who are at a healthy weight.*	42% of adults aged 20 years and older in 1988–94	34.9% of adults aged 20 years and older in 1999–2002	NHANES, CDC, NCHS	60%
19–2	Reduce the proportion of adults who are obese.†	23% of adults aged 20 years and older in 1988–94	30.4% of adults aged 20 years and older in 1999–2002	NHANES, CDC, NCHS	15%
19–3	Reduce the proportion of children and adolescents who are overweight or obese.‡	11% of children 6–19 years old in 1988–94	16.0% of children 6–19 years old in 1999–2002	NHANES, CDC, NCHS	5%
		11% of children 6–11 years old in 1988–94	15.8% of children 6–11 years old in 1999–2002	NHANES, CDC, NCHS	5%
		11% of children 12–19 years old in 1988–94	16.1% of children 12–19 years old in 1999–2002	NHANES, CDC, NCHS	5%

*Healthy weight is defined as a BMI of 18.5 to 24.9.

†Obesity is defined as a BMI ≥ 30.0.

‡*Weight Classifications in Children:* Overweight in Children: BMI-for-age ≥ 95th percentile; At Risk for Overweight: BMI-for-age = 85th to 95th percentile.

Source: U.S. Department of Health and Human Services, *Healthy People 2010: Understanding and Improving Health,* 2nd ed. (Washington, D.C.: U.S. Government Printing Office), November 2000; and Public Health Service, U.S. Department of Health and Human Services, *HP 2010 Progress Review: Nutrition and Overweight,* January 21, 2004.

TABLE 8-4 ***Healthy People 2010*** **Objectives Compared to Current Data on Obesity and Overweight**

Recently, the state of Arkansas has set a national example of measuring students' BMI. In 2003, as part of a statewide multifaceted legislative initiative, Arkansas required that every public school student have a BMI assessment performed and reported confidentially to parents. The legislation also required schools to provide parents with information about the health risks their child could develop as a result of being overweight.[131]

Awareness Building, Education, and Research

The primary cabinet-level departments involved in federal obesity policy in the United States include the Department of Health and Human Services and the U.S. Department of Agriculture. The Federal Trade Commission, an independent agency, is also involved in obesity prevention efforts.

DEPARTMENT OF HEALTH AND HUMAN SERVICES (DHHS)

Most of the agencies and offices within DHHS are involved in obesity-related programs, including the Centers for Disease Control and Prevention, the Centers for Medicare and Medicaid Services, the Food and Drug Administration, the National Institutes of Health, the Health Resources and Services Administration, the Office of Women's Health, the Administration on Aging, the Head Start Bureau, and the Indian Health Service.[132] Details about programs in CDC and NIH follow, and Table 8-5 describes obesity-related programs and policies in other DHHS agencies and offices.

See Chapter 7 for organization charts of the Department of Health and Human Services and the Department of Agriculture.

TABLE 8-5 Obesity-Related Programs and Policies in Selected Agencies at the U.S. Department of Health and Human Services

Food and Drug Administration	*Obesity Working Group (OWG)*—A 2004 report titled "Counting Calories" focused on six components regarding caloric exchange as a means for controlling weight: Food labeling; Enforcement; Educational partnerships; Restaurants; Therapeutics; and Research.
Administration on Aging (AOA)	*You Can! Steps to Healthier Aging* program promotes physical activity and sound nutrition in elderly populations. *National Policy and Resource Center on Nutrition, Physical Activity, and Aging* at Florida International University promotes healthy aging. *Eat Better & Move More* program is a community-based initiative located in 10 communities across the United States.
Health Resources and Services Administration (HRSA)	The Maternal and Child Health Bureau (MCHB) coordinates several obesity-related programs, such as *Bright Futures in Practice: Physical Activity,* a series of exercise guidelines for children and adults, and *Bright Futures in Practice: Nutrition,* a guidebook that discusses health and nutrition. HRSA also supports the *National Adolescent Health Information Center (NAHIC)* at the University of California.
Head Start Bureau and the Indian Health Service (IHS)	Head Start has also teamed up with the IHS to develop the *Healthy Children, Healthy Families* program and the *Healthy Communities: A Focus on Diabetes and Obesity Prevention* program. Several tribal Head Start pilots have been selected for training and technical assistance and community-specific interventions designed to reduce the obesity rate in American Indian and Alaska Native populations.

Centers for Disease Control and Prevention (CDC)

The National Center for Chronic Disease Prevention and Health Promotion (NCCDPHP) at the CDC has been leading the agency's obesity-related prevention and health promotion efforts. Major federal-level initiatives targeted toward obesity prevention, such as the *Steps to a HealthierUS, Healthy Lifestyles and Disease Prevention (Small Steps),* and the *VERB* media campaign, are administered by NCCDPHP. The *Steps to a HealthierUS* focuses on community-based health initiatives related to obesity. The program, which directly funds efforts at the city and community levels, was launched in 2003 by former DHHS Secretary Tommy Thompson and is based on the goals outlined in the *President's HealthierUS Initiative.*[133] Communities receiving grants are required to address asthma, obesity, and diabetes prevention; to make special efforts to reach underserved populations; to share information and best practices with other communities; to encourage public–private partnerships; and to evaluate program outcomes against a series of goals.

Healthy Lifestyles and Disease Prevention (Small Steps) uses multimedia public service announcements, such as an interactive website, and television advertisements to try to further the goals of the *Steps to a HealthierUS.*[134] See the website at www.smallstep.gov.

VERB is a multi-ethnic media campaign based on social marketing principles and behavior change models with the goal of increasing and maintaining physical activity in youth aged 9 to 13 ("tweens").[135] *VERB* also targets parents to encourage promotion of physical activity. Although preliminary results of the *VERB* campaign are positive, it is too early to tell whether media can effectively increase physical activity among this population of youth.[136] Nonetheless, a recent IOM report noted that "a broad multimedia campaign focused on obesity prevention offers the best possibility of reaching a sizeable and broad audience, on a continuing basis, to generate support for policy changes and

TABLE 8-6 Examples of CDC-Funded State-Level Nutrition and Physical Activity Programs to Prevent Obesity and Other Chronic Diseases

STATE	WEBSITE
Colorado	www.cdphe.state.co.us/pp/COPAN/Obesity.html
Montana	www.dphhs.mt.gov/index.shtml
North Carolina	www.eatsmartmovemorenc.com/
Oregon	http://oregon.gov/DHS/ph/pan/index.shtml
Pennsylvania	www.panaonline.org
Texas	www.tdh.state.tx.us/phn/obesity-plan.pdf
Florida	www.doh.state.fl.us/Family/obesity/index.html
Kentucky	www.fitky.org
Maryland	www.fha.state.md.us/fha/cphs/npa/index.html
Rhode Island	www.health.ri.gov/disease/obesity/index.php
West Virginia	www.wvdhhr.org/bph/oehp/hp/obesity/default.htm

provide needed information to parents, children and youth."[137] CDC and its partners have also developed the *HHS Blueprint for Action on Breastfeeding,* which establishes a comprehensive national breastfeeding policy.[138] Breastfeeding may be protective against obesity and may increase the acceptability of fruits and vegetables among infants.[139]

The CDC also funds *State-Level Nutrition and Physical Activity Programs to Prevent Obesity and Other Chronic Diseases.* Targeted goals of the program include (1) increased consumption of fruits and vegetables, (2) increased physical activity, (3) promotion of breastfeeding, (4) reducing television watching, and (5) balancing energy intake and expenditure.[140] As part of the program, some states received funding to gather data, engage partners, and develop a state plan to address overweight and obesity; examples of websites for these programs and state plans are shown in Table 8-6.

Within NCCDPHP, divisions that implement and oversee obesity-prevention programs include the *Division of Adolescent and School Health (DASH)* and the *Division of Nutrition and Physical Activity (DNPA).* The most notable obesity-related program administered by DASH is the *Coordinated School Health Program (CSHP),* which promotes healthful behavior in schools by focusing on an integrated model involving eight components: health education, physical education, health services, nutrition services, counseling and social services, healthful school environments, health promotion for staff, and family and community involvement. (See the Programs in Action feature on page 264.) [141] DASH also developed and distributes the *School Health Index for Physical Activity and Healthy Eating,* a self-assessment guide for schools to measure progress in physical activity and nutrition initiatives against a series of benchmarks and goals.[142] DNPA supports a wide variety of obesity-related programs at the community level, which can be divided into four different categories: prevention, applied research, tracking of health behaviors, and health communication.

The CDC also funds two major research initiatives that encompass obesity prevention: the *Health Protection Research Initiative* and the *Prevention Research Centers.* The *Health Protection Research Initiative* is a new CDC program to generate research that can be used in outreach efforts to employers to inform them about the benefits of wellness programs and the cost-effectiveness of a healthy workplace.[143] The CDC also funds 33 *Prevention Research Centers,* which investigate ways to prevent and control chronic diseases, including obesity, at the community level.[144] Funding helps to support infrastructure and community-based research projects, especially those that target **translational research.**

Translational research The adaptation of more highly controlled, experimental health promotion interventions that are effective to less controlled, but more generalizable, community-based conditions.

National Institutes of Health (NIH)

NIH serves a dual function in the fight against obesity: It seeks to further obesity prevention awareness as well as working on research and treatment measures. The Obesity Research Task Force, founded in 2003, is an interagency committee that strives to increase funding for obesity-related research through interagency proposals and coordination. The Obesity Research Task Force developed the *Strategic Plan for NIH Obesity Research,* which calls for intensifying obesity efforts along several fronts, including[145]

1. Behavioral and environmental approaches to modifying lifestyle to prevent or treat obesity
2. Pharmacological, surgical, and other medical approaches to prevent or treat obesity effectively and safely
3. Examining the link between obesity and diseases such as Type 2 diabetes, heart disease, and certain cancers
4. Research on special populations at high risk for obesity, including children, ethnic minorities, women, and older adults
5. Translating basic science results into clinical research and then into community intervention studies
6. Disseminating research results to the public and to health professionals

NIH also oversees several different government institutes that administer programs focused on obesity-related education and research. Examples of obesity-related programs from several NIH agencies, including the National Heart, Lung and Blood Institute, the National Cancer Institute, the National Institute of Diabetes and Digestive and Kidney Diseases, and the National Institute of Environmental Health Sciences, are highlighted below.

The *National Heart, Lung, and Blood Institute (NHLBI)* introduced the *Obesity Education Initiative (OEI)* in January 1991 to help reduce the prevalence of overweight and obesity in people, while increasing physical activity. Obesity-related goals are furthered by a two-part strategy, one part focused on high-risk audiences and the other on the general population.[146] High-risk interventions include targeting specific ethnic groups with culturally appropriate messages, as exemplified in the *Pathways* outreach program targeted at American Indian and African-American communities. The general-population approach involves communicating messages through programs such as *Hearts N' Parks,* which encourages heart-healthy eating and exercise.[147]

The *National Cancer Institute (NCI)* promotes healthful lifestyles that lead to lower cancer incidences.[148] NCI directs a leading government nutritional public education campaign, *5 A Day for Better Health*, which promotes fruit and vegetable consumption as an essential component of a healthful lifestyle.[149] The program, started in 1991, has evolved into one of the most visible nutrition education campaigns in the United States.[150]

The Weight-control Information Network (WIN) is located within the *National Institute of Diabetes and Digestive and Kidney Diseases (NIDDK).* The WIN Network provides science-based materials on obesity, weight maintenance, and nutrition and is also involved in educational outreach to high-risk populations, including the *Sisters Together: Move More, Eat Better* program, which promotes healthful nutrition and exercise among women within targeted African-American communities.[151]

Recently, the *National Institute of Environmental Health Sciences* has begun to encourage research in two specific areas examining the link between obesity and the built environment. The first area deals with understanding the role of the built environment in causing and/or exacerbating obesity and related comorbidities; the second deals with developing, implementing, and evaluating prevention/intervention strategies that affect the built environment in order to reduce the prevalence of overweight and obesity.

Projects in this area seek to enhance our understanding of the roles played by regional planning, housing, transportation, media, access to healthful foods and availability of public spaces, including green spaces (community gardens, walking paths, and the like) as determinants of dietary practices and physical activity.[152]

UNITED STATES DEPARTMENT OF AGRICULTURE (USDA)

USDA is responsible for a range of food and nutrition programs that affect obesity via (1) nutritional advice and guidance, (2) food labeling regulations, (3) food and obesity education campaigns, (4) distribution of food products to schools, and (5) oversight and protection of the nation's agricultural and dairy markets. USDA's division of Food, Nutrition, and Consumer Services (FNCS), includes two departments that target obesity-related programs and policies: the Food and Nutrition Service (FNS) and the Center for Nutrition Policy and Promotion (CNPP).

The Food and Nutrition Service administers nutrition assistance programs to needy and eligible populations through food assistance, school lunch, and school-based educational programs.[153] FNS developed and disseminates *Team Nutrition*, a program that provides educational materials for children aged 4 to 18 and offers advice on how to maintain a healthful weight.[154] The *HealthierUS* School Challenge, which is an extension of the president's *HealthierUS* initiative, is designed to build upon the FNS *Team Nutrition* program. This initiative recognizes schools that achieve the goal of meeting voluntary nutrition and physical activity standards set by the FNS.[155]

Chapter 5 provides a detailed discussion of the USDA and DHHS food and nutrition assistance programs.

The Center for Nutrition Policy and Promotion (CNPP) develops nutrition education information and works to disseminate research findings via outreach materials to target populations.[156] CNPP developed the *USDA Healthy Eating Index (HEI)*, which is a tool for measuring overall diet quality that enables users to compare their diets with USDA recommendations in the Dietary Guidelines for Americans and the MyPyramid food guidance system. The HEI is an important tool for monitoring diet quality, given that research has indicated that a low HEI score is associated with overweight and obesity.[157] Additionally, along with FDA and other government agencies, CNPP is asking manufacturers to promote use of nutritional labels on food.[158]

FEDERAL TRADE COMMISSION (FTC)

The FTC has an important role to play in ensuring that the marketplace is receptive to healthful lifestyles and nutrition. More specifically, the FTC oversees (1) claims of health effects and labeling of food, (2) disclosure of caloric information, and (3) deceptive marketing of foods and food-related products.[159] The FTC has compiled a set of obesity-related consumer information publications to guide the general public in making better diet and health choices and in not falling for "too good to be true" claims on topics such as weight loss.[160] The FTC has also published an educational guide, *Red Flag,* to help media outlets avoid publicizing fraudulent weight-loss claims.[161] In order to deal with such claims, the FTC has relied primarily on federal district court complaints against companies accused of perpetrating fraud and has filed cases against the worst offenders.

See this chapter's Professional Focus feature for tips on how to evaluate popular weight-loss diets.

RECENT LEGISLATIVE EFFORTS

Multiple pieces of legislation have been introduced into the U.S. Congress with the goal of reducing the obesity epidemic through increased educational efforts. Table 8-7 details selected bills introduced during the 108th Congress in 2004. It is expected that similar bills will be introduced in subsequent legislative sessions. Details about legislation that

TABLE 8-7 Examples of Obesity-Related Legislation in the 108th Congress (2004)

BILL TITLE	DESCRIPTION OF BILL
Improved Nutrition and Physical Activity (IMPACT) Act (S.1172, HR 716)	• Provides grants to train health professionals to treat and prevent obesity • Authorizes block grants to state and local entities for obesity programs
Obesity Prevention Act (HR 2227)	• Requires physical activity, nutrition education, and availability of fruits, vegetables, whole grains, and low-fat dairy products in schools • Establishes a Commission on Obesity Treatment and Prevention
Prevention of Childhood Obesity Act (S.2894)	• Authorizes a grant program at CDC to provide funds (1) to school districts to promote nutrition and physical activity in schools and (2) to nonprofits for after-school programs
Obesity Reduction Act (S.2551)	• Establishes a Congressional Council to develop model obesity prevention plans for elementary and middle schools
Healthy Lifestyle and Prevention (HELP) America Act (S.2558)	• Restores authority of the secretary of agriculture to regulate the sale of junk foods in schools • Creates standards and incentives to provide for sidewalks, bike lanes, and intersections • Provides tax credits to businesses that offer employee health programs • Requires nutritional information on menus of chain restaurants • Restores the rule-making authority of the FTC to issue restrictions on advertising to children
Menu Education and Labeling Act (HR 3444)	• Amends the Federal Food, Drug and Cosmetic Act to ensure that consumers receive information about the nutritional content of restaurant foods
Child Nutrition and WIC Reauthorization Act (S.2097), now PL 108-265	• Establishes Local School Wellness Policies and funding efforts to provide technical assistance and best practices to schools and states. Policies include: Goals for nutrition education, physical activity, and other activities to promote student wellness, and nutrition guidelines for all foods available on each school campus
Personal Responsibility in Food Consumption Act (HR 339)	• Grants the food industry immunity from "claims of injury relating to a person's weight gain, obesity or any health condition associated with weight gain or obesity"
Commonsense Consumption Act (S.1428)	• Protects restaurants, processors, distributors, advertisers, and others from civil liability for an individual's weight gain or related health problems caused by the consumption of specific foods

includes mandates related to nutrition and/or physical activity can be found at the CDC legislative website or at the policy website maintained by the Center for Science in the Public Interest (see the list of Internet addresses at the end of this chapter).

Regulating Environments

Much emphasis has been placed on the "toxic environment" or "obesogenic" environment in the development of obesity. It is logical to conclude that regulation of environmental factors, such as food availability and opportunities for physical activity, can influence diet and exercise habits, which in turn lead to the decrease or increase of obesity rates in a population. From an economic perspective, increased rates of obesity lead to higher costs to society in a number of ways, including direct health care costs (for example, increase in Type 2 diabetes rates) or work productivity losses. Thus, because society bears some of the financial burden of the obesity problem, this is a population-level issue, which can justify population-level approaches, such as regulation.[162] Systematic change of environments has greater potential than education alone to affect overall diet and physical activity patterns in populations.[163] A number of proposed options for regulating food, school, and built environments are presented in the following sections.

THE FOOD ENVIRONMENT

Regulatory interventions that could help create more healthful food environments include mandatory food product labeling at restaurants and restrictions on food advertising. The Food and Drug Administration has the authority to regulate nutritional labeling of processed foods in the United States; however, the federal Food, Drug, and Cosmetic Act exempts restaurants, which means they are not required to disclose nutrition information. Exempting the restaurant industry from disclosing the content and nutritional value of food makes it difficult for consumers to estimate the energy content of the food they consume away from home.[164] Americans spend almost half of their total food budget (46 percent) on eating outside the home.[165] Thus, mandating point-of-sale nutrition information in restaurants would enable consumers to make more informed dietary decisions. It might also encourage restaurants to modify their ingredients and menus to provide a greater number of healthful food and beverage options.[166]

A recent Institute of Medicine (IOM) report, *Preventing Childhood Obesity: Health in the Balance,* noted that more than half of television advertisements directed at children promote foods and beverages such as candy, fast food, snack foods, soft drinks, and sweetened breakfast cereals that are high in calories and fat and are low in fiber and other essential nutrients.[167] And though the IOM report concluded that there is not enough evidence of food, beverage, and entertainment advertising's adverse impacts on children to call for a ban on all such advertising, it did issue several recommendations, including the development of guidelines for the advertising and marketing of foods, beverages, and sedentary entertainment directed at children and youth, with attention to product placement, promotion, and content.[168]

Many countries have mechanisms to regulate marketing to children, including regulations specific to marketing food.[169] For example, Sweden and Norway ban all marketing to children under the age of 12, and the province of Québec, Canada, bans marketing to children under 13.[170] The impact of such advertising bans has not been evaluated, with the exception of a 1990 study showing that Québec children were less likely to consume sugary cereals than their English-speaking Canadian counterparts who were exposed to cereal advertisements. Such bans, however, are difficult to evaluate because children are exposed to many different types of advertising and promotional techniques. Therefore, it is hard to tell whether advertising bans have played a role in improving children's diets.[171]

THE SCHOOL ENVIRONMENT

Because children spend a significant percentage of their formative years in school, a healthful school environment is an important venue for shaping a child's future eating and physical activity habits. A recent Institute of Medicine report recommends that "schools should provide a consistent environment that is conducive to healthful eating behaviors and regular physical activity."[172]

Food is typically available for sale in most schools in two ways: (1) the USDA-regulated national school lunch program (NSLP), breakfast program, and after school snack programs, and (2) "competitive foods," which include food sold from snack shops, school stores, and vending machines and in à la carte lines in the cafeteria, as well as through bake sales, fundraisers, and other school activities. State education departments receive subsidies from the USDA for school meals if the programs follow national nutritional

Chapter 11 provides details about the USDA child nutrition programs and discusses strategies for building healthful school environments and the USDA nutrition-related initiatives to meet the *HealthierUS* School Challenge.

guidelines and offer free or reduced-price meals to children from low-income households.[173] In 2001, the USDA issued a report to Congress highlighting concerns about the impact of the sale of competitive foods and of the accompanying decrease in student participation—and thus in funding—on the overall viability of the NSLP.[174] Another concern noted by the USDA in its report to Congress is that when children are taught in the classroom about good nutrition but are surrounded by snack vending machines, snack bars, school stores, and à la carte foods, a high percentage of which may be low-nutrient, energy-dense foods (such as sweetened soft drinks, other sugar-sweetened drinks, and fried potatoes), they receive a mixed message.[175] The school setting offers a unique opportunity to extend health information in the classroom to a "learning laboratory" setting in the environment, but mixed messages in the school environment can contradict nutrition education lessons in the classroom.[176] Consistent messages about nutrition and physical activity can be facilitated and reinforced through parental and community involvement via School Health Advisory Councils (SHACs) and through participation in school activities, parent–teacher organizations, and local board of education meetings.[177]

As we have noted, one of the biggest challenges with food choices offered in the school environment is the issue of "competitive foods." Federal regulations restrict the sale during meal times of only a small subset of competitive foods. These include "foods of minimal nutritional value," such as hard candy, chewing gum, and soft drinks.[178] However, federal guidelines do not prohibit selling these foods *outside of the cafeteria area* at any time during the day. Other competitive foods that are not regulated by the federal government include hamburgers, potato chips, french fries, and pizza, which are often sold in competition with the NSLP.[179]

National efforts to regulate the school meal program include Title I, "Healthier Kids and Schools," of the *HELP America Act* (see Table 8-7). Several states have begun to regulate food and beverage availability in schools, including attempts to regulate foods available from the NSLP as well as those available from outside vendors.

Another key regulatory opportunity at the state level lies in improving and encouraging physical and nutrition education programs. Physical education programs often fail to meet the recommendations for daily physical education for children.[180] Physical education standards often are not enforced at the state or local level because schools and districts have so many other mandated curriculum requirements.[181] The National Center for Education Statistics reported that only 50 percent of schools were required by state or district mandates to provide nutrition education to children in kindergarten through grade 8.[182] This figure was even lower for grades 9 through 12: Only 40 percent of ninth and tenth grades and 20 percent of eleventh and twelfth grades were required to provide nutrition education. Ultimately, failure to fund nutrition and physical education could result in increased levels of obesity and, therefore, increased state and federal health care spending.[183]

At the national level, the Child Nutrition and WIC Reauthorization Act (PL 108-265) may assist in school-based efforts for obesity prevention (see Table 8-7). Key provisions that target nutrition education efforts include, among others,[184]

- Providing opportunities for states to receive additional funds to establish *Team Nutrition Networks* to promote nutrition education and active lifestyles.
- Creating new approaches to improve the nutrition environment in schools by establishing *Local School Wellness Policies,* which include goals for nutrition education, physical activity, and other school-based activities designed to promote student wellness, and nutrition guidelines for all foods available on each school campus during the school day.

Several states require that all school districts establish and maintain School Health Advisory Councils (SHACs) or School Health Councils (SHCs) to establish health-related goals, to make policy recommendations, and to help coordinate implementation of school health programs. These SHACs can serve as vehicles to implement Local School Wellness Policies as mandated in the Child Nutrition and WIC Reauthorization Act.[185]

THE BUILT ENVIRONMENT

As noted earlier, the "built environment" encompasses a variety of community design elements such as street layout, zoning, transportation options, stairs, public and green spaces (walking paths, parks/recreation areas, playgrounds, community gardens), and business areas (farmers' markets, supermarkets, and grocery stores).[186] Various investigators have found an association between obesity and urban sprawl, as well as other aspects of the built environment (for example, lack of grocery stores in inner cities).[187]

The Institute of Medicine recommends that local governments, private developers, and community groups expand opportunities for physical activity, including recreational facilities, parks, playgrounds, sidewalks, and safe streets and neighborhoods, especially for populations at high risk of childhood obesity.[188] Regulatory options to enhance opportunities for physical activity and to increase access to affordable, nutritious food through the built environment include mixed-use zoning, improved bicycling and walking opportunities, and supermarket development in underserved areas.

One way in which the built environment can be regulated to enhance physical activity and to increase access to fresh, nutritious food is through *mixed-use zoning*. Land use in local municipalities is determined through zoning, which can further hinder opportunities for physical activity. Zoning regulations that locate commercial and residential areas together can promote physical activity by making it possible to walk or bike from home to work, school, shopping, and entertainment. A second regulatory option for enhancing the built environment is through *improved bicycling and walking opportunities*. Safety needs to be considered when increasing these opportunities, because residents may be reluctant to walk or bike in areas with high rates of pedestrian accidents.[189] Other environmental factors that encourage biking and walking are crosswalks, sidewalks, bike paths, and sufficient light and shade.[190] Programs that identify and create bicycle routes to schools and provide safety education have been found to increase physical activity.[191]

Another policy to improve the built environment is to *relocate supermarkets in urban and rural areas* that currently have few stores and high unemployment rates. Efforts to promote supermarket development in low-income areas in Philadelphia could be a model for providing families with access to fresh, nutritious foods and for subsequently stimulating economic growth. In 2002, the city of Philadelphia created a City Council Food Marketing Task Force, which included representatives from the supermarket industry, local government, and the nonprofit sector.[192] The Task Force members agreed to several policy recommendations to stimulate more supermarket development in the city of Philadelphia, which included implementing market assessment techniques to highlight unmet demand for supermarkets in neighborhoods, giving priority to assembling land for supermarkets, reducing regulatory barriers to supermarket investment, arranging transportation to supermarkets for shoppers who do not have access to a full-service supermarket, and providing public incentives to help locate new supermarkets in underserved locations. In spring of 2004, Pennsylvania enacted legislation authorizing $100 million for the establishment of supermarkets in low-income areas.[193]

Private Enforcement and Litigation

Legislative action can effectively provide for enforcement of behaviors that have been shown to affect public health, and legislation has been successfully used in preventing environmental contamination, regulating automobile safety, and controlling tobacco use. However, new legislative initiatives have been seeking to *restrict* the ability to use this strategy by limiting the jurisdiction of the courts in these cases. These efforts serve to emphasize the role of personal responsibility by limiting consumer access to the courts and to increase reliance on traditional public health authorities.[194] Examples of these types of legislation include (1) the Personal Responsibility in Food Consumption Act (H.R. 339), or the "Cheeseburger Bill," which sought to grant the food industry immunity from "claims of injury relating to a person's weight gain, obesity, or any health condition associated with weight gain or obesity,"[195] and (2) the Commonsense Consumption Act (S. 1428), which protects restaurants, food processors, distributors, advertisers, and others from civil liability for an individual's weight gain or related health problems caused by consumption of specific foods.[196]

One advantage of obesity-related lawsuits is that they have attracted increased public attention to the obesity epidemic and its health consequences. As a result, some food companies are responding with pledges to modify their products and marketing practices in the interest of obesity reduction.[197] Thus, from a public health point of view, the threat of a lawsuit can also be an impetus to change.[198]

Pricing Policies

More recently, researchers have begun to explore the connection among agricultural subsidies, economic polices, and the obesity epidemic.[199] The majority of agricultural subsidies in the United States are targeted to a small number of agricultural crops such as corn, wheat, and soybeans.[200] A large percentage of these commodity crops is used as either animal feed for meat production and/or to provide ingredients for highly processed foods such as the high-fructose corn syrup used in soft drinks. Academicians and policy analysts have begun to point out that from a public health perspective, it would make more sense for the United States government to subsidize the growing of foods such as fruits and vegetables so that food costs would be lower for more healthful foods compared to some more highly processed foods.[201] Consider the following argument:

> There are lots of subsidies for the two things we should be limiting in our diet, which are fat and sugar, and there are not a lot of subsidies for broccoli and Brussels sprouts. . . . What would happen if we took away the subsidies on the sugar and fat? Probably not much. They might go up a little bit, but the cost of food is not the cost of the final products. But if we're trying to do something political that might make a difference, try subsidizing fruit and vegetable growers so the cost is comparatively lower for better foods.[202]

Taxing is another type of pricing policy measure that legislatures can employ to influence consumers' buying practices. Tax incentives could be used to encourage more healthful dietary and physical activity behaviors, by (1) encouraging employers to promote worksite wellness programs, and (2) encouraging real estate developers to convert unused or abandoned spaces into physical activity–oriented facilities or to include green space and accessible sidewalks in their plans for residential development.[203] The National Governors Association's Center for Best Practices and the World Health Organization have noted that taxes on less nutritious foods are tools that can be used to influence consumer food-buying behavior.[204] Federal and state governments currently impose taxes on alcohol and tobacco; such taxes raise revenue but also promote public health and discourage consumption of these products.

Although states and cities that choose to levy taxes on less nutritious foods, including soft drinks, candy, chewing gum, and snack foods (such as potato chips), may not appreciably alter food consumption patterns, these tax revenues could be used to help fund healthful eating and nutrition education campaigns.[205] Opponents argue that because these taxes are levied on the purchase of foods that all income groups consume, they disproportionately affect low-income people, who spend a greater percentage of their total income on foods.[206]

Others have argued that such a tax penalizes the wrong target, because it affects consumers and not manufacturers, who may be more to blame for the preponderance of the many low-nutrient-density foods.[207] Public opinion is also divided on the issue of a "junk food" tax.

Societal-Level Solutions

The World Health Organization (WHO) report *Obesity—Preventing and Managing the Global Epidemic* first highlighted obesity as a worldwide problem that now affects most countries.[208] More recently, WHO has declared overweight one of the top ten health risks in the world.[209]

Countries in Asia, the Middle East, and Latin America are already experiencing a double burden of undernutrition and nutritional disease, such as diabetes and heart disease, caused by increasing rates of obesity as well as poverty.[210] The rise in obesity on a global scale means that health systems (and thus government budgets) will face an ever-growing financial burden from chronic disease unless effective obesity prevention and treatment strategies are implemented.[211]

Examples of social and environmental trends that may be contributing to the global obesity epidemic include increased use of motor transport; increased traffic hazards for walkers and cyclists; fewer opportunities for recreational physical activity; greater quantities of food available; more frequent and widespread food-purchasing opportunities; rising use of soft drinks to replace water; multiple television channels available around the clock; and globalization of markets, which favors energy-dense foods of low nutritional value.[212]

A primary goal of public health initiatives addressing the global obesity epidemic is to increase the awareness of people in sectors outside the health care field (such as culture and education, commerce and trade, development, planning, and transport) of the potentially adverse effects of their various actions on the ability of people to maintain energy balance. Table 8-8 outlines a range of societal-level solutions that can be implemented for obesity prevention at the population level.[213]

TABLE 8-8 **Potential Societal-Level Solutions for Obesity Prevention**

SETTING OR SECTOR	POTENTIAL SOCIETAL INTERVENTION
National governments	• Provide economic incentives for supply of "healthy" foods and disincentives for supply of "unhealthy" foods
Food supply	• Produce, distribute, and promote food products that are low in dietary fat and energy • Introduce economic incentives for supermarkets to locate in low-income neighborhoods • Introduce new and improved labeling schemes (covering fat and energy) that do not mislead the consumer
Media	• Regulate television advertising aimed at children • Incorporate positive behavior change messages into television programs and popular magazines
Non-governmental/ international organizations	• Develop and implement healthful eating, physical activity, and obesity prevention programs • Provide training in obesity prevention for doctors and other health care workers
Education sites	• Introduce nutrition standards for school meals • Provide classes in food preparation and cooking • Increase range of physical activities offered at school
Worksites	• Provide healthful food and drink options in staff restaurants • Empower employees to integrate physical activity into work day
Neighborhoods, homes, and families	• Set up community gardening programs, farmers' markets, and food cooperatives • Increase the "walkability" of city centers and residential areas • Set up walking programs in shopping malls and parks, and open safe-cycling routes

Source: Adapted from S. Kumanyika and coauthors, Obesity prevention: The case for action, *International Journal of Obesity* 26 (2002): 425–36.

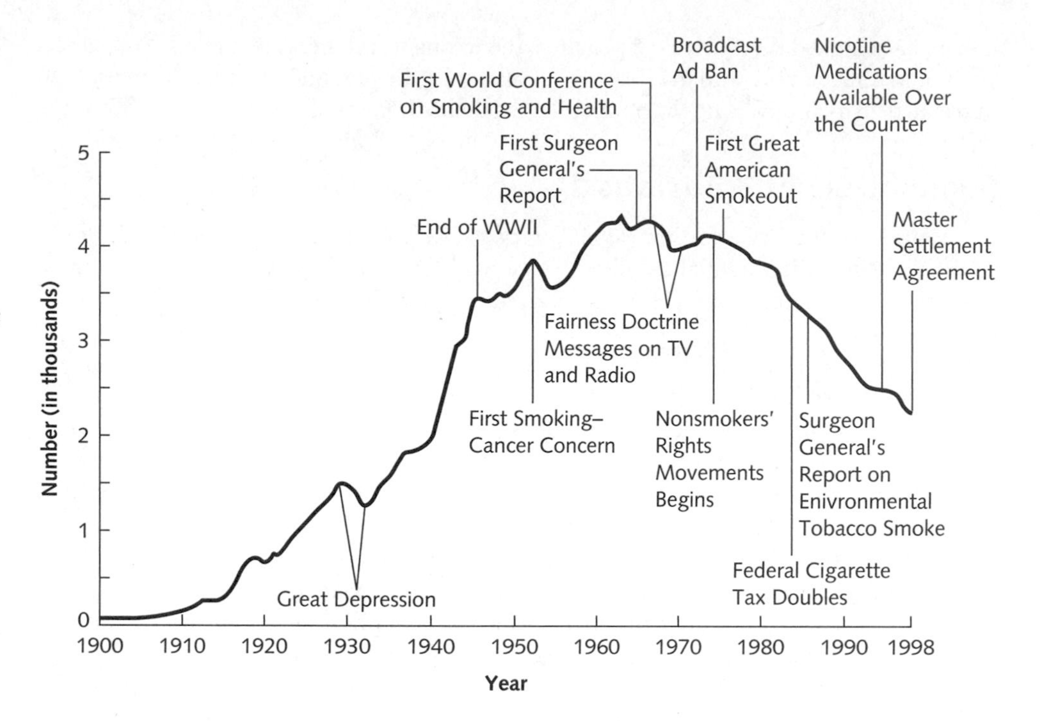

FIGURE 8-3 Annual Adult Per Capita Cigarette Consumption and Major Smoking and Health Events—United States, 1900–1998

Source: United States Department of Agriculture and "Reducing Tobacco Use: A Report of the Surgeon General"; accessed at http://www.cdc.gov/tobacco/sgr/sgr_2000/index.htm.

Where Do We Go from Here?

Awareness of obesity and overweight as a significant public health issue is in its beginning stages. The rapid increases in prevalence of obesity and overweight are fairly recent, and standards for child overweight have been available only since the early 2000s. With a multifactorial problem such as obesity, multiple approaches are necessary, and change may occur slowly. The progress of the obesity epidemic can be likened to another public health problem—smoking. As illustrated in Figure 8-3, smoking rates rose from the early 1900s through the 1960s. Awareness of the association between smoking and cancer arose during the 1950s, and public health initiatives were put into place as a result. These initiatives resulted in societal changes that occurred over a 30-year period. As these public health approaches began to be implemented, smoking rates reached a plateau and then started to drop. It can be expected that a similar pattern will emerge with the obesity epidemic. Public health approaches and societal change must begin now, but it is likely to be several years before the surveillance data show decreases in the prevalence of obesity and obesity-related diseases. A public health approach to address the global obesity epidemic must apply the same kind of multifaceted and coordinated approach that reduced tobacco use in order to change individual behavior patterns and effectively address environmental barriers to physical activity and healthful food choices.[214]

In general, environmental changes will follow a strong lead from policy and/or social change. However, before this can occur, a paradigm shift is needed—one that recognizes "obesogenic environments" as the main drivers of the obesity epidemic and that actively recruits sectors outside the field of health and nutrition (culture and education, commerce and trade, development, planning, and transport) as essential allies in tackling the obesity epidemic. Similarly, sectors such as local governments, schools, the food industry, and the media (through their marketing and advertising practices) need a paradigm shift to recognize how much they can contribute to reversing the obesity epidemic.[215]

TABLE 8-9 Summary of Recommendations from the Institute of Medicine's *Preventing Childhood Obesity*

1. *National Priority*	Government at all levels should provide coordinated leadership for the prevention of obesity in children and youth.
2. *Industry*	Industry should make prevention of obesity in children and youth a priority by developing and promoting products, opportunities, and information that will encourage healthful eating behaviors and regular physical activity.
3. *Nutrition Labeling*	Nutrition labeling should be clear and useful so that parents and youth can make informed product comparisons and decisions to achieve and maintain energy balance at a healthy weight.
4. *Advertising and Marketing*	Industry should develop and strictly adhere to marketing and advertising guidelines that minimize the risk of obesity in children and youth.
5. *Multi-Media and Public Relations Campaign*	The Department of Health and Human Services should develop and evaluate a long-term national multi-media and public relations campaign focused on obesity prevention in children and youth.
6. *Community Programs*	Local governments, public health agencies, schools, and community organizations should collaboratively develop and promote programs that encourage healthful eating behaviors and regular physical activity.
7. *Built Environment*	Local governments, private developers, and community groups should expand opportunities for physical activity, including recreational facilities, parks, playgrounds, sidewalks, bike paths, routes for walking or bicycling to school, and safe streets and neighborhoods.
8. *Health Care*	Physicians, nurses, and other clinicians should engage in the prevention of childhood obesity.
9. *Schools*	Schools should provide a consistent environment that is conducive to healthful eating and regular physical activity.
10. *Home*	Parents should promote healthful eating behaviors and regular physical activity for their children.

Source: Adapted from Institute of Medicine, *Preventing Childhood Obesity: Health in the Balance.* (Washington, DC: The National Academies Press, 2005).

What can the community nutritionist do? As we have seen, the first step is to build an awareness of the magnitude of the problem, especially awareness of the chronic diseases associated with obesity and their ultimate financial costs to the taxpayer. A second step is to put into effect policies and practices to change both individual behavioral factors and environmental factors. Because it is difficult to target behavioral factors without a supportive environment, it is essential to bring together other interested stakeholders through community coalitions or groups devoted to a common goal. Legislators at all levels (local, state, and national) need to be made aware of this issue and to be included in these efforts. Funding for research should be increased, with special allocations for new and innovative pilot programs, studies on changing environmental factors, and evaluation of obesity-related policies and their impact on the population. Finally, solutions for prevention of overweight and obesity need to be creative—perhaps by involving nontraditional partners or targeting less obvious determinants of obesity (such as parenting skills).

How can the community nutritionist get started? Current recommendations for obesity prevention, such as the Institute of Medicine's *Preventing Childhood Obesity: Health in the Balance* report (see the recommendations listed in Table 8-9), or the recommendations in *The Surgeon General's Call to Action to Prevent and Decrease Overweight and Obesity,*

(*Text discussion continues on page 266.*)

Several good resources for the community nutritionist, including background information, model legislation, and other resources promoting healthful nutrition and physical activity, can be found at the Center for Science in the Public Interest Policy Options website (www.cspinet.org/nutritionpolicy/policy_options.html), as well as at the other Internet sites listed at the end of this chapter.

Coordinated School Health Programs

Because the majority (95 percent) of children attend schools,[1] and most schools have nutrition resources and opportunities for physical activity, school systems are an excellent avenue for obesity-prevention programs. Schools can target obesity prevention through behaviorally based classroom education programs that target individual and cognitive factors, as well as environmental influences such as physical activity (through physical education classes) and diet (through cafeterias and vending sales of foods).

The Division of Adolescent and School Health (DASH) at the Centers for Disease Control and Prevention (CDC) has introduced and supported a model for health promotion programs at the school level. This model, known as a *Coordinated School Health Program* (see Figure 8-4), views the school in a multidimensional fashion, in which all components at the school level work together to maintain consistent, healthful messages.[2] Thus, messages about health are delivered to the students through different modalities that reinforce the concepts and appeal to all types of learning styles. Probably the best example of a coordinated school health program that addresses both nutrition and physical activity is the *Coordinated Approach To Child Health,* or *CATCH,* program.

FIGURE 8-4 CDC Coordinated School Health Model
The *Coordinated School Health Model* views the school in a multidimensional fashion, in which all components at the school level work together to maintain consistent, healthful messages that are reinforced using different modalities and appeal to all types of learning styles.

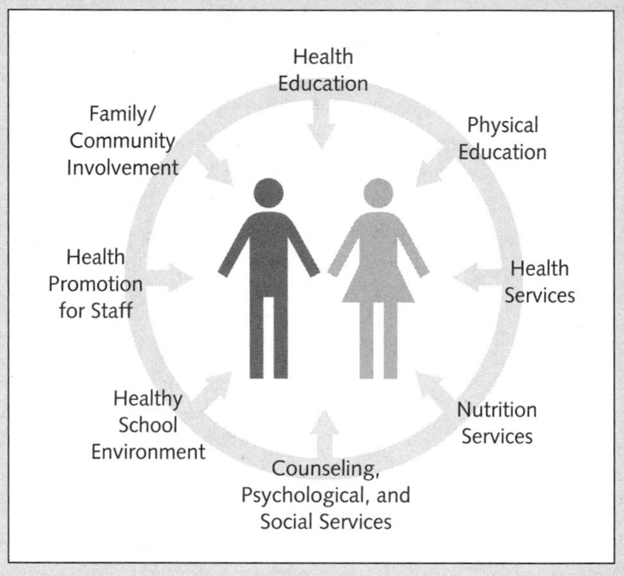

Source: Centers for Disease Control and Prevention (CDC). Division of Adolescent and School Health, *School Health Defined: Coordinated School Health Program;* available at www.cdc.gov/HealthyYouth/CSHP/index.htm.

Goals and Objectives

The overall goal of the Coordinated Approach To Child Health (CATCH) program is to create healthy children and healthy school environments. The specific aims of the program are

1. To encourage students to consume a diet that is low in fat (30 percent of energy) and saturated fat (10 percent of energy) and higher in fruits and vegetables;
2. To encourage students to participate in increased amounts of moderate to vigorous physical activity (MVPA), or activity that makes your heart beat fast and makes you breathe hard;
3. To increase MVPA in schools to 50 percent of the physical education class;
4. To provide food in school cafeterias that is lower in fat and saturated fat; and
5. To encourage parental participation in the school health program.

CATCH was originally developed and evaluated as the Child and Adolescent Trial for Cardiovascular Health (CATCH) from 1991 to 1994.[3] The dissemination phase of CATCH was conducted beginning in 1996, shortly after the main trial. The goal of dissemination is to promote the philosophy, materials, and methods of coordinated school health to school teachers and staff throughout Texas.[4]

Priority Population

CATCH targets several population groups. These include elementary school children and their parents, elementary school teachers, and school administration and staff.

The main trial of CATCH included a cohort of 5,106 third-grade students enrolled in 96 schools in 4 sites (San Diego, California; New Orleans, Louisiana; Minneapolis, Minnesota; and Austin, Texas).[5] The current dissemination phase of CATCH includes training school teachers and staff to implement the program.

Rationale for the Intervention

The original CATCH was funded by NHLBI to determine whether a school-based curriculum could affect cardiovascular risk factors, such as blood cholesterol levels, diet, and physical activity. Because children's diets were high in fat and saturated fat, and it has been shown that health behaviors track from childhood into adulthood,[6] it was reasoned that changes in children's diets and physical activity habits would benefit them in the future as well as in the present.

Since CATCH had shown promise in changing diet and physical activity patterns in children, the Diabetes Council of the Texas State Department of Health Services (formerly the Texas Department of Health) began to fund dissemination of CATCH to schools in Texas as a program that targets behaviors that are precursors to chronic diseases such as obesity, Type 2 diabetes, and cardiovascular disease.

Methodology

The main CATCH study was a randomized clinical trial in which each of 96 schools at 4 sites (24 schools per site) was assigned to one of three conditions:

- Control (usual health program) ($n = 40$),
- School-based program ($n = 28$), and
- School-based program plus family component ($n = 28$).

Control schools implemented their usual health program, while intervention schools implemented behaviorally based classroom curricula for grades 3–5, a physical education component, and a cafeteria program. Schools with a family-based component also had a series of lessons designed to be done at home with parents or guardians, as well as family-based health fairs at the schools.

The dissemination phase of CATCH was conducted after the main trial, with funding from the Texas Department of State Health Services.[7] CATCH was packaged into a set for schools, and the name was changed from "Child and Adolescent Trial for Cardiovascular Health" to "Coordinated Approach To Child Health" to better reflect the implementation of the program instead of the randomized clinical trial. Initially, *opinion leaders* (people who influence other people's attitudes about a program) and *change agents* (people who can influence decisions to implement a program) were contacted and familiarized with the program. As these people became better acquainted with the program, they began to implement it in their own schools and school districts. Because these people were leaders in their organizations and communities, they began to influence others, who adopted the program or who suggested legislative efforts to promote CATCH-type programs more widely. Partnerships between groups with the common goal of promoting school-based physical activity and nutrition programs to decrease obesity and related risk factors began; these partnerships evolved into a coalition. These dissemination efforts led to recognition of the program, especially in local communities and among Texas legislators. As a result, Texas Senate Bill 19 (now Texas Education Code 38.013) was passed in 2001. This legislation mandated daily physical education for kindergarten through grade 5 (through grade 6 if the school is K-6), formation of school health advisory councils (SHACs) for nutrition and physical activity, and implementation of coordinated school health programs in all elementary schools by 2007. Continued visits, training sessions, and presentations by CATCH staff further served to publicize the program.

CATCH dissemination was measured using both quantitative and qualitative methods to determine the reach of the program and subsequent implementation. These methods included enumeration of the schools that purchased CATCH materials or attended a training session; surveys mailed to training participants; and observations of physical education classes.

Results

The CATCH program resulted in significant changes in self-reported diet and physical activity levels of the children.[8] These changes were maintained for 3 years without additional intervention.[9] Although changes in rates of child overweight were not found in the CATCH main trial, it should be noted that decreasing the rate of child obesity was not a targeted goal of CATCH. During the time period for the main CATCH trial (1991–94), rates of child overweight were not a significant public health priority, largely because the increase in the prevalence of child overweight was just beginning to be evident. A more recent follow-up evaluation of CATCH as implemented in El Paso, Texas, shows a leveling in the prevalence of child overweight and at risk of overweight in elementary school children after 3 years of the program, compared to children in control schools.[10] Dissemination rates have been significant as well: As of October 2004, more than 1,600 schools had adopted part of the CATCH curriculum, and more than 700 schools had been trained in coordinated school health.

Lessons Learned

The main CATCH trial demonstrated that it is possible to implement a school-based health promotion program to change child and adolescent diet and physical activity patterns; however, these changes in diet and physical activity do not necessarily result in changes in related physiologic risk factors. More recent data indicate that implementation of CATCH in schools is associated with significant decreases in BMI over time.[11] These results may be reflective of the

Continued

Programs in Action

Coordinated School Health Programs—*continued*

increased rate of child overweight in today's society. With a larger population of overweight children, it may be that diet and physical activity behaviors have significantly changed over time, and thus, the changes targeted by CATCH may be more evident now compared to the early 1990s, when child overweight was not so prevalent.

CATCH is an excellent example of *translational research,* in which studies that are rigorously evaluated under controlled conditions and show promising results are "translated" into community-based interventions that are implemented in real-life situations.[12] CATCH was evaluated in a randomized clinical trial in which all the schools and measures were strictly monitored and differences between the schools were minimized. In the dissemination phase of CATCH, the program is individualized to each school setting. This means there is somewhat less control over implementation of the program, but a greater probability that the changes will be maintained over time. The goal of translational research is to maintain the effectiveness of the program while still providing for implementation in a variety of real-life settings.

The dissemination phase of CATCH presents new challenges and new learning opportunities. Implementing CATCH involved partnering with different organizations and groups to advance common goals, packaging the materials in an easy-to-follow program, and extensive training sessions and networking that highlight the compatibility and flexibility of the program.* Although it is evident that this strategy can produce widespread dissemination and effective program implementation, further evaluation and monitoring of student-related outcomes, such as effects on the prevalence of overweight, will determine the ultimate success of this approach.

References

1. National Center for Health Statistics, "Participation in Education, Enrollment by Age"; available at http://www.nces.ed.gov/programs/coe/2004/section1/table.asp?tableID=97.
2. Centers for Disease Control and Prevention (CDC), Division of Adolescent and School Health, National

provide an outline of activities and areas to target.[216] Here are some examples of immediate steps that a community nutritionist can take:

- Developing and promoting community awareness campaigns to highlight health risks associated with obesity
- Organizing a community coalition or partnership to address obesity-related issues
- Working with schools in the formation and maintenance of Local School Wellness Policies or Coordinated School Health Programs
- Implementing effective behaviorally based interventions through community-based clinics and organizations, schools, worksites, faith-based communities, and WIC centers
- Encouraging health care professionals to measure and track BMI and to provide counseling and referrals for patients who exceed current standards for overweight and obesity
- Providing opportunities for increased access to physical activity and healthful foods, such as encouraging funding for walking trails and safe routes to schools; developing standards for foods sold in schools and at public venues; and encouraging healthful catering guidelines for school, community, and worksite functions
- Writing or educating local municipal officials, state legislators, and/or members of Congress about upcoming obesity-related legislation

The development of a solution to the obesity epidemic is a complex issue that will likely involve a combination of societal and individual-level approaches. It may well be that the solution includes either factors that have not been previously targeted on a large-scale basis, or a combination of effective approaches within one comprehensive program. Although the task of decreasing the epidemic of obesity seems formidable, it is also a challenge that will provide increasing employment and personal opportunities, especially for the field of nutrition.

Center for Chronic Disease Prevention and Health Promotion, "School Health Defined: Coordinated School Health Program"; available at www.cdc.gov/HealthyYouth/CSHP/index.htm.
3. R. V. Luepker and coauthors, Outcomes of a field trial to improve children's dietary patterns and physical activity: The Child and Adolescent Trial for Cardiovascular Health, *Journal of the American Medical Association* 275 (1996): 768–776; and C. L. Perry and coauthors, School-based cardiovascular health promotion: The Child and Adolescent Trial for Cardiovascular Health (CATCH), *Journal of School Health* 60 (1990): 406–13.
4. D. M. Hoelscher and coauthors, Dissemination and adoption of the Child and Adolescent Trial for Cardiovascular Health (CATCH): A case study in Texas, *Journal of Public Health Management and Practice* 7 (2001): 90–100.
5. R. V. Luepker and coauthors, 1996.
6. S. H. Kelder and coauthors, Longitudinal tracking of adolescent smoking, physical activity, and food choice behaviors, *American Journal of Public Health* 84 (1994): 1121–6.
7. D. M. Hoelscher and coauthors, 2001.
8. R. V. Lupeker and coauthors, 1996.
9. P. R. Nader and coauthors, Three-year maintenance of improved diet and physical activity, *Archives of Pediatric and Adolescent Medicine* 153 (1999): 695–704.
10. K. J. Coleman and coauthors, Impacting epidemic increases in child risk for overweight in low-income schools: The El Paso Coordinated Approach to Child Health (El Paso CATCH), *Archives of Pediatric and Adolescent Medicine* 159 (2005): 217–24.
11. Ibid.
12. R. E. Glasgow, E. Lichtenstein, and A. C. Marcus, Why don't we see more translation of health promotion research to practice? Rethinking the efficacy-to-effectiveness transition, *American Journal of Public Health* 93 (2003): 1261–7.

*Implementation of a Coordinated School Health program such as CATCH involved partnering with such groups as the state education agency (health and physical education, child nutrition services), state department of agriculture, health education centers, community health agencies, parent-teacher associations, pediatric and state medical groups, the Centers for Disease Control and Prevention, and state chapters of the American Heart Association and American Cancer Society.

Internet Resources

Data on Obesity and Overweight

National Health and Nutrition Examination Survey (NHANES)	**www.cdc.gov/nchs/nhanes.htm**
Behavioral Risk Factor Surveillance System	**www.cdc.gov/brfss**
Youth Risk Behavior Surveillance System	**www.cdc.gov/HealthyYouth/yrbs/index.htm**
Arkansas Center for Health Improvement	**www.achi.net**

General Information on Obesity/Overweight

Dietary Guidelines 2005	**www.healthierus.gov/dietaryguidelines**
CDC Obesity	**www.cdc.gov/nccdphp/dnpa/obesity/index.htm**
University of California Berkeley Center for Weight and Health	**http://nature.berkeley.edu/cwh**
CDC Obesity Trends	**www.cdc.gov/nccdphp/dnpa/obesity/trend/index.htm**
American Dietetic Association	**www.eatright.org**
Information from the National Institute of Diabetes and Digestive and Kidney Diseases	**www.niddk.nih.gov/health/nutrit/nutrit.htm**
NIDDK Weight-Control Information Network (WIN)	**http://win.niddk.nih.gov/index.htm**

Healthy Weight Journal	**www.healthyweight.net**
American Obesity Association	**www.obesity.org**
USDA Food and Nutrition Information Center, Weight Control and Obesity	**www.nal.usda.gov/fnic/etext/000060.html**

Measurement of Obesity/Overweight and Food Intake

School Physical Activity and Nutrition (SPAN) survey	**www.sph.uth.tmc.edu/hnc/SPAN/SPAN%20Home.htm**
BMI calculator	**www.cdc.gov/nccdphp/dnpa/bmi/calc-bmi.htm**
Food composition	**www.nal.usda.gov/fnic/foodcomp**
Calorie Control Council	**www.caloriecontrol.org**

Programs

Coordinated Approach to Child Health	**www.sph.uth.tmc.edu/catch/**
Planet Health	**www.hsph.harvard.edu/prc/proj_planet.html**
Stanford's Student Media Awareness to Reduce Television (SMART)	**http://hprc.stanford.edu/pages/store/itemDetail.asp?169**
HUGS International, Inc. for teens and adults	**www.hugs.com**
Team Nutrition	**www.fns.usda.gov/tn**
National Weight Control Registry	**www.uchsc.edu/nutrition/WyattJortberg/nwcr.htm**
NIH Strategic Plan for Obesity Research	**http://obesityresearch.nih.gov/about/strategic-plan.htm**
Steps to a HealthierUS Initiative	**www.healthierus.gov/steps/index.html**
Active Living Research (funded by Robert Wood Johnson Foundation)	**www.activelivingresearch.org/**
America on the Move	**www.americaonthemove.org/**
Action for Healthy Kids	**www.actionforhealthykids.org/**

Legislation on Physical Activity and Nutrition

CDC information on nutrition and physical activity legislation	**http://apps.nccd.cdc.gov/DNPALeg/**
Center for Science in the Public Interest Policy Information	**www.cspinet.org**
Health Policy Tracking Service	**www.hpts.org**
National Alliance for Nutrition and Activity	**www.cspinet.org/nutritionpolicy/nana.html**
Obesity Report Card and Map	**www.ubalt.edu/experts/obesity/**

Recommendations for Obesity Prevention

The Surgeon General's Call to Action to Prevent and Decrease Overweight and Obesity	**www.surgeongeneral.gov/topics/obesity/**
Institute of Medicine *Preventing Childhood Obesity: Health in the Balance*	**www.iom.edu/project.asp?id=5867**
WHO *Obesity—Preventing and Managing the Global Epidemic*	**www.who.int/nut/documents/obesity_executive_summary.pdf**
WHO *Diet, Nutrition and the Prevention of Chronic Diseases*	**www.who.int/nut/documents/trs_916.pdf**

Case *Study*

Worksite Health Promotion Program for Prevention of Overweight

Scenario

You are a consultant nutritionist who has recently been contacted by the headquarters of a large manufacturing plant in your city. This company has about 300 employees in one location, with large open spaces around the company and an on-site cafeteria. In addition, there are several break rooms that contain vending machines. The employees are shift workers, 75 percent are blue-collar, and the majority are Latino and African-American.

Six months ago, a group from your local university came to the company and measured heights and weights of the employees as part of a larger study. After calculating BMIs, the university researchers found that the majority (65 percent) of employees were either overweight or obese. The company president had recently had a heart attack and was appalled at the high rate of overweight among company employees, so he decided to take action. He has hired you to put together a one-year worksite health promotion program that targets obesity prevention for the employees. He has told you that he is willing to change company policies regarding food and physical activity and that you have a budget to develop some infrastructure and implement an intervention. Because the company president is investing a great deal of effort and resources in the program, he expects to see some success over time. On the basis of your previous experience in designing, implementing, and evaluating worksite-based programs, you know that you will need to obtain "buy-in" from the stakeholders (people who will be implementing and participating in the intervention), as well as appropriate goals for the program.

Learning Objectives

1. Identify program outcomes based on *Healthy People 2010* objectives.
2. Identify individual and environmental determinants of obesity in the company.
3. Determine steps for implementation of a new worksite health promotion program.
4. Outline a worksite health promotion program that targets obesity prevention and treatment in this company.

Foundation: Acquisition of Knowledge and Skills

1. Find the healthy weight goals for adults from the objectives of *Healthy People 2010* in this chapter.
2. List seven benchmarks of success for worksite health promotion from the Wellness Council of America (Welcoa) at www.welcoa.org.
3. Review previous worksite health promotion programs for weight loss and obesity prevention.
4. Access the ADA position paper on *The Role of Dietetics Professionals in Health Promotion and Disease Prevention* at www.eatright.org/Public/NutritionInformation/92_adar1102.cfm.
5. Behaviorally based programs have been found to be the most effective, so review basic behavioral theories (see Theory at a Glance at www.cancer.gov/aboutnci/oc/theory-at-a-glance.) These theories are also discussed in Chapter 15.

Step 1: *Identify Relevant Information and Uncertainties*

1. Identify relevant determinants of overweight and/or obesity in this population. Be sure that the determinants you target can be changed through a worksite health promotion program.
2. Determine appropriate weight-loss goals for this type of population and this type of program.
3. List different strategies for a worksite health promotion program that can be implemented at this particular site.
4. Determine key stakeholders for implementation of the program and new company policies. (In other words, whom do you need to persuade to implement the program?)

Step 2: *Interpret Information*

1. Determine which behavioral theory or theories could be used in a worksite setting with the strategies you proposed.
2. Formulate a plan to bring together key stakeholders for implementation of the program and convince them that they should implement the program.

3. List specific aims for behavioral objectives (diet and physical activity behaviors) that will lead to prevention of weight gain or to weight loss in this population.
4. Outline several strategies for the proposed worksite health promotion program. Be sure to include strategies for both nutrition and physical activity.

Step 3: *Draw and Implement Conclusions*

1. Develop a proposal for the company president that includes an account of any previous work that has been done in this area, specific goals or objectives for the program, a list of company employees who will help you implement the program, an outline of the program that includes specific strategies to be implemented, and a timetable for the program.

Step 4: *Engage in Continuous Improvement*

1. Create an evaluation plan to determine whether you have made a significant difference in the overweight/obesity problem. Remember to measure your primary outcome (such as body weight or BMI) as well as behavioral outcomes such as diet and physical activity.
2. What barriers do you anticipate during the implementation of this health promotion program? How do you intend to address these barriers?
3. What can you do to be sure that the program is institutionalized—that is, continues to be implemented—in the company after the initial year?

Recommended Reading

1. D. J. Hennrikus and R. W. Jeffery, Worksite intervention for weight control: A review of the literature [Review] [60 refs], *American Journal of Health Promotion* 10(6) (1996): 471–98.
2. R. W. Jeffery and coauthors, Long–term maintenance of weight loss: Current status, *Health Psychology* 19 (2000): 5–16.
3. K. M. McTigue and coauthors, Screening and interventions for obesity in adults: Summary of the evidence for the U.S. Preventive Services Task Force, *Annals of Internal Medicine* 139 (2003): 933–49.

Professional Focus

Diet Confusion: Weighing the Evidence

Lose weight while you sleep! Lose 30 pounds in just 20 days! Eat the foods you love and lose weight! You will never be hungry! Do these claims look familiar to you? With the recent focus on the increase in obesity in the United States and the world, there are burgeoning efforts to promote diet books, products, and programs. The truth is that although most diets can provide a weight loss in the short term, few people can lose weight and keep it off permanently, and some of these claims might actually be harmful. Dieting is big business in the United States. In 2004, one nationally representative survey found that one-third of American adults are on a diet, and this number has increased from 24 percent in 2000.[1]

How Do Diets Work?

Diets work because people limit their food consumption. Excess weight is the consequence of an energy imbalance, caused by overconsumption of food or decreased physical activity relative to individual requirements. Limiting of dietary intake can occur through elimination or restriction of certain food groups, such as carbohydrates; portion control through prepackaged meals, snacks or drinks; alteration of meal patterns or content; and control of food intake through point systems or monitoring. A comparison of the approximate caloric content and macronutrient distribution of several types of diets is provided in Table 8-10.

What Are Some Common Diets?

Although the current diet fad can change quickly, certain types of diets have appeared during the past few years. Here are some of the most common:[2]

- ***Dr. Atkins New Diet Revolution.*** In this diet, consumption of high-fat meats, cheeses, and fats is encouraged, while consumption of carbohydrates (such as fruit, breads, and cereals) is severely limited. The underlying premise of the diet is that elimination of these foods will produce a "benign dietary ketoacidosis," which leads to a decrease in hunger and slows excessive food consumption. Ketosis can be accompanied by bad breath, nausea, headaches, and fatigue. High intakes of protein may exacerbate gout and kidney disease, and high intakes of saturated fat can increase blood cholesterol levels.

- ***The Zone Diet.*** This rigid eating plan separates foods into "macronutrient blocks."
- ***The South Beach Diet.*** This regimen is a more healthful version of the Atkins high-protein, low-carbohydrate diet, with incorporation of lower-fat protein sources such as chicken and fish, whole grains, and vegetables and fruits. The plan does limit some foods, such as carrots, bananas, pineapple, and watermelon, and the first phase of the diet is more restrictive than later phases.
- ***Weight Watchers.*** Dieters may use a list of core foods or a point system to select and eat foods to reduce caloric intake and lose weight. In Weight Watchers, no food is forbidden, but all must be balanced with other choices.
- ***Dr. Ornish Eat More, Weigh Less.*** Weight loss is based on consuming a very-low-fat diet (10 percent of its kilocalories from fat), with little meat, oils, nuts, butter, dairy (except non-fat), sweets, or alcohol. The original Ornish plan included diet together with exercise and stress reduction.
- ***Eat Right for Your Blood Type.*** This diet is based on the claim that your blood type determines the types of foods that you should eat and how your body absorbs nutrients. For example, people with type O blood should consume meat, seafood, fruits and vegetables, but less wheat and beans. There is no scientific basis for this claim.
- ***Dr. Phil's Ultimate Weight Solution.*** The book describing this diet focuses on "Keys to Weight Loss Freedom," but these concepts do not include defined meal plans or recipes. The diet promotes seafood, poultry, meat, low-fat dairy, whole grains, fruits, vegetables, and some oils. Supplements and weight-loss bars and shakes are also promoted.
- ***The New Glucose Revolution.*** This eating plan encourages consumption of low-glycemic foods, such as beans, pasta, most fruits, vegetables, low-fat dairy, and meats. Unfortunately, the glycemic index is not always a reliable measure of increases in blood glucose, because it can vary with the food itself and with other foods consumed at the same time. In addition, it is dependent on the amount of food consumed.

A recent study published in the *Journal of the American Medical Association* evaluated four of these diets (the Atkins, Ornish, Weight Watchers, and Zone diets) and found that all four modestly reduced body weight and some cardiac risk factors at one year.[3] Adherence to each diet for the 12-month period varied, ranging from 50 percent for the Ornish diet and 53 percent for the Atkins diet to 65 percent for both the Weight Watchers and Zone diets. The subjects who had the best adherence to the diets had the best results, and cardiac risk factors were more closely associated with weight loss than with diet type. In general, the subjects had more difficulty following the more restrictive diets (the Ornish and Atkins diets). Although this is just one study with small sample sizes, it does suggest that there are many ways to lose weight, that people find it difficult to adhere to very restrictive diets for a long time, and that we need to find methods of keeping people motivated to stay on any new eating plan.

How Can You Evaluate a Diet to Determine Whether It Is Healthful?

Frequently, community nutritionists are asked to provide guidance on various diets or diet plans. What can you do to determine whether a particular diet plan is useful to a consumer? Use the checklist that follows.[4]

1. *Does the weight-loss program systematically eliminate one group of foods from a person's eating pattern?* For example, are all carbohydrates systematically eliminated from a person's diet? Are dairy products eliminated? In general, a diet that eliminates a certain food group is probably lacking in important nutrients and dietary variety, and it will be difficult for a person to adhere to that eating plan.
2. *Does the weight-loss program encourage specific supplements or foods that can be purchased only from selected distributors?* These supplements or foods often contain ingredients that may be harmful or unproven.
3. *Does the weight-loss program tout magic or miracle foods or products that burn fat?* The only way to burn fat is to increase your physical activity levels or decrease the amount of total food that you consume. You cannot "burn" fat with sauna belts, body wraps, thigh-reducing creams or similar products. If you consume more than you expend or if you lower your physical activity level and keep your food intake the same, your body will store the extra calories as fat.
4. *Does the weight-loss program promote bizarre quantities of only one food or one type of food?* Some diets include eating only one food each day, or unlimited amounts of certain foods, such as grapefruit or cabbage soup. Such advice runs counter to everything we know about the broad spectrum of human nutritional needs.

TABLE 8-10 Comparison of Diet Programs/Eating Plans to Typical American Diet

TYPE OF DIET	EXAMPLE	GENERAL DIETARY CHARACTERISTICS*	COMMENTS
Typical American diet		CHO: 50% PRO: 15% Fat: 35% Average of 2,200 kcal/d	• Low in fruits and vegetables, dairy and whole grains • High in saturated fat and unrefined carbohydrates
Balanced-nutrient, moderate-calorie approach	DASH Diet or Diet based on MyPyramid food guide; Commercial plans such as Diet Center, Jenny Craig, Nutri/System, Physician's Weight Loss, Shapedown Pediatric Program, Weight Watchers, Setpoint Diet, Volumetrics	CHO: 55–60% PRO: 15–20% Fat: 20–30% Usually 1,200 to 1,700 kcal/d	• Based on set pattern of selections from food lists using regular grocery store foods or prepackaged foods supplemented by fresh food items • Low in saturated fat and ample in fruits, vegetables, and fiber • Recommend reasonable weight-loss goal of 0.5 to 2.0 lb/week • Prepackaged plans may limit food choices • Most recommend exercise plan • Many encourage dietary record-keeping • Some offer weight-maintenance plans/support
Very low-fat, high-carbohydrate approach	Ornish Diet (Eat More, Weigh Less), Pritikin Diet, T-Factor Diet, Choose to Lose, Fit or Fat	CHO: $\geq$ 65% PRO: 10–20% Fat: $\leq$ 10–19% Limited intake of animal protein, nuts, seeds, other fats	• Long-term compliance with some plans may be difficult because of low level of fat • Can be low in calcium • Some plans restrict healthful foods (seafood, low-fat dairy, poultry) • Some encourage exercise and stress-management techniques
Low-carbohydrate, high-protein, high-fat approach	Atkins New Diet Revolution, Protein Power, Stillman Diet (The Doctor's Quick Weight Loss Diet), the Carbohydrate Addict's Diet, Scarsdale Diet	CHO: $\leq$ 20% PRO: 25–40% Fat: $\geq$ 55–65% Strictly limits CHO to less than 100–125 g/d	• Promote quick weight loss (much is water loss rather than fat loss) • Ketosis causes loss of appetite • Can be too high in saturated fat • Low in carbohydrates, vitamins, minerals, and fiber • Not practical for long-term because of rigid diet or restricted food choices

*CHO = Carbohydrate, PRO = Protein.

5. *Does the weight-loss program have rigid menus?* If a diet has specific meal plans and times to eat, it will be difficult to incorporate individual taste preferences. People are unique, so no one diet plan will work for everyone. A person who loves Thai food will not succeed on a diet if there is no way to incorporate Thai food into his or her eating plan.
6. *Does the weight-loss program promote specific food combinations?* Some diets provide combinations of foods that should or should not be eaten at the same time. These food combinations have no basis in fact and needlessly restrict the dieter's options for reasonable dietary intake and food choices.
7. *Does the weight-loss program promise a weight loss of more than 2 pounds per week for an extended period of time?* If so, the initial weight loss will probably be due to water loss. A more realistic diet plan will aim for a weight loss of 0.5 to 2.0 pounds per week.
8. *Does the weight-loss program provide a warning to people with diabetes, high blood pressure, or other health conditions?* People with preexisting health conditions need to consult a physician or other health care provider before beginning any diet. Elimination of certain food groups or eating excessive amounts of certain foods can exacerbate these problems, as well as interfere with the efficacy of certain medications.

TABLE 8-10 Comparison of Diet Programs/Eating Plans to Typical American Diet—*continued*

TYPE OF DIET	EXAMPLE	GENERAL DIETARY CHARACTERISTICS*	COMMENTS
Moderate-carbohydrate, high-protein, moderate-fat approach	The Zone Diet, Sugar Busters, South Beach Diet	CHO: 40–50% PRO: 25–40% Fat: 30–40%	• Diet rigid and difficult to maintain • Enough CHO to avoid ketosis • Low in carbohydrates; can be low in vitamins and minerals
Novelty diets	Immune Power Diet, Rotation Diet, Cabbage Soup Diet, Beverly Hills Diet, Dr. Phil	Most promote certain foods, or combinations of foods, or nutrients as having unique (magical) qualities	• No scientific basis for recommendations
Very low-calorie diets	Health Management Resources (HMR), Medifast, Optifast	Less than 800 kcal/d	• Requires medical supervision • For clients with BMI $\geq$ 30 or BMI $\geq$ 27 with other risk factors; may be difficult to transition to regular meals
Weight-loss online diets	Cyberdiet, DietWatch, eDiets, Nutrio.com	Meal plans and other tools available online	• Recommend reasonable weight-loss goal of 0.5 to 2.0 lb/week • Most encourage exercise • Some offer weight-maintenance plans/support

Source: Adapted from M. Boyle and S. L. Anderson, *Personal Nutrition,* 5th ed. (Belmont, CA: Wadsworth/Thomson Publishing, 2004), pp. 276–277; Weighing the Diet Books, *Nutrition Action Newsletter,* January/February 2004: 3–8; M. Freedman and coauthors, Popular diets: A scientific review, *Obesity Research* 9 (2001): 1S–39S; and A guide to rating the weight-loss websites, *Tufts University Health and Nutrition Letter,* May 2001, pp. 1–4.

9. *Does the weight-loss program encourage or promote increased physical activity?* Although people can lose weight by limiting food intake alone, research has shown that the most successful weight-loss plans include lifestyle changes, such as exercise.
10. *Does the weight-loss program encourage an intake that is very low in calories (below 800 kcal/d)* without supervision of medical experts? Very low-calorie diets are designed to be used for persons with severe obesity or obesity with other health-related problems. Because the energy level is so low, the diet must be supplemented with vitamins and minerals. In addition, the patient must be strictly observed for any adverse health effects. Finally, the person needs dietary counseling to handle "real" food choices before the end of the diet, or weight gain can quickly ensue.

What Can You Do?

Some of your clients may believe that weight loss is a lost cause, but don't give up! There are several strategies and diets that have been proved successful. The strategies supported by the most evidence are detailed in a recent analysis by the USDA and backed up by data from the National Weight Control Registry, a study that examines people who have lost at least 30 pounds and have maintained that loss for at least a year.[5]

In your practice as a community nutritionist, there are several steps that you can take to prepare yourself for dealing with the public—and with fad diets:

1. Be familiar with the current fad diets. Before you can answer questions, you need to be familiar with the latest diet craze. Study the diet—these food plans often include scientifically based statements intermingled with inaccuracies, so you have to know the literature to refute any incorrect claims.
2. Recommend appropriate weight-loss strategies and programs. A recent evidence-based review indicates that most weight loss is associated with consumption of diets that contain about 1400 to 1500 calories per day, so it is essential to control energy intake for any weight-loss plan.[6] Weight Watchers has been cited as a good option in many recent studies for the variety of foods offered and for principles based on scientific evidence. Internet-based programs may be good for people who like to keep records and need support but cannot attend group sessions. The DASH diet has been found to significantly affect hypertension and other chronic disease outcomes, and it is free on the NIH website. It is interesting to note that about half of the people in the National Weight Control Registry (NWCR) lost weight without any formal program, indicating that the more individualized a program, the more likely it is for people to adhere to it for longer periods of time. Finally, it should be noted that most successful weight-loss attempts include regular exercise of some type.
3. Refer the public to websites that list resources for determining whether a diet is a fad. The following websites contain good information or handouts that the public can use to determine whether following a diet will be harmful or not: (1) the Federal Trade Commission (FTC) website (www.ftc.gov), including Weighing the Evidence in Diet Ads at www.ftc.gov/bcp/menu-home .htm, and (2) the American Heart Association *Fad Diets* at www.americanheart.org/presenter.jhtml?identifier =4584.
4. Report fraudulent or deceptive weight-loss claims. Any weight-loss claims that are distributed via the Internet, television, or print media can be reported at www.ftc.gov or by calling 1-877-FTC-HELP (1-877-382-4357).

Websites

American Dietetic Association
www.eatright.org
DASH (Dietary Approaches to Stop Hypertension) diet
www.nhlbi.nih.gov/health/public/heart/hbp/dash
eDiets
www.ediets.com
Weight Watchers
www.weightwatchers.com
Jenny Craig
www.jennycraig.com
Diet Center
www.dietcenter.com
DietWatch
www.dietwatch.com
Cyberdiet
www.cyberdiet.com/reg/index.html
Nutrisystem
www.nutrisystem.com
Nutrio.com
http://nutrio.com/servlet/nutrio
South Beach Diet online
www.southbeachdiet.com/public/
Health Management Resources
www.yourbetterhealth.com
Weight-control Information Network on choosing a safe and successful diet
http://win.niddk.nih.gov/publications/choosing.htm
USDA Food and Nutrition Information Center on weight control and obesity
www.nal.usda.gov/fnic/etext/000060.html
Wheat Council "Setting the Record Straight"
www.wheatfoods.org
National Weight Control Registry
www.uchsc.edu/nutrition/WyattJortberg/nwcr.htm
Learning tool for fad diets:
Go to **wemarket4u.net/fatfoe** to see an ad for FatFoe™ Eggplant Extract, and click on the "order now" button.
Aim for a Healthy Weight
www.nhlbi.nih.gov/health/public/heart/obesity/lose_wt/index.htm
Obesity, Physical Activity, and Interactive Web Applications
www.nhlbi.nih.gov/health/public/heart/index.htm#obesity
Weight Loss and Nutrition Myths
http://win.niddk.nih.gov/publications/myths.htm

References

1. Calorie Control Council National Consumer Survey, 2004; available at http://www.caloriecontrol.org/trndstat.html.
2. Adapted from Weighing the diet books, *Nutrition Action Newsletter,* January/February 2004: 3–8; M. Freedman and coauthors, Popular diets: A scientific review, *Obesity Research* 9 (2001): 1S–39S; M. L. Dansinger and coauthors, Comparison of the Atkins, Ornish, Weight Watchers, and Zone diets for weight loss and heart disease risk reduction, *Journal of the American Medical Association* 293 (2005); 43–53; V. S. Retelny, Fad diet review, available at www.foodandhealth.com, 2005, accessed January 2005; and Wheat Council, "Setting the Record Straight" at www.wheatfoods.org/.
3. M. L. Dansinger and coauthors, 2005.
4. Adapted from American Heart Association, "Quick Weight Loss or Fad Diets," available at www.americanheart.org/presenter.jhtml?identifier=4584; Weighing the diet books, *Nutrition Action Newsletter,* January/February 2004: 3–8; and Weight-control Information Network on choosing a safe and successful diet, available at http://win.niddk.nih.gov/publications/choosing.htm.
5. M. Freedman and coauthors, Popular diets: A scientific review, *Obesity Research* 9 (2001): 1S–39S; and National Weight Control Registry; available at www.uchsc.edu/nutrition/WyattJortberg/nwcr.htm.
6. M. Freedman and coauthors, 2001.

Chapter *Nine*

Health Care Systems and Policy

Learning *Objectives*

After you have read and studied this chapter, you will be able to:

- Describe factors affecting the cost and delivery of health care.
- Explain why health promotion is a major component of the rhetoric about health care reform at the national level.
- Differentiate between traditional systems of health care and managed forms of health care.
- Describe eligibility requirements for and services provided to recipients of Medicare and Medicaid.
- Identify consumer trends affecting health care.
- State the value of using medical nutrition therapy protocols to document client outcomes in various health care settings.

This chapter addresses such issues as health care policy and delivery systems, current reimbursement issues, policies, and regulations, and outcomes-based research, which are Commission on Accreditation for Dietetics Education (CADE) *Foundation Knowledge and Skills* requirements for dietetics education.

Chapter *Outline*

Something To Think About...

The enjoyment of the highest attainable standard of health is one of the fundamental rights of every human being without distinction of race, religion, political belief, economic or social condition. . . . Governments have a responsibility for the health of their peoples which can be fulfilled only by the provision of adequate health and social measures.

– PREAMBLE TO THE CONSTITUTION OF THE WORLD HEALTH ORGANIZATION

Introduction

Prevention of disease makes sense, especially in light of the cost of health care. Health care expenditures in the United States continue to increase. In 1965 these costs totaled $42 billion, but in 2002, Americans spent almost $1.6 trillion for health care.[1] This hefty sum represents nearly 15 percent of the gross domestic product (GDP)—compared with 9 percent in 1980.[2] By the year 2010, health care costs are expected to reach $2.6 trillion.[3] Compared with other industrialized nations, U.S. per capita health spending exceeds that of other countries by significant margins.[4]

A strange paradox exists today in U.S. health care: It treats preventable illness rather than investing in prevention. A former secretary of health and human services observed that prevention "must become a national obsession."[5] He went on to say that health promotion and disease prevention offer perhaps the best opportunity to reduce the ever-increasing portion of resources spent treating preventable illness and functional impairment. Likewise, in the U.S. surgeon general's remarks before the Joint Economic Committee of Congress on October 1, 2003, Richard H. Carmona, MD, stated, "There is no greater imperative in American health care than switching from a treatment-oriented society to a prevention-oriented society. Right now we've got it backwards. We wait years and years, doing nothing about unhealthy eating habits and lack of physical activity until people get sick. Then we spend billions of dollars on costly treatments, often when it is already too late to make meaningful improvements to their quality of life or lifespan."[6]

Public policy is now attempting to direct the medical system toward health promotion, disease prevention, and the efficient use of scarce resources. *Healthy People 2010,* the U.S. health agenda designed to help reduce preventable disease, has two basic goals: to increase quality and years of healthy life and to eliminate health disparities among different population segments in the United States.[7] The American Dietetic Association (ADA) agrees with this paradigm shift and asserts that "health promotion and disease prevention endeavors are the best population strategies for reducing the current burden of chronic disease . . . ,"[8] rather than health care assuming only a curative or treatment role.

The ADA continues to maintain that all citizens of the United States should have access to preventive and therapeutic health care.[9] Many studies show that early detection and intervention, immunization, and behavior change could significantly reduce many of the leading causes of death and disability.[10] By investing in health maintenance through health promotion and disease prevention, many of the economic and social costs of disease and injury could be avoided. Good health could be preserved at a reduced cost if we made the "front end" (i.e., prevention) the primary concern, rather than waiting to devote substantial resources to illness and disability after they strike.[11] Yet the current health care system generally provides only limited reimbursement for prevention activities

Figure 1-1 in Chapter 1 compares the leading causes of death in the United States for 1900 and 2002.

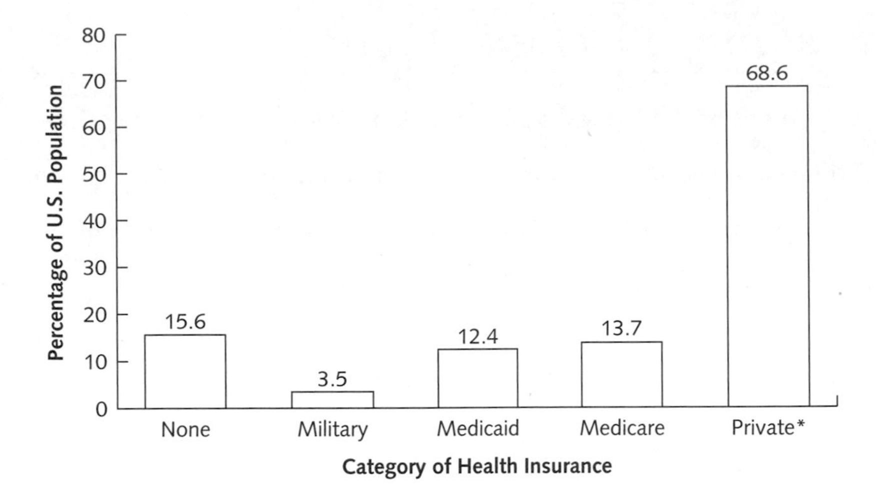

FIGURE 9-1 Categories of Health Insurance and Percentage of U.S. Population Enrolled
Categories of health insurance and percentage of U.S. population that has each type. Of 288 million people in the United States, 243 million had health insurance in 2003. Percentages do not add to 100 because a person can be covered by more than one type of health insurance during the year.

*About 88 percent of people covered by private insurance are covered through an employer.

Source: U.S. Census Bureau.

and/or intervention for conditions such as obesity, cardiovascular disease, osteoporosis, and other chronic conditions that contribute to increasing health care costs.[12] It seems that the most logical avenue, in light of the prevalence of chronic diseases among Americans, is chronic disease prevention. Not only are chronic diseases the leading causes of death in the United States,[13] but they also can limit everyday activities and alter the ability of community members to lead independent lives.

Chronic disease is not just an adulthood problem. Data from 1999 through 2001 indicated that 7.9 percent of boys and 4.5 percent of girls under 18 years of age had limitation of activity caused by one or more chronic health conditions.[14] For adults, 6.1 percent, 12.9 percent, and 20.5 percent of those aged 18–44 years, 45–54 years, and 55–64 years, respectively, had limitation of activity due to a chronic condition or conditions.[15] Among adults between 45 and 64 years, heart and other circulatory conditions, as well as arthritis and other musculoskeletal conditions, were the leading disorders that limited their activity.

This chapter introduces you to the challenges facing health care. One question, for example, is how we can balance the traditional medical model of health care with a wellness/preventive-medicine model. Other issues discussed here are resource allocation and cost containment, social justice and adequate access to health care resources, program accountability and quality in health care, and funding for health promotion and disease prevention.

An Overview of the Health Care Industry

The pluralistic system of health care in the United States includes many parts: private insurance, group insurance, Medicare, Medicaid, workers' compensation, the Veterans Health Administration medical care system, the Department of Defense hospitals and clinics, the Public Health Service's Indian Health Service, state and local public health programs, and the Department of Justice's Federal Bureau of Prisons. Currently, the system is structured around the provision of health insurance. In 2003, 84.4 percent of the U.S. population were insured, and 15.6 percent were not.[16] Some choose not to have health insurance because they can pay for their health care; however, many Americans are forced to make this choice by poverty. The uninsured will be discussed later in the chapter.

In the United States, there are two general categories of **health insurance:** private and public health insurance.[17] Approximately 68.6 percent of the U.S. population have private insurance, and 29.6 percent are covered by governmental health insurance (see Figure 9-1).[18]

Health insurance Protection against the financial burdens associated with health care services and assurance of access to the health care system.

Private Insurance

As we have noted, more Americans carry private insurance than are covered under a governmental health program. The following sections discuss a variety of plans within this privatized system.

TRADITIONAL FEE-FOR-SERVICE PLANS

Private insurance can be in the form of traditional fee-for-service insurance or **group contract** insurance. The traditional fee-for-service plans include a billing system in which the provider of care charges a fee for each service rendered. This type of insurance is provided by both commercial insurance companies and not-for-profit organizations, such as Blue Cross and Blue Shield and independent employee health plans. Traditional fee-for-service plans account for only about 10 percent of insurance coverage today. Critics of fee-for-service plans claim that they encourage physicians to provide more services than are necessary.[19] Proponents of fee-for-service systems prefer the greater flexibility and unrestricted access to physicians, tests, hospitals, and treatments.

Group contract A health insurance contract that is made with an employer or other entity and covers a group of persons identified as individuals by reference to their relationship to the entity.

Managed-care system An approach to paying for health care in which insurers try to limit the use of health services, reduce costs, or both. These health plans are subject to utilization review (UR). That review aims to prevent unnecessary treatment by requiring enrollees to obtain approval for nonemergency hospital care, denying payment for wasteful treatment, and monitoring severely ill patients to ensure that they get cost-effective care.

Health maintenance organization (HMO) A prepaid plan that both finances and delivers health care. HMOs enroll patients as members, charge a fixed fee per year, and provide all medical services deemed necessary. Enrollees generally must use the plan's providers or face financial penalties.

Preferred provider organization (PPO) A group of providers, usually hospitals and doctors, who contract with private indemnity (fee-for-service) insurance companies to provide medical care for a discounted fee. PPOs are subject to peer review and strict use controls in exchange for a consistent volume of patients and speedy turnaround on claim payments.

GROUP CONTRACT INSURANCE

In the latter part of the twenieth century, the nation's health care system went through a major transition from the traditional unmanaged fee-for-service system to a predominantly **managed-care system,** represented by **health maintenance organizations (HMOs)** and **preferred provider organizations (PPOs).** All are prepaid group practice plans that offer health care services through groups of medical practitioners. The presumed goal of managed care is improved quality of care with decreased costs. In 2002, managed-care plans had almost 76.1 million Americans enrolled, including those with public insurance in a managed-care plan (12.8 million and 5.4 million Medicaid and Medicare beneficiaries, respectively).[20] Figure 9-2 shows HMO enrollments since the mid-seventies. The number of HMOs peaked in 1999 and is now declining.

Considering job-based coverage, in 1988 only 27 percent of employees were enrolled in a managed-care plan, and this figure increased to 54 percent, 73 percent, and 86 percent in 1993, 1996, and 1998, respectively. In 2004, 95 percent of employees were enrolled, the majority of workers with job-based coverage belonging to a PPO plan (55 percent), followed by HMOs (25 percent), and POS plans (15 percent; see below).[21] By law, employers with 25 or more employees must offer their employees HMO membership as an alternative to traditional health insurance plans.

In HMOs, physicians practice as a group, sharing facilities and medical records. The physicians may either be salaried or provide contractual services. There are five general models of HMOs:

1. *Staff model:* The HMO owns and operates its own facility; is equipped for laboratory, pharmacy, and X-ray services; and hires its own physicians and other health care providers.
2. *Group model:* The HMO contracts with one or more multispecialty group practices that contract to provide health care services exclusively to its members.
3. *Network model:* Much like the group model, the HMO contracts with multiple group practices, hospitals, and other providers to provide services to its members, but in a nonexclusive arrangement.
4. *Independent practice association (IPA):* A decentralized model—or HMO without walls—in which the HMO contracts with individual physicians to care for plan members in their own private offices for a discounted fee. The physicians are free to contract with more than one plan and may provide care on a fee-for-service basis as well.

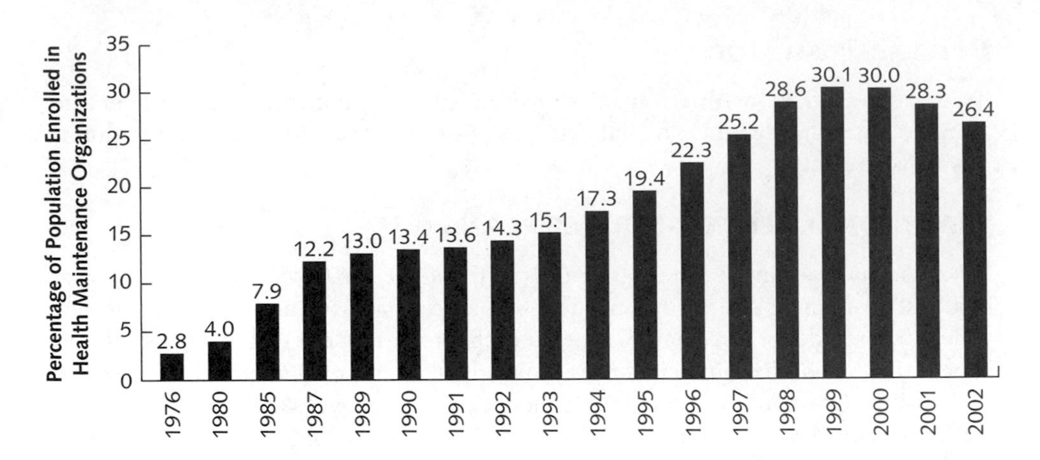

FIGURE 9-2 HMO Enrollments, 1976–2002
HMO enrollments peaked in 1999.
Source: National Center for Health Statistics, *Health, United States, 2003* (Hyattsville, MD: 2003). Illustration adapted from Public Agenda website at www.publicagenda.org.

5. *Point-of-service (POS) plan:* Like IPAs, these popular "open HMOs" allow members the option of using health care providers in a plan's network at a reduced cost, or going to health care providers not in the network—at a higher cost to themselves.

HMOs typically provide comprehensive services across the continuum of care. In some HMO programs, the provider receives a **capitation** payment, usually a specific amount per enrollee per month, to provide a defined group of health care services (see Figure 9-3). Dietitians may be included under specialists or as part of the primary care provider portion, depending on the contractual agreement of the HMO.[22]

FIGURE 9-3 A Sample of How a Capitation Payment Is Used
Source: Adapted from American Dietetic Association, *Medical Nutrition Therapy across the Continuum of Care* (Chicago, IL: American Dietetic Association, 1998), p. 2.

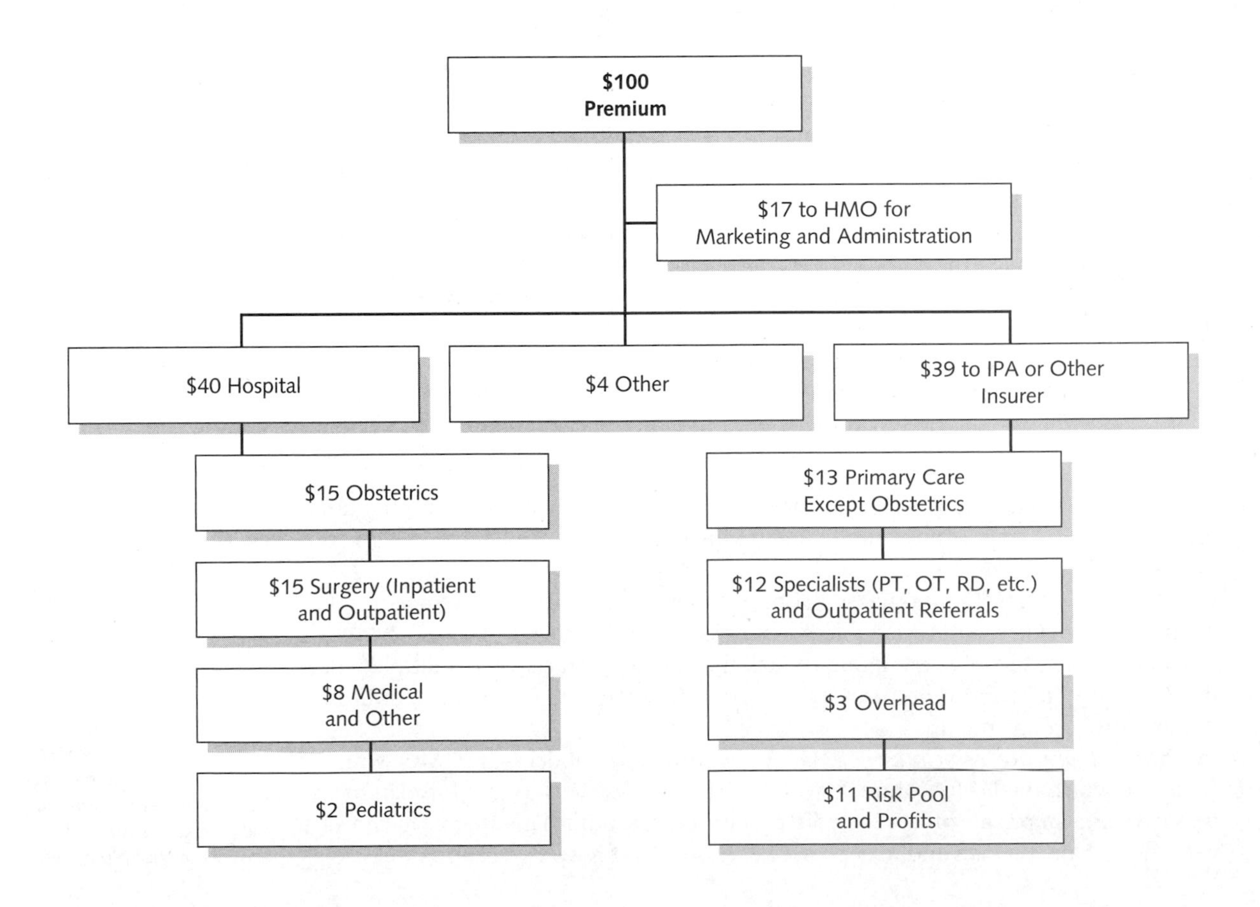

"It's a get-well card from your hospitalization insurance company."

The HMO idea—a fixed cost to the consumer, with health care insurer and health care provider as one and the same—is viewed as a more cost-effective way of practicing medicine than the traditional fee-for-service systems. Because HMOs make money by keeping you healthy, they have a greater stake in your wellness than most fee-for-service doctors.[23] Prepaid group health plans emphasize health promotion, because they provide health care services at a preset cost. By keeping people healthy, HMOs avoid the need for lengthy hospitalizations and costly services. Enrollees of HMOs are hospitalized less frequently than patients of fee-for-service physicians.[24]

Public Insurance

The Centers for Medicare and Medicaid Services (CMS) is the federal agency responsible for administering the Medicare, Medicaid, SCHIP (State Children's Health Insurance), and several other health-related programs, including HIPAA (the Health Insurance Portability and Accountability Act of 1996) and CLIA (Clinical Laboratory and Improvement Amendments). The two major public health insurance plans in the United States are **Medicare** and **Medicaid.** A comparison of their features is provided in Table 9-1.

Capitation A predetermined fee paid per enrollee per month to the participating health care provider.

Medicare A federally run entitlement program through which people age 65 years or older and people in certain other eligible categories receive health insurance.

Medicaid A federally aided, state-administered entitlement program that provides medical benefits for certain low-income persons in need of health and medical care.

Centers for Medicare and Medicaid Services (CMS) A federal agency that establishes guidelines and monitors Medicare, Medicaid, SCHIP, HIPAA, and CLIA.

Workers' compensation, which pays benefits to workers who have been injured on the job, is another public-sector health benefit program. The State Children's Health Insurance Program (SCHIP) provides health coverage to uninsured children whose families earn too much money to qualify for Medicaid but too little to afford private coverage.[25] Health care services are also provided by the Department of Veterans Affairs (VA), the Public Health Service (including the Indian Health Service), the Department of Defense (including the Civilian Health and Medical Program of the Uniformed Services, or CHAMPUS), public hospitals and community health centers, and state and local public health programs.[26]

THE MEDICARE PROGRAM

In 2003, over 39 million individuals were enrolled in Medicare. This program was established in 1965 by Title XVIII of the Social Security Act and is administered by the **Centers for Medicare and Medicaid Services (CMS)** of the Department of Health and Human Services. The Social Security Administration provides information about program eligibility and handles enrollment.[27] Medicare is designed to assist

- People 65 years of age or older;
- People of any age with end-stage renal disease;

	MEDICARE	MEDICAID ADMINISTRATION
Administration	Social Security Office Centers for Medicare and Medicaid Services	Local welfare office Varies within state, territory, or the District of Columbia
Financing	Trust funds from Social Security; contributions from insured	Taxes from federal, state, and local sources
Eligibility	People 65 years of age and older, people with end-stage renal disease, people eligible for Social Security/Railroad Retirement Board disability programs for 24 months; Medicare-covered government employees, possibly others	Individuals with low incomes, people 65 or older, the blind, persons with disabilities, all pregnant women and infants with family incomes below 133% of poverty level, possibly others
Benefits*	Same in all states **Hospital insurance (Part A)** *helps* pay for inpatient hospital care, skilled nursing facility care, home health care, hospice care. **Medical insurance (Part B)** *helps* pay for physicians' services, outpatient hospital services, home health visits, diagnostic X-ray, laboratory, and other tests; necessary ambulance services, other medical services and supplies, outpatient physical or occupational therapy and speech pathology; partial coverage of mental health treatment, kidney dialysis, medical nutrition therapy services for people with diabetes or kidney disease, and certain preventive services.†	Varies from state to state **Hospital services:** inpatient and outpatient hospital services, other laboratory and X-ray services, physician services, screening, diagnosis, and treatment of children, home health care services **Medical services:** many states pay for dental care, health clinic services, eye care and glasses, prescribed medications, and other diagnostic, rehabilitative, and preventive services, including nutrition services
Typical Exclusions	Regular dental care and dentures, routine physical exams and related tests, eyeglasses, hearing aids and examinations to prescribe and fit them, most prescription drugs, nursing home care (except skilled nursing care), custodial care, immunizations (except for pneumonia, influenza, and hepatitis B), cosmetic surgery	Varies from state to state
Premium Costs (2005)	Part A: none if eligible, or $206–$375/month Part B: $78.20/month	None (federal government contributes 50% to 80% to states to cover eligible persons)

*Medicare beneficiaries who have both Part A and Part B can choose to get their benefits through a variety of risk-based plans (e.g., HMOs, PPOs), known as the Medicare Advantage Plan, which may expand coverage. An additional premium may apply.

†Certain recipients are covered for bone mass measurements, colorectal cancer screening, diabetes self-management training and supplies, glaucoma screening, mammogram screening, Pap test and pelvic examination, prostate cancer screening, and certain vaccinations. The Medicare Moderization Act of 2003 expanded coverage. For more information, visit www.medicare.gov.

Source: Adapted from U.S. Department of Health and Human Services, 2002 *Guide to Health Insurance for People with Medicare* (Washington, D.C.: U.S. Department of Health and Human Services, 2002); and Centers for Medicare and Medicaid Services, *Your Medicare Benefits* (Baltimore, MD: U.S. Department of Health and Human Services, 2004).

TABLE 9-1 A Comparison of Medicare and Medicaid Services

- People eligible for Social Security or Railroad Retirement Board disability benefits for 24 months;
- Individuals who are receiving or are eligible to receive retirement benefits from Social Security or Railroad Retirement Boards; and
- People who had Medicare-covered government employment.

To obtain Medicare benefits, recipients are offered the Original Medicare Plan or a Medicare Advantage Plan. Basically, Medicare consists of two separate parts: hospital insurance (Part A) and medical insurance (Part B). No monthly premium is required for

Medicare Part A if a person or his or her spouse is entitled to benefits under either Social Security or the Railroad Retirement System or has worked a sufficient period of time in federal, state, or local government to be insured, because premiums were paid through payroll taxes while the individual or spouse was working.[28] Those who do not meet these qualifications (40 or more quarters of Medicare-covered employment) may purchase Part A coverage if they are at least age 65 and meet certain requirements.[29] Specifically, for 2005, for those working 30–39 quarters, the premium is $206 per month, whereas those having less than 30 quarters pay a monthly premium of $375. For Medicare Part B, the premium is $78.20 per month. The Department of Health and Human Services announces these premiums annually.

Medicare Part A

Medicare Part A provides hospital insurance benefits that include inpatient hospital care, care at a skilled nursing facility, and some home health care. Deductible and **coinsurance** fees apply. For inpatient care, the deductible for the first 60 days of care in 2005 is $912; coinsurance amounts are $228 per day for days 61–90 of a hospital stay, and $456 per day for days 91–150. Beyond 150 days, Medicare pays for hospital charges. Inpatient days exceeding 90 count toward an individual's lifetime reserve days, which cannot exceed 60 days over one's life. For skilled nursing care, there is a coinsurance fee of $114 per day for days 21–100 annually. Hospital inpatient charges are reimbursed according to a **prospective payment system (PPS)** known as **diagnosis-related groups (DRGs)** (discussed in detail later in this chapter). Since 1983, the government has shifted a larger portion of health care costs to Medicare beneficiaries through larger **deductibles,** greater use of services with coinsurance, and use of services not covered by Medicare.

Medicare Part B

Medicare Part B is an optional medical insurance program financed through premiums paid by enrollees and contributions from federal funds; it provides supplementary medical insurance benefits for eligible physician services, outpatient hospital services, certain home health services, and durable medical equipment. In addition to the monthly premium, there is a $110 deductible per year (2005),[30] and a 20 percent copay applies for each service. As of 2002, Medicare pays qualified dietitians and nutrition professionals who enroll in the Medicare program as providers, regardless of whether they provide medical nutrition therapy (MNT) services in an independent practice setting, hospital outpatient department, or any other setting, except for patients in an inpatient stay in a hospital or skilled nursing facility.[31] Enrolled Medicare MNT providers are able to bill Medicare for MNT services provided to Medicare beneficiaries with Type 1 diabetes, Type 2 diabetes, gestational diabetes, nondialysis kidney disease, and post-kidney-transplant status using specified codes. A physician's referral for MNT is required.

Coverage Gaps

The two most notable gaps in Medicare coverage have been prescription drug coverage and skilled nursing/long-term institutional care. Traditionally, most prescription drugs are not covered at all under the Medicare program. As we noted in the discussion of Medicare Part A, only 100 days of skilled nursing/long-term care are covered annually. Thereafter, patients or their families must either pay the costs themselves or "spend down" their assets in order to reduce their net worth and be eligible for Medicaid coverage of long-term care. However, in December 2003, President George W. Bush signed into law the Medicare Prescription Drug, Improvement, and Modernization Act of 2003 (Medicare Modernization Act).[32] The Medicare Moderization Act provides optional

Coinsurance A cost-sharing arrangement in which the insured assumes a portion of the costs of covered services.

Prospective payment system (PPS) A payment system under which hospitals are paid a fixed sum per case according to a schedule of diagnosis-related groups.

Diagnosis-related groups (DRGs) A method of classifying patients' illnesses according to principal diagnosis and treatment requirements, for the purpose of establishing payment rates.

Deductibles The amount of expense that must be incurred by a person who is insured before an insurer will assume any liability for all or part of the remaining cost of covered services.

coverage to Medicare recipients, including drug discount cards/prescription drug plans and additional preventive benefits (wellness physical exam, cardiovascular disease blood screening, and diabetes screening for those at risk), in addition to those preventive benefits already covered (cancer screening, bone mass measurements, and vaccinations).[33]

For those in the Original Medicare Plan, a Medigap policy may be purchased if the individual participates in both Medicare Part A and Part B. This is a supplemental insurance policy sold by private insurance companies to help pay the deductible, coinsurance fees, prescription drug costs, and certain services not covered by Medicare. Because the Original Medicare Plan is a fee-for-service plan, a fee is typically charged for each health care service or supply received.[34]

Another option for individuals is to receive their Medicare benefits through a Medicare Advantage Plan. These plans must cover at least the same benefits covered by Medicare Part A and Part B; however, the costs may vary through these Medicare Managed-Care Plans, Medicare Preferred Provider Organization Plans, Medicare Private Fee-for-Service Plans, or Medicare Specialty Plans.[35] To join, the individual must have Medicare Part A and Part B. The Part B premium is still paid ($78.20 monthly in 2005), and a monthly premium may have to be paid to the Medicare Advantage Plan provider.

For additional benefits, Medicare recipients often explore other options. They may continue insurance coverage through a current or former employer. Individuals may also choose to purchase nursing home or long-term care policies, which pay cash amounts for each day of covered nursing home or at-home care. Finally, individuals may qualify for full Medicaid (see the next section) benefits or at least to receive some state assistance in paying Medicare costs.

THE MEDICAID PROGRAM

Medicaid, an entitlement program insuring almost 36 million individuals,[36] was established as a joint state and federal program, the federal government paying 50 percent or more of the costs depending on a state's per capita income. ("States" include states, U.S. territories, and the District of Columbia.) It was established in 1965 by Title XIX of the Social Security Act. Medicaid provides assistance with medical care for

- eligible persons with low incomes;
- certain pregnant women and children with low incomes;
- older adults, the blind, and people with disabilities; and
- members of families with dependent children in which one parent is absent, incapacitated, or unemployed.

Poverty guidelines are published annually by the Department of Health and Human Services. The 2005 guidelines are summarized in Table 5-1 in Chapter 5. Updates to the poverty guidelines (poverty line) can be found at http://aspe.hhs.gov/poverty/index.shtml.

The individual states administer the program and define eligibility, benefits and services, and payment schedules. Typically, one must meet three criteria: income, categorical, and resource. Income must often be below—sometimes significantly below—the federal poverty guidelines.

Because states administer the program, an individual may qualify for Medicaid in one state but not in another. Generally, those eligible for Medicaid include the following:

- Those eligible for Temporary Assistance for Needy Families (TANF) or Supplemental Security Income (SSI)
- Children under 6 years old living in a household at or below 133 percent of the poverty guidelines and all children born after September 30, 1983, who are under age 19 and living in a household at or below the poverty guidelines
- Pregnant women (eligible only for services related to pregnancy/complications, delivery, and postpartum care)

- Recipients of adoption or foster care assistance under Title IV of the Social Security Act
- Special protected groups, including individuals who lose cash assistance as a consequence of work or increased Social Security benefits
- Medicare beneficiaries with low incomes

To meet the categorical requirements, one must be a member of a family with dependent children or be an older adult, blind, or a person with a disability. The resource test sets a maximum allowable amount for liquid resources and other assets. Income and asset eligibility standards vary widely among the states, territories, and the District of Columbia.

Medicaid covers inpatient and outpatient hospital services; physician, pediatric/family nurse practitioner, and nurse–midwife services; selected health center and rural health clinic services; prenatal care and family planning services/supplies; vaccines for children and other services for those under 21 years, laboratory and X-ray services; and skilled nursing home and home health services, among others. Some states include other benefits, such as prescription drug coverage and dental services, but there is significant variability among states.[37] In many states, Medicaid programs cover certain forms of nutrition services provided by dietitians.[38]

Medicaid covers less than half of those below the poverty line.[39] The American Medical Association has recommended that Medicaid be expanded to provide acute-care coverage for all persons below the poverty line.[40] Although this would increase the cost of services provided, it would improve access to health services and potentially decrease health care costs in the long run.

THE STATE CHILDREN'S HEALTH INSURANCE PROGRAM

President William J. Clinton signed into law the Balanced Budget Act of 1997, which included Title XXI, the **State Children's Health Insurance Program (SCHIP).** SCHIP was the largest single expansion of health insurance coverage for children in more than 30 years.[41] At the time, nearly 11 million American children—1 in 7—were uninsured. In fact, from 1988 to 1998 the proportion of children insured through Medicaid increased from 15.6 percent to 19.8 percent, while the percentage of children without health insurance increased from 13.1 percent to 15.4 percent. This increase was attributed to fewer children being covered by an employer-sponsored health insurance.[42] The SCHIP initiative was designed to reach these children, many of whom were part of working families with incomes too high to qualify for Medicaid but too low to afford private health insurance. For example, in 2004, in most states, uninsured children under the age of 19, whose families earned up to $36,200 per year (family of four) were eligible.[43] States are able to use part of their federal funds to expand outreach and ensure that all children eligible for Medicaid and the new SCHIP program are enrolled. The initiative is a partnership between the federal and state governments that will help provide children with health coverage. Because Medicaid allows states flexibility in determining eligibility, states currently cover children whose family incomes range generally from below the poverty guidelines to as high as 300 percent of the poverty guidelines. Funds for the program became available to states in 1997. States receive federal matching funds only for actual expenditures to insure children. In 2003, almost 6 million children were covered by SCHIP.[44] Figure 9-4 summarizes the enrollment in SCHIP since 1999.

State Children's Health Insurance Program (SCHIP) Created under Title XXI of the Social Security Act; expands health coverage to uninsured children whose families earn too much income to qualify for Medicaid but too little to afford private coverage.

Under the program, states have flexibility in targeting eligible uninsured children. States may choose to expand their Medicaid programs, design new child health insurance programs, or create a combination of both. States choosing a new children's health insurance program may offer one of the following benchmark plans: the standard Blue Cross/Blue Shield Preferred Provider Option offered by the Federal Employees Health

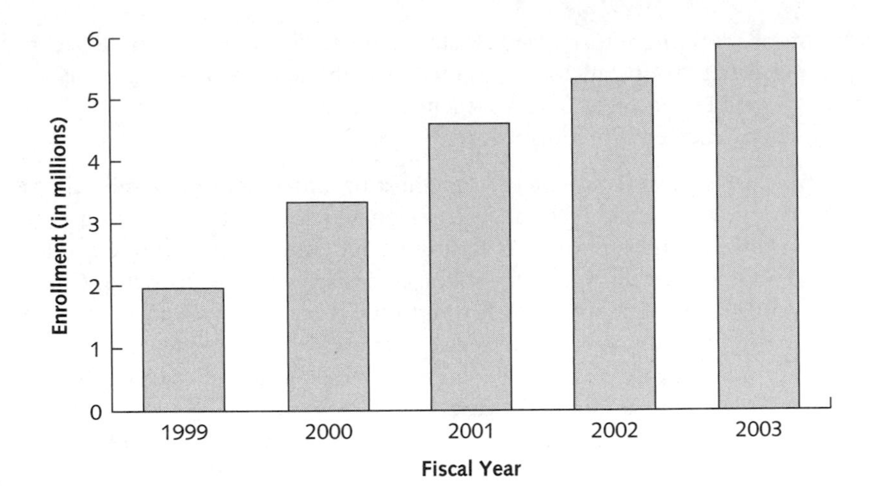

FIGURE 9-4 SCHIP Enrollment, 1999–2003
SCHIP enrollment has increased steadily since the program's inception.

Source: Centers for Medicare and Medicaid Services, State Children's Health Insurance Program (SCHIP) site portal, Enrollment Reports; available at www.cms.hhs.gov/SCHIP/enrollment.

Insure Kids Now is a website with eligibility and contact information for each state, each territory, and the District of Columbia. It also contains information about local and national outreach activities, including school-based outreach. Visit www.insurekidsnow.gov.

Benefit Program; a health benefit plan offered by the state to its employees; or the HMO benefit plan with the largest commercial enrollment in the state. A state may also choose to offer the "equivalent" of one of the benchmark plans. If a state chooses this option, its plan's value must be at least equal to the benchmark plan's, and it must include inpatient and outpatient hospital services, physicians' surgical and medical services, laboratory and X-ray services, and well baby/child care services including immunizations. In addition, a benchmark-equivalent plan must include benefits similar to the benchmark plan coverage of prescription drugs, mental health services, vision care, and hearing-related care. States choosing the Medicaid option must offer the full benefit package.

The Uninsured

In theory, health care coverage is available to virtually all U.S. citizens through one of four routes: Medicare for the elderly and people with disabilities, Medicaid for low-income women and children and some low-income men and people with certain disabilities, employer-subsidized coverage at the workplace, or self-purchased coverage for those ineligible for the previous three.[45] Yet an estimated 45 million people (15.6 percent of the population) with no insurance coverage at all live in the United States, and perhaps an even larger number of people have coverage that is inadequate for any major illness.[46]

In 1987, 12.9 percent of the population were uninsured, and the proportion of the uninsured continued to increase until it peaked in 1998 at 16.3 percent. The current rate represents an increase from a rate that fell to 14.2 percent in 2000. Although the 2003 data represent a percentage increase over 2002 in those without coverage, the actual number of people insured actually increased.

Who, then, are the uninsured? Statistics show that they are not the elderly, who have Medicare, or the very poor, who have Medicaid. Instead, those who lack coverage are primarily people in the middle (for example, the working poor and those who work for small businesses). More than half are in families with incomes below 200 percent of the poverty guidelines.[47] They also include the self-employed, those who work part-time, seasonal workers, the unemployed, full-time workers whose employers offer unaffordable insurance or none at all, and early retirees—aged 55 through 64—who retired from companies that either offered no health insurance or have since dropped it.[48] These persons are classified further as the employed uninsured and the nonworking uninsured. In 2001, included among the uninsured were 9.2 million children.[49]

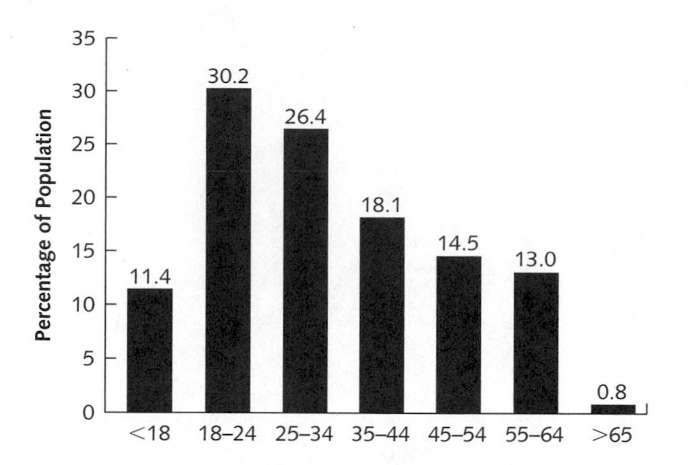

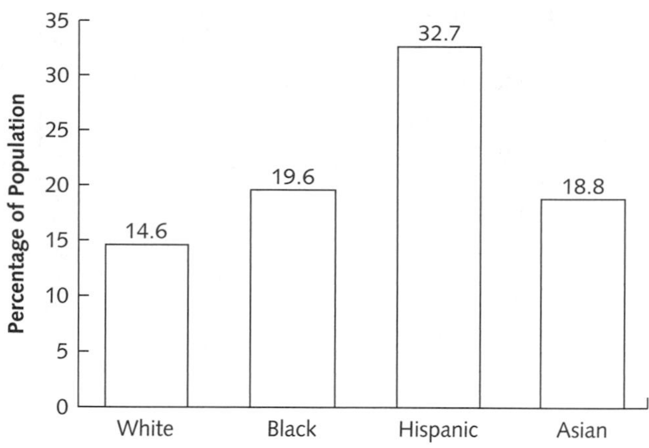

FIGURE 9-5 Percentage of Persons in the United States without Health Care Coverage, by Age, and Race or Ethnic Origin,* 2003

Two-thirds of uninsured persons are under age 35, and those of Hispanic origin account for the greatest proportion of the uninsured.

*In 2001 and earlier, Current Population Surveys asked respondents to report only one race (White, White Not of Hispanic Origin, Black, or Asian and Pacific Islander). Persons of Hispanic origin may be of any race. Being Hispanic was reported by 11.8 percent of White, 2.7 percent of Black, 26.5 percent of American Indian and Alaska Native, and 10.0 percent of Hawaiian, and Other Pacific Islander households reporting only one race.

Source: U.S. Census Bureau, Historical Health Insurance Tables; available at www.census.gov/hhes/hlthins/historic/index.html.

Looking more closely reveals that those covered by employment-based health insurance fell to 60.4 percent.[50] Health insurance premiums increased 59 percent from 2001 to 2004, with employee contributions increasing 49 to 57 percent over that same period.[51]

Considering children, in 2001, 9.2 million children under 19 years (12.1 percent) were uninsured.[52] The proportion of children without insurance did not change (11.4 percent of all children) from 2002 to 2003; however, impoverished children continued to be the children most likely not to be insured (19.2 percent). Considering race, regardless of age, in 2003, 19.6 percent, 18.8 percent, and 32.7 percent of blacks, Asians, and Hispanics, respectively, did not have insurance; these figures were unchanged from 2002 (see Figure 9-5). For 2001–2003, 27.5 percent of American Indian and Alaska natives did not have health insurance; this percentage was also unchanged.[53]

When those without health insurance do get sick, they often wind up using the most expensive treatment available—hospital emergency room care—or they delay getting treatment and later require more expensive and prolonged medical services. These costs are shifted to the people who are insured.

All community members, including the employed and nonworking uninsured, the homeless, and others, should be able to obtain medical care when it is needed. However, cost was a barrier for health care access for 1 in 7 individuals living in the United States in 2003. Between 2001 and 2003, however, access to needed medical care improved. During this time, the percentage of those who had no insurance and a low income fell from 16.4 percent to 13.2 percent. In fact, the unmet medical needs of children from low-income households decreased to such a degree that income-related differences in access to health care for children disappeared.[54]

Demographic Trends and Health Care

Between 1946 and 1964, 78 million babies were born in the United States; these individuals—the baby boomers—now make up one-third of the population.[55] By the year 2030, the baby boom will become a senior boom, with 21 percent of the population over 65 years of age.[56] Not only will the elderly be greater in number, but they may require care for a greater number of years, placing a heavier burden on the long-term care system (see Figure 9-6). Because older Americans consume a disproportionate amount of medical care, the demand for such care, including pharmaceutical products and services, can be expected to rise.[57]

FIGURE 9-6 Number of Elderly Needing Long-Term Care, 1990 and 2030

The number of individuals needing skilled nursing care is expected to rise as the baby boomers age.

Source: A Call for Action: Final Report of the Pepper Commission (Washington, D.C.: U.S. Government Printing Office, 1990).

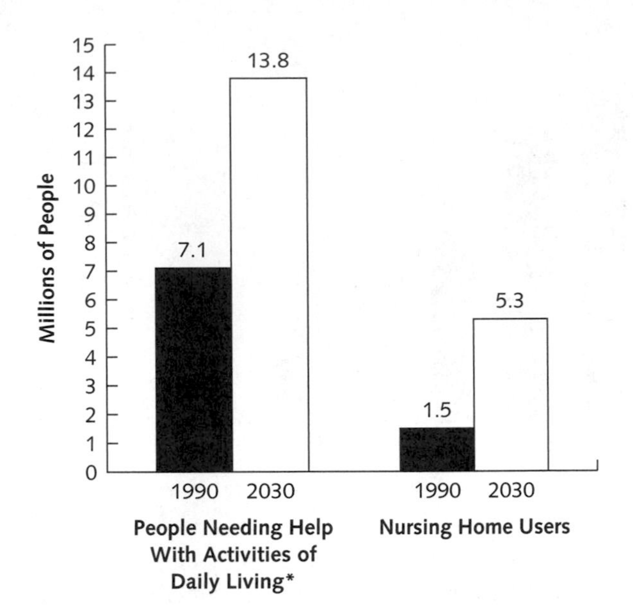

*Activities of daily living (ADL) include activities such as bathing, dressing, toileting, continence, and feeding.

Racial and geographical factors in the population are also important to the shape of the future. In some parts of the United States, particularly the Southwest, the Hispanic population will dramatically increase. To the extent that such a population may exhibit differing utilization patterns for medical services or pharmaceuticals, such changes may significantly affect the marketplace. Geographical demographics will also be important, especially if the population drift from the Northeast to the Southwest and the Sun Belt continues.[58]

The Need for Health Care Reform

To determine the rating of a particular health care system, one must examine three crucial variables: cost, quality, and access.[59] At the zero end of the scale is no health care system. As we have seen, millions of Americans cannot afford to buy into or gain meaningful, ongoing access to any health care at all. At the other end of the scale is high-quality, reasonably priced, accessible health care. On such a scale, how does the U.S. health care system rate?

Before you respond, consider the following scenario:[60] Imagine that you are the decision maker in a large corporation, and I approach you and try to sell you a product. I say that I want to sell you a key piece of equipment that meets the following specifications:

- It will cost you $3,200 per employee per year.
- It will consume up to half of each profit dollar and will rise in price by 15 to 30 percent annually.
- There is a tremendous unexplained variation in this product depending on who uses it.
- There is no way to measure its quality in terms of appropriateness, reliability, or outcome.
- And you'll just have to take my word for it when I tell you that we adhere to the highest professional standards.

Would you buy this product? Many believe the current U.S. health care system fits this description. Not only is it expensive, but we don't necessarily know what we are paying for or whether what we are paying for is worth it.[61]

Ursula Markus/Photo Researchers

By the year 2030, the number of elderly needing nursing home care will more than triple.

The term *health care reform* refers to the efforts undertaken to ensure that everyone in the United States has access to affordable, quality health care. Among the challenges for health care reform are how to make health care accessible to everyone, how to contain costs, how to provide nursing home care to those who need it, and how to ensure that Medicare and Medicaid can serve all who are eligible.

As we will see, cost, access, and quality are interrelated; manipulating one has an astounding impact on the others. For example, some people argue that we should abandon free enterprise and turn the system over to the government, as has been done in other countries, including Canada. Critics of government-run health care systems say they appear promising at first but soon bog down in bureaucracy, unable to keep pace with advances in medical technology. Some point to the Canadians who travel to the United States to purchase treatment out of their own pockets rather than wait in line.[62] Is there a way to extend the scope of the system without sacrificing quality?

Health care policy makers are studying alternative models of delivery and financing in hopes of applying to the United States approaches that have been successful in other nations.[63] The U.S. health care system appears both to have higher costs (see Figure 9-7) and to offer less access than the systems of other industrialized nations. During the last two decades, U.S. health care trends differed from those in other nations in a variety of ways, most notably in rising costs and erosion of access to health care services. Per capita health spending in the United States exceeds that of other industrialized countries by huge margins.[64] The following sections consider each of these issues.

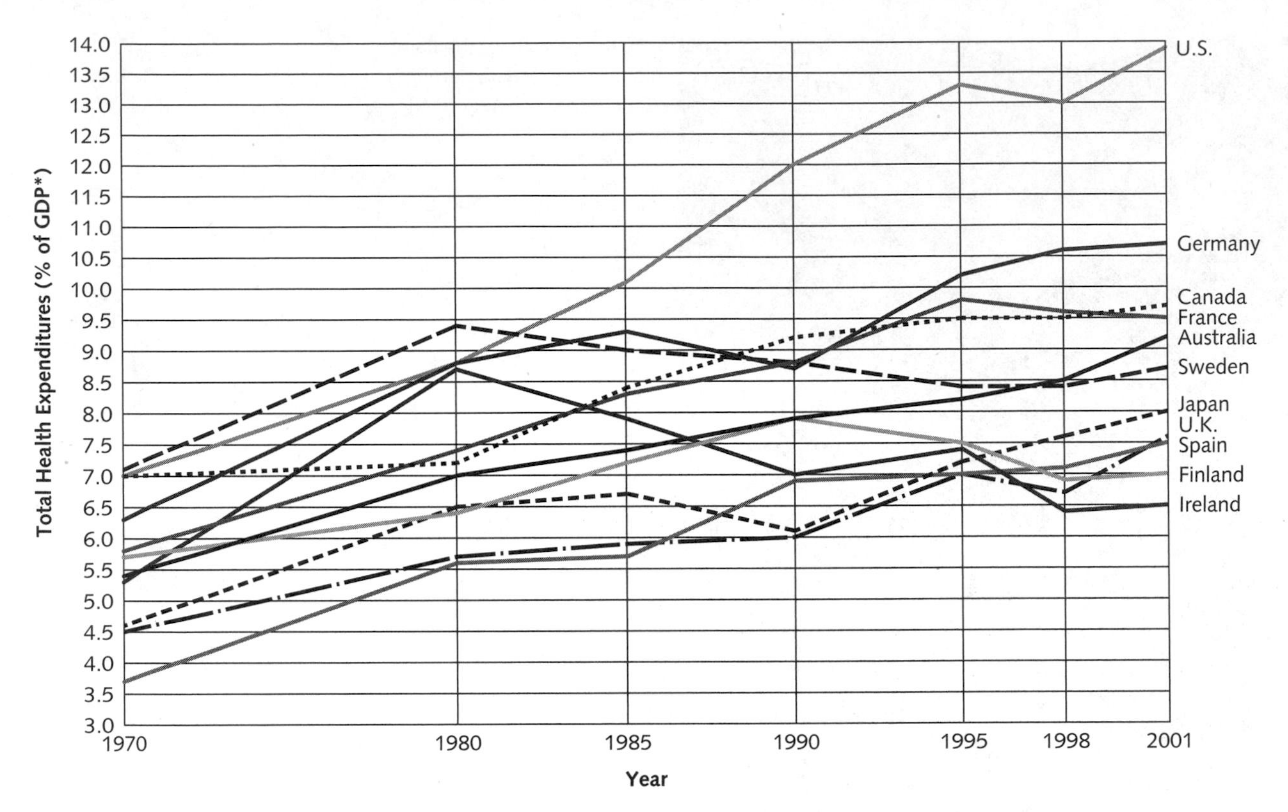

*Gross domestic product (GDP) represents the total value of a nation's output, income, or expenditures produced within its borders. GDP is more specific than gross national product (GNP), the total retail market value of all goods and services.

FIGURE 9-7 Total Health Expenditures as a Percentage of Gross Domestic Product (GDP), 1970–2001

Source: National Center for Health Statistics, *Health, United States, 2001* (Hyattsville, MD: National Center for Health Statistics, 2001); health data from Organization of Economic Cooperation and Development.

Copayment The portion of the charge for medical services that the patient must pay.

Third-party reimbursement involves three parties in the process of paying for medical services. The *first party* (patient) receives a service from the *second party* (physician, hospital, or other health care provider). The *third party* (insurance company or government) pays the bill.

THE HIGH COST OF HEALTH CARE

Health care inflation is well established. As noted in the introduction to this chapter, in 2002 Americans spent almost $1.6 trillion for health care.[65] Figure 9-8 tracks the rise in U.S. health care costs since 1960. In fact, since that year, health care expenditures have increased over 800 percent.

The level of health care activity is expected to grow. This growth is a result of various factors, including an aging population, increased demand (fostered in part by more consumer awareness of health issues), and continuing advances in medicine, which make it possible to do more for people than ever before.[66]

A major contributor to health care expenditures in the United States is the administrative cost of the insurance process itself. Yet another factor contributing to the cost of our health care is the practice of defensive medicine and the associated phenomenon of ever-rising professional liability costs, which have skyrocketed. Some say that we have become a litigious society. For example, an OB-GYN reported that in 2002 his premium was $23,000 and that it then increased to $47,000 in 2003 and finally to $84,000 in 2004.[67] Patient safety and the legal process remain at the forefront of the medical malpractice crisis, and in order to help curtail the cost of liability insurance, reforms are necessary.[68]

Efforts at Cost Containment

Efforts to curb soaring health care costs cover a broad spectrum: slowing hospital construction, modifying hospital and physician reimbursement mechanisms, reducing the length of hospital stays, increasing **copayments** and deductibles for insured employees and Medicare recipients, changing eligibility requirements for Medicaid, reducing unnecessary

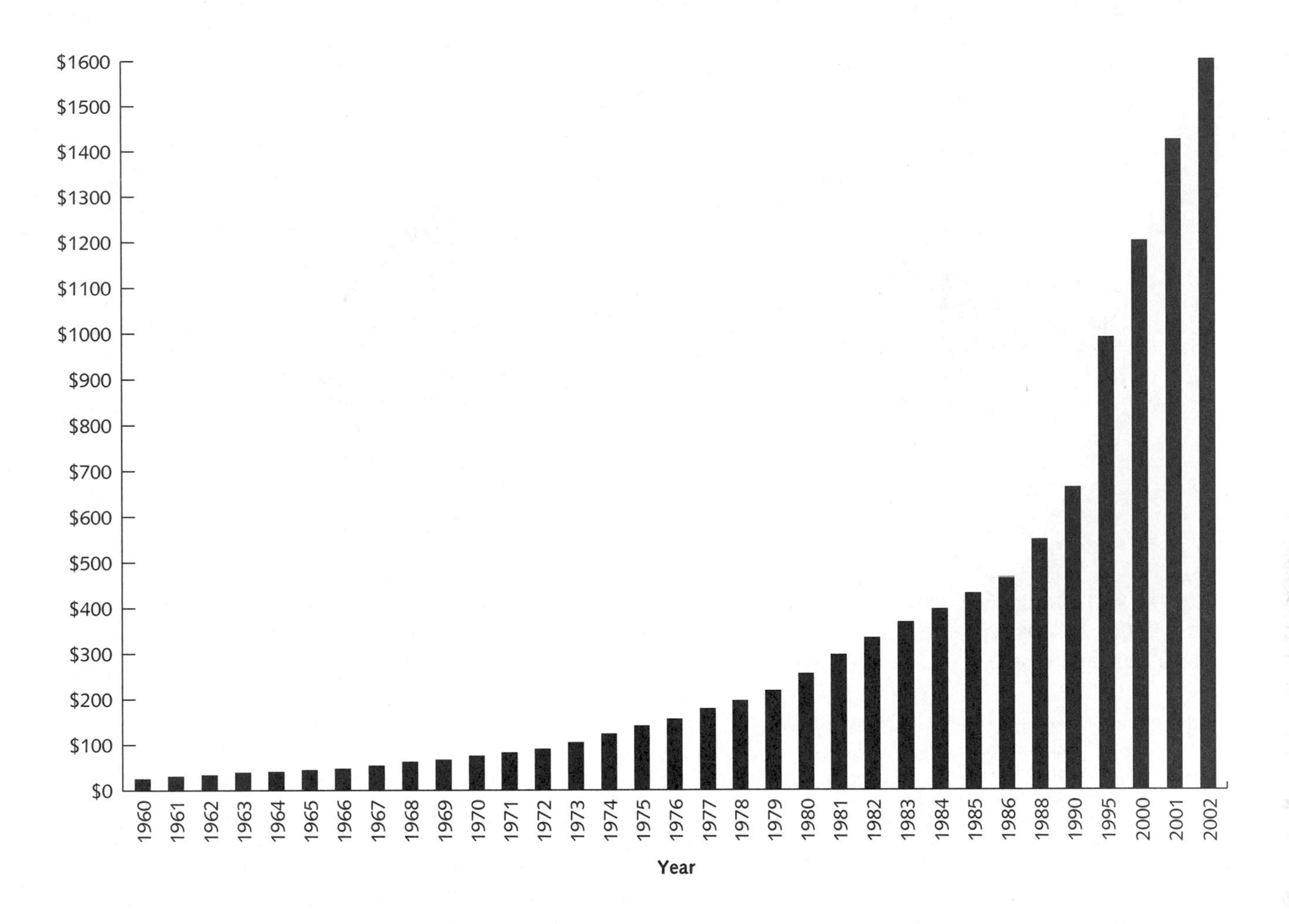

FIGURE 9-8 National Health Expenditures (billions of dollars), 1960–2002

Source: Adapted from *Source Book of Health Insurance Data* (Washington, D.C.: Health Insurance Association of America, 2003).

surgery by requiring patients to obtain second opinions, restricting access to new technology, encouraging alternative delivery systems, and emphasizing prevention.[69] Use of generic drugs has also been utilized to help contain the costs of health care.[70]

The recent cost containment effort in the United States is actually a fierce competition among *third-party* payers (government, insurance companies, and employers) to control their own costs. This effort has been characterized by three trends:[71]

1. There is a movement away from traditional fee-for-service health care to newer models of managed care, evident in the enrollments in HMOs and PPOs.
2. As more and more of their profits are siphoned off into health care coverage, companies are increasingly attempting to manage the health care of their employees themselves to reduce expenditures. In an effort to avoid **cost shifting,** many businesses are moving to **self-insured health plans,** thereby determining which benefits are covered and assuming the risks involved.[72]
3. The payers (government, insurance companies, and employers) are actively setting **reimbursement** restrictions and limitations.

Cost shifting A much-criticized aspect of the existing health care system in which hospitals and other providers bill indemnity (fee-for-service) insurers at higher rates to recover the costs of charity care and to make up for discounts given to HMOs, PPOs, Medicare, and Medicaid.

The largest components of national health care expenditures are hospital care (31 percent) and physician and clinical services (25 percent), as illustrated in Figure 9-9.[73] Therefore, efforts to contain costs have largely been aimed at these providers.

One example of cost containment is the prospective payment system (PPS) that the federal government implemented as a result of the 1983 Social Security Act Amendments.

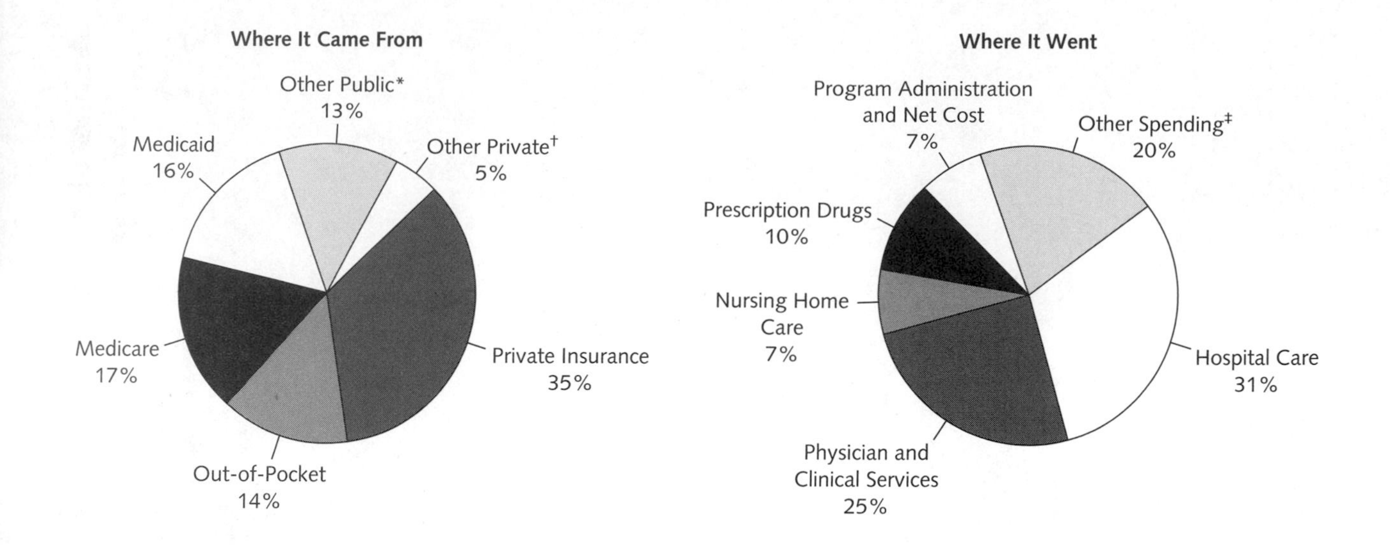

FIGURE 9-9 The Nation's Health Dollar, 2002: Where It Came from and Where It Went

*"Other Public" includes SCHIP and Medicaid SCHIP expansion, workers' compensation, public health activity, Department of Defense/Department of Veterans Affairs/Indian Health Service health programs, and state and local hospital subsidies, and school health.

†"Other Private" includes industrial in-plant, privately funded construction, and nonpatient revenues, including philanthropy.

‡"Other Spending" includes dentist services, other professional services, home health, durable medical products, over-the-counter medicines and sundries, public health, research, and construction.

Source: Centers for Medicare and Medicaid Services. Office of the Actuary. National Health Statistics Group, 2004 [Cowan and coauthors, National health expenditures, 2002, *Health Care Financing Review* 25 (Summer 2004): 143–166].

The purpose of the PPS was to change the behavior of health care providers by changing the incentives under which care is provided and reimbursed. Prospective payment means knowing the amount of payment in advance. The PPS uses diagnosis related groups (DRGs) as a basis for reimbursement. Patients are classified according to their **principal diagnosis,** secondary diagnosis, sex, age, and surgical procedures.

The DRG approach is based on a system of classifying hospital admissions. The system begins with the ninth edition of *International Classification of Diseases: Clinical Modifications,* abbreviated **ICD-9-CM,** which contains approximately 10,000 possible reasons for a hospital admission, organized into 23 major categories. The 23 categories are subdivided into 490 DRGs. Tables compute average cost per discharge by state, region (rural or urban), hospital bed size, and other factors.[74] All DRGs have been assigned a *relative weight* that reflects the cost of caring for a patient in the particular category. Table 9-2 shows a sample payment based on DRGs. Note that a patient with a complication or comorbidity (for example, with malnutrition) is assigned a higher relative weight, reflecting the need for more intensive services.

One consequence of the PPS has been an increased focus on outpatient services as opposed to more costly inpatient care. This trend has significant implications for dietitians. It will continue to bring increased opportunities for consulting in outpatient settings, such as hospital outpatient clinics and **home health agencies** and for private-practice counseling and consulting in physician or other health care provider offices, HMOs, health and fitness facilities, weight loss programs, community health centers and clinics, and group patient education classes.

EQUITY AND ACCESS AS ISSUES IN HEALTH CARE

Is health care a basic right? The majority of people in the United States (54 percent) say that providing health insurance to the uninsured should be a top legislative priority.[75] In reality, as deVise has observed, health care may be more of a privilege than a right:

> If you are either very poor, blind, disabled, over 65, male, female, white, or live in a middle- or upper-class neighborhood in a large urban center, you belong to a privileged class of health care recipients, and your chances of survival are good. . . . But, if you are none of these, if you are only average poor, under 65, female, black, or live in a low-income urban

TABLE 9-2 Sample Payment Based on DRGs, with and without Complication/Comorbid Condition

NAME OF DRG	DRG NUMBER	MEDICARE RELATIVE WEIGHT		BASE RATE		PAYMENT AMOUNT
Respiratory infections and inflammations without complication/ comorbid condition	080	1.0404	×	$4,000	=	$4,162
Respiratory infections and inflammations with complication/ comorbid condition	079	1.8144	×	$4,000	=	$7,258

Source: D. D'Abate Cicenas, Increasing Medicare reimbursement through improved DRG coding, *Reimbursement and Insurance Coverage for Nutrition Services* (Chicago: American Dietetic Association, 1991), p. 53. Used with permission of Ross Products Division, Abbott Laboratories, Columbus, OH 43216. © 1990 Ross Products Division, Abbott Laboratories.

> neighborhood, small town, or rural area, you are a disenfranchised citizen as far as health care rights go, and your chances of survival are not good.[76]

In 1983, a presidential commission studying ethical issues in medicine reported, "Society has a moral obligation to ensure that everyone has access to adequate [health] care without being subjected to excessive burdens."[77] Proponents of this view argue that just as the federal government provides for defense, postal delivery, and certain other services, it should provide at least a *minimal* amount of basic health care.[78]

This debate leads to another question: Access to what? What *is* an acceptable level of health care? The states that have considered or passed health care plans for their uninsured have aimed at providing "basic" or "minimum" health care benefits unlike the "comprehensive benefits" offered through the national health plans of other industrialized countries.

Providing comprehensive benefits, of course, do not necessarily mean providing unlimited care. The right to health care in Britain, Germany, and Canada does not mean the right to all treatments. Although most services provided in these countries are covered, the extent to which services are offered varies substantially across countries. Equity in health care in reality means a commitment to providing some common, adequate level of care. As yet, however, no country has explicitly determined what this level is.[79]

In countries with universal access, referral systems tend to restrict access to high-technology services while maintaining comprehensive coverage of services. This is different from the approach in the United States of providing open access to technological services but restricting the type and quantity of services that are covered under the various insurance plans.[80]

Racial and Ethnic Disparities in Health

Even though significant improvements in the health of racial and ethnic minorities have been reported, health disparities persist among different populations.[81] A recent report on racial and ethnic disparities presents national trends in race- and ethnicity-specific rates for 17 health status indicators during the 1990s. All racial and ethnic groups experienced improvements in rates for 10 of the 17 indicators. At the same time, the report shows that despite these overall improvements, in some areas the disparities for ethnic and racial minorities remained the same or even increased.

The report is part of the *Healthy People* initiative—an effort to set health goals for each decade and then measure progress toward achieving them.[82] The indicators reflect various aspects of health and include infant mortality, teen births, prenatal care, and low birthweight, as well as death rates for all causes and for heart disease, stroke, lung and

Self-insured health plan A health plan whereby the risk for medical costs of employees is assumed by the employer rather than an insurance company or managed-care plan.

Reimbursement Payment made by a third party (for instance, government or private or commercial insurance).

Principal diagnosis The condition chiefly responsible for the patient's need for services. The principal diagnosis determines the payment the hospital or other provider receives from Medicare.

Secondary diagnoses are also referred to as comorbidities. A comorbid condition is present at the time of admission to the hospital but is not the primary reason for treating that patient. For example, if a patient is admitted for a cholecystectomy and has diabetes mellitus, diabetes mellitus is a comorbidity.

ICD-9-CM (*International Classification of Diseases—Clinical Modifications,* 9th ed.) Codes used by health care providers on billing forms to classify diseases/diagnoses.

Home health agency An agency that provides home health care. To be certified under Medicare, a home health agency must provide skilled nursing services and at least one additional therapeutic service (physical, speech, or occupational therapy; medical social services; or home health aide services) in the home.

breast cancer, suicide, homicide, motor vehicle crashes, and work-related injuries. Infectious diseases such as tuberculosis and syphilis are also included. The percent of children in poverty and the percent of the population living in communities with poor air quality round out the set of measures developed to allow comparisons among national, state, and local areas on a broad set of health indicators.

All racial and ethnic groups experienced improvement in rates for ten of the indicators: prenatal care; infant mortality; teen births; death rates for heart disease, homicide, motor vehicle crashes, and work-related injuries; the tuberculosis case rate; the syphilis case rate; and poor air quality.[83] For five more indicators—total death rate and death rates for stroke, lung cancer, breast cancer, and suicide—there was improvement in rates for all groups except American Indians and Alaska natives. The percent of children under 18 years old living in poverty improved for all groups except Asian or Pacific Islanders, and the percent of low-birthweight infants improved only for black non-Hispanics.

One of the goals of the *Healthy People* initiative is to reduce disparities in health. However, for about half of the indicators, the disparities improved only slightly, and disparities actually widened substantially for deaths due to work-related injuries, to motor vehicle crashes, and to suicide. "In many ways, Americans of all ages and in every racial and ethnic group have better health today," former Surgeon General David Satcher said. "But our work isn't done until all infants have the same chance to thrive, all mothers have equal access to prenatal care, and all Americans are equally protected from cancer, heart disease, and stroke." Whereas the goals of *Healthy People 2000* aimed at reducing disparities, the *Healthy People 2010* plan aims at the elimination of disparities in health among all population groups, with special emphasis on six areas: infant mortality, child and adult immunizations, HIV/AIDS, cardiovascular disease, breast and cervical cancer screening and management, and diabetes complications.[84]

Health Care Reform in the United States

Practically all industrialized countries except the United States have national health care programs.[85] Coverage is generally universal (everyone is eligible regardless of health status) and uniform (everyone is entitled to the same benefits). Costs are paid entirely from tax revenues or by some combination of individual and employer premiums and government subsidization.

The concept of government-sponsored comprehensive health care is not new to the United States.[86] In 1934, President Franklin D. Roosevelt strongly supported national health insurance (NHI) and almost pushed to have it included with old age and unemployment insurance in the Social Security Act of 1935. Fearing that NHI might jeopardize passage of the Social Security Act, however, he decided to drop the proposal. As a result of World War II and the passage of the Hill-Burton Act of 1946, federal monies were diverted away from NHI and used for construction of new hospitals. Two decades later, the nation shifted its focus from NHI to providing for those without private insurance. Consequently, through the efforts of Presidents John F. Kennedy and Lyndon B. Johnson, Congress enacted the Social Security Amendments of 1965, which created Medicare (Title XVIII) and Medicaid (Title XIX).

Now, four decades later, increased health care costs and decreased patient satisfaction with the health care available in the United States have prompted consideration of a new approach to health care. Rather than proposing the comprehensive reform of health care, many suggest incremental reforms on broad issues, such as health insurance reform, physician malpractice reform, and incentives to induce businesses to include health promotion initiatives in their insurance plans.[87] Changes in medical education are also being discussed. Because medicine has its roots in the treatment of acute disease, the greatest emphasis is on training physicians in treating patients with chronic diseases, the most prevalent problem in health care today, through a coordinated management team, including nutrition professionals.[88] Again, reform undoubtedly needs to include an increased focus on prevention of chronic disease.

Health care reform for the United States raises a formidable list of issues, including overall cost containment, universal access, emphasis on prevention, and reduction in administrative superstructure and costs.[89] These issues require difficult decisions. Consider the following questions:[90]

- Whom should be covered?
- How can coverage be increased to reach all people?
- What services should be considered in basic health care packages?
- Should health care cover both acute problems and prevention?
- Who should decide what constitutes preventive services?
- Who will pay for this coverage—consumers, employers, government?
- Where will government get the money to pay for it?
- How can health care costs be reduced or contained?
- What are the advantages and disadvantages of managed competition versus single-payer systems?

While the government remains undecided on what kind of health care system is needed and on how to pay for it, health care reform is evolving at an accelerating rate without legislation. The health care industry's determined efforts to curb the growth of costs while increasing access to services has transformed the traditional approach to health care in the United States into one emphasizing a managed-care approach.[91]

Nutrition as a Component of Health Care Reform

Community health care systems must include provision of nutrition services to preserve health and prevent disease. Realizing the importance of nutrition in overall health, even in the face of a nontraditional health care approach such as managed care, the ADA believes that medical nutrition therapy is an essential component of disease management and health care and that qualified nutrition professionals must provide it.[92] Gro Harlem Brundtland, MD, MPH, the director-general of the World Health Organization, has said, "Nutrition is a cornerstone that affects and defines the health of all people, rich and poor. It paves the way for us to grow, develop, work, play, resist infection and aspire to realization of our fullest potential as individuals and societies. . . . Putting first things first, we must . . . realize that resources allocated to preventing and eliminating disease will be effective only if the underlying causes of malnutrition—and their consequences—are successfully addressed. This is the 'gold standard': health and human rights. It makes for both good science and good sense, economically and ethically. . . ."[93]

One cannot have good health without proper nutrition. Conversely, poor nutrition contributes substantially to infant mortality, retarded growth and development of children, premature death, illness, and disability in adults, and frailty in the elderly, causing unnecessary pain and suffering, reduced productivity in the workplace, and increased health care costs.[94]

Many believe that nutrition services are the cornerstone of cost-effective prevention and are essential to halting the spiraling cost of health care. The ADA has urged that provision of nutrition services be included in any health care reform legislation.[95]

In addition, health care reform legislation needs to recognize the registered dietitian as the nutrition expert of the health care team, with a scope of practice that includes the following:[96]

- *Nutrition assessment* for the purpose of determining individual and community needs and recommendations of appropriate nutrient intake to maintain, recover, or improve health

- *Nutrition counseling and education* of individuals, families, community groups, and health professionals
- *Research and development* of appropriate nutrition practice guidelines
- *Administration* through *management* of time, finances, personnel, protocols, and programs
- *Consultation* with patients, clients, and other health professionals
- *Evaluation* of the effectiveness of nutrition counseling/education and community nutrition programs

COST-EFFECTIVENESS OF NUTRITION SERVICES

The **cost-effectiveness** of nutrition services has been well documented.[97] The ADA encourages all its practitioners to document the cost-effectiveness of nutrition services. Community nutritionists need to compete successfully for a fair share of the health care dollar. To do so, they must document the demand for and effectiveneess of nutrition services so that they can market those services to health care officials, providers, payers, and the public.

Obviously, no payer in the health care system wants additional costs. For a new technology or service, including nutrition services, to be a reimbursable benefit, it must prove its cost-effectiveness. Only services that have a proven impact on the quality of patient care will be funded. As Simko and Conklin have said, no expenditure of resources is justified for a service that fails to achieve its intended outcome.[98]

Cost-effectiveness An approach to evaluation that takes into account both costs and outcomes of intervention for a specific purpose. The analysis is especially useful for comparing alternative methods of intervention.

Cost-effectiveness studies compare the costs of providing health care against a desirable change in patient health outcomes (for example, a reduction in serum cholesterol in a patient with hypercholesterolemia).[99] Figure 9-10 shows a model for testing the costs and benefits of nutrition services. As this model shows, effective nutrition therapy can produce economic benefits as a result of altered food habits and risk factors.

FIGURE 9-10 Benefits of Nutrition Intervention

Source: Adapted from M. Mason and coauthors, Requisites of advocacy: Philosophy, research, documentation. Phase II of the costs and benefits of nutritional care, *Journal of the American Dietetic Association* 80 (1982): 213.

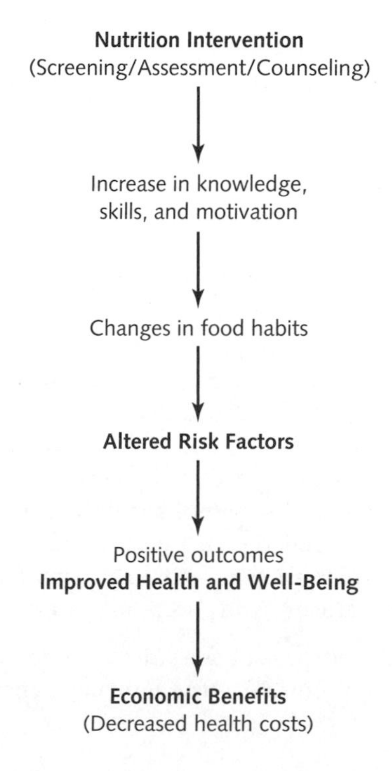

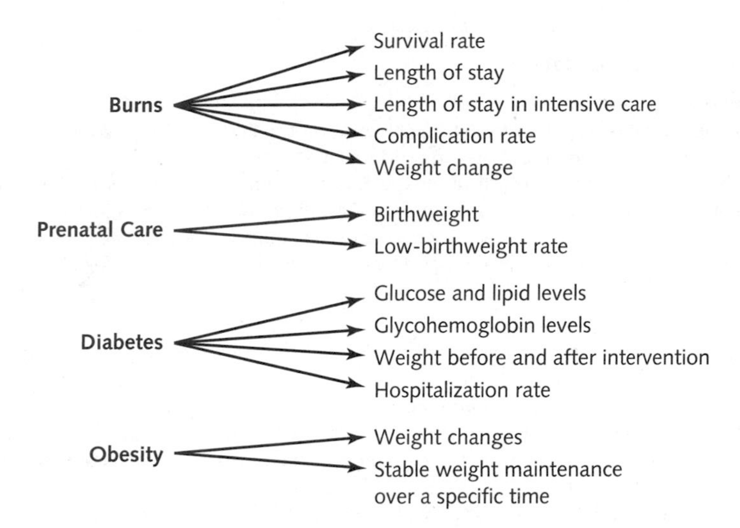

FIGURE 9-11 Measurable Outcomes of Nutrition Intervention

Source: R. Gould, The next rung on the ladder: Achieving and expanding reimbursement for nutrition services. © The American Dietetic Association. Reprinted by permission from *Journal of the American Dietetic Association* 91 (1991): 1383.

In an effort to enhance the quality, efficiency, and effectiveness of the health care system, policy makers are urging physicians and other health professionals to develop **practice guidelines** or **protocols** that clearly specify appropriate care and acceptable limits of care for each disease state or condition. Care delivered according to a protocol has been linked with positive **outcomes** for the patient or client.[100] Examples of outcomes include measures of control (serum lipid profiles, glycolated hemoglobin), quality of life, dietary intake, and patient satisfaction. The ADA has developed a variety of client protocols that define the minimum number of office visits and activities required for successful nutrition intervention and the outcomes that can be expected from the dietetics professional implementing the protocol and, more recently, a series of evidence-based practice guidelines.[101] Nutrition protocols serve as frameworks to help practitioners in the assessment, development, and evaluation of nutrition interventions. Developing standardized protocols of care (practice guidelines) for nutrition intervention is considered a must for achieving payment for nutrition services and expanding current levels of third-party reimbursement.[102]

Protocols have been developed for a variety of conditions, including enteral and parenteral feeding support, gestational diabetes, AIDS, hyperlipidemia, hypertension, irritable bowel syndrome, diabetes mellitus, oncology, congestive heart failure, pressure ulcer management, pre-end stage renal disease, chronic obstructive pulmonary disease, high-risk prenatal care, weight management, and anorexia and bulimia nervosa.

Documentation of specific *outcomes* of nutrition intervention—clinical data, laboratory measures, anthropometric measures, and dietary intake data—is also necessary. Figure 9-11 shows examples of outcome measures of nutrition intervention in burn injury, prenatal care, diabetes, and obesity.[103] When determining the outcomes of a given intervention, remember to ask the following questions: (1) Does the nutrition intervention make a difference in terms of disease-specific indicators? (2) Is the nutrition intervention worth the cost? Table 9-3 lists steps for developing protocols that enhance the effectiveness of nutrition services and make it easier to evaluate their quality and effectiveness.[104] When one is developing protocols for a clinical practice setting, using an evidence-based approach will undoubtedly yield the best evidence to answer practice-questions related to the protocol.[105]

The contribution of nutrition to preventing disease, prolonging life, and promoting health is well recognized. Accumulated evidence shows that when nutrition services are integrated into health care, diet and nutritional status change with the following results:[106]

- The birthweight of infants born to high-risk mothers improves.
- The prevalence of iron-deficiency anemia is reduced.
- Serum cholesterol and the risk of heart attacks are reduced.
- Glucose tolerance in persons with diabetes improves.
- Blood pressure in hypertensive patients is lowered.

As demonstrated by the research highlighted on page 299, the benefits of providing nutrition services far outweigh the costs of providing those services.

Practice guidelines Guidelines to be used by doctors, hospitals, and other health professionals for treating various conditions in order to ensure the most cost-effective care.

Protocol Detailed guidelines for care that are specific to the disease or condition and the type of patient.

Outcome An end result of the health care process; a measurable change in the patient's state of health or functioning.

TABLE 9-3 Development of Patient Care Protocols

1. Define the patient population.

- Select a prevalent problem (diagnosis).
- Select a population or problem for which evidence (in the literature or in your own experience) indicates that nutrition intervention can produce clinical outcomes.
- Define the population as specifically as possible, based on medical diagnosis, state of diagnosis if relevant (e.g., new diagnosis versus long-standing diagnosis of diabetes), and other relevant factors (medication, age, other medical conditions). Example: hypercholesterolemic patients (cholesterol $\geq$ 200 mg/dL) who are either newly diagnosed or first-time referrals.

2. Define the treatment.

- Base standards for treatment on accepted methods of nutritional management (e.g., the National Cholesterol Education Program guidelines or the National Institutes of Health Consensus Panel recommendations on diabetes management).
- Emphasize individualization of specific dietary patterns within limits of treatment standards.
- Specify expected length of treatment (e.g., 3-month trial of diet) and expected minimum number of visits.
- Specify expected length of each visit, as well as length of time between visits.
- Outline specific activities or topics to be addressed during the course of treatment and provide appropriate educational materials and tools.

3. Identify expected outcomes.

- Include as appropriate:

  Anthropometric measurements (weight, body mass index, skinfolds, etc.)
  Laboratory values (fasting blood sugar, hemoglobin A_{1c}, cholesterol, lipid profile, etc.)
  Clinical data (medication change, blood pressure change, etc.)
  Dietary intake (change in eating behaviors, intake of specific nutrients)
  Functional status (ability to perform activities of daily living [ADLs], ability to work, ability to live independently, etc.)
  Psychosocial outcomes (perceived quality of life, etc.)
  Economic outcomes (cost-effectiveness, cost–benefit analysis of reducing hospital stay, reimbursement by third-party payers for services provided, etc.)

- Specify appropriate points in time when outcomes should be measured.
- If possible, specify expected magnitude of change and associated timeline.
- Identify expected and measurable intermediate outcomes (e.g., positive changes in knowledge, behavior, decision making, involvement in self-care).

Source: Adapted from M. K. Fox, Defining appropriate nutrition care. © 1991, The American Dietetic Association. Reprinted by permission from *Reimbursement and Insurance Coverage for Nutrition Services,* 1991, p. 93.

Medical Nutrition Therapy and Medicare Reform

The ADA believes that reimbursement for nutrition services through both Medicare and Medicaid is wholly inadequate. Often, the elderly choose not to seek appropriate nutrition services because Medicare's coverage is limited and they are unable to pay for the services themselves.[107] The **Nutrition Screening Initiative** recommends that nutrition screening for the elderly be included in the U.S. health care system and that Medicare cover and reimburse nutrition assessment and treatment for those found to be at nutritional risk. Such action is considered crucial to lowering health care costs, because malnourished persons have longer hospital stays and higher hospital costs than persons without malnutrition. This chapter's Programs in Action feature demonstrates how nutrition screening can help lower nutrition risk among frail, homebound older adults.

Nutrition Screening Initiative A program of the American Academy of Family Physicians, the ADA, and the National Council on Aging, formed in 1990 as a multifaceted effort to promote nutrition and improved nutritional care for the elderly.

Medical Nutrition Therapy Providing Return on Investment

Research demonstrates the cost-effectiveness of medical nutrition therapy.

- Oxford Health Plan* saved $10 for every $1 spent on nutrition counseling for at-risk elderly patients. Monthly costs for Medicare claims alone tumbled from $66,000 before the nutrition program to $45,000 afterwards. As a result, the health plan continued use of nutrition screenings.
- The Lewin Group† documented an 8.6 percent reduction in hospital utilization and a 16.9 percent reduction in physician visits associated with medical nutrition therapy for patients with cardiovascular disease.
- The Lewin Group† additionally documented a 9.5 percent reduction in hospital utilization and a 23.5 percent reduction in physician visits when medical nutrition therapy was provided to persons with diabetes mellitus.
- The University of California Irvine‡ demonstrated that lipid drug eligibility was obviated in 34 of 67 subjects; the estimated annual cost savings from the avoidance of lipid medication was $60,652.
- Pfizer Corporation§ projected $728,772 in annual savings from reduced cardiac claims of their employees from an on-site nutrition/exercise intervention program.
- The U.S. Department of Defense‖ saved 3.1 million dollars in the first year of a nutrition therapy program utilizing registered dietitians counseling 636,222 patients with cardiovascular disease, diabetes, and renal disease.

*Oxford Health Plan's pilot nutrition screening program applied to Medicare population in New York, between 1991 and 1993.

†Johnson, Rachel, "The Lewin Group—What does it tell us, and why does it matter?" *Journal of the American Dietetic Association* (1999) 99: 426–27.

‡Sikland, G., et al., "Medical nutrition therapy lowers serum cholesterol and saves medication costs in Medicare populations with hypercholesterolemia," *Journal of the American Dietetic Association* (1998) 98: 889–94.

§Pfizer Corp., Lipid Intervention Program, http://healthproject.stanford.edu/koop/pfizer99/documentation.html.

‖"The Cost of Covering Medical Nutrition Therapy Services Under TRICARE: Benefits Costs, Cost Avoidance and Savings," final report prepared by the Lewin Group, Inc. for the Department of Defense Health Affairs, 11/15/98.

Source: American Dietetic Association, *Medical Nutrition Therapy Works,* April 2001; available at www.eatright.org.

Since 1992, the legislative priority of ADA has been the inclusion of **medical nutrition therapy** as a covered benefit in health care delivery.[108] Since the failure of the 1993 Health Security Act and other health care reform bills to pass, the ADA Health Care Reform Team has been focused on securing a mechanism for nutrition reimbursement under existing federal insurance programs. Medicare was amended in 2000 to cover nutrition therapy as an outpatient benefit under Part B of the Medicare program.[109] Chapter 7 provides a discussion of the legislative process that led to Medicare's reimbursement of MNT for certain Medicare recipients.

An ADA-financed independent study projected the cost of extending coverage of medical nutrition therapy to all Medicare beneficiaries under Medicare Part B to be less than $370 million over 7 years, when savings are considered. Savings would be greater than costs after the third year of enactment (see Figure 9-12).[110] For example, if coverage had begun in 1998, in 2001 an additional cost to Medicare Part B of $389 million would have been offset by a reduction in cost to Part A of $401 million, resulting in a net savings

Medical nutrition therapy The range of specific medical nutrition therapies for various conditions is determined following a complete assessment of the client's nutritional status. Medical nutrition therapy includes dietary modifications and nutrition counseling, as well as more complex methods of nutrition support using specialized nutrition therapies (for example, nutritional supplements and enteral and parenteral feedings).

Programs in Action

Evaluating Nutrition Risk in Older Adults

The elderly are the largest demographic group likely to be at risk for malnutrition in the United States. In one survey, over two-thirds of participants in the Elderly Nutrition Program of the Older Americans Act were found to be at moderate to high nutritional risk.[1]

A 1990 survey conducted for the Nutrition Screening Initiative (NSI) found that a large number of older adults met NSI criteria for nutrition risk—skipping meals, poverty, social isolation, disability, illness, chronic medication use, and advanced age. Screening and assessment are recommended by NSI as the first steps in improving nutrition status in the elderly. Nutrition screening is a low-cost way to identify and work with or refer individuals at potential nutrition risk. Appropriate medical nutrition therapy can enhance mobility and alertness, reduce medical complications, and promote faster wound healing. Although nutrition services in the home may not be covered by Medicaid, many states use Medicaid waivers, provided for under the Social Security Act, to allocate funds for in-home services not ordinarily covered by Medicaid.[2]

The NSI DETERMINE Checklist is a nutrition screening tool to help identify warning signs of potential nutrition problems. It includes ten simple questions, to be answered "yes" or "no," about factors that could influence nutritional status. The checklist can be completed and scored by professionals, paraprofessionals, or members of the public, including family members.

Campaign Long-Term Nutrition Risk Reduction demonstrates how nutrition screening and case management can help lower nutrition risk among frail, homebound older adults.

"Determine Your Nutritional Health" Checklist[3]

1. I have an illness or condition that made me change the kind or amount of food I eat.
2. I eat fewer than two meals each day.
3. I eat few fruits or vegetables or milk products.
4. I have three or more alcoholic drinks almost every day.
5. I have tooth or mouth problems that make it hard for me to eat.
6. I don't always have enough money to buy the food I need.
7. I eat alone most of the time.
8. I take three or more different prescribed or over-the-counter medicines a day.
9. Without wanting to do so, I have lost or gained 10 pounds in the last 6 months.
10. I am not always physically able to shop, cook, and/or feed myself.

Campaign Long-Term Nutrition Risk Reduction

Nutrition and home care professionals employed by United Home Care Services, Inc. (UHCS), a home care service

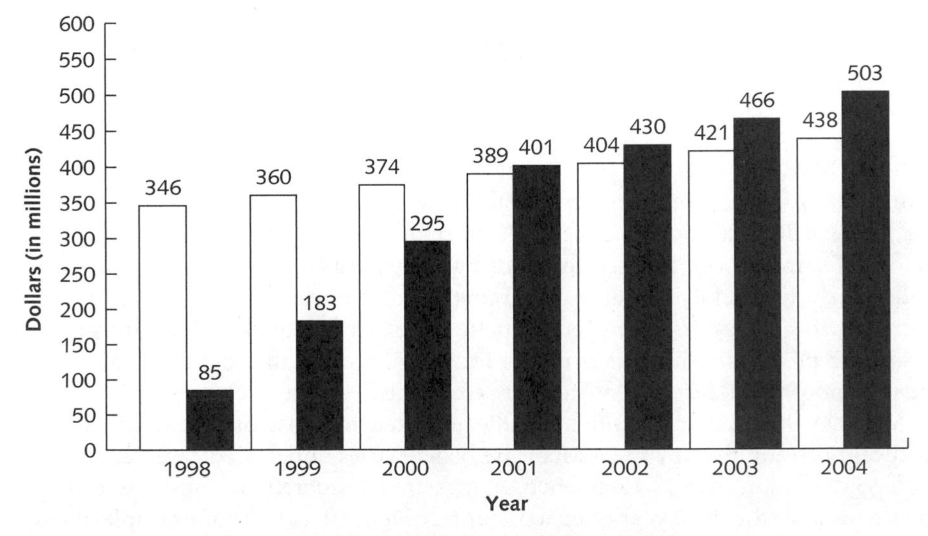

FIGURE 9-12 Saving Medicare Millions

Source: Lewin Group, 1997.

provider in Florida, used nutrition screening checklists and assessment tools to establish long-term medical nutrition therapy (MNT) care plans for its elderly clients.

Goals and Objectives

This campaign aimed to improve the nutritional status of frail, homebound older adults receiving home services under the Medicaid Waiver Program by nutritionally screening clients, providing home-based medical nutrition therapy where indicated, using a coordinated case management approach to determine need for further services, and evaluating the effectiveness of home-based medical nutrition therapy.

Methodology

Frail, homebound older adults who were clients of UHCS residing in Dade County, Florida, were contacted by a case manager for the purpose of completing a "Nutrition Screening Checklist." This checklist was adapted from the screening checklist developed by the NSI. Clients identified as "at risk" after screening were referred to a dietetic technician for an initial home visit and in-depth nutritional assessment. The case manager and dietetic technician then met with a UHCS-registered dietitian to review assessment results, determine risk, and establish a care plan for medical nutrition therapy. The dietitian carried out the care plan through monthly visits. The case manager was kept informed via progress notes and case narrative tools. At the time of discharge from the medical nutrition therapy care plan, the in-depth nutritional assessment was repeated and compared to the pretherapy assessment.

Results

Semiannual samplings of 20 percent of all discharged patients demonstrated that approximately 89 percent of clients surveyed lowered their nutrition risk scores after receiving home-based medical nutrition therapy.

References

1. M. Ponza, J. C. Ohls, and B. E. Millen, *Serving Elders at Risk: The Older Americans Act Nutrition Programs, National Evaluation of the Elderly Nutrition Program, 1993–1995* (Washington, D.C.: Mathematica Policy Research, 1996).
2. Position of the American Dietetic Association: Nutrition across the spectrum of aging, *Journal of the American Dietetic Association* 105 (2005): 616–33.
3. Nutrition Screening Initiative.

Source: Community Nutritionary (White Plains, NY: Dannon Institute, Spring 1999). Used with permission. For more information about the *Awards for Excellence in Community Nutrition,* go to www.dannon-institute.org.

of $11 million. The savings to the Medicare program come from fewer hospital admissions and fewer complications requiring a physician's visit. The data used in the study were particularly significant for persons with diabetes and cardiovascular disease. Spending for diabetes and cardiovascular disease accounts for about 60 percent of annual Medicare spending.[111] In the long run, the program would save more in medical expenses than it costs to operate.

In summary, medical nutrition therapy is an integral component of cost-effective medical treatment. It can reduce health costs by improving patient outcomes and reducing recovery time. The ADA believes that the coverage of appropriate medical nutrition therapy, when medically necessary, should be included in any basic health care benefit package.

Go to www.eatright.org/Member/Files/fedregmntfinalregs1101.pdf for the *Federal Register*'s Final Rule for Medicare Part B Medical Nutrition Therapy Benefit.

On the Horizon: Changes in Health Care and Its Delivery

The future of health care in the United States will be shaped by current trends in society at large and in the field of health care, as well as by the choices we make for health care reform. The paradigm shift from sickness to wellness will undoubtedly be one of the strongest factors affecting health care.

To a greater extent than most of us are willing to accept, today's disorders of overweight, heart disease, cancer, blood pressure, and diabetes are by and large preventable. In this light, true health insurance is not what one carries on a plastic card but what one does for oneself.

– L. Power in G. Edlin and E. Golanty, *Health and Wellness* (3rd ed.)

For several decades, the dominant paradigm has been the medical model. During the 1970s, people were guided by the philosophy that the health care system would do everything possible in terms of curative and treatment services to make them well. In the 1990s, people viewed wellness as a function of prevention and began to accept responsibility for their own health. The focus on the pursuit of health, marked by an increased interest in nutrition, fitness, and health promotion, was reflected in the growth of corporate "wellness" programs and a widening choice of health care practitioners.

Our health care system still contains a number of barriers to focusing on prevention of poor health habits, however. The biomedical approach to illness underestimates and underemphasizes behavioral and lifestyle influences on disease. Physicians think in terms of treating or correcting conditions rather than preventing them. To transform health care, medical education will also undoubtedly need to change.[112]

The challenge for the next decade is to change the U.S. approach to health care from a system based on treatment of acute conditions to one based on disease prevention and health promotion. Physicians, public health workers, registered dietitians and other health practitioners, health educators, and community health organizations have joined ranks to emphasize health promotion and disease prevention as a more economical route to good health than the more costly procedures necessitated by sickness and disease.[113] According to physician Andrew Weil, "The medical facility of the future should look like a cross between a health spa and a hospital. Not only should it offer a smorgasbord of therapeutic options, its main focus should be on educating patients on how to stay well once they leave."[114]

The future offers much that is positive for the profession of dietetics. The public's thinking about health and nutrition has matured, interest in positive health is growing steadily, and demand for health promotion products and services is increasing. Yet to be achieved, however, are the effective provision and allocation of resources such as nutrition services as part of preventive care. To accomplish this, a coordinated strategy for health care, political will, and active collaboration of both health care professionals and consumers of health care services will be required.

As noted earlier, health care reform is a difficult undertaking. It involves more than cost containment and universal access. Community nutritionists need to educate the payers of health care about the inherent value of including nutrition services in their policies.[115] In arguing for reimbursable nutrition services, highlight its benefits:[116]

- Nutrition services are attractive, progressive health benefits that are relatively inexpensive compared with other types of benefits.
- Nutrition services benefits enhance the insurance product (the employee benefit package).
- Nutrition services have a preventive component (they help keep employees healthy).
- Nutrition services benefits attract healthy subscribers.
- Nutrition services are manageable—they can be easily documented.
- Nutrition services help patients become more self-reliant by helping them fight disease, avoid hospitalization, and reduce the use of other, more expensive medical therapies.
- Nutrition care speeds recovery.

Health care reform for the United States is certain, but the exact nature of the reform will continue to evolve. Undoubtedly, the changes required of the health care system will transform it slowly over time. Remember that change at the federal level begins with local advocacy and that the prescription for success is persistence.

Internet Resources

Agency for Health Care Research and Quality *An arm of the U.S. Department of Health and Human Services.*	**www.ahcpr.gov**
America's Health Insurance Plans *A trade group for HMOs and PPOs.*	**www.ahip.org**
American Association of Retired Persons *Offers approval for select managed-care plans.*	**www.aarp.org**
American Dietetic Association *Updates on medical nutrition therapy information and resources.*	**www.eatright.org/Member/83_12954.cfm**
Centers for Medicare and Medicaid Services *The federal agency that administers Medicare, Medicaid, SCHIP, HIPAA, and CLIA.*	**www.cms.hhs.gov**
CDC's Office of Minority Health *Minority health resources and training from the Centers for Disease Control and Prevention.*	**www.cdc.gov/omh/default.htm**
Department of Health and Human Services *The principal agency for protecting the health of all Americans. Provides essential human services, especially for those who are least able to help themselves.*	**www.hhs.gov**
Health Pages *Issues report cards on major managed-care plans.*	**www.thehealthpages.com**
Healthfinder *Serves to organize the mass of health and nutrition information available from federal and state agencies.*	**www.healthfinder.gov**
Intelihealth *Offers a wide collection of consumer health information from the National Heart, Lung, and Blood Institute, the National Institutes of Health, the National Library of Medicine, and others.*	**www.intelihealth.com**
Joint Commission on Accreditation of Healthcare Organizations (JCAHO) *The primary accreditation organization that evaluates hospitals and outpatient clinics.*	**www.jcaho.org**
Medicare *Official U.S. government site for people with Medicare.*	**www.medicare.gov**
National Committee for Quality Assurance (NCQA) *Evaluates managed-care plans in terms of patient records, complaints, equipment, and personnel.*	**www.ncqa.org**
National Institutes of Health *Comprises 27 institutes and centers. Mission is to uncover new knowledge that will lead to better health for everyone.*	**www.nih.gov**
Office of Minority Health *Useful publications—such as "Closing the Gap"—and links to related sites.*	**www.omhrc.gov**
Social Security Administration *The best site for information on the Medicare and Medicaid programs.*	**www.ssa.gov**

Insurance Access

By Alice Fornari, EdD, RD, and Alessandra Sarcona, MS, RD

Scenario

You are a registered dietitian (RD) working in an ambulatory clinic that provides nutrition counseling to Medicaid clients. Nutrition services are not currently provided to Medicare clients, because Medicare has historically not covered this service. An elderly client is referred to you for counseling regarding his elevated blood glucose levels, which remain high despite medications. Upon opening the client's record, you realize the client is on Medicare, not Medicaid. You ask the client to verify this information, and it is correct. You ask the client whether he can pay for your services. He cannot because of his fixed income, from Social Security. You have on your bulletin board the memorandum from the clinic administrator reiterating the policy that no patients can be seen on clinic time if they are not able to pay for services and do not receive insurance reimbursement. You remember reading in the *Journal of the American Dietetic Association* (January 2002) about medical nutrition therapy (MNT) insurance coverage for Medicare clients diagnosed with diabetes mellitus. Although you want to investigate further this Medicare coverage of MNT, you cannot assume for this visit that the client will be covered for your services under Medicare.

Learning Outcomes

- Identify individuals who have access to health care/nutrition services through government and/or private insurance.
- Relate allocation of resources to health care issues, specifically considering the access to health care of clients with limited resources.
- Communicate to clients and colleagues policies and procedures for obtaining health care resources as an agency or individual provider.
- Integrate and monitor strategies that support MNT services and desired nutrition outcomes.

Foundation: Acquisition of Knowledge and Skills

1. Define terms specific to private and public insurance.
2. List comparative information about Medicare and Medicaid (see Table 9-1 on page 282).
3. Access and review the article Becoming a Medicare provider: Systems for success, *Journal of the American Dietetic Association* 101 (2001): 1412. As a member of ADA, you can search the journal online at www.eatright.org.
4. Access and review the ADA Position Paper on Nutrition Services in Managed Care (www.eatright.org).

Step 1: *Identify Relevant Information and Uncertainties*

1. Explain the issue facing the nutritionist.
2. Identify a range of solutions focused on enabling the client to obtain nutrition counseling by a qualified nutrition professional.
3. Identify and analyze the information related to health care resources that would be useful to help resolve the issue.

Step 2: *Interpret Information*

1. Compare and contrast the arguments related to possible solutions on the basis of supporting evidence.
2. As you analyze the problem, how might you compensate for your own biases?

Step 3: *Draw and Implement Conclusions*

1. Within the clinic structure, document in writing the steps you would need to pursue to ensure that the Medicare clients covered under the new MNT regulations are referred to you for nutrition services.

Step 4: *Engage in Continuous Improvement*

1. To support your solution to the issue, develop an outcomes assessment plan to document that your intervention is cost-effective and beneficial to the client in terms of disease management and reducing risk for complications. Refer to www.eatright.org/member/PolicyInitiatives/83.cfm to identify criteria supporting the benefits of MNT specific to the management of diabetes mellitus.

Ethics and the Nutrition Professional

Life is full of paradoxes. For example, even though health promotion and disease prevention are paramount in halting the alarming escalation in the cost of health care in this country, the United States spends less than 5 percent of all dollars directed toward health care on public health and disease prevention.[1] Similarly, although one goal of nutrition is to apply scientific knowledge to feed all people adequately, every fifth child in the United States is vulnerable to hunger. The United States spends more on health care than other nations, but certain health disparities between racial and ethnic groups exist, particularly in pregnancy outcomes, infant mortality rate, nutritional status, life expectancy, and food insecurity. And finally, one issue that arises in connection with developing countries is whether it is fundamentally wrong that so much preventable sickness and death occur in the world.

As a community nutritionist, how do you address such issues? This Professional Focus feature reviews some of the ethical questions in the field of health promotion that are related specifically to community nutritionists. Its intent is not to arrive at a conclusion or present solutions to ethical dilemmas; rather, it seeks to present the issues for your consideration and stresses the need for moral sensitivity in the planning and implementation of community nutrition programs. As Aristotle said, "We are what we repeatedly do." Moral sensitivity and characteristics such as honesty, integrity, loyalty, and candor are developed through practice.[2]

What Is Ethics?

Philosophers throughout history have struggled with questions of how to live and work ethically. Ethics is a philosophical discipline that attempts to determine what is morally good and bad, right and wrong. Ethics helps decision makers seek criteria to evaluate different moral stances.[3]

As a community nutritionist, you may wonder what ethics has to do with your professional activities. Certainly, as a community nutritionist, you will not often confront such media issues as euthanasia, abortion, capital punishment, insider trading, maternal surrogacy, infanticide, the withdrawal of nutrition support for terminally ill patients, or the right to die. Nevertheless, situations arise in community settings that will force you to make ethical decisions.

Community nutritionists working with the media or food industry must consider the accuracy of product descriptions and claims, as well as words and images that can mislead the public.[4] As a manager, the community nutritionist may face ethical dilemmas in allocating resources. In setting priorities, she may have to make some decisions that necessitate ethicical judgments. If she believes that all eligible clients have the right to receive optimal nutritional care, then how should she decide which clients will actually receive that care? (In this case, optimal nutritional care might mean a homebound elderly person's receipt of home-delivered meals.) The community nutritionist involved in research exercises honesty, accuracy, and integrity in conducting studies and publishing the results. Consider the impact of using falsified data in determining nutrition policy for funding a new or existing nutrition program.

Codes of Ethics

Simple answers to ethical questions are elusive, but many health care organizations and professional associations have established codes of ethics to provide guidance in resolving ethical dilemmas.[5] Codes of ethics are written to guide decision making in areas of moral conflict; outline the obligations of the practitioner to self, client, society, and the profession; and "assist in protecting the nutritional health, safety, and welfare of the public."[6] The American Dietetic Association (ADA) published its first code of ethics in 1942.[7] The most recent code (presented in the next section) became effective in 1999 and applies to all ADA members and credentialed practitioners.

A code of ethics for nutritionists and other professionals working in international situations is likewise critical, as C. E. Taylor notes: "Needs are so obvious that the temptation is great to rush in with programs that seem reasonable; but international work is full of surprises. Each new activity needs to be carefully tested."[8] Consider the story of the monkey and the fish:

> After a dam burst, a flood raged through an African countryside. A monkey, standing in safety on the riverbank, watched a fish swim into its view. "I will save this poor fish from drowning," thought the monkey. And, swinging from a tree branch, he scooped up the fish and carried it, gasping, to land. "Throw me back," pleaded the fish. Reluctantly, the monkey agreed, scratching his head in bewilderment at the fish's lack of appreciation for the aid he had so selflessly offered.[9]

Code of Ethics for the Dietetics Profession

Preamble

The American Dietetic Association and its credentialing agency, the Commission on Dietetic Registration, believe it is in the best interests of the profession and the public they serve that a Code of Ethics provide guidance to dietetics practitioners in their professional practice and conduct. Dietetics practitioners have voluntarily developed a Code of Ethics to reflect the values and ethical principles guiding the dietetics profession and to outline commitments and obligations of the dietetics practitioner to self, client, society, and the profession.

The Ethics Code applies in its entirety to members of the American Dietetic Association who are Registered Dietitians (RDs) or Dietetic Technicians, Registered (DTRs). Except for sections dealing solely with credential, the Code applies to all American Dietetic Association members who are not RDs or DTRs. Except for aspects dealing solely with membership, the Code applies to all RDs and DTRs who are not ADA members. All of the aforementioned are referred to in the Code as "dietetics practitioners."

The purpose of the Commission on Dietetic Registration is to assist in protecting the nutritional health, safety, and welfare of the public by establishing and enforcing qualifications for dietetic registration and for issuing voluntary credentials to individuals who have attained those qualifications. The Commission has adopted this Code to apply to individuals who hold these credentials.

Principles

The dietetics practitioner:

1. Conducts himself/herself with honesty, integrity, and fairness.
2. Practices dietetics based on scientific principles and current information.
3. Presents substantiated information and interprets controversial information without personal bias, recognizing that legitimate differences of opinion exist.
4. Assumes responsibility and accountability for personal competence in practice, continually striving to increase professional knowledge and skills and to apply them in practice.
5. Recognizes and exercises professional judgment within the limits of his/her qualifications and collaborates with others, seeks counsel, or makes referrals as appropriate.
6. Provides sufficient information to enable clients and others to make their own informed decisions.
7. Protects confidential information and makes full disclosure about any limitations on his/her ability to guarantee full confidentiality.
8. Provides professional services with objectivity and with respect for the unique needs and values of individuals.
9. Provides professional services in a manner that is sensitive to cultural differences and does not discriminate against others on the basis of race, ethnicity, creed, religion, disability, sex, age, sexual orientation, or national origin.
10. Does not engage in sexual harassment in connection with professional practice.
11. Provides objective evaluations of performance for employees and coworkers, candidates for employment, students, professional association memberships, awards, or scholarships; makes all reasonable effort to avoid bias in any kind of professional evaluation of others.
12. Is alert to situations that might cause a conflict of interest or have the appearance of a conflict and provides full disclosure when a real or potential conflict of interest arises.
13. Who wishes to inform the public and colleagues of his/her services, does so by using factual information; does not advertise in a false or misleading manner.
14. Promotes or endorses products in a manner that is neither false nor misleading.
15. Permits the use of his/her name for the purpose of certifying that dietetics services have been rendered only if he/she has provided or supervised the provision of those services.
16. Accurately presents professional qualifications and credentials:
    - Uses Commission on Dietetic Registration awarded credentials: "RD" or "Registered Dietitian"; "DTR" or "Dietetic Technician, Registered"; "CSP" or "Certified Specialist in Pediatric Nutrition"; "CSR" or "Certified Specialist in Renal Nutrition"; and "FADA" or "Fellow of the American Dietetic Association" only when the credential is current and authorized by the Commission on Dietetic Registration. The dietetics practitioner provides accurate information and complies with all requirements of the Commission on Dietetic Registration program in which he/she is seeking initial or continued credentials from the Commission on Dietetic Registration.
    - Is subject to disciplinary action for aiding another person in violating any Commission on Dietetic Registration requirements or aiding another person in representing himself/herself as Commission on Dietetic Registration credentialed when he/she is not.
17. Voluntarily withdraws from professional practice under the following circumstances:
    - Has engaged in any substance abuse that could affect his/her practice.
    - Has been adjudged by a court to be mentally incompetent.
    - Has an emotional or mental disability that affects his/her practice in a manner that could harm the client or others.
18. Complies with all applicable laws and regulations concerning the profession and is subject to disciplinary action under the following circumstances:
    - Has been convicted of a crime under the laws of the United States which is a felony or a misdemeanor, an essential element of which is dishonesty, and which is related to the practice of the profession.
    - Has been disciplined by a state, and at least one of the grounds for the discipline is the same or substantially equivalent to these principles.
    - Has committed an act of misfeasance or malfeasance which is directly related to the practice of the profession as determined by a court of competent jurisdiction, a licensing board, or an agency of a governmental body.
19. Supports and promotes high standards of professional practice; accepts the obligation to protect clients, the public, and the profession by upholding the Code of Ethics for the Profession of Dietetics and by reporting alleged violations of the Code through the defined review process of the American Dietetic Association and its credentialing agency, the Commission on Dietetic Registration.

Source: From the American Dietetic Association, Code of Ethics for the Profession of Dietetics, *Journal of the American Dietetic Association* 99 (1999): 109–113.

Guiding Principles

Four basic principles are used in ethical decision making and in developing guidelines for professional practice:[10] (1) autonomy—respecting the individual's rights of self-determination, independence, and privacy; (2) beneficence—protecting clients from harm and maximizing possible benefits; (3) nonmaleficence—the obligation not to inflict harm intentionally; and (4) justice—striving for fairness in one's actions and equality in the allocation of resources.

To determine whether an issue in the community setting raises an ethical question, consider these ethical principles expressed as questions:[11]

1. Does the nutritional program, message, product, or service foster or deter the individual's ability to act freely? (The ADA code of ethics items 3, 6, 13, and 14 address this principle of autonomy.)
2. Does the nutritional program, message, product, or service help people or harm them? (Items 2, 4, and 5, and 18 in the ADA code of ethics address these principles of beneficence and nonmaleficence.)
3. Does the nutritional program, message, product, or service unfairly or arbitrarily discriminate among persons or groups? (Items 1, 10, and 11 in the ADA code of ethics address this principle of justice.)

Consumers are eager to know about nutrition. The principle of beneficence moves us, as nutrition educators, to provide consumers with truthful and convincing information based on current scientific knowledge. In this way, we protect them from fraudulent misinformation and also motivate them to change their diet accordingly.[12] The principle of autonomy motivates us to provide consumers with factual information that includes both the weaknesses and the strengths of the scientific data supporting a given behavior, service, or product; with this information, individuals can exercise their right to make an informed choice or decision.

Health Promotion and Ethics

The purpose of health promotion is to motivate people to adopt and maintain healthful practices in order to prevent illness and functional impairment. Many hold that by investing in health promotion and disease prevention activities, we can avoid the much greater economic and social costs of disease and disability. The challenge today is to provide the public with the opportunity to benefit from appropriate nutrition knowledge and services. However, this challenge raises a hidden moral issue worthy of consideration. At what point is scientific knowledge sufficiently documented to warrant translating it into dietary messages to the public? What responsibilities and rights do we have to alter individual lifestyles in our effort to promote public health? As health promoters, we are sometimes criticized for taking a paternalistic approach with a "We know better than you" attitude. Moral sensitivity demands that we respect the dignity of persons—their right to make their own choices. For example, we may carefully and creatively design messages for older women at risk of osteoporosis, encouraging them to use dairy products in their daily diet, but our target audience has the right to resist our efforts and choose not to do so. An adequate calcium intake is a good thing, but life offers many other good things as well.

Health promotion for a number of issues (for example, cigarette smoking and drinking and driving) necessitates a paternalistic approach that both restricts private liberties and promotes group virtues such as beneficence and concern for the common good. Such paternalism is for the most part considered legitimate and reflects the view that the good of each of us is not the same thing as the good of all of us together.[13]

Community nutritionists, as health promoters, call attention to other health risks—a diet high in saturated fat or low in fiber, commercial advertising of empty-calorie foods to children, and nutrition fraud in the marketplace, among others. Should governments, therefore, move from taxing cigarettes and alcohol to taxing companies that manufacture high-fat confections or cereals high in sugar? Because obesity and a sedentary lifestyle are associated with a number of chronic diseases (for instance, hypertension, coronary artery disease, and diabetes) and with increased health care costs, should persons who eat too many calories or too much fat and those who fail to exercise regularly be taxed to discourage these lifestyles and raise revenues for health care? Should we fine pregnant women who smoke or drink? In other words, to what extent should society tolerate and bear the burden for the health risks that individuals choose to take? Such are the ethical dilemmas facing those who work in health promotion. The ethical conflict is how to achieve the goal of protecting and promoting public health while ensuring an individual's freedom of choice.

Ethical Decision Making

Analytical skills are necessary to resolve ethical dilemmas. One must objectively evaluate the individual circumstances of each situation, gather relevant data, consider possible alternatives, consult experts as necessary, and take appropriate actions to accomplish the greatest good for the greatest number. The particular action chosen must adhere to the general ethical principles of autonomy, beneficence, nonmaleficence, and justice.[14]

As community nutritionists, you are certain to face ethical dilemmas in both your professional and your personal lives.

In closing this section, we leave you with a set of questions.[15] In the months to come, consider your responses to these questions in light of your moral sensitivity regarding these situations, your understanding of ethical principles, and your discussions with other professionals experienced in making ethical decisions.

- Is it right to save lives by immunization, nutrition, oral rehydration therapy, or chemotherapy when those who are saved face a life of despair?
- Do the United States, Canada, and other developed countries have a moral obligation toward the less developed countries?
- When limited resources necessitate setting priorities in program planning, who should receive benefits? Infants? Children? Pregnant women? Working people? Elderly persons?
- What are the ethical limits to promotional activities of multinational corporations? Is it acceptable to market infant formula or soft drinks in developing countries? Should we ban television advertising of empty-calorie foods to children?
- In setting program priorities, are there situations in which one ethnic group should be favored over another?
- Do we have the right to ask individuals to adjust and, in some cases, to abandon their ethnic and cultural customs, traditions, or cuisines in the interests of improved health?

References

1. D. R. Smith, Public health and the winds of change, *Public Health Reports* 113 (1998): 160–61.
2. S. L. Anderson, Dietitians' practices and attitudes regarding the Code of Ethics for the Profession of Dietetics, *Journal of the American Dietetic Association* 93 (1993): 88–91.
3. S. Tamborini-Martin and K. V. Hanley, The importance of being ethical, *Health Progress* 70 (1989): 24.
4. J. N. Neville and R. Chernoff, Professional ethics: Everyone's issues, *Journal of the American Dietetic Association* 88 (1988): 1286.
5. J. Sobal, Research ethics in nutrition education, *Journal of Nutrition Education* 24 (1992): 234–38.
6. American Dietetic Association, Code of ethics for the profession of dietetics, *Journal of the American Dietetic Association* 99 (1999): 109–13.
7. Neville and Chernoff, Professional ethics, p. 1287.
8. C. E. Taylor, Ethics for an international health profession, *Science* 153 (1966): 716–20, as cited by P. F. Basch, *Textbook of International Health* (New York: Oxford University Press, 1990), pp. 407–8.
9. C. Levine, Ethics, justice, and international health, *Hastings Center Report* 4 (1977): 5–6.
10. M. Barry, Ethical considerations of human investigations in developing countries: The AIDS dilemma, *New England Journal of Medicine* 319 (1988): 1083–85.
11. M. W. Kreuter, M. J. Parsons, and M. P. McMurry, Moral sensitivity in health promotion, *Health Education* (November/December 1982): 11–13.
12. K. McNutt, Ethics: A cop or a counselor? *Nutrition Today* 26 (1991): 36–39.
13. D. Beauchamp, Lifestyle, public health and paternalism, in *Ethical Dilemmas in Health Promotion*, ed. S. Doxiadis (New York: Wiley, 1987), pp. 69–81.
14. A. Fornari, Approaches to ethical decision making, *Journal of the American Dietetic Association* 102 (2002): 865–66.
15. Adapted from Basch, *Textbook of International Health.*

Community Nutritionists in Action: Delivering Programs

Section *Two*

Serena M. sits on the edge of her bed and surveys the baby things spread around the room. A pearly white crib stands near the window on her left, and an old chest of drawers, newly painted yellow, fills the space on her right. On the floor next to her husband's crumpled jeans and T-shirt are a stuffed rabbit, a stack of disposable diapers, two rubber squeeze toys, a box of Q-tips, and an airplane mobile still in its box. She wonders whether she has everything she needs for the new baby—her first—due in just a few weeks.

Serena is excited about this change in her life and a little scared, too. Money is tight. She and her husband, Todd, live with his parents because they cannot afford their own apartment. They can't even afford a car, which is why Todd works at the auto body shop just four blocks over. She tries not to let her fear show, even though she frets daily about life with her in-laws and the prospect of being a mother. "I'm nearly 17," Serena says to herself. "My mother was about this age when she had my brother." As she gets up to take clothes out of the dryer, she wonders whether her mother ever felt scared, anxious, and alone.

Here is a young woman, in many respects a child herself, who is expecting her first baby. Does she know how to care for an infant? Is the family's income sufficient to cover the costs of providing for a child? Does this family need food assistance? Is there a way for this young woman to have her baby and continue in school? Is Serena eating properly during her pregnancy? Does she plan to breastfeed the baby? How can the health and nutritional outcome for this baby and its mother be improved?

This real-life scenario reflects the many activities undertaken by community nutritionists: identifying a nutritional problem in the community (a significant number of pregnant teenagers give birth to low-birthweight infants), selecting a target population (low-income pregnant teenagers), asking questions about why the problem developed (What do pregnant teenagers know about nutrition during pregnancy? Do pregnant teenagers use health services for prenatal counseling?), and figuring out how best to address the problem (programs that promote prenatal counseling and breastfeeding and enhance parenting skills among teenagers). The desired outcome is ultimately to reduce the number of low-birthweight babies born to teenagers and to improve infant health.

This section reviews the major food assistance and nutrition programs of the federal government and other noteworthy community-based programs. These programs help "protect" people like Serena and her family from environmental and social conditions that place them at risk for disease and poor health. They offer some security to people who have little money to spend on food, and they help control the problems associated with malnutrition. The section is divided along life-cycle lines, with chapters on mothers and infants, children and adolescents, and adults. A chapter on world hunger and food insecurity is included because this area is increasingly important in a shrinking global community.

We encourage you to draw on the material you have already learned as you study the chapters that follow. By now, you can appreciate the complexities of national nutrition monitoring, poverty, food insecurity, and community needs assessment. As you review the programs described in these chapters, consider how you would change the delivery of community-based nutrition programs and the policies that influence them.

Chapter *Ten*

Mothers and Infants: Nutrition Assessment, Services, and Programs

Learning *Objectives*

After you have read and studied this chapter, you will be able to:

- List the recommendations for maternal weight gain during pregnancy.
- Explain the relationship of maternal weight gain to infant birthweight.
- Identify nutritional factors and lifestyle practices that increase health risk during pregnancy.
- Describe the benefits of breastfeeding.
- Describe the purpose, eligibility requirements, and benefits of the federal nutrition programs available to assist low-income women and their children.
- Identify the common nutrition-related problems of infancy.
- Describe current recommendations for feeding during infancy.

This chapter addresses such issues as nutrient requirements across the life span, economics and nutrition, and availability of food and nutrition programs in the community, which are Commission on Accreditation for Dietetics Education (CADE) *Foundation Knowledge and Skills* requirements for dietetics education.

Chapter *Outline*

Something To Think About...

All aspects of a society, whether it is relatively static or in a stage of dramatic upheaval, affect the health of every member of the family and of the community, and especially of its most physiologically and culturally vulnerable groups—mothers and young children.

– Cicely D. Williams
– Derrick B. Jelliffe

Introduction

The effects of nutrition extend from one generation to the next, and this is particularly evident during pregnancy. Research has demonstrated that the poor nutrition of a woman during her early pregnancy can impair the health of her *grandchild*—and can do so even after that child has become an adult.[1] For example, if a mother's nutrient stores are inadequate early in pregnancy when the placenta is developing, the fetus will develop poorly, no matter how well the mother eats later. After getting such a poor start on life, the female child may grow up poorly equipped to support a normal pregnancy, and she, too, may bear a poorly developed infant.

Infants born of malnourished mothers are more likely than healthy women's infants to become ill, to have birth defects, and to suffer retarded mental or physical development.[2] Malnutrition in the prenatal and early postnatal periods also affects learning ability and behavior. Impaired intrauterine growth may also "program" the fetus for chronic diseases, such as coronary heart disease, hypertension, or Type 2 diabetes, in adult life. According to the fetal origins hypothesis, if a woman's nutrient intake is under- or oversupplied—particularly at critical phases of fetal development—long-term alterations may occur in tissue function.[3] For example, if a woman's energy intake is low during the third trimester of pregnancy, pancreatic cell development in the fetus may be hindered, resulting in impaired glucose tolerance and increased risk of developing diabetes later in life.[4] Clearly, it is critical to provide the best nutrition possible at the early stages of life. This chapter focuses on the nutrition and health recommendations for pregnancy, lactation, and infancy and examines the nutrition programs and services that target pregnant women and their infants.

TABLE 10-1
A Comparison of Infant Mortality Rates Worldwide, 2004

Country	IMR
Denmark	3.0
Japan	3.0
Norway	3.0
Sweden	3.0
Austria	4.0
Belgium	4.0
Finland	4.0
France	4.0
Germany	4.0
Italy	4.0
Spain	4.0
Switzerland	4.0
Canada	5.0
Israel	5.0
Netherlands	5.0
New Zealand	5.0
United Kingdom	5.0
Australia	6.0
Ireland	6.0
United States	7.0

Source: Adapted from UNICEF, *The State of the World's Children* 2005 (Oxford: Oxford University Press, 2004).

Trends in Maternal and Infant Health

The health of a nation is often judged by the health status of its mothers and infants. One of the best indicators of a nation's health, according to epidemiologists, is the **infant mortality rate (IMR).** Although the United States spends more money on health care than most other countries, its IMR of 7.0 is considerably higher than that of several industrialized countries'—for example, 3.0 for Denmark, Japan, and Sweden (see Table 10-1). Furthermore, important measures of increased risk of morbidity and death, such as incidence of low birthweight and fetal alcohol syndrome, have increased since 1990, rather than showing any recent improvement.[5] In addition, disparities in IMRs persist between ethnic groups and between poor and nonpoor infants, as shown in Figure 10-1. The IMR of 13.8 for black infants remains more than twice as high as the IMR of 5.8 for white infants. Although infant mortality rates among blacks and whites have been in steady decline throughout the last few decades, the discrepancy between the rates among blacks

Healthy People 2010

Reduce fetal and infant deaths.

Infant mortality rate (IMR) Infant deaths under one year of age, expressed as the number of such deaths per 1,000 live births.

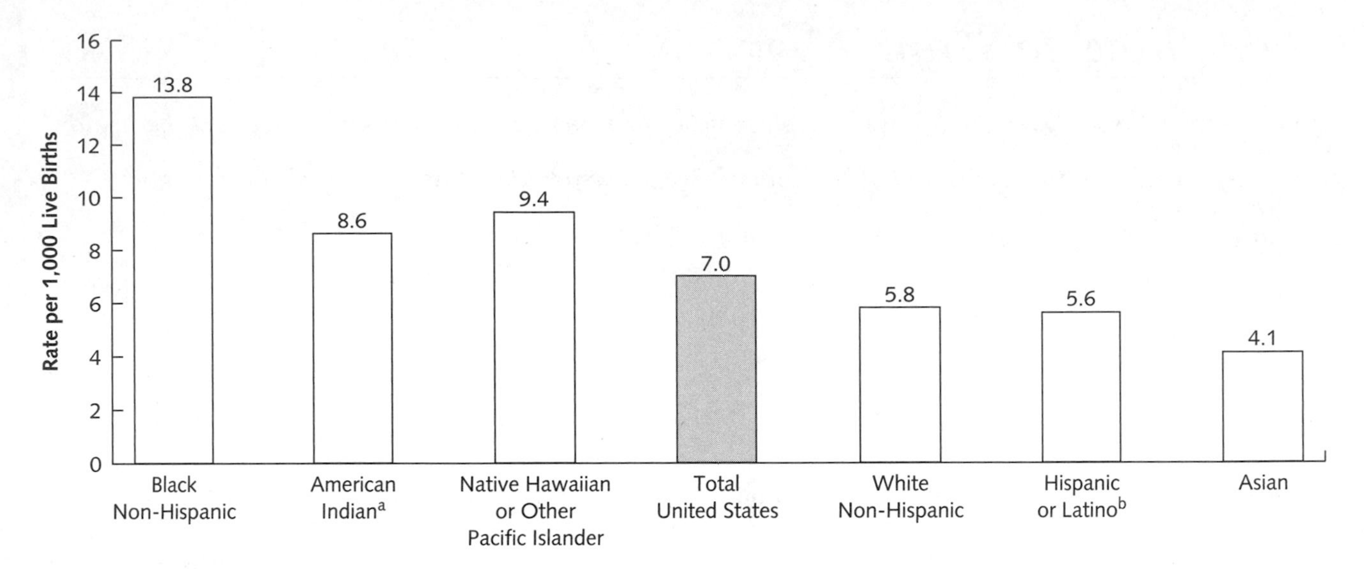

FIGURE 10-1 Infant Mortality Rates by Race and Ethnicity, 2002

Source: National Vital Statistics Reports (Hyattsville, MD: National Center for Health Statistics), 2004; available at www.cdc.gov/nchs.

and rates among whites has remained largely unchanged. The failure to further improve the IMR in the United States has been attributed to the number of infants born with low birthweights. In 2003 the infant mortality rate ranged from a low of 4.8 in New Hampshire and Massachusetts to a high of 10.6 in Mississippi and 11.3 in the District of Columbia.[6] The leading causes of death among infants are birth defects, preterm delivery and **low birthweight (LBW),** sudden infant death syndrome (SIDS), and maternal complications during pregnancy.

If the pregnant woman does not receive adequate nourishment and does not gain the recommended amount of weight, she may give birth to a baby of low birthweight (LBW). Not all small babies are unhealthy, but birthweight and length of gestation are the primary indicators of the infant's future health status. An LBW baby is more likely to experience complications during delivery than a normal-weight baby and has a statistically greater chance of having physical and mental birth defects, of contracting diseases, and of dying during the first year of life. Low birthweight in full-term infants is a major contributing factor to infant mortality. Clearly, a key to reducing infant mortality is reducing the incidence of LBW babies. Doing so requires that several factors be addressed: poverty, minority status, lack of access to health care, inability to pay for health care, poor nutrition, low level of educational achievement, unsanitary living conditions, and unhealthful habits such as smoking, drinking, and drug use. Figure 10-2 depicts the percentage of LBW infants by race in the United States. In 2002 the percentage of low-birthweight births ranged from a low of 5.8 percent in Alaska and Oregon to a high of 11.2 percent in Mississippi and 11.6 percent in the District of Columbia.[7]

Improving the health and nutrition status of pregnant women and infants remains a national challenge. Although **maternal mortality rates** have decreased significantly during the last five decades from a high of 83.3 per 100,000 live births in the 1950s to a low of 7.2 for white women in 2001, black women still have more than a three times greater risk of dying than white women.[8] The *Healthy People 2000* objective for maternal mortality of no more than 3.3 maternal deaths per 100,000 live births was not achieved during the twentieth century; substantial improvements are needed to meet the same objective for *Healthy People 2010.*

Healthy People 2010

Reduce low birthweight (LBW) and very low birthweight (VLBW).

Low birthweight (LBW) A birthweight of $5\frac{1}{2}$ pounds (2,500 grams) or less, used as a predictor of poor health in the newborn and as a probable indicator of poor nutritional status of the mother during and/or before pregnancy. ***Very low birthweight*** is defined as less than 1,500 grams, or 3 pounds 4 ounces.

Maternal mortality rate Women's deaths assigned to causes related to pregnancy, expressed as the number of such deaths per 100,000 live births.

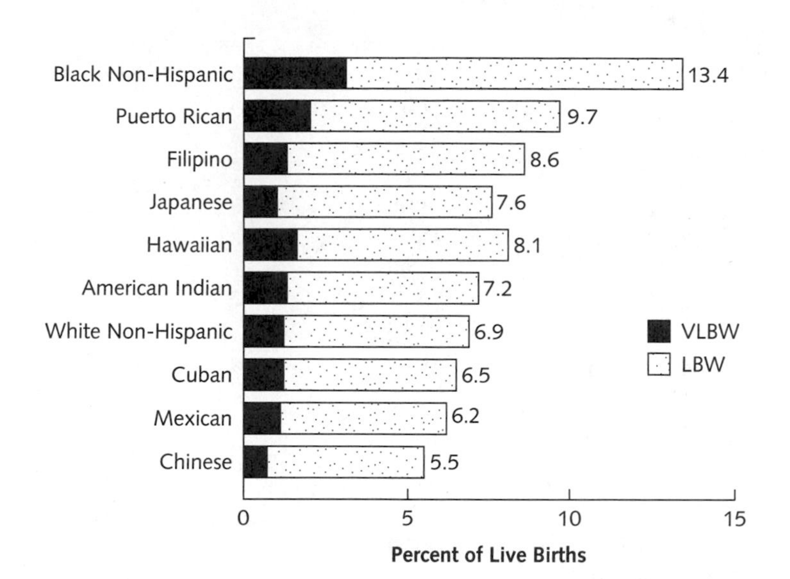

FIGURE 10-2 Percentage of Low-Birthweight and Very-Low-Birthweight Infants by Race and Hispanic Origin of Mother in the United States, 2002

Source: National Center for Health Statistics, *National Vital Statistics Report*, Vol. 52, No. 10; available at www.cdc.gov/nchs.

National Goals for Maternal and Infant Health

Within the past two decades a number of research reports, such as *Healthy People 2000* and *Healthy People 2010,* have established goals and recommendations designed to improve the nutrition and health status of mothers and infants.[9] *Healthy People 2010* takes a broad scope, encompassing maternal and infant health, as well as child health, birth defects, and developmental disabilities, and includes 23 objectives in 8 categories (sample objectives are shown in Table 10-2). Objectives focusing on mortality include infant, fetal, and maternal deaths, perinatal deaths, deaths of children 5 to 14 years of age, and deaths of adolescents and young adults. Objectives addressing risk factors include the areas of preterm births and infant sleep position. A focus on developmental disabilities is included, with new objectives on the incidence of several specific conditions and folate intake among women of childbearing age.

In order for the United States to achieve further reductions in infant mortality and eliminate racial and ethnic differences in pregnancy outcomes, health care professionals must focus on changing the behaviors—both protective and risky—that affect pregnancy outcomes.[10] For example, health problems and behaviors such as smoking, substance abuse, and poor nutrition need to be addressed in preconception screening and counseling.[11] Examples include daily folate consumption (a protective factor) and alcohol use (a risk factor). Presently, only 21 percent of women of childbearing age are consuming the recommended 400 micrograms of folate daily. Binge drinking among pregnant women increased fourfold during the 1990s.[12]

The use of timely prenatal care can also help to mitigate risks by identifying women who are at high risk of high blood pressure or other maternal complications. Other actions, such as breastfeeding, newborn screening, and primary care in infancy, can significantly improve infants' health and chances of survival. For example, the incidence of sudden infant death syndrome (SIDS), the leading cause of postneonatal mortality in the United States, decreased from 79 per 100,000 live births in 1996 to 56 per 100,000 in 2001. The *Healthy People 2010* target is 25. This improvement reflects the success of the "Back to Sleep" education campaign that encourages parents to place healthy infants on their backs to sleep.[13] Approximately 66 percent of infants are now put to sleep on their backs, compared to 35 percent in 1996. The *Healthy People 2010* target is 70 percent.

Healthy People 2010

Reduce maternal deaths.

Healthy People 2010

- Increase abstinence from cigarettes among pregnant women.
- Increase abstinence from alcohol among pregnant women.
- Reduce the occurrence of fetal alcohol syndrome (FAS).

Healthy People 2010

- Increase the proportion of pregnant women who receive early and adequate prenatal care.
- Reduce deaths from sudden infant death syndrome (SIDS).
- Increase the percentage of healthy full-term infants who are put down to sleep on their backs.

HEALTHY PEOPLE (HP) *2010* OBJECTIVE NUMBER	HP 2010 OBJECTIVE	BASELINE (Year)	PROGRESS REVIEW (2002)	HP 2010 TARGET
16.1c	Reduce Infant mortality to no more than 4.5 per 1,000 live births.	7.2 (1998)	7.0	4.5
	a. Black	13.8 (1998)	13.8	4.5
	b. American Indian/Alaska Native	9.3 (1998)	8.6	4.5
	c. Hispanics	5.8 (1998)	5.6	4.5
16.4	Reduce the maternal mortality rate to no more than 3.3 per 100,000 live births.	9.9 (1999)	8.9 24.9	3.3
	a. Among black women	25.4 (1999)		3.3
16.18	Reduce the incidence of fetal alcohol syndrome to no more than 0.12 per 1,000 live births.	0.67 (1993)	NA*	0.12
16.10 a-b	Reduce low birthweight (LBW) to an incidence of no more than 5% of live births and very low birthweight (VLBW) to no more than 0.9% of live births.	7.6% (1998) 1.4 (1998)	7.8% 1.5%	5% 0.9%
	a. LBW among black infants	13.0% (1998)	13.3%	5%
	b. VLBW among black infants	3.1% (1998)	3.1%	0.9%
16.12	Increase to at least 85% the proportion of mothers who achieve the minimum recommended weight gain during their pregnancies.	75% (1990)	NA	85%
16.19 a-b	Increase to at least 75% the proportion of mothers who breastfeed their babies in the early postpartum period and to at least 50% the proportion who continue breastfeeding until their babies are 5 to 6 months old.	Early postpartum period: 64% (1998) At 6 months: 29% (1998)	70% 33%	75% 50%

*NA = Data are not available.

TABLE 10-2 Progress Review for Meeting the *Healthy People 2010* Objectives to Improve Maternal, Infant, and Child Health

HEALTHY PEOPLE 2010 PROGRESS REVIEW

According to the most recent *Healthy People 2010* Review, progress toward the *Healthy People 2010* maternal and infant objectives has been uneven (see Table 10-2).[14] Improvement has occurred in the following areas: (1) There has been a decline in infant mortality rates for Hispanics, whites, and American Indians. (2) The incidence (new cases) of spina bifida and other neural tube defects decreased from 6 per 10,000 live births in 1996 to 4.8 per 10,000 in 2000. The target is 3 per 10,000. The decrease was due in large part to increased consumption of folic acid from fortified foods or dietary supplements, as evidenced by the rise in the median red blood cell folate level in nonpregnant

HEALTHY PEOPLE (HP) *2010* OBJECTIVE NUMBER	HP 2010 OBJECTIVE	BASELINE (Year)	PROGRESS REVIEW (2002)	HP 2010 TARGET
16.17 a-d	Increase abstinence from tobacco use by pregnant women to at least 99%, increase abstinence from alcohol use by pregnant women to at least 94%, and increase abstinence from illicit drug use by pregnant women to 100 percent.	Tobacco: 87% (1998)	89%	99%
		Alcohol: 86% (1996–1997)	90.9%	94%
		Illicit Drugs: 98% (1996–1997)	NA	100%
16.6	Increase to at least 90% the proportion of all pregnant women who receive prenatal care in the first trimester of pregnancy.	83% (1998)	84%	90%
16.15	Reduce the incidence of spina bifida and other neural-tube defects to 3 per 10,000 live births.	6 per 10,000 live births (1996)	4.8	3 per 10,000 live births
19.11	Increase the proportion of pregnant and lactating women who meet dietary recommendations for calcium.	39% (1988–1994)	NA	75%
19.12c	Reduce iron deficiency to less than 7% among women aged 12 to 49 years.	11% (1988–1994)	12%	7%
19.13	Reduce anemia among low-income pregnant females in their third trimester.	29% (1996)	33% (2001)	20%
16.16a	Increase the proportion of nonpregnant women aged 15 to 44 years who consume 400 μg of folic acid each day from fortified foods or dietary supplements.	21% (1991–1994)	NA	80%

TABLE 10-2 Progress Review for Meeting the *Healthy People 2010* Objectives to Improve Maternal, Infant, and Child Health—*continued*

women aged 15 to 44 years. (3) There has been an increase in breastfeeding by women in all racial and ethnic groups. And (4) cigarette smoking during pregnancy has continued to decline. In 2001, 12 percent of women smoked during pregnancy, compared to 20 percent of pregnant women in 1990.[15] However, either no progress or movement in the *wrong* direction has occurred in the areas of maternal death for African-American women, iron deficiency in women aged 12 to 49 years, fetal alcohol syndrome, and low birthweight. The proportion of pregnant women who receive prenatal care beginning in the first trimester has shown little change in recent years. Long-standing racial and ethnic disparities among the recipients of prenatal care also remain much the same. In 2001, 83 percent of all pregnant women received such care, but for

Healthy People 2010

- Reduce the occurrence of spina bifida and other neural-tube defects (NTDs).
- Increase the proportion of pregnancies begun with an optimum folate level.

Healthy People 2010

- Reduce iron deficiency among pregnant females.
- Reduce anemia among low-income pregnant females in their third trimester.

black women the proportion was only 74 percent and for American Indian/Alaska Native women only 69 percent. The *Healthy People 2010* target is 90 percent. The proportion of pregnant women aged 15 to 44 years who had abstained from alcohol in the month preceding the survey was 90.9 percent in 2002, with little variance by race or ethnicity. The target is 94 percent.

Improving the health of mothers and infants remains a national priority.[16] As a follow-up to the *Healthy People 2000* initiative, and in order to continue to achieve gains in maternal and infant health in the current decade, an interagency workgroup, along with the surgeon general, has identified several key areas on which to focus attention:[17]

- Ensure the capacity for tracking the effects of welfare reform on the availability and utilization of prenatal care, particularly for immigrant populations.
- Seek to streamline eligibility requirements for maternal and infant programs so as to increase participation.
- Ensure that maternal and infant health care programs are culturally sensitive.
- Direct additional research toward determining the cause of black women's increased risk of dying from maternal complications.
- Increase research on genetic, environmental, and behavioral factors that have an influence on pregnancy outcomes for mother and child.
- Develop new interventions to reduce alcohol consumption during pregnancy, especially binge drinking.
- Increase dissemination of information to health care providers about the benefits of daily folate intake before, during, and after pregnancy.
- Help ensure that maternal and infant health care programs are tailored to the communities they serve by strengthening the nonmedical aspects of the programs, such as transportation and child care.

Healthy Mothers

A number of factors contribute to maternal and infant health. Genetic, environmental, and behavioral factors affect risk and the outcome of pregnancy.[18] A woman's nutrition prior to and throughout pregnancy is crucial both to her health and to the growth, development, and health of the infant she conceives. Ideally, a woman starts pregnancy at a healthful weight, with filled nutrient stores and the firmly established habit of eating a balanced and varied diet (see Table 10-3). In this section, we discuss maternal weight gain, adolescent pregnancy, and nutrition assessment in pregnancy.

TABLE 10-3 Food Guide for Pregnant and Lactating Women

	NUMBER OF SERVINGS*	
FOOD	Nonpregnant Women	Pregnant or Lactating Women
Breads/cereals	6 to 11	7 to 11 (7+)
Vegetables	3 to 5	4 to 5 (5+)
Fruits	2 to 4	3 to 4 (4+)
Meat/meat alternates	2 to 3	3 (3+)
Milk/milk products	2	3 to 4 (4+)

*Numbers in parentheses indicate numbers of servings recommended for the pregnant teenager.

TABLE 10-4 Recommended Weight Gain for Pregnant Women Based on Body Mass Index (BMI)

BMI	WEIGHT CATEGORY	RECOMMENDED GAIN (Pounds)*
<18.5	Underweight	28–40
18.5–24.9	Normal weight	25–35
25.0–29.9	Overweight	15–25
≥30.0	Obesity	≥15

*Teens should strive to gain the maximum pounds in their ranges; short women (less than 62 inches tall) should strive for the minimum. Weight gain varies widely, and these values are suggested only as guidelines for identifying individuals whose weights may be too high or low for health.

Source: Adapted from NHLBI, *Clinical Guidelines on the Identification, Evaluation, and Treatment of Overweight and Obesity in Adults: The Evidence Report,* June 1998, and National Academy of Sciences, Food and Nutrition Board, *Nutrition During Pregnancy* (Washington, D.C.: National Academy Press, 1990).

Maternal Weight Gain

Normal weight gain and adequate nutrition support the health of the mother and the development of the fetus. The National Academy of Sciences recommendations for weight gain take into account a mother's prepregnancy weight-for-height or body mass index (BMI), as shown in Table 10-4. The committee recommends that a woman who begins pregnancy at a healthful weight should gain between 25 and 35 pounds. Women pregnant with twins need to gain 35 to 45 pounds; women pregnant with triplets need to gain 45 to 55 pounds. An underweight woman needs to gain between 28 and 40 pounds, an overweight woman between 15 and 25 pounds.

Healthy People 2010

Increase the proportion of mothers who achieve a recommended weight gain during their pregnancies.

Approximately three-quarters of married women who deliver at full term gain the recommended weight during pregnancy. Two groups of women who continue to gain less than the recommended level of weight during pregnancy—pregnant teenagers and African-American women—also are at particularly high risk for having LBW infants and other adverse pregnancy outcomes.[19]

Low weight gain in pregnancy is associated with increased risk of delivering an LBW infant; these infants have high mortality rates.[20] Excessive weight gain in pregnancy increases the risk of complications during labor and delivery, as well as postpartum obesity. Obese women also have an increased risk for complications during pregnancy, including hypertension and gestational diabetes.

Weight gain should be lowest during the first trimester—2 to 4 pounds for the trimester—followed by a steady gain of about a pound per week thereafter, as shown in Figure 10-3. If a woman gains more than the recommended amount of weight early in pregnancy, she should not try to diet in the last weeks. Dieting during pregnancy is not recommended. A *sudden* large weight gain, however, may indicate the onset of pregnancy-induced hypertension, and a woman who experiences this type of weight gain should see her health care provider. See Table 10-5 for tips on counseling pregnant women regarding healthful weight gains.

Adolescent Pregnancy

More than 800,000 teenagers become pregnant in the United States each year—1 out of every 20 babies is born to a teenager—and more than a tenth of these mothers are under age 15.[21] The complex social, emotional, and physical factors involved make teen pregnancy one of the most challenging situations for nutrition counseling. According to a paper from the American Dietetic Association, pregnant adolescents are nutritionally at risk and require intervention early and throughout pregnancy.[22] Medical and nutritional

Healthy People 2010

Reduce pregnancies among adolescent females.

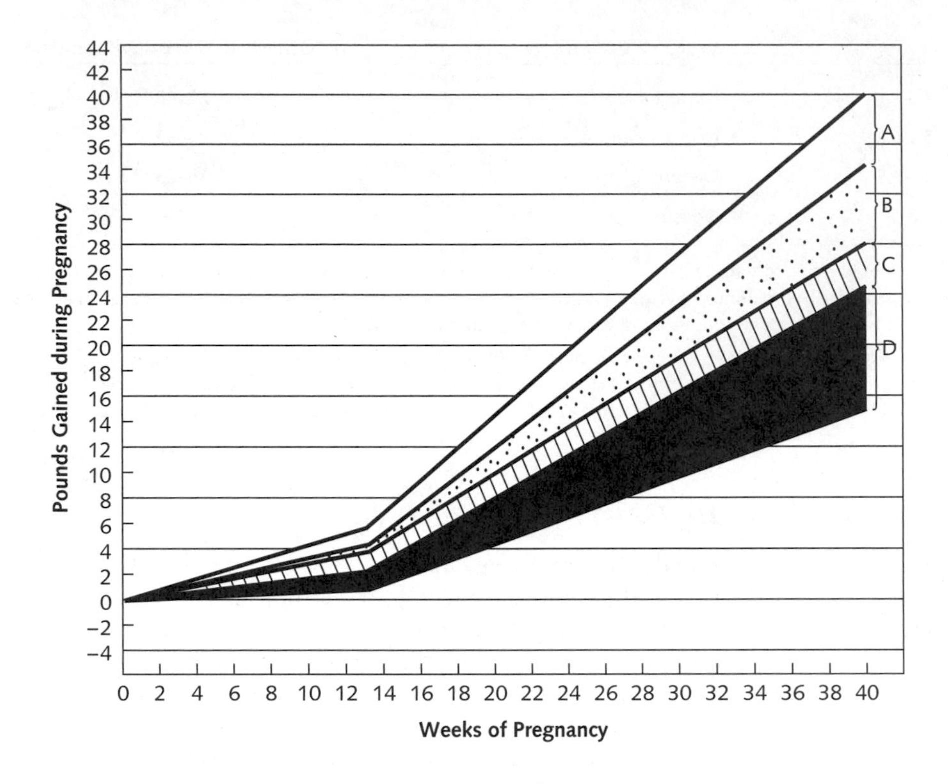

FIGURE 10-3 Desirable Weight Gains during Pregnancy

The woman who is of normal weight prior to pregnancy should gain in the B–C range, 25 to 35 pounds, during the pregnancy. The underweight woman should gain in the A–B range, 28 to 40 pounds. The woman who is overweight prior to pregnancy should gain in the D range, 15 to 25 pounds.

Source: Courtesy of National Dairy Council, *Great Beginnings: The Weighting Game Graph* (Rosemont, IL: National Dairy Council, 1991); adapted from Food and Nutrition Board, Institute of Medicine, National Academy of Sciences, *Nutrition during Pregnancy* (Washington, D.C.: National Academy Press, 1990).

risks are particularly high when the teenager is within 2 years of menarche (usually 15 years of age or younger).[23] Risks include higher rates of pregnancy-related hypertension, iron-deficiency anemia, premature birth, stillbirths, LBW infants, and prolonged labor in pregnant teens than in older women. In addition, mothers under 15 years of age bear more babies who die within the first year than do any other age group.

The increased energy and nutrient demands of pregnancy place adolescent girls, who are already at risk for nutritional problems, at even greater risk. To support the needs of both mother and infant, adolescents are encouraged to strive for pregnancy weight gains at the upper end of the ranges recommended for pregnant women (see Table 10-4). Those who gain between 30 and 35 pounds during pregnancy have lower risks of delivering LBW infants. Adequate nutrition can substantially improve the course and outcome of adolescent pregnancy.[24]

Nutrition Assessment in Pregnancy

Nutrition assessment and monitoring during pregnancy can be divided into three categories: preconception care, the initial prenatal visit, and subsequent prenatal visits. Because a woman's nutrition status and lifestyle habits prior to pregnancy can influence the outcome of pregnancy, these are important factors to consider prior to pregnancy. Nutritional risk factors that may be present at the start of pregnancy are listed in Table 10-6. Preconception care should ideally be available to all women. It should include nutrition assessment, nutrition counseling, and appropriate supplementation and referral to correct nutritional problems existing prior to conception. Prepregnancy weight-for-height should be categorized using BMI so that appropriate weight gain can be recommended for pregnancy (refer to Table 10-4).

TABLE 10-5 Points to Consider for Optimal Weight Gain in Pregnancy

What to Look for If Weight Gain Is Slow or If Weight Loss Occurs:

- Is there a measurement or recording error?
- Is the overall pattern acceptable? Was a lack of gain preceded by a higher than expected gain?
- Was there evidence of edema at the last visit, and is it resolved?
- Is nausea, vomiting, or diarrhea a problem?
- Is there a problem with access to food?
- Have psychosocial problems led to poor appetite?
- Does the woman resist weight gain? Is she restricting her energy intake? Does she have an eating disorder?
- If the slow weight gain appears to be a result of self-imposed restriction, does she understand the relationship between her weight gain and her infant's growth and health?
- Is she smoking? How much?
- Is she using alcohol or drugs (especially cocaine or amphetamines)?
- Does her energy expenditure exceed her energy intake?
- Does she have an infection or illness that requires treatment?

What to Look for If Weight Gain is Very Rapid:

- Is there a measurement or recording error?
- Is the overall pattern acceptable? Was the gain preceded by weight loss or a lower than expected gain?
- Is there evidence of edema?
- Has the woman stopped smoking recently? The advantages of smoking cessation offset any disadvantages associated with gaining some extra weight.
- Are twins a possibility?
- Are there signs of gestational diabetes?
- Has there been a dramatic decrease in physical activity without an accompanying decrease in food intake?
- Has the woman greatly increased her food intake? Obtain a diet recall, making special note of high-fat foods. However, rapid weight gain is often accompanied by normal eating patterns, which should be continued. If intake of high-fat or high-sugar foods is excessive, encourage substitutions.
- If serious overeating is occurring, explore why. Does stress, depression, an eating disorder or boredom play a factor? Is there need for special support or a referral?

Source: B. Worthington Roberts and S. R. Williams, *Nutrition Throughout the Life Cycle,* 3rd ed. (St. Louis, MO: Times Mirror/Mosby, 1996).

Assessment of pregnant teens follows the general pattern for assessment of older pregnant women (see Table 10-7). Assessment issues include acceptance of the pregnancy, food resources and food preparation facilities, body image, living situation, relationship with the father of the infant, peer relationships, nutrition status, prenatal care, nutrition attitude and knowledge, preparation for child feeding, financial resources, continuation of education, day care, and knowledge of and attitudes toward different methods of feeding infants.[25] See Table 10-8 for sample questions to include in a nutrition interview with a pregnant woman.

The nutrition status of all women should be assessed at their initial prenatal visit. This assessment should include[26]

TABLE 10-6 Nutritional Risk Factors in Pregnancy

PERSONAL CHARACTERISTICS (Cannot Change)	PERSONAL HABITS, LIVING SITUATION (Seek to Change)	CHRONIC PREEXISTING MATERNAL MEDICAL PROBLEMS (Screen, Treat)	OBSTETRIC HISTORY (Screen, Prevent)	CURRENT/POTENTIAL PREGNANCY-INDUCED PROBLEMS (Screen, Prevent, Treat)
Age (years)	Low socioeconomic status	Hypertension	Low birthweight	Anemia
Adolescent (<15)	Smoking	Type 1 diabetes	Macrosomia	Iron deficiency
Older woman (>35)	Alcohol/drug use	Type 2 diabetes	Stillbirth	Folate deficiency
Family history	Poor diet	Heart disease	Abortion	Pregnancy-induced hypertension (PIH)
Diabetes	Eating disorders	Pulmonary disease	Fetal anomalies	
Heart disease	Malnutrition	Renal disease	High parity	Gestational diabetes
Hypertension	Obesity	Maternal PKU	Multipara	Weight gain
PKU	Underweight	AIDS		Inadequate
				Excessive
History of poor obstetric outcome	Sedentary lifestyle			
	Restricted diet			

Source: Adapted with permission from B. Worthington-Roberts and S. R. Williams, *Nutrition in Pregnancy and Lactation,* 5th ed. (St. Louis, MO: Times Mirror/Mosby, 1993) p. 240.

- *Dietary measures.* Diet history, including food habits, attitudes, and folklore; allergies; use of vitamin and mineral supplements; and lifestyles (for example, substance abuse or existence of pica).
- *Clinical measures.* Obstetric history, including outcome of previous pregnancies, interval between pregnancies, and history of problems during the course of previous pregnancies (PIH, gestational diabetes, iron-deficiency anemia, or pattern of inadequate or excessive weight gain).
- *Anthropometric measures.* Measurement of weight-for-height; the weight should be recorded and plotted on a weight gain grid (refer to Figure 10-3). Skinfold measures are not recommended for routine assessment.
- *Laboratory values.* Screening for anemia by hematocrit and/or hemoglobin. A urine analysis for ketones, glucose, and protein spillage may be ordered for women with gestational diabetes or preeclampsia.

During each subsequent prenatal visit, weight gain should be monitored and the pattern of weight gain evaluated. Screening for anemia should be repeated on at least one other occasion during pregnancy. Routine assessment of dietary practices is recommended for all women so that their need for an improved diet or vitamin or mineral supplementation can be evaluated. The nutritional status of women in high-risk categories should be reevaluated at each visit.

Poor dietary practices should be improved by appropriate interventions. These may include general nutrition education, individualized diet counseling, and referral to food assistance programs (for instance, the Special Supplemental Nutrition Program for Women, Infants, and Children [WIC] and the Food Stamp Program) or to programs that promote improved food acquisition or preparation practices (such as the Expanded Food and Nutrition Education Program—EFNEP).

Healthy Babies

The growth of infants directly reflects their nutritional well-being and is the major indicator of their nutritional status. A baby grows more rapidly during the first year of life than ever again; its birthweight doubles during the first 4 to 6 months and triples by the end of

TABLE 10-7 Protocol for Nutrition Assessment in Pregnancy

Initial Evaluation

Review clinical data:
- Height and weight
- Gynecological age (adolescents)
- Physical signs of health
- Expected delivery date

Review laboratory data:
- Hematocrit or hemoglobin
- Urinalysis

Assess attitude about and acceptance of prepregnancy weight and feelings about weight gain during pregnancy.

Assess intake patterns using dietary methodology best suited to client and professional.

Make preliminary assessment of food resources and refer to supportive agencies if necessary.

Check for nausea and vomiting and suggest possible remedy.

Discuss supplemental vitamins and minerals.

Make initial plan that sets priorities for issues.

Determine client's understanding of relation between nutrition and health.

Second Visit

Check on referrals to other agencies.

Discuss results of initial evaluation and suggest any changes necessary in dietary patterns (use printed materials as appropriate).

Do any further investigations when necessary:
- Laboratory studies
  - For specific diagnosis of anemia:
    - Protoporphyrin heme or serum ferritin
    - Serum or red cell folate
    - Serum vitamin B_{12}

Further probing of dietary habits if necessary

Monitor weight gain; discuss projected weight gain for following visit and total for gestation.

Assess and address issues affecting nutrition status in order of priority for the individual:
- Activity level
- Appetite changes
- Pica, food cravings, and aversions
- Allergies/food intolerances
- Supplementation practices

Subsequent Visits

Monitor and support appropriate weight gain; include discussion of fitness and encourage safe exercise.

Continue to address issues affecting nutrition status.

Check for heartburn, small food-intake capacity, and elimination problems; suggest dietary interventions.

Begin preliminary discussion and comparison of advantages/disadvantages of breastfeeding and formula feeding.

Final Prenatal Visit(s)

Discuss infant feeding.

If breastfeeding is chosen, provide preliminary guidance about breastfeeding practices.

If formula feeding is chosen, discuss product selection and preparation; define important details about feeding techniques.

Postpartum Visits

Help client understand safe methods of managing weight following delivery.

Review infant feeding practices and infant growth; provide assistance when problems are identified.

Source: Adapted from L. K. Mahan and J. M. Rees, *Nutrition in Adolescence* (St. Louis, MO: Times Mirror/Mosby, 1984).

TABLE 10-8 Questions for a Client-Focused Nutrition Interview in Pregnancy

What you eat and some of the lifestyle choices you make can affect your nutrition and health now and in the future. Your nutrition can also have an important effect on your baby's health. Please answer these questions by circling the answers that apply to you.

Eating Behavior

1. Are you frequently bothered by any of the following? (Circle all that apply): Nausea Vomiting Heartburn Constipation		
2. Do you skip meals at least 3 times a week?	No	Yes
3. Do you try to limit the amount or kind of food you eat to control your weight?	No	Yes
4. Are you on a special diet now?	No	Yes
5. Do you avoid any foods now for health or religious reasons?	No	Yes

Food Resources

6. Do you have a working stove?	No	Yes
Do you have a working refrigerator?	No	Yes
7. Do you sometimes run out of food before you are able to buy more?	No	Yes
8. Can you afford to eat the way you should?	No	Yes
9. Are you receiving any food assistance now?	No	Yes
(Circle all that apply): Food stamps School breakfast School lunch WIC Donated food Commodity foods		
10. Do you feel you need help in obtaining food?	No	Yes

Food and Drink

11. Which of these did you drink yesterday?
(Circle all that apply):

Soft drinks	Coffee	Tea	Fruit drink
Orange juice	Grapefruit juice	Other juices	Milk
Kool-Aid	Beer	Wine	Alcoholic drinks
Water	Other beverages (list) ________		

the first year. Adequate nutrition during infancy is critical to support this rapid rate of growth and development. Clearly, from the point of view of nutrition, the first year is the most important year of a person's life. This section provides a brief overview of nutrient requirements, current recommendations and health objectives for feeding healthy infants, and the relationship between infant feeding and selected pediatric nutrition issues.

Nutrient Needs and Growth Status in Infancy

The infant's rapid growth and metabolism demand an adequate supply of all essential nutrients. Because of their small size, infants need smaller total amounts of these nutrients than adults do, but relative to body weight, infants need over twice as much of many of the nutrients. After 6 months, energy needs increase less rapidly as the baby's growth rate begins to slow down, but some of the energy saved by slower growth is spent on increased activity.

ANTHROPOMETRIC MEASURES IN INFANCY

Anthropometric measurements routinely obtained in the examination of infants include length, weight, and head circumference. These measures assess physical size and growth.

Infants should be weighed nude, using a table model beam scale that allows the infant to lie or sit. Length should be measured in the recumbent position on a measuring

TABLE 10-8 Questions for a Client-Focused Nutrition Interview in Pregnancy—*continued*

Food and Drink—*continued*

12. Which of these foods did you eat yesterday?
    (Circle all that apply):

Cheese	Pizza	Macaroni and cheese	
Yogurt	Cereal with milk		
Other foods made with cheese (such as tacos, enchiladas, lasagna, cheeseburgers)			
Corn	Potatoes	Sweet potatoes	Green salad
Carrots	Collard greens	Spinach	Turnip greens
Broccoli	Green beans	Green peas	Other vegetables
Apples	Bananas	Berries	Grapefruit
Melon	Oranges	Peaches	Other fruit
Meat	Fish	Chicken	Eggs
Peanut butter	Nuts	Seeds	Dried beans
Cold cuts	Hot dog	Bacon	Sausage
Cake	Cookies	Doughnut	Pastry
Chips	French fries		
Other deep-fried foods, such as fried chicken or egg rolls			
Bread	Rolls	Rice	Cereal
Noodles	Spaghetti	Tortillas	

Were any of these whole grain?	No	Yes
13. Is the way you ate yesterday the way you usually eat?	No	Yes

Lifestyle

14. Do you exercise for at least 30 minutes on a regular basis (3 times a week or more)?	No	Yes
15. Do you ever smoke cigarettes or use smokeless tobacco?	No	Yes
16. Do you ever drink beer, wine, liquor, or any other alcoholic beverages?	No	Yes

17. Which of these do you take?
    (Circle all that apply):
    Prescribed drugs or medications
    Over-the-counter products (such as aspirin, Tylenol, antacids, or vitamins)
    Street drugs (such as marijuana, speed, downers, crack, or heroin)

Source: Reprinted with permission from *Nutrition During Pregnancy and Lactation: An Implementation Guide*, Copyright 1992 by the National Academy of Sciences. Courtesy of the National Academy Press, Washington, D.C.

board that has a fixed headboard and a movable footboard attached at right angles to the surface.

Head circumference measures can confirm that growth is proceeding normally or help detect protein-energy malnutrition (PEM) and evaluate the extent of its impact on brain size. To measure head circumference, a nonstretchable tape is placed around the largest part of the infant's head: just above the eyebrow ridges, just above the point where the ears attach, and around the occipital prominence at the back of the head.[27] The head circumference percentile should be similar to the infant's weight and length percentiles.

Appendix A provides an example of how to plot measures on a growth chart; the CDC Growth Charts can be downloaded from www.cdc.gov/growthcharts.

The National Center for Health Statistics (NCHS) growth charts are used to analyze measures of growth status in infants. A single plotting is used to assess how an infant ranks in comparison to other infants of the same age and sex in the United States. Excessive weight-for-length (above the 95th percentile) indicates overweight. An infant whose weight-for-length falls below the 5th percentile is classified as exhibiting **failure to thrive,** and those below the 10th percentile are suspect for failure to thrive. These infants require further evaluation and care. Under normal conditions, the growth rate usually varies within 2 percentiles. Greater variation indicates the possibility of inadequate nutrition and needs to be evaluated further.

Failure to thrive Inadequate weight gain of infants.

Breastfeeding Recommendations

Breastfeeding offers both emotional and physical health advantages.[28] Emotional bonding is facilitated by many events and behaviors of mother and infant during the early months and years; one of the first can be breastfeeding.

During the first 2 or 3 days of lactation, the breasts produce colostrum, a premilk substance containing antibodies and white cells from the mother's blood. Colostrum and breast milk both contain the bifidus factor that favors the growth of the "friendly" bacteria *Lactobacillus bifidus* in the infant's digestive tract so that other, harmful bacteria cannot grow there. Breast milk also contains the powerful antibacterial agent lactoferrin and other factors—including several enzymes, several hormones, and lipids—that help protect the infant against infection.

Breast milk is also tailor-made to meet the nutrient needs of the young infant. Breastfed infants usually require no nutrient supplements except vitamin D and fluoride. At 4 to 6 months, infants may require an iron supplement, depending on their dietary intake.

Breastfeeding provides other benefits as well. It protects against allergy development during the vulnerable first few weeks, the act of suckling favors normal tooth and jaw alignment, and breastfed babies are less likely to be obese because they are less likely to be overfed. A woman who wants to breastfeed can derive satisfaction from all these advantages. These attributes, along with the convenience and lower cost of breastfeeding, have led many organizations and medical experts to encourage breastfeeding for all normal full-term infants.[29] Breastfeeding in the United States declined after World War II to a low of about 25 percent of infants in 1970. Breastfeeding rates more than doubled in the late 1970s and early 1980s and then declined slightly in the late 1980s. By 2002, 70 percent of mothers initiated breastfeeding, 33 percent were breastfeeding babies who were 6 months old, and 20 percent were still breastfeeding babies who were 1 year old (see Figure 10-4). The *Healthy People 2010* goal for breastfeeding is to increase the incidence

FIGURE 10-4 Percentage of Mothers Breastfeeding Their Infants, 1970–2002 with 2002 Breakdown by Race

Source: 2002 Breastfeeding Trends Report (Cleveland, OH: Ross Products Division, Abbott Laboratories).

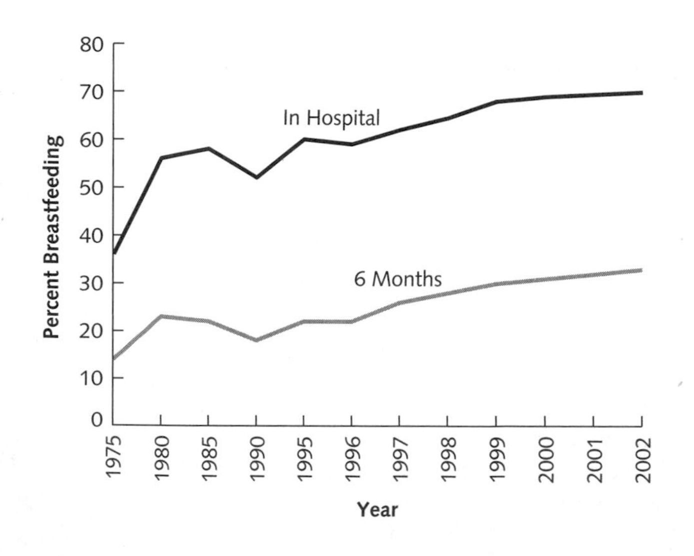

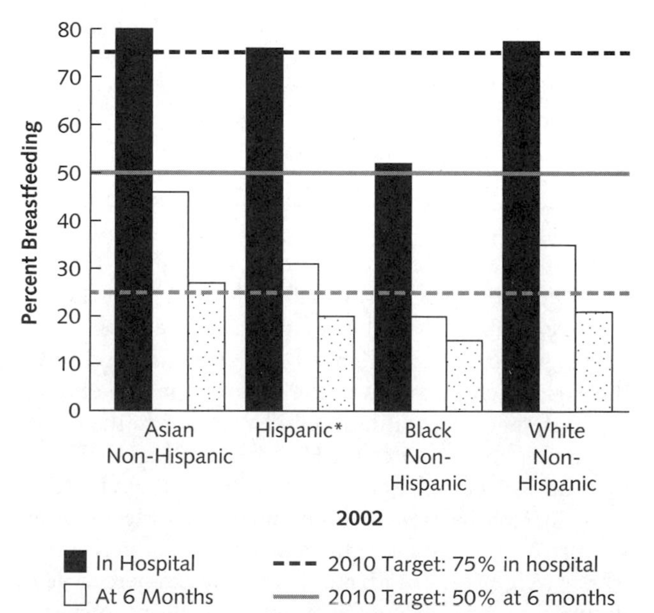

of breastfeeding to 75 percent at discharge from the hospital, 50 percent at 6 months and 25 percent at 1 year.[30] In 2002, the highest percentages at each of the three stages were recorded for Asians and the lowest for blacks, with whites and Hispanics in between (see Figure 10-4). Analysis of data from the Ross Laboratories Mothers Survey indicates that breastfeeding rates continue to be the highest among women who are older, well educated, relatively affluent, and/or living in the western United States.[31]

BREASTFEEDING PROMOTION

A number of barriers to achieving the nation's health objective of increasing the incidence of breastfeeding have been identified. These include lack of knowledge, an absence of work policies and facilities that support lactating women (for example, extended maternity leave, part-time employment, facilities for pumping breast milk or breastfeeding, and on-site child care); the portrayal of bottle feeding rather than breastfeeding as the norm in American society; and the lack of breastfeeding incentives and support for low-income women.[32] Health care professionals and hospitals can provide support by offering encouragement and supplying accurate information on breastfeeding. Table 10-9 lists ten steps recommended by the World Health Organization that hospitals and health professionals can take to promote breastfeeding among women.[33]

Loving Support Makes Breastfeeding Work is the WIC breastfeeding promotion campaign, which is national in scope and is being implemented at the state agency level. The goals of the campaign are to encourage WIC participants to initiate and continue breastfeeding; increase referrals to WIC for breastfeeding support; increase general public acceptance and support of breastfeeding; and provide technical assistance to WIC state and local agency professionals in the promotion of breastfeeding.[34]

One example of a successful approach to increasing breastfeeding rates in low-income, urban populations is the peer counseling method promoted by the La Leche League

TABLE 10-9 Baby-Friendly Hospitals: Ten Steps to Successful Breastfeeding

To Promote Breastfeeding, Every Maternity Facility Should:

- Develop a written breastfeeding policy that is routinely communicated to all health care staff.
- Train all health care staff in the skills necessary to implement the breastfeeding policy.
- Inform all pregnant women about the benefits and management of breastfeeding.
- Help mothers initiate breastfeeding within 1/2 hour of birth.
- Show mothers how to breastfeed and how to maintain lactation, even if they need to be separated from their infants.
- Give newborn infants no food or drink other than breast milk, unless medically indicated.
- Practice rooming-in, allowing mothers and infants to remain together 24 hours a day.
- Encourage breastfeeding on demand.
- Give no artificial nipples or pacifiers to breastfeeding infants.*
- Foster the establishment of breastfeeding support groups and refer mothers to them at discharge from the facility.

*Compared with nonusers, infants who use pacifiers breastfeed less frequently and stop breastfeeding at a younger age. C. G. Victora and coauthors. Pacifier use and short breastfeeding duration: Cause, consequence, or coincidence? *Pediatrics* 99 (1997): 445–53.

Source: United Nations Children's Fund and World Health Organization, *Barriers and Solutions to the Global Ten Steps to Successful Breastfeeding, 1994.*

International.[35] The WIC program has initiated breastfeeding promotion projects among low-income women using the peer counselor method. With this approach, peer support counselors are trained to provide culturally appropriate interventions for initiating and maintaining breastfeeding in their communities. The peer counselor is paired with an expectant or new mother for individual assistance and informal discussions in their neighborhood. Results of similar grassroots approaches to breastfeeding promotion are encouraging; both the rate and the duration of breastfeeding among the women in the programs have shown significant increases.[36]

Focus groups can be useful in designing breastfeeding promotion projects. For example, the project "Best Start: Breastfeeding for Healthy Mothers, Healthy Babies" was organized by a coalition of nutrition and public health officials concerned about declining rates of breastfeeding in the southeastern United States.[37] Focus groups were especially helpful in exploring topics related to breastfeeding among participants who were ambivalent or undecided. Research findings pointed to the need for a carefully coordinated campaign that utilized a combination of strategies to improve the image of breastfeeding and help women overcome the barriers they perceived to breastfeeding. On the basis of information derived from the focus group interviews, guidelines were formulated for program development (see page 327 for the guidelines).

For more information about the *Babies Were Born to Be Breastfed Campaign*, contact the National Women's Health Information Center's free breastfeeding helpline at 1-800-994-WOMAN (9662) or visit its website at www.4women.gov.

In 2004, the U.S. Department of Health and Human Services' Office on Women's Health and the Advertising Council launched a new national campaign that encourages first-time mothers to breastfeed exclusively for 6 months. "Babies were born to be breastfed" is the campaign slogan. The campaign seeks to build confidence while highlighting the consequences of not breastfeeding, citing studies showing that babies who are not breastfed exclusively for the first 6 months may be more likely to develop allergies in childhood, to have more colds, ear infections, and other respiratory illnesses, and to make more visits to the doctor. The campaign includes television, radio, newspaper, magazine, and outdoor public service announcements (PSAs) that communicate the importance of breastfeeding.

Other Recommendations on Feeding Infants

Like the breastfeeding mother, the mother who offers formula to her baby has reasons for making her choice, and her feelings should be honored. Infant formulas are manufactured to approximate the nutrient composition of breast milk. The immunologic protection of breast milk, however, cannot be duplicated.

Many mothers breastfeed at first and then wean within the first 1 to 6 months. Whole cow's milk is not recommended during the first year of life, according to the American Academy of Pediatrics (AAP).[38] Therefore, when a woman chooses to wean her infant during the first 6 months of life, it is imperative that she shift to *formula,* not to plain milk of any kind—whole, low-fat, or nonfat. Only formula contains enough iron (to name but one of many factors) to support normal development in the baby's first months of life. National and international standards have been set for the nutrient content of infant formulas.

For infants with special problems, many variations of infant formulas exist. Special formulas based on soy protein are available for infants allergic to milk protein, and formulas with the lactose replaced can be used for infants with lactose intolerance.

For most infants, breast milk and/or iron-fortified formula provide all the nutrients required for the first 6 months after birth. After that, breast milk and/or iron-fortified infant formula should remain an infant's primary beverage until the end of the first year of life, but solid foods should also be given to provide needed nutrients and to expose infants to a wide variety of flavors and textures, as well as to encourage the mastery of feeding skills. Introducing a variety of flavors and foods before age 2 may increase acceptance

Using Focus Group Findings

Using comments made by women participating in focus groups, planners of the Best Start breastfeeding promotion project formulated the following guidelines.

- **Campaign tone.** The tone should be emotional to reflect the strong feelings women attach to their aspirations for their children and themselves as mothers.
- **Message design.** Educational messages should be succinct and easily understood to counteract the mistaken belief that breastfeeding is complicated and requires major lifestyle changes. Promotional messages should emphasize confidence and the pride breastfeeding mothers gain from nursing.
- **Spokespersons.** Celebrities are not perceived as credible sources of advice on infant feeding. Also, most focus group respondents find it difficult to identify with the wealthy women featured on many of the pamphlets, posters, and other breastfeeding promotion materials used in health departments. Whenever possible, women featured in the materials should be of the same economic, ethnic, and age groups as those targeted or should not have a clear class affiliation. Visual images in print and broadcast materials should communicate modernity and confidence.
- **Educational approaches.** Educational strategies and materials need to be redesigned so that they no longer reinforce women's perceptions of breastfeeding as difficult. The emphasis on being healthy and relaxed and following special dietary guidelines needs to be replaced with reassurance that most women produce sufficient quantities of highly nutritious breast milk despite variations in diet, stress levels, and health status.
- **Professional training.** Motivational and training materials are needed to counter professionals' mistaken belief that their economically disadvantaged clients are not interested in breastfeeding and do not value health professionals' advice. Counseling strategies need to be redesigned to address the special needs of low-income women. Of special concern are recommendations for overcoming clients' lack of confidence and enabling them to realize their aspirations as women and mothers.
- **Program activities/components.** A variety of mutually reinforcing activities are needed to reach social network members who influence women's decisions about infant feeding and who create hospital, home, and community environments conducive to lactation.

Source: Adapted from C. Bryant and D. Bailey, The use of focus group research in program development. *Best Start*—the national breastfeeding promotion program for economically disadvantaged minorities and teenagers—was started in 1990.

---

of a wider variety of flavors and foods in later childhood.[39] Table 10-10 shows a suggested pattern for feeding infants.

Primary Nutrition-Related Problems of Infancy

Iron deficiency and food allergies are two of the most significant nutrition-related problems of infants.

IRON DEFICIENCY

Iron deficiency remains a prevalent nutritional problem in infancy, although it has declined in recent years largely because of the increasing use of iron-fortified formulas.[40] The use of cow's milk earlier than recommended in infancy can cause iron deficiency

(*Text discussion continues on page 330.*)

TABLE 10-10 Infant Feeding Guide for Healthy Infants*

FOODS†	BIRTH TO 3 MONTHS‡	4–6 MONTHS‡	6–8 MONTHS‡
Human Milk	Breastfeed about 10–12 feedings per 24 hours in the 1st month	Breastfeed about 7–9 feedings in 24 hours	Breastfeed about 4–6 feedings in 24 hours
or	About 8–10 feedings per 24 hours in the 2nd and 3rd months		
Iron-Fortified Infant Formula	0–1 month: 18–24 oz 1–2 months: 22–28 oz 2–3 months: 24–32 oz	4–5 months: 25–40 oz 5–6 months: 25–45 oz	24–32 oz Offer cup
Cereal and Breads	NONE	• Iron fortified infant cereal, by spoon Start with 2–4 tbsp rice cereal, mix with human milk, formula or water, feed twice a day	• All varieties of plain, boxed infant cereal, (2–4 tbsp) feed twice a day • Slowly introduce crackers, dry unsweetened cereals, zwieback and toast (1 serving, feed twice a day)
Fruit Juices	NONE	NONE	• 100% fruit juice with Vitamin C Offer in child-size cup, 1–2 oz, twice a day
Vegetables	NONE	NONE	• Mashed winter squash, sweet peas, green beans, carrots, and spinach 2 tbsp, twice a day
Fruits	NONE	NONE	• Fresh or cooked mashed banana, applesauce, jarred fruits 2 tbsp, twice a day
Protein Foods	NONE	NONE	• Meats or poultry, plain, chopped, jarred 1–2 tbsp, twice a day • Plain yogurt, 1–2 tbsp/day
Water	NONE	NONE	2–4 oz, twice a day

*Every infant is different. Consult with your health care provider to make sure your infant is getting what he or she needs. The American Academy of Pediatrics recommends exclusive breastfeeding for the first 6 months, and that breastfeeding continue for at least 12 months.

†Do not give honey for the first year. Avoid foods high in sugar and/or fat, such as corn syrup, karo syrup, sweetened drinks, puddings, cookies, cakes, candy, mixed dinners, bacon, lunch meats, french fries, or creamed vegetables.

FOODS[†]	8–10 MONTHS[‡]	10–12 MONTHS[‡]
Human Milk	Breastfeed 4 or more feedings in 24 hours	Breastfeed 3 or more feedings in 24 hours
Iron-Fortified Infant Formula	24–32 oz Offer cup	16–24 oz Offer cup
Cereal and Breads	• All varieties of plain, boxed infant cereal 2-3 tbsp, twice a day • Soft breads such as plain bagels, rolls and muffins, or unsweetened dry cereal; 2–3 small servings	• Unsweetened dry cereal, toast, crackers, bread, bagels,rolls, plain muffins, rice and noodles; 2–3 small servings
Fruit Juices	• 100% fruit juice with Vitamin C Offer in child-size cup 1–2 oz, twice a day	• 100% fruit juice with Vitamin C Offer in child-size cup 2 oz, twice a day
Vegetables	• Cooked, mashed vegetables • Soft, bite-size pieces 3–4 tbsp, twice a day	• Cooked, mashed vegetables • Soft, bite-size pieces 1/4 cup, twice a day
Fruits	• Peeled, soft, fresh fruits, or fruits canned in water or juice, such as bananas, pears and peaches • Soft, bite-size pieces, no seeds 3–4 tbsp, twice a day	• All peeled, soft, fresh fruits such as bananas, pears and peaches • Canned fruit in water or juice • Soft, bite-size pieces, no seeds 1/4 cup, twice a day
Protein Foods	• Well-cooked, bite-sized pieces of meat, poultry or fish; mild cheese • Cooked beans, egg yolk, cottage cheese 2–3 tbsp a day	• Strips of tender lean meats, chicken, fish, ground or chopped meats, and cheese strips 1 oz or 1/4 cup, twice a day
Water	2–4 oz, twice a day	2–4 oz, twice a day

AVOID FOODS THAT CAN CAUSE CHOKING

Firm, smooth, or slippery foods:
- Hot dogs, sausages, or toddler hot dogs
- Peanuts and other nuts
- Hard candy, jelly beans
- Whole beans
- Whole grapes, berries, cherries, melon balls, or cherry and grape tomatoes
- Whole pieces of canned fruit

Small, dry, or hard foods:
- Popcorn
- Peanuts, nuts and seeds
- Whole grain kernels
- Small pieces of raw carrots or other raw or partially cooked hard vegetables or fruits
- Pretzels
- Cooked or raw whole kernel corn
- Potato and corn chips

Sticky or tough foods:
- Peanut butter or other nut or seed butters
- Raisins and other dried fruit
- Tough meat or large chunks of meat
- Marshmallows
- Chewing gum
- Caramels or other chewy candy

Acceptable Finger Foods
- Small pieces of ripe soft peeled banana, peach, or pear
- Small strips of toast or bread
- Cooked macaroni
- Thin slices of mild cheese
- Soft cooked chopped vegetables such as string beans or potatoes
- Teething biscuits
- Soft moist finely chopped meats

[‡]Feeding skills: *0–3 months:* strong extrusion reflex. *3–6 months:* extrusion reflex diminishes, and the ability to swallow nonliquid foods develops; chewing action begins. *6–8 months:* able to feed self finger foods, and begins to drink from cup. *8–10 months:* begins to hold own bottle; reaches for and grabs food and spoon. *10–12 months:* begins to master spoon.

Sources: Adapted from New Jersey WIC: Infant Feeding Guide for Healthy Infants, 2003, as adapted from the Massachusetts and Maine WIC Programs. Reprinted by permission of the New Jersey Department of Health and Senior Services; and *Feeding Infants: A Guide for Use in Child Nutrition Programs,* Food and Nutrition Service, United States Department of Agriculture, FNS-258, 2002.

Programs in Action

Using Peer Counselors to Change Culturally Based Behaviors

For many years, the Special Supplemental Nutrition Program for Women, Infants, and Children (WIC) clinics and organizations such as La Leche League have used peer counselors to help women successfully breastfeed their infants. Incidence of breastfeeding among rural, low-income women in Iowa was far greater among women in a peer counseling group (82 percent) than among women in a control group (31 percent), and women in the peer counseling group also were more likely to continue breastfeeding.[1] Breastfeeding decisions among a group of low-income women in Georgia were influenced more by the women's social support networks than by the attitudes of health professionals.[2] These programs, along with the case that follows, demonstrate that a peer counselor who has breastfed a baby, has been trained in lactation support and counseling, and understands a pregnant woman's view of breastfeeding can be very influential in motivating women to initiate and continue breastfeeding.

Goals and Objectives

The goals of Best Beginnings of Forsyth County were to increase the number of women who breastfeed their newborn infants and to increase the number of women who continue to breastfeed past the first few weeks of their baby's life. The program objectives were to develop the curriculum for the didactic and experiential components of the training and to add group discussions and breastfeeding classes.

Target Audience

The Best Beginnings breastfeeding education program was designed for pregnant women who applied for support under the WIC program in Forsyth County. Approximately one-third of those women were Hispanic.

Rationale for the Intervention

Best Beginnings of Forsyth County was initiated to increase breastfeeding and duration of breastfeeding among women in low-income settings. Prior to intervention, only about 10 percent of mothers in the target audience initiated breastfeeding, and almost no women continued breastfeeding past six weeks. In 1991, the Forsyth County WIC Program decided to lay the groundwork for a peer counselor program to try to increase the rate of breastfeeding among WIC participants.

Methodology

In 1992, the Forsyth County WIC program received a grant to begin a breastfeeding support program. Peer counselor training began in 1992 with three women, all of whom had been enrolled in the WIC program and each of whom had breastfed a baby for more than six months. Peer counselor training included 20 hours of didactic instruction on lactation

because of its poor iron content and its potential to cause gastrointestinal blood loss in susceptible infants.[41] Other factors contributing to iron deficiency in infancy include breastfeeding for more than six months without providing supplemental iron, feeding infant formula not fortified with iron, the infant's rapid rate of growth, low birthweight, and low socioeconomic status. To prevent iron deficiency, the AAP recommends that infants be fed breast milk or iron-fortified formula for the first year of life, with appropriate foods added as shown in Table 10-10.

FOOD ALLERGIES

Genetics is probably the most significant factor affecting an infant's susceptibility to food allergies.[42] Nevertheless, food allergies are much less prevalent in breastfed babies than in formula-fed infants. At-risk infants can be identified from elevated cord blood levels of immunoglobulin E (IgE) or by a family history. Breast milk is recommended for infants allergic to cow's milk protein and is preferable to soy or goat's milk formulas, because infants are sometimes intolerant of these proteins as well. To reduce the risk of food sensitivity or allergic reactions to other foods, new foods should be introduced singly to facilitate prompt detection of allergies. For example, if a cereal causes and counseling, a written test, and practical instruction on counseling. Course information included advantages of and myths about breastfeeding, breast physiology, mechanics of breastfeeding, common problems, and high-risk situations. On completion of the training course, peer counselors took a competency exam and completed a six-week internship.

Peer counselors encouraged breastfeeding among WIC participants through individual counseling, classes, and telephone support. They increased public awareness of breastfeeding by taking part in health fairs, baby fairs, and the county fair.

Results

The Best Beginnings program of breastfeeding education through peer counselors has been highly successful. Within six months after the start of the program in 1992, the breastfeeding initiation rate had increased from 10 percent to 26 percent. In fiscal year 1997–1998, almost 50 percent of Forsyth County women enrolled in WIC had initiated breastfeeding, compared to 38 percent of WIC participants statewide. The percentage of participants who continued breastfeeding for more than six weeks increased from virtually zero to over 27 percent in fiscal year 1996–1997, compared to 19 percent of WIC participants statewide.

Forsyth County was recognized in 1997 and 1998 for having the highest breastfeeding initiation rate among urban counties in North Carolina.

Lessons Learned

The enthusiasm of the peer counselors was important to the success of Forsyth County's breastfeeding support program. In addition, peer counselor programs can have a positive influence on the peer counselors themselves. Of the peer counselors who no longer work with Best Beginnings, a large percentage left to complete their education. Most now work full-time and no longer depend on federally subsidized programs.[3]

References

1. E. Schafer, M. K. Vogel, S. Viegas, and C. Hausafus, Volunteer peer counselors increase breastfeeding duration among rural low-income women, *Birth* 25 (1998): 101–6.
2. A. S. Humphreys, N. J. Thompson, and K. R. Miner, Intention to breastfeed in low-income pregnant women: The role of social support and previous experience, *Birth* 25 (1998): 169–74.
3. Statement by Sheila Britt-Smith, M.Ed., R.D., WIC Breastfeeding Coordinator, Forsyth County Department of Public Health.

Source: Community Nutritionary (White Plains, NY: Dannon Institute, Fall 2000). Used with permission. For more information about the *Awards for Excellence in Community Nutrition,* go to www.dannon-institute.org.

irritability due to skin rash, digestive upset, or respiratory discomfort, discontinue its use before going on to the next food. At least two days should elapse between the introduction of one new food and that of another, to allow time for clinical symptoms to appear.[43] For infants with a strong family history of food allergies, solid foods should not be given before 6 months of age, and eggs, milk, wheat, soy, peanuts, tree nuts, shellfish, fish, and foods containing these major food allergens should not be given before 1 year of age.[44]

Domestic Maternal and Infant Nutrition Programs

Nutrition plays a vital role in the outcome of pregnancy as well as in the growth and development of infants. A stable base of essential programs and services is required to meet maternal and infant health care needs. This section describes the nutrition programs and related services available to meet the demands of pregnancy and infancy, two of the most vulnerable periods of the life cycle (see Table 10-11).

TABLE 10-11 Federal Nutrition and Health Care Programs That Assist Mothers and Their Infants

PROGRAM	PARTICIPANTS	BENEFITS
Food Stamps	Anyone with income < 130% of poverty guidelines†	Increased ability to purchase food
WIC*	Pregnant women, postpartum and lactating women, infants, and children up to 5 years of age	Vouchers to purchase healthful foods or direct food supplements, nutrition education, and referral to health services
FMNP	Persons eligible for WIC	Increased fruit and vegetable consumption
Commodity Supplemental Foods	Pregnant and postpartum women, infants, and children < 6 years of age with incomes ≤ 185% of poverty guidelines	Monthly food package of fruits, vegetables, meats, infant formula, beans, and other available foods
EFNEP	Persons with incomes ≤ 125% of poverty guidelines who have children under 19 years of age	Nutrition education
Medicaid	Anyone with income < 133% of poverty guidelines	Complete health care
Healthy Start (Medicaid)	Pregnant women with incomes < 185% of poverty guidelines; certain high risk pregnancies	Prenatal and postpartum care
EPSDT (Health Chek)	Infants, children, and adolescents up to 18 years of age	Health screening: dental checks, health education, hearing, vision

*WIC = Special Supplemental Nutrition Program for Women, Infants, and Children; FMNP = Farmers' Market Nutrition Program; EFNEP = Expanded Food and Nutrition Education Program; EPSDT = Early Periodic Screening, Diagnosis, and Treatment.

†See Table 5-1 in Chapter 5 for 2005 Poverty Guidelines.

The WIC Program

In 1969, the White House Conference on Food, Nutrition, and Health recommended that special attention be given to the nutritional needs of pregnant and breastfeeding women, infants, and preschool children. As a result, the Special Supplemental Nutrition Program for Women, Infants, and Children (WIC) was authorized in 1972 by PL 92-433 as an amendment to the Child Nutrition Act of 1966. The legislation states that the WIC program is to "serve as an adjunct to good health care, during critical times of growth and development." To encourage earlier and more frequent utilization of health services, federal regulations mandate that local agencies may qualify as WIC sponsors only if they can make health care services available to WIC enrollees.[45]

The WIC program is based on two assumptions. One is that inadequate nutritional intakes and health behaviors of low-income women, infants, and children make them vulnerable to adverse health outcomes. The other is that nutrition intervention at critical periods of growth and development will prevent health problems and improve the health status of participants.

WIC is federally funded but is administered by the states. Cash grants are made to authorized agencies of each state and to officially recognized American Indian tribes or councils, which then provide WIC services through local service sites. Priority for the creation of local programs is given to areas whose populations need benefits most, judging on the basis of high rates of infant mortality, low birthweight, and low income. WIC began as a two-year pilot project for each of fiscal years 1973 and 1974 to provide supplemental foods to infants, children up to age 5, and pregnant, breastfeeding, and nonbreastfeeding postpartum women who qualify financially and are considered by competent professionals to be at nutritional risk because of inadequate nutrition and inadequate income.[46] Competent professionals include physicians, nutritionists, nurses, and other health officials. Financial eligibility is determined by income (between 100 percent and 185 percent of the poverty guidelines or below) or by participation in the Food Stamp Program or Medicaid. Nutritional risks, determined by a health professional, may include one of three types: medically based risks (anemia, underweight, obesity, maternal age, history of high-risk pregnancies, HIV infection); diet-based risks (inadequate dietary pattern, gastrointestinal disorders, renal or cardiorespiratory disorders); or conditions that make the applicant predisposed to medically based or diet-based risks, such as alcoholism or drug addiction (see Table 10-12).

TABLE 10-12 **Nutritional Risk Criteria for the Special Supplemental Nutrition Program for Women, Infants, and Children**

To be considered for the WIC program, an applicant must exhibit at least one of the nutritional risk factors listed below.

Women

Conditions Complicating the Prenatal and/or Postpartum Periods:
- low hematocrit or hemoglobin
- insufficient or excessive prenatal weight gain
- insufficient or excessive pregravid or postpartum weight
- excessive use of alcohol, drugs, or tobacco
- inadequate dietary status (as assessed by WIC standards)

General Obstetrical Risks:
- younger than 18
- multifetal gestation
- closely spaced pregnancies
- high parity and young age

History of:
- preterm delivery
- low-birthweight infant
- infant with congenital defect
- miscarriage or stillbirth
- neonatal death
- gestational diabetes

Nutrition-Related Risk Conditions:
- eating disorders
- gastrointestinal disorders
- chronic or pregnancy-induced hypertension
- thyroid disorders
- diabetes
- infectious diseases
- depression
- cancer
- renal disease
- homelessness
- migrancy

Infants and Children

Low-birthweight or preterm infants

Abnormal pattern of growth: short stature, underweight, overweight

Failure to thrive

Inadequate growth

Low head circumference

Inadequate dietary status (as assessed by WIC standards)

Low hematocrit or hemoglobin (after 6 months of age)

Elevated blood lead levels

Food allergies/intolerances (as specified by WIC)

Fetal alcohol syndrome

Inborn errors of metabolism

Source: USDA, Nutrition Program Facts: WIC, December 2004; and www.fns.usda.gov/wic.

The WIC program plays both a remedial and a preventive role. Services provided include the following:

- Food packages or checks or vouchers to purchase specific foods each month. **WIC foods** include iron-fortified infant formula and infant cereal, iron-fortified breakfast cereal, vitamin C–rich fruit or vegetable juice, eggs, milk, cheese, peanut butter, dried beans and peas, tuna fish, and carrots. WIC foods are intended to supplement participants' food intakes; each of the WIC foods is rich in one or more of the nutrients protein, calcium, iron, vitamin A, and vitamin C, which tend to be low in the diets of the population that WIC serves.
- Nutrition education, including individual nutrition counseling and group nutrition classes
- Referral to health care services, including breastfeeding support, immunizations, prenatal care, family planning, and substance abuse treatment

Sample WIC Food Package*

Children (1–5 years) or Pregnant and Breastfeeding Women†

9 qt juice (frozen, reconstituted)
36 oz cereal (hot or cold)
24 qt milk‡
2.5 dozen eggs
1 lb dried beans/peas, *or*
18 oz peanut butter

*An enhanced food package is available for women whose infants do not receive formula from the WIC program (includes addition of cheese, tuna, and carrots).
†Maximum monthly allowance.
‡28 qt for women; cheese may be substituted for a portion of the milk.

Providing nutritious supplemental foods to needy pregnant women is expected to improve the outcome of pregnancy. For infants and children, the food supplements are intended to reduce the incidence of anemia and to improve physical and mental development. Some states distribute food directly, but most provide vouchers or checks that participants can use at participating grocery stores. The food voucher lists the quantities of specific foods, including brand names, that can be purchased with the voucher or check.

Recently, a review of the current WIC food package was undertaken by a committee of the Institute of Medicine.[47] As a first step toward determining whether changes are needed to strengthen the nutritional quality of the WIC food packages, the committee evaluated the dietary intakes of the WIC-eligible population. In its preliminary report, *Proposed Criteria for Selecting the WIC Food Packages*, the committee proposed criteria for making changes in the current WIC food packages as listed in the margin.

Proposed Criteria for WIC Food Packages

The WIC food packages:

- Reduce the prevalence of inadequate or excessive nutrient intakes
- Follow Dietary Guidelines for Americans and dietary recommendations for infants
- Support breastfeeding
- Are suitable for persons who may have limited transportation options, storage, and cooking facilities
- Are readily acceptable and available, and take into account cultural food preferences.

The combination of supplementary food, nutrition education, and preventive health care distinguishes WIC from other federal food assistance programs. Some of the potential impacts of program participation are summarized in Figure 10-5. Program benefits include improved dietary quality, more efficient food purchasing, better use of health services, and improved maternal, fetal, and child health and development.[48]

WIC WORKS

WIC has been described as one of the most efficient programs undertaken by the federal government. Over the years, the program has expanded significantly; the 1.9 million women and children served in 1980 at a cost of $725 million had grown by 2004 to more than 7.9 million women, infants, and children served at a cost of $4.9 billion. The dramatic growth of WIC since 1974 (see Figure 10-6) has prompted policy makers to focus attention on quantifying the program's benefits.[49] To date, the program's proven benefits include the following:

WIC Foods In the 2004 WIC reauthorization, the definition of WIC foods was enhanced by expanding the definition beyond addressing specific nutritional deficiencies to focus also on "foods that promote the health of the population served, as indicated by relevant nutrition science, public health concerns, and cultural eating patterns."

- A General Accounting Office review of previous WIC studies concluded that WIC reduces the incidence of low birthweight ($< 2{,}500$ grams) and very low birthweight ($< 1{,}500$ grams) by 25 percent and 44 percent, respectively.[50]
- The 1986 National WIC Evaluation released by the USDA found that WIC has contributed to a reduction of 20 to 33 percent in fetal mortality and that the head size of infants whose mothers participated in the WIC program during pregnancy increased measurably.[51]
- A USDA study of the health records of more than 100,000 mothers and babies found that the program increased birthweight significantly.

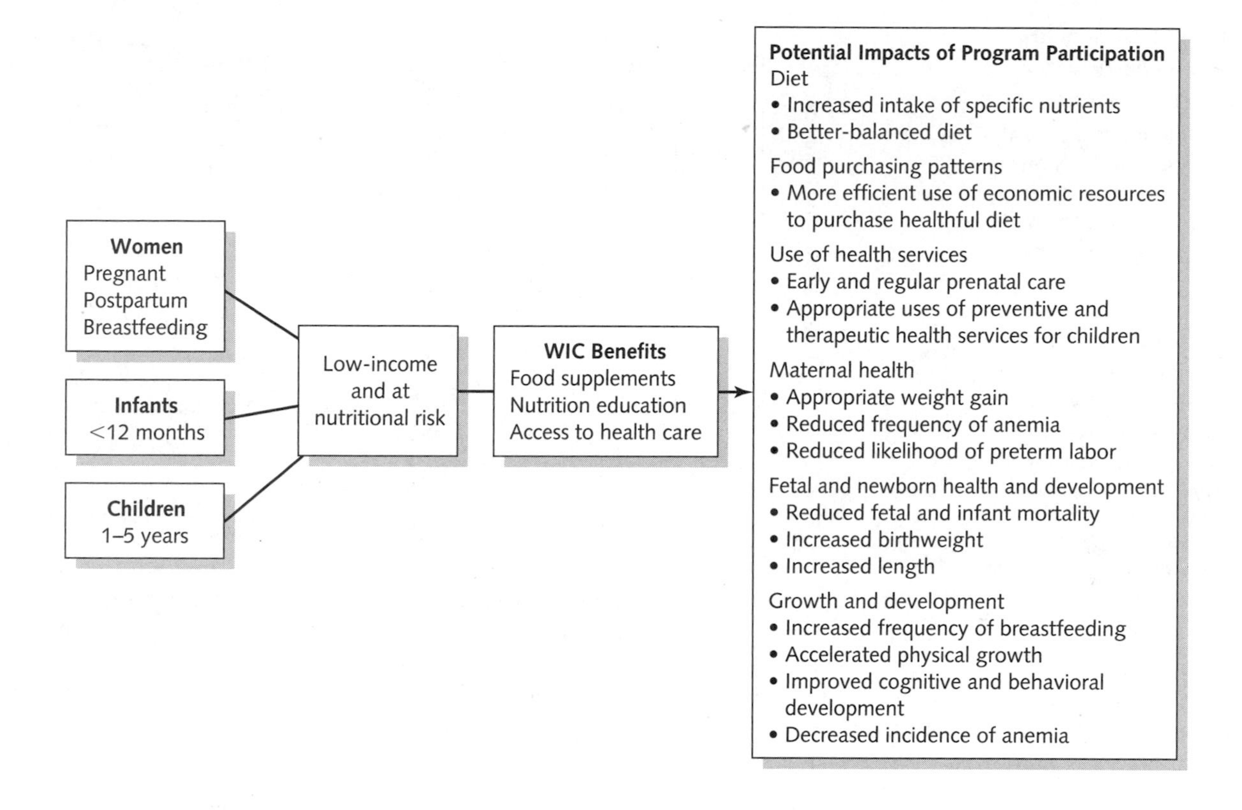

FIGURE 10-5 WIC Benefits and Potential Program Impacts

Source: D. Rush, *The National WIC Evaluation* (Washington, D.C. U.S. Department of Agriculture, 1987).

- Women who participate in WIC have longer pregnancies leading to fewer premature births. This not only benefits the infants but saves millions of dollars in Medicaid costs that would otherwise have been incurred for neonatal intensive care. It is estimated that every dollar spent on WIC for pregnant women can save as much as $4.21 in Medicaid costs.[52] (See the cost–benefit analysis of the WIC program provided in the box on page 337.)
- A study conducted in five states—Florida, North Carolina, Minnesota, Texas, and South Carolina—found that each dollar spent on WIC participants prenatally resulted in a $1.77 to $3.90 savings per participant in Medicaid costs.[53]
- A Yale University School of Medicine study found a remarkable decrease in the prevalence of anemia among low-income children in New Haven since the early 1970s. The researchers concluded, "The marked improvement can most probably be attributed to the nutritional supplementation with iron-fortified foods provided by the WIC program."[54]
- WIC also appears to lead to better mental performance. WIC children whose mothers participated in WIC during pregnancy had better scores than nonparticipants on vocabulary and memory tests.
- WIC participation can lead to improved breastfeeding rates.[55]
- WIC participation is associated with regular use of health care services; WIC children are more likely than not to receive some form of immunization against infectious diseases.
- Data show that the WIC program makes a significant contribution to reducing food insecurity among first-time program participants and suggest the need to consider food insecurity as a risk criterion for the WIC Program.[56]
- The WIC Program may contribute to the adequacy of iron intake among low-income infants and children.[57]

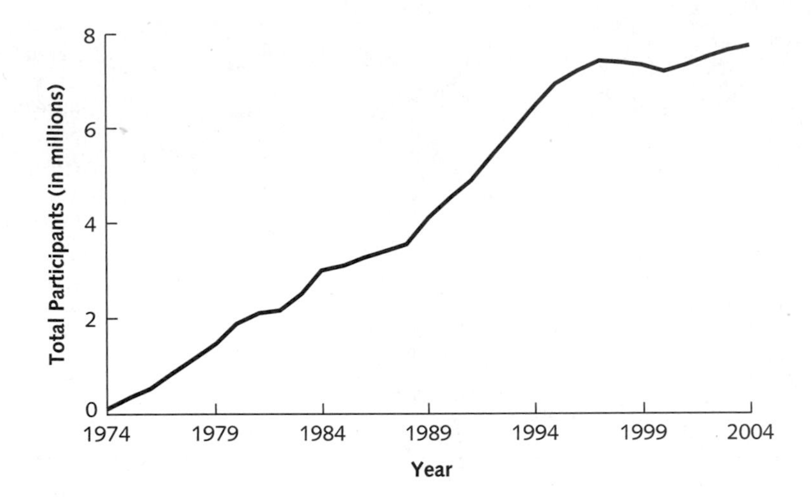

FIGURE 10-6 Annual Growth in Participation in the WIC Program
The dramatic growth of WIC since 1974 includes a diverse population of participants today, including approximately 38 percent Hispanic, 35 percent white, 20 percent black, 3.5 percent Asian/Pacific Islander, and 1.4 percent American Indian/Alaska Native.

Source: S. Bartlett and coauthors, WIC participant and program characteristics, 2003 (Alexandria, VA: Food and Nutrition Service), 2003; and Institute of Medicine, *Proposed Criteria for Selecting the WIC Food Packages: A Preliminary Report of the Committee to Review the WIC Food Packages* (Washington, D.C.: National Academy Press, 2004).

Unlike most of the other food assistance programs, WIC is not an entitlement program and therefore can serve only as many people as its annual appropriation from Congress permits.[58] Under federal regulations, once a local agency has reached its maximum caseload, vacancies must be filled in the following order of priority to ensure that program resources are allocated to those at greatest nutritional risk.

1. Pregnant women, breastfeeding women, and infants determined to be at nutritional risk by a blood test, anthropometric measures, or other documentation of a nutrition-related medical condition
2. Infants, up to 6 months of age, whose mothers were at nutritional risk during pregnancy
3. Children at nutritional risk as determined by a blood test, anthropometric measures, or documentation of a nutrition-related medical condition
4. Pregnant or breastfeeding women and infants at nutritional risk because of an inadequate dietary pattern
5. Children at nutritional risk because of an inadequate dietary pattern
6. Nonbreastfeeding, postpartum women at nutritional risk
7. Persons certified for WIC solely because of homelessness or migrancy

Because of the caps placed on allocated federal funds, WIC currently reaches about 80 percent of eligible persons. Barriers to participation in WIC include limited government funding, lack of transportation, insufficient time or money to travel to the clinic, insufficient outreach to potential WIC recipients, the absence or expense of child care, inconvenient clinic hours, and understaffing of WIC facilities.

Overall, research evaluating WIC's effectiveness finds positive outcomes for WIC participants compared to nonparticipants.[59] The following measures would help lower the barriers to program participation and improve the nutrition status and health care of women, infants, and children even further.

- Increase congressional funding for the program to enable more eligible women, infants, and children to receive WIC benefits.
- Improve the coordination between WIC and programs providing or financing maternal and child health care services.
- Implement outreach activities to increase access to WIC program benefits by all who are eligible.[60]

A Cost–Benefit Analysis of the WIC Program

A cost–benefit analysis helps managers gauge how well a program is meeting its clients' needs and provides data that can be used to influence policy makers. Let's consider an evaluation of prenatal participation in the Special Supplemental Nutrition Program for Women, Infants, and Children (WIC). This cost–benefit analysis involved the following steps:

1. **Identify the primary client.** The primary client was the Division of Maternal and Child Health, Department of Environment, Health, and Natural Resources, Raleigh, North Carolina.
2. **Specify the purpose of the evaluation.** The cost–benefit analysis was designed to assess the effect of participation in the prenatal WIC program on low birthweight and on Medicaid costs for the medical care of infants born in 1988.
3. **Specify the objectives of the evaluation.** The objectives of the evaluation of WIC prenatal participation in North Carolina in 1988 were to determine
   - The birthweight of all infants born to WIC mothers
   - The birthweight of a random sample of infants born to non-WIC mothers
   - The cost per client of participating in the WIC program
   - The type and cost of hospital claims for newborn care paid by Medicaid
4. **Calculate a dollar value for each benefit of the program.** The direct benefits to participants in the WIC program included food and nutrition counseling. Indirect benefits included the birth of infants weighing more than 2,500 grams and reduced costs to Medicaid for the medical management of newborn care. (The dollar value assigned to these benefits was not reported in the published analysis.)
5. **Calculate the costs associated with the program.** Both personnel and material costs were considered. In this analysis, the direct costs included the total value of food vouchers redeemed through the WIC program; administrative costs, estimated at an average of $170 per woman; and the newborn medical care costs paid by Medicaid. Medicaid covered such costs for newborn care as physician services, medications, and inpatient or outpatient care. The direct costs are summarized in Table 10-13, which shows the Medicaid costs for infants whose mothers received WIC prenatal care, the Medicaid costs for infants whose mothers did not participate in WIC, and the estimated WIC program costs.
6. **Calculate the total cost per client.** The average cost of the program was $179 per white woman, $164 per black woman, and $170 overall (both groups taken together).
7. **Calculate the benefit-to-cost ratio and/or the net savings.** The benefit-to-cost ratio, also shown in Table 10-13, was calculated by subtracting the Medicaid costs for the WIC group (column 1) from the Medicaid costs for the non-WIC group (column 2) and dividing by the WIC costs (column 3). The net savings was the actual dollar difference between the total benefits and the total costs. The estimated net savings (column 1 subtracted from column 2) was $343 for whites and $615 for blacks.

TABLE 10-13 Average Costs of Medicaid and Average Costs of WIC Services

	MEDICAID COSTS			
	WIC	Non-WIC	WIC COSTS*	BENEFIT-TO-COST RATIO
White	$1,778	$2,121	$179	1.92
Black	1,902	2,517	164	3.75
Total	1,856	2,350	170	2.91

*WIC costs include administrative and food costs.

Source: Copyright The American Dietetic Association. Reprinted by permission from *Journal of the American Dietetic Association*, Vol. 93: 1993, p. 166.

The cost–benefit analysis revealed that the savings in Medicaid costs outweighed the costs of WIC services. The benefit-to-cost ratio was 1.92 for white women and 3.75 for black women, which means that for each dollar spent on WIC, Medicaid saved $1.92 on whites and $3.75 on blacks. In addition, women who received WIC prenatal care gave birth to fewer infants with low and very low birthweights than women who did not participate in WIC.

A cost–benefit analysis undertaken by another state with a different sample of women and infants would probably produce slightly different results in program costs and in the benefit-to-cost ratio. Even so, fiscal evaluations can be used to convince policy makers that money allocated for the WIC program is well spent and that cost savings can be achieved with nutrition intervention.

Source: The description of the cost–benefit analysis of the WIC program was adapted from P. A. Buescher and coauthors, Prenatal WIC participation can reduce low birthweight and newborn medical costs: A cost–benefit analysis of WIC participation in North Carolina, *Journal of the American Dietetic Association* 93 (1993): 163–66. The procedure for conducting a cost–benefit analysis was adapted from the Ross Roundtable Report, *Benefits of Nutrition Services: A Costing and Marketing Approach* (Columbus, OH: Ross Laboratories, 1987), pp. 24–29.

Other Nutrition Programs of the U.S. Department of Agriculture

Several programs of the U.S. Department of Agriculture (USDA) directly or indirectly provide nutrition support during pregnancy and infancy.

FOOD STAMP PROGRAM

The Food Stamp Program (FSP) is designed to improve the diets of people with low incomes by providing benefits to cover part or all of their household's food budget. Chapter 5 provided an overview of this program. As explained previously, program participants can use the food stamp benefits to buy food in any retail store that has been approved by the Food and Nutrition Service to accept them. For many economically disadvantaged women, the FSP is the major means by which they are able to purchase adequate diets for their families. The Select Panel for the Promotion of Child Health reported that food stamp users purchase more nutritious foods per dollar spent on food than eligible households that do not participate in the program.[61]

WIC FARMERS' MARKET NUTRITION PROGRAM

The WIC Farmers' Market Nutrition Program (FMNP) was created to provide fresh, nutritious fruits and vegetables to WIC participants and to expand the awareness and use of farmers' markets by consumers. Eligible recipients receive FMNP coupons. A study reported by the Food Research and Action Center showed that women in the WIC program who received FMNP coupons increased both fruit and vegetable consumption by about 5 percent compared with WIC women who did not receive coupons. The FMNP participants also patronized farmers' markets more than nonrecipients, even after they were no longer eligible for WIC.[62]

COMMODITY SUPPLEMENTAL FOOD PROGRAM

The Commodity Supplemental Food Program (CSFP) is a direct food distribution program providing supplemental foods and nutrition education. The CSFP provides supplemental foods to infants and children and to pregnant, postpartum, and breastfeeding women with low incomes who are vulnerable to malnutrition and live in approved project areas. Recipients may not participate in both WIC and CSFP. The USDA purchases the foods for distribution through state agencies on a monthly basis.

EXPANDED FOOD AND NUTRITION EDUCATION PROGRAM (EFNEP)

The Expanded Food and Nutrition Education Program (EFNEP) is a federally funded program designed specifically for nutrition education. The EFNEP is directed at low-income families and is administered by the USDA Extension Service. It was authorized in 1968 to provide food and nutrition education to homemakers with young children. The program is implemented by trained nutrition aides from the local community under the supervision of county "cooperative extension educators." These paraprofessionals work to develop a one-to-one relationship with disadvantaged homemakers enrolled in the program. In recent years, multimedia strategies have been tested to expand the scope of nutrition education for this population.[63]

Nutrition Programs of the U.S. Department of Health and Human Services

The U.S. Department of Health and Human Services (DHHS) also sponsors several programs that are concerned with health and nutrition status during pregnancy and infancy.

TITLE V MATERNAL AND CHILD HEALTH PROGRAM

Enacted in 1935, Title V of the Social Security Act is the only federal program concerned exclusively with the health of mothers, infants, and children. It provides federal support to the states to enhance their ability to "promote, improve, and deliver" maternal, infant, and child health (MCH) care and programs for children with special health care needs (CSHCN), especially in rural areas and regions experiencing severe economic stress. The aim of Congress in passing this legislation was to improve the health of mothers, infants, and children in areas where the need was greatest.

The states are allocated Title V MCH funds to be used for (1) services and programs to reduce infant mortality and improve child and maternal health and for (2) services, programs, and facilities to locate, diagnose, and treat children who have special health care needs (for example, chronic medical conditions) or are at risk of physical or developmental disabilities.[64] The Title V MCH program provides for nutrition assessment, dietary counseling, nutrition education, and referral to food assistance programs for infants, preschool and school-aged children, children with special health care needs, adolescents, and women of childbearing age. Title V helps create healthy communities by working with local groups to identify and address their local health needs, ranging from teen pregnancy to low immunization rates. It also supports training in nutrition for health and nutrition professionals who are involved in developing nutrition services.

The Title V MCH program is administered federally by the Bureau of Maternal and Child Health and Resources Development in the Health Resources and Services Administration of the Public Health Service. Administration of the MCH program is the responsibility of the MCH unit within each state's health agency. Most of the CSHCN programs are also administered through state health agencies; some, however, are delivered through other state agencies, such as welfare departments, social service agencies, or state universities. Under the law, each state is required to operate a "program of projects" in each of five areas: (1) maternity and infant care, (2) intensive infant care, (3) family planning, (4) health care for children and youth, and (5) dental care for children.[65]

States have varying degrees of control over local use of Title V funds. Most state MCH funds are used to support well-child checkups, immunization programs, vision and hearing screenings, and school health services, as well as other programs. CSHCN program funds are used to provide direct services to children with special health care needs through local clinics and/or fee-for-service arrangements with physicians in private practice. Regardless of the method of delivering CSHCN programs, a multidisciplinary approach to providing health care is used by nearly all states.[66]

The Select Panel for the Promotion of Child Health reported that Title V program efforts have resulted in significant improvements in maternal and child health. The program is believed to have contributed to the decline in infant and maternal mortality, to the reduction of disability in children with handicaps, and to a general improvement in the health status of children.[67]

MEDICAID AND EPSDT

Congress created the Medicaid program in 1965 to ensure financial access to health care for the economically disadvantaged. It was enacted through Title XIX of the Social Security Act. The Medicaid program is a state-administered entitlement program built on the welfare model. It is constructed as a medical assistance program that reimburses providers for specific services delivered to eligible recipients. Under amendments enacted in 1967, the states are required to provide Early Periodic Screening, Diagnosis, and Treatment (EPSDT) as a mandatory Medicaid service. As outlined by Congress, the purpose of EPSDT is to improve the health status of children from low-income families by providing health services not typically found under the current Medicaid program. EPSDT requires an assessment of the nutrition status of eligible children and their referral for treatment.[68] Whereas Medicaid is mainly a provider payment program for medical services, EPSDT regulations stipulate that states must develop protocols for identifying eligible children, informing them of the EPSDT program, and ensuring that referral, preventive, and treatment services are made available to participants.

The federal administration of Medicaid is the responsibility of the Center for Medicare and Medicaid Services (CMS) within the DHHS. The CMS's Office of Child Health administers EPSDT at the federal level. Medicaid and EPSDT have been credited with increasing the access of low-income women, infants, and children to health care services. EPSDT, by virtue of its aggressive preventive strategy, has been effective in improving child health.[69] However, strict eligibility requirements and federal statutory policies have limited the ability of these programs to reach their full potential.[70]

COMMUNITY HEALTH CENTERS

The Community Health Centers program was initiated by the Office of Economic Opportunity in 1966 and authorized by the Public Health Service Act. It is designed to provide health services and related training in medically underserved areas. The primary program focus is on comprehensive primary care services through community health centers, including migrant health centers, Appalachian Health Demonstration projects, the Rural Health Initiative project, and the Urban Health Initiative project.[71] Preventive services are also offered through Community Health Centers, including well-child care, nutrition assessment, and health education. The program is administered federally by the Bureau of Community Health Services within the Health Resources and Services Administration of the Public Health Service.

THE HEALTHY START PROGRAM

The fact that African Americans and other minorities continue to have increased rates of infant mortality and low birthweight constitutes a major public health problem. In 1991, the Health Resources and Services Administration (HRSA) of the U.S. Department of Health and Human Services (DHHS) funded 15 urban and rural sites in communities with infant mortality rates that were 1.5 to 2.5 times the national average.[72] The Healthy Start Program's goal is to identify and develop community-based approaches to reducing infant mortality and improving the health of low-income women, infants, children, and their families. Healthy Start projects address multiple issues, such as providing adequate prenatal care, promoting positive prenatal health behaviors, and reducing barriers to accessing health care services. Healthy Start specializes in outreach and home visits and focuses on getting women into prenatal care as early as possible. Improving the low-birthweight rate requires improvements in the practices and behavior of women while pregnant. Significant savings in health care costs can accrue from enabling mothers to add a few ounces to a baby's weight before birth. An increase of 250 grams (about 0.5 pound) in birthweight saves an average of $12,000 to $16,000 in first-year medical expenses. Prenatal interventions—such as the Healthy Start Program—that result in a normal-weight birth (over 2,500 grams, or 5.5 pounds) can save $59,700 in medical expenses in the infant's first year. Presently, there are 96 federally funded Healthy Start projects. Over 90 percent of all Healthy Start families are African American, Hispanic, or Native American.

Looking Ahead: Improving the Health of Mothers and Infants

Many of the existing health care programs do not themselves offer nutrition counseling or education. These programs frequently refer their clients to the WIC program for nutrition services. The heavy caseloads in many WIC programs, however, limit the amount of personalized nutrition counseling they can provide. Clearly, more must be done to ensure that quality nutrition counseling is available and accessible for pregnant and lactating women and their infants.

A few states have been successful in providing reimbursable nutrition counseling services to maternal and child health programs. The South Carolina High Risk Channeling Project provides nutrition services, reimbursable by Medicaid, to all participants. The Kentucky Department for Health Services uses some of its MCH Block Grant funds to hire public health nutritionists specifically to provide nutrition counseling for high-risk clients in the local maternal and child health care programs.[73]

Some voluntary health organizations, such as local chapters of the La Leche League International and the March of Dimes Birth Defects Foundation, offer classes that are helpful to particular groups or can provide appropriate nutrition education materials helpful to the community nutritionist working with mother–child populations (see the list of Internet Resources).

The increasing numbers of working women, including those who are planning families or have young infants, along with the growth in worksite health promotion programs, have implications for community nutritionists in providing nutrition education and related services to this population. Some worksites have included components designed for pregnant and lactating women (for example, breastfeeding promotion and prenatal education programs) in their overall health promotion programs.[74]

Where adolescents are concerned, a variety of comprehensive community programs exist to help the pregnant teen with her educational, social, medical, and nutritional needs. Most of these programs include three components: (1) early and consistent prenatal care, (2) continuing education on a classroom basis, and (3) counseling on an individual or group basis.[75] The importance of these programs in improving maternal and child health is well recognized. Because teen pregnancy is usually unplanned, efforts should be made to improve the nutrition and health status of all adolescents through nutrition education and counseling in the classroom as well as in physicians' offices.

To ensure that all pregnant women have access to satisfactory prenatal services in the future, efforts must be made to convince policy makers of the importance of the following recommendations:[76]

- Food supplementation and nutrition education should be available to all pregnant women with low incomes.
- Additional federal funds should be appropriated to make WIC available to all pregnant low-income women.
- Nutrition counseling and education should be provided to all pregnant women whose care is financed by Medicaid.
- State Medicaid programs should be required to include nutrition counseling and education as reimbursable services.
- Federal and state funds should be provided to health department clinics and community health centers to allow for employment of public health nutritionists to offer nutrition counseling to all pregnant women who use these facilities.
- Health insurance policies should include prenatal nutrition counseling as a reimbursable service for all pregnant women living in the United States.

Internet Resources

American Academy of Pediatrics	**www.aap.org**
Bright Futures	**www.brightfutures.org**
Maternal and Child Health Library	**www.mchlibrary.info/databases**
Infant Feeding Action Coalition (INFACT) Canada	**www.infactcanada.ca**
Feeding Infants: A Guide for Use in the Child Nutrition Programs	**www.fns.usda.gov/tn/Resources/feeding_infants.pdf**
La Leche League	**www.lalecheleague.org**
March of Dimes	**www.marchofdimes.com**
Maternal and Child Health Bureau	**www.mchb.hrsa.gov**
National Center for Education in Maternal and Child Health	**www.ncemch.org**
Mayo Clinic Healthy Living Centers	**www.mayoclinic.com**
National Breastfeeding Promotion Project	**www.social-marketing.org/success/cs-nationalwic.html**
Nutrition for Limited Resource Groups	**http://nirc.cas.psu.edu/limitres.cfm**
National Women's Health Information Center (DHHS)	**www.4women.gov**
Pregnancy Nutrition Surveillance System	**www.cdc.gov/pednss/**
Promotion of Mother's Milk, Inc.	**www.promom.org**
Center for Food Safety and Applied Nutrition	**http://vm.cfsan.fda.gov/~dms/wh-preg.html**
WIC Breastfeeding Promotion	**www.nal.usda.gov/wicworks/Learning_Center/index.html#breastfeeding**
WIC Learning Center	**www.nal.usda.gov/wicworks**
WIC Program	**www.fns.usda.gov/wic**
WIC Infant Feeding Guide for Healthy Infants	**www.nal.usda.gov/wicworks/Sharing_Center/NJ/infant%20feeding%20guide.pdf**
WIC Program Studies	**www.fns.usda.gov/oane/MENU/Published/WIC/WIC.HTM**
WIC Works Educational Materials Database	**www.nal.usda.gov/wicworks/Databases/about_database.html**
WIC Innovative Practices: Profiles of 20 Programs	**www.ers.usda.gov/publications/efan04007**
WISEWOMAN	**www.cdc.gov/wisewoman/**

Case *Study*

Promotion of Breastfeeding

By Alice Fornari, EdD, RD, and Alessandra Sarcona, MS, RD

Scenario

As a lactation specialist (international board-certified lactation consultant) and registered dietitian, you have been doing consulting work for physicians specializing in obstetrics and gynecology. Most of your counseling has been geared to middle- and upper-class pregnant and lactating women. You have created nutrition education pamphlets outlining the benefits of breastfeeding, as well as lactation management to promote continued breastfeeding. Most of your visits were to patients' homes. You also held group sessions focusing on nutrition and lactation for pregnant mothers enrolled in birthing preparation classes.

The community hospital in your area has set up a Medicaid program called Healthy Start for pregnant women with incomes less than 185 percent of the poverty guidelines for prenatal and postpartum care and has requested your services to set up a breastfeeding promotion program for program participants. The population is predominantly African American (70 percent), Hispanic (22 percent), and Caucasian (8 percent).

Learning Outcomes

- Describe the benefits of breastfeeding and identify the barriers to breastfeeding.
- Identify the tools for advocating breastfeeding using peer counselors to change culturally based behaviors.
- Distinguish cultural differences that affect delivery of nutrition education.

Foundation: Acquisition of Knowledge and Skills

1. Review the following American Dietetic Association position paper entitled Breaking the Barriers to Breastfeeding, which is available from *Journal of the American Dietetic Association,* 101 (2001): 1213 or at www.eatright.org/Public/GovernmentAffairs/92_8236.cfm. Then list the benefits of breastfeeding and the barriers to breastfeeding. (*Note*: You can search for updates to any of the ADA Position Papers at www.eatright.org.)
2. Referring to your text, describe the rationale for using peer counselors to change culturally based behaviors.

Step 1: *Identify Relevant Information and Uncertainties*

1. In review of the breastfeeding trends in the United States, compare the Healthy Start population to the findings in the ADA position paper.
2. Identify the barriers to breastfeeding that may exist in the Healthy Start population.

Step 2: *Interpret Information*

1. Compare and contrast your experience with counseling pregnant women on lactation in a physician's referral practice to the present situation with Medicaid recipients in a Healthy Start program; take into account any bias you may have toward breastfeeding.
2. Outline the tools for advocating a breastfeeding peer counselor program utilized at WIC and La Leche that may be useful for this population; go to the website www.nal.usda.gov/wicworks/Topics/index.html and click on breastfeeding campaigns. Then review the National WIC Breastfeeding Promotion Project and the website at www.lalecheleague.org/ed/PeerCounsel.html to review the peer counselor program.

Step 3: *Draw and Implement Conclusions*

1. Judging on the basis of your interpretations in the questions posed in Step 2, what types of revisions would you make to your nutrition education materials, counseling, and group sessions in order to help minimize barriers to successful breastfeeding in the Healthy Start setting?

Step 4: *Engage in Continuous Improvement*

1. As you continue to work with breastfeeding promotion at Healthy Start, what steps would you take to encourage breastfeeding up to 1 year?

Chapter *Eleven*

Children and Adolescents: Nutrition Issues, Services, and Programs

Learning *Objectives*

After you have read and studied this chapter, you will be able to:

- Describe three nutritional problems currently experienced by U.S. children and adolescents.
- Specify four *Healthy People 2010* nutrition objectives for children and adolescents.
- Discuss four nutrition assistance programs aimed at improving the health and nutritional status of children, including their purposes and types of assistance offered.
- Describe factors that increase the likelihood of obesity in children.

This chapter addresses such issues as the influence of age, growth, and normal development on nutrition requirements; availability of food and nutrition programs in the community; and current information technologies, which are Commission on Accreditation for Dietetics Education (CADE) *Foundation Knowledge and Skills* requirements for dietetics education.

Chapter *Outline*

Something To Think About...

Children are one-third of our population and all of our future.

– Select Panel for the Promotion of Child Health, 1981

Introduction

Good health is fundamental to the growth, development, and well-being of all children and adolescents and ultimately helps to protect them from chronic diseases as adults. Widespread immunization, improved sanitation, public education on nutrition and health, and the discovery of antibiotics have dramatically reduced the rates of child morbidity and mortality due to infectious diseases. Unfortunately, the status of this group today is far from satisfactory, and new perils have arisen in the past few decades: motor vehicle accidents; violence due to suicide, homicide, and abuse; sexually transmitted diseases; substance abuse; exposure to environmental pollutants; and an alarming increase in the prevalence of overweight and obese children and adolescents.[1] Despite advances in clinical and preventive medicine, children and adolescents in the United States have significant health and nutritional concerns that deserve attention.

This chapter reviews the eating patterns of children and adolescents, the factors that influence these patterns, and current nutrition-related problems of children and adolescents. It also examines the nutrition programs that target children and teenagers and the challenges of operating and improving the child nutrition programs. These programs all have the objective of improving child and adolescent nutrition and, ultimately, enhancing health. For our purposes, children are generally categorized as ages 1 to 11 years and adolescents as ages 12 to 19 years. Other age categories appear in this chapter, however, because the literature is not entirely consistent on the ages that constitute childhood and adolescence.

Healthy People 2010 National Nutrition Objectives

In 2000, the U.S. Department of Health and Human Services (DHHS) outlined a set of health and nutrition objectives for children and adolescents in its publication *Healthy People 2010: Understanding and Improving Health.*[2] The publication lists the priority health-related areas for children and adolescents. Physical activity and fitness, nutrition, and dental health were among the priority concerns. The *Healthy People 2010* initiative builds on the outcomes of the earlier *Healthy People 2000* initiative. Progress was made in reaching some of the goals for the year 2000. The prevalence of growth retardation among low-income children, for example, decreased for all races combined and for some Hispanics and Asian/Pacific Islanders. The percent of elementary and secondary schools offering low-fat choices for breakfast and lunch increased considerably, although only about one in five schools offered lunches that met goals for total fat and saturated fat content by the year 2000.[3]

Other objectives moved away from the year 2000 targets. The prevalence of overweight among children and adolescents of all ethnic groups has increased substantially.[4] The proportion of students in grades 9 to 12 who participated in daily physical education declined from 42 percent in 1991 to 29 percent in 1999. Both the incidence and the prevalence of diabetes increased for the population as a whole and among the special population groups for whom there are data: American Indians, Alaska Natives, Mexican Americans, and African Americans. No progress was made in reducing the prevalence of iron deficiency among young children overall, although the prevalence did decline for low-income children. The objective to increase the population's consumption of calcium-rich foods also moved away from its target—only about 1 in 10 females aged 11 to 24 years consumed the recommended number of servings by the late 1990s.

Healthy People 2010 includes separate targets for fruits, vegetables, and grains in order to be consistent with the current set of *Dietary Guidelines for Americans*. Objectives are included for foods eaten at school and for nutrition education in schools. Table 11-1 shows several of the major *Healthy People 2010* objectives for children and adolescents, as well as any progress made thus far in achieving the objectives.

Healthy People 2010

Increase the proportion of middle schools, junior high schools, and senior high schools that provide school health education to prevent health problems in several areas (including unhealthful dietary patterns and inadequate physical activity).

Healthy People 2010

Increase the proportion of adolescents who engage in moderate physical activity for at least 30 minutes on 5 or more days per week. Increase the proportion of adolescents who engage in vigorous physical activity that promotes cardiorespiratory fitness 3 or more days per week for 20 or more minutes per occasion. Increase the proportion of the nation's public and private schools that require daily school physical education for all students.

Healthy People 2010

Reduce the proportion of children and adolescents who are overweight or obese.

Healthy People 2010 Progress Review

The most recent progress review for the *Healthy People 2010* initiative shows that overall, the data for the *Healthy People 2010* objective on the weight status of children reflect a trend for the worse.[5] The proportion of children and adolescents aged 6 to 19 years who are overweight increased from 11 percent in the late 1980s to 16 percent in 2002. Less than 5 percent of children and adolescents, male or female, were overweight in the late 1960s. The *Healthy People 2010* target is 5 percent.

The three objectives for fruit, vegetable, and grain consumption have shown little or no progress in this decade. The average number of daily servings of fruit consumed by people 2 years of age and older changed little, going from 1.6 in 1994–1996 to 1.5 in 1999–2000. Two to four servings are recommended. Vegetable intake also showed little change: an average of 3.4 daily servings in 1994–1996, compared with 3.3 in 1999–2000. Three to five servings are recommended, with at least one-third being dark green or orange vegetables. In 1999–2000, only 8 percent of vegetable servings consumed by children aged 2 to 19 years were dark green or orange, whereas fried potatoes constituted about one-half.[6]

The progress review of *Healthy People 2010* suggests that significant efforts should be made to bring about further progress toward achievement of the objectives that promote healthful weights and food choices in children and adolescents. The following steps are recommended.[7]

- Because most postpubertal children regain any lost weight within a year, promote the initiation of behavioral therapy for overweight children before the onset of puberty. Such therapy should include the teaching of social skills to counter taunting and to maintain and develop friendships.
- Educate children and their families about the health benefits of being physically active at any size and the possible added benefit of thereby attaining modest weight reduction.
- Encourage schools to find health-promoting ways to offset potential revenues lost from vending machine sales in the event that such machines are banned in some jurisdictions.
- Demonstrate to schools that regularly scheduled periods set aside for physical education during the school day can boost academic achievement by students, rather than detract from it.
- Develop and implement strategies to increase physical activity among children with disabilities. This step should include the use of media that feature disabled adults who can serve as role models of physical fitness for youth who have disabilities.

HEALTHY PEOPLE (HP) *2010* OBJECTIVE NO.	HP 2010 OBJECTIVE	BASELINE (Year)	PROGRESS REVIEW (2002)	HP 2010 TARGET
19.2	Reduce prevalence of iron deficiency among young children:			
	a. Aged 1 to 2 years	9%	7%	5%
	b. Aged 3 to 4 years	4% (1988–1994)	—[†]	1%
19.3	Reduce the proportion of children and adolescents who are overweight or obese.	11% (1988–1994)	16%	5%
19.4	Reduce growth retardation among low-income children aged 5 and younger to less than 5%.	8% (1997)	8%	5%
19.18	Increase the proportion of persons aged 2 years and older who consume less than 10% of calories from saturated fat.	Females 2 to 11 years: 23%; 12 to 19 years: 34%; Males 2 to 11 years: 23–25%; 12 to 19 years: 27% (1994–1996)	NC*	75%
19.9	Increase the proportion of persons aged 2 years and older who consume no more than 30% of calories from total fat.	Females 2 to 19 years: 34–36%; Males 2 to 19 years: 30–33% (1994–1996)	NC	75%
19.11	Increase the proportion of persons aged 2 years and older who meet dietary recommendations for calcium.	Females 2 to 8 years: 79%; 9 to 19 years: 19%; Males 2 to 8 years: 89%; 9 to 19 years: 52% (1988–1994)	—	75%
22.6	Increase the proportion of adolescents who engage in moderate physical activity for at least 30 minutes on 5 or more of the previous 7 days.	27% (1999)	26%	35%
22.7	Increase the proportion of adolescents who engage in vigorous physical activity that promotes cardiorespiratory fitness 3 or more days per week for 20 or more minutes per occasion.	65% (1999)	65%	85%
22.9	Increase the proportion of adolescents who participate in daily school physical education.	29% (1999)	32%	50%

*NC = Little or no change.

[†]— = Cannot assess; limited data.

Source: National Center for Health Statistics, Department of Health and Human Services, *Healthy People 2010 Progress Review* (Hyattsville, MD: Public Health Service, 2004); available at www.cdc.gov/nchs.

TABLE 11-1 ***Healthy People 2010*** **Objectives to Improve Child and Adolescent Health**

What Are Children and Adolescents Eating?

Since the late 1980s, identifying nutritional problems in children's diets and developing initiatives to help improve what children eat have received considerable attention from government agencies, nonprofit groups, and the medical community. Studies from a number of organizations have shown that children are failing to meet the recommended nutrition guidelines by not consuming enough fruits and vegetables and by eating too many foods high in fat and added sugars.[8]

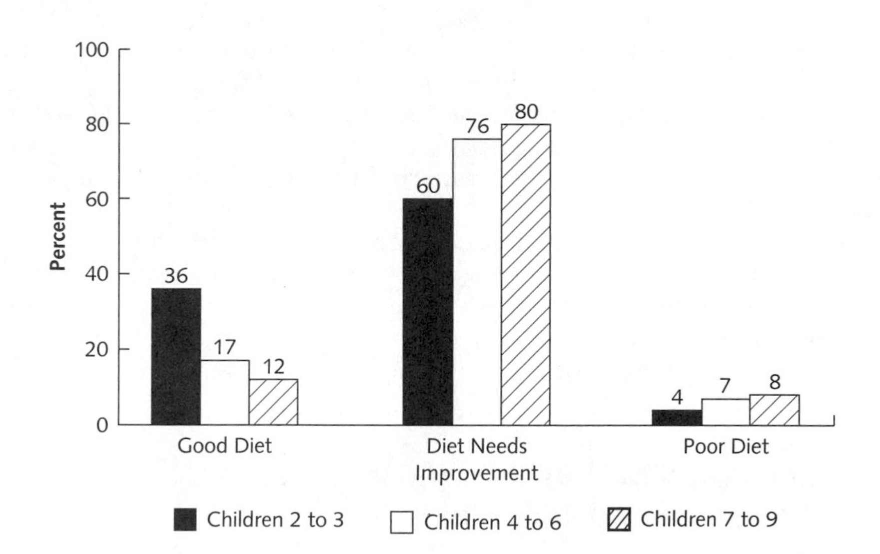

FIGURE 11-1 A Healthy Eating Report Card for Children Aged 2 to 9

Source: Center for Nutrition Policy and Promotion, USDA, 2001.

Children's eating habits have changed over the past two decades. Dietary data, collected in large nationwide surveys to determine trends in nutrient intakes, have shown that most children and adolescents have either a poor diet or one that needs improvement. The USDA's Center for Nutrition Policy and Promotion uses the **Healthy Eating Index (HEI)** as an indicator of diet quality.[9] It provides an overall picture of the variety and quantity of foods people choose to eat and of their compliance with specific dietary recommendations. The HEI measures ten components, each representing a different aspect of a healthful diet, such as fat consumption and fruit and vegetable intake.

As shown in Figure 11-1, the dietary quality of most children aged 2 to 9 is less than optimal. Moreover, children who live in poor families, compared to those who do not live in poverty, are more likely to have a diet rated as poor or needs improvement. Dietary quality continues to decline from childhood to adolescence, especially with the decreased consumption of vegetables, fruits, and milk, and the increased intake of soft drinks.

As indicated by the Healthy Eating Index (HEI), children aged 7 to 9 have a lower-quality diet than younger children, and the lower quality is associated with a decline in their fruit and sodium HEI scores—perhaps because as children get older, they consume more fast food and salty snacks.[10] Additionally, most children do not meet the recommended intake of vegetables or meat.

By not following the current nutrition recommendations, children are missing important daily nutrients, as shown in the results of the Continuing Survey of Food Intakes by Individuals (CSFII). In the latest survey, children of all ages, races, and ethnic groups were at risk of inadequate intakes of magnesium, zinc, and vitamins A and E.[11] Less than 15 percent of teenage girls aged 14 to 18 met recommended goals for calcium, and magnesium, and only 50 percent met the standards for iron. Females between the ages of 9 and 13 also had a tendency to have low intakes of these minerals.

More than two-thirds of children consumed well above the recommended levels of total fat and saturated fat.[12] For all children, the percent of total fat from milk and eggs decreased, whereas the percentage of fat from cheese and snacks increased.[13]

The CSFII showed that beverage choices for all ages changed from whole milk to lower-fat milk, soft drinks, and fruit and fruit-flavored drinks. These changes were especially pronounced for adolescents. For example, a teenage boy's consumption of soda rose from an average of 2 daily servings in the late 1980s to 3 servings a decade later.[14] On any given day, a majority of all children consume soft drinks, which are the second leading energy source

Healthy Eating Index (HEI) A summary measure of the quality of one's diet. The HEI provides an overall picture of how well one's diet conforms to the nutrition recommendations contained in the Dietary Guidelines for Americans and the Food Guide Pyramid. The index factors in such dietary practices as consumption of total fat, saturated fat, cholesterol, and sodium, and the variety of foods in the diet. Check out the interactive HEI at www.cnpp.usda.gov.

for children ages 2 to 18. These soft drinks add from 19 teaspoons of sugar for females aged 6 to 8 years, to 36 teaspoons, or three-fourths of a cup of sugar, for males aged 14 to 18 years.[15] The primary energy source for children and youth comes from a high consumption of grain products, found in dough-based dishes such as pizza, pastas, and Mexican food.[16] These foods not only are high in calories but are often high in total fat and sodium and low in fiber.

Over the past 20 years, the portion sizes of commonly consumed foods, such as soft drinks and hamburgers, have increased. Large food portions that provide more calories than smaller portions may be contributing to the increasing prevalence of overweight in children and teens. For example, children 3 to 5 years of age consumed 25 percent more of an entrée when they were presented with portions that were double an age-appropriate standard size.[17]

Influences on Child and Adolescent Eating Patterns and Behaviors

Despite the importance of healthful eating habits during the growing years, children and youth are not meeting the current nutrition recommendations for health. As children grow into teens, unhealthful eating patterns become more pronounced. The reasons for this decline in good eating habits can be correlated with growing independence from parents, eating away from home, concern with physical appearance and body weight, the need for peer acceptance, and busy schedules.

The traditional pattern of the family gathered around the kitchen table has changed, and fewer families are eating meals together. The increasing popularity of fast food, dining out, and take-out meals, and the increasing demand for convenience and prepared foods, are related to shifts in family structure and work schedules. As the number of single parents and the number of women in the workforce increase, it is likely that dining out will continue to increase. Nearly half of family food expenditures are spent on food and beverages outside the home, and more than one-third of the dollars spent on food away from home are spent on fast foods.[18]

When dining in environments other than home, the majority of adolescents do not make food choices primarily in terms of health and nutrition. Instead, taste, appearance of the food, hunger, and price are the primary factors influencing food selections.[19] Dining out is more prevalent among older teens, who have the greatest freedom, mobility, and income.

More than 52 percent of "dining out" takes place at schools, followed by quick-serve restaurants, vending machines, convenience stores, and worksites.[20] Children and teens consume between 35 and 40 percent of their energy intake at school.[21] As children age, participation in the National School Lunch Program declines from 66 percent of elementary school students to 55 percent of middle school students, and only 40 percent of high school students participate.[22] Teenagers tire of standing in line for the traditional meal program that is regulated in accordance with federal nutrition standards; instead, they quickly purchase foods sold in vending machines and à la carte areas in the school. USDA nutrition standards for school meals do not apply to foods sold in these areas, and an abundance of high-fat, high-sugar items are often for sale. Approximately 78 percent of middle schools and high schools have vending machines present in the cafeterias and hallways, and almost 95 percent of these schools offer à la carte foods.[23]

Quick-serve restaurants provide almost one-third of the meals adolescents eat away from home. An average teen visits a quick-serve restaurant at least twice a week and spends more than $5 a visit, for a nationwide total of over $13 billion annually.[24] These restaurants typically offer foods high in total fat, saturated fat, and sodium, and low in fiber, iron, and calcium. Adolescents also spend $5.2 billion each year on after-school snacks in convenience stores.[25] Quick-serve outlets and convenience stores are often located near school buildings, making them easily accessible food sources. Furthermore, because many quick-serve restaurants and convenience stores employ adolescents, these worksites influence dietary habits through food discounts and free meals that are often provided as part of employee benefits.

Adolescent lifestyles show that convenience and peer pressure strongly influence food choices. As we have noted, adolescents frequently turn to vending machines, quick-serve restaurants, and convenience stores for foods that can be prepared and served quickly. They also want to sleep longer in the morning instead of taking time to eat breakfast. Breakfast is the most commonly missed meal, and studies show that 24 percent of adolescent girls and 20 percent of boys skip breakfast.[26] Adolescents also spend a substantial amount of time socializing with friends while eating. Friends help to define acceptable behavior for the peer group, including eating patterns and habits.

In a media-saturated world, adolescents are the target of more intense and specialized marketing efforts than ever before.[27] They often influence their parents' decisions on where the family dines, what types of foods are eaten at home, and what brands are purchased. And they represent a growing demographic projected to increase to 43 million in 2020 from 36 million today.[28]

Television is the favorite advertising medium of the food industry. Quick-serve restaurants spend more than 95 percent of their advertising budgets on television.[29] The average child aged 8 to 18 years spends approximately 6 hours and 45 minutes per day using media, including television, Internet, radio, video games, magazines, phones, and compact discs.[30] In these same households, 47 percent report that the television is on most of the time, and 65 percent indicate that the television is on during meals.[31] African-American children aged 8 to 18 have the highest total daily media exposure, followed by Hispanic children and then Caucasian children.[32]

Studies show that there is a correlation between television viewing and eating patterns and habits.[33] Excessive television viewing may play a role in the increasing rates of overweight among children, in part because they often watch television, sitting and burning few calories, while consuming high-calorie snacks. Mass media, especially television and magazines, also influence how adolescent girls and boys view their bodies by showing unrealistic body ideals. This is especially true for girls who try to become alarmingly thin, while imitating the models and actresses they see in the media. Between 50 and 75 percent of adolescent girls are dissatisfied with their weight and body image.[34]

Another form of advertising to children and youths is found in an unlikely place—public school. Schools have allowed advertisers to promote many products, including foods and beverages, directly to children during the school day in order to receive additional monies that contribute to relieving tight budgets. These advertisers are building brand loyalty at an early age by using ads in school buses, on scoreboards, and in hallways and by providing corporation-sponsored curricula, books, and posters in the classrooms.

Weighing In on the Problem of Childhood Obesity

Eating practices influence a child's physical growth, cognitive development, and overall health. As we noted earlier, the lack of good nutrition habits, coupled with physical inactivity, has led to an epidemic of overweight children and adolescents during the past two decades.[35] During the past two decades, the percentage of overweight children has nearly doubled, and the percentage of overweight adolescents has almost tripled. The same conditions associated with overweight adults, such as Type 2 diabetes, high blood lipids, and hypertension, are now appearing in young children and teens with greater frequency.

See Chapter 8 for more about factors contributing to, and proposed solutions to, the epidemic of child obesity.

Refer to Appendix A for information on assessing children's BMI and growth status. See also www.cdc.gov/growthcharts.

There are two terms used to refer to children and adolescents whose excess body weight could pose medical risks: **overweight** (BMI-for-age $\geq$ 95th percentile) and **at risk of overweight** (BMI-for-age = 85th to $<$ 95th percentile). BMI changes with age and gender, so as children and adolescents increase in age, BMI also increases.

Dietary intake and nutritional status affect the maturation rate of children and youth. Early maturation, characterized in adolescents by a skeletal age greater than 3 months in advance of chronological age, is associated with increased fatness in adulthood and an increase in abdominal fat.[36] Overweight children tend to be taller, to have advanced bone ages, and to mature earlier than children who are not overweight.[37]

CHILDHOOD OBESITY AND THE EARLY DEVELOPMENT OF CHRONIC DISEASES

Childhood obesity is associated with hyperinsulinemia, hypertriglyceridemia, and reduced HDL-cholesterol concentrations. Overweight children are at risk for cardiovascular disease, insulin resistance and Type 2 diabetes, sleep apnea, gallbladder disease, psychosocial dysfunction, and other serious health problems. A variety of orthopedic complications in overweight children can affect the feet, legs, and hips because the tensile strength of bone and cartilage has not fully developed to carry substantial quantities of excess weight. Hospital costs associated with childhood obesity increased from $35 million to $127 million in the last two decades.[38] Moreover, overweight children and adolescents are more likely to become overweight and obese adults.[39]

The health consequences related to excess weight can begin early in life, especially with respect to cardiovascular disease. Nearly 60 percent of overweight children have been shown to have at least one cardiovascular risk factor, compared to 10 percent of those with a BMI less than the 85th percentile.[40] Furthermore, 25 percent of overweight children had two or more risk factors. Hypertension occurs with low frequency in children. However, in a study of nearly 7,000 children, persistent elevated blood pressure occurred nine times more frequently in those who were overweight.[41]

Diabetes mellitus, the fifth deadliest disease in the United States, is a chronic disease with no cure. The risk of developing Type 1 diabetes is higher than the risk of all other severe chronic diseases of childhood.[42] The peak incidence of Type 1 diabetes usually occurs between 10 and 12 years of age in girls and between 12 and 14 years of age in boys. A recent study found that the prevalence of being overweight at the onset of Type 1 diabetes had tripled from the 1980s to the 1990s. Type 2 diabetes, a form of diabetes that is generally diagnosed among adults, has been rapidly increasing in children and teens.[43] Children diagnosed with Type 2 diabetes are generally overweight, have a strong family history of the disease, are older than 10 years of age, and tend to be members of certain racial and ethnic groups (for example, African American, Hispanic, and Native American). The complications of diabetes include heart disease, stroke, vision loss and blindness, amputation, and kidney disease.[44]

Social stigmatization and low self-esteem are other childhood consequences of excess weight. Overweight children often become targets of early discrimination. For example, children aged 6 to 10 associate obesity with negative characteristics, such as laziness and sloppiness.[45] When asked with whom they would like to be friends, 10- to 11-year olds rank overweight children lowest.[46] These same children may choose younger friends, who are less inclined to discriminate and be judgmental about the older child's weight. Furthermore, overweight children and youth often experience psychological stress from teasing by peers that can lead to poor body image and low self-esteem. A negative self-image in adolescence often persists into adulthood and may have a long-term impact on achievement and income-earning potential.[47]

Overweight BMI-for-age $\geq$ 95th percentile.

At risk of overweight BMI-for-age = 85th to $<$ 95th percentile.

Genetic susceptibility to obesity, lifestyle, family eating patterns, lack of positive role models, and inactivity all contribute to overweight and obesity in this population. According to one recent study, the strongest predictors of overweight and overfatness in fifth-graders were their prior status in third grade.[48] Second to prior obesity, the strongest predictor of subsequent obesity in children is television viewing.[49]

These data on the morbidity and mortality associated with overweight and obesity demonstrate the importance of the prevention of weight gain, as well as the role of obesity treatment in maintaining and improving health and quality of life. Prior to leaving office, U.S. Surgeon General David Satcher called on communities, schools, health care providers, and the media to address the epidemic of overweight and obesity among U.S. children with a multifaceted strategy addressing the health, social, physical, educational, and environmental issues relevant to effectively preventing and treating this escalating public health problem.[50] Through widespread action on the part of all Americans, *The Surgeon General's Call to Action to Prevent and Decrease Overweight and Obesity* "aims to catalyze a process that will reduce the prevalence of overweight and obesity on a nationwide scale."[51]

The Surgeon General's Call to Action to Prevent and Decrease Overweight and Obesity is available at www.surgeongeneral.gov/topics/obesity.

Other Nutrition-Related Problems of Children and Adolescents

The physical, developmental, and social changes that occur during childhood and the teen years can markedly affect eating behaviors and may have long-term health implications. Childhood is a critical time in human development. Children typically grow taller by 2 to 3 inches and heavier by 5 or more pounds each year between the age of one and adolescence.[52] They master fine motor skills (including those related to eating and drinking), become increasingly independent, and learn to express themselves appropriately. Adolescence is a time of change. Between the ages of about 10 and 18 years in girls and between about 12 and 20 years in boys, there are marked changes in physical, intellectual, and emotional growth and development. Many aspects of the maturation process are influenced by dietary intake and nutritional status. Failure to meet nutrition needs during the early years can potentially affect growth, delay sexual maturation, and lead to chronic illnesses as an adult.[53]

Most children and adolescents in the United States are perceived as "healthy." Nevertheless, many experience a variety of health and nutritional problems, some related to their risk-seeking behaviors and inability to deal with abstract notions such as "good health" and the link between current behaviors and long-term health. In general, the nutritional health of U.S. children is better today than ever before. Overt nutrient deficiencies, with the exception of iron deficiency, are not the public health problems they once were, although low calcium intakes during the peak bone-building years increase the risk of osteoporosis in later life, especially among females. Specific nutrition-related problems among children and adolescents include undernutrition, iron-deficiency anemia, high blood cholesterol levels, dental caries, eating disorders, and overweight and obesity.

UNDERNUTRITION

Undernutrition is a problem for some children in the United States, especially those from low-income and migrant families or certain ethnic and racial minority groups (for example, African Americans and Asians).[54] Children in foster care, many of whom live in poverty, and homeless children are also at risk for undernutrition.[55] Some groups of adolescents are at risk for reduced energy and food intakes. Adolescents from low-income families and those who have run away from home or who abuse alcohol or drugs are at risk nutritionally. African-American and Hispanic teenagers are nearly twice as likely as white teenagers to live in poverty. Juveniles who live on the street tend to have a host of health problems, including substance abuse and malnutrition. Irregular meal patterns, combined with a high rate of use of substances such as alcohol, marijuana, cocaine, and amphetamines, contribute to low nutrient and energy intakes among street youth.[56]

For more about hunger and food insecurity, see Chapter 5.

In 2003, 11.2 percent of U.S. households (36.3 million people) were food insecure, meaning that at some time during the year, these households were unable to acquire enough food for all household members because of insufficient resources.[57] About 11 million U.S. children—including 5.1 million children under the age of 6—lived in poverty in 2002.[58] After a slight decline in low-income families during the nineties, the percentage of children living in low-income families started to rise again in 2002.[59] The prevalence of hunger in children was six times higher in single-mother families; three times more prevalent among racial and ethnic minorities, particularly African Americans and Hispanics, and ten times more prevalent in households with incomes below 185 percent of the poverty threshold.[60]

The widespread practice of dieting among adolescents, especially girls, makes them at risk for undernutrition. In a survey of 7- to 13-year-old children, almost half were concerned about their weight, more than one-third had dieted, and almost one-tenth demonstrated signs of anorexia nervosa. As girls increase in age, concerns about weight also increase. Thirteen percent of the 17,354 females in grades 7 through 12 who were interviewed in the Minnesota Adolescent Survey reported being chronic dieters, which is defined as always on a diet or having been on a diet for 10 of the previous 12 months.[61] In 1999, 59 percent of high school girls and 26 percent of high school boys nationwide reported trying to lose weight during the previous 30 days.[62] The results of the National Youth Risk Behavior Survey found adolescent girls engaging in extreme weight-loss behaviors, such as vomiting, using laxatives, or taking diet pills.[63]

IRON DEFICIENCY AND IRON-DEFICIENCY ANEMIA

Iron deficiency is one of the most common nutritional deficiencies, not only in the United States but throughout the world. Iron-deficiency anemia results from depletion of total body iron and impaired hemoglobin production. It is associated with diminished cognitive function, behavior changes, delayed growth and development, and impaired immune function in children.[64]

The *Healthy People 2010* target is to reduce iron deficiency to less than 5 percent for children aged 1 to 2 years.[65] Current prevalence estimates of impaired iron status indicate that 7 percent of children aged 1 to 2 years are iron deficient—down from 9 percent in 1988–1994 and that 3 percent have iron-deficiency anemia. Roughly 4 percent of children aged 3 to 4 years are iron deficient. The year 2010 target is 1 percent.

The incidence of iron-deficiency anemia has decreased in recent years, thanks to iron fortification of infant formula and cereal and increased emphasis on breastfeeding. However, the prevalence of iron deficiency is higher among children living at or below the poverty threshold than among other children.[66] Data from the Pediatric Nutrition Surveillance System survey (PedNSS), which monitors the general health and nutrition status of low-income U.S. children who participate in the public health programs, reveal a higher prevalence of anemia in the PedNSS population and reflect the greater risk for iron-deficiency anemia among low-income children.[67] The problem still exists in 10 percent of minority children 1 to 3 years of age.[68] African-American, Mexican-American, and Southeast Asian children are more likely to be iron deficient than children of other ethnic groups.[69]

Adolescents have special iron needs. In boys, the requirements for absorbed iron increase from about 1.0 to 2.5 milligrams/day.[70] This increase reflects the expanding blood volume and rise in hemoglobin concentration that accompany sexual maturation. The total daily iron loss of menstruating women is about 1.4 milligrams.[71] Whereas most males have an adequate iron intake during adolescence, many females do not. Data from the CSFII indicate that females between 12 and 19 years of age have iron intakes below what is recommended.

BLOOD LEAD LEVEL

Concern about the link between high blood lead levels and lowered intelligence in children led to a large-scale public health effort to eliminate or reduce lead in gasoline, food cans, drinking water, and house paint. Today, small children are most likely to be exposed to lead if they live in houses built before 1950 that were painted with lead-based paints. Such children may eat paint chips containing lead or other lead-contaminated objects, thus increasing their blood lead levels. An elevated blood lead level is defined as a blood lead level high enough to warrant further medical evaluation for the possibility of adverse mental, behavioral, physical, or biochemical effects. The Centers for Disease Control and Prevention recommends that children with blood lead levels exceeding 10 μg/dL be evaluated and referred for treatment.

Healthy People 2010

- Eliminate elevated blood lead levels in children.
- Increase the proportion of persons living in pre-1950s housing whose homes have been tested for the presence of lead-based paint.

Overall, the percentage of children aged 1 to 5 years with elevated blood lead levels decreased more than 80 percent between 1976 and 1994. Even so, more than 400,000 children in the United States—most of them under the age of 6—have blood lead levels $\geq$ 10 μg/dL. Children who live in poor families are more likely to have high blood lead levels than children of the same age living in high-income families. African-American children are more likely than children of other racial and ethnic groups to have high blood lead levels. Continued efforts are needed to reduce household lead hazards and prevent the exposure of children to lead.[72]

DENTAL CARIES

Although dental caries are largely preventable, this remains the most common chronic disease of children aged 5 to 17 years.[73] By age 5, 60 percent of all children have had tooth decay, and more than 90 percent of 18-year-olds have experienced decay.[74] Children in low-income households, and especially those who are American Indian, African American, or Hispanic, have three times the risk of tooth decay because they lack access to or encounter barriers to accessing dental services.[75]

School-based sealant programs are an excellent means of reaching low-income children with preventive dental care. At no cost to children 6 to 14 years of age, some schools offer oral hygiene instruction, fluoride rinses, and dental sealants for permanent molars.[76] There is typically a 60 percent decrease in tooth decay on teeth after sealant application.[77]

Fortunately, the incidence of dental caries in children has decreased by as much as 30 to 50 percent over the last two decades, owing in part to fluoridation of public drinking water, improved dental hygiene, and the use of fluoride in toothpastes and mouthwashes. Twenty states and the District of Columbia met the *Healthy People 2000* target for fluoridation.[78] Children living in communities with water fluoridation experienced 30 to 50 percent fewer cavities than those who lived where there was no fluoridated water.[79] In the United States, about 100 million people still are not receiving the benefits of community water fluoridation.[80] Community coalitions are essential in educating citizens about the benefits of water fluoridation.

HIGH BLOOD CHOLESTEROL

There is considerable evidence that atherosclerosis begins in childhood and that this process is related to high blood cholesterol levels. In fact, data from the Bogalusa Heart Study, a long-term epidemiologic study of cardiovascular disease risk factors in a white and African-American population of children and young adults, found that elevated LDL-cholesterol concentrations in children persist into adulthood and increase the risk of cardiovascular disease.[81]

The Expert Panel on Blood Cholesterol Levels in Children and Adolescents of the National Cholesterol Education Program (NCEP) classifies a total blood cholesterol level of $\geq$ 200 mg/dL or an LDL-cholesterol level of $\geq$ 130 mg/dL as high when associated with a family or parental history of hypercholesterolemia. The panel recommends that children with high blood cholesterol be evaluated further and receive medical nutrition therapy if appropriate. Drug therapy is recommended only for children over 10 years of age whose blood cholesterol level has not responded to an adequate trial of medical nutrition therapy lasting from 6 months to 1 year. When compared with children in other countries, children and adolescents in the United States have higher blood cholesterol levels and higher dietary intakes of saturated fat and cholesterol.[82] In fact, teenagers have many of the same risk factors for high blood cholesterol as adults: family history of coronary heart disease; diets high in total fat, saturated fat, and cholesterol; hypertension; low activity levels; and smoking.

EATING DISORDERS

Eating disorders have become serious health problems in recent years. The most common eating disorders are anorexia nervosa and bulimia nervosa. A constellation of individual, familial, sociocultural, and biological factors contribute to these disorders, which threaten physical health and psychological well-being. Some individuals are more predisposed to develop an eating disorder than others. For example, about 90 percent of people with eating disorders are female. Caucasians are more likely than African Americans and other minority groups to develop an eating disorder.[83] African-American females have greater acceptance of increased body weight and are less preoccupied with social consequences of obesity.[84] Finally, most individuals who develop eating disorders are adolescents or young adults who typically begin experiencing food-related and self-image problems between the ages of 14 and 30 years. Because these syndromes are surrounded by secrecy, their prevalence is not known with certainty, although it has increased dramatically within the past two decades.

Anorexia nervosa and bulimia nervosa may affect about 3 percent of all teenage girls in the United States. These types of eating disorders are often seen in adolescent athletes, many of whom compete in sports such as gymnastics, wrestling, distance running, diving, horse racing, and swimming that demand a rigid control of body weight.[85]

Children with Special Health Care Needs

Children with special needs from infancy through adolescence are served by public school systems and early-intervention services. Several terms and classifications have been used in describing the population with special needs, including developmental disabilities, developmental social needs, handicapping conditions, chronic disorders, and chronic illnesses.[86]

An estimated 17 percent of U.S. children aged 18 years or younger experience some form of disability that limits their participation in school, play, and social activities.[87]

A person with a disability is any person who has a physical, developmental, behavioral, or emotional impairment that substantially limits one or more major life activities, has a record of such impairment, or is regarded as having such impairment.[88] Diseases and conditions included in this definition are cerebral palsy, epilepsy, metabolic diseases, and mental retardation. Children with special health care needs are at increased nutritional risk because of feeding problems, metabolic aberrations, drug/nutrient interactions, decreased mobility, and alterations in growth patterns. In addition to experiencing a range of health problems, such as loss of hearing and vision, and poor fitness, 70 to 90 percent of children with special health care needs also have unique nutritional requirements.[89] An interdisciplinary approach to managing children with special health care needs is recommended. The interdisciplinary team may include a physician, nurse, psychologist, dentist, dietitian, school administrator, food service director, special education staff, and occupational, physical, and speech therapists. The child and his or her parents or caregivers are important team members whose insights and observations help the team determine which foods and feeding strategies work.[90]

The interdisciplinary team evaluation of the health and nutritional status of children with special needs includes the dietary history, medical history, anthropometric measurements, clinical assessment, and feeding assessment.[91] The nutritional assessment forms the basis for the development of a nutrition plan for the toddler or school-aged child and his or her family. The nutrition plan describes the child's food preferences, the family's beliefs and values about foods, and the child's mealtime behavior, nutritional needs, and ability to feed himself or herself. It includes feeding goals for the child—for example, "David will learn to chew solid food" or "Karen will learn to hold a spoon in her right hand"—and provides detailed information about the child's special diet and any equipment required during feeding. The primary feeding goal is to help the child learn to feed herself or himself.

Fortunately, nutrition care for these children has improved because of legislation, increased involvement of various agencies, improved delivery of community-based programs, and better home health care. In 1986, Congress passed the Education of the Handicapped Act Amendments (PL 99-457), which mandate the provision of comprehensive nutrition services to children witih special needs who are 3 to 5 years of age, using a community-based approach that focuses on the family. The legislation also recognizes nutritionists as the health professionals qualified to provide developmental services to children with special health care needs.

In recent years, there has also been increased emphasis on ensuring that children with special health care needs who participate in school feeding programs are able to receive substitutions for food items because of their disabilities. This is mandated by USDA's nondiscrimination regulations and the National School Lunch and Breakfast policies. The regulations require that a physician's statement be provided, listing the disability, such as diabetic diet, PKU, or lactose-free, along with the reason for meal modification and the specific substitution that is required. Additionally, nutrition goals should be included in the child's or adolescent's Individualized Education Plan. This IEP is a written statement that contains the program of special education and related services to be provided to a child with a disability.[92]

One program that gives children and youth with special needs an opportunity for year-round training and competition is Special Olympics. The nutrition section of Special Olympics, Health Promoter, offers locally based, ongoing, health promotion programs with the goal of making good nutrition and physical fitness routine for athletes with special needs. *Healthy People 2010* lists goals for individuals with special health care needs to accomplish, such as more physical activity, controlling weight, better nutrition, and improved access to health care and clinical preventive services, oral health, and mental health.[93]

Programs in Action

Nutrition Education Strategies for Preadolescent Girls

Studies have documented the high prevalence of unhealthful dieting behaviors among adolescent girls.[1,2] These behaviors may have a negative impact on nutritional intake and psychological well-being. Of even greater concern are the increasing rates of unhealthful dieting behaviors and body dissatisfaction among girls of elementary school age. Television shows, commercials, magazines, and movies filled with images of thin models fuel body image problems.

In spite of the high prevalence rates of unhealthful dieting behaviors, disordered eating, and body dissatisfaction among youth, few primary prevention programs—programs that prevent a problem from occurring at all—have been both implemented and evaluated. Several have been school based, with the school serving as a catalyst for community outreach interventions that are directed toward groups of individuals,[3,4] namely preadolescent and adolescent girls. It has been suggested that children and adolescents can learn at school about the dangers of unsafe weight-loss methods and about safe ways to maintain a healthful weight.[5] Other possible avenues for reaching preadolescent and adolescent girls include religious institutions, community centers, and clubs such as the Boys and Girls Club and the YMCA and YWCA. These institutions are well established and trusted, so they may be better able to reach the target population and attract participants.

The "Free to Be Me" program directed its attention toward social and environmental factors (the media and the family), personal factors (body image), and behavioral factors (dieting behaviors). It also worked within the framework of an established, widely accepted community program, the Girl Scouts, as a catalyst to connect with the target population and create trust in the program.

"Free to Be Me" is a Girl Scout badge program that helps young girls feel good about their bodies. It was designed to decrease unhealthful weight control behaviors in preadolescent girls. Its primary emphasis was on improving body image by taking an in-depth look at what young girls see in the media and on helping fifth- and sixth-grade Girl Scouts accept a wide variety of body shapes and sizes.

Goals and Objectives

The goal of the project was to assess the feasibility and short-term impact of "Free to Be Me," a program aimed at preventing unhealthful dieting and promoting a healthful body image among preadolescent girls. Objectives for participants upon completion of the six-session intervention were to decrease incidence of unhealthful weight control behaviors, to increase knowledge of media influences on body image and food choices, to increase the interest in healthful eating and body image, to improve participants' ability to critically evaluate media messages, and to improve overall body image.

Methodology

"Free to Be Me" included six 90-minute sessions that were presented during consecutive biweekly Girl Scout meetings. The girls completed activities on body development, the media's impact on body image and self-esteem, and combating negative images. For example, they critically analyzed media

The History of Child Nutrition Programs in Schools

Federal programs addressing the nutritional needs of children and adolescents have existed for more than 150 years. The Children's Bureau, at one time part of the U.S. Department of Labor, issued dietary advice to parents and teachers and conducted nutrition surveys of low-income children. School feeding programs, augmented by the financial support of local school districts, philanthropic organizations, and private donors, likewise began in the early 1900s. Federal involvement in school food programs increased during the 1930s with the passage of an amendment to the Agricultural Act of 1933, which established a fund to purchase surplus agricultural commodities for donation to needy families and child nutrition programs, including school lunch programs.

As a result of testimony from the surgeon general that "70 percent of the boys who had poor nutrition 10 to 12 years ago were rejected by the draft," legislation was introduced messages and looked for alternative positive messages. After analyzing media messages, the girls were encouraged to write letters to the media about positive body image. Following completion of the program, they were awarded a "Free to Be Me" badge designed specifically for the program.

A 3-hour training session taught troop leaders to teach the program. Each troop leader received a detailed handbook, along with materials and supplies. A registered dietitian served as project coordinator and worked closely with troop leaders. Parental involvement included receiving weekly mailings, assisting with take-home activities, preparing healthful snacks, and viewing end-of-session skits.

Results

A group-randomized, controlled study was designed to evaluate program effectiveness using 12 troops that participated in "Free to Be Me" and 13 nonparticipating troops as controls. The evaluation focused on program feasibility and short-term effectiveness, to be assessed on the basis of changes exhibited upon program completion and at a 3-month follow-up. The program had a significant influence on media-related attitudes and behaviors, including internalization of sociocultural ideals, self-efficacy to affect weight-related social norms, and print media habits. A modest program impact on body-related knowledge and attitudes (that is, on acceptance of body size, knowledge about puberty, and perceived weight status) was apparent immediately after the intervention, but not at follow-up. Significant changes were not noted for dieting behaviors. Nevertheless, satisfaction with the program was high among girls, parents, and leaders.

Lessons Learned

Community nutrition programs may be successfully implemented within Girl Scout troops. Intervention programs for young adolescent girls have the potential to promote a positive body image and prevent unhealthful dieting behaviors.

References

1. M. W. Felts, A. Parrillo, T. Chenier, and P. Dunn, Adolescents' perceptions of relative weight and self-reported weight-loss activities: Analysis of YRBSS national data, *Journal of Adolescent Health* 18 (1996): 20–6.
2. J. C. Rosen and J. Gross, Prevalence of weight reducing and weight gaining in adolescent girls and boys, *Health Psychology* 6 (1987): 131–47.
3. D. Neumark-Sztianer, R. Butler, and H. Palti, Eating disturbances among adolescent girls: Evaluation of a school-based primary prevention program, *Journal of Nutrition Education* 27 (1995): 24–31.
4. N. Piran, On the move from tertiary to secondary and primary prevention: Working with an elite dance school. In *Preventing Eating Disorders: A Handbook of Interventions and Special Challenges,* ed. N. Piran and C. Steiner-Adair (Philadelphia: Brunner/Mazel, 1999).
5. Guidelines for school health programs to promote lifelong healthy eating, *Journal of School Health* (1997) 67: 9–26.

Source: Adapted from *Community Nutritionary* (White Plains, NY: Dannon Institute, Spring 2000). Used with permission. For more information about the *Awards for Excellence in Community Nutrition,* go to www.dannon-institute.org.

to give the school lunch program permanent status and to authorize the appropriations necessary to keep it running in the future. In 1946, President Harry Truman signed into law the National School Lunch Act (as amended, PL 79-396) with the following policy objectives:

> It is hereby declared to be the policy of Congress, as a measure of national security, to safeguard the health and well-being of the Nation's children and to encourage the domestic consumption of nutritious agricultural commodities and other food, by assisting the states, through grants-in-aid and other means, in providing an adequate supply of foods and other facilities for the establishment, maintenance, operation, and expansion of non-profit school lunch programs.[94]

To receive cash and commodity assistance under the statute, states had to operate school lunch programs on a nonprofit basis, serve free or reduced-price lunches for needy children, and provide lunches that met certain federal standards. Twenty years later, in 1966, the Child Nutrition Act was passed. It expanded federal efforts to improve child nutrition by establishing numerous programs of year-round food assistance to children of all ages. Many of the programs and policies developed during the 1960s and 1970s still exist today, although most have been modified over the years. This section describes the major domestic nutrition programs for children and adolescents.

Nutrition Programs of the U.S. Department of Agriculture

The Food and Nutrition Service was established in 1969 to administer the food assistance programs of the U.S. Department of Agriculture (USDA). The primary aim of this agency is to make food assistance available to people who need it. Other goals include improving the eating habits of U.S. children and stabilizing farm prices through the distribution of surplus foods. In this section, we examine five of the agency's largest food assistance programs for children: the National School Lunch Program, the School Breakfast Program, the Summer Feeding Program, the After School Snack Program, and the Commodity Distribution Program. Table 11-2 gives a brief description of the programs designed specifically for children, and Table 11-3 outlines programs that benefit children by assisting their families. These programs are also described in Chapters 5 and 10.

THE NATIONAL SCHOOL LUNCH PROGRAM

The National School Lunch Program (NSLP) operates in more than 99,800 public and nonprofit private school and residential child care institutions. In 2004, it provided nutritionally balanced, low-cost or free lunches to more than 28 million school children each day, at a cost of more than $6.6 billion.[95] Public school districts and independent private, nonprofit schools that participate in the NSLP are entitled to receive reimbursement dollars and donated commodities for each meal that is served. In return, they must offer free or reduced-price lunches to eligible children and meet specific nutrition guidelines (see Table 11-4 on page 363).

In accordance with the *Dietary Guidelines for Americans*, school lunches must provide no more than 30 percent of an individual's calories from fat, and 10 percent or less from saturated fat. Regulations also require that lunches provide one-third of the DRI for calories, protein, calcium, iron, vitamin A, and vitamin C for the applicable age or grade groups. Even though school lunches must meet federal nutrition requirements, local school food authorities make decisions about what specific foods to serve and how they are prepared.

Any child at a participating school may purchase a meal through the NSLP. Children from families with incomes at or below 130 percent of the poverty guidelines are eligible for free meals. Those with incomes between 130 and 185 percent of the poverty guidelines are eligible for reduced-price meals, where students can be charged no more than $0.40. Children from families with incomes greater than 185 percent of the poverty guidelines pay full price, although their meals are still federally subsidized by a small amount. Table 11-5 on page 363 lists the income eligibility guidelines for the Child Nutrition Programs. Local districts set their own prices for paid meals but must still operate as a nonprofit program. In addition to cash reimbursements listed in Table 11-6 on page 364, schools are entitled to receive commodity foods for each lunch meal served. In 2005, the commodity value for each lunch served was 17.25 cents.

There has been a strong effort by the School Nutrition Association and other education organizations to seek legislation that would eliminate the reduced-price category by raising "free" eligibility to the "reduced" limit of 185 percent of the poverty guidelines and, thereby, allowing these children a free meal. Children classified as eligible only for reduced-price

PROGRAM	PURPOSE(S)	TYPE OF ASSISTANCE	ELIGIBILITY REQUIREMENTS OF PROGRAM PARTICIPANTS
National School Lunch Program	Assist states in making the school lunch program available to students and encourage the domestic consumption of nutritious agricultural commodities. Public and nonprofit private schools of high school grade and under, public and private nonprofit residential child care institutions (except Job Corps Centers), residential summer camps that participate in the Summer Food Service Program for Children, and private foster homes are eligible to participate.	Formula grants*	1. All students attending schools where the lunch program is operating may participate. 2. Lunch is served free to students who are determined by local school authorities to live in households with incomes at or below 130 percent of the federal poverty guidelines. 3. Lunch is served at a reduced price to students who live in households with incomes between 130 percent and 185 percent of the poverty guidelines. 4. Students from families with incomes over 185 percent of the poverty guidelines pay full price for lunch.
School Breakfast Program	Assist states in providing a nutritious, nonprofit breakfast for school students. Eligible schools and residential child care facilities are the same as for the National School Lunch Program.	Formula grants	Eligibility requirements are the same as for the National School Lunch Program.
After School Snack Program	Assists school-based after-school programs in providing healthful snacks to children.	Available through NSLP	Sites can qualify to serve all children free of charge based on the percentage of children receiving free and reduced-price meals at the school.
Special Milk Program for Children	Provide subsidies to schools and institutions to encourage the consumption of fluid milk by children. Any public or private nonprofit school or child care institution (for example, nursery school, child care center) of high school grade or under (except Job Corps Centers) may participate on request if it does not participate in a meal service program authorized under the National School Lunch Act or the Child Nutrition Act of 1966.	Formula grants	All students attending schools and Institutions in which the program is operating may participate.
Summer Food Service Program for Children	Assist states in conducting nonprofit food service programs for low-income children during the summer months and at other approved times, when area schools are closed for vacation.	Formula grants	Homeless children and children attending public or private nonprofit schools and residential camps or participating in the National Youth Sports Program can receive free meals.

*Formula grants are a type of funding mechanism in which the funding agency distributes funds to states on the basis of a "formula" that takes into account a variety of factors, such as the number of breakfasts or lunches served to eligible children, the number of breakfasts or lunches served free or at a reduced price, and the national average payment for the program.

Source: Catalog of Federal Domestic Assistance Programs, June 2004; available at www.cfda.gov/public.

TABLE 11-2 USDA Food Assistance Programs Specifically for Children

meals are often from the "working poor"—households that are ineligible for free meals yet do not have enough money to pay for reduced-priced meals. The Child Nutrition and WIC Reauthorization Act of 2004 (Public Law 108-265) included a pilot test for five states to eliminate the reduced-priced category and evaluate its effect on program participation. The National School Lunch Program is the only federal nutrition program that provides different levels of benefits based on a family's income and household size.

TABLE 11-3 Other USDA Food Assistance Programs That Benefit Children*

PROGRAM	PURPOSE(S)
Child and Adult Care Food Program	Assist states in initiating, maintaining, and expanding nonprofit food service programs for children and elderly or impaired adults in nonresidential day care facilities. After-school programs operated by community groups may also serve snacks to teenagers aged 12–18 years in low-income areas.
Commodity Supplemental Food Program	Improve the health and nutritional status of low-income pregnant, postpartum, and breastfeeding women, infants, and children up to 6 years of age, and elderly persons through the donation of supplemental foods
Emergency Food Assistance Program (Food Commodities)	Make food commodities available to states for distribution to needy persons such as the unemployed, welfare recipients, and low-income individuals
Food Commodities for Soup Kitchens	Improve the diets of the homeless
Food Distribution (popularly called the Food Donation Program)	Improve the diets of school and preschool children and other groups and increase the market for domestically produced foods acquired under surplus removal or price support operations
Food Distribution Program on Indian Reservations	Improve the diets of needy persons in households on or near Indian reservations and increase the market for domestically produced foods acquired under surplus removal or price support operations
Food Stamp Program	Improve the diets of low-income households by increasing their ability to purchase foods
Homeless Children Nutrition Program	Assist state, city, county, and local governments, other public institutions, and private nonprofit organizations in providing food service throughout the year to homeless children under the age of 6 in emergency shelters
Nutrition Assistance for Puerto Rico	Improve the diets of needy persons and families living in Puerto Rico through a cash grant alternative to the Food Stamp Program
Special Supplemental Nutrition Program for Women, Infants, and Children (WIC)	Provide, at no cost, supplemental nutritious foods, nutrition education, and referrals to health care to low-income pregnant, breastfeeding, and postpartum women, infants, and children to age 5 who are determined to be at nutritional risk
WIC Farmers' Market Nutrition Program	Provide fresh, nutritious unprepared foods such as fruits and vegetables from farmers' markets to low-income women, infants, and children; expand the awareness and use of farmers' markets; and increase sales at farmers' markets

*Refer to Table 5-5 on pages 135–140 for a complete description of the programs shown in this table.

Source: Catalog of Federal Domestic Assistance Programs, June 2004.

School food service operators must take great care to protect the confidentiality of a student's eligibility status and not to discriminate against any child who qualifies for a free or reduced-price meal. The names of children who receive free or reduced-price lunches cannot be "published, posted, or announced in any manner to other children," nor can children who participate in the program be required to use a separate lunchroom, lunchtime, serving line, cafeteria entrance, or medium of exchange.

THE SCHOOL BREAKFAST PROGRAM

The 1966 Child Nutrition Act established funding for a pilot breakfast program, which was made permanent in 1975. By 2004, the program had grown annually to 8.8 million children participating at a federal cost of more than $1.7 billion.[96] Even though a school

FOOD GROUP	PRESCHOOL* (Age)		GRADE SCHOOL THROUGH HIGH SCHOOL* (Grade)		
	1 to 2	3 to 4	K to 3	4 to 6	7 to 12
Meat or Meat Alternate					
1 serving:					
Lean meat, poultry, or fish	1 oz	1½ oz	1½ oz	2 oz	3 oz
Cheese	1 oz	1½ oz	1½ oz	2 oz	3 oz
Large egg(s)	1/2	3/4	3/4	1	1½
Cooked dry beans or peas	1/4 c	3/8 c	3/8 c	1/2 c	3/4 c
Peanut butter	2 tbs	3 tbs	3 tbs	4 tbs	6 tbs
Yogurt	1/2 c	3/4 c	3/4 c	1 c	1½ c
Peanuts, soynuts, tree nuts, or seeds†	1/2 oz	3/4 oz	3/4 oz	1 oz	1½ oz
Vegetable and/or Fruit					
2 or more servings, both to total	1/2 c	1/2 c	1/2 c	3/4 c	3/4 c
Bread or Bread Alternate‡					
Servings	5 per week	8 per week	8 per week	8 per week	10 per week
Milk					
1 serving of fluid milk	3/4 c	3/4 c	1 c	1 c	1 c

*The quantities listed represent per-lunch minimums for each age and grade except those for the oldest group, which are recommendations. Schools unable to serve the recommended quantities for grades 7 to 12 must provide at least the amount shown for grades 4 to 6.

†These meat alternates may be used to meet no more than half of the meat or meat alternate requirement; therefore, they must be used in a meal with another meat or meat alternate.

‡Schools must serve daily at least 1/2 serving of bread or bread alternate to the youngest age group and at least 1 serving to older children.

Source: U.S. Department of Agriculture, National School Lunch Program, Food Program Facts, 2003.

TABLE 11-4 National School Lunch Patterns for Different Ages

may offer the NSLP, it does not have to offer the breakfast program. This program, operated in the same manner as the School Lunch Program, provides a nutritious breakfast to all students attending a school where the program is offered.

Breakfast is served free, or at a reduced price of no more than $0.30, to students from households with incomes at or below the income eligibility guidelines. Breakfasts served to students paying full price are also subsidized. Schools may qualify for higher, "severe-need"

TABLE 11-5 Annual Income Eligibility Guidelines for the Federal Child Nutrition Programs, 2005–2006

HOUSEHOLD SIZE	FEDERAL POVERTY GUIDELINES 100% OF POVERTY*	FREE MEALS 130% OF POVERTY	REDUCED-PRICE MEALS 185% OF POVERTY
1	$ 9,570	$ 12,441	$ 17,705
2	12,830	16,679	23,736
3	16,090	20,917	29,767
4	19,350	25,155	35,798
5	22,610	29,393	41,829
6	25,870	33,631	47,860
7	29,130	37,869	53,891
8	32,390	42,107	59,922
Each additional, +	$ 3,260	$ 4,238	$ 6,031

*Guidelines are adjusted annually. Income guidelines are higher in Alaska and Hawaii; see Table 5-1 on page 118.

Sources: Federal Register, Vol. 70, No. 52, 3/18/05, pp. 13160–63; United States Department of Agriculture.

TABLE 11-6 2004–2005 Reimbursement Rates for Sponsors of the National School Lunch, School Breakfast, and After School Snack Programs*

National School Lunch Program	Reimbursement Rate per Meal or Snack
Paid	$0.21
Reduced-Price	$1.84
Free	$2.24
School Breakfast Program	
Paid	$0.23
Reduced-Price	$0.93
Free	$1.23
After School Snack Program	
Paid	$0.05
Reduced-Price	$0.30
Free	$0.61

*Payment rates are higher in Alaska and Hawaii to reflect the higher cost of providing meals in those states.

reimbursements if 40 percent of their lunch participation is served free or at a reduced price. About 65 percent of breakfasts served in 2003 received severe-need payments. To receive federal reimbursement, participating schools must follow standard meal patterns in which breakfasts provide one-fourth of the DRI values for protein, calcium, iron, vitamin A, vitamin C, and calories. The breakfast can be either hot or cold, but it must include milk, fruit or juice, and either two servings of bread, two meats, or a meat and bread (see Table 11-7). Unlike school lunches, a commodity value is not applied to breakfasts served to children.

The number of children who participate in the school breakfast program, including those who qualify for a free or reduced-price meal, is far below the number of children who participate in the school lunch program. State agencies and local school boards are exploring alternative ideas to encourage more children to take advantage of this important program. To encourage greater participation, states such as Minnesota and Maryland have developed initiatives that support district breakfast programs with additional funding. Minnesota's Fast Break to Learning program, started in 1999, offers breakfasts to all students in participating schools at no charge.[97] These schools receive state funding that helps to compensate for the loss of dollars from students who would have normally been charged for a breakfast. In 130 schools that participate in Maryland's Meals for Achievement project, children are served a free breakfast in the classroom everyday. In both of these examples, additional schools can apply to offer free breakfast, provided that funding is available.[98]

TABLE 11-7 Foods Required for Breakfasts Provided under the School Breakfast Program

- 1/2 pt of fluid milk as a beverage, on cereal, or both—one serving; AND
- 1/2 c serving fruit OR 1/2 c full-strength fruit or vegetable juice—one serving; AND
- bread or bread alternate (one slice whole-grain or enriched bread, or an equivalent serving of cornbread, biscuits, rolls, muffins, etc.; or 3/4 c or 1 oz serving of cereal)—two servings; OR
- meat or meat alternate (one serving of protein-rich foods such as an egg; or a 1-oz serving of meat, poultry, fish, or cheese; or 2 tbsp of peanut butter)—two servings; OR
- one bread AND one meat

Source: Fact Sheets on the Federal Food Assistance Programs (Washington, D.C.: Food Research and Action Center, June 2004). Used with permission from the Food Research and Action Center.

Fourteen states provide financial incentives to school districts that offer a breakfast program and 22 states mandate that districts offer school breakfast. Recognizing the important effect that breakfast has on a child's ability to learn in the classroom, school food service operators are using a number of approaches to increase participation, such as serving in the classroom, providing "grab 'n go" breakfasts, or offering breakfast after the first class. Some districts are offering breakfast free of charge to *all* students, regardless of whether the child's eligibility status is "free," "reduced-price," or "paid," and without additional state and federal funding.

THE AFTER SCHOOL SNACK PROGRAM

In 1998, Congress expanded the National School Lunch Program to include reimbursement for snacks served to children, through the age of 18, in after school educational and enrichment programs.[99] After school snacks are provided to children on the same income eligibility basis as school lunches and must meet federal nutrition standards to qualify for reimbursement. However, programs that operate in schools where 50 percent or more of the enrolled students are eligible for free and reduced-priced meals, called "area eligible" sites, may receive the "free" rate of reimbursement for all snacks served to children.

THE SUMMER FOOD SERVICE PROGRAM FOR CHILDREN

The Summer Food Service Program (SFSP) is an entitlement program created by Congress in 1968 to ensure that children would have access to nutritious meals when school was not in session during summer vacation. During the 2003 Summer Food Service Program, close to 30,000 sites served 117 million children, for a federal expenditure of $256.6 million.[100]

For the program to operate in a community, it must have a local sponsor that either contracts for food services or prepares its own food. Many school districts act as both the sponsor and the provider. For example, a district may provide food not only in its own locations but also to other school districts and to local parks and recreation programs. Programs operate as either open or enrolled sites:

- "*Open sites*" are located in areas where 50 percent or more of the children come from families whose household income is below 185 percent of the federal poverty guidelines. Any child or teen who lives in the area of an open site, even though she or he is not enrolled in the site's program, may participate free of charge in the SFSP. Sponsors of open-site programs may be reimbursed for all meals served, regardless of the income of each individual child, provided that they meet specified nutrition requirements.
- "*Enrolled sites*" are those where 50 percent of the children attending the program come from families whose income is greater than 185 percent of the federal poverty guidelines. Sponsors of enrolled sites, unlike those of open sites, must document the incomes of participating children in order to claim reimbursement.

The SFSP has faced a lack of sponsors, often because of the labor-intensive accounting and application procedures required by the program, as well as insufficient funding. To increase the number of sponsors, the Seamless Summer Feeding Waiver, begun in 2002, gives school districts the option of claiming meals under the National School Lunch Program instead of the SFSP. Another initiative to increase participation was launched as a result of the 2002 Farm Bill. Fourteen states were eligible to participate in a pilot study where the cost of meals served by sponsors was reimbursed at the maximum allowable rate, which reduced the amount of administrative record keeping.[101] Under the Child Nutrition Reauthorization Act of 2004, the pilot was made permanent and extended to six more states.

THE FOOD DISTRIBUTION PROGRAM

The food distribution division of the USDA's Food and Nutrition Service coordinates the distribution of commodities to public and private nonprofit schools that provide meals to students. This program purchases food only of U.S. origin with funds provided through direct appropriations from Congress and price supports. During 2003, the USDA purchased more than $849 million worth of commodities, totaling over 1.2 billion pounds.[102]

School districts are given an entitlement dollar value, based on the number of lunches served during the previous year, which is used to determine the amount of commodity food items that will be allocated to the district. Schools received 17.25 cents worth of commodity foods per meal for the 2005 school year, but the entitlement is adjusted annually to reflect changes in the Price Index of Foods Used in Schools and Institutions.

Two types of commodities are provided to nutrition programs. The amount of entitlement commodities provided (such as vegetables, meats, and cheeses) is based on the number of lunches served multiplied by the annual per-meal commodity rate. Bonus commodities, such as apricots, salmon, and fruit cups, are available through surpluses of agricultural products.

Nutrition Programs of the U.S. Department of Health and Human Services

Recall from Chapter 10 that several DHHS programs include a child care component: the Title V Maternal and Child Health Program; Medicaid and the Early and Periodic Screening, Diagnosis, and Treatment (EPSDT) programs; and the primary-care programs of Community Health Centers. Another DHHS program that benefits children is the Head Start Program.

Initiated in 1965 by the Office of Economic Opportunity, Head Start was authorized by the Economic Opportunity and Community Partnership Act of 1967 (PL 93-644, as amended in 1974). The program is coordinated by the Administration for Children and Families within the DHHS. It is one of the most successful federal government programs for child development. Head Start provides children from low-income families with comprehensive education, social, health, and nutrition services. Eligible children range in age from birth to the age at which they begin to attend school. Parental involvement in the program planning and operation is emphasized. Head Start projects provide meals and snacks as well as nutrition assessment and education for children and their parents. The nutrition services are meant to complement the health and education components of the program. In FY2003, about 910,000 preschool children from low-income families were enrolled. The Select Panel for the Promotion of Child Health reported that Head Start has been shown to improve children's health. Children participating in Head Start have a lower incidence of anemia, receive more immunizations, and have better nutrition and improved overall health than children who do not participate.[103]

Initiated in 1995, the Early Head Start Program expands the benefits of Head Start's early childhood development and family support services to low-income families with children under age 3 and to pregnant women. In 2003, there were more than 650 Early Head Start programs serving 62,000 infants and toddlers.[104]

Impact of Child Nutrition Programs on Children's Diets

Child Nutrition Program food service directors use federal dietary guidelines to plan menus for the National School Lunch and Breakfast Programs. Because of these standards, school meal programs have promoted healthful eating habits and contributed to the quality of children's overall diets. However, despite progress that has enhanced the nutrition quality of school meals, results of research conducted in the 1990s indicated that school meals were failing to meet certain key nutritional goals.[105]

The USDA's Continuing Survey of Food Intakes by Individuals (CSFII) assessed dietary intake in more than 5,000 children aged 6 to 18. One objective was to examine relationships between their participation in school meal programs and their dietary intake. Researchers reported that children who ate both breakfast and lunch at school on any given day received over 50 percent of their daily food energy from these meals.[106] Participation was also associated with higher intakes of total fat, saturated fat, and sodium, and with lower intakes of added sugars, than the intakes of those who did not participate. Students who participate in both the school breakfast and lunch programs come much closer to meeting the "5 a day" goal for fruits and vegetables, consume more grains, and drink more milk, while eating fewer sweet and salty snacks and drinking fewer sweetened beverages, compared to those students not participating in the school meal programs. Participation in only the NSLP was associated with higher intakes of nutrients, such as the B vitamins, calcium, magnesium, and zinc, than the intakes of those who did not eat a school lunch.[107] Children who ate a school lunch got 15 percent of their lunch calories from saturated fat and 13 percent from added sugars. Those who did not participate received only 11 percent of their lunch calories from fat but got 23 percent from added sugars, possibly from soda consumption. School lunch participants drank about three times as much milk, but only half as much soda, as children who did not participate in the program. School breakfast participants had higher intakes of many vitamins and minerals, and had no increase in saturated fat and added sugars, compared to those who skipped breakfast at school and at home.

Even though school meal programs appeared to have a positive effect on children's consumption of milk, fruit, vegetables, and some vitamins and minerals, there was evidence that school meals contributed too much fat in menu items such as pizza, chicken sandwiches, french fries, and baked goods. In light of these findings, the USDA launched a reform of the NSLP in 1994, starting with public hearings and a proposed rule. In 1995, a final rule was adopted that directed schools to upgrade the nutritional value of their meals. Several elements of this reform are collectively referred to as the *School Meals Initiative for Healthy Children,* or SMI.[108] The regulations for SMI allow food service operators the flexibility to choose one of four USDA menu planning systems, provided that menus served over a school week adhere to the Dietary Guidelines for Americans and meet specified nutrient standards for calories, total fat, saturated fat, protein, carbohydrate, calcium, iron, and vitamins A and C. The USDA requires that states perform an SMI compliance review of school programs to determine whether menus are meeting nutrient standards.

By 2000, nearly two-thirds of all school districts said they had "fully implemented" their chosen approach to menu planning.[109] To meet the nutrition standards, menus included more fruits, vegetables, whole grains, and low-fat and reduced-fat foods. These items, often served as larger portions to meet calorie requirements for specific ages, did not contribute to increasing fat percentages. More vegetarian and vegan options are now being offered, as well as salad bars and a variety of prepared salads. The creative use of packaging for food items helped to market healthful choices to students. In addition to complying with federal regulations, school food service directors throughout the country have also implemented a number of creative concepts to improve the quality and nutritional value of school meals.

Factors Discouraging Participation in Child Nutrition Programs

The environment in some schools discourages students from eating meals provided by the NSLP and SBP and encourages food choices and eating habits that are not consistent with the Dietary Guidelines for Americans.[110] Efforts to promote healthful eating habits and provide nutrition instruction may be contradicted in school settings where the sale of food and beverages, many with low nutritional appeal, is promoted in snack bars, school stores, and vending machines. Even less healthful à la carte items sold by the school cafeteria may "compete" with school meals for a student's meal money.

By federal statute, competitive foods are any foods sold in competition with USDA school meals and are categorized as "foods of minimal nutritional value." These foods provide less than 5 percent of the DRI values for eight specified nutrients per serving: protein, niacin, riboflavin, thiamin, calcium, iron, and vitamins A and C. Examples of foods in this category include soft drinks, chewing gum, candies, and candy-coated popcorn. By law, these foods may not be sold in food service areas during the meal-serving period.

However, in many schools, the sale of competitive foods in vending machines outside the cafeteria represents additional income that can be spent for discretionary purposes by the school.[111] Many school districts negotiate exclusive contracts, or "pouring rights," to receive guaranteed profit margins, up-front cash incentives, and other nonmonetary incentives, such as scoreboards and drinks for meetings. Many of these lucrative packages are worth millions of dollars and help supplement the budgets of financially strapped school districts. In exchange, the school exclusively sells and promotes the major beverage company's products. Funds generated from these competitive sales accrue to school organizations, departments, or programs other than child nutrition programs.

These exclusive contracts can hurt the financial success of child nutrition programs that are expected to be self-supporting and to operate as a business within the school system. Such programs must generate enough annual revenue to pay for food, supplies and equipment needed to prepare and serve meals, the cost of salaries and benefits for personnel, indirect costs, and other expenses incurred in operating a business. The primary source of funds that support local child nutrition programs is reimbursement provided by the USDA through participation in the NSLP and SBP. However, federal funds provide, on average, only half the revenues needed to operate these programs. Students who pay for school meals are the second largest source of revenue. The remaining revenue is generated from after school snack programs, summer feeding programs, contract feeding, and catering. More and more child nutrition programs are struggling to remain self-sufficient, because food, salary, and benefit expenditures increase far more quickly than revenue from reimbursement dollars and student meal prices.

School administrators are looking for ways to increase classroom time within the existing school day. Many secondary schools have moved to "block schedules" that give students more class time but allow for just one 50-minute lunch period during which the food service staff feeds an entire school. These types of schedules, which result in short lunch periods and long lunch lines, force students either to skip lunch or to select foods they can eat quickly from snack bars or vending machines. There are no federal guidelines that specify a time frame that schools must follow when providing meals to students. The law merely states that adequate time should be given, and this allows every school administrator to interpret "adequate" differently.

Building Healthful School Environments

Many states, and many school districts, are developing policies that limit the sale of competitive foods and less healthful food choices. In 2004, 21 states introduced bills that would affect vending sales and the sale of other competitive foods in schools. States such as California and Texas have already implemented sweeping changes to nutrition policies in schools. For example, the 2004 Texas Public School Nutrition Policy mandated changes at all school levels, such as eliminating deep-fat frying; restricting portion sizes on chips, milk, fruit drinks, and certain snacks and sweets; limiting fats and sugar; and offering fruits and vegetables daily at every point of service. In all grades, the portion size for french fries, a school lunch "staple," was reduced to 3 ounces, and students are allowed to purchase only one serving.[112]

Healthy People 2010

Increase the proportion of children and adolescents aged 6 to 19 years whose intakes of meals and snacks at school contribute to good overall dietary quality.

The surgeon general's call to action to prevent and decrease overweight and obesity outlines the following actions for creating healthful school environments:[113]

- Provide age-appropriate nutrition and health education to help students develop lifelong healthful lifestyle habits.
- Ensure that meals offered through school breakfast and lunch programs meet healthful standards.
- Adopt policies that require all foods and beverages available on school campuses and school events to contribute toward eating patterns that are consistent with the Dietary Guidelines for Americans.
- Provide food options that are low in fat, calories, and added sugars.
- Ensure that healthful snacks and foods are provided in vending machines, school stores, and other venues.
- Prohibit access to vending machines that compete with healthful school meals in elementary schools; restrict access in middle, junior, and high schools.
- Provide adequate time for students to eat school meals, and schedule lunch periods at reasonable hours around midday.

In response to the epidemic of overweight children that has resulted from unhealthful eating behavior and a lack of physical activity, legislators passed one of the most significant provisions of the Child Nutrition and WIC Reauthorization Act of 2004 (Public Law 108-265). This law requires every school district that participates in the NSLP to adopt a board-approved wellness policy. The Wellness Policy must set goals for nutrition education, physical activity, and other school-based activities designed to promote student wellness. Schools must also establish nutrition standards for all foods that are available on each school campus during the school day, with the objective of promoting student health and reducing childhood obesity. The law also requires that school districts establish a wellness task force that includes parents, students, school board members, school administrators, members of the public, and school food service staff members to create these policies.

Nutrition Education Programs

Nutrition education strategies aimed at children or their caregivers are found in both the public and private sectors. Their goal is to improve eating patterns among children. Numerous government and commercial sites on the Internet provide databases of educational materials and resources and, in some cases, offer interactive games and puzzles for children. Refer to this chapter's Internet Resources on page 377 for a list of government and commercial websites related to children and adolescents. Public-sector and private-sector nutrition education initiatives are described below (see page 371).

Action for Healthy Kids

Action for Healthy Kids (AFHK) is an integrated, national–state initiative that addresses childhood obesity by focusing on changes in the school environment, with leadership provided by former Surgeon General David Satcher. AFHK includes a partnership of 40 national organizations, industry groups and government agencies representing education, physical activity, and health and nutrition—such as the Association for Supervision and Curriculum Development, the National Association of State Boards of Education, the National Association for Sport and Physical Education, the American Dietetic Association, the American Academy of Pediatrics, the United States Department of Agriculture, and the United States Department of Education.

In the short term, AFHK is working to increase the number of health-promoting schools that support sound nutrition and physical activity in order to slow the rate of increasing overweight among American children. In the long term, AFHK aims to play a key role in preventing childhood overweight nationwide. To achieve these goals, Action for Healthy Kids has three main objectives:

- Improving schoolchildren's eating habits by increasing access to nutritious food and beverages on school grounds, while decreasing access to high-calorie, low-nutrient options, as well as by integrating nutrition education into the curriculum for all schoolchildren.
- Increasing schoolchildren's physical activity through physical education courses, recess, the integration of physical activity into academic classes, and after school and co-curricular fitness programs.
- Educating administrators, educators, students, and parents about the role of sound nutrition and physical activity in academic achievement.

Fifty-one *Action for Healthy Kids* state teams are implementing a variety of creative interventions at the grassroots level to promote sound nutrition and physical activity throughout the school environment. Each team has developed an action plan that is appropriate for its own state's educational system, culture and resources. For instance:

- The Alabama team conducted a school vending machine survey with 1,400 school principals. The team used the results to help develop a "Guide to Healthy Vending," which it distributed to all principals in the state.
- The Delaware team is working to provide staff training on how to test students' physical fitness in order to ensure that 50 percent of students show an improvement on physical fitness tests.
- The Indiana team is providing all superintendents with a position paper summarizing the relationship between recess and academic performance, and a resource kit of selected before school and after school activities and nutrition programs.
- The Texas team is working to ensure that the majority of its school districts have a school health advisory council responsible for making recommendations and monitoring nutrition and physical activity programs within the district.
- The Kansas team has awarded 13 mini-grants to school teams that have developed strategies for improving choices of healthful foods and beverages on campus and for increasing physical activity during and after school.

Check out the www.actionforhealthykids.org website for the latest information and resources to improve nutrition and physical activity in schools. The AFHK website provides a one-stop source for information on AFHK's national and state team initiatives, statistics on nutrition and physical activity in schools, and a variety of resources for school-based change. Additionally, AFHK has collaborated with CDC and others to develop a wellness policy tool that is available as a resource for schools creating their wellness policy.

Source: Action for Healthy Kids, 2005, www.actionforhealthykids.org.

Bob Daemmrich/StockBoston

Successful nutrition education activities help children and teenagers focus on their interests and on the relationship of good eating and physical activity to health.

Nutrition Education in the Public Sector

Schools can play a key role in reversing the trend of obesity and lack of physical activity by offering comprehensive nutrition education programs.[114] Schools are an ideal setting for nutrition education for the following reasons:

- More than 95 percent of all children and adolescents are enrolled in school.
- More than 50 percent eat at least one of three meals at school.
- Professionally prepared teachers and staff can provide nutrition education that teaches students to resist social pressures for unhealthful eating.

Severe drops in funding from year to year have decreased the capacity of the USDA, state agencies, and local sponsors to deliver effective nutrition education to children. The USDA's Nutrition Education and Training (NET) program, begun in 1977 with $27 million, has not been adequately funded since the 1980s and was not reauthorized during 2004. Even though the number of students participating in child nutrition programs has continued to grow, these same programs have received limited funding for nutrition education.

Despite the lack of significant and consistent federal funding for national nutrition education initiatives, successful nutrition projects have been implemented. To improve fruit and vegetable consumption of children, the 2002 Farm Bill appropriated $6 million for schools in Iowa, Indiana, Michigan, Ohio, and the Zuni Indian Tribal Organization in New Mexico to purchase fresh and dried fruit and fresh vegetables to be available free of charge to children through the school day.[115] The project has shown several positive benefits. Students reported enjoying the healthful snack option and being involved in preparing the fruits and vegetables for distribution. School administrators commented that the USDA offered flexibility: Each school was allowed to implement the program with the approach that best met the needs of the school. Food service staff reported that not only lunch participation increased but also the consumption of fruits and vegetables at mealtime. The pilot program showed that given the resources, schools can create an environment where healthful snack options can be a reality.

Programs in Action

Empowering Teens to Make Better Nutrition Decisions

Unhealthful eating and lack of physical activity are major contributors to adulthood morbidity and mortality in the United States.[1] These habits are prevalent among youth also. By the time children graduate from high school, more than 70 percent do not eat enough fruits and vegetables, 84 percent eat too much fat, and nearly one-third do not engage in regular vigorous physical activity.[2] Fast food is increasingly common in high schools,[3] and fast-food advertising is prevalent.[4] These elements contribute to the fact that more than 16 percent of young people are considered overweight. The incidence of obesity has significant public health implications today and for years to come.

"Food on the Run" recognized the importance of empowering teens to make better decisions about their diet, activity, and health. The project was born out of collaboration among ten California communities that recognized the lack of nutrition education materials and programs for high school students.

Participating communities worked with California Project LEAN, a program of the Public Health Institute and the California Department of Health Services, to develop the framework for "Food on the Run."

Goals and Objectives

"Food on the Run" sought to improve the health of high school students through the promotion of accurate nutrition information in the classroom and increased availability of healthful food options on campus. Its primary objectives were

- To create a high school youth advocacy model that motivates students to advocate for more healthful food and physical activity options in their communities.
- To advance locally identified policy and environmental changes that increase the number and promotion of healthful food items and physical activity options on participating school campuses.
- To motivate students to make more healthful food choices and to become physically active.

Target Audience

Program participants were low-income students in high schools where at least 40 percent of the students were eligible for free and reduced-price meals. During the 1998–1999 school year, the 28 "Food on the Run" schools reached 11 percent of California's low-income high school students.

Rationale for the Intervention

In general, high school students need to improve their eating habits and level of physical activity. It is believed that these students will be more motivated to change if they play an integral role in the formulation of health program strategies and messages.

"Food on the Run" uses the spectrum of prevention as a basis for its intervention.[5] This framework states that the following components are necessary to effect change at the individual and community levels:

- Strengthening individual knowledge and skills
- Promoting community education
- Educating providers
- Fostering coalitions and networks

FIGURE 11-2 The Team Nutrition Logo

The 2004 Child Nutrition Reauthorization made the Fresh Fruit and Vegetable pilot program a permanent program, increased funding to $9 million, and added four more states (Mississippi, Washington, Pennsylvania, and North Carolina) and two tribal organizations in Arizona and South Dakota.

Two other successful USDA nutrition education programs, the Expanded Food and Nutrition Education Program (EFNEP) and Team Nutrition, target the improvement of children's health. EFNEP, a cooperative extension program, began in 1968 to assist low-income youth and families in acquiring knowledge, skills, and attitudes that contribute to their nutritional well-being and to the improvement of the total family's diet. Team Nutrition, begun in 1995, focuses attention on the important role that school meals, nutrition education, and a healthful school environment play in teaching students the importance of good dietary habits and physical activity.[116] The Team Nutrition logo is shown in Figure 11-2. Team Nutrition has three behavior-focused strategies:

1. Provide training to school food service professionals that enable them to prepare and serve nutritious meals.

- Changing organizational practices
- Influencing policy and legislation

Methodology

Each "Food on the Run" school, in conjunction with students, set its own nutrition and physical activity policy agenda. During the 1998–1999 school year, each school worked with a coalition of local organizers, health providers, and private industry to build its program. Components included the recruitment and training of 10–20 high school student advocates, the implementation of at least seven school-based activities, and at least two activities to increase parent awareness and involvement. Specific activities included taste tests of low-fat foods, presentations to school boards, setting up of a sports equipment checkout table at lunch, initiation of a cafeteria salad bar, and a change from reduced fat (2 percent) to low-fat (1 percent) milk in school cafeterias.

California Project LEAN supported "Food on the Run" communities with training, resources, media tools, research, and development of food and physical activity messages.

Results

Program success was evaluated with student surveys and an assessment of the school environment. The environment assessment described the eating and physical activity environment at participating high schools using pre- and post-test measures. During the 1998–1999 school year, statistically significant increases ($p = .05$) were observed for physical activity knowledge (6 percent) and attitude (4 percent); nutrition knowledge (5 percent), attitude (5 percent), and behavior (9 percent); healthy eating options (5.7 out of a possible 11 points); healthy eating promotional efforts on school campuses (2.3 out of a possible 5 points); and physical activity options available to students at the schools (3.3 out of a possible 6 points).

Lessons Learned

Contrary to popular belief, high school students want opportunities to "eat healthy" and be more physically active. Student involvement is the key to offering healthful foods that sell and physical activity classes that are full. When high school students are involved in the formulation of nutrition and physical activity messages and policy strategies, behavior change can occur.

References

1. J. M. McGinnis and W. H. Foege, Actual causes of death in the United States, *Journal of the American Medical Association* 270 (1993): 2207–12.
2. *Physical Activity and Good Nutrition: Essential Elements for Good Health,* Centers for Disease Control and Prevention, 1999.
3. *ASFSA 1999 School Lunch Trend Survey,* The American School Food Service Association, 1999.
4. *California High School Fast Food Survey: Findings & Recommendations* (Berkeley, CA: The Public Health Institute, February 2000).
5. L. Cohen and S. L. Swift, The spectrum of prevention: Developing a comprehensive approach to injury prevention, *Injury Prevention* 5 (1999): 203–7.

Source: Adapted from *Community Nutritionary* (White Plains, NY: Dannon Institute, Spring 2001). Used with permission. For more information about the *Awards for Excellence in Community Nutrition,* go to www.dannon-institute.org. The California Project LEAN program is available at www.caprojectlean.org.

2. Promote nutrition education in schools that teaches and encourages students to make healthful lifestyle choices.
3. Build school and community support for healthful school environments that promote positive nutrition messages throughout the school environment.

There are many government and nonprofit organizations at the national, state, and local levels that are also implementing initiatives that aim to reduce overweight and increase physical activity among youth. One such program is the Centers for Disease Control and Prevention's Coordinated School Health Program (CSHP), which combines health education and promotion, disease prevention, and access to health and social services in an integrated, comprehensive manner. To reach beyond the classroom and cafeteria, CSHP also provides opportunities for parental and community involvement in promoting healthful behaviors to students.

An excellent example of a coordinated school health program that addresses both nutrition and physical activity is the *Coordinated Approach To Child Health,* or *CATCH,* program described in Chapter 8's Programs in Action feature.

The following are some examples of programs that effectively promote healthful eating to children using a variety of methods.

FIGURE 11-3 A Nutrition Education Activity from the USDA's YourSELF Program for Seventh and Eighth Graders

Source: U.S. Department of Agriculture, *YourSELF Middle School Nutrition. Education Kit,* 1998; available at www.fns.usda.gov/tn/Resources/yourself.html.

Let's Eat!

Treat your brain to this puzzle. When you're through, give your brain a break: eat something!

```
T D I N H R B E A N S D I O A E A B
C E R E A L A U T N B M I D A T N E
E R E G G S W C O C N I T R P P O T
P I Z Z A F Y T M H T L I O P O S F
E R N W A O E W A E W K Y T L R E E
S O C P A S T A T D I E O L E K T A
D H E E A I A A O D E Z G I R C E S
H U G R A P E S E E A H U O A H B E
M U F F I N D U S R H E R Y O O P E
Y B R O C C O L I C N A T H N P A R
D N I T I O M T E H E G T B H V N E
P I A E C R A C K E R S B U F E C C
K I W I D A I U H E U L T R S C A R
O O S D A N E E T S A H P R Y S K I
S O R H R G E F T E C P N I R E E I
D G O E H E L A S A G N A T I N S D
P E P P E R S T E W N A S O A E E S
T S P E A R O N V E O N S I A O E D
```

How many words can you find? Hidden within this puzzle are:

THREE MILK GROUP FOODS
THREE MEAT GROUP FOODS
FIVE FRUIT GROUP FOODS
THREE VEGETABLE GROUP FOODS
FIVE BREAD GROUP FOODS
THREE COMBINATION FOODS

YourSELF

YourSELF is one of the first federal initiatives on nutrition and physical activity. It is designed to speak directly to adolescents in the seventh and eighth grades. Released in 1998 by the USDA, the *YourSELF* Middle School Nutrition Education Kit contains materials for health education, consumer science, or family living classes that teach students to make healthful eating choices and become physically active. The kit includes the *YourSELF* magazine, a student workbook, a teacher's guide, reproducible masters, and ideas for linking classroom and cafeteria activities.[117] A portion of one of the *YourSELF* handouts is shown in Figure 11-3.

EAT SMART. PLAY HARD.

This national FNS nutrition education and promotion campaign is designed to convey motivational messages about healthful eating and physical activity.[118] The campaign uses Power Panther™ as the primary communication vehicle for delivering messages about nutrition and physical activity to children and their caregivers. The target audience for this campaign is children aged 2 to 18 years who are participating, or are eligible to participate, in FNS nutrition assistance programs. Campaign messages are based on the Dietary

Guidelines for Americans and focus on four basic themes: eating breakfast, healthful snacking, achieving balance, and physical activity.

5 A DAY THE COLOR WAY

This is the Produce for Better Health Foundation's national campaign to promote the increased consumption of fruits and vegetables by adults and children.[119] This national campaign includes a specific program for elementary schools called "There's a Rainbow on My Plate," which promotes a colorful diet of fruits and vegetables. The campaign includes curriculum guides for teachers and techniques to promote healthful eating in the school food service program. The campaign also targets the produce sections of supermarkets.

VERB. IT'S WHAT YOU DO

This is a national multicultural campaign coordinated by the U.S. Department of Health and Human Services and the Centers for Disease Control and Prevention.[120] Using social marketing strategies to influence the behavior of young people aged 9 to 13, or "tweens," the *VERB* campaign encourages physical activity every day. The campaign combines paid advertising, marketing strategies, and partnership efforts to reach tweens. There are also *VERB* materials and Web resources for organizations that have support staff working with tweens.

POWERFUL BONES, POWERFUL GIRLS

This is a national bone health campaign that partners the National Osteoporosis Foundation with the Centers for Disease Control and Prevention and the Department of Health and Human Services.[121] It is a multiyear campaign to promote optimal bone health in girls 9 to 12 years of age and thus to reduce their risk of osteoporosis later in life. The goal is to encourage calcium consumption and physical activity for building and maintaining strong bones. The campaign also targets adults who influence tweens, including parents, teachers, and health care professionals.

Nutrition Education in the Private Sector

Some voluntary, nonprofit health organizations have developed nutrition education programs or materials for children or their caregivers. The American Heart Association (AHA), for example, has a variety of Schoolsite Program modules for children in kindergarten through grade 12. Each module comes with a teacher's guide, learning and kick-off activities, a videotape, and handouts for students, all geared to the appropriate grade level.[122] HeartPower! Online is the American Heart Association's curriculum-based program for teaching preschool, elementary, and middle school students about the heart and how to keep it healthy for a lifetime. It includes lesson plans and activity sheets about nutrition, physical activity, living tobacco-free, and knowing how the heart works.

The American Cancer Society developed "Changing the Course," a comprehensive nutrition education program for students in kindergarten through grade 12. Changing the Course uses a video, together with handouts and overheads, to help students develop good eating habits and reduce cancer risk by eating more fruits and vegetables and less fat.[123] A variety of books, pamphlets, and brochures written for children or their caregivers are available from the American Dietetic Association, the American Academy of Pediatrics, and other organizations.

The Kids Café Program is sponsored by ConAgra Foods through the ConAgra Feeding Children Better Foundation in partnership with America's Second Harvest. Since the program's inception in 1989, Kids Cafés have become one of the nation's largest free meal service programs for children.[124] The primary goal of the Kids Café Program is to provide free and prepared food and nutrition education to hungry children. Kids Cafés across the country achieve this goal by utilizing existing community resources, such as Boys and Girls Clubs, community recreation centers, or schools—places where children already congregate naturally. Today, there are 141 America's Second Harvest affiliate food banks and food-rescue organizations operating more than 1,200 Kids Café sites, which serve more than 12 million meals per year to children in 41 states and Washington, D.C. In addition to providing hot meals to hungry kids, some Kids Café programs also offer a safe and welcoming place where, under the supervision of caring staff, a child can get involved in educational, recreational, and social activities that draw on existing community programs and often include family members.

Keeping Children and Adolescents Healthy

Programs and services designed to keep children and adolescents healthy can have a lasting effect on the nation's public health. Healthy children and adolescents mean healthy communities. Children and youth who learn good eating habits, exercise regularly, refrain from smoking, learn to manage stress, and develop a strong sense of self are less likely to turn to drugs and alcohol to solve problems and more likely to know how to live constructively.

Successful, effective programs or services for children and youth recognize the stresses of life in the early twenty-first century and the mixed messages in the media about products and values. Programs and services founded on basic health promotion principles must consider the urgent health issues facing today's young people: suicide, child abuse, teen pregnancy, sexually transmitted diseases, eating disorders, overweight, substance abuse, and others. Positive nutrition messages support and expand on other health promotion concepts.

What types of programs work? Health educators have found that programming for children and adolescents succeeds when it is fun and informative. Programs work best when they are geared to a specific health or nutritional objective, such as weight loss, improved eating patterns, or increased activity levels. For instance, Wings of America, a youth program for Amerian Indians based in Santa Fe, New Mexico, has sparked a revival of running among American Indian children and teenagers. Wings teams help these children develop pride, self-esteem, and cultural identity, in addition to promoting fitness.[125] Involving children and adolescents in the planning and implementation of a program increases its effectiveness, as does using peer support to help the participants make decisions about their health. Effective programs also employ trained staff who work well with these populations. The youth development approach offers a strategy for linking young people meaningfully with their communities and involving them in designing and implementing programs and services.[126]

In the final analysis, developing programs and services that meet the needs of our nation's children and youth means recognizing that as today's children grow, they have to perform increasingly complex tasks in a world of constant technological and environmental change. Improving the health of today's young people will enhance the quality of their lives in the immediate future and enlarge their potential for contributing to our nation's future as adults. In promoting the health and well-being of children and youth, we are recognizing "that children matter for themselves, that childhood has its own intrinsic value, and that society has an obligation to enhance the lives of children today."[127]

Internet Resources

Government and University Websites

Center for Disease Control and Prevention	**www.cdc.gov/growthcharts**
Administration for Children and Families	**www.acf.dhhs.gov**
Adolescence Directory Online	**http://education.indiana.edu/cas/adol/adol.html**
Child Care Nutrition Resource System	**www.nal.usda.gov/childcare**
Child Nutrition Programs	**www.fns.usda.gov/fns**
Harvard Eating Disorders Center (HEDC)	**www.hedc.org**
Head Start Bureau	**www.acf.dhhs.gov/programs/hsb**
Lead Program at CDC	**www.cdc.gov/nceh/lead**
National Center for Education in Maternal and Child Health	**www.ncemch.org**
National Clearinghouse on Families & Youth	**www.ncfy.com**
National Institute of Child Health and Development	**www.nichd.nih.gov**
President's Council on Physical Fitness and Sports	**www.fitness.gov**
School Nutrition Dietary Assessment Study II: Summary of Findings	**www.fns.usda.gov/oane/MENU/Published/CNP/FILES/SNDAIIfindsum.htm**
School Nutrition Program (Nutrient Analysis Protocols)	**www.nalusda.gov/fnic/schoolmeals/resources/naproto.html**
USDA Food Guide Pyramid (For Young Children)	**www.usda.gov/cnpp**
USDA's School Meals Initiative for Healthy Children/Team Nutrition	**http://schoolmeals.nal.usda.gov**
Team Nutrition Booklist	**www.tn.fcs.msue.msu.edu/booklist.html**
Action For Healthy Kids (AFHK)	**www.actionforhealthykids.org**
U.S. Department of Housing and Urban Development Office of Lead Hazard Control	**www.hud.gov/offices/lead**
U.S. Government Health Information Site	**www.healthfinder.gov**

Websites Related to Disabilities

Cornucopia of Disability Information (CODI)	**http://codi.buffalo.edu/**
National Council on Disability	**www.ncd.gov**
National Dissemination Center for Children with Disabilities	**www.nichcy.org**

Other Websites

Academy for Eating Disorders	**www.aedweb.org**
American Academy of Pediatrics	**www.aap.org**
School Nutrition Association	**www.schoolnutrition.org**
Canadian Pediatric Society	**www.cps.ca**
Childhood Obesity Prevention Partnership	**www.kidnetic.com**
Food Research and Action Center	**www.frac.org**
International Life Sciences Institute's Take10!™ Program	**www.take10.net**
I Am Your Child Program	**www.iamyourchild.org**
Kids Café	**www.secondharvest.org**
Kids Count	**www.aecf.org/kidscount**
Kids Food Cyberclub	**www.kidfood.org**
KidsHealth	**www.kidshealth.org**
Produce for Better Health Foundation	**www.5aday.org**
Team Nutrition's Local Wellness Policy Web Page	**www.fns.usda.gov/tn/Healthy/wellnesspolicy.html**
Children's Nutrition Research Center at Baylor College of Medicine	**www.kidsnutrition.org**
Fantastic Food Challenge	**http://commtechlab.msu.edu/products/foodchallenge.html**
VERB. It's What You Do	**www.verbnow.com**
Powerful Bones, Powerful Girls	**www.cdc.gov/powerfulbones/index2.html**
Clueless in the Mall	**http://calcium.tamu.edu**
Dairy Council of California	**www.dairycouncilofca.org**
Dole 5 a Day: Just for Kids	**www.dole5aday.com**
HHS Pages for Kids	**www.hhs.gov/kids**
KidsHealth.org/Hey Teens	**www.kidshealth.org/teen/index.html**
Why Milk?	**www.whymilk.com**
HeartPower!	**www.americanheart.org**

Note: For updates and links to these and other nutrition-related sites go to our website at www.wadsworth.com/nutrition.

Case *Study*

The Child Nutrition Program

By Alice Fornari, EdD, RD, and Alessandra Sarcona, MS, RD

Scenario

A school district in a suburban community has offered you the position of director of the Child Nutrition Program. The school district consists of one high school, two middle schools, and four elementary schools. You have accepted the position, seeing it as an exciting challenge after your 15 years of experience directing a hospital food service. The district has expressed interest in developing a nutrition and health intervention program for the students as part of the Child Nutrition Program.

After your month of "settling in" to the job, you begin to look at the goal of developing a nutrition and health intervention program for the students. You have done a nutrient analysis of the school lunch menu that revealed that 40 percent of the calories in the average lunch are derived from fat. Twenty percent of the student body receives free and reduced-price lunches. Government commodities are provided but are not fully utilized. A large percentage of the sales of the lunch program—and therefore a large percentage of its revenues—comes from snack items. Recently, the school nurse has released statistics on the weight status of the students. She outlined the percentage of students with a BMI greater than the 85th percentile to determine overweight trends: 12 percent of the high school students, 20 percent of the middle school students, and 14 percent of the elementary school students fell in this category. Another relevant issue is that the middle school physical education curriculum has been reduced from daily to two days per week because of budget constraints and the need for more classroom time in order to meet rigorous academic requirements.

Learning Outcomes

- Identify the *Healthy People 2010* objectives for children and adolescents.
- Identify USDA nutrition guidelines for Child Nutrition Program meals.
- Outline the process for implementing a school-based nutrition/health intervention program.

Foundation: Acquisition of Knowledge and Skills

1. Acquire from the text the *Healthy People 2010* objectives for children and adolescents.
2. From the website www.fns.usda.gov/cnd/menu/menu.planning.approaches.for.lunches.doc, review the guidelines for menu planning for the national school lunch program for healthful school meals.
3. From the website www.fns.usda.gov/tn/healthy/call2act.pdf, outline the objectives of TEAM Nutrition and review the "Ten Keys to Promote Healthy Eating in Schools."

Step 1: *Identify Relevant Information and Uncertainties*

1. To set up a pilot program for a specific target group, sort the pieces of information presented in the case to determine which school in the district has the most risk factors.
2. Identify conditions in the school and aspects of the lunch menu that are affecting the population determined to be at greatest risk; list a range of possible ways to address these issues.

Step 2: *Interpret Information*

1. Identify the strengths and weaknesses of incorporating TEAM Nutrition for your target school; from the TEAM Nutrition website, review some projects/activities other schools have implemented that might be feasible for your school.
2. Devise a plan to develop a task force from appropriate groups within the school to support the implementation of TEAM Nutrition; include a written letter.

Step 3: *Draw and Implement Conclusions*

1. Outline a time line for implementing TEAM Nutrition.

Step 4: *Engage in Continuous Improvement*

1. Outline the variables that would need to be monitored in the first phase to evaluate a desired outcome for your intervention program.
2. From the TEAM Nutrition website, investigate the availability of grants. Develop a fact sheet for the task force that would support applying for a grant.

Chapter *Twelve*

Growing Older: Nutrition Assessment, Services, and Programs

Learning *Objectives*

After you have read and studied this chapter, you will be able to:

- Describe the potential impact of the graying of America on health care services.
- List national goals for health promotion for adults.
- Identify factors influencing the nutritional status of older adults.
- Describe the components of a nutrition assessment of older adults.
- Describe the purpose and function of the Nutrition Screening Initiative.
- Describe community nutrition programs that are intended to provide nutrition assistance to older adults.

This chapter addresses such issues as the influence of age on nutrient requirements, screening individuals for nutritional risk, collecting pertinent information for nutrition assessments, availability of food and nutrition programs in the community, and current information technologies, which are Commission on Accreditation for Dietetics Education (CADE) *Foundation Knowledge and Skills* requirements for dietetics education.

Chapter *Outline*

Something To Think About...

How far you go in life depends on your being tender with the young, compassionate with the aged, sympathetic with the striving, and tolerant of the weak and the strong. Because someday in life you will have been all of these.

– George Washington Carver

Introduction

Americans are living longer than ever before, and the *average* age at death (life expectancy) has changed dramatically. A child born in 2002 could expect to live 77 years, about 30 years longer than a child born in 1900. Among women, life expectancy from birth is 80 years for white females and 75 years for black females.[1] For men, life expectancy from birth is 75 years for white males and 68 years for black males. On the other hand, the *maximum* age at which people die—that is, the *maximum life span*—has changed less dramatically. It seems that the aging phenomenon cuts off life at a rather fixed point in time.

Technically, the life span is the oldest documented age to which a member of a given species is known to have survived. For humans, this is about 125 years.

To what extent is aging inevitable? Apparently, aging is an inevitable, natural process programmed into our genes at conception. Nevertheless, we can adopt lifestyle habits, such as consuming a healthful diet, exercising, and paying attention to our work and recreational environments, that will slow the process within the natural limits set by heredity. Life expectancy has also risen as a result of better prenatal and postnatal care and improved means of combating disease in older adults. For example, the death rate from heart disease began to decline in the 1960s and continues to fall today. Over half of the drop is attributed to a decline in smoking and in the numbers of people with high blood pressure or high blood cholesterol. Clearly, good nutrition can retard and ease the aging process in many significant ways. However, no potions, foods, or pills will prolong youth. People who claim to have found the fountain of youth have been selling its waters for centuries, but products purporting to prevent aging profit only the sellers, not the buyers.

One approach to the prevention of aging has been to study other cultures in the hope of finding an extremely long-lived people and then learning their secrets of long life. The views of the experts can best be summed up by saying that disease can *shorten* people's lives and that poor nutrition practices make diseases more likely to occur, but poor health is not an inevitable consequence of aging. Thus, by postponing and slowing disease processes, optimal nutrition can help to prolong life up to the maximum life span—but cannot extend it further.[2] This chapter focuses on the diseases that seem to come with age, their risk factors, and the nutrition assessment of older adults; it also examines the programs and services that target older adults for health promotion and disease prevention. We begin with a look at the demographic trends characteristic of this segment of the population and the national nutrition and health goals for improving the quality of life of Americans as they age.

Elderly people in Okinawa, Japan have among the lowest mortality rates in the world from a multitude of chronic diseases of aging. The Okinawa Centenarian Study seeks to uncover the genetic and lifestyle factors responsible for this successful aging phenomenon. See http://okinawaprogram.com.

Demographic Trends and Aging

The number of elderly (aged 65 years and older) in the United States will double by 2030 to more than 71.5 million people. In 2003, nearly 36 million people were age 65 and over and accounted for just over 12 percent of the population. This proportion is expected to

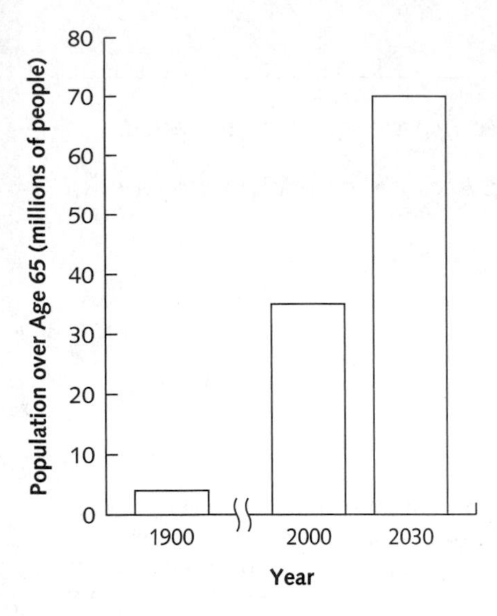

FIGURE 12-1 The Aging of the Population

In 1900, 4 percent of the U.S. population (3.1 million people) were over 65 years of age; in 2000, 12.4 percent (35 million people) were over 65; by 2030, 20 percent (about 71.5 million people) will have reached age 65.

Source: Data from *A Profile of Older Americans, 2003*; available at www.aoa.gov.

rise to approximately 14 percent in 2010 and to 20 percent by 2030. Nearly 12 percent of the population will be over age 74 by 2030.[3] The increased growth in the elderly population in the United States is illustrated in Figure 12-1.

Both the baby boom that took place between 1946 and 1964 and improved life expectancy are important contributors to the growing elderly population in the United States. Baby boomers will increase the numbers of the older middle-aged (ages 46 through 63) until 2011, when they will begin to swell the ranks of the retired population. The growth rate of the older population is expected to slow after 2030, when the last baby boomers turn 65 years old. By 2030, minority groups will represent 26.4 percent of the population of older adults (up from 17.2 percent in 2002), with the largest increases in Hispanics and Asians.

Policy makers and others concerned with meeting the health needs of older adults are alert to the implications of these demographic changes, because the elderly tend to consume a large amount of total health care and long-term care resources. Consider that persons aged 65 years and older represented just over 12 percent of the total population in 2003, yet they accounted for more than 30 percent of the country's health care costs—utilizing more hospital services and consuming more than 40 percent of all prescription drugs.[4] With the "graying of America," these health care demands can only increase. As Joseph A. Califano, a former secretary of health and human services, testified before the Committee on a National Research Agenda on Aging,

> The aging of America will challenge all our political, retirement, and social service systems. As never before, it will test our commitment to decent human values. Nowhere is the aging of America freighted with more risk and opportunity than in the area of health care.[5]

Healthy Adults

Researchers and marketing analysts have struggled to find a useful way to segment—and, therefore, target—the older adult market. One frequently used segmentation divides older adults into the "mature," aged 55–64; the "young-old," aged 65–74; and the "old-old," over age 74.

The most important goal of health promotion and disease prevention for adults as they age is maintaining health and functional independence. Many of the health problems associated with the later years are preventable or can be controlled. An individual's current health profile is substantially determined by behavioral risk factors. As shown in Table 12-1, the leading causes of death for adults of various ages include heart disease, cancer, stroke, chronic lung disease, diabetes, injuries, and liver disease; all have been associated with behavioral risk factors. Thus, many adults today would benefit from changes in their lifestyle behaviors. For example, changing certain risk behaviors into healthful ones can improve the quality of life for older persons and lessen their risk of disability. Improvements in diet and nutritional status and weight control can enhance the health of older adults, as well as help control risk factors for disease in middle-aged and younger adults.[6]

For the past four decades, health professionals have provided the American public with health-related messages to lower risk of chronic diseases such as cardiovascular disease, cancer, diabetes, osteoporosis, and obesity.[7] These messages have encouraged such things as avoidance or cessation of cigarette smoking, maintenance of a healthful body weight, reduced intake of saturated fats and cholesterol, regular physical activity, regular health screenings for blood pressure and cholesterol, cancer-related screenings, and consumption of a diet rich in whole grains, fruits, vegetables, and fiber.

TABLE 12-1 The Leading Causes and Numbers of Deaths among Adults, Aged 25–44, 45–64, and 65+; All Races, Both Sexes, 2002

	AGE GROUPS		
RANK	**25–44**	**45–64**	**65+**
1	Unintentional injury: 27,454	Cancer: 143,416	Heart disease: 577,353
2	Cancer: 20,008	Heart disease: 100,378	Cancer: 392,145
3	Heart disease: 16,155	Unintentional injury: 21,578	Cerebrovascular: 143,780
4	Suicide: 11,501	Cerebrovascular: 15,869	Chronic low respiratory disease: 109,158
5	HIV: 7,531	Diabetes mellitus: 15,452	Influenza and pneumonia: 59,235
6	Homicide: 7,505	Chronic low respiratory disease: 14,720	Alzheimer's disease: 58,205
7	Liver disease: 3,476	Liver disease: 13,131	Diabetes mellitus: 54,717
8	Cerebrovascular: 2,934	Suicide: 9,517	Nephritis: 34,389
9	Diabetes mellitus: 2,747	HIV: 5,729	Unintentional injury: 32,973
10	Influenza and pneumonia: 1,294	Septicemia: 5,416	Septicemia: 26,688

Sources: Office of Statistics and Programming, National Center for Injury Prevention and Control, CDC; National Center for Health Statistics (NCHS) Vital Statistics System.

National Goals for Health Promotion

Primary *Healthy People 2010* focus areas for older adults include reducing the prevalence of, and the overall number of people who suffer from, diseases such as arthritis, osteoporosis, cancer, diabetes, and kidney disease. Other age-specific goals include increasing the number of older persons receiving pneumonia and influenza vaccinations and colorectal cancer screenings, along with increasing daily physical activity and cardiovascular health.[8] Table 12-2 shows the current status of the U.S. adult population on various "healthy lifestyle" habits and presents the key national health objectives for the year 2010 that focus on improving the health of adults as they age.

Overall, the progress review data on the *Healthy People 2010* objectives for the weight status of adults reflect a trend for the worse.[9] The proportion of adults aged 20 years and older who are at a healthful weight—that is, who have a body mass index (BMI) in the range of 18.5 to 24.9—decreased from 42 percent in 1988–1994 to 34.9 percent in 1999–2002. The target is 60 percent. The age-adjusted proportion of adults aged 20 years and older who are obese (they have a BMI of 30.0 or more) increased from 23 percent in the survey period 1988–1994 to 30.4 percent in 1999–2002.

As illustrated in Figure 12-2, the three objectives for fruit, vegetable, and grain consumption have shown little or no progress in this decade. The average number of daily servings of fruit consumed changed little, going from 1.6 in 1994–1996 to 1.5 in 1999–2002. Two to four servings are recommended. Vegetable intake also showed little change: an average of 3.4 daily servings in 1994–1996 compared with 3.3 in 1999–2000. Three to five servings are recommended, with at least one-third being dark green or orange vegetables. In 1999–2002, only 11 percent of vegetable servings consumed by adults 20 years of age and older were dark green or orange, whereas fried potatoes constituted 22 percent. In 1999–2002, the proportion of all grain products consumed that were whole grain was 13 percent for adults. The target for all is that one-half be whole grain.

There has also been little or no change since the past decade in the status of most objectives for physical activity and fitness. Since the 1997 baselines, modest improvements have been recorded for two objectives: (1) A smaller proportion of the adult population (aged 18 years and older) reports pursuing no leisure-time physical activity, and (2) a larger proportion of adults perform physical activities that enhance muscular strength and endurance.

TABLE 12-2 ***Healthy People 2010*** **Objectives to Improve the Health of Adults and Progress Review of** ***Healthy People 2010*** **Objectives**

HEALTHY PEOPLE 2010 OBJECTIVE	BASELINE (Year)	CURRENT STATUS (Year)	HP 2010 TARGET
19.1 Increase the proportion of adults who are at a healthy weight.	42% (1998–1994)	34.9% (1999–2002)	60%
19.2 Reduce the percentage of adults who are obese.	23% (1998–1994)	30.4% (1999–2002)	15%
19.5 Increase the proportion of persons aged 2 years and older who consume at least two daily servings of fruit.	28% (1994–1996)	NC*	75%
19.6 Increase the proportion of persons aged 2 years and older who consume at least three daily servings of vegetables, with at least one-third being dark green or deep yellow vegetables.	3% (1994–1996)	NC	50%
19.7 Increase the proportion of persons aged 2 years and older who consume at least six daily servings of grain products, with at least three being whole grains.	7% (1994–1996)	NC	50%
19.8 Increase the proportion of persons aged 2 years and older who consume less than 10% of calories from saturated fat.	36% (1994–1996)	NC	75%
19.9 Increase the proportion of persons aged 2 years and older who consume no more than 30% of calories from total fat.	33% (1994–1996)	NC	75%
19.10 Increase the proportion of persons aged 2 years and older who consume 2,400 milligrams or less of sodium daily.	21% (1988–1994)	—[†]	65%
19.11 Increase the proportion of persons aged 2 years and older who meet dietary recommendations for calcium.	45% (1988–1994)	—	75%
19.17 Increase the proportion of physician office visits made by patients with a diagnosis of cardiovascular disease, diabetes, or hyperlipidemia that include counseling or education related to diet and nutrition.	42% (1997)	—	75%
19.18 Increase food security among U.S. households and in so doing reduce hunger.	88% (1995)	NC	94%
22.2 Increase the proportion of adults who engage regularly, preferably daily, in moderate physical activity for at least 30 minutes per day.	32% (1997)	33% (2003)	50%
3.1 Reduce the overall cancer death rate.	202.7 deaths per 100,000 population (1999)	201 (2000)	160/100,000
5.3 Reduce the overall rate of diabetes per 1,000 population.	40 cases/1,000 (1997)	45/1,000 (2000)	25/1,000
7.5 Increase the proportion of worksites with 50 or more employees that offer a comprehensive employee health promotion program to their employees.	34% (1999)	—	75%
12.1 Reduce coronary heart disease deaths.	208 deaths per 100,000 population (1998)	196 (2000)	166/100,000
12.14 Reduce the proportion of adults with high total blood cholesterol levels ($\geq$ 240 mg/dL).	21% (1988–1994)	18.3% (2000)	17%

*NC = Little or no change.

[†]— = Cannot assess, limited data.

Source: National Center for Health Statistics, Department of Health and Human Services, *Healthy People 2010 Progress Review* (Hyattsville, MD: Public Health Service, 2004); *Healthy People 2010;* available at www.cdc.gov/nchs.

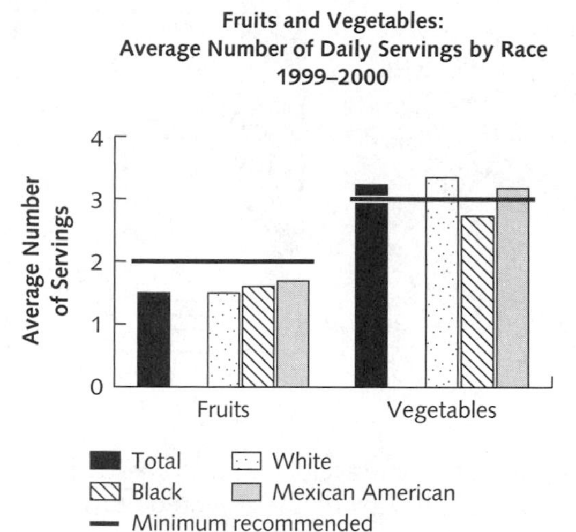

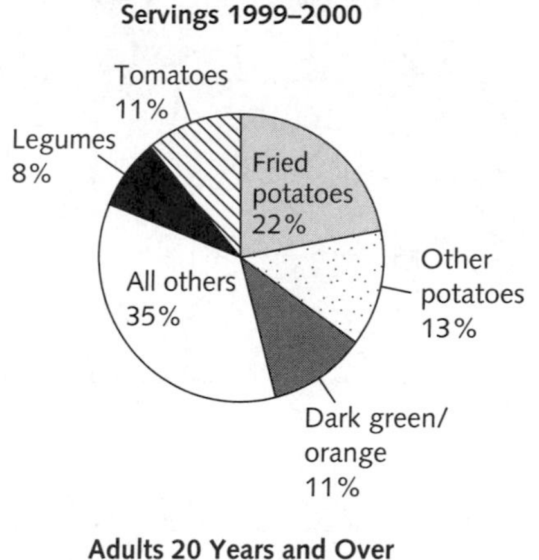

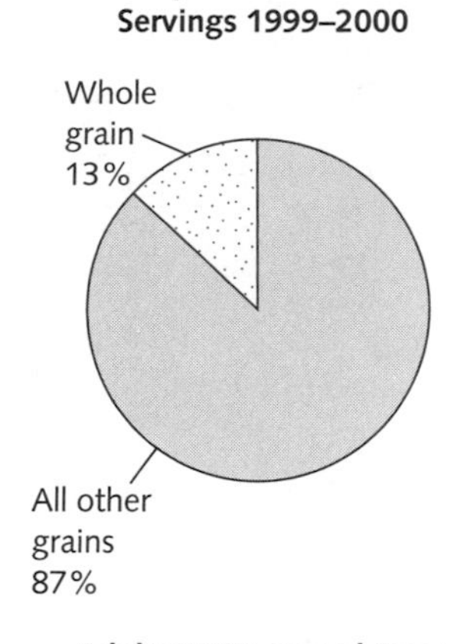

FIGURE 12-2 ***Healthy People 2010*** **in Review: Fruits, Vegetables, and Whole Grains**

Source: Healthy People 2010 Progress Review: Nutrition and Overweight, January 21, 2004.

In 2002, 38 percent of adults engaged in no leisure-time physical activity, compared with 40 percent in 1997. Females were 14 percent more likely to be in this category in 2002 than males. Among five racial and ethnic groups, Hispanics had the highest rate of being sedentary during leisure time; whites had the lowest. Data for 2003 show that 33 percent of adults aged 18 years and older engaged regularly in moderate physical activity (at least 30 minutes of moderate activity at least 5 times a week) compared to 32 percent in 1997.

Approximately 35 percent of men 65 years of age and older and 26 percent of women of that age were physically active at least 5 times a week for 30 minutes per time or were vigorously active for 20 minutes at least 3 times per week. The *Healthy People 2010* target is 50 percent. In 2002, as in 1997, 23 percent of adults engaged in vigorous physical activity that promoted cardiovascular fitness for 20 or more minutes on at least 3 occasions per week. The target is 30 percent.

As a result of the most recent review of the *Healthy People 2010* objectives, health professionals from a number of government agenices suggested steps to bring about further progress in achieving the *HP 2010* objectives that promote healthful weights and food choices.

- Educate the public about the health benefits of being physically active at any size and the possible added benefit of thereby attaining modest weight reduction.
- Promote partnerships with community planners to design neighborhoods that encourage and support increased opportunities for physical activity in appropriate and safe locations.
- Educate the public on how to use the nutrition facts panel on food products and on how to select appropriate portion sizes for healthful diets.
- Offer incentives for worksites to provide safe, convenient, and affordable venues for employees to engage in physical activity.
- Also recommended as effective were "point-of-decision" prompts, such as motivational signs that encourage people to use the stairs instead of elevators.

Understanding Baby Boomers

Baby boomers, or the approximately 77 million individuals who were born between 1946 and 1964, represent almost one-third of the U.S. population. By virtue of their large numbers, baby boomers are a driving force for current and future trends. An understanding of

Steve Dickenson, Copley News. Reprinted with permission.

their preferences, character, lifestyle, and location is and will continue to be critical to health promotion programs and services. In 1991, the first of this generation turned 45, and in 2030, the last of the baby boomers will turn 65.

Although there are several subsets of baby boomers, some general characteristics can be noted:[10]

- Boomers have the power to change the marketplace. Because of their numbers and affluence, they are able to drive trends, especially as they age. Of importance to community nutritionists, these consumers are generally concerned about what is healthful and convenient.
- Boomers make decisions based on personal beliefs and want to be empowered. They prefer health programs that offer information and options in a learner-involved format, such as a supermarket tour.
- Boomers are constantly pressed for time as they juggle careers, child care, elder care, home responsibilities, and leisure activities. Programs need to be practical and convenient and should be presented in an understandable format.
- Boomers look for value and quality in their investments and are becoming thriftier with age. They seek information on how to relate market choices to value.
- Boomers will not age gracefully. Programs must be upbeat and dynamic for these on-the-go consumers.
- Boomers like nostalgia. Nostalgia can be used to reinforce nutrition messages—for example, modifying traditional family recipes and holiday menus to reflect current nutritional advice.

Nutrition Education Programs

Chapter 17 offers strategies for designing nutrition education programs for adult learners.

Nutrition education strategies aimed at adults are found in both the public and private sectors. They strive to increase nutrition knowledge and skills and to improve eating patterns among adults of all ages. Nutrition education for health promotion is generally based on the Dietary Guidelines for Americans and other food guidance systems. The nutritional goals of these educational tools are to help consumers select diets that provide an appropriate amount of energy to maintain a healthful weight; meet the recommended intakes for all nutrients; are varied in types of fat and moderate in total fat, caloric sweeteners, sodium, cholesterol, and alcohol; and are adequate in complex carbohydrate and fiber.

Public nutrition education programs include the Expanded Food and Nutrition Education Program, described in Chapter 10, and the Food and Drug Administration (FDA) and U.S. Department of Agriculture (USDA) public education campaign on the food label in cooperation with other federal, state, and local agencies.[11] For example, the Michigan State University Cooperative Extension offers a workbook for culturally diverse low-income consumers on how to use the food label to manage fat intake.

A primary challenge facing nutrition educators is to improve nutrition education strategies to reduce the major risk factors for coronary heart disease and cancer, the leading causes of death among adults. The National Heart, Lung, and Blood Institute initiated the guidelines of the National Cholesterol Education Program (NCEP) to help prevent heart disease by reducing saturated fat, *trans* fat, and cholesterol in the American diet.[12] The National Cancer Institute (NCI) designed its *5 a Day for Better Health* program to increase per capita fruit and vegetable consumption. The NCI's 5 a Day program promotes a simple nutrition message that physicians, nurses, community nutritionists, and other health care professionals can reinforce to their clients: *Eat five or more servings of fruits and vegetables every day for better health.*[13]

A number of trade and professional organizations are likewise directing some of their nutrition education strategies to help adults understand the role of nutrition in health promotion and disease prevention. Brochures, booklets, newsletters, videos, and interactive websites that offer simple ways to trim dietary fat, interpret nutrition labels, implement the Dietary Guidelines, or improve shopping skills are available from the American Dietetic Association, the American Association for Retired Persons, the National Dairy Council, General Mills, and the Produce Marketing Association, among others.[14] For example, the International Food Information Council's *Weight Loss: Finding a Weight Loss Program that Works for You* is a booklet that helps consumers choose safe and effective weight-loss methods.[15]

In many communities, food retailers and food service establishments provide point-of-purchase information and literature to their customers.[16] Other communities offer seminars and grocery store tours to help consumers understand food labels.[17]

Health Promotion Programs

Evidence continues to indicate that adults of all ages need to modify their current eating patterns and other behaviors to reduce the risk of chronic diseases. However, changes in behavior can be very difficult to make. An important characteristic of community nutrition interventions is that they can reach people in many different contexts of their daily lives. Supportive social environments can help individuals change their behavior.[18] For this reason, community- and employer-based programs for health promotion are expanding and are facilitating lifestyle changes. For example, many employers now provide worksite health promotion programs offering classes and activities for smoking cessation, weight loss, and stress management. These programs vary widely—some are simple and inexpensive (distribution of health information pamphlets), whereas others are more complex (comprehensive risk factor screening and intensive follow-up counseling). In general, worksite health promotion efforts can be classified under four main areas:[19]

Chapter 15 discusses popular theories of behavior change and strategies for designing community nutrition interventions.

- *Policies.* Smoking, alcohol and other drugs.
- *Screenings.* Health risk/health status, cancer, high blood pressure, and cholesterol.
- *Information or activities.* Individual counseling, group classes, workshops, lectures, special events, and resource materials such as posters, brochures, pamphlets, and videos. Topics typically covered include cancer, high blood pressure, cholesterol, smoking, exercise and fitness, nutrition, weight control, and stress management.
- *Facilities or services.* Nutrition, physical fitness, alcohol and other drugs, and stress management.

Worksite Health Promotion Programs

As the evidence mounts that worksite health promotion (WHP) cuts costs and produces a healthier workforce, more employers are giving WHP programs greater attention. Much of this newfound respect is probably due to the impressive savings reported by 12 companies in *Fortune* magazine. Here are some of the highlights:

- *Aetna:* Five state-of-the-art centers kept exercisers' health care costs $282 lower than those of nonexercisers.
- *L.L. Bean:* Thanks to a healthy workforce, annual insurance premiums were half the national average.
- *Dow Corporation:* On-the-job injury strains dropped 90 percent.
- *Johnson & Johnson:* Health screening saved $13 million a year in absenteeism and health care costs.
- *Quaker Oats:* Because of an integrated health management approach, health insurance premiums were nearly a third less than the national average.
- *Steelcase:* Personal health counselors motivated high-risk employees to reduce major risk factors, generating an estimated $20 million in savings over 10 years.
- *Union Pacific:* Reduced hypertension and smoking saves more than $3 million a year.

Source: D. H. Chenoweth, *Worksite Health Promotion* (Champaign, IL: Human Kinetics, 1998), p. 12. © D. H. Chenoweth.

The national objectives for improving the health of the nation include the target that by the year 2010, at least 75 percent of all worksites with 50 or more employees will offer a comprehensive employee health promotion program to their employees.

According to the U.S. Department of Health and Human Services (DHHS), two out of three worksites with at least 50 employees offer some form of health promotion programming.[20] Successful nutrition promotion programs range from introducing heart-healthy menus into company cafeterias to reducing blood cholesterol levels through screening and intervention.[21] At the world headquarters for Coca-Cola in Atlanta, Georgia, the *HealthWorks Program* focuses on comprehensive health promotion. The nutrition theme is carried through group weight-loss classes, cooking demonstrations and taste tests, and individualized diet instruction available to all employees, spouses, children, and retirees.[22]

Current efforts to help young adults identify their familial risk factors for chronic disease conditions and programs that tout the benefits of lifelong healthful eating and regular physical activity should enable the older adults of tomorrow to enjoy a productive and satisfying life well into advanced age.

Aging and Nutritional Status

Growing old is often associated with frailty, sickness, and a loss of vitality. Although the aging in our society do experience chronic illness and associated disabilities, this population is very heterogeneous: Older people vary greatly in their social, economic, and lifestyle situations, functional capacity, and physical condition.[23] Each person ages at a different rate, sometimes making chronological age different from biological age. Most older persons live at home, are fully independent, and have lives of good quality.[24] Only 4.5 percent of older adults live in nursing homes.

TABLE 12–3 The Effects of Aging on Biological Function

ORGAN SYSTEM	AGING EFFECT
Skin	Dryness, wrinkling, mottled pigmentation, loss of elasticity, dilation of capillaries
Head and neck	Macular degeneration, hearing loss
Cardiovascular	Thickening heart wall and valves, decreased cardiac output, increased collagen rigidity, decreased elasticity of blood vessels with calcification
Pulmonary	Stiffening of tissue, decreased vital capacity, decreased maximum oxygen consumption, decreased breathing capacity
Renal	Decreased size, decreased GFR*, decreased renal blood flow, decreased renal concentrating ability
Endocrine	Altered circulating hormone levels and actions
Gastrointestinal (GI tract)	Altered perception of taste and smell, altered GI motility, decreased muscle strength, decreased digestive secretions, decreased absorption (calcium, iron, vitamin B_{12}, vitamin D)
Nervous	Decreased sensory perception, decreased muscle response to stimuli, decreased cognition and memory, loss of brain cells
Musculoskeletal	Progressive loss of skeletal muscle, degeneration of joints, decalcification of bone

*GFR = Glomerular filtration rate.

Source: Adapted from G. L. Jensen, M. McGee, and J. Binkley, Nutrition in the elderly, *Gastroenterol Clin North Am.* 2001; 30: 314, and American Dietetic Association, *Nutrition Care of the Older Adult*, 2nd edition (Chicago: American Dietetic Association, 2004), p. 65.

Primary Nutrition-Related Problems of Aging

Although aging is not completely understood, we know that it involves progressive changes in every body tissue and organ: the brain, heart, lungs, digestive tract, and bones (see Table 12-3). After age 35, functional capacity declines in almost every organ system. Such changes affect nutritional status. Some, including oral problems, interfere with nutrient intake; others affect absorption, storage, and utilization of nutrients; and still others increase the excretion of, and need for, specific nutrients. Examples of various conditions associated with aging that can affect nutritional status include sensory impairments, altered endocrine, gastrointestinal, and cardiovascular functions, and changes in the renal and musculoskeletal systems. Both genetic and environmental factors contribute to these declines.[25] Many of the changes are inevitable, but a healthful lifestyle that combines moderation with adequate intakes of all essential nutrients can forestall degeneration and improve the quality of life into the later years.

To meet nutritional needs, the diets of older adults should include:

- At least 3 servings of low-fat dairy products or suitable alternatives.
- 2 or more servings from the meat group that feature dried beans, fish, eggs, and lean cuts of meat and poultry.
- At least 5 servings of fruits and vegetables featuring foods that are richly colored (dark green, orange, red, or yellow).
- At least 6 servings of nutrient-dense, fiber-rich whole grains and fortified cereals.
- At least 8 glasses of water per day to prevent dehydration.

Older adults face the challenge of choosing a nutrient-dense diet. Although caloric needs may decrease with age, the need for certain nutrients such as calcium, vitamin D, vitamin C, vitamin B_{12}, and vitamin B_6 may actually increase with the effects of aging.[26] For example, as many as 30 percent of persons older than 50 years may experience reduced stomach acidity, which can interfere with their ability to absorb vitamin B_{12}, calcium, and iron effectively from foods. In addition, there may be increased needs for vitamin D in older adults who have low intakes of fortified dairy products and have less ability to make the vitamin when their skin is exposed to sunlight. Assessing the diet quality of older adults is important for identifying issues relevant to their health and nutritional status. According to the Healthy Eating Index, a composite measure of overall

Programs in Action

Worksite Wellness Works for Firefighters

The ecological perspective on health promotion consists of five layers of factors: intrapersonal, interpersonal, institutional, community, and public policy.[1] Police and fire stations are both institutions and communities for their employees, officers, and firefighters. Worksite nutrition education in this type of environment has been shown to successfully improve nutrition status and eating behaviors.[2] Employer commitment to a worksite education program is said to be important to the success of worksite health programs.[3] The "5 a Day Nutrition Education Program" offers an example of a successful program that was endorsed by a municipal fire department and targeted to firefighters.

Goals and Objectives

The goal of the 5 a Day Nutrition Education Program was to develop and implement a comprehensive nutrition education program to improve the nutrition behaviors and fruit and vegetable intake of firefighters in the City of St. Paul, Minnesota. Program objectives included determining the level of fruit and vegetable intake among St. Paul firefighters, determining readiness to change as it pertained to fruit and vegetable consumption, identifying barriers to increasing intake, increasing awareness of the health benefits of eating fruits and vegetables, increasing firefighter confidence on how to add fruits and vegetables to their diet, and providing a supportive environment.

Target Audience

The target audience was 340 firefighters, predominantly male, between the ages of 25 and 50, employed by the City of St. Paul. The firefighters were stationed in companies of 4 to 12 firefighters at 18 stations in and around St. Paul.

Rationale for the Intervention

In 1999, the City of St. Paul's Risk Management Division, in partnership with their managed-care carrier (HealthPartners), the Fire and Safety Services Department, and the firefighters union Local 21, joined together to create a health promotion program to improve the health of the firefighters. HealthPartners provided a medical claim analysis of the firefighter population, which identified coronary heart disease as the number-one disease with preventable costs. Poor nutrition and physical inactivity, two modifiable risk factors of interest to representatives of the firefighters, were chosen as initial areas for health improvement.

A preliminary focus group revealed that firefighters did not perceive that they were at risk for heart disease, that they were resistant to reducing meat portions, that they were unfamiliar with the health benefits of fruits and vegetables, and that they would be willing to eat more available and affordable fruit.

Methodology

The 5 a Day Nutrition Education Program included three components: awareness building, behavior, and environmental interventions. To build awareness, a "Try for 5"

dietary quality based on the Dietary Guidelines for Americans and the Food Guide Pyramid, only 15 percent of the population aged 45 to 64 years consumes a "good" diet, and it is the dairy and fruit groups that need the most improvement.[27] Figure 12-3 on page 392 summarizes the overall dietary quality of independent, free-living middle-aged and elderly adults. General nutrition guidelines for older adults are listed in the margin on page 389.[28] Clearly, an adequate intake of nutrients and fiber from a variety of foods throughout life, together with moderate intakes of calories and fat and regular physical activity, helps immensely to promote good health in the later years.

As a person gets older, the chances of suffering a chronic illness or functional impairment are greater. In the United States, approximately 80 percent of all persons 65 years of age and older have at least one chronic condition, and 50 percent have at least two chronic conditions.[29] Among the diseases that befall some people in later life are heart disease, hypertension, cancer, diverticulosis, osteoporosis, dementia, diabetes, and gum disease. Almost 50 percent of people over age 65 have high blood pressure, and approximately 30 percent have heart disease. Other chronic conditions contributing to **disability** include arthritis, strokes, and disorders of vision and hearing (see Table 12-4 on page 393).

Disability Any restriction on or impairment in performing an activity in the manner or within the range considered normal for a human being.

calendar, which included tips on adding fruits and vegetables, was sent to the firefighters' homes. A directory of local farmers' markets was included in this mailing. Prizes were offered to firefighters who tried at least five tips. Additionally, colorful posters were hung in all participating fire stations. Eighteen 1-hour interactive nutrition workshops offered to all firefighters constituted the behavior phase. In these workshops, firefighters prepared and tasted recipes and competed in teams for prizes. A supportive environment was created by placing free fruit baskets in each participating station after completion of the workshops and by providing a coupon to be used to purchase fruits and vegetables to refill the baskets.

Results

The "Try for 5" home mailing was sent to 340 firefighters. Over a period of 3 days, 275 firefighters attended the 5 a Day nutrition workshops. Most (84 percent) of the participating firefighters stated that they planned to share the workshop information with a friend or family member. A majority (85 percent) said that they intended to eat more fruits and vegetables as a result of participating in the workshop. Nearly all (97–99 percent) of the participants said that they understood the benefits of eating fruits and vegetables, could identify ways to eat more fruits and vegetables, and had the confidence to add fruits and vegetables to their diet. The participants rated the workshop's interactive games and recipe preparation segments as good or excellent. At the beginning of the 5 a Day program, firefighters reported eating an average of 4.4 servings of fruits and vegetables each day. Six months later this had increased to 4.8 servings each day.

Lessons Learned

The success of this program was a result of the collaboration among and support of the many key stakeholders in the program. Resources and expertise contributed by multiple partners reduced the burden of development costs and time on any single partner. Shared vision, interest, expertise, and resources can positively affect employee health.

References

1. K. R. McLeroy and coauthors, An ecological perspective on health promotion programs, *Health Education Quarterly* 15 (1998): 355–77.
2. M. E. Briley, D. H. Montgomery, and J. Blewett, Worksite nutrition education can lower total cholesterol levels and promote weight loss among police department employees, *Journal of the American Dietetic Association* 92 (1992): 1382–84.
3. P. R. McCarthy and coauthors, What works best for worksite cholesterol education? Answers from targeted focus groups, *Journal of the American Dietetic Association* 92 (1992): 978–81.

Source: Adapted from *Community Nutritionary* (White Plains, NY: Dannon Institute, Spring 2001). Used with permission. For more information about the *Awards for Excellence in Community Nutrition,* go to www.dannon-institute.org.

Dementia (especially Alzheimer's disease) is a major contributor to disability and placement in nursing homes for those over age 75.[30] Malnutrition can occur secondary to these conditions, as noted in Table 12-5 on page 394. Many of these conditions require special diets that can further compromise nutritional status in the older adult. Also, there are differences in disease prevalence among racial and ethnic groups. For this reason, nutrition interventions designed to reduce disease risks must be sensitive to ethnic or cultural differences and preferences. Minority-group elderly are more likely to have malnutrition secondary to chronic diseases such as heart disease, renal disease, diabetes mellitus, obesity, and certain cancers. Many times, they have less access to health care and have decreased quality of life and increased mortality.[31]

Polypharmacy, or the use of multiple drugs, is problematic for many older adults. The average older person fills more than 30 prescriptions a year and often takes 3 or more drugs at a time.[32] Cardiac drugs are most widely used by the elderly, followed by drugs to treat arthritis, psychic disorders, and respiratory and gastrointestinal conditions. Many older adults also use over-the-counter medications such as antacids, as well as nutritional herbal remedies and supplements.[33] Long-term use of a variety of medications and supplements

Polypharmacy The taking of three or more medications regularly; occurs in one-third of those over 65 years of age.

FIGURE 12-3 Dietary Quality Ratings of Persons Age 45 or Older, by Age Group and Poverty Status*

Source: Healthy People 2010 Progress Review: Nutrition and Overweight, January 21, 2004.

*Dietary quality was measured by the Healthy Eating Index (HEI). The HEI is a summary measure of people's overall diet quality. The HEI is expressed as one score on a scale of 1 to 100 but consists of the sum of ten components. Each component score can range from 0 to 10. Components 1–5 measure the degree to which a person's diet conforms to the recommended number of servings for the five major food groups: Grains, Vegetables, Fruits, Milk, and Meat. A high score for these components is reached by maximizing consumption of recommended amounts. Components 6–9 measure compliance of total fat and saturated fat intake with the *Dietary Guidelines for Americans* and of cholesterol and sodium with the *Daily Values* listed on the Nutrition Facts label. A high score is reached by consuming at or below recommended amounts. The last component evaluates variety in the diet. A person consuming eight or more different foods each day will score 10 points. The HEI is available at www.usda.gov/cnpp.

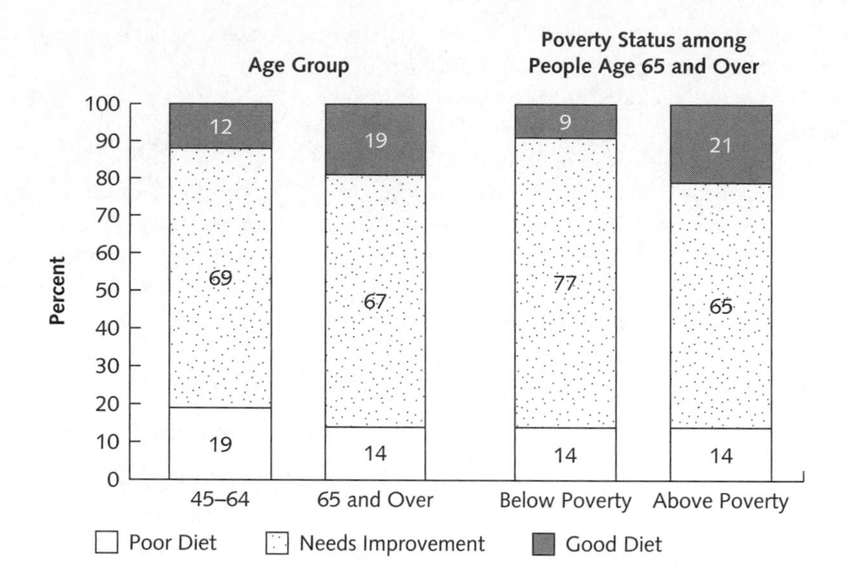

increases the risk of drug–nutrient and drug–drug interactions. Individuals with impaired nutritional status and poor dietary intakes are at the highest risk.

Individually or in combination, the social, economic, psychological, cultural, and environmental factors associated with aging may interact with the physiological changes and further affect nutrition status in older adults. These factors are listed in the margin on page 393.

Nutrition Policy Recommendations for Health Promotion for Older Adults

As we noted in Chapter 1, efforts at health promotion and disease prevention are conducted at several levels and are designed to help adults of all ages change their eating patterns and other behaviors to reduce the risk of chronic disease. Optimal nutrition contributes to healthy aging. As primary prevention, nutrition helps promote health and affects the quality of life in older adults. As secondary prevention, nutrition can lessen risks from chronic diseases; as tertiary prevention, medical nutrition therapy slows disease progression and reduces symptoms.[34] The American Dietetic Association (ADA) has designated aging as a priority area for its current public policy advocacy program and supports efforts to help older adults better manage their health. Several major nutrition policy recommendations for promoting the health of older adults are related to dietary guidance and nutrition services.[35]

1. Dietary guidance for older adults should:
   - Target messages to the special concerns of this population.
   - Emphasize the need for balance between food intake to both meet nutrient needs and maintain a healthful weight.
   - Promote the consumption of more fruits, vegetables, and whole-grain products, as well as the choice of lean meats and low-fat or fat-free dairy foods.
   - Address drug–nutrient interactions.
   - Become integrated into the training of physicians, dietitians, and other health professionals.
2. Nutrition services should be:
   - Structured to include nutrition assessment and guidance.

TABLE 12-4 Distribution of Most Common Chronic Conditions That Cause Disability in Older Persons

CHRONIC CONDITIONS	TOTAL PERCENT
Hypertension	49.2
Hearing impairment	38.5
Arthritis	36.1
Heart disease	31.1
Vision disease	22.9
Any cancer	20.0
Visual impairment	18.0
Diabetes	16.0
Orthopedic impairment	15.8
Chronic sinusitis	15.1
Cerebrovascular disease	9.0
Asthma	8.0
Emphysema/chronic bronchitis	6.0

Source: National Center for Health Statistics (2003).

- Incorporated into institution-, community-, or home-based health care programs for older adults.
- Tailored to the needs of older persons who are homebound, live in isolation, or are chronically ill.
- Provided by credentialed nutrition professionals.

3. Food manufacturers should:
   - Develop easy-to-prepare, tasty food products that are nutrient dense.
   - Use food labels set in larger type that provide sufficient information to guide older consumers.
4. The federal government should:
   - Disseminate information about model programs that have successfully delivered nutrition education and services to older adults.
   - Require its sponsored programs to address the calorie and nutrient needs of older participants.
   - Promote research and surveillance programs to improve nutritional status.
   - Develop a system of appropriate funding for nutrition services delivered to older adults.

The ADA recommends that nutrition services (including nutrition assessment and monitoring), therapeutic interventions as needed, and nutrition counseling and education be included throughout the continuum of health care services for older adults.[36] A discussion of the guidelines and tools for nutrition screening and assessment of older adults follows in the next three sections.

RISK FACTORS INFLUENCING THE NUTRITIONAL STATUS OF OLDER ADULTS

Physiological
- Dietary intake
- Lack of appetite
- Inactivity/immobility
- Poor taste and smell
- Alcohol or drug abuse
- Chronic disease
- Polypharmacy
- Physical disability
- Oral health problems

Psychological
- Loneliness
- Dementia
- Cognitive impairment
- Depression
- Loss of spouse
- Social isolation

Environmental
- Inadequate housing
- Inadequate cooking facilities
- Lack of transportation

Socioeconomic
- Cultural beliefs
- Poverty
- Limited education
- Literacy level
- Limited access to health care
- Institutionalization

Evaluation of Nutritional Status

Up to one-quarter of all elderly patients and one-half of all hospitalized elderly may be suffering from malnutrition.[37] About 3.6 million elderly adults (10.4 percent) had incomes below the poverty threshold in 2002; another 2.2 million were classified as "near-poor,"

TABLE 12-5 Malnutrition That Is Secondary to Disease, Physiologic State, or Medication Use

DISEASE OR CONDITION	EFFECTS ON NUTRITIONAL STATUS
Atherosclerosis	May increase difficulties in regulating fluid balances if caused by congestive heart failure. If the individual is incapacitated, energy needs decrease.
Cancer	Weight loss, lack of appetite, and secondary malnutrition are common.
Dental and oral disease	May alter the ability to chew and thus reduce dietary intake. Increased likelihood of choking and aspiration.
Depression and dementia	Increased or decreased food intakes are common. A person with dementia may have decreased ability to get food, or the appetite may be very small or very great. Judgment and balance in meal planning are generally absent.
Diabetes mellitus (Type 1)	If untreated, increased risk of undernutrition results; increased risk of other diet-related diseases, such as hyperlipidemia.
Diabetes mellitus (Type 2)	Increased risk of other diet-related diseases such as hyperlipidemia; weight loss is needed if obesity is present.
End-stage kidney disease	Alters fluid and electrolyte needs; uremia may alter appetite and increase risk of malnutrition.
Gastrointestinal disorders	Increased risk of malabsorption of nutrients and consequent undernutrition.
High blood pressure	Hyper- or hypokalemia can be increased by dietary means; weight gain may exacerbate high blood pressure.
Osteoarthritis	Makes motion difficult, including those activities related to meal preparation. Predisposes people to a sedentary lifestyle and may give rise to obesity.
Osteoporosis	Limits the ability to purchase and prepare food if mobility is affected. If severe scoliosis is present, the appetite may be altered.
Smoking	Smoking may alter weight status. Alters serum levels of some nutrients such as ascorbic acid and carotenes. Chronic smoking gives rise to emphysema and chronic obstructive pulmonary diseases (COPD), which make it difficult to eat owing to breathing problems.
Stroke	May alter abilities in the cognitive and motor realms related to food and eating. If the individual is incapacitated, his or her energy needs decrease.

Source: Adapted from Institute of Medicine. *The Second Fifty Years: Promoting Health and Preventing Disability* (Washington, D.C.: National Academy Press, 1992), pp. 168–69.

Activities of daily living (ADLs) Bathing, dressing, grooming, transferring from bed or chair, going to the bathroom (toileting), and feeding oneself.

Instrumental activities of daily living (IADLs) Food preparation, use of telephone, housekeeping, laundry, use of transportation, responsibility for medication, managing money, and shopping.

with incomes up to 125 percent of the poverty threshold. In addition, a national survey found that about 30 percent of all noninstitutionalized people over the age of 65 live alone, 45 percent take multiple prescription drugs that can interfere with appetite and nutrient absorption, 27 percent have difficulty in performing one or more **activities of daily living (ADLs),** and an additional 13 percent have difficulties with **instrumental activities of daily living** (for example, preparing meals and taking medications)—all factors that place older persons at nutritional risk. Older persons who have problems with the activities of daily living are known as the frail elderly. Because they depend on others to perform these essential activities, they are likely to be at risk for malnutrition.[38] Identifying older adults at nutritional risk is an important first step in maintaining quality of life and functional status.

TABLE 12-6 Major Indicators of Poor Nutritional Status in Older Adults

- Significant weight loss over time:
  5.0% or more of prior body weight in 1 month.
  7.5% or more of body weight in 3 months.
  10.0% or more of body weight in 6 months or involuntary weight loss.
- Significantly low or high weight for height: 20% below or above desirable weight for height.
- Significant reduction in serum albumin: serum albumin of less than 3.5 g/dL.
- Significant change in functional status: change from "independent" to "dependent" in two of the ADLs or one of the nutrition-related IADLs.*
- Significant and sustained inappropriate food intake:
  Failure to consume the recommended minimum from one or more food groups, or a sufficient variety of foods for a period of 3 months or more. Excessive consumption of fat, saturated fat, and/or alcohol.
- Significant reduction in mid-arm circumference: to less than 10th percentile (NHANES standards).
- Significant increase or decrease in triceps skinfolds: to less than 10th percentile or more than 95th percentile (NHANES standards).
- Significant obesity: body mass index over 30 or triceps skinfolds above 95th percentile (NHANES standards).
- Other nutrition-related disorders: presence of osteoporosis, folic acid or vitamin B_{12} deficiency.

*IADLs are instrumental activities of daily living (e.g., meal preparation and financial management).

Source: The Nutrition Screening Initiative, a project of the American Academy of Family Physicians, The American Dietetic Association, and the National Council on Aging, Inc.

NUTRITION SCREENING

The American Dietetic Association, the American Academy of Family Physicians, and the National Council on Aging have collaborated since 1990 in an effort to promote nutrition screening and early intervention as part of routine health care. Its focus is on the elderly, one of the largest population groups in the United States at risk of poor nutrition.

The Nutrition Screening Initiative has identified a number of specific risk factors and indicators of poor nutrition status in older adults (see Table 12-6). Some of the risk factors shown in the margin on page 393 increase the risks for dietary inadequacy, excess, or imbalance and involve social, economic, and psychological factors rather than physical health problems. Others indicate risks of malnutrition secondary to disease rather than those caused primarily by lack of food.[39]

A key premise of the Nutrition Screening Initiative is that nutritional status is a "vital sign"—as vital to health assessment as blood pressure and pulse rate.[40] The Nutrition Screening Initiative has developed a 10-question self-assessment "checklist" that can be distributed at any office or agency for the elderly (see Figure 12-4). Individuals identify factors placing them at risk to arrive at their score. This checklist addresses disease, eating status, tooth loss or mouth pain, economic hardship, reduced social contact, multiple medications, involuntary weight loss or gain, and need for assistance with self-care. The word DETERMINE is used as a mnemonic device with the checklist and helps provide basic nutrition information (see Figure 12-5). Each letter in DETERMINE stands for a risk factor.[41] As Figure 12-6 on page 398 shows, people identified as being at risk should be followed up with more in-depth screening and assessment of nutritional status (Level I or Level II screens) by a health professional (see Table 12-7 on page 399).

The Level I screen is a basic nutrition screen designed for social service and health professionals to use to identify older adults who may need medical attention or nutrition services. This screen includes determinations of height and weight and provides

FIGURE 12-4 Checklist to Determine Your Nutritional Health

Source: Reprinted with permission by the Nutrition Screening Initiative, a project of the American Academy of Family Physicians, The American Dietetic Association, and the National Council on Aging, Inc., and funded in part by a grant from Ross Laboratories, a division of Abbott Laboratories.

The Warning Signs of poor nutritional health are often overlooked. Use this checklist to find out if you or someone you know is at nutritional risk.

DETERMINE YOUR NUTRITIONAL HEALTH

Read the statements below. Circle the number in the yes column for those that apply to you or someone you know. For each yes answer, score the number in the box. Total your nutritional score.

	YES
I have an illness or condition that made me change the kind and/or amount of food I eat.	2
I eat fewer than 2 meals per day.	3
I eat few fruits or vegetables, or milk products.	2
I have 3 or more drinks of beer, liquor or wine almost every day.	2
I have tooth or mouth problems that make it hard for me to eat.	2
I don't always have enough money to buy the food I need.	4
I eat alone most of the time.	1
I take 3 or more different prescribed or over-the-counter drugs a day.	1
Without wanting to, I have lost or gained 10 pounds in the last 6 months.	2
I am not always physically able to shop, cook and/or feed myself.	2
TOTAL	

Total Your Nutritional Score. If it's —

0-2 **Good!** Recheck your nutritional score in 6 months.

3-5 **You are at moderate nutritional risk.** See what can be done to improve your eating habits and lifestyle. Your office on aging, senior nutrition program, senior citizens center or health department can help. Recheck your nutritional score in 3 months.

6 or more **You are at high nutritional risk.** Bring this checklist the next time you see your doctor, dietitian or other qualified health or social service professional. Talk with them about any problems you may have. Ask for help to improve your nutritional health.

These materials developed and distributed by the Nutrition Screening Initiative, a project of:

AMERICAN ACADEMY OF FAMILY PHYSICIANS

THE AMERICAN DIETETIC ASSOCIATION

NATIONAL COUNCIL ON THE AGING, INC.

Remember that warning signs suggest risk, but do not represent diagnosis of any condition. Turn the page to learn more about the Warnings Signs of poor nutritional health.

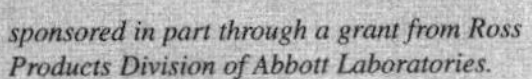

sponsored in part through a grant from Ross Products Division of Abbott Laboratories.

The Nutrition Screening Initiative • 1010 Wisconsin Avenue, NW • Suite 800 • Washington, DC 20007

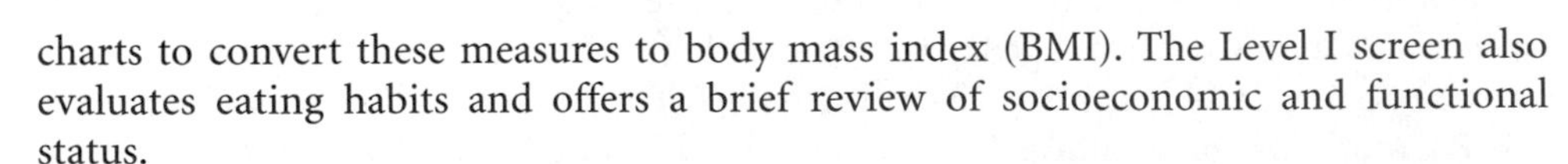

charts to convert these measures to body mass index (BMI). The Level I screen also evaluates eating habits and offers a brief review of socioeconomic and functional status.

The Level II screen provides more specific diagnostic information on nutritional status. It is designed for health and medical professionals to use with older adults who have a potentially serious medical or nutritional problem. This in-depth screening tool focuses on the components of nutrition assessment: anthropometric indicators (for instance, weight, height, and body composition), clinical indicators (such as oral health and general physical exam), biochemical indicators (for example, serum albumin, serum cholesterol, hemoglobin, and blood glucose), and dietary indicators (including dietary history). The Level II screen also assesses chronic medication use and the living environment of the individual (assistance, facilities, support systems, safety), cognitive status, emotional status, and functional status.[42]

The Nutrition Checklist is based on the Warning Signs described below. Use the word DETERMINE to remind you of the Warning Signs.

DISEASE
Any disease, illness or chronic condition which causes you to change the way you eat, or makes it hard for you to eat, puts your nutritional health at risk. Four out of five adults have chronic diseases that are affected by diet. Confusion or memory loss that keeps getting worse is estimated to affect one out of five or more of older adults. This can make it hard to remember what, when or if you've eaten. Feeling sad or depressed, which happens to about one in eight older adults, can cause big changes in appetite, digestion, energy level, weight and well-being.

EATING POORLY
Eating too little and eating too much both lead to poor health. Eating the same foods day after day or not eating fruit, vegetables, and milk products daily will also cause poor nutritional health. One in five adults skip meals daily. Only 13% of adults eat the minimum amount of fruit and vegetables needed. One in four older adults drink too much alcohol. Many health problems become worse if you drink more than one or two alcoholic beverages per day.

TOOTH LOSS/ MOUTH PAIN
A healthy mouth, teeth and gums are needed to eat. Missing, loose or rotten teeth or dentures which don't fit well or cause mouth sores make it hard to eat.

ECONOMIC HARDSHIP
As many as 40% of older Americans have incomes of less than $6,000 per year. Having less--or choosing to spend less--than $25-30 per week for food makes it very hard to get the foods you need to stay healthy.

REDUCED SOCIAL CONTACT
One-third of all older people live alone. Being with people daily has a positive effect on morale, well-being and eating.

MULTIPLE MEDICINES
Many older Americans must take medicines for health problems. Almost half of older Americans take multiple medicines daily. Growing old may change the way we respond to drugs. The more medicines you take, the greater the chance for side effects such as increased or decreased appetite, change in taste, constipation, weakness, drowsiness, diarrhea, nausea, and others. Vitamins or minerals when taken in large doses act like drugs and can cause harm. Alert your doctor to everything you take.

INVOLUNTARY WEIGHT LOSS/GAIN
Losing or gaining a lot of weight when you are not trying to do so is an important warning sign that must not be ignored. Being overweight or underweight also increases your chance of poor health.

NEEDS ASSISTANCE IN SELF CARE
Although most older people are able to eat, one of every five have trouble walking, shopping, buying and cooking food, especially as they get older.

ELDER YEARS ABOVE AGE 80
Most older people lead full and productive lives. But as age increases, risk of frailty and health problems increase. Checking your nutritional health regularly makes good sense.

The Nutrition Screening Initiative • 1010 Wisconsin Avenue, NW • Suite 800 • Washington, DC 20007
The Nutrition Screening Initiative is funded in part by a grant from Ross Products Division of Abbott Laboratories.

FIGURE 12-5 The Warning Signals Nutrition Checklist

Source: Reprinted with permission by the Nutrition Screening Initiative, a project of the American Academy of Family Physicians, The American Dietetic Association, and the National Council on Aging, Inc., and funded in part by a grant from Ross Laboratories, a division of Abbott Laboratories.

NUTRITION ASSESSMENT

Periodic nutrition assessment is useful for identifying and tracking elderly persons at nutritional risk. The components of nutrition assessment for the elderly are listed in Table 12-8 on page 400, along with indicators to determine risk. A geriatric nutrition assessment includes the elements discussed in the list that follows:[43]

- *Anthropometric measures.* Height, weight, and skinfold measures are affected by aging. Height decreases over time because of changes in the integrity of the skeletal system as a result of bone loss. Measurements of height are sometimes difficult to obtain as a consequence of poor posture or the inability to stand erect unassisted. In such cases, a recumbent anthropometric measure such as knee-to-heel height can be used instead.[44]
- *Clinical assessment.* The clinical assessment should evaluate the condition of hair, skin, nails, musculature, eyes, mucosa, and other physical attributes. An oral examination

FIGURE 12-6 A Practical Approach to Nutrition Screening

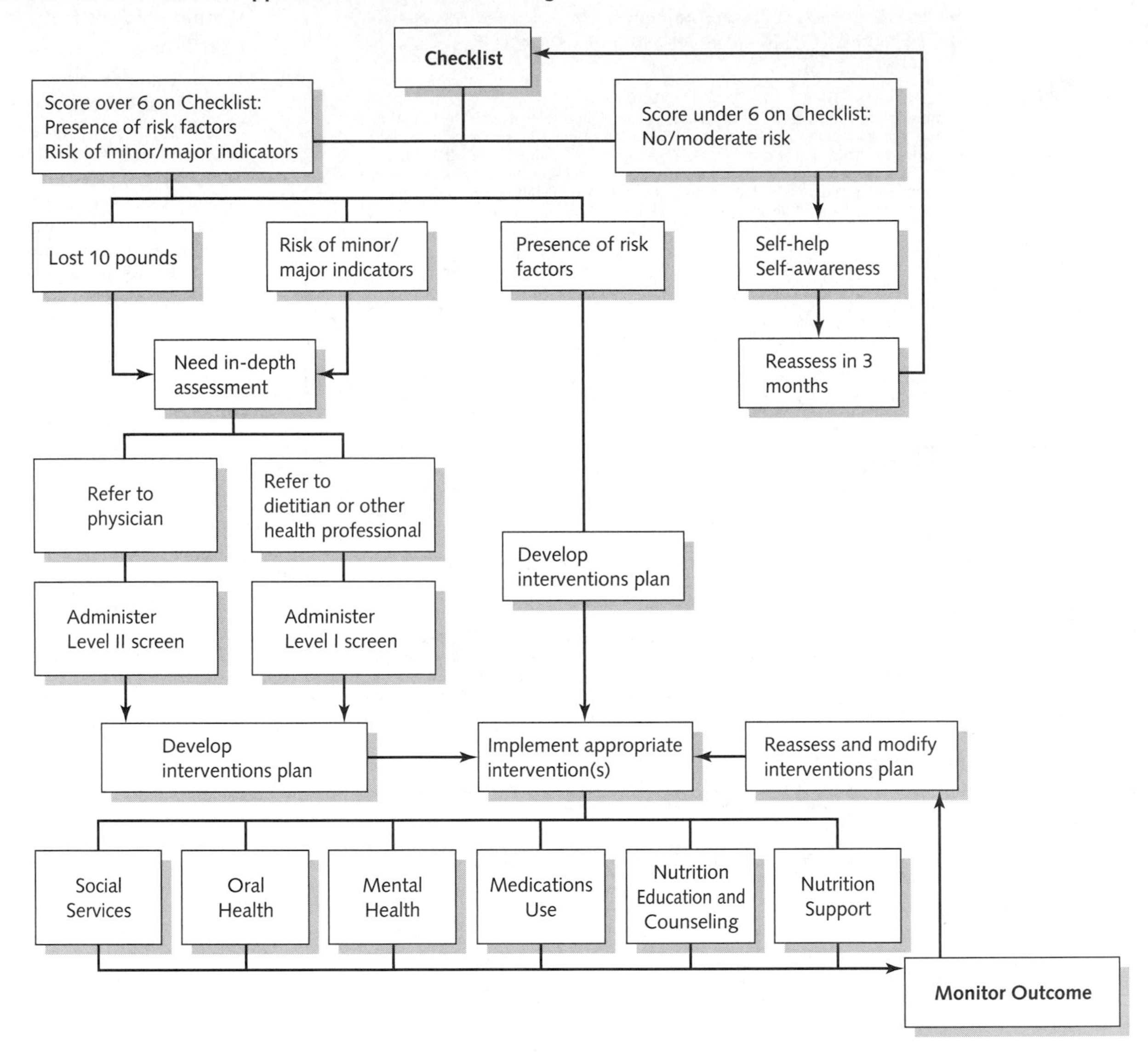

Risk Factors

- Inappropriate food intake
- Poverty
- Social isolation
- Dependence/disability
- Acute/chronic diseases or conditions
- Chronic medications use
- Advanced age (80+)

Major Indicators

- Weight loss of 10 lb or more
- Under-/overweight
- Serum albumin below 3.5 g/dL
- Change in functional status
- Inappropriate food intake
- Mid-arm muscle circumference <10th percentile
- Triceps skinfold <10th percentile or >95th percentile
- Obesity
- Nutrition-related disorders
  - Osteoporosis
  - Osteomalacia
  - Folate deficiency
  - B_{12} deficiency

Minor Indicators

- Alcoholism
- Cognitive impairment
- Chronic renal insufficiency
- Multiple concurrent medications
- Malabsorption syndromes
- Anorexia, nausea, dysphagia
- Change in bowel habits
- Fatigue, apathy, memory loss
- Poor oral/dental status
- Dehydration
- Poorly healing wounds
- Loss of subcutaneous fat or muscle mass
- Fluid retention
- Reduced iron, ascorbic acid, zinc

Source: Reprinted with permission from Nutrition Screening Initiative, Incorporating Nutrition Screening and Interventions into Medical Practice, 1994, p. 28.

TABLE 12-7 Nutrition Screening Initiative: Level I and Level II Screens

	LEVEL I SCREEN	LEVEL II SCREEN
Primary User	Social workers and health care professionals	Physicians and other qualified health care professionals
Data Evaluation	Height Weight Dietary data Daily food intake Living environment Functional status	Height Weight Dietary data Daily food intake Living environment Functional status Laboratory and anthropometric data Clinical features Mental/cognitive status Medication use

Source: Adapted from Nutrition Screening Initiative, *Nutrition Screening Manual for Professionals Caring for Older Americans* (Washington, D.C.: Nutrition Screening Initiative, 1991).

is useful for determining the condition of the mouth and teeth, the need for dentures or the condition of existing dentures, and oral lesions.[45] An assessment of the client's ability to chew, swallow, and self-feed is recommended.

- *Biochemical assessment.* Biochemical parameters are affected by the aging process, as well as by polypharmacy, chronic disease, and hydration status. However, serial measures of blood parameters can be useful in evaluating nutritional risk. Serum albumin is generally used to assess visceral protein status in the elderly. Low serum albumin levels are associated with increased morbidity and mortality in the elderly. Measurements of serum cholesterol, hemoglobin, blood glucose, and antigen-recall skin tests are also included.
- *Dietary assessment.* A detailed record of current food consumption and a history of changes in eating habits over time are needed to assess diet adequacy. The evaluation should detect persons who avoid certain food groups, adhere to unusual dietary practices, or consume excessive or insufficient amounts of essential nutrients. The adequacy of fluid intake should be assessed as well.
- *Functional assessment.* Functional assessment measures changes in the basic functions necessary to maintain independent living. Activities of daily living (ADLs) are self-care activities (such as bathing, dressing, and feeding). Instrumental ADLs, or IADLs, which require a higher level of functioning, include activities such as meal preparation, financial management, and housekeeping. Many of these activities (for example, shopping, cooking, and self-feeding) are closely related to adequate nutrition status.
- *Medication assessment.* Note the types and doses of various prescription and over-the-counter drugs. Evaluate the individual's drug intake for possible nutrient–drug interactions that could affect the absorption and metabolism of, and requirements for, specific nutrients. Also identify any drugs that may depress the appetite or alter the perception of taste.
- *Social assessment.* Financial resources, living arrangements, and social support network, including availability of caregivers, should be evaluated as part of the nutrition assessment because these factors can directly affect a person's nutritional status. Poverty (annual income of less than $9,570 per person in 2005) and social isolation particularly impair the nutritional status of many older adults, as noted perceptively by a professor of psychiatry:

> It is not what the older person eats but with whom that will be the deciding factor in proper care for him. The oft-repeated complaint of the older patient that he has little incentive to prepare food for only himself is not merely a statement of fact but also a rebuke to the

TABLE 12-8 Nutrition Assessment for Older Adults

NUTRITION ASSESSMENT COMPONENT	INDICATOR OF RISK
Anthropometric Measurements	
Weight Height	Measured weight-for-height 80% or below or 120% or above midpoint of medium frame (Metropolitan Life Tables) Reported usual body weight more than 5% less than actual weight Involuntary weight loss or gain of more than 10 lb in past 6 months
Body mass index (BMI)	Underweight: BMI < 18 Overweight: BMI ≥ 25
Body composition	Mid-arm muscle circumference 20% or more below NHANES standards Triceps skinfold 40% or less of, or 190% or more above, NHANES standards Waist circumference > 40" for men > 35" for women
Clinical Assessment	
Individual/family medical history	Use of tobacco, alcohol, drugs Chronic illness or disability High blood pressure > 140/90 mm Hg
Oral cavity exam	Lesions in mouth, bleeding gums, mobile teeth, ill-fitting dentures, need for dentures
Physical examination	Signs and symptoms of possible diet-related problems
Biochemical Assessment	
Total blood cholesterol	Elevated (high risk): ≥ 240 mg/dL > 200 mg/dL plus risk factors
LDL-C	Elevated: > 130 mg/dL
HDL-C	Low: < 40 mg/dL
Hemoglobin	Risk of anemia: Males: 45–64 yr: 13.2 g/dL; 65+ yr: 13.6 g/dL Females: 45–64 yr: 11.8 g/dL; 65+ yr: 11.9 g/dL
Serum albumin	Low: < 3.5 g/dL
Blood glucose	Outside normal range of 70–110 mg/100 ml
Dietary Assessment	
Food diary (3 or more days) Food frequency 24-hour recall	Deficient or excessive intakes of calories and/or nutrients
Functional Assessment	
Activities of daily living (ADLs) Instrumental activities of daily living (IADLs)	A change from independence to dependence (needs assistance most of the time) with 2 ADLs or 1 nutrition-related IADL
Medication Assessment	Use of 3 or more prescribed medications; daily use of over-the-counter medications (e.g., aspirin, antacids) and/or nutritional or herbal supplements; nutrition quackery; drug–nutrient interactions

Source: Adapted with permission by the Nutrition Screening initiative, a project of the American Academy of Family Physicians, The American Dietetic Association, and the National Council on Aging, Inc. and funded in part by a grant from Ross Laboratories, a division of Abbott Laboratories.

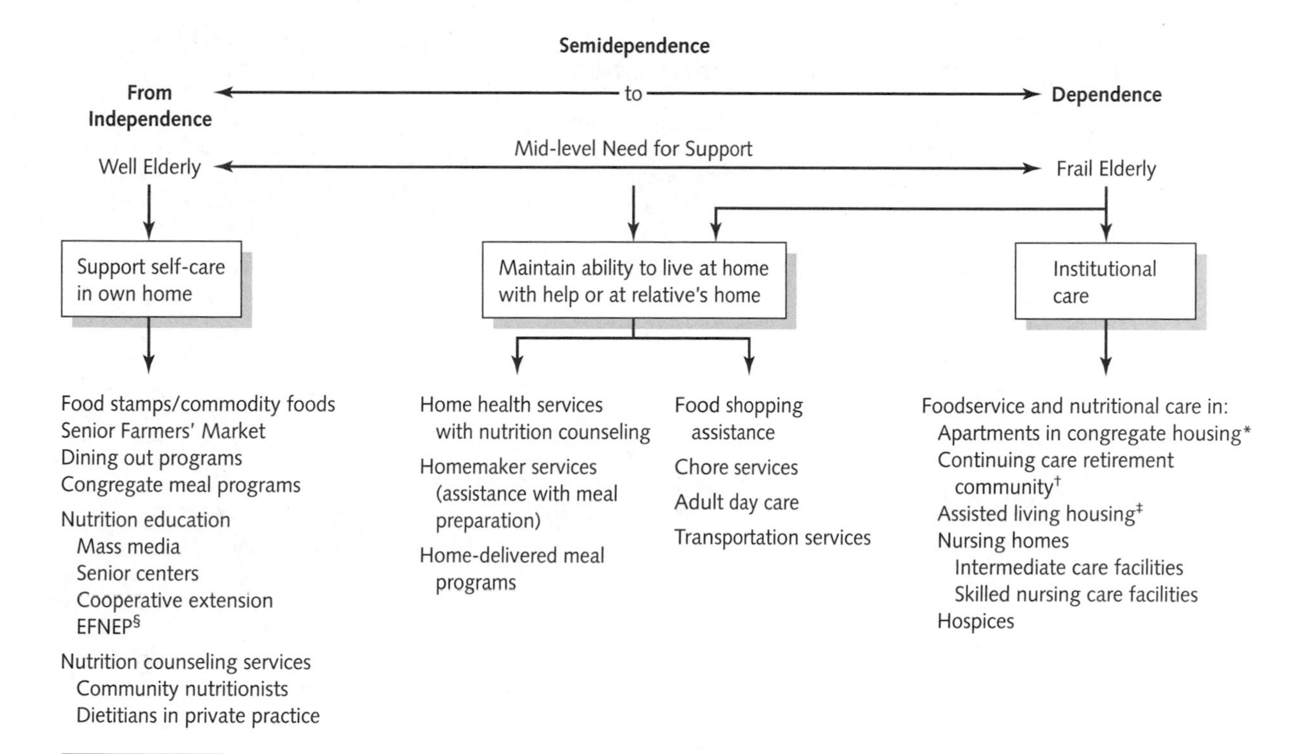

**Apartments in congregate housing*: allows older persons to maintain a private apartment but makes supportive services easily available (transportation, common dining room, other personal services).

†*Continuing-care retirement community*: offers full spectrum of services as needed—from meal service only to assisted-living services to skilled nursing care.

‡*Assisted-living housing*: offers elderly people more supportive services and supervision than congregate housing (for example, help with ADLs, meals provided in common dining room, emergency assistance available 24-hours/day).

§ EFNEP = Expanded Food and Nutrition Education Program.

FIGURE 12-7 Overview of Community Nutrition Programs for Older Adults Community-based services can be located by contacting the Eldercare Locator at 800-677-1116 or www.eldercare.gov.

Source: Adapted from H. T. Philips and S. H. Gaylord, eds., *Aging and Public Health.* Copyright Springer Publishing Company, Inc., New York 10012. Used by permission.

questioner for failing to perceive his isolation and aloneness and to realize that food . . . for one's self lacks the condiment of another's presence which can transform the simplest fare to the ceremonial act with all its shared meaning.[46]

Community-Based Programs and Services

Until the early 1970s, nutrition services for older adults, with the exception of food stamps, were found primarily in hospitals and long-term care facilities. Efforts were then made to expand services from the hospitals to include communities and homes.

In response to the socioeconomic problems that trouble many older adults and may lead to malnutrition—low income, inadequate facilities for preparing food, lack of transportation, and inability to afford dental care, among others—federal, state, and local agencies have mandated nutrition programs for the elderly.[47] Older adults' need for nutrition services depends on their level of independence, which can be depicted as a continuum (see Figure 12-7). Currently, community nutrition programs support the functional independence of older individuals in ambulatory care centers, adult day care centers, hospices, and home settings. Community nutritionists need to be familiar with organizations and programs providing nutrition and other health-related services to older adults.[48]

See Table 5-5 on page 135 for more information about federal nutrition assistance and nutrition education programs for older adults.

TABLE 12-9 Nutrition Programs for Older Adults

PROGRAM	TYPE OF INTERVENTION	FUNDING SOURCE	ELIGIBLE/AVAILABLE SERVICES
Elderly Nutrition Program	Congregate and home-delivered meals, special diets	DHHS AoA	Meals; transportation; shopping assistance; limited nutrition education, information, referral, and attention to needs of homebound elderly
Food Stamp	Income subsidy	USDA	Electronic benefits transfer (EBT) for food purchases
Food Stamp Nutrition Education	Nutrition education	USDA CES	Provide information about making healthful food choices. Materials for older adults available at www.nal.usda.gov/fnic/foodstamp/index.html
Adult Day Care Food Program	Meal program, supervised day care	USDA	Meals and snacks to participating day care programs
Senior Farmers' Market Nutrition Program (SFMNP)	Income subsidy	USDA	Provide low-income seniors with coupons that can be exchanged for eligible foods at farmers' markets, roadside stands, and community-supported agriculture programs.
Medicare/Medicaid	Third-party payment system	DHHS CMS SSA	Covers medical and related services provided by participating hospitals, HMOs, private medical practices, ambulatory centers, rehabilitation and skilled nursing facilities, home health agencies, and hospice programs. Eligible nutrition services vary depending on the setting of care and the deemed medical necessity.

AoA = U.S. Department of Health and Human Services, Administration on Aging.
DHHS = U.S. Department of Health and Human Services.
CES = Cooperative Extension Service.
CMS = U.S. Department of Health and Human Services, Centers for Medicare and Medicaid Services.
SSA = Social Security Administration.
USDA = U.S. Department of Agriculture.

Source: Food and Nutrition Service, *Food Program Facts* (Washington, D.C.: United States Department of Agriculture, 2004).

A summary of the nutrition programs for older adults is provided in Table 12-9. Check out the resources available online from the organizations listed at the end of the chapter.

General Assistance Programs

The National Council on Aging has created www.benefitscheckup.org to help older adults quickly identify federal and state assistance programs that may improve the quality of their lives.

The Supplemental Security Income (SSI) Program improves the financial plight of the very poor directly by increasing a person's or family's income to the defined poverty threshold. This sometimes helps older people retain their independence.

Another system of financial support to older Americans is the third-party reimbursement system, which we discussed in Chapter 9. Third-party payers (for example, Medicare, Medicaid, and Blue Cross/Blue Shield) sometimes reimburse the costs of health-related services, including such nutrition services as nutrition screening, assessment, and counseling and enteral or parenteral nutrition support. Generally, the nutrition service must be deemed "medically necessary." Whether a service is reimbursable and the extent of reimbursement vary from state to state and from case to case.

Social work agencies can provide older adults with information about appropriate nutrition resources in the community, such as congregate meal sites and home-delivered meals programs. On physician referral, Home Health Services, offered through local private and public organizations, provide home health aides to assist older people with shopping, housekeeping, and food preparation.[49]

Nutrition Programs of the U.S. Department of Agriculture

The Food Stamp Program was not designed specifically for older people, but it can nevertheless help older adults in need of financial assistance. The Food Stamp Program, which is administered by the USDA, enables qualifying people to obtain an Electronic Benefits Transfer (EBT) debit-type card that they can use to buy food at authorized grocery stores or at Senior Farmers' Markets. Currently, about 9 percent of participants in the Food Stamp Program are elderly (aged 60 or over), and only 30 percent of the eligible elderly participate in the program.[50] Reasons for nonparticipation by the elderly include the "stigma" of receiving assistance, confusing paperwork, and a lack of public information about eligibility requirements. In an effort to reach more eligible older adults, California and Wisconsin provide those over 65 years of age with a cash equivalent to their entitled food stamp benefit as part of their SSI check.

Visit the Food Resource and Action Center (FRAC) at www.frac.org/html/news/fsp/fselderlycenter.htm for information on how to better connect older persons with the Food Stamp Program.

The USDA also sponsors meal and snack programs for the Adult Day Care Centers operating in many communities through its Child and Adult Day Care Program. Adult day care facilities care for seniors while their care providers are away from the home. USDA's Commodity Supplemental Food Program is also available in a limited number of states and can provide monthly food packages to persons 60 years of age and older.

Nutrition Programs of the U.S. Department of Health and Human Services

The Older Americans Act (OAA) of 1965, which authorizes and funds the Administration on Aging and all of its programs, was amended in 1972 (PL 92-258) to establish and fund the federal Elderly Nutrition Program (ENP). The ENP is authorized to provide grants to promote the delivery of nutrition services in local communities: (1) Under Title III, Grants for State and Community Programs on Aging, grants are made to the 655 Area Agencies on Aging, and (2) under Title VI (added in 1978), Grants for Native Americans, grants are made to 233 Tribal Organizations representing American Indians, Alaska Natives, and Native Hawaiians. These grants are used to fund local congregate and home-delivered meals programs. The following section provides an overview of the congregate and home-delivered meals programs.

THE ELDERLY NUTRITION PROGRAM

With the graying of America, increased attention is being given to delivering cost-effective nutrition and health-related services to older persons in the community. The Elderly Nutrition Program (ENP) is intended to improve older people's nutritional status and

TABLE 12-10 Sample Title III Meal Pattern

FOOD TYPE	RECOMMENDED PORTION SIZE
Meat or meat alternative	3 oz, cooked portion
Vegetables and fruits	Two 1/2 c portions*
Enriched white or whole-grain bread or alternative	1 serving (one slice bread or equivalent)
Butter or margarine	1 tsp
Milk	8 oz milk or calcium equivalent
Dessert	1 serving

*A vitamin C–rich fruit or vegetable is to be served each day; a vitamin A–rich fruit or vegetable is to be served at least three times per week.

Source: U.S. Department of Health and Human Services.

enable them to avoid medical problems, continue living in communities of their own choice, and stay out of institutions. Its specific goals are to provide

- Low-cost, nutritious meals
- Opportunities for social interaction
- Nutrition screening and assessment
- Nutrition education and shopping assistance
- Counseling and referral to other social and rehabilitation services
- Transportation services

The dietary reference intakes (DRI) are listed on the inside front cover of this text.

The current ENP legislation makes meals available at least 5 days a week, supplying about a third of the dietary reference intakes (DRI) for one meal, two-thirds of the DRI for two meals, and 100 percent of the DRI when three meals are served (see Table 12-10). States must also ensure that meals comply with the Dietary Guidelines for Americans and take into account local cultural preferences. Menu planning is done with the advice of a registered dietitian or someone with comparable experience.[51] Meals can be breakfast, lunch, or dinner, depending on the needs of participants. There is no cost for meals, but participants sometimes make voluntary contributions.

One aspect of ENP is the Congregate Meals Program. Administrators try to select sites for congregate meals that will be accessible to as many of the eligible elderly as possible. The congregate meal sites are often community centers, senior centers, faith-based facilities, schools, adult day care facilities, or elderly housing complexes. Through the Home-Delivered Meals (HDM) Program, meals are delivered to those who are homebound either permanently or temporarily. The home-delivery program—often referred to as "Meals on Wheels"—ensures nutrition, but its recipients miss out on the social benefits of coming to congregate meal sites; every effort is made to persuade them to come to the shared meals, if they can. The home-delivered meals can be hot, cold, frozen, dried, or canned. Breakfast, lunch, dinner, or some combination of meals may be provided 5 to 7 days per week, where feasible. The DHHS's Administration on Aging administers these feeding programs, whereas the states, usually in conjunction with local county and city agencies, have responsibility for their daily operation and administration.

All persons over 60 years of age and their spouses (regardless of age) are eligible to receive meals from these programs, regardless of their income level. However, priority is given to low-income minority populations, older persons with the greatest economic or social need, and extremely old individuals.[52] Because American Indians, Alaska Natives, and Native Hawaiians tend to have lower life expectancies and higher rates of illness at younger ages, Title VI allows tribal organizations to set the age at which older people can participate in the program. Since 1972, these programs have grown significantly, accounting for annual federal funding of about $566 million in 2003. In total, the ENP provided about 250.4 million meals

TABLE 12-11 Ten Nutritional Risk Factors Diagnostic for the Need for Assistance among the Elderly

1. Consumption of fewer than 8 main meals (hot or cold) per week.
2. Drinking of very little milk (less than half a pint per day).
3. Little or no intake of fruits or vegetables.
4. Wastage of food, even if supplied hot and ready to eat.
5. Long periods of the day without food or beverages.
6. Depression or loneliness.
7. Unexpected weight change (gain or loss).
8. Shopping difficulties.
9. Low income.
10. Presence of disabilities (including alcoholism).

Source: L. Davies, Nutrition and the elderly: Identifying those at risk, *Proceedings of the Nutrition Society* 43 (1984): 299. © 1984, reprinted with the permission of Cambridge University Press.

to about 2.74 million seniors.[53] For every federal dollar spent, the ENP collects more than $2 from state, local, and private donations. Funding assists with food purchasing and preparation, facilities, and transportation for persons otherwise unable to participate.

Current evaluations of home-delivered meals programs ask whether the most needy elderly are going unserved and who should be given priority for receiving food assistance—those who lack access to food because of social or economic disabilities, or those with medical disabilities.[54] Criteria of nutritional risk are needed in order to assign priority status among the elderly experiencing food insecurity. One such assessment tool was designed in Great Britain so that social workers would be able to identify nutritional risk factors related to poverty, frailty, and loss of coping skills among the homebound elderly (see Table 12-11). In the United States, eligibility criteria for home-delivered meals vary with whether the program is a federally, state-, or locally operated program.

In an effort to reduce the cost of providing home-delivered meals, some states have initiated "luncheon clubs." These clubs have permitted several seniors living in close proximity to one another and receiving home-delivered meals to congregate in neighbors' homes. Only one meal delivery stop is therefore required, and the seniors benefit from the social interaction.

A two-year congressionally mandated evaluation of the programs for congregate and home-delivered meals generally shows that the programs improve the dietary intake and nutritional status of their clients (see Figure 12-8).[55] Participants generally have greater diversity in their diets and higher intakes of essential nutrients, and they are less likely than nonparticipants to report food insecurity.[56] Other benefits come as a result of screening and the referrals generated by such programs. Additional benefits are derived from the activities associated with the congregate meals services: nutrition counseling, physical activity programs, adult education, and other classes and activities. Participants benefit, too, from the opportunity for increased socializing.

However, despite these positive outcomes, deficiencies in the meals programs are noted in the lack of regular provision of weekend or evening meals to those who cannot get food or cannot cook. In addition, 41 percent of Title III ENP service providers have waiting lists for home-delivered meals, suggesting a significant unmet need for these meals.[57]

The special needs of the homebound elderly need to be given greater priority.[58] The dietary intake of the homebound elderly might improve if more than one meal per day were provided, if the meal furnished to the client were prepared with greater percentages of the DRI, and if meals were provided 7 days a week rather than 5.[59]

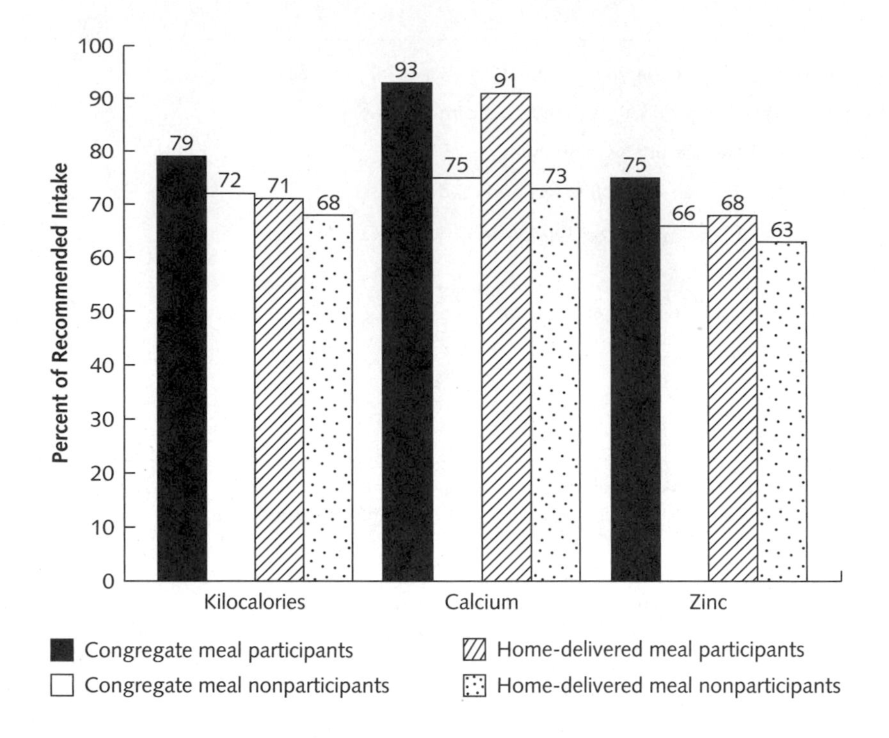

FIGURE 12-8 Intakes of Selected Nutrients by Participants and Nonparticipants in the Congregate Meal and Home-Delivered Meal Programs

Meal participants have a higher calorie intake and reach a higher percentage of the recommended intakes for most nutrients than nonparticipants similar in age and socioeconomic status. Homebound elderly have lower nutrient intakes than elderly persons who can leave their home.

Source: Data from M. Ponza, J. Ohls, and B. Millen, *Serving Elders at Risk: The Older Americans Act Nutrition Programs: National Evaluation of the Elderly Nutrition Program 1993–1995, Vol. I, Title III Evaluation Findings* (Princeton, NJ: Mathematica Policy Research, 1996).

Although Title III nutrition programs are required to provide nutrition education to their clients at least twice a year, these efforts are usually limited to the congregate meal sites. With the exception of a limited amount of printed material, virtually no education is provided to the staff who purchase and prepare these meals. Greater emphasis on nutrition education for both the helper and the client receiving home-delivered meals is warranted.

The social atmosphere at congregate meal sites can be as valuable as the foods served.

Ryan McVay/Getty Images

One such effort—the Senior Nutrition Awareness Project (SNAP) of Rhode Island—in partnership with the Meals on Wheels program, provides home-delivered meal participants with easy-to-read nutrition tips and information via a brief monthly newsletter called *Nutrition to Go*.[60] The newsletter is delivered to participants by the meal drivers each month and includes low-cost recipe ideas, self-assessment quizzes, and nutrition information on a monthly topic. As effective as it has been, the Elderly Nutrition Program faces certain challenges on the horizon:[61]

- Changing demographics are likely to increase the demand for program services—particularly for home-delivered meals. The number of persons 85 years and older is expected to double by 2030; because of disabling conditions, this group is less likely to live independently.[62]
- Changes in the present health care system will affect the ENP as more people are discharged early and in need of community health services.
- Depending on changes in public policy and funding, the ENP may be challenged to meet increased demand at a time of decreasing federal funding.

Private-Sector Nutrition Assistance Programs

In some communities, food banks enable older people on limited incomes to buy good food for less money. A food bank project buys industry "irregulars"—products that have been mislabeled, underweighted, redesigned, or mispackaged and would therefore ordinarily be thrown away. Nothing is wrong with this food; the industry can claim it, for tax purposes, as a donation, and the buyer (often a food-preparing site) can obtain the food for a small handling fee and make it available at a greatly reduced price.

The *Meals On Wheels Association of America* (MOWAA) provides leadership and professional training to those who provide congregate and home-delivered meals, develops partnerships, and offers grant opportunities to its members (for example, volunteer coordinators or nutrition directors at congregate meal and home-delivered meal programs).[63] MOWAA helps to reach older adults in communities not fully serviced by the Elderly Nutrition Program. In some communities, MOWAA member programs provide weekend and holiday meals in addition to ENP's typical five luncheon meals.

Nutrition Education and Health Promotion Programs for Older Adults

Nationwide about one-third of all seniors live in rural areas—communities with populations of 2,500 or less.[64] Few nutrition education efforts for these seniors are available, although some new programs have been designed to target this large audience. In Florida, the Area Agency on Aging of Central Florida has teamed with Florida Cooperative Extension to provide food labeling educational materials for the elderly and training programs for volunteers working with the elderly.

The Harvest Health at Home—Eating for the Second Fifty Years (HHH) project in North Dakota reaches rural communities through a series of newsletters.[65] The focus is on improving eating behaviors, with an emphasis on preventive nutrition messages such as decreasing fat intake and increasing fiber in the diet. Part of the intervention's success is due to the practical suggestions included in the newsletters (such as lists comparing brand names); such measures are generally valuable and effective with elders. The HHH intervention has also been successful in collaborating with other health professionals; for example, the newsletter publicizes the schedule of routine screenings sponsored by local health departments and clinics.

Programs in Action

Bringing Food and Nutrition Services to Homebound Seniors

Mobile health services have been highly successful in improving the well-being of individuals who are physically isolated from traditional medical and health settings. A mobile health unit in Virginia was created to reach rural elderly residents who could not obtain conventional health services because of illness, transportation problems, or financial factors.[1] Participants reached by the mobile health unit demonstrated increased participation in cancer screenings and increased immunization rates.

The case that follows, the Mobile Market Program, offers a creative solution for reaching homebound elderly whose major limitation to food access is transportation. By bringing a traveling market to their clients, volunteers in the Mobile Market Program empower the elderly and homebound to overcome their isolation and take more responsibility for their own health and well-being.

Goals and Objectives

Mobile Meals of Toledo, Ohio, devised the Mobile Market, a grocery store on wheels, to improve the independence and quality of life for the elderly, convalescing, chronically ill, disabled, and homebound by giving them the opportunity to shop easily for their groceries. Ongoing objectives include providing shopping opportunities to those who lack them and encouraging clients to progress from receiving home-delivered meals to using the Mobile Market Program.

Target Audience

The Mobile Market serves seniors, the physically and mentally challenged, the homebound, and residents of center-city neighborhoods with limited availability of supermarkets. The Market visits residents of 50 housing facilities for seniors and the disabled on a weekly basis.

Rationale for the Intervention

Mobile Meals of Toledo's existing Home-delivered Meal Program had a waiting list in some parts of the city. The organization realized that some clients needed assistance only in obtaining groceries, rather than in actual meal preparation and delivery. Thus, moving clients from Mobile Meals to the Market Program would reduce the waiting time for those who needed the full meal program. Additionally, many clients who had been receiving home-delivered meals were able to prepare their own meals but lacked shopping options and transportation. With the services of the Mobile Market available, agency caseworkers could better serve their neediest clients.

Focus group interviews show that seniors are interested in changing their eating behavior.[66] Including practical activities in programs can help motivate these changes. For example, at the White Crane Senior Center in Chicago, a combination health care and wellness center founded by seniors, monthly cooking classes provide an opportunity to modify and taste new recipes and try new foods.

The supermarket can serve as a forum to promote healthful diets to older persons. At some supermarkets, dietitians interact with older consumers through store tours for people on special diets and in-store cooking classes showing how to prepare meals with foods that help lower the risks for chronic diseases. Retailers now recognize the value of providing customers with point-of-purchase information about nutrition and health. Supermarket programs for older adults can increase product sales, attract new customers, and contribute to customer loyalty. Many quality nutrition programs exist that can be implemented with a minimum of cost and effort. The Food Marketing Institute's *To Your Health!* program offers a supermarket kit that provides program planners with tools and ideas for promoting healthful eating and activity habits for older shoppers.[67]

Whatever the setting, nutrition programs for seniors can be designed for cost-effectiveness by considering the following elements that have been found to contribute to the cost-effectiveness of past interventions involving seniors.[68]

- Begin nutrition programs with a personalized approach, such as a self-assessment of nutritional status or behaviors and subsequent comparison with recommendations.
- Use a behavioral approach that combines self-assessment with self-management techniques (for example, goal setting or social support).

Methodology

The Mobile Market is a grocery store on wheels that visits 50 housing facilities for seniors and the disabled on a weekly basis to provide residents with the opportunity to shop for their groceries. The Market carries more than 1,200 grocery items, including meats, fresh produce, dairy products, baked goods, and other items typically carried by grocery stores, such as stationery, cleaning supplies, and reading materials. The Market is small—37 feet long—to enable shoppers to focus easily and not become overwhelmed. Shoppers who are unable to visit the Mobile Market during its regularly scheduled stop can phone in orders for delivery. A voucher program allows the clients of mental health agencies to purchase groceries without using cash. The Mobile Market helps promote health screenings and other community health programs to its clients and arranges for flu and pneumonia vaccinations.

The Mobile Market has several partners in the community. The program is partially funded by United Way of Greater Toledo and the Area Office on Aging of Northwestern Ohio. The Market also receives a Community Development Block Grant. A local grocery store provides technical advice and sells the Market merchandise at cost to provide significant cost savings to its customers, who are frequently those most in need yet least able to pay for services.

Results

Approximately 650 to 700 clients purchased groceries from the Mobile Market on a weekly basis in 1999. Caseworkers reported that they made fewer home visits because they knew many of their clients' dietary needs were being met. Site managers acknowledged the positive effect of the opportunity to socialize offered by the Mobile Market; it was the only social outlet of the week for many homebound residents. As part of its vaccination program, the Mobile Market administered over 600 flu and pneumonia immunizations.

Lessons Learned

The Mobile Market Program is particularly beneficial to rural areas and inner-city neighborhoods that have limited options for grocery shopping. It also has led to a decrease in the number of people on waiting lists for meal programs.

Reference

1. B. B. Alexy and C. Elnitsky, Rural Mobile Health Unit: Outcomes, *Public Health Nursing* 15 (1998): 3–11.

Source: Adapted from *Community Nutritionary* (White Plains, NY: Dannon Institute, Fall 2000). Used with permission. For more information about the *Awards for Excellence in Community Nutrition,* go to www.dannon-institute.org.

- Allow for active participation in the program (for example, hands-on cooking classes and small-group discussions).
- Pay attention to motivators and reinforcements (for example, ease of food preparation, and opportunities for social interaction).
- Empower participants by enhancing personal choice and self-control of health-related behaviors.
- Target specific subgroups of older adults. Needs and interests differ by age, income, and health status.
- Be sensitive to age-related physical changes. For example, consider the visual and hearing capabilities of the audience.

Finally, remember to plan for the evaluation of the program. Clarify and document program outcomes in order to measure the impact of the intervention. Who benefits? How does the impact vary by type of person served? Is the program cost-effective?

Chapter 14 provides steps for evaluating program elements and effectiveness.

For adults over 50, health promotion efforts seek to preserve independence, productivity, and personal fulfillment.[69] The premise of health promotion is that individuals can enjoy benefits from healthful behaviors at any age. To this end, most states now offer community wellness centers for seniors that include services at all levels of prevention. Resource people for these efforts at the local level include public health nutritionists employed by county health departments, registered dietitians working with local nursing homes and community hospitals as consultants, and county cooperative extension educators. In addition, some community groups and churches offer support and self-help groups.

AGING WELL

Stress Busters

- Relax
- Go for a walk
- Breathe deeply
- Think positively

Emotional Well-Being

- Reduce stress
- Learn relaxation techniques
- Cultivate a garden
- Seek out laughter
- Take time for spiritual growth
- Adopt and love a pet
- Take time off

Social Health

- Be socially active
- Volunteer for a special cause
- Make new friends
- Enroll in lifelong learning
- Be active in your community

Nutritional Health

- Choose nutrient-dense foods
- Eat 5 to 9 fruits and vegetables every day
- Drink plenty of water
- Keep fat intake at a healthful level
- Get adequate fiber

Physical Health

- Be physically active
- Get adequate sleep
- Challenge your mental skills
- Do aerobic and strength-training exercises at least 3 times a week
- Stretch for flexibility

Lifelong Habits for Successful Aging

- Cherish your personal values and goals
- Develop good communication skills
- Balance diet and exercise to maintain a healthful weight
- Practice preventive health care
- Develop skills and hobbies to enjoy for a lifetime
- Manage time
- Learn from mistakes
- Nurture relationships with family and friends
- Enjoy, respect, and protect nature
- Accept change as inevitable
- Plan ahead for financial security

FIGURE 12-9 The Aging Well Pyramid

The time to prepare for old age is early in life. Practice the items found at the base of the pyramid to achieve an optimal sense of well-being. Use the inner four compartments of the pyramid to create a balance among all aspects of your life: nutrition, physical activity, social health, and emotional well-being. Use the tip of the pyramid to manage everyday stresses such as traffic gridlock, exams, and work deadlines.

Looking Ahead: And Then We Were Old

As a nation, we tend to value the future more than the present, putting off enjoying today so that we will have money, prestige, or time to have fun tomorrow. The elderly feel this loss of future. The present is their time for leisure and enjoyment, but often they have no experience in using leisure time.

The solution is to begin to prepare for old age early in life, both psychologically and nutritionally (see Figure 12-9). Preparation for this period should, of course, include financial planning, but other lifelong habits should be developed as well. Each adult needs to learn to reach out to others to forestall the loneliness that will otherwise ensue. Adults need to develop some skills or activities that they can continue into their later years—volunteer work with organizations, reading, games, hobbies, or intellectual pursuits—and that will give meaning to their lives. Each adult needs to develop the habit of adjusting to change, especially when it comes without consent, so that it will not be seen as a loss of control over one's life. The goal is to arrive at maturity with as healthy a mind and body as possible; this means cultivating good nutritional status and maintaining a program of daily exercise.

In general, the ability of the elderly to function well varies from person to person and depends on several factors. The "life advantages" listed below seem to contribute to good physical and mental health in later years.[70]

- Genetic potential for extended longevity. Some persons seem to have inherited a reduced susceptibility to degenerative diseases.
- A continued desire for new knowledge and new experiences. Some studies suggest that "active" minds, ever involved in learning new things, may be more resistant to decline.
- Socialization, intimacy, and family integrity. Older persons thrive in situations where love, understanding, shared responsibility, and mutual respect are nurtured.
- Adherence to a nutritious diet, combined with avoidance of excesses of food energy, fat, cholesterol, and sodium. A balanced diet with adequate intakes of all essential nutrients has a positive impact on health and weight management.
- Avoidance of substance abuse.
- Acceptable living arrangements.
- Financial independence.
- Access to health care, including a family physician, health clinic, public health nursing service providing home health care, dentist, podiatrist, physical therapist, pharmacist, and community nutritionist.

Everyone knows older people who have maintained many contacts—through relatives, church, synagogue, or fraternal orders—and have not allowed themselves to drift into isolation. Upon analysis, you will find that their favorable environment came about through a lifetime of effort. These people spent their entire lives reaching out to others and practicing the art of weaving others into their own lives. Likewise, a lifetime of effort is required for good nutritional status in the later years. A person who has eaten a wide variety of foods, maintained a healthful weight, and remained physically active will be best able to withstand the assaults of change.

Internet Resources

Government Sites

HealthierUS.gov	**www.healthierus.gov**
Steps to a Healthier US Initiative	**www.healthierus.gov/steps/index.html**
Small Step Campaign	**www.smallstep.gov**
Administration on Aging (AOA)	**www.aoa.dhhs.gov**
Centers for Disease Control and Prevention	**www.cdc.gov/nchs/agingact.htm**
Eldercare Locator	**www.eldercare.gov**
Federal Interagency Forum on Aging-Related Statistics	**www.agingstats.gov**
First Gov for Seniors	**www.seniors.gov**
Food Safety.gov	**www.foodsafety.gov/~fsg/fsgsr.html**
Medicare Program	**www.medicare.gov**
National Center for Health Statistics	**www.cdc.gov/nchs**
National Heart, Lung, and Blood Institute	**www.nhlbi.nih.gov/index.htm**

National Institute on Aging	**www.nia.nih.gov/**
National Women's Health Information Center	**www.4women.gov**
Office of Minority Health	**www.omhrc.gov**
USDA Food and Nutrition Service (FNS) Nutrition Assistance Programs	**www.fns.usda.gov/fns/**
United States Senate Special Committee on Aging	**http://aging.senate.gov/public/**
Weight Control Information Network	**http://win.niddk.nih.gov/index.htm**
Women's Health Initiative	**www.nih.gov**
Organizations	
Alzheimer's Disease Education and Referral Center (ADEAR)	**www.alzheimers.org**
Nutrition Screening Initiative (American Academy of Family Physicians)	**www.aafp.org/nsi**
American Association of Retired Persons (AARP)	**www.aarp.org/health/healthguide/**
American Diabetes Association	**www.diabetes.org**
American Federation of Aging Research (AFAR)	**www.afar.org**
American Geriatrics Society	**www.americangeriatrics.org**
American Heart Association	**www.americanheart.org**
American Institute for Cancer Research	**www.aicr.org**
Gerontological Society of America	**www.geron.org**
Home- and Community-Based Services	**www.hcbs.org**
Leadership Council of Aging Organizations (LCAO)	**www.lcao.org**
Meals on Wheels (MOW) Association of America	**www.mowaa.org**
National Association of Area Agencies on Aging (N4A)	**www.n4a.org**
National Association of Child and Adult Care Food Programs	**www.cacfp.org**
National Association of State Units on Aging	**www.nasua.org**
National Council on the Aging	**www.ncoa.org**
National Council on Aging's Benefits Check Up	**www.benefitscheckup.org**
National Osteoporosis Foundation	**www.nof.org**
North American Menopause Society	**www.menopause.org**
Universities	
Florida International University's National Policy and Resource Center on Nutrition and Aging	**www.fiu.edu/~nutreldr**
Emory University's MedWeb—Geriatrics	**www.medweb.emory.edu/MedWeb**
University of Michigan's Health and Retirement Study (HRS)	**http://hrsonline.isr.umich.edu/**
Tufts University Human Nutrition Research Center on Aging	**www.hnrc.tufts.edu**
Commercial	
AgeNet Eldercare Network	**www.agenet.com**
Elder Care Online	**www.ec-online.net**
ElderNet	**www.eldernet.com**

Case *Study*

Postmenopausal Nutrition and Disease Prevention Program

By Alice Fornari, EdD, RD, and Alessandra Sarcona, MS, RD

Scenario

A physician (gynecological specialist) wants to create a nutrition program for the patients in his private practice who are postmenopausal or are at a target age of approximately 50–65 years old. The physician is leaving you, as the consulting RD, responsible for identifying the nutrition topics that would best meet the needs of this population. The physician states that many of his patients have been asking him about soy, calcium, heart-healthy eating, and weight management. The target group consists of middle-income to upper-middle-income females, predominantly Caucasian. The physician's office has a conference room that could be utilized for nutrition lectures.

Learning Outcomes

- Identify the national goals for health promotion and disease prevention for adults.
- Recognize the leading nutrition-related diseases and causes of death of women.
- Select appropriate nutrition strategies for disease prevention among postmenopausal women.

Foundation: Acquisition of Knowledge and Skills

1. Review the *Healthy People 2010* objectives for adults presented in the chapter.
2. From the Centers for Disease Control and Prevention (CDC) website, www.cdc.gov, list the top killers/conditions of women, or go to www.nhlbi.nih.gov/whi/factsht.htm and click on *Why WHI*; list the diseases that the Women's Health Initiative is researching and give an explanation for why these conditions are being studied.

Step 1: *Identify Relevant Information and Uncertainties*

1. Research the Women's Health Initiative (WHI), www.nhlbi.nih.gov/whi/factsht.htm, and extract from this initiative the nutrition information that could be useful for your program.
2. Besides reading the physician's report, how would you go about determining what nutrition issues take precedence among the women in the physician's private practice and among those in your target group, aged 50–65?

Step 2: *Interpret Information*

1. On the basis of the nutrition issues identified, what resources would you use to determine the legitimacy of nutrition topics relative to health issues and disease prevention? For example, how would you explore the impact of soy on the chronic diseases affecting women?

Step 3: *Draw and Implement Conclusions*

1. Based on the information gathered from the CDC and WHI and on the target group's nutrition issues and resources used, prioritize a list of nutrition topics that would be appropriate to discuss in a program with five $1\frac{1}{2}$-hour sessions. From your list, create an outline of a program proposal to present to the physician that includes a rationale for having a nutrition program, a list of the program's nutrition topics, and information and activities that will be utilized to assist the participants in learning about these topics.

Step 4: *Engage in Continuous Improvement*

1. Identify variables that could be used to monitor the impact of the program on women's health.

Chapter *Thirteen*

World Hunger and Food Insecurity: Challenges and Opportunities

Learning *Objectives*

After you have read and studied this chapter, you will be able to:

- Describe the current status of world food insecurity.
- List causes of world food insecurity.
- Give reasons why women and children are particularly at risk with regard to hunger.
- Describe the purpose and goals of recent international food policy initiatives.
- Describe the global public health issues related to world food insecurity that will continue to challenge policy makers and program designers in the twenty-first century.
- List actions that individuals might take to eliminate world food insecurity.

This chapter addresses such issues as the influence of socioeconomic and cultural factors on food behavior; food availability and access for the individual, family, and community; and using current information technologies, which are Commission on Accreditation for Dietetics Education (CADE) *Foundation Knowledge and Skills* requirements for dietetics education.

Chapter *Outline*

Something To Think About...

The world does not exist in a series of separately functioning compartments. The last lines from the chapter on poverty in the State of the World Report sum this up succinctly: "Sheets of rain washing off denuded watersheds flood exclusive neighborhoods as surely as slums. Potentially valuable medicines lost with the extinction of rain forest species are as unavailable to the rich in their private hospitals as they are to the poor in rural clinics. And the carbon dioxide released as landless migrants burn plots in the Amazon or the Congo warms the globe as surely as do the fumes from automobiles and factory smoke-stacks in Los Angeles or Milan."

Poverty, food security, environment, nutrition—they are all inescapably interlinked. Those of you who have the knowledge to embrace issues of nutrition education also have the power to ameliorate the human predicament. And there is no objective in this world more worth pursuing than improving the human condition.

– Stephen Lewis

Introduction

All people need food. Regardless of race, religion, sex, or nationality, our bodies experience similarly the effects of food insecurity and its companion malnutrition—listlessness, weakness, failure to thrive, stunted growth, mental retardation, muscle wastage, scurvy, anemia, rickets, osteoporosis, goiter, tooth decay, blindness, and a host of other effects, including death.[1] Apathy and shortened attention span are two of a number of behavioral symptoms that are often mistaken for laziness, lack of intelligence, or mental illness in undernourished people.[2] Malnutrition is the biggest risk factor for illness worldwide.[3] Some 2 billion people, mostly women and children, are deficient in one or more of three major micronutrients: iron, iodine, and vitamin A.[4] For the year 2025, the projected United Nations figure for total world population is approximately 7.8 billion, thus further stretching the world's food resources.[5]

Food security consists of access by all people at all times to enough food for an active and healthy life. Food security has two aspects: ensuring that adequate food supplies are available and ensuring that households whose members suffer from undernutrition have the ability to acquire food, either by producing it themselves or by being able to purchase it.

Mapping Poverty and Undernutrition

Food insecurity was once viewed as a problem of overpopulation and inadequate food production, but now many people recognize it as a problem of poverty. Food is available but not accessible to the poor who have neither land nor money. In 1978, Robert McNamara, then president of the **World Bank,** gave what stands as the classic description of absolute poverty: "A condition of life so limited by malnutrition, illiteracy, disease, squalid surroundings, high infant mortality, and low life expectancy as to be beneath any reasonable definition of human decency" (see Table 13-1).[6]

World Bank A group of international financial institutions owned by the governments of more than 150 nations. The bank provides loans for economic development.

REGION	GNI*	INFANT MORTALITY RATE (IMR)	LIFE EXPECTANCY	LITERACY RATE	SAFE WATER SUPPLY (%)	UNDER-5 MORTALITY RATE (U5MR)
General Differences†						
Least developed countries	$304	98	49	52	58	155
Other developing countries	$1,255	60	62	74	79	87
Industrialized countries	$28,337	5	78	98	100	6
Regional Differences						
		IMR		Undernourished Children (%)		U5MR
Sub-Saharan Africa		104		29		175
South Asia		67		46		92
East Asia and Pacific		31		17		40
Latin America and Caribbean		27		7		32
Middle East and North Africa		45		14		56

*GNI = Gross national income.

†Notice that the poorer nations have higher infant and under-5 mortality rates, shorter life expectancies, and lower literacy rates than richer nations. In short, the quality of life suffers from poverty.

Source: Data from UNICEF, *The State of the World's Children 2005* (New York, NY: UNICEF, 2005).

TABLE 13-1 The Gap between Developed and Developing Countries

The Food and Agriculture Organization of the United Nations (FAO) estimates that of the more than 6 billion people in the world, at least 852 million people—18 percent of the developing world's population—suffer from chronic, severe undernutrition, consuming too little food each day to meet even minimum energy requirements (see Figure 13-1).[7] Approximately 1.2 billion people live in poverty. Living standards declined during the past two decades, partly because of accelerated rates of population growth and environmental decline, but also as a result of lower export earnings, rising inflation, and higher interest

FIGURE 13-1 A Regional Distribution of the Proportion and Number of the World's Chronically Undernourished Populations, 2000–2002

Source: Food and Agriculture Organization of the United Nations, *Assessment of the World Food Security Situation* (Rome: FAO, 2004).

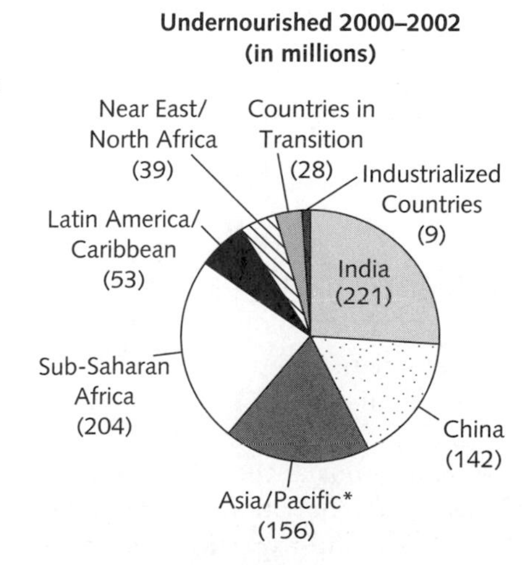

Proportion Undernourished, by Region

Countries in transition
Near East/ North Africa
Latin America/ Caribbean
Asia/Pacific
Sub-Saharan Africa
1990–1992*
2000–2002
0 10 20 30 40
Percent Undernourished

*1993–1995 for countries in transition

rates on foreign debts.[8] In other words, the poor earned less and paid more. According to the FAO, widespread chronic hunger is most likely to be found in developing countries that can neither produce enough food to feed their populations fully nor earn enough foreign exchange to import food to cover their food deficits.[9]

To qualify as chronically and severely undernourished by FAO standards, a person must consume fewer than the calories required to perform the basic physiological functions and light physical activity. This minimum is usually in the range of 1,700 to 1,960 kilocalories/day.

See the World Food Programme's interactive hunger map for presentations on hunger hot spots and poverty around the world. Go to www.wfp.org/ and click on Hunger Map.

Those who live with chronic poverty often face unsafe drinking water, intestinal parasites, insufficient food, a low-protein diet, stunted growth, low birthweights, illiteracy, disease, shortened life spans, and death. In *Quiet Violence: View from a Bangladesh Village,* Hartman and Boyce provide a good introduction to life in the villages of the developing world. The lives of these villagers are more difficult than anything we have ever known, and yet their hopes and dreams are not unlike our own. They exhibit resourcefulness, hard work, and dignity in the midst of circumstances that require a persistence and personal strength that most of us will never need to call upon in our lifetimes. Hari, one of the landless laborers in the village, reflects on his life just days before his death: "Between the mortar and the pestle, the chili cannot last. We poor are like chilies—each year we are ground down, and soon there will be nothing left."[10]

Poverty is much more than an economic condition and exists for many reasons, including overpopulation, the greed of others, unemployment, and the lack of productive resources such as land, tools, and credit.[11] Consequently, if we are to provide adequate nutrition for all the earth's hungry people, we must transform the economic, political, and social structures that both limit food production, distribution, and consumption and create a gap between rich and poor.[12]

Malnutrition and Health Worldwide

Nearly 30 percent of the world's population experience some form of malnutrition.[13] Some 10.6 million children under the age of 5 years die each year from the parasitic and infectious diseases associated with poverty.[14] At least 75 percent of all child deaths are caused by neonatal disorders and a few treatable infectious diseases, such as diarrhea, pneumonia, malaria, and measles (see Figure 13-2). These diseases interact with poor nutrition to form a vicious cycle in which the outcome for many is death, as shown in Figure 13-3.

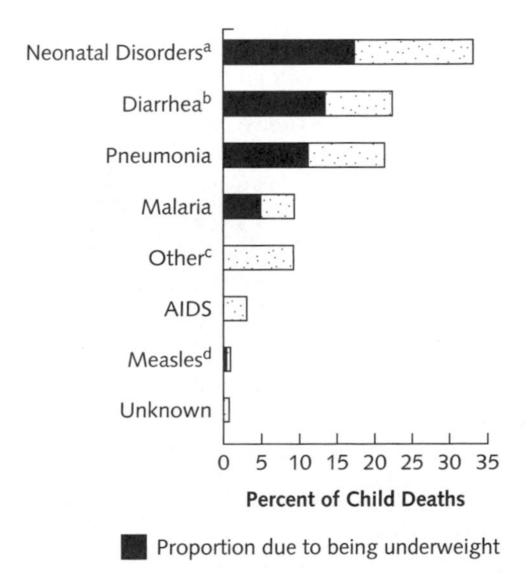

FIGURE 13-2 Top Child Killers: Causes of Child Mortality, Percentages, 2005

[a]Birth asphyxia, neonatal sepsis, congential anomalies, birth trauma, prematurity.
[b]Adequate sanitation is crucial to reducing under-5 mortality and morbidity rates, yet 2.4 billion people lack access.
[c]Accidents, other
[d]Of all the vaccine-preventable diseases (diphtheria, tuberculosis, measles, pertussis, tetanus, hepatitis B, meningitis, yellow fever), measles kills the most children.

Source: UNICEF, 2005.

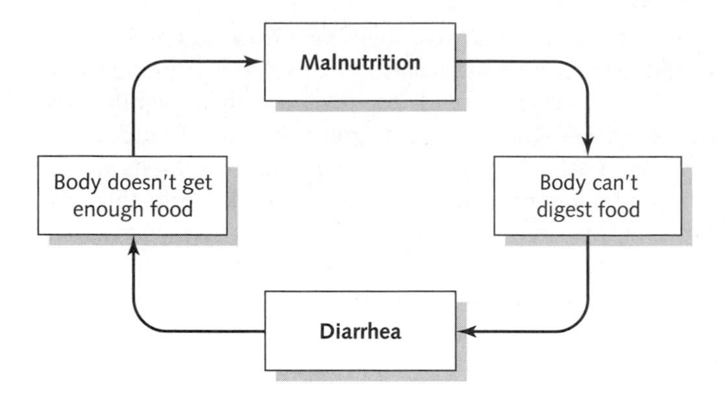

FIGURE 13-3 The Vicious Circle of Malnutrition

Source: UNICEF.

Hungry people receive such small quantities of food that they develop multiple nutrient deficiencies. Their undernutrition may result from lack of food energy or from a lack of both food energy and protein. Such distinctions can easily be made on paper, and the extremes are evident in individuals, but for the most part, the differences blur.

More than 150 million children in developing countries suffer from malnutrition (see Figure 13-4).[15] **Protein–energy malnutrition (PEM)** is the most widespread form of malnutrition in the world today. Children who are thin for their height may be suffering from acute PEM (recent severe food restriction), whereas children who are short for their age may be suffering from chronic PEM (long-term food deprivation). PEM includes **kwashiorkor,** a protein-deficiency disease; **marasmus,** a deficiency disease caused by inadequate food intake; and the states in which these two extremes overlap. Children suffering from PEM are likely to develop infections, nutrient deficiencies, and diarrhea (see Table 13-2).

Protein–energy malnutrition (PEM) The world's most widespread malnutrition problem; includes **kwashiorkor** (a deficiency disease caused by inadequate protein intake), **marasmus** (a deficiency disease caused by inadequate food intake—starvation), and the states in which they overlap.

Worldwide, three micronutrient deficiencies are of particular concern: vitamin A deficiency, the world's most common cause of preventable child blindness and vision impairment; iron-deficiency anemia; and iodine deficiency, which causes high levels of goiter and child retardation.[16]

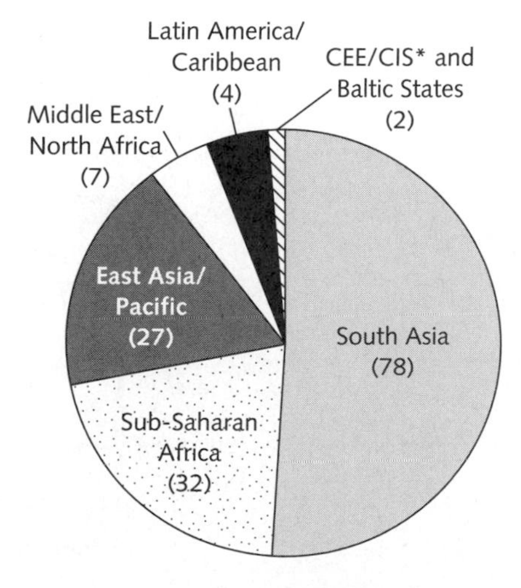

FIGURE 13-4 Number of Malnourished Children (in millions)

In developing countries, 150 million children are still malnourished. Indicators of child malnutrition include:

Stunting (short-for-age): Children with a low *height-for-age* measure (measure of linear growth); a measure of chronic malnutrition.

Wasting (thin-for-height): Children with a low *weight-for-height* measure; a measure of acute malnutrition.

Underweight: Children with a low *weight-for-age* (a synthesis of height-for-age and weight-for-height); reflects both stunting and wasting.

*CEE/CIS = Central and Eastern Europe and the Commonwealth of Independent States.

Source: UNICEF, 2005.

TABLE 13-2 Characteristic Features of Marasmus and Kwashiorkor

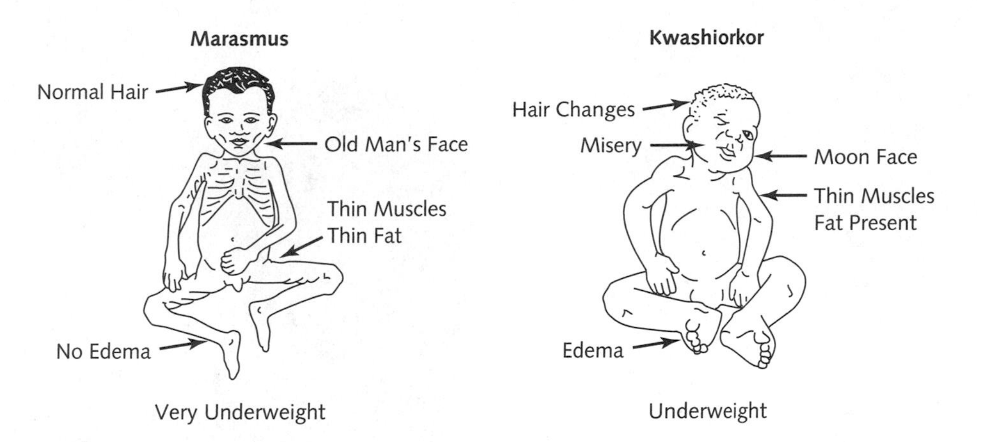

FEATURES	MARASMUS	KWASHIORKOR
Essential features		
Edema	None*	Lower legs, sometimes face, or generalized*
Wasting	Gross loss of subcutaneous fat, "all skin and bone"*	Less obvious; sometimes fat, blubbery
Muscle wasting	Severe*	Sometimes
Growth retardation in terms of body weight	Severe*	Less than in marasmus
Mental changes	Usually none	Usually present
Variable features		
Appetite	Usually good	Usually poor
Diarrhea	Often (past or present)	Often (past or present)
Skin changes	Usually none	Often, diffuse depigmentation; occasional, "flaky-paint"* or "enamel" dermatosis
Hair changes	Texture may be modified but usually no dispigmentation	Often sparse—straight and silky, dispigmentation—greyish or reddish
Moon face	None	Often
Hepatic enlargement	None	Frequent, although it is not observed in some areas
Biochemistry pathology		
Serum albumin	Normal or slightly decreased	Low*
Urinary urea per gram of creatinine	Normal or decreased	Low*
Urinary hydroxyproline index	Low	Low
Serum-free amino acid ratio	Normal	Elevated*
Anemia	May be observed	Common; iron or folate deficiency may be associated
Liver biopsy	Normal or atrophic*	Fatty infiltration*

*The most characteristic or useful distinguishing features.

Source: C. D. Williams, N. Baumslag, and D. B. Jelliffe, *Mother and Child Health: Delivering the Services,* 3rd ed. (New York: Oxford University Press, 1994), p. 156.

Infants can be the first to show the signs of undernutrition due to their high nutrient needs. No famine, no flood, no earthquake, no war has ever claimed the lives of 200,000 children in a single week. Yet UNICEF estimates that malnutrition and disease claim more than that number of child victims under the age of 5 years *every week* at the rate of one every three seconds.

Louise Gubb/The Image Works

This chapter's Programs in Action highlights the international activities that target vitamin A deficiency.

- *Vitamin A deficiency.* More than 100 million children are affected by vitamin A deficiency (VAD). Of these, an estimated 500,000 children become partially or totally blind as a result of insufficient vitamin A in the diet. Vitamin A deficiency is also associated with other forms of malnutrition, infection, diarrhea, and a high rate of mortality.
- *Iron deficiency.* Iron-deficiency anemia is estimated to affect some 2 billion people. Iron deficiency in infancy and early childhood is associated with decreased cognitive abilities and resistance to disease.
- *Iodine deficiency.* Iodine deficiency, the major preventable cause of mental retardation worldwide, is a risk factor for both physical and mental retardation in about 1 billion people. About 655 million people worldwide—especially in mountainous regions—are estimated to have goiter, and more than 16 million suffer overt cretinism.

The malnutrition that comes from living with food insecurity is one of the major factors influencing life expectancy. According to the *2005 State of the World's Children Report,* life expectancy at birth is 77 years in the United States and about 79 years in Canada.[17] Worldwide, life expectancy averages about 63 years, but in sub-Saharan Africa it is approximately 46 years. In countries heavily affected by HIV/AIDS—life expectancy ranges from a high of 39 years, in Botswana, to the lowest of all—only 33 years—in the small African countries of Zambia and Zimbabwe.

Hunger and malnutrition can be found in people of all ages, sexes, and nationalities. Even so, these problems hit some groups harder than others.

EFFECTS OF MALNUTRITION ON THOSE MOST VULNERABLE

When nutrient needs are high (as in times of rapid growth), the risk of undernutrition increases. If family food is limited, pregnant and lactating women, infants, and children are the first to show the signs of undernutrition. The effects of food insecurity can be devastating to this group of the population.

As we noted in Chapter 10, to support normal fetal growth and development, women must gain adequate weight during pregnancy. Healthy women in developed countries gain an average of about 27 pounds. Studies among poor women reveal that they have an average weight gain of only 11 to 15 pounds. In some areas, women may have caloric deficits of up to 42 percent and do not gain any weight at all during pregnancy.[18] As a consequence, they give birth to babies with low birthweights.

Birthweight is a potent indicator of an infant's future health status. A low-birthweight baby (less than $5\frac{1}{2}$ pounds, or 2,500 grams) is more likely to experience complications during delivery than a baby of normal weight and has a statistically greater-than-normal chance of exhibiting stunted physical and cognitive growth during childhood, contracting diseases, and dying early in life. More than 20 million LBW babies are born in the developing world each year.[19] Low birthweight contributes to more than half of the deaths of children under 5 years of age. Low-birthweight infants suffering undernutrition after their births incur even greater risks. They are more likely to get sick, to fail to obtain nourishment by sucking, and to be unable to win their mothers' attention by energetic, vigorous cries and other healthy behavior. They can become apathetic, neglected babies, which compounds the original malnutrition problems.

Almost one-third of all children in developing countries are stunted—suffering from chronic undernutrition. If stunting occurs during the first five years of life, the physical and cognitive impairments are usually irreversible.[20] Mortality statistics reflect the hazards to those most vulnerable. The **infant mortality rate** ranges from about 8 (Costa Rica) to over 166 (Sierra Leone) in the poorest of the developing countries. The death rate for children from 1 to 5 years of age is no more favorable; it ranges from 20 to 30 times higher in developing countries than in developed countries.[21] Maternal mortality rates (listed in the margin) are equally shocking.

The lifetime chance of a woman dying in pregnancy or childbirth:
In least developed countries: 1 in 16
In industrialized countries: 1 in 4,085

Maternal Mortality Rates
- Industrialized countries: 12
- Central and Eastern Europe: 55
- East Asia and Pacific: 140
- Latin America and Caribbean: 190
- Middle East and North Africa: 360
- South Asia: 430
- Sub-Saharan Africa: 1,100

UNICEF regards the **under-5 mortality rate (U5MR)** as the single best indicator of children's overall health and well-being (see Table 13-1).[22] UNICEF argues that this rate reflects the overall resources a country directs at children:

> The U5MR reflects the nutritional health and the health knowledge of mothers; the level of immunization and ORT (oral rehydration therapy) use; the availability of maternal and health services (including prenatal care); income and food availability in the family; the availability of clean water and safe sanitation; and the overall safety of the child's environment.[23]

THE ECONOMIC BURDEN OF MALNUTRITION AND HUNGER

The "grim tally of human lives cut short or scarred by disability leaves no doubt that hunger is morally unacceptable," states the FAO in the latest edition of *The State of Food Insecurity in the World*.[24] However, calculating the economic costs of hunger shows that it is also unaffordable—not only to those affected by hunger but also to the future development of the countries in which they live.

The burden of hunger includes both direct and indirect costs. The most obvious are the direct health-related expenses associated with maternal complications in pregnancy and with the poor health of low-birthweight babies and malnourished children, who are at increased risk of conditions such as diarrhea, measles, malaria, and pneumonia, as well as chronic diseases. A very rough estimate indicates that these direct costs add up to about $30 billion per year.[25]

The indirect costs of hunger include lost productivity and income caused by problems associated with chronic hunger: premature death, disability, absenteeism in the workplace, and reduced educational and occupational opportunities.[26] The FAO estimates these indirect costs to be hundreds of billions of dollars. Studies that measure the impact of malnutrition on physical and mental development have established correlations with reduced

Infant mortality rate (IMR) Infant deaths under 1 year of age, expressed as a rate per 1,000 live births.

UNICEF The United Nations International Children's Emergency Fund, now referred to as the United Nations Children's Fund.

Under-5 mortality rate (U5MR) The number of children who die before the age of 5 for every 1,000 live births.

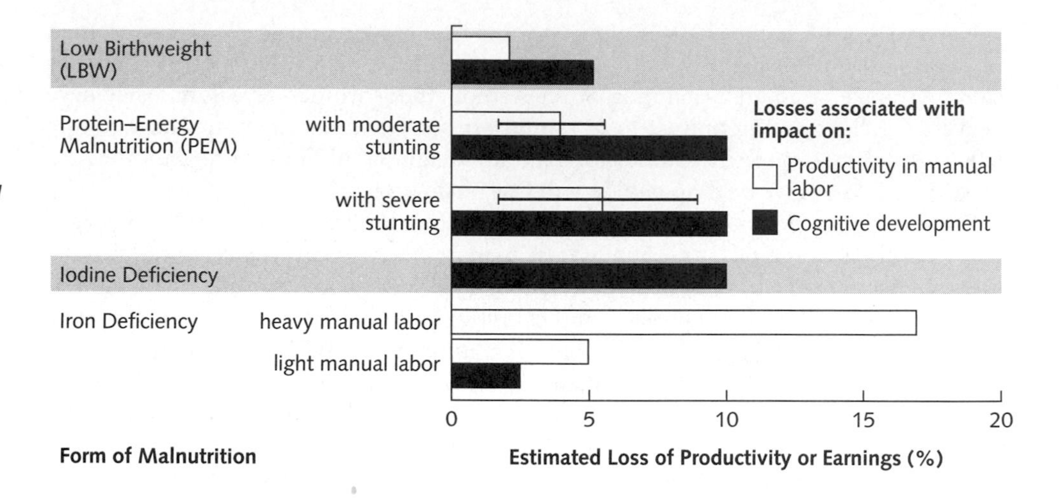

FIGURE 13-5 Impact of Various Forms of Malnutrition on Productivity and Lifetime Earnings

Source: FAO, *The State of Food Insecurity in the World 2004.*

productivity and earnings, as shown in Figure 13-5. Every child whose physical and mental development is stunted by hunger and malnutrition stands to lose 5 to 10 percent or more in lifetime earnings.

This $24 billion—needed to meet basic human needs each year—is less than the amount spent annually in Europe on cigarettes ($50 billion) and the amount spent on beer each year in the United States ($31 billion).

Every year that hunger persists at current levels will cost developing countries future productivity of $500 billion from lives lost to disease or disability.[27] On the other hand, UNICEF estimates that most child malnutrition in the developing world could be eliminated with the expenditure of an additional $24 billion a year (see the margin note to put this sum of money in perspective). This amount would cover the cost of the resources needed to control the major childhood diseases, halve the rate of child malnutrition, bring clean water and safe sanitation to all communities, make family planning services universally available, and provide almost every child with at least a basic education.[28]

Food Insecurity in Developing Countries

World hunger is more extreme than domestic hunger. In fact, most people would find it hard to imagine the severity of poverty in the developing world:

> Many hundreds of millions of people in the poorest countries are preoccupied solely with survival and elementary needs. For them, work is frequently not available or pay is low, and conditions barely tolerable. Homes are constructed of impermanent materials and have neither piped water nor sanitation. Electricity is a luxury. Health services are thinly spread, and in rural areas only rarely within walking distance. Permanent insecurity is the condition of the poor. . . . In the wealthy countries, ordinary men and women face genuine economic problems. . . . But they rarely face anything resembling the total deprivation found in the poor countries.[29]

Figure 13-6 shows some of world hunger's many causes. World hunger is a problem of supply and demand, inappropriate technology, environmental abuse, demographic distribution, unequal access to resources, extremes in dietary patterns, and unjust economic systems. Two generalizations and an important question are suggested by Figure 13-6:

- The underlying causes of global hunger and poverty are complex and interrelated.
- Hunger is a product of poverty resulting from the ways in which governments and businesses manage national and international economies.
- The question, then, is "Why are people poor?"

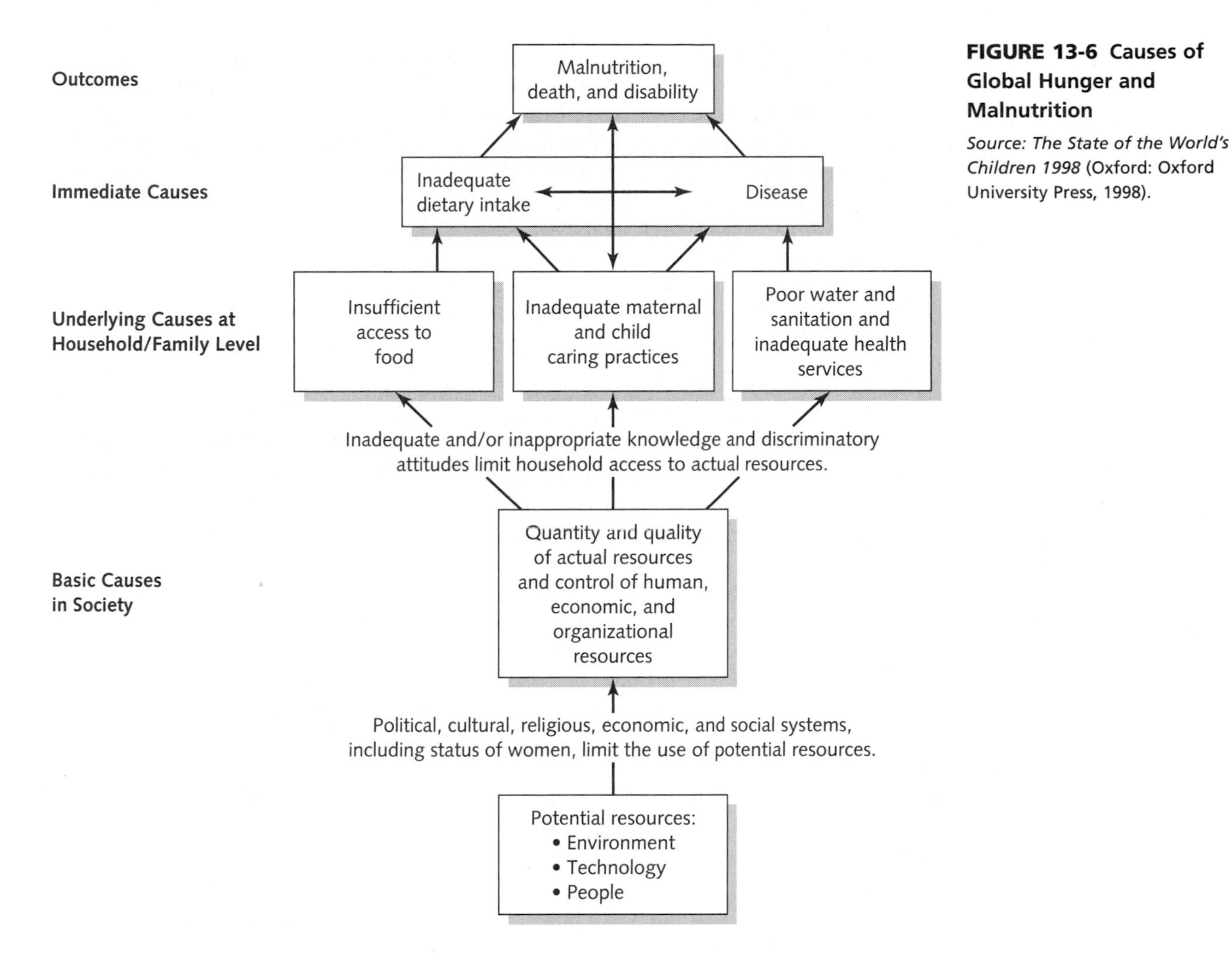

FIGURE 13-6 Causes of Global Hunger and Malnutrition

Source: The State of the World's Children 1998 (Oxford: Oxford University Press, 1998).

Poverty contributes to hunger in many important ways. Oftentimes, people who are poor are powerless to change their situation because they have little access to vital resources such as education, training, food, health services, and other vehicles of change. The roots of hunger and poverty, like those of many other current problems, can be found in numerous historical and natural developments, including colonialism, economic institutions, corporate systems, population pressure, resource distribution, and agricultural technology.

The Role of Colonialism

The colonial era led to hunger and malnutrition for millions of people in developing countries. Although no longer called colonialism, much of this same activity still continues today. The African experience provides a good example of the colonial process. Britain, the Netherlands, Germany, France, and other nations originally colonized the African continent largely to gain a source of raw materials for industrial use. Accordingly, the colonial powers created a governing infrastructure designed merely to move Africa's minerals, metals, cash crops, and wealth to Europe. They provided few opportunities for education, disrupted traditional family structures and community organization, and greatly diminished the ability of the African people to produce their own food.

Before the Europeans arrived, small landowners throughout much of Africa had cleared forests to grow beans, grains, or vegetables for their own use. With colonialism, wealthy Africans and foreign investors took over the fertile farmland and forced the rural poor onto marginal lands that could produce adequate food only with irrigation and fertilizer, which were beyond the means of the poor. The fertile lands were used to grow cotton, sesame, sugar, cocoa, coffee, tea, tobacco, and livestock for export. As more raw materials were exported, more food and manufactured products had to be imported. Imported goods cost money—also beyond the reach of the poor.

Per capita production of grains for food use has declined for the last 25 years in Africa, while sugar cane production has doubled and tea production has quadrupled. The country of Chad recently harvested a record cotton crop in the same year that it experienced an epidemic of famine. Sixty percent of the gross national product of Ghana, Sudan, Somalia, Ethiopia, Zambia, and Malawi is derived from cash crops that finance both luxury imported goods for the minority and international debts. Huge amounts of soybeans and grains are fed to livestock to produce protein foods that the poor cannot afford to purchase.

International Trade and Debt

Over the years, developing countries have seen the prices of imported fuels and manufactured items rise much faster than the prices they receive for their export goods, such as bananas, coffee, and various raw materials, on the international market. The combination of high import costs and low export profits often pushes a developing country into accelerating international debt that sometimes leads to bankruptcy.

Debt and trade are closely related to the progress a country can make toward achieving adequate diets for its people. As import prices increase relative to export prices, a country's money "moves abroad" to pay for the imports. With more and more of its money abroad, the country is forced to borrow money, usually at high interest rates, to continue functioning at home. Many of its financial resources must then go to pay the interest on the borrowed money, thus draining the economy further. Creditor nations may not demand much, or any, capital back, but they do require that interest be paid each year, and the interest can consume most of a country's gross national product. Large and growing debts can slow or halt a nation's attempt to deal effectively with its problems of local food insecurity. As more and more of its financial resources are used to pay interest on the country's trade debts, less and less money is available to deal with food insecurity at home.[30] Each year, the debt crisis worsens and leads to further problems with hunger.

Since 1996, 27 countries have benefited from some debt relief under the *Heavily Indebted Poor Countries (HIPC) Initiative*—a collaborative effort between the World Bank and other financial institutions.[31] This initiative has identified one of the key steps in solving debtor countries' hunger problems. Once the tremendous financial drain caused by their international debt is eliminated, the countries can choose to allocate more financial resources for the tasks of developing the infrastructure, agriculture, and other types of development that would lead to less hunger, such as investments in education, health, nutrition, water supply, and sanitation. Enhancing market access to help countries diversify and expand trade is also necessary. Trade policies in rich countries remain highly discriminatory against developing country exports.[32]

The Role of Multinational Corporations

National economic policies in the developing countries based on the export of cash crops, such as coffee, often have a negative effect on household food security. The competition between cash crops and food crops for farmland provides a classic example of the plight of the poor. Typically, the fertile farmlands are controlled by large landowners and **multinational corporations** that hire indigenous people for below-subsistence wages to grow crops to be exported for profit, leaving little fertile land for the local farmers to grow food. The local people work hard cultivating cash crops for others, not food crops for themselves. The money they earn is not even enough to buy the products they help produce. As a result, imported foods—bananas, beef, cocoa, coconuts, coffee, pineapples, sugar, tea, winter tomatoes, and the like—fill the grocery stores of developed countries, while the poor who labored to grow these foods have less food and fewer resources than when they farmed the land for their own use. Additional cropland is diverted for non-food cash crops such as tobacco, rubber, and cotton. These practices have also had an adverse effect on many U.S. farmers. The foreign cash crops often undersell the same U.S.-grown produce. The U.S. farmers cannot compete with these lower-priced imported foods, so they may be forced out of business.

Export-oriented agriculture thus consumes the labor, land, capital, and technology that is needed to help local families produce their own food. For example, the resources used to produce bananas for export could be reallocated to provide food for the local people. Some have suggested that the developed countries could help alleviate the world food problem not by *giving* more food aid to the poor countries, but by *taking* less food away from them.[33]

Countless examples can be cited to illustrate how natural resources are diverted from producing food for domestic consumption to producing luxury crops for those who can afford them:

- Africa is a net *exporter* of barley, beans, peanuts, fresh vegetables, and cattle (not to mention luxury crops such as coffee and cocoa), yet 40 percent of Africans cannot obtain sufficient food on a day-to-day basis, and Africa has a high incidence of protein-energy malnutrition among young children.[34]
- Over half of the U.S. supply of several winter and early spring vegetables comes from Mexico, where infant deaths associated with poor nutrition are common.
- Half of the agricultural land in Central America produces food for export, while in several Central American countries the poorest 50 percent of the population eat only half the protein they need.[35]

Besides diverting acreage from the traditional staples of the local diet, some multinational corporations also contribute to hunger through their marketing techniques. Their advertisements lead many consumers with limited incomes to associate products such as cola beverages, cigarettes, infant formulas, and snack foods with Western culture and prosperity. A poor family's nutritional status suffers when its tight budget is pinched further by the purchase of such goods.

The United Nations has commissioned several studies in the hope of establishing an international code of conduct for multinational corporations.[36] These corporations could increase the credit and capital available to the developing world; and these resources, if properly used, could help to eliminate food insecurity. The multinational corporations also possess the scientific knowledge and organizational skills needed to help develop improved food and agricultural systems. Experience shows, however, that sustained outside pressure may have to be applied to some of these corporations to help ensure that human needs do not become subordinate to political and financial gains.

Multinational corporations Transnational companies (TNC) with direct investments and/or operative facilities in more than one country. U.S. oil and food companies are examples.

The Role of Overpopulation

The current world population is approximately 6 billion, and the United Nations projects 7.8 billion by the year 2025. The earth may not be able to support this many people adequately.[37] The world's present population is certainly of concern, as is the projected

World Population Growth

1950 2.5 billion
1975 4.1 billion
2000 6.0 billion
2025 7.8 billion
2050 9.0 billion

The transition of population growth rates from a slow-growth stage (high birth rates and high death rates) through a rapid-growth stage (high birth rates and low death rates) to a low-growth stage (low birth rates and low death rates) is known as the demographic transition.

increase in that population. Nevertheless, population is only one aspect of the world food problem. Poverty seems to be at the root of both problems—hunger and overpopulation.

Three major factors affect population growth: birth rates, death rates, and standards of living. Low-income countries have high birth rates, high death rates, and low standards of living. When people's standard of living rises, giving them better access to health care, family planning, and education, the death rate falls. In time, the birth rate also falls. As the standard of living continues to improve, the family earns sufficient income to risk having smaller numbers of children. A family depends on its children to cultivate the land, secure food and water, and provide for the adults in their old age. Under conditions of ongoing poverty, parents will choose to have many children to ensure that some will survive to adulthood. Children represent the "social security" of the poor. Improvements in economic status help relieve the need for this "insurance" and so help reduce the birth rate. The relationships between the infant mortality rate and the population growth rate reveal that hunger and poverty reflect both the level of national development and the people's sense of security.[38]

In many countries where economic growth has occurred and all groups share resources relatively equally, the rate of population growth has decreased. Examples include Costa Rica, Sri Lanka, Taiwan, and Malaysia. In countries where economic growth has occurred but the resources are unevenly distributed, population growth has remained high. Examples include Brazil, Mexico, the Philippines, and Thailand, where a large family continues to be a major economic asset for the poor.[39]

As the world's population continues to grow, it threatens the world's capacity to produce adequate food in the future. The activity of billions of human beings on the earth's limited surface is seriously and adversely affecting our planet: wiping out many varieties of plant life, using up our freshwater supplies, and destroying the protective ozone layer that shields life from the sun's damaging rays—in short, overtaxing the earth's ability to support life.[40] Population control is one of the most pressing needs of this time in history. Until the nations of the world resolve the population problem, they must all deal with its effects and make efforts to support the life of the populations that currently exist.

Distribution of Resources

Land reform—giving people a meaningful opportunity to produce food for local consumption, for example—can combine with population control to increase everyone's assets. However, in much of the developing world, control over land and other assets is highly inequitable. Resources are distributed unequally not only between the rich and the poor within nations, but also between rich and poor nations. But if the wealthy nations simply give aid to the poor nations, the poor nations will be weakened further. Instead, the wealthy nations must foster self-reliance in the poor nations. Doing so will initially require some economic sacrifices by the wealthy nations but will ultimately benefit large numbers of hungry people.

Developing nations must be allowed to increase their agricultural productivity. Much is involved, but to put it simply, poor nations must gain greater access to five things simultaneously: land, capital, water, technology, and knowledge.[41] Increasing poor people's access to assets, including credit, is also essential to ending hunger (see Figure 13-7). A number of microcredit initiatives, such as the Grameen Bank in Bangladesh, offer small loans to help very poor women generate income through small-scale projects such as basket weaving and chicken raising. As the microenterprise income raises the women out of poverty, nutritional benefits can be seen. For example, their children have increased arm circumferences and their daughters are more likely to be enrolled in school.[42] Equally important, each nation must make improving the condition of all its people a

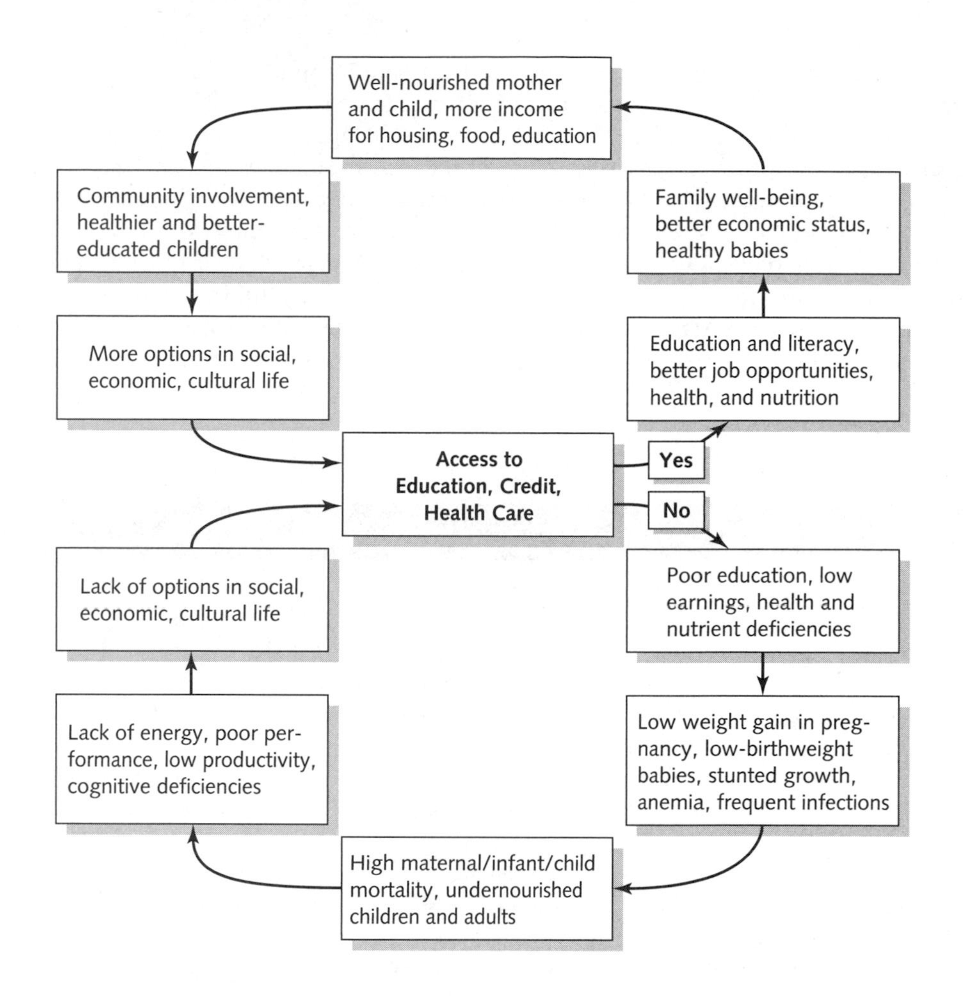

FIGURE 13-7 Breaking Out of the Cycle of Despair

Source: Adapted from Bread for the World Institute, *Hunger 1996: Causes of Hunger* (Silver Spring, MD: Bread for the World Institute, 1995), p. 91.

political priority. International food aid may be required temporarily during the development period, but eventually this aid will be less and less necessary.

Agricultural Technology

Governments can learn from recent history the importance of developing local agricultural technology. A major effort made in the 1960s and 1970s—the **green revolution**—demonstrated both the potential for increased grain production in Asia and the necessity of considering local conditions. The industrial world made an effort to bring its agricultural technology to the developing countries, but the high-yielding strains of wheat and rice that were selected required irrigation, chemical fertilizers, and pesticides—all costly and beyond the economic means of many of the farmers in the developing world.

International research centers need to examine the conditions of developing countries and orient their research toward **appropriate technology**—labor-intensive rather than energy-intensive agricultural methods. Instead of transplanting industrial technology to the developing countries, small, efficient farms and local structures for marketing, credit, transportation, food storage, and agricultural education should be developed.

For example, labor-intensive technology, such as the use of manual grinders for grains, is appropriate in some places because it makes the best use of human, financial, and natural resources. A manual grinder can process 20 pounds of grain per hour, replacing the

Green revolution The development and widespread adoption of high-yielding strains of wheat and rice in developing countries. The term *green revolution* is also used to describe almost any package of modern agricultural technology delivered to developing countries.

Appropriate technology A technology that utilizes locally abundant resources in preference to locally scarce resources. Developing countries usually have a large labor force and little capital; the appropriate technology would therefore be labor-intensive.

Biotechnology The use of biological systems or living organisms to make or modify products. Includes traditional methods used in making products such as wine, beer, yogurt, and cheese; cross breeding to enhance crop production; and modification of living plants, animals, and fish through the manipulation of genes (the latter is called genetic engineering).

mortar and pestle, which can grind a maximum of only 3 pounds in the same time.[43] The specific technology that is appropriate for use varies from situation to situation.

Biotechnology may result in the development of drought-tolerant crop varieties with increased yield and resistance to pests and plant diseases. Some researchers believe that a form of plant biotechnology known as **genetic engineering** may also help mitigate problems of malnutrition by enhancing the nutritional content of staple foods, such as rice high in beta-carotene and iron.[44] However, more research is needed to determine the long-term environmental and health effects of the large-scale utilization of genetically modified crops.[45] The accompanying box highlights some of the potential benefits and risks associated with genetically altered crops, and gives examples of genetically modified crops that are currently available.[46]

Risks vs. Benefits: The Debate over Genetically Modified (GM) Crops

Potential Benefits

- **Increased nutritional value of staple foods:** Genes are being inserted into rice to make it produce beta-carotene, which the body converts into vitamin A. This experimental transgenic "golden rice" has the potential to reduce vitamin A deficiency, a leading cause of blindness and a significant factor in many child deaths.
- **Reduced environmental impact:** Scientists are developing trees with modified cell lignin content. When used to make pulp and paper, the modified wood requires less processing with harsh chemicals.
- **Increased fish yield:** Researchers have modified the gene that governs growth hormones in tilapia, a farmed fish, offering the prospect of increased yield and greater availability of fish protein in local diets.
- **Increased nutrient absorption by livestock:** Animal feed under development will improve animals' absorption of phosphorus. This reduces the phosphorus in animal waste, which pollutes groundwater.
- **Tolerance of poor environmental conditions:** Scientists are working to produce transgenic crops that are drought-resistant or salt-tolerant, allowing the crops to be grown on marginal land.

Potential Risks

- **Inadequate controls:** Although safety regimes are being improved, control over GMO releases is not completely effective. In 2000, for example, a maize variety cleared only for animal consumption was found in food products.
- **Transfer of allergens:** Allergens can be transferred inadvertently from an existing to a target organism, and new allergens can be created. For example, when a Brazil nut gene was transferred to soybean, tests found that a known allergen had also been transferred. However, the danger was detected in testing, and the soybean was not released.
- **Unpredictability:** GM crops may have unforeseen effects on farming systems—for example, by taking more resources from the soil or using more water than normal crops.
- **Undesired gene movement:** Genes brought into a species artificially may cross accidentally to an unintended species. For example, resistance to herbicide could spread from a GM crop into weeds, which could then become herbicide-resistant themselves.
- **Environmental hazards:** GM fish might alter the composition of natural fish populations if they escape into the wild. For example, fish that have been genetically modified to eat more in order to grow faster might invade new territories and displace native fish populations.

Some GMOs Currently Available

GMO* SPECIES	GENETIC MODIFICATION	SOURCE OF GENE	PURPOSE OF GENETIC MODIFICATION
Maize	Insect resistance	*Bacillus thuringiensis*	Reduced insect damage
Soybean	Herbicide tolerance	*Streptomyces* spp.	Greater weed control
Cotton	Insect resistance	*Bacillus thuringiensis*	Reduced insect damage
Rice	Pro-Vitamin A	*Erwinia*	Increase vitamin A supply
		Daffodil	

*GMO = Genetically modified organism; full biotechnology glossary available at www.fao.org/biotech/gloss.htm.

Source: The Food and Agriculture Organization of the United Nations; available at www.fao.org.

A Need for Sustainable Development

Environmental concerns must be taken more seriously as well. The amount of land available for crop production is as important as the condition of the soil and the availability of water. Soil erosion is now accelerating on every continent at a rate that threatens the world's ability to continue feeding itself. Erosion of soil has always occurred; it is a natural process. But in the past, processes that build up the soil—such as the growth of trees—have compensated for erosion.

Where forests have already been converted to farmland and there are no trees, farmers can practice crop rotation, thus alternating soil-devouring crops with soil-building crops. An acre of soil planted one year in corn, the next in wheat, and the next in clover loses 2.7 tons of topsoil each year, but if it is planted only in corn three years in a row, it loses 19.7 tons a year. When farmers must choose whether to make three times as much money planting corn year after year or to rotate crops and earn less money, many choose the short-term profits. Ruin may not follow immediately, but it will follow.[47]

There is a growing recognition that governments need to encourage and support efforts at sustainable development. **Sustainable development** is defined as the successful management of agricultural resources to satisfy changing human needs while maintaining or enhancing the natural resource base and avoiding environmental degradation.

Consider that

> Poverty drives ecological deterioration when desperate people overexploit their resource space, sacrificing the future to salvage the present. The cruel logic of short-term needs forces landless families to raise plots in the rain forest, plow steep slopes, and shorten fallow periods. Ecological decline, in turn, perpetuates poverty as degraded ecosystems offer diminishing yields to their poor inhabitants. A self-defeating spiral of economic deprivation and ecological degradation takes hold.[48]

Sustainable development entails the reduction of poverty and food insecurity in environmentally friendly ways. It includes the following four interrelated objectives:[49]

1. Expand economic opportunities for low-income people, to increase their income and productivity in ways that are economically, environmentally, and socially viable in the long term.
2. Meet basic human needs for food, clean water, shelter, health care services, and education.

Genetic engineering The use of biotechnology to alter the genes of a plant in an effort to create a new plant with different traits; in some cases, a plant's gene(s) may be deleted or altered, or a gene(s) may be introduced from different organisms or species; a form of biotechnology. The foods or crops produced are called genetically modified (GM) or genetically engineered (GE).

Sustainable development Development that meets the needs of the present without compromising the ability of future generations to meet their own needs.

3. Protect and enhance the natural environment by managing natural resources in a way that respects the needs of present and future generations.
4. Promote democratic participation by all people in economic and political decisions that affect their lives.

People-Centered Development

The more developed countries, with the exceptions of New Zealand and Australia, are located geographically to the *north* of most of the developing countries.

We have used the word *developing* in this chapter to classify certain countries, but what is development? According to Oxfam America, a nonprofit international agency that funds self-help development and disaster relief projects worldwide, development enables people to meet their essential needs, extends beyond food aid and emergency relief, reverses the process of impoverishment, enhances democracy, and makes possible a balance between populations and resources. It also improves the well-being and status of women, respects local cultures, sustains the natural environment, measures progress in human (not just monetary) terms, and involves change, not just charity. Finally, development requires the empowerment of the poor and promotes the interests of the majority of people worldwide, in the global *North* as well as the *South*.[50]

Development cannot be measured by gross national product or the quantity of the community harvest. Instead, development should serve the people and requires ongoing community involvement and participation in project development and implementation. In Tanzania, the Iringa Nutrition Program involves community members in the assessment and analysis of problems and decisions about appropriate actions.[51] The program is designed to increase people's awareness of malnutrition and thus improve their capacity to take action. Fundamental to the program are the United Nations' child survival activities; these include the regular quarterly weighing of children under 5 years of age in the villages, with discussion of the results by the village health committees. From this analysis, a set of appropriate interventions for solving problems can be identified. Recent evaluations indicate that this process has contributed to significant decreases in infant and child malnutrition and mortality.[52]

The cornerstone of true development was best expressed decades ago by Mahatma Gandhi: "Whenever you are in doubt . . . apply the following test. Recall the face of the poorest and the weakest man whom you may have seen, and ask yourself if the step you contemplate is going to be of any use to him. Will he gain anything by it? Will it restore him to a control over his own life and destiny?"[53]

Nutrition and Development

The United Nations views a healthful, nutritious diet as a basic human right—one that the FAO and WHO are pledged to secure. However, achieving improved nutritional well-being worldwide requires broad action on many issues, including the following:[54]

- Ensuring that the poor and malnourished have adequate access to food
- Preventing and controlling infectious diseases by providing clean water, basic sanitation, and effective health care
- Promoting healthful diets and lifestyles
- Protecting consumers through improved food quality and safety
- Preventing micronutrient deficiencies
- Assessing, analyzing, and global monitoring of the nutritional status of populations at risk
- Incorporating nutrition objectives into development policies and programs

Nutrition and health are now seen as instruments or tools of economic development as well as goals. The inclusion of nutrition objectives in growth and development policies holds the promise of increasing the productivity and earning power of people worldwide. Well-nourished people are more productive, are sick less often, and earn higher incomes.[55] In 1996 the World Food Summit, convened by the Food and Agriculture Organization, served to focus the world's attention on the continuing problem of food insecurity.[56] World leaders from 186 nations adopted the overall goal of reducing by half the number of undernourished people by no later than the year 2015.[57]

More recently, the world's political leaders met at the Millennium Summit of the United Nations to set goals that call for a dramatic reduction in poverty and marked improvements in the lives of the world's poor by the year 2015. The Millenium Development Goals set targets for progress in eight areas: poverty and hunger, primary education, women's equality, child mortality, maternal health, disease, environment, and a global partnership for development (see Table 13-3). The Summit leaders concluded that although sustainable development to overcome chronic undernutrition is difficult both economically and politically, the payoff is high: more people contributing to the economic, social, and cultural life of their communities, nations, and the world.[58]

The World Health Organization's Global Database on National Nutrition Policies and Programs (GDNNPP) monitors and evaluates the progress in implementing the World Declaration and Plan of Action; available at www.who.int/nut/db_pol.htm.

Agenda for Action

Although the problem of world hunger may seem overwhelming, it can be broken down into many small, local problems. Significant strides can then be made toward solving them at the local level. Even if the problem of poverty itself is not immediately or fully solved, progress is possible. Lessons can be learned from the 30 countries—representing about half of the developing world's population—that are on track to meet the World Food Summit goal of cutting extreme poverty rates in half by 2015.[59] Several of these countries applied a twin-track approach: strengthening social safety nets, while at the same time addressing underlying causes of food insecurity with new initiatives to stimulate agricultural production, increase education and employment opportunities, and reduce poverty.[60] The twin-track approach is illustrated in Figure 13-8. Brazil's Zero Hunger Program demonstrates the success of the twin-track program: By buying food for school lunch and other food assistance programs directly from local farmers, it improves dietary intakes of children, increases food availability for families, increases the income of farmers, and contributes to national food security.[61]

Focus on Children

Children are the group most strongly affected by poverty, malnutrition, and food insecurity and its related effects on the environment.[62] However, there is hopeful news for children in developing countries. **GOBI,** a child survival plan set forth by UNICEF, has made outstanding progress in cutting the number of hunger-related child deaths. GOBI is an acronym formed from four simple, but profoundly important, elements of UNICEF's "Child Survival" campaign: growth charts, oral rehydration therapy (ORT), breast milk, and immunization.

GROWTH MONITORING

The use of growth monitoring to determine the adequacy of child feeding requires a worldwide education campaign. A mother can learn to weigh her child every month and chart the child's growth on specially designed paper. She can learn to detect the early stages of hidden malnutrition that can leave a child irreparably retarded in mind and body. Then at least she will know she needs to take steps to remedy the malnutrition—if she can.

GOBI An acronym formed from the elements of UNICEF's Child Survival campaign—Growth charts, Oral rehydration therapy, Breast milk, and Immunization.

FACTOR	GOAL	TARGETS, 2015	PROGRESS, 1990–2004
Poverty	*Eradicate extreme poverty and hunger*	Reduce by half the proportion of people living on less than a dollar a day. Reduce by half the proportion of people who suffer from hunger.	**Mixed.** On current trends and projections, this goal and its related targets will be achieved in aggregate terms, mostly owing to strong economic growth in China and India. However, most sub-Saharan African countries will in all likelihood miss these targets.
Primary education	*Achieve universal primary education*	Ensure that all boys and girls complete a full course of primary schooling.	**Mixed.** Several regions are on target to meet this goal, including Central and Eastern Europe and the Commonwealth of Independent States (CEE/CIS) and Latin America and the Caribbean. East Asia and the Pacific have almost met the target a full decade ahead of schedule. Shortfalls appear likely across sub-Saharan Africa.
Gender equality	*Promote gender equality and empower women*	Eliminate gender disparity in education by 2015.	**Insufficient.** Despite significant progress toward gender parity in primary schools, shortfalls are still likely in about one-third of developing countries at the primary level and in over 40 percent of countries at the secondary level.
Child survival	*Reduce child mortality*	Reduce by two-thirds the mortality rate among children under five.	**Seriously off track.** The fourth MDG is commonly regarded as the furthest from being achieved. Only one region—Latin America and the Caribbean—is on track, although substantial progress has been made in several East Asian countries.
Families and women	*Improve maternal health*	Reduce by three-quarters the maternal mortality ratio.	**Seriously off track.** Only 17 percent of countries, accounting for 32 percent of the developing world's population, are on track.
Health	*Combat HIV/AIDS, malaria, and other diseases*	Halt and begin to reverse the spread of HIV/AIDS. Halt and begin to reverse the incidence of malaria and other major diseases.	**Seriously off track.** HIV prevalence is rising in many countries. Although prevalence rates are highest in southern Africa, the rate of increase is sharpest in Europe and Central Asia, and absolute numbers are large in China and India. Malaria is proving difficult to contain, and the global incidence of tuberculosis is rising.
Water and sanitation	*Ensure environmental sustainability*	Reduce by half the proportion of people without sustainable access to safe drinking water and basic sanitation.	**Mixed.** The world is on track to meet the target for drinking water, as global access to improved drinking water sources increased from 77 percent in 1990 to 83 percent in 2002. However, progress in sub-Saharan Africa has fallen short. Sanitation remains an even greater challenge: On current trends, the target will be missed by a margin of more than half a billion people.
Policy changes in developed countries regarding development aid, trade, and debt	*Develop a global partnership for development*	Deal comprehensively with the debt problems of developing countries. Develop further an open, nondiscriminatory trading and financial system.	Some countries have benefited from debt relief under the Heavily Indebted Poor Countries (HIPC) Initiative, but much more needs to be done. Lower trade barriers are needed to improve market access by developing countries.

Source: UNICEF, *The State of the World's Children, 2005.*

TABLE 13-3 Progress and Setbacks in Efforts to Meet the Millennium Development Goals (MDG)

ORAL REHYDRATION THERAPY (ORT)

Most children who die of malnutrition do not starve to death—they die because their health has been compromised by dehydration from infections causing diarrhea. Until recently, there was no easy way of stopping the infection–diarrhea cycle and saving their lives; now, the spread of **oral rehydration therapy (ORT)** is preventing an estimated

FIGURE 13-8 A Twin-Track Strategy to Escape Poverty and Eliminate Hunger
The Food and Agriculture Organization (FAO) believes that the World Food Summit goal of "reducing the number of undernourished people by half by no later than 2015" is both attainable and affordable. FAO recommends a twin-track strategy that addresses both the *causes* and the *consequences* of poverty and hunger. *Track One* includes interventions for improving personal income and food availability for the poor by strengthening income-generating opportunities (for example, by improving the productivity of small farmers). *Track Two* supplies a safety net to provide direct assistance to the most vulnerable groups, including pregnant and lactating women, infants and small children, school children, unemployed urban youth, and the elderly, disabled, and sick, including people living with HIV/AIDS.

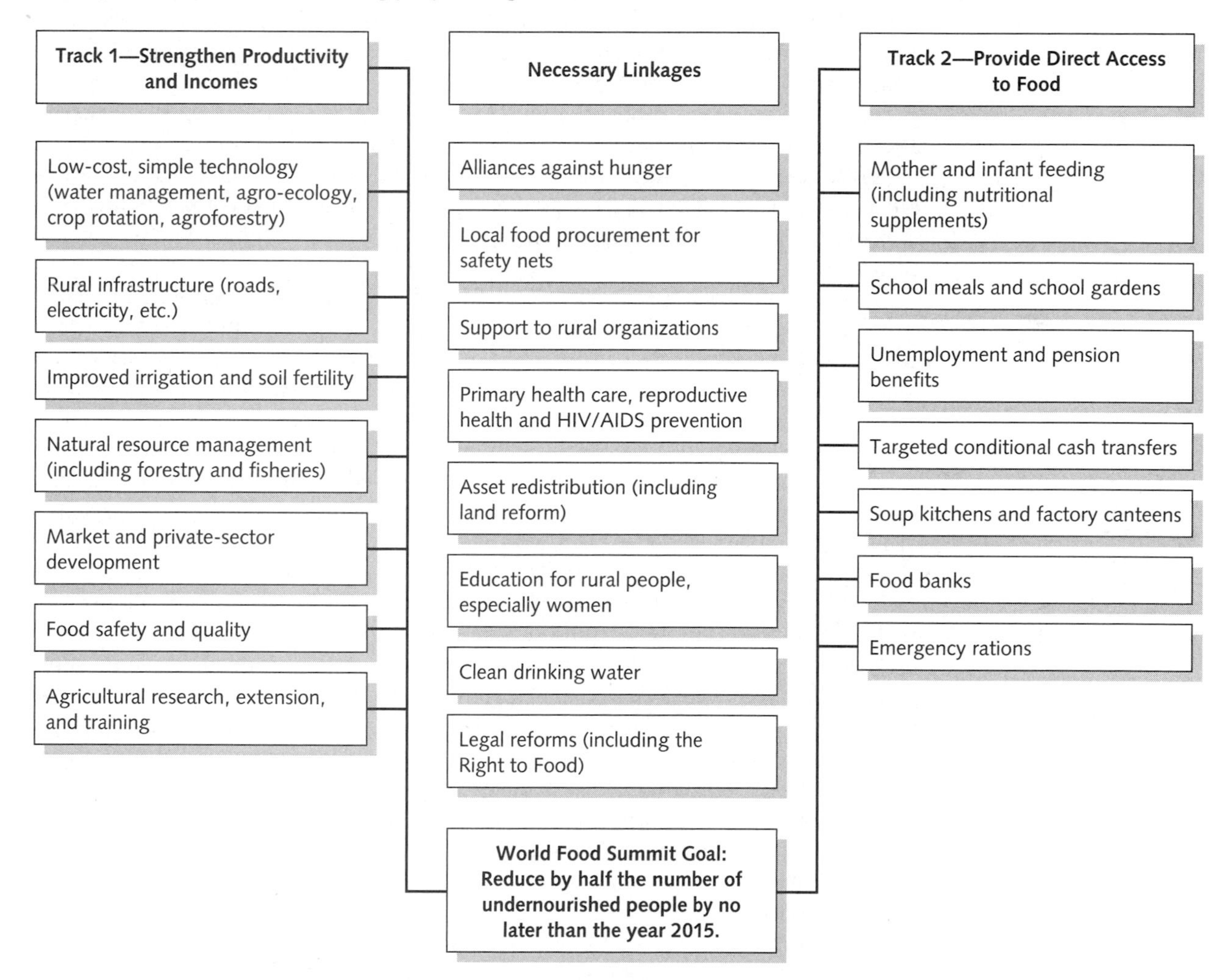

Source: Adapted from FAO, *The State of Food Insecurity in the World 2004*, p. 32.

1 million dehydration deaths each year.[63] ORT involves the administration of a simple solution that mothers can make themselves, using locally available ingredients; the solution increases a body's ability to absorb fluids 25-fold.[64] International development groups also provide mothers with packets of premeasured salt and sugar to be mixed with boiled water in rural and urban areas. A safe, sanitary supply of drinking water is a prerequisite for the success of the ORT program. Contaminated water perpetuates the infection–diarrhea cycle.

Oral rehydration therapy (ORT) The treatment of dehydration (usually due to diarrhea caused by infectious disease) with an oral solution; as developed by UNICEF, ORT is intended to enable a mother to mix a simple solution for her child from substances that she has at home.

PROMOTION OF BREASTFEEDING

Until the middle of the twentieth century, in most developing countries, babies were breastfed for their first year of life—with supplements of other milk and cereal gruel added to their diets after the first several months.[65] Despite improved overall rates of breastfeeding during the 1990s, fewer than half of all infants are now being exclusively breastfed for up to 4 months. The global recommendation now is for exclusive breastfeeding for the first 6 months, joined thereafter by timely complementary, or semi-solid, foods along with continued breastfeeding up to 2 years of age.[66] A number of factors contributed to this unfortunate decline, including the aggressive promotion and sale of infant formula to new mothers; the encouragement of bottle feeding by health care practitioners, who send mothers home from the hospital with free samples of formula after delivery of the newborn; and the global pattern of urbanization and accompanying loss of cultural ties supporting breastfeeding, combined with more women working outside the home.[67] Overall, the World Health Organization (WHO) estimates that improved breastfeeding practices could save the lives of more than 1.5 million children each year.[68]

In 1981, the World Health Assembly—comprising the health ministers of nearly all countries—adopted the International Code of Marketing of Breastmilk Substitutes. The code stipulates that health facilities must never be involved in the promotion of infant formula and that free samples should not be provided to new mothers. In 1992, WHO and UNICEF began a Baby-Friendly Hospital Initiative to help transform hospitals into centers that promote and support good infant feeding practices.

Table 10-9 on page 325 provides a list of Ten Steps to Successful Breastfeeding from the WHO Baby-Friendly Hospital Initiative.

The Baby-Friendly Hospital Initiative has fostered an increased awareness of the importance of breastfeeding worldwide and of exclusive breastfeeding in the early months of life; as a consequence, the average duration of breastfeeding is beginning to increase globally.[69] Today, there are more than 15,000 Baby-Friendly Hospitals in 134 countries.

The finding that HIV can be transmitted through breast milk has complicated infant feeding recommendations. New guidelines released by the Joint United Nations Programme on HIV/AIDS, WHO, and UNICEF call for urgent action to educate, counsel, and support HIV-positive women regarding safe infant feeding practices.[70] In developing countries with people at increased risk of other infectious diseases and malnutrition leading to high infant mortality rates, the mortality risk associated with formula feeding may outweigh the risks of acquiring HIV infection.[71] Recent reports indicate that transmission of HIV may be lower among exclusively breastfed 3-month-old infants of HIV-positive women than among such infants who were partially breastfed. However, confirmation of any protective effect of exclusive breastfeeding on the risk of mother-to-infant transmission of HIV is urgently needed.[72]

Replacing breast milk with infant formula in environments and economic circumstances that make it impossible to feed formula safely may lead to infant undernutrition. Breast milk, the recommended food for infants, is sterile and contains antibodies that enhance an infant's resistance to disease. In the absence of sterilization and refrigeration, formula in bottles is an ideal breeding ground for bacteria. More than 1 billion people in developing countries do not have access to safe drinking water, and 2.4 billion lack adequate sanitation.[73] Thus, feeding infants formula prepared with contaminated water often causes infections leading to diarrhea, dehydration, and failure to absorb nutrients. In countries where poor sanitation is prevalent, breastfeeding should take priority over feeding formula. Studies indicate that a bottle-fed infant living in poverty is up to 14 times as likely to die from diarrhea-related causes and 4 times more likely to die of pneumonia than an exclusively breastfed baby.[74]

The promotion of breastfeeding among mothers in developing countries has many benefits. Breast milk is hygienic, readily available, and nutritionally sound, and it provides infants with immunologic protection specific to their environment. In the developing world, its advantages over formula feeding can mean the difference between life and death.

TIMELY AND APPROPRIATE COMPLEMENTARY FEEDING

Breastfeeding permits infants in many developing countries to achieve weight and height gains equal to those of children in developed countries until about 6 months of age, but then the majority of these children fall behind in growth and development because inadequate complementary foods are added to their diets. Even if infants are protected by breastfeeding at first, they must be weaned. The **weaning period** is one of the most dangerous times for children in developing countries for a number of reasons. For one thing, newly weaned infants often receive nutrient-poor diluted cereals or starchy root crops. For another, infants' foods are often prepared with contaminated water, making infection almost inevitable. Attitudes toward food may also affect nutrition. In some areas of India, for example, a child may be forbidden to eat curds and fruit because they are "cold" or bananas because they "cause convulsions."[75]

An important contributor to children's malnutrition in developing countries is the high bulk and low energy content of the available foods. The diet may be based on grains, such as wheat, rice, millet, sorghum, and corn, as well as starchy root crops, such as the cassava, sweet potato, plantain, and banana. These may be supplemented with legumes (peas or beans), but rarely with animal proteins. Infants have small stomachs, and most cannot eat enough of these staples (grains or root crops) to meet their daily energy and protein requirements. They need to be fed more nutrient-dense foods during the weaning period. The most promising weaning foods are usually concentrated mixtures of grain and locally available **pulses**—that is, peas or beans—which are both nourishing and inexpensive.[76] Mothers are advised to continue breastfeeding up to 2 years or beyond while they introduce safe, appropriate, and adequate complementary foods.

IMMUNIZATIONS

As for immunizations (the *I* of GOBI), they could prevent most of the 2 million deaths each year from measles, diphtheria, tetanus, whooping cough, poliomyelitis, and tuberculosis. However, adequate protein nutrition is necessary for vaccinations to be effective; otherwise, the body may use the vaccine itself as a source of protein. It used to be difficult to keep vaccines stable in their long journeys from laboratory to remote villages. Now, however, the discovery of a new measles vaccine that does not require refrigeration has made universal measles immunization for young children possible, and many countries are reporting coverage rates of 80 to 90 percent. The immunization achievements of the last 2 decades are credited with preventing approximately 3 million deaths a year and with protecting many millions more from disease, malnutrition, blindness, deafness, and polio.[77] Still, more than 30 million children in the world are unimmunized because vaccines are unavailable, because health services are poorly provided or inaccessible, or because families are uninformed or misinformed about when and why to bring their children for immunization.[78]

Making the World Fit for Children

The first World Summit for Children in history was convened by UNICEF in September 1990, bringing together representatives of 159 nations for the purpose of making a renewed commitment to improving the plight of the world's children. Significantly, *nutrition* was mentioned for the first time in world history as an internationally recognized human right.[79] The overall goal of ending child deaths and malnutrition was broken

Weaning period The time during which an infant's diet is changed from breast milk to other nourishment.

Pulses A term used for legumes, especially those that serve as staples in the diets of developing countries.

down into specific targets in a Plan of Action agreed upon by the countries in attendance. An immediate result of this summit was an increase in the number of governments actively adopting the child survival strategies of UNICEF and WHO. UNICEF's goals for nutrition and food security included

- A 50 percent reduction in the 1990 levels of moderate to severe malnutrition among children under 5 years old.
- A 50 percent reduction in the 1990 levels of low-birthweight infants.
- The virtual elimination of blindness and other consequences of vitamin A deficiency.

Strategies devised to achieve these goals included universal immunization; oral rehydration therapy; a massive effort to promote breastfeeding as the ideal food for at least the first 6 months of an infant's life; an attack on malnutrition involving nutrition surveillance that focuses on growth monitoring and weighing of infants at least once every month for the first 18 months of life; and nutrition and literacy education that will empower women in developing countries and lead to a reduction in nutrition-related diseases among vulnerable children.[80]

PROGRESS TOWARD MEETING THE WORLD SUMMIT FOR CHILDREN GOALS

The number of malnourished children in the developing world declined from around 174 million in the early 1990s to 150 million today.[81] Half of all malnourished children live in South Asia and more than one fifth in sub-Saharan Africa. The actual number of malnourished children in sub-Saharan Africa has actually increased over the decade, partly because of the lack of progress and the increase in overall population size.

Countries can be grouped by their level of commitment to reducing malnutrition. In Costa Rica, Chile, Argentina, Ecuador, Uruguay, Egypt and Malaysia, fewer than 5 percent of the population are malnourished. Some more populous countries—China, India, and Bangladesh—have designed effective nutrition strategies but lack sufficient resources for implementing them. Other countries, such as Ethiopia, Nigeria, and Nepal, are without strategies because of a lack of political consensus on the causes of the malnutrition problem. Finally, some countries (such as Angola, Haiti, Rwanda, and Sudan) were unlikely to meet the Summit goals because of war or civil strife.

Large-scale efforts in certain areas of Africa and India show promise for reducing the malnutrition rates for children in those regions. Despite economic hardship, Tanzania's Iringa nutrition program, mentioned earlier, has more than halved the rate of severe malnutrition in 3 years. Some progress has been made on vitamin and mineral deficiency goals. Many countries, including Bangladesh, Brazil, India, Malawi, and the Philippines, now use immunization systems to deliver vitamin A capsules. However, there are still 3 million deaths each year from diseases that could have been prevented with vaccination. Nearly 80 developing countries have now banned the free or subsidized distribution of infant formula to new mothers in hospitals and clinics. Although oral rehydration therapy now saves the lives of 1 million children annually, another 3 million children still die of dehydration from diarrheal disease each year. More improvement is needed. The progress that has been made toward meeting the World Summit for Children goals is shown in Table 13-4.[82]

Focus on Women

Women are more susceptible than men to food insecurity and undernutrition for a number of reasons. In addition to their increased nutrient needs during childbearing years, many women in developing countries are responsible, even during their pregnancies, for most of the physical labor required to procure food for their families. The poor nutrition of some women results both from their family's lack of access to food and from unequal distribution of food within the family itself. A woman will feed her husband, children, and other family members first, eating only whatever is left. Furthermore, each time she becomes pregnant, her body's nutrient reserves are drained.

Social beliefs may also limit women's food intakes. In the Indian Punjab, the director of a program aimed at relieving undernutrition in local villages found that an undernutrition rate of 10 to 15 percent persisted even after a major effort was made to provide supplementary foods to families. The majority of those affected were young girls, who were unable to demand their share and were regarded by other family members as not deserving a fair share.[83]

Women make up 50 percent of the world's population. With their children, they represent the majority of those living in poverty. Thus any solution to the problems of poverty and hunger is incomplete—indeed, is hopeless—if it fails to address the role of women in developing countries.[84]

In many countries, over 90 percent of the population live in rural areas. Women living in rural poverty endure oppressive conditions. They are often overworked and underfed, yet they are expected to carry most of the burden of their family's survival. In many cultures, they are the last to get food, although they spend long hours each day procuring water and firewood and pounding grain by hand. In many countries, women in rural areas not only are the primary food producers but also are responsible for child care and food preparation. Often they have to work as harvesters on other people's lands as well. Husbands are frequently absent from their homes—not by choice, but because the changing global economy has forced many men to leave home to find paying jobs. They have gone to look for work in the cities or to find employment growing export crops on distant commercial farms.

Development projects are often large in scale and highly technological, but they frequently overlook women's needs. Typically, only men have access to education and training programs. Yet women play a vital role, biologically and socially, in the well-being of their nation's people. Their nutrition during pregnancy and lactation determines the future health of their children. If women are weakened by malnutrition themselves or ignorant about how to feed their families, the consequences ripple outward to affect many other individuals. The importance of women in these countries is increasingly being appreciated, and many countries now offer development programs with women in mind.

Seven basic strategies are at the heart of women's programs:

- Removing barriers to financial credit so that women can obtain loans for raw materials and equipment to enhance their role in food production
- Providing access to time-saving technologies—seed grinders, for example
- Providing appropriate training to make women self-reliant
- Teaching management and marketing skills to help women avoid exploitation
- Making health and day care services available to provide a healthful environment for the women's children
- Forming women's support groups to foster strength through cooperative efforts
- Providing information and technology to promote planned pregnancies

The recognition of women's needs by some development organizations is an encouraging trend in the efforts to contend with the world hunger crisis. The following examples from Sierra Leone and Ghana illustrate how women's development programs work:[85]

> Balu Kamara is a farmer in Sierra Leone in West Africa, where farming is difficult, particularly for women. There women have little money and must take out loans to buy seed rice and to pay for the use of oxen. The price of rice is so low, though, that at the end of the

WORLD SUMMIT GOAL	PROGRESS MADE	UNFINISHED BUSINESS
Infant and under-5 mortality: reduction by one-third in infant mortality and U5MR	• More than 60 countries achieved the U5MR goal. • At the global level, U5MR declined by 11 percent.	• U5MR rates increased in 14 countries (9 of them in sub-Saharan Africa) and were unchanged in 11 others. • Serious disparities remain in U5MR within countries: by income level, urban vs. rural, and among minority groups.
Polio: global eradication by 2000	• More than 175 countries are polio-free.	• Polio is still endemic in 20 countries.
Routine immunization: maintenance of a high level of immunization coverage	• Sustained routine immunization coverage is at 75 percent for three doses of combined diphtheria/pertussis/tetanus vaccine (DPT3).	• Less than 50 percent of children under 1 year of age in sub-Saharan Africa receive DPT3.
Measles: reduction by 95 percent in measles deaths and 90 percent in measles cases by 1995 as a major step to global eradication in the longer run	• Worldwide reported measles incidence declined by almost 40 percent between 1990 and 1999.	• In 14 countries, measles vaccination coverage is less than 50 percent.
Deaths due to diarrhea: reduction by 50 percent	• This goal was achieved globally, according to WHO estimates.	• Diarrhea remains one of the major causes of death among children.
Acute respiratory infections (ARI): reduction of ARI deaths by one-third in children under 5	• ARI case management has improved at the health center level.	• ARI remains one of the greatest causes of death among children.
Malnutrition reduction by half of severe and moderate malnutrition among under-5 children	• Malnutrition declined by 17 percent in developing countries. South America achieved the goal with a 60 percent reduction in underweight prevalence.	• 150 million children are still malnourished, more than three-fourths of them in Asia. The absolute number of malnourished children has increased in Africa.
Low birthweight: reduction of the rate of low birthweight (less than 2.5 kilograms) to less than 10 percent	• To date, 100 developing countries have low-birthweight levels under 10 percent.	• Over 9 million newborns in South Asia and over 3 million newborns in sub-Saharan Africa each year are of low birthweight.
Vitamin A deficiency: virtual elimination by the year 2000	• More than 40 countries reach the large majority of their children (over 70 percent) with at least one high-dose vitamin A supplement a year. UNICEF estimates that as many as 300,000 child deaths a year have been prevented in this way.	• In the least developed countries, 40 percent of children are not receiving even one high-dose vitamin A supplement—and the majority of those who get one dose do not receive the required second dose.

TABLE 13-4 Progress Made toward Meeting the World Summit for Children Goals

growing season the women do not earn enough money to repay their loans. Yet, as the economy worsens, it is up to the women to carry the burdens; it is up to the women to stretch what resources are available to feed their families regardless of hardships.

Balu is the leader of the Farm Women's Club, a basket cooperative the women formed to make and sell baskets so they could pay their debts and continue farming. Finding time to weave baskets is difficult. Yet the women and their cooperative are succeeding. On the value of the Farm Women's Club, Balu says, "We have access to credit and a cash income. We have the opportunity to learn improved methods of agriculture and marketing and to increase our belief in ourselves and ease our families through the hungry season."

Gari (processed cassava) is becoming increasingly popular in Ghana because of the shortage of many other food items and because, once prepared, it is easy to cook. But it is very time-consuming to prepare gari—peeling and grating the fresh cassava, fermenting it over several days, squeezing the water from the fermented cassava, and finally roasting it over a wood fire.

WORLD SUMMIT GOAL	PROGRESS MADE	UNFINISHED BUSINESS
Iodine-deficiency disorders: virtual elimination	• Some 72 percent of households in the developing world are using iodized salt, compared to less than 20 percent at the decade's beginning. As a result, 70 million newborns are protected yearly from significant loss in learning ability.	• There are still 35 countries where less than half the households consume iodized salt.
Breastfeeding: empowerment of all women to breastfeed their children exclusively for 4 to 6 months and to continue breastfeeding, with complementary food, well into the second year of life	• Exclusive breastfeeding rates increased over the decade. • Gains were also made in timely complementary feeding.	• Only about half of all infants are exclusively breastfed for at least the first 4 months of life.
Growth monitoring: growth promotion and regular growth monitoring of children to be institutionalized in all countries by the end of the 1990s	• A majority of developing countries have implemented growth-monitoring and promotion activities.	• Growth-monitoring information is often not used as a basis for community, family, or government action.
Household food security: dissemination of knowledge and supporting services to increase food production	• The number of people in developing countries lacking sufficient calories in their diets has decreased marginally.	• In sub-Saharan Africa, about one-third of the people lack sufficient food.
Water: universal access to safe drinking water	• 900 million people obtained access to improved water.	• Some 1 billion people still lack access.
Sanitation: universal access to sanitary means of excreta disposal	• 987 million additional people gained access to decent sanitation facilities.	• 2.4 billion people, including half of all Asians, lack access. • 80 percent of those lacking sanitation live in rural areas.
Gender disparities in education: reduction of current disparities between boys and girls	• The primary school enrollment gap between girls and boys has been halved from 6 percent to 3 percent.	• The gender gap has not narrowed sufficiently over the decade in sub-Saharan Africa.
Adult literacy: reduction of adult illiteracy rate to at least half of its 1990 rate, with emphasis on female literacy	• Adult illiteracy has declined from 25 percent to 20 percent.	• The absolute number of illiterate adults has remained at nearly 900 million over the last decade. • Illiteracy is increasingly concentrated among women, especially in South Asia and sub-Saharan Africa.

Source: Adapted from UNICEF, *We the Children: Meeting the Promises of the World Summit for Children* (New York: UNICEF, June 2001); and *State of the World's Children 2005;* available at www.unicef.org.

TABLE 13-4 Progress Made toward Meeting the World Summit for Children Goals—*continued*

To help village women in the Volta Region increase their income through gari processing, an improved technology was introduced with the help of the National Council on Women and Development. The process involves a special mechanical grater, a pressing machine to squeeze the water from the grated cassava, and a large enamel pan for roasting. This pan holds ten times the volume of the traditional cassava pot. The system was developed locally, with advice from the women themselves contributing to the success of the project.

Before, the women produced 50 gari bags every week. Now they are able to produce 5,000 to 6,000 bags a week. However, this increased output of gari can be maintained only with a higher yield of cassava in the area. Therefore, a male cassava growers' association has been formed to step up cassava production, and a tractor has been acquired by the women's cooperative to put more land under cassava cultivation.

International Nutrition Programs

Nutrition programs in developing countries vary considerably. A sample job description for a nutrition specialist working with the FAO is shown in Figure 13-9. International nutrition programs include both large-scale and small-scale operations, may be supported with private or public funds, and may focus on emergency relief or long-term development.[86] In developing countries, emphasis has generally been placed on four types of nutrition interventions:

1. Breastfeeding promotion programs with guidance on preparing appropriate weaning foods

FIGURE 13-9 Sample Job Description for a Nutrition Specialist with the Food and Agriculture Organization

Source: Adapted from Vacancy Announcement (No. 130-ESN) of the Food and Agriculture Organization of the United Nations. Used with permission.

Position Title: Nutrition Officer (Nutrition Information)
Responsible To: Senior Officer (Nutrition Assessment)
Level Grade: P–4

Responsibilities:
Under the general direction of the Chief of the Nutrition Planning, Assessment and Evaluation Service, Food Policy and Nutrition Division, and the supervision of the Senior Officer (Nutrition Assessment), this specialist is responsible for the development of activities aimed at improving information necessary for assessing and analyzing nutritional status of populations at global, national, and local levels, especially in the drought-prone African countries.

Job Duties:
In accordance with departmental policies and procedures, this specialist carries out the following duties and responsibilities:

- Assists member countries in the design and implementation of data collection and analysis activities to meet the needs of policy makers and planners concerned with food and nutrition issues.
- In cooperation with other organization units, strengthens the skills and technical capacity of member countries, particularly within the agricultural sector, to assess, analyze, and monitor the food and nutrition situation of their own population.
- Provides data and background information for Expert Committees and Councils.
- Provides consultations on energy and nutrient requirements, dietary assessment methodology, and foodways.
- Prepares relevant reports for the organization and its committees.
- Performs other related duties.

Qualifications and Experience—Essential
The specialist should have the following educational qualifications and experience:

- A university degree in Nutrition or Biological Science with postgraduate qualification in Nutrition.
- Seven years of progressively responsible professional experience, including the planning and implementation of nutrition data collection, processing, and analysis.
- Experience in the management and analysis of nutrition-related data using advanced statistical applications on microcomputers.

Qualifications and Experience—Desirable

- Experience in international work and research.
- Relevant work experience in Africa.
- Experience in the organization of seminars, training courses, or conferences.

Skills

- Working knowledge of English and French (level C).
- Demonstrated ability to write related technical reports. Familiarity with project design techniques.
- Courtesy, tact, and ability to establish and maintain effective work relationships with people of different national and cultural backgrounds.
- Experience using word-processing equipment and software.

2. Nutrition education programs typically focusing either on infant and child feeding guidelines and practices and child survival activities or on the incorporation of nutrition education into primary school curricula and teacher training programs
3. Food fortification and/or the distribution of nutrient supplements (e.g., vitamin A capsules) and the identification of local food sources of nutrients in short supply
4. Special feeding programs designed to provide particularly vulnerable groups with nutritious supplemental foods[87]

In many countries, there is mounting evidence of grass roots progress in improving agricultural, water, education, and health services, especially for children.[88] Experiences in Sierra Leone and Nepal are encouraging examples.

In Sierra Leone, a food product was developed from rice, sesame (benniseed), and peanuts that were hand pounded and cooked to make a flour meal. The local children not only found it tasty, but whereas they had been malnourished before, they thrived when this product supplemented their diets. The village women formed a cooperative to reduce the drudgery of preparing the food and rotated the work on a weekly or monthly basis.[89] The government also established a manufacturing plan to produce and market the mixture at subsidized prices. The success of the venture can be directly attributed to the involvement of the local people in identifying the problem and devising a solution that met their needs.

A similar success story unfolded in Nepal. A supplementary food made from soybeans, corn, and wheat, mixed in a ratio of 2 to 1, yielded a concentrated "superflour" of high biological value suitable for young children. A nutrition rehabilitation center tested this superflour by giving undernourished children and their mothers two cereal-based meals a day and giving the children three additional small meals of superflour porridge daily. Within 10 days, the undernourished children had gained weight, lost their edema, and recovered their appetites and social alertness. The mothers, who saw the remarkable recoveries of their children, were motivated to learn how to make the tasty supplementary food and incorporate it into their local foodstuffs and customs.

These two examples offer hope, but the real issue of poverty remains to be addressed. One-shot intervention programs offering nutrition education, food distribution, food fortification, and the like are not enough. It is difficult to describe the misery a mother feels when she has received education about nutrition but cannot grow or purchase the foods her family needs. She now knows *why* her child is sick and dying but is unable to *apply* her new knowledge.

Looking Ahead: The Global Challenges

In the developing world, the health care crisis revolves around the daily struggle for survival and the growing disparity between the haves and the have-nots. For the poor, the struggle for safe water, adequate nutrition, and access to basic health care leaves no energy or resources for other concerns—and most have little hope of winning the battle. For most of us, living in the developed world, health care reform means we are assured of meeting our own and our family's ongoing health care needs, including access to the latest miracle drug or the availability of a bone marrow transplant. But as one physician from the hospital ship *M/V Anastasis* remarked on visiting Ghana in West Africa, "We are foolhardy to believe that we can be a healthy society when the world around us is languishing with diseases and poverty that we could alleviate."[90]

In addition, we face many new challenges if we are to meet the World Health Organization's main social target of "health for all in the twenty-first century." Consider the following list of issues that we must deal with, among others:

- *The pandemic of HIV/AIDS.* By the end of 2003, an estimated 38 million people in the world were living with HIV/AIDS. More than 95 percent of them are living in developing countries,[91] and approximately 2.1 million are children less than 15 years of age.[92] The majority of these children were born to mothers with HIV, acquiring the virus near the time of birth or during breastfeeding. More than 15 million children worldwide have been orphaned by the epidemic, and 80 percent of these orphans live in sub-Saharan Africa.[93] Children orphaned by the rampant HIV/AIDS pandemic are more likely than other children to be malnourished and unschooled. Poverty itself increases the risk of infection, because AIDS education is hampered by less access to health services and mass media, and by lower levels of literacy, among the poor in many countries.[94] Additionally, the loss of adult lives to AIDS deprives families and communities of healthy workers. Not only do these lost workers make up the most economically productive age group, but they also are the very same household members who previously had the responsibility to care for their elders.

- *The trend toward urbanization.*[95] According to FAO estimates, most of the increase in the world's population between 2000 and 2030 will occur in urban areas.[96] Urbanization has been a factor in the worldwide decline in breastfeeding; the increased consumption of fats, sugars, meat, and wheat; and the outbreak of cholera in several countries due to contamination of the urban water supply. As nations continue the rural-to-urban transition, the incidence of chronic diarrhea from polluted water and foodborne contamination remains a major public health problem in urban slums.

- *Rapid population growth.* The earth's population will increase from 6 to 7.8 billion by 2025—most of these people will be born into poor families in developing countries. Agricultural production will need to increase to feed these people. The number of nonagricultural jobs must also increase to support those not working in agriculture.

- *Destruction of the global environment.* The earth's capacity to sustain life is being impaired by many complex interrelated developments, including overconsumption, industrial pollution, overgrazing, and deforestation. This destruction of natural resources threatens the health and well-being of today's people and future generations as well.

Personal Action: Opportunity Knocks

The problems addressed in this chapter may appear to be so great that they can be approached only through worldwide political decisions. Indeed, the members of the International Conference on Nutrition stressed that intensive worldwide efforts were needed to overcome hunger and malnutrition and to foster self-reliant development. To this end, many individuals and groups are working to improve the future well-being of the world and its people through a number of national and international organizations. Check out the resources available online from the organizations listed in the Internet Resources section that follows.

Consider working with others who have similar interests, follow current hunger-related legislation, and call for change by writing and telephoning local and national political representatives and expressing concern about hunger-related issues. Encourage your church or synagogue or mosque to support both overseas work and domestic outreach efforts to feed the hungry; support these efforts with monetary contributions.[97]

Individuals can help change the world through their personal choices.[98] Our choices have an impact on the way the rest of the world's people live and die. The world food problem derives in part from the demands we in the developed world place on the world's finite natural resources. In a sense, we contribute to the world food problem. People in affluent nations have the freedom and means to choose their lifestyles; people in poor nations do not. We can find ways to reduce our consumption of the world's resources by using only what we need.

It is ironic that whereas other societies cannot secure enough clean water for people to drink, our society produces bottles of soda that contain one calorie of artificial sweetener in 12 ounces of water that cost 800 calories to produce. In the United States, billions of dollars are spent annually to lower calorie consumption, while more than 850 million people in the rest of the world can rarely find an adequate number of calories to consume.[99] Thus, choosing a diet at the level of necessity, rather than excess, would reduce the resource demands made by our industrial agriculture. In fact, those who study the future are convinced that the hope of the world lies in "the widespread simplification of life that is vital to the well-being of the entire human family."[100] Personal lifestyles do matter, for a society is nothing more than the sum of its individuals. As we go, so goes our world.

George McGovern, former U.S. senator and U.S. ambassador to the UN Agencies on Food and Agriculture in Rome, calls us all to action with the following words—excerpted from his book *The Third Freedom: Ending Hunger in Our Time:*

> Hunger is a political condition. The earth has enough knowledge and resources to eradicate this ancient scourge. Hunger has plagued the world for thousands of years. But ending it is a greater moral imperative now than ever before, because for the first time humanity has the instruments in hand to defeat this cruel enemy at a very reasonable cost.
>
> What will it cost if we don't end the hunger that now afflicts so many of our fellow humans? The World Bank has concluded that each year malnutrition causes the loss of 46 million years of productive life, at a cost of $16 billion annually, several times the cost of ending hunger and turning this loss into productive gain.
>
> Of course it is impossible to evaluate with dollars the real cost of hunger. What is the value of a human life? The twentieth century was the most violent in human history. With nearly 150 million people killed by war. But in just the last half of that century nearly three times as many died of malnutrition or related causes. How does one put a dollar figure on this terrible toll silently collected by the Grim Reaper? What is the cost of 800 million hungry people dragging through shortened and miserable lives, unable to study, work, play, or otherwise function normally because of the ever-present drain of hunger and malnutrition on body, mind, and spirit? What is the cost of millions of young mothers breaking under the despair of watching their children waste away and die from malnutrition? This is a problem we can resolve at a fraction of the cost of ignoring it. We need to be about that task now. I give you my word that anyone who looks honestly at world hunger and measures the cost of ending it for all time will conclude that this is a bargain well worth seizing. More often than not, those who look at the problem and the cost of its solution will wonder why humanity didn't resolve it long ago.[101]

Internet Resources

Bread for the World Institute	**www.bread.org**
CARE	**www.care.org**
Catholic Charities USA	**www.catholiccharitiesusa.org**
Church World Service	**www.churchworldservice.org**
Food and Agriculture Organization	**www.fao.org/sd**
Freedom from Hunger	**www.freefromhunger.org**
Global Health Affairs, U.S. Department of Health and Human Services	**www.globalhealth.gov**
Global Health Council	**www.globalhealth.org/**
Hunger Web	**http://nutrition.tufts.edu/academic/hungerweb/**
International Council for Control of Iodine Deficiency Disorders	**www.people.virginia.edu/~jtd/iccidd/mi/idd.htm**
International Food Policy Research Institute	**www.ifpri.org**
International Fund for Agricultural Development (IFAD)	**www.ifad.org**
International Nutritional Anemia Consultative Group (INACG)	**http://inacg.ilsi.org**
International Vitamin A Consultative Group (IVACG)	**http://ivacg.ilsi.org**
Micronutrient Initiative	**www.mn-net.org**
Oxfam America	**www.oxfamamerica.org**
Pan American Health Organization	**www.paho.org**
RESULTS	**www.results.org**
The Hunger Project	**www.thp.org**
UNAIDS	**www.unaids.org**
UNICEF	**www.unicef.org**
United Nations Development Programme (UNDP)	**www.undp.org**
United Nations System Standing Committee on Nutrition	**http://ceb.unsystem.org/hlcp/scn.htm**
United States Agency for International Development	**www.usaid.gov/our_work/global_health/mch**
U.S. National Committee for World Food Day	**www.worldfooddayusa.org**
The World Bank	**www.worldbank.org/data**
World Food Programme	**www.wfp.org**
World Health Organization	**www.who.int/en**
WHO Global Database on Child Growth and Malnutrition	**www.who.int/nut**

Case *Study*

UNICEF's Child Survival Campaign

By Alice Fornari, EdD, RD, and Alessandra Sarcona, MS, RD

Eight dietetic interns, including you, are embarking on a journey to Guatemala for a 4-week community rotation. Each semester for the past 3 years, students from your university have been visiting the same village through Friends World and have done constructive projects such as building a small schoolhouse and a church and cultivating a garden. This is the first year the dietetic interns will participate with a focus on nutrition. The Dietetic Internship Director will also attend to supervise; she is fluent in Spanish (the primary language of Guatemala), as are four of the eight interns. The goal of this rotation is to provide nutrition education on UNICEF's Child Survival campaign called GOBI (an acronym for the following elements of the UNICEF campaign: *growth charts, oral rehydration therapy, breast milk,* and *immunization*). UNICEF has initiated this program in many developing countries and will be setting up GOBI within the year in the Guatemalan village you will be visiting. Another goal of your rotation is to identify data that could be useful for a nutritional assessment and monitoring tool for the children of this village.

Learning Outcomes

- Describe the current status of malnutrition and health worldwide.
- Identify the components of UNICEF's Child Survival Campaign.
- Recognize barriers to meeting nutrition and health goals in developing countries.
- Identify data that would be useful as part of an assessment specific to the GOBI program.

Foundation: Acquisition of Knowledge and Skills

1. Identify at least three organizations working toward the goal of ending world hunger.
2. Describe the current status of malnutrition and health worldwide.
3. Access the World Summit for Children Goals for the Year 2000 at www.action.org/goals.html.
4. Describe the basic principles of UNICEF's GOBI program from your text and at www.rehydrate.org/index.html. Also review the three "Fs": Female education, Family spacing, and Food supplements, that are the focus of the GOBI-FFF Program at www.rehydrate.org/facts/gobi_fff.htm.

Step 1: *Identify Relevant Information and Uncertainties*

1. Which goals from the World Summit for Children are related to the GOBI program?
2. From the Web, note progress made by UNICEF's Child Survival Campaign (GOBI) in other developing countries. Go to: www.rehydrate.org/facts/progress_nutrition.htm
3. List some major contributors to children's malnutrition in developing countries.

Step 2: *Interpret Information*

1. Create a list of data that will be useful to include in a nutritional assessment of the children in this village.
2. Describe some barriers that you and your group may encounter in delivering your nutrition messages relative to the GOBI program to the local population.

Step 3: *Draw and Implement Conclusions*

1. Select an element of GOBI (growth charts, oral rehydration therapy, breast milk, or immunization) that you and your fellow interns would view as a priority during your 4 weeks in the Guatemalan village. Include your rationale for choosing this element.
2. Describe intervention strategies that you might utilize with the Guatemalan community for the element of GOBI that you selected.

Step 4: *Engage in Continuous Improvement*

1. Establish a plan for monitoring the outcomes of the GOBI element you selected that would be consistent with one of the World Summit for Children Goals.

Vitamin A Field Support Projects

At least 100 million children under the age of 5 suffer from vitamin A deficiency (VAD) worldwide.[1] Millions more children consume inadequate amounts of vitamin A–rich foods and are thus at risk for VAD. Manifestations of VAD range from mild xerophthalmia (night blindness and/or Bitot's spots) to dryness of the conjunctiva and cornea and, in severe cases, to melting of the cornea and blindness. In Asia alone, 200,000 children are blinded each year, and two-thirds of these children die soon afterward. Recent epidemiologic research has identified a relationship between marginal VAD, documented by reduced levels of circulating serum retinol, and higher mortality and morbidity rates from infectious diseases in children.[2] Researchers have confirmed that even mild VAD significantly increases the death rate among children aged 6 months to 6 years. In particular, VAD significantly increases the severity of and risk for diarrheal disease, measles, and pneumonia—three of the main health threats facing children in the developing world.[3] Evidence from Africa suggests that vitamin A supplements can substantially reduce mortality and complications among children with measles, presumably by protecting epithelial tissue and ensuring the proper maintenance and functioning of the immune system.[4]

The most common factor contributing to the magnitude of VAD worldwide is the chronic inadequate dietary intake of vitamin A. Other contributors include poor nutritional status of mothers during pregnancy and lactation, low prevalence of breastfeeding, delayed or inappropriate introduction of complementary foods, high incidence of infection (such as diarrhea, acute respiratory infection, and measles), low levels of maternal education, drought, civil strife, poverty, and ecological deprivation in some regions resulting in limited availability of vitamin A–rich foods.[5] For example, the production and consumption of vitamin A–rich foods in Africa (dark green, leafy vegetables, orange-colored fruits and tubers, and red palm oil) are influenced strongly by seasonal trends and cultural practices.[6]

The U.S. Agency for International Development (USAID) support for vitamin A and other micronutrients began 35 years ago with the fortification of nonfat dry milk for U.S. food donation programs. Support from Congress eventually led to funding of the Vitamin A Technical Assistance Program at Helen Keller International in 1988 and the Vitamin A Field Support Project (VITAL) from 1989 to 1994. USAID support for vitamin A and other micronutrients increased throughout the 1990s—launching the OMNI Project, combining support for vitamin A, iron, and iodine nutrition programs worldwide. Building on the experience of the VITAL and OMNI projects and the increased commitment by Congress to combat vitamin A deficiency, USAID in 1998 developed a comprehensive global initiative for micronutrient programs—the Micronutrient Operational Strategies and Technologies (MOST) project.[7] MOST is currently working in 11 countries: Ghana, Madagascar, Uganda, and Zambia in Africa; Bangladesh, India, Morocco, Nepal, and the Philippines in Asia and the near East; and El Salvador and Nicaragua in Latin America.

MOST is a $45 million program administered by USAID for the promotion of activities designed to improve the micronutrient status of at-risk populations throughout the world, especially for vitamin A. Like other programs designed to eradicate and prevent VAD in developing countries, MOST seeks an appropriate balance among dietary diversification, food fortification, and supplementation to deliver micronutrients to at-risk populations in an effective yet affordable way.

Dietary Diversification

MOST strategies include stimulating the production and consumption of vitamin A–rich foods through agricultural production, home gardening, food preservation, nutrition education, and social marketing. For example, MOST promotes the consumption of papayas, an excellent source of vitamin A, by pregnant women in the South Pacific. In some regions, foods containing vitamin A—especially vegetables and fruits—are readily available but are underutilized by vulnerable groups, particularly weaning-age children and pregnant and lactating women, because of traditional customs and beliefs. Consequently, dietary diversification programs in these areas focus on intensive nutrition education and social marketing campaigns to foster necessary community understanding, motivation, and participation.

Home gardens play a critical role in alleviating VAD in many communities. They provide a regular and secure supply of household food.

USAID also sponsors several VAD projects aimed at increasing vitamin A consumption by improving food processing techniques. Solar drying has been introduced as the appropriate technology for the preservation of mangoes, papayas, sweet potatoes, pumpkins, green leaves, and other vitamin A–rich foods in several countries. When solar-dried, these foods retain both flavor and carotene content and can alleviate seasonal variation in the availability of food. For example, the MANGOCOM Project in Senegal attempts to improve the vitamin A intake of weaning-age children by promoting dried mangoes, produced by women's cottage industries, as finger foods and fruit purees for toddlers.[8]

Food Fortification

Fortifying food is generally a large-scale undertaking and will be effective only if the target groups can buy and will consume the fortified product. In several developed countries, commonly consumed foods (such as margarine and milk) have been successfully fortified with vitamin A. Sugar is fortified with the vitamin in several regions of Central America, notably through VAD projects in Guatemala, El Salvador, and Honduras.[9] Pilot trials with vitamin A–fortified rice are under way in Brazil. A pilot program in the Philippines is currently testing vitamin A–fortified margarine as an alternative means of improving the vitamin A status of children.

Distribution of Vitamin A Supplements

Most commonly, VAD intervention programs periodically distribute vitamin A supplements in the form of high-dose capsules or oral dispensers. UNICEF typically donates the vitamin A capsules and helps organize the distribution efforts.* Often, supplements are delivered in conjunction with ongoing local health services, primary health care programs (for example, maternal–child health projects), or national vaccination campaigns. UNICEF estimates that as many as 300,000 child deaths are prevented each year by vitamin A supplementation.

Vitamin A supplementation programs to date have reported a number of operational obstacles: low priority given to distribution of the supplements by primary health care workers, a lack of community demand for vitamin A, and a lack of awareness among policy makers of the critical nature of vitamin A nutritional status. However, similar obstacles have been overcome by immunization programs, which now reach 80 percent of targeted children worldwide.[10] For this reason, WHO and UNICEF have encouraged that vitamin A supplementation be integrated into existing immunization programs: "The provision of vitamin A supplementation through immunization services could dramatically expand the coverage of children in late infancy, giving a boost to vitamin A status before the critical period of weaning." Since immunization programs primarily target infants under 12 months of age, similar coverage would need to be provided by other means to children from 1 to 6 years old.

Supplementation is considered a temporary measure for the control and prevention of VAD. More lasting solutions include cultivating vitamin A–rich foods, fortifying foods with vitamin A, promoting improved food habits, eliminating poverty, and improving sanitation worldwide (see Table 13-5).[11] Therefore, countries pursuing the high coverage achieved by integrating vitamin A supplement distribution into immunization programs are encouraged to allocate resources to these alternative efforts as well.

Fortunately, the problem of VAD is not insurmountable. The numerous VAD projects worldwide have demonstrated three principles critical to successful VAD intervention efforts. First, successful interventions include preliminary formative research (such as focus group interviews), target audience segmentation, pre-testing, and evaluation in program planning. Secondly, support by policy makers and participation by local

TABLE 13-5 Examples of Cost-Effective Interventions to Reduce Micronutrient Malnutrition

INTERVENTION	EXAMPLES
Food/Menu-based	Modification in quantity and/or quality and diversity of menus Improved bioavailability through modified preservation/preparation procedures Household food-to-food fortification
Fortification	Sugar, salt, and other condiments Oils and margarine Cereals and flours
Supplementation	Periodic high doses Frequent low doses
Public health measures	Breastfeeding promotion Immunizations Parasite control Sanitation/safe water
Social and economic developmental measures	Female literacy Family spacing Income generation

Source: Adapted from B. A. Underwood, Micronutrient malnutrition, *Nutrition Today* 33 (1998): 125; and B. A. Underwood, From research to global reality: The micronutrient story, *Journal of Nutrition* 128 (1998): 145–51.

community members is critical to sustaining the program. Finally, multiple channels of communication are recommended (mass media, traditional forms of media, and personal communications).

In the final analysis, individual countries will need to choose the most practical and cost-effective mix of VAD interventions given local customs, resources, and needs. Numerous international agencies, including WHO, UNICEF, the FAO, the World Bank, and USAID continue their commitment to support country efforts to meet the World Summit for Children's goal of virtually eliminating VAD.

References

1. *Nutrition: What Are the Challenges?*, UNICEF, 2005; The State of the World's Children 1998: A UNICEF Report, *Nutrition Reviews* 56 (1998): 115–23.
2. K. P. West and coauthors, Efficacy of vitamin A in reducing preschool child mortality in Nepal, *Lancet* 338 (1991): 67–71.
3. The State of the World's Children, 2005.
4. Vitamin A deficiency in Africa, *Vital News* 3(2) (1992): 1–10.
5. Vitamin A deficiency in Asia, *Vital News* 3(3) (1992): 1–11.
6. Ibid., p. 4.
7. The discussion of the MOST program was adapted from USAID, available at www.usaid.gov.
8. J. Rankins, MANGOCOM: A nutrition social marketing module for field use, *Journal of Nutrition Education* 24 (1992): 192–94.
9. Vitamin A deficiency in Latin America and the Caribbean, *Vital News* 3(1) (1992): 8.
10. The discussion of vitamin A supplementation and immunization programs was adapted from Linking vitamin A activities to primary health care programs, *Vital News* 2(1) (1991): 1–7.
11. M. G. Herrera and coauthors, Vitamin A supplementation and child survival, *Lancet* 340 (1992): 267–71; B. A. Underwood and P. Arthur, The contribution of vitamin A to public health, *The FASEB Journal* 10 (1996): 1040–48; and B. A. Underwood, Scientific research: Essential, but is it enough to combat world food insecurities? *Journal of Nutrition* 133 (2003): 1434S–1437S.

*The recommended dosing schedule for a vitamin A capsule distribution program is 200,000 IU given twice a year. The dose for children less than 1 year of age and for those who are significantly underweight is 100,000 IU twice a year. (A. Gadomski and C. Kjolhede, *Vitamin A Deficiency and Childhood Morbidity and Mortality* [Baltimore, MD: Johns Hopkins University Publications, 1988].)

Community Nutritionists in Action: Planning Nutrition Interventions

Section *Three*

Leon T. can't decide what to do next. "I could go downstairs and finish the pull-toy for young Kevin," he muses. Leon remembers leaving a pile of sawdust and wood chips scattered about the floor down there. "Or I could read the new *Reader's Digest*, just arrived today." Leon pats his shirt pocket for his glasses. Even with the magazine's large print, he can't read it very well without his glasses. Nope, they aren't in his shirt pocket, and they aren't in his pants pocket either, or on the end table. "I can't keep track of anything these days." He stands and starts a search of the easy chair and sofa. No luck.

He moves to the dining table, where he picks among the dirty dishes, magazine flyers, newspapers, and mail. His frustration mounts quickly when he realizes he can't remember what he's looking for. "Ah, look what I found—my pills!" Momentarily satisfied, he goes to the kitchen to take his medications. He's being treated for high blood pressure, underactive thyroid, and angina. He's suddenly confused, though, knowing there was something he was aiming to do. What was it?

Things just haven't been the same for Leon since his oldest, dearest friend Bill moved to Memphis a year ago to be close to his only daughter. Leon and Bill played golf regularly and met for dinner several nights a week. Now Leon doesn't seem to have the energy for much of anything, and he no longer enjoys puttering around his woodworking shop or reading. At 76, Leon lives alone and likes it that way. But how much longer can he live independently? Does he need assistance with meals, medications, and doctor's visits? Are some of his symptoms due to depression or some other medical condition? Does he have family living nearby, and if so, what is the nature of his relationship with them?

These are the types of questions the community nutritionist asks when assessing Leon's needs. The community nutritionist aims to determine whether existing nutrition services and programs can improve his quality of life and help him continue to live independently.

This section describes how to use the results of a community needs assessment by reviewing several important questions: *Who* has a nutritional problem that is not being met? *How* did this problem develop? *What* programs and services exist to alleviate this problem? *Why* do existing services fail to help the people who experience this problem? The answers to these and other questions help community nutritionists understand the many factors that influence the health and nutritional status of a particular group.

In this section you'll learn how to lay out a plan for designing a program or intervention, write goals and objectives, work with culturally diverse groups, choose nutrition messages, market your program, evaluate its impact, and locate funding to cover program costs. As in Section One, the thread running through these discussions is entrepreneurship and how its essential principles—creativity and innovation—help community nutritionists "do more with less." The ultimate goal of community nutrition is to design programs that improve people's health and nutritional status.

Chapter *Fourteen*

Program Planning for Success

Learning *Objectives*

After you have read and studied this chapter, you will be able to:

- Describe six factors that can trigger program planning.
- Describe seven steps in designing, implementing, and evaluating nutrition programs.
- Discuss three reasons for conducting evaluations of programs.
- Discuss three major principles to consider when preparing an evaluation report.

This chapter addresses such issues as needs assessment; program planning, monitoring, and evaluation; and current information technologies, which are Commission on Accreditation for Dietetics Education (CADE) *Foundation Knowledge and Skills* requirements for dietetics education.

Chapter *Outline*

Something To Think About...

One should get as many new ideas as possible.

– Isabel Archer in Henry James's *The Portrait of a Lady*

Introduction

In the early 1990s, the state of Arizona foresaw a gap in services. Data obtained from the National Health Interview Survey on Child Health showed that at least two-thirds of children between the ages of one week and five years were placed in child care. Some of these children had special health care needs. Arizona's own data predicted that by the year 2000, more than 24,000 children would have developmental delays and other special needs in Arizona and that many of these children would participate in child care programs. The problem was that only limited training in providing quality nutrition services was available for child care workers.

The Arizona Department of Health Services was determined to find a solution to the problem. It organized a team to develop a course that would make child care workers more aware of the challenges involved in feeding children with special health care needs.[1] The team was composed of health professionals with expertise in pediatric medicine, occupational therapy, speech therapy, child care programs, and community nutrition. A parent of a child with special needs was included on the team. The Arizona Department of Health Services, Office of Nutrition Services, obtained funding for the project through a Maternal and Child Health Improvement Project Field Grant and linked with the Office of Women's and Children's Health to develop the program.

The community nutritionists assigned to the project had a leadership role in designing and evaluating the course. They reviewed research studies and other technical material related to feeding children with special needs, determined who would use the course materials, assessed the knowledge and skill levels of child care workers, and determined what types of nutritional problems child care workers were likely to encounter in day care and other settings. They evaluated the course and made adjustments in course materials. The end product was Project CHANCE, a course designed and tested by health professionals to meet the training needs of child care workers.[2]

This chapter presents an overview of program planning—the process of designing a program to meet a nutritional need or fill a gap in services. It describes the process of program evaluation, explaining why evaluation is important, who conducts the evaluation, how evaluation findings are used, and how to prepare an evaluation report. Topics such as developing the intervention and marketing plan, choosing nutrition messages, managing personnel and data, and identifying funding sources are discussed in the remaining chapters of Section Three.

Factors That Trigger Program Planning

The decision to develop a nutrition program or modify an existing program is usually made in response to some precipitating event. Perhaps the community needs assessment revealed that some elderly people were not taking advantage of the community's Meals-on-Wheels program or that a large number of children living in an inner-city neighborhood had low intakes of vitamins A and C. In other cases, as shown in Figure 14-1, the

FIGURE 14-1 Factors That Trigger Program Planning

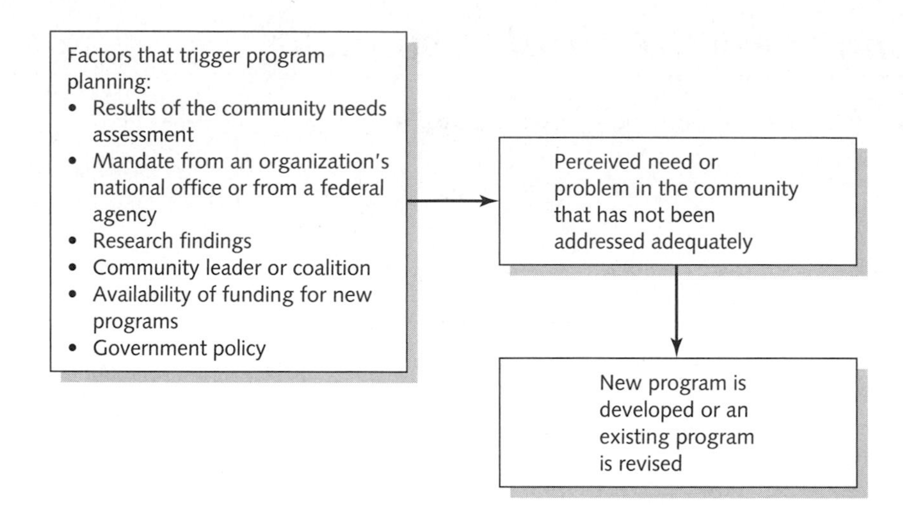

stimulus for program planning may be a mandate handed down from an organization's national office. For example, when the national office of the American Heart Association (AHA) chose nutrition and physical fitness as the organization's national health promotion initiatives for one year, state AHA offices then determined whether their existing programs met the organization's mandate in these areas.

Research findings sometimes trigger the planning process. The report of the National Cholesterol Education Program (NCEP) on the detection, evaluation, and treatment of high blood cholesterol in adults, first published in 1988, described the findings of major studies that linked high blood cholesterol levels to coronary heart disease (CHD) risk. Two additional NCEP reports described population strategies for reducing blood total cholesterol. These reports led some hospitals and municipal health departments to review their programs for reducing CHD risk within their communities and, in some cases, to develop new programs to promote cardiovascular health.[3] Findings from the Bogalusa Heart Study, an epidemiologic study begun in Louisiana in 1974, led to development of the Heart Smart Family Health Promotion program, a school-based program that targeted high-risk elementary schoolchildren and their families. The program was designed to involve parents in improving the eating and activity patterns of schoolchildren.[4]

The concerns of a well-known community leader or coalition may stimulate program planning, as when a community activist helps state agencies plan alcohol and drug treatment programs under the welfare reform law. Government policy and the availability of new funding can also spur program planning. When the U.S. Department of Agriculture increased its funding for the WIC Farmers' Market Program, the National Commission on Small Farms called for expansion of the program to every state—an action that continues to motivate states where the program has not been implemented to offer the program to eligible mothers and children.[5] Regardless of the impetus, the community nutritionist considers developing a program when there is a nutritional or health problem in the community that has not been resolved adequately.

Steps in Program Planning

Mission statement A broad statement or declaration of an organization's purpose or reason for being.

The community nutritionist reviews her organization's mission statement before developing or modifying a program. A **mission statement** is a broad declaration of the organization's purpose and a guideline for future decisions. It provides an identity and proclaims,

		Discussed in ...
Step 1	Review the results of the community needs assessment.	This chapter
Step 2	Define program goals and objectives.	This chapter
Step 3	Develop a program plan.	
	• Design the intervention.	Chapter 15
	• Design the nutrition education component.	Chapter 17
	• Develop the marketing plan.	Chapter 18
Step 4	Define the management system.	Chapter 19
Step 5	Identify funding sources.	Chapter 20
Step 6	Implement the program.	Chapter 19
Step 7	Evaluate program elements and effectiveness.	This chapter and Chapters 15–19

FIGURE 14-2 Steps in Program Planning

"This is what this organization is all about."[6] The community nutritionist ensures that all programs fulfill her organization's mandate. If the match between the mission statement and the program concept is good, then she has a reasonable level of confidence about gaining senior management's support for it. If the match is not good, it will be difficult to justify the resources, time, and expense of a new program and to secure funding for it. Consider, for example, the FirstRate Spa and Health Resort mentioned in Chapter 1 (see Table 1-6 on page 20). Its mission statement reads, "The FirstRate Spa and Health Resort works to enhance the health and fitness of its members by promoting physical activity, healthy eating, and self-care." The director of health promotion will have difficulty obtaining internal support for a new program whose participants are not spa clients.

The mission statement of the American Dietetic Association appears on page 577 in Chapter 19.

The program planning process consists of several steps, as shown in Figure 14-2. The first step is to review the results of the community needs assessment. In Step 2, goals and objectives that specify the expected outcomes of the program are defined. Step 3 is to develop a program plan that describes the intervention, the appropriate nutrition messages for the target population, and how the program will be marketed. In Steps 4 and 5, decisions are made about the management system, budgeting, and potential funding sources for program activities. The program is implemented in Step 6 and evaluated in Step 7. Evaluation focuses on program elements such as the nutrition education and marketing materials and the program's effectiveness—in other words, evaluation activities are designed to determine whether the program accomplished what it was designed to accomplish. Finally, after the program's effectiveness has been verified, colleagues, community leaders, and the community at large are notified of its success.

Chapter 2 describes how to conduct a community needs assessment.

Step 1: Review the Results of the Community Needs Assessment

The community needs assessment provides information about the target population's nutritional problem or need. It is a major impetus for program planning. Consider the needs assessment undertaken by Project MANA, a program that provides emergency food relief to Hispanics, Caucasians, and migrant workers in the Incline Village area of Nevada. Project MANA staff were alerted to a problem with the Thanksgiving basket's food items, which traditionally included turkey, stuffing, and cranberry sauce. Some Spanish-speaking families were not familiar with cranberry sauce and did not know how to cook with it. The outcome of the needs assessment was to change an existing program so that families were offered a choice, the traditional Thanksgiving basket or a new basket containing chicken, salsa, and tortillas.[7]

When the community assessment identifies a gap in services, a new program may be developed to fill the gap. A needs assessment of the food and nutrition situation of low-income Hispanic children living in inner-city Hartford, Connecticut, found that about 53 percent of caregivers did not breastfeed their infants. The assessment report called for the development of culturally sensitive campaigns that promote breastfeeding and inform caregivers about the appropriate times for introducing weaning foods in this population. "*Lactancia, Herencia y Orgullo*" ("Breastfeeding, Heritage and Pride"), an education program based at the Hispanic Health Council, was designed to increase breastfeeding among low-income Hispanic women.[8]

See Table 2-1 on page 39 and Table 3-1 on page 65 to review the needs assessment results and objectives for Case Study 1, on women and coronary heart disease, and Case Study 2, on the nutritional status of elderly persons living at home.

Some results of the community needs assessments for Case Studies 1 and 2, first described in Chapter 2, are shown in Tables 14-1 and 14-2. The tables summarize key findings of the assessments and indicate areas where interventions are needed. (The tables are not meant to be comprehensive, and other information could have been presented.) The results of the needs assessment for Case Study 1 indicate a low level of awareness of CHD risk factors among women and a typical pattern of low activity levels and high intakes of saturated fat—factors that contribute to CHD risk and could serve as points for intervention. The results for Case Study 2 reveal several nutritional problems among the city's elderly population. Note the language and organization of the summaries.

TABLE 14-1 Results of the Community Needs Assessment for Case Study 1: Women and Coronary Heart Disease (CHD)

Results of Needs Assessment

The city of Jeffers and its four adjoining municipalities have a population of 612,000. The city's economic base is light manufacturing and service industries. Many residents have moved from traditional ethnic neighborhoods near downtown to the independent "bedroom" communities outside the city.

It is an ethnically diverse city. A majority (46 percent) of the population is non-Hispanic white, but there are several major ethnic groups: blacks (22 percent), Hispanics (17 percent), Portuguese (7 percent), and Asians (6 percent).

A survey of women aged 18–72 years, conducted at five medical clinics in the major metropolitan area, found that about two-thirds (64 percent) do not know that CHD is the leading cause of death among women. Three out of four cannot identify two major CHD risk factors. About one-third (32 percent) of women smoke cigarettes—a figure higher than the state average of 27 percent.

Although women claim to be eating a healthful diet and exercising regularly, 33 percent of all women surveyed are obese (BMI $\geq$ 30.0). This figure is similar to the state average of 32 percent, but higher than the national average of 30.4 percent. About 50 percent of black women and 42 percent of Hispanic women are obese.

Women's mean intake of saturated fat is 14 percent of total energy. This intake is above the recommended intake level of $<$ 10 percent of total energy. (Mean intake of saturated fat reported in the Continuing Survey of Food Intakes of Individuals was 12 percent of total energy.)

Weight-management programs and classes are available through private health/fitness clubs and dietitians in private practice. Nearly two-thirds (64 percent) of women report that they are trying to lose weight, but only 8 percent of women participate in organized classes and programs.

The local affiliate of the heart association offers some programming for the general public. Brochures describing the leading risk factors for CHD are available at a majority of hospitals and medical clinics. Only 14 percent of women were aware that some local restaurants feature "heart smart" meals on their menus.

Women indicate that they want to do more to reduce their CHD risk and improve their health. No current programs are designed specifically to help women reduce their CHD risk.

TABLE 14-2 Results of the Community Needs Assessment for Case Study 2: Nutritional Status of Independent Elderly Persons (> 75 years)

Results of Needs Assessment

The city of Jeffers has a population of 434,000. This is 71 percent of the total metropolitan area population (that is, Jeffers plus its four adjoining municipalities). The number of elderly persons > 75 years living in the city of Jeffers is 60,760 (14 percent), an increase of 12 percent since 1999. More than half (56 percent) of these elderly people are living independently, typically in their own homes or apartments.

The overall nutritional status of the 34,025 independent elderly is fairly good, but several key nutritional problems emerged:

- About 2 percent or 680 elderly persons, mostly black and Hispanic women, had serum vitamin A concentrations < 20 μg/dL. At these low levels, impairment of immune function and dark adaptation and development of ocular lesions are likely.
- Twenty percent of white women and 38 percent of white men had low hemoglobin concentrations. Among non-Hispanic black men, 62 percent had low hemoglobin levels. (CDC criteria for low hemoglobin are < 13.5 g/dL in men and < 12.0 g/dL in women.)
- One in four elderly men and one in five elderly women had serum LDL-cholesterol concentrations ≥ 160 mg/dL.
- About two-thirds (64 percent) of independent elderly persons had hypertension.
- More than half (56 percent) reported using laxatives for the relief of constipation.
- About 12 percent of independent elderly persons reported having difficulty performing two or more personal care activities (for example, bathing, dressing, using the toilet, getting into and out of bed or chair, eating).

Fewer than one in five (18 percent) use community-based services such as personal care services, homemaker services, and adult day care.

The numbers of independent elderly persons living in Jeffers who participate in federal assistance programs are shown here:

Program	Number (%) of Participants
Food Stamps	3,240 (32%)
Social Security	8,912 (88%)
Medicare/Medicaid	7,190 (71%)
Supplemental Security Income	1,215 (12%)
Veterans Benefits	810 (8%)

Delivery of health care and nutritional services has traditionally occurred through the state Department of Social Services.

They were written for a broad audience, which is likely to include policy makers, community leaders, and the general public.

Step 2: Define Program Goals and Objectives

The next step in the program planning process is to define the program goals and objectives. Goals are broad statements of desired changes or outcomes. They provide a general direction for the program. Objectives are specific, measurable actions to be completed within a specified time frame. An objective has four components: (1) the action or activity to be undertaken, (2) the target population, (3) an indication of how success will be measured or evaluated, and (4) the time frame in which the objective will be met. There are three types of objectives:[9]

▸ **Outcome objectives**—These are measurable changes in a health or nutritional outcome, such as an increase in knowledge of folate-rich food sources, a decrease in blood total cholesterol concentration, an increase in serum ferritin concentration, or a change in functional status. An outcome objective might state, "By the year 2010, increase calcium intakes so that at least 75 percent of females aged 9 to 19 years consume recommended calcium intakes" or "By 2010, reduce iron deficiency to less than 5 percent among low-income children under 2 years."[10]

▶ **Process objectives**—These are measurable activities carried out by the community nutritionist and other team members in implementing the program. They specify the manner in which the outcome objectives will be achieved. A process objective might state, "Each community nutritionist will conduct two nutrition lectures per week over the course of the three-month program."

▶ **Structure objectives**—These are measurable activities surrounding the budget, staffing patterns, management systems, use of the organization's resources, and coordination of program activities. A structure objective might read, "On the last day of each month for the next 12 months, each community nutritionist will submit an itemized statement of expenses related to conducting the program."

The community nutritionists who participated in Case Study 1 developed two broad goals and several outcome objectives (see the following list) for a program designed to help women in the city of Jeffers reduce their CHD risk.

CASE STUDY 1: WOMEN AND CHD RISK

Goals	Outcome Objectives
Reduce death from CHD among women	1. Decrease the number of CHD deaths among women to a rate of no more than 115 per 100,000 within 5 years. 2. Decrease the number of women who smoke by 33 percent within 2 years. 3. Decrease the mean saturated fat intake of women by 3 percent (from their current intake of 14 percent of total energy) within 1 year.
Increase women's awareness of CHD	1. Within 1 year, increase by 50 percent the number of women who can specify CHD as the leading cause of death in the city of Jeffers. 2. Increase the number of women who can specify two major CHD risk factors by 25 percent within 1 year.

These goals and objectives are the framework on which the program elements such as the type of intervention, the nutrition messages, and the marketing campaign are built. The goals and outcome objectives are the basis for determining whether the program was effective—that is, whether the program was successful in raising women's awareness of CHD risk factors and in reducing death from CHD among women.[11]

Step 3: Develop a Program Plan

Using the goals and objectives as a guide, the community nutritionist develops a program plan, which consists of a description of the proposed intervention, the nutrition education component, and the marketing plan. Other factors may also be described in the program plan: the number of clients expected to use the program; the staff, equipment, and material resources required to administer the program; the facilities available

Patrik Giardino/CORBIS

Program planning is best done in teams composed of people with different skill sets, ideas, opinions, and perspectives. The more extensive the program planning, the greater the chance that the program will succeed.

for staff offices and teaching rooms; and the level of staff training required prior to implementing the program. The community nutritionist plans how the program will work after asking many questions: What criteria determine a client's eligibility for the program? What federal, state, and/or local regulations must be considered in administering the program? What educational materials are needed to convey important nutrition messages? What provisions should be made for follow-up? Who will be responsible for implementing the program and assessing its effectiveness? How much money is required to administer the program? Well-designed programs are scrutinized before being made available to the target population.

The program plan is usually developed after reviewing existing programs and talking to colleagues and other professionals who have worked with similar programs or with the target population. One easy way to network with colleagues and stay informed about new programs and services is to join one or more listservs. Listservs are types of Internet mailing lists to which people subscribe, much like a magazine or newspaper.[12] They are convenient, electronic bulletin or message boards for exchanging ideas and information. Some popular listservs for community nutritionists are given in the list of Internet Resources on page 471.

Step 4: Develop a Management System

In this context, the term *management* refers to two types of structures needed to implement the program: personnel and data systems. The personnel structure consists of the employees responsible for overseeing the program and determining whether it meets its objectives. The structure of the data management system is the manner in which data about clients, their use of the program, and the outcome measures are recorded and analyzed.

An important part of program planning is calculating the "management" costs of the program. Both direct costs (such as the salaries and wages of program personnel, materials needed, travel expenses, and equipment) and indirect costs (such as office space rental, utilities, and janitorial services) must be determined to identify the true cost of a program.

Step 5: Identify Funding Sources

Community nutritionists in nonprofit organizations and government agencies face many challenges in securing funding for all aspects of a program. Money may be available in the current year's budget for staff time to develop the program's format, choose nutrition education messages, and plan a marketing campaign, but there may not be enough money to print educational materials or to allocate personnel to pretesting a survey instrument. At this point in the program planning process, the community nutritionist reviews the program elements (for example, educational materials and marketing campaign) and considers whether outside funding in the form of cash grants or in-kind contributions from partners can be found. He or she identifies the area where financial support is needed, reviews possible funding sources, and prepares and submits a grant application for funding. The grant-writing process requires the community nutritionist to be clear about the purpose of the funding request, demonstrate a specific need, and explain how the funds will be used to enhance the program's effectiveness. Chapter 20 provides details for the grant-writing and grant-management processes.

Step 6: Implement the Program

This is the action phase of the program planning process. **Implementation** is "the set of activities directed toward putting a program into effect."[13] The format of the program has been finalized, educational materials printed, the marketing plan prepared, and staff trained. Now it is time to put the program into operation—to link the program goals with the plan of action.

Implementing the program as conceived is challenging, and glitches in program delivery are inevitable. Perhaps no one on the team thought about modifying handouts for Portuguese-speaking clients who participate in the program, or no one was aware of cultural barriers to teenage girls participating in an inner city's after-school fitness program. The key to successful implementation is to observe all aspects of program delivery and consider ways in which delivery can be improved.

Step 7: Evaluate Program Elements and Effectiveness

Evaluation is the use of scientific methods to judge and improve the planning, monitoring, effectiveness, and efficiency of health, nutrition, and other human service programs. The purpose of program evaluation is to gather information for making decisions about redistributing resources, changing program delivery, or continuing a program.[14] It takes the guesswork out of planning and implementing programs and occurs throughout the

Cartoon by John Chase.

Implementation The set of activities directed toward putting a program into effect.

Evaluation The measurable determination of the value or degree of success in achieving specific objectives.

program planning process. Some Internet resources that are useful for the program evaluation process are included in the list that begins on page 471. The next sections describe the purposes and uses of evaluations.

WHY EVALUATION IS NECESSARY

Although the immediate purpose of program evaluation is to help managers make decisions about the short- and long-term operation of their programs, evaluations also serve to inform the community at large about a program's success or failure. When a community-based nutrition program succeeds, nutritionists in other locations across the country want to know how the lessons can be used in their communities. Likewise, when a program fails, community nutritionists want to examine its flaws and figure out how to avoid them in the future.

However beautiful the strategy, you should occasionally look at the results.
– Winston Churchill

Program evaluations force community nutritionists to determine whether they are progressing toward their initial goals and whether these goals are still appropriate. Evaluations may be used for administrative purposes: to determine whether some elements of a program should be changed, to identify ways in which interventions can be improved, to pinpoint weaknesses in program content, to meet certain accountability requirements of the funding agency or senior management, to ensure that program resources (such as supplies, equipment, personnel, and facilities) are being used properly, or to conduct a cost–benefit analysis. They may be undertaken to test innovative approaches to a nutrition or public health problem, to fulfill policy or planning purposes, or to support the advocacy of one program over another. Finally, evaluations may be undertaken to determine whether objectives have been met or whether priorities need to be changed.[15] The finding that a program is not accomplishing its objectives signals a need to consider whether the program is worthwhile or whether its goals can be accomplished in some other fashion.[16] Consult Table 14-3 for a list of reasons for undertaking evaluations.[17]

HOW EVALUATION FINDINGS ARE USED

Evaluation findings have many uses. Sometimes they are used to influence an executive or politician who has the authority to distribute resources and shape public policy. For example, in preparing its position on medical nutrition therapy, the American Dietetic Association (ADA) reviewed the literature on the economic benefits of nutrition services in acute, outpatient, home, long-term, and preventive care and in the care of pregnant women, infants, children, and older adults. Its evaluation of the impact of nutrition services led to the development of a platform on the benefits of nutrition services in health care. The platform was used in the ADA's grassroots lobbying campaign to inform members of Congress and state legislatures about the value of medical nutrition therapy.[18]

Chapter 7 describes the complexities of the legislative process and lists opportunities for influencing public policy at the grassroots level (see Table 7-2 on page 213).

Evaluation findings alert managers and policy makers to the need for expanding or refining programs. For instance, in the United States, the Food Research and Action Center (FRAC) conducts an annual evaluation of the efforts of each state and the District of Columbia to provide nutritious summer meals to children of low-income families. Evidence of a decline in participation in any state is a signal that this state should develop innovative ways of increasing children's participation in the program.

Generally, evaluation findings are used at two levels. Of course, they are applied to an immediate problem and hence are used by managers and program staff who are focused on that problem. But they are also used to shape policies and services beyond the scope of the original problem. In other words, evaluations have many different audiences, some of whom may be directly involved in the program, whereas others are not involved at all or may be concerned with the program at some future date.

TABLE 14-3 Reasons for Undertaking Evaluations

Evaluation to Improve Your Program

- To improve methods of placing clients in various activity programs
- To measure the effect of your program or the extent of client progress in your program
- To assess the adequacy of program goals
- To identify weaknesses in the program content
- To measure staff effectiveness
- To identify effective instructional, leadership, or facilitation techniques
- To measure the effectiveness of resources (such as materials, supplies, equipment, or facilities)

Evaluation to Justify Your Program or to Show Accountability

- To justify the budget or expenditures
- To show the need for increased funds
- To justify staff, resources, facilities, etc.
- To justify program goals and procedures
- To account for program practices
- To compare program outcomes against program standards

Evaluation to Document Your Program in General

- To record client attendance and progress
- To document the nature of client involvement and interaction
- To record data on clients who drop out and those who complete the program
- To document the major program accomplishments
- To list program weaknesses expressed by staff or others
- To list leader/therapist functions and activities
- To describe the context or atmosphere of the treatment setting
- To file supportive statements and testimonies about the program

Source: A. D. Grotelueschen and coauthors, *An Evaluation Planner* (Urbana: Office for the Study of Continuing Professional Education, University of Illinois at Urbana-Champaign, 1974). Permission to reprint granted by Arden D. Grotelueschen.

WHO CONDUCTS THE EVALUATION?

Evaluations may be carried out by program staff, other agency staff, or outside consultants. The evaluator may be intimately familiar with all aspects of a program because he or she manages or is involved with it, or the evaluator may have a limited knowledge of the program. Regardless, the evaluator is responsible for all aspects of the evaluation, from negotiating the evaluation focus to collecting data and preparing the final report. Because evaluations often occur in a politically charged atmosphere, where program stakeholders fret about the evaluation's outcome and its ramifications, the evaluator must be sensitive to this environment. In the final analysis, the evaluator must be able to recognize what has to be done and must remain objective about the evaluation and its findings.[19]

THE PROGRAM EVALUATION PROCESS

The purpose and scope of an evaluation depend on the questions being asked about the program. An evaluation may focus on one particular program element—for example, determining whether screening every client for high blood cholesterol concentration is cost-effective or whether a significant portion of the target population is aware of the nutrition messages appearing in posters placed in city buses. Or the evaluation may be comprehensive and examine the design of the program, how it is delivered, and whether it is being used properly. Evaluation occurs across all areas of program planning, from design to implementation, and no one method is always most effective for carrying out an evaluation. Rather, each evaluation must be tailored to the organization or department in which it is conducted.

When evaluating their programs, community nutritionists begin by asking the following questions:[20]

- Did the intervention reach the target population?
- For which participants was the program most effective?
- For which participants was the program least effective?
- Was the intervention implemented according to the original program plan?
- Was the program effective—that is, did it accomplish what it was supposed to accomplish?
- How much does the program cost?
- What are the program's costs relative to its effectiveness and benefits?

The answers to these and other questions help the community nutritionist design a better, more effective program and formulate recommendations for colleagues and community leaders. Recommendations can be made about the suitability of program goals and objectives, whether the objectives should be changed, whether the program can be applied successfully in another setting or among a different group of participants, and how the program should be changed to improve delivery.[21]

EVALUATION AS A PLANNING TOOL

Evaluation is fundamental to every step of community assessment and program planning. Recall from Chapter 2 that the community needs assessment itself is an evaluation of a population's health or nutritional status. It is designed to find answers to questions about who in the community has a nutritional problem, how the problem developed, what programs and services exist to address the problem, and what can be done to alleviate the problem. The evaluation tools of the community needs assessment are the health risk appraisal, focus group discussions, screenings, surveys, interviews, and direct assessments of nutritional status, as we saw in Chapter 3. During program planning, evaluation occurs at every step. In the design stage, managers develop goals and objectives for the program to determine its impact and effectiveness.

They conduct formative evaluation to achieve a good fit between the program and the target population's needs, to develop appropriate nutrition messages for the target population (Chapter 17), and to design a marketing plan for the program's target market (Chapter 18). In this section, we describe evaluations that occur during the implementation of the program and when the program is completed. Managers use evaluation findings to plan changes in programs, interventions, and staff activities.

Formative Evaluation

Evaluation occurs right from the beginning of the design phase. It is often necessary and prudent to pilot-test certain design elements during the development phase. This process, called **formative evaluation,** helps pinpoint and eliminate any kinks in the proposed delivery system or intervention before the program is implemented fully. Formative evaluation can be used to assess educational materials in terms of the appropriateness of language used, accuracy and completeness of the contents, and readability.[22]

Formative evaluation The process of testing and assessing certain elements of a program before it is implemented fully.

For example, program planners in an inner-city district of Philadelphia found that most pregnant women don't consider breastfeeding their newborns and that those who do breastfeed do so for only a few weeks. They decided to conduct a formative evaluation before they agreed on the final intervention design. They surveyed potential program clients about their family's support for breastfeeding. The planners learned that few new mothers had a close female relative living nearby who could help explain how to breastfeed an infant and that many new mothers in the younger age groups had boyfriends or husbands who had negative attitudes about breastfeeding. The results of this formative evaluation led the planners to add two pieces to the program intervention: (1) the provision of an experienced mother to befriend the new mother prior to the infant's birth and (2) sessions with husbands and boyfriends to help change their attitudes about breastfeeding.

Process Evaluation

Monitoring how a program operates helps managers answer questions and make decisions about what services to provide, how to provide them, and for whom. This **process evaluation** involves examining program activities in terms of (1) the age, sex, race, occupation, or other demographic variables of the target population; (2) the program's organization, funding, and staffing; and (3) its location and timing.[23] Process evaluation focuses on program *activities* rather than on outcomes.

Process evaluation is gaining recognition as a tool to help managers make good decisions. Through process evaluation, managers can systematically exclude various explanations that may arise for a given outcome. If the program appears to have had no effect, process evaluation can determine which, if any, of the following problems was the reason.[24]

- The program was not properly implemented (in other words, program staff were not fully effective in implementing the program).
- The program could not be implemented properly in some participants (which suggests that compliance with the program protocol was a problem for some participants).
- Some participants had difficulty accessing the program.

Alternatively, if the program has had a beneficial effect, process evaluation can determine whether that effect was due, in fact, to the program or to one of the following:

- The greater receptivity of some participants or target groups compared with others
- Competing interventions

Process evaluation focuses on how a program is delivered. In the course of conducting a process evaluation, the evaluator examines the target population to determine how individuals were attracted to the program and to what extent they participated. In addition, the program is evaluated for bias in terms of how participants were served—that is, whether the target group received too much or too little coverage by the program. Such information can be obtained by examining the records kept on program participants and by conducting surveys of the target population and the program participants. The participant records should reveal which services were used and how often. Surveys help define the characteristics of clients who used the program, compared with those who dropped out or refused to use the program.

Process evaluation deals with activities that are planned to occur. In Case Study 1, an intervention strategy was designed to increase women's awareness of coronary heart disease (CHD) risk factors and to reduce their deaths from CHD. Several process objectives were developed, including these three:

- Provide the eight sessions of the "Heartworks for Women" program over a 12-week period at each participating worksite.
- Administer a CHD risk factor knowledge test to each participant at the beginning and end of the program.

Process evaluation A measure of program activities or efforts—that is, of how a program is implemented.

- Obtain an estimate of saturated fat intake using a 24-hour dietary recall completed by each participant at the beginning and end of the program.

We might examine the third process objective listed above—to obtain an estimate of usual saturated fat intake using a 24-hour recall of all participants at the beginning and end of the program—and consider how the planned activity compared with the actual results. If 66 employees entered the program and 24-hour recalls were obtained on 57 of them as planned, we can calculate the percentage of activities attained, using the following formula:

$$\frac{\text{Actual activities}}{\text{Planned activities}} \times 100 = \text{Percentage of activities attained}$$

$$\frac{57}{66} \times 100 = 86.4 \text{ percent}$$

Thus 86.4 percent of the planned 24-hour recalls were actually obtained during the specified time period. A number of questions arise immediately: Why weren't recalls obtained from all participants? Was there a problem scheduling employees to see the dietitian and, if so, why? Were the instructions to participants unclear? The information obtained through process evaluation signals the program manager that additional planning is required to improve the program's delivery.

Another example of process evaluation comes from the Child and Adolescent Trial for Cardiovascular Health (CATCH), an intervention that targeted the school environment, staff, students, and the students' families in four centers located in California, Louisiana, Minnesota, and Texas. CATCH was designed to assess the effectiveness of school food service programs, physical education, classroom instruction, and family activities to reduce CHD risk in elementary schoolchildren. Process evaluation was used to assess the level of standardization of teacher instruction, the number of CATCH classroom activities completed out of the total number expected for each school session, the number of promotional activities sponsored by the food service operation, and the number of support visits made by CATCH staff to schools over the course of the year. Process evaluation determined that the fourth-grade classes had the lowest rate of completion of classroom activities, whereas third-grade classes had the highest rate of completion. The number of cafeteria promotional events ranged from about 6 to 14, and the number of support visits by CATCH staff ranged from 2.5 to 19.7. The overall findings of the process evaluation provided information that helped intervention planners improve the intervention.[25]

See the Programs in Action feature in Chapter 8 for more about the CATCH program.

Impact Evaluation

Impact evaluation is used to determine whether and to what extent a program or an intervention accomplished its stated goals. It describes the specific effect of program activities on the target population. In the "Heartworks for Women" program, the skills-building intervention in Case Study 1, the impact of the program would be the knowledge about CHD risk, and about major sources of dietary saturated fat, acquired by women who participated in the program. The impact evaluation would assess whether women had learned the key risk factors for heart disease and whether they could describe the types of fat in the diet, among other things. In the CATCH intervention, described previously, one unexpected finding of impact evaluation was that students in classes where teachers modified the CATCH sessions to suit their needs learned more about diet, health, and heart disease than students in classes headed by teachers who did not modify their sessions. Perhaps these teachers were more motivated, confident, and creative and had better communication skills than teachers who made no changes in the lesson plans.

Impact evaluation The process of determining whether the program's methods and activities resulted in the desired immediate changes in the client.

The finding led the intervention planners to consider changing the model on which the intervention was based.[26]

Impact evaluation focuses on immediate indicators of a program's success. Depending on the program's goals and objectives, it might examine variables such as beliefs, attitudes, decision-making skills, self-esteem, self-efficacy, and knowledge.[27]

Outcome Evaluation

The purpose of **outcome evaluation** (also referred to as summative evaluation) is to determine whether the program or intervention had an effect on the target population's health status, food intake, morbidity, mortality, or other outcomes. Outcome evaluations are a challenging managerial activity, for they require technical skills in survey design and analysis. The problems associated with outcome evaluations arise from the difficulty of determining whether a particular effect was "caused" by the intervention and was not due to some extraneous factor. It is possible that factors beyond the control of the program staff influenced the outcome significantly. Such confounding factors might include secular trends within the community, the occurrence of unexpected events such as a natural catastrophe, or certain characteristics of clients (such as their tending to "self-select" for the program).[28]

Case Study 1 included a skills-building program, the "Heartworks for Women" program, and nutrition messages directed at women in the community at large. Did the intervention strategy accomplish its goals and objectives? The intervention manager plans outcome evaluations that target the three time frames (1, 2, and 5 years) specified in the objectives. (Refer to the program objectives described on page 456.) The first evaluation will be undertaken 12 months after the startup of the intervention and program. Data will be collected from women who enrolled in the "Heartworks for Women" program and from women in the broader community. The purpose of the evaluation will be to determine whether the nutrition and smoking messages delivered through the intervention resulted in a behavior change in the target population. In other words, did women who participated in the "Heartworks for Women" program reduce their intake of saturated fat by 3 percent? Did the number of women who smoke decrease by 33 percent? The findings of the outcome evaluation will be used to modify the intervention strategy, nutrition messages, marketing plan, and other program elements.

Outcome evaluation, like impact evaluation, is tied to the program's goals and objectives. It is designed to account for a program's accomplishments and long-term effectiveness in terms of a health change in the target population. Outcome evaluation measures are associated with factors relevant to the particular program. Such measures can include serum ferritin levels, percent body fat, calcium intake, stroke prevalence, blood pressure, and use of home food services, depending on the nature of the program.

Outcome evaluation The process of measuring a program's effectiveness in changing one or more aspects of nutritional or health status.

Structure evaluation The process of determining adequacy of the internal processes and resources needed to deliver a program, including personnel (staff training) and environmental factors (supply of instructional materials, adequacy of facility and equipment).

Structure Evaluation

Here, structure consists of personnel and environmental factors related to program delivery such as the training of personnel; the adequacy of the facility; the use of equipment such as laptop computers, overhead projectors, and skinfold calipers; and the storage of participant records.[29] In preparing a **structure evaluation** of the "Heartworks for Women" program in Case Study 1, the senior manager specifies several structure objectives, three of which follow.

- The 12-month operating budget for the program is $274,500.
- Monthly operating budgets for the next 12 months will not show a variance exceeding 0.02 percent of the total operating budget.

TABLE 14-4 Costs and Benefits Considered in a Cost–Benefit Analysis of the "Heartworks for Women" Program

TYPE	COSTS	BENEFITS
Direct	Personnel—salaries Utilities (telephone, Internet) Travel (to/from worksites) Office supplies Postage Equipment Instructional materials Design Printing Promotion and advertising	Revenue from program
Indirect	Personnel—benefits Rental of office space Utilities Equipment depreciation Maintenance and repairs Janitorial services	Increased exposure for the program and the organization within the community Reduced use of health care services by employees of participating worksites

- In-house educational materials will be used at a monthly rate that does not exceed 10 percent of the total stored supply.

These and other structure objectives provide targets that regulate the resources needed to deliver the program. The third objective, for example, helps staff members plan for the use of educational materials. It guides session instructors in monitoring their use of the materials, and it helps the staff member who coordinates the supply and storage area maintain an adequate supply of materials for the program. The manager uses the findings from structure evaluation to make changes in the internal processes that support the staff and program activities.

Fiscal or Efficiency Evaluation

The purpose of **fiscal evaluation** is to determine how program outcomes compare with their costs. There are two types of efficiency evaluations: cost–benefit analysis and cost-effectiveness analysis.

In performing a cost–benefit analysis, managers estimate both the tangible and intangible benefits of a program and the direct and indirect costs of implementing that program, as summarized in Table 14-4 for the "Heartworks for Women" program. Once these have been specified, they must be translated into a common measure, usually a monetary unit. In other words, the cost–benefit analysis examines the program outcomes in terms of money saved or reduced costs. A prenatal program that costs $200 per participant to produce and results in reduced medical costs of $800 can be expressed as a cost–benefit ratio of 1:4. For every one dollar required to produce the program, there is a four-dollar savings in medical costs.[30]

Refer to Chapter 10 for a discussion of a cost–benefit analysis of the Special Supplemental Nutrition Program for Women, Infants, and Children (WIC).

The second type of fiscal evaluation is cost-effectiveness analysis. Unlike cost–benefit analysis, which reduces a program's benefits and costs to a common monetary unit, cost-effectiveness analysis relates the effectiveness of reaching the program's goals to the monetary value of the resources going into the program. With this type of evaluation, similar programs can be compared to one another or ranked in order of their cost per program goal. A cost-effectiveness analysis would be done, for example, to determine which of two methods of intervention—individual dietary counseling or group nutrition education classes—produces a desired outcome for less cost.

Fiscal or efficiency evaluation The process of determining a program's benefits relative to its cost.

Programs in Action

Feast with the Beasts™

Community institutions can be desirable nutrition education partners for schools. Institutions have access to staffing and funding to augment education efforts, particularly in financially strapped districts. They may approach the education process with a new and refreshing perspective, different resources, and creative thoughts. Partnerships between schools and community institutions are most successful when the two groups work toward a common education goal while continuing to pursue their individual goals and objectives.

School districts in underserved areas face a particular challenge in setting goals for nutrition education programs. Nutrition problems—specifically iron-deficiency anemia, obesity, and elevated heart disease risk factors—are prevalent among underserved populations.[1] The goals of nutrition education, therefore, become not only mastery of food and nutrition basics but also health improvement and disease prevention.

Institutions often define goals and objectives through the strategic planning process—discussed in Chapter 19. The focus of strategic planning is on formulating objectives; assessing past, current, and future conditions and events; evaluating the organization's strengths and weaknesses; and making decisions about the appropriate course of action. In contrast to strategic planning, program planning covers a shorter time frame and serves to support the strategic plan and overall mission. The values, behaviors, and perceived needs of community members should be identified and understood.[2] An institutional partner in a nutrition education program will want to identify nutrition-related problems, along with opportunities to reduce those problems.

According to social cognitive theory, learning results from interactions among the individual, other people, and the environment.[3] The individual actively observes another person's behavior in a particular environment and sees the results of that behavior. Community institutions may be a source of "environments" from which the student can learn. In the case that follows, the learning environment is an area zoo, and students learn from observing the feeding of animals.

The "Feast with the Beasts" program exemplifies a win–win partnership between an institutional setting (a zoo) and area schools. Students learn about nutrition in an entertaining and engaging way, while the zoo fulfills its mission to educate the public about the importance of conserving the earth's environment and animal species.

In 1995, the Brookfield (Illinois) Zoo began its Feast with the Beasts program as a way to provide underserved and disadvantaged children with nutrition education and healthful lunches. Children whose teachers agree to incorporate Feast with the Beasts into their curriculum receive classroom lessons in nutrition and conservation and then visit the zoo on a field trip. To date, the program has served 23,000 children.

Goals and Objectives

The zoo's goals are to generate increased awareness of the natural world, to encourage scientific inquiry, to promote understanding of the relationship of humankind with the natural world, and to develop sensitivity and a responsible attitude toward animals and their habitats. Student-specific goals are to help students identify basic principles of nutrition, acquire skills for choosing a nutritious diet, eat healthfully for life, and care about animals and their habitats.

COMMUNICATING EVALUATION FINDINGS

Because the primary purpose of evaluations is to provide information for decision making, you do not want your findings to go unused. If they are stored away in a filing cabinet or dumped into the "circular file," a program may continue missing the mark or a success story may go unnoticed. With careful planning and work, you can ensure that the main findings of your evaluation get the attention they deserve.

Even as you begin the evaluation, you should be thinking about the final report! As the evaluation progresses, make notes of how problems were handled and which documents or materials were used. Retain copies of survey instruments and computer printouts as reference items or for use in the report's appendix. When you begin preparing your report, keep these three rules in mind:[31]

- Communicate the information to the appropriate potential users.
- Ensure that the report addresses the issues that the users perceive to be important.
- Be sure that the report is delivered in time to be useful and in a form that the intended users can easily understand.

Methodology

A team of professionals, including educators, animal specialists, and a registered dietitian, developed Feast with the Beasts. Teachers qualify for the free program if more than half the students in their school participate in the government-subsidized lunch program. They receive a teachers' guide to use during the five days before the zoo field trip. It contains a pre-test, lessons, and activities related to animal and human nutrition.

On the day of the field trip, teachers take their students to see the five species of animals they have studied in class. A meeting with the zoo dietitian and her assistant includes a free lunch and hands-on activities.

Lessons following the zoo visit include two days of activities and a posttest. Students bring home a completed activity book to share with their families. Teachers complete a survey and document student scores on the pre-tests and post-tests.

Results

The program's success was evaluated in 1996 through teacher surveys and student pre-test and post-test scores. A total of 33 teacher surveys from 19 schools were returned. On a scale of 1 to 5, where 5 is the highest score, teachers rated Feast with the Beasts 4.9 on content and relevance, 4.7 on format, and 4.8 on organization, ease of use, and reception by students. Pre-visit materials received a mean score of 4.8, and in-zoo and post-visit components received a mean score of 4.6. More than 70 percent of teachers have repeated the program every year since 1995.

Student post-test scores were significantly higher than pre-test scores—83 percent and 59 percent, respectively. These results demonstrate that students were able to identify and choose more healthful foods more often after participating in Feast with the Beasts.

Lessons Learned

- A unique aspect of the Feast with the Beasts program is its use of the link between human nutrition and animal nutrition.
- The program was able to capture the students' undivided attention by showing, for example, that dolphins need vitamins just as humans do.
- The Program supports the *Healthy People 2010* goal of eliminating health disparities by providing a fun atmosphere in which underserved children, their parents, and their teachers could compare animal and human nutrition to learn the importance of healthful eating habits.[4]

References

1. Guidelines for school health programs to promote lifelong healthy eating, *Journal of School Health* 67 (1997): 9–26.
2. M. Lipton, Demystifying the development of organizational vision, *Sloan Management Review* 37 (1996): 83–92.
3. K. Glanz and B. K. Rimer, *Theory at a Glance: A Guide for Health Promotion Practice* (Bethesda, MD: U.S. Department of Health and Human Services, Public Health Service, National Institutes of Health Publication (NIH) 97-3896: September 1997).
4. *Healthy People 2010: National Health Promotion and Disease Prevention Objectives* (Washington, D.C.: U.S. Department of Health and Human Services, November 2000).

Source: Adapted from *Community Nutritionary* (White Plains, NY: Dannon Institute, Fall 1999). Used with permission. For more information about the *Awards for Excellence in Community Nutrition,* go to www.dannon-institute.org.

With these issues in mind, prepare your report, which may be either informal (for example, a short memorandum) or formal (such as a full report). Even if the report is only a three-page memorandum, it should be concise and understandable and should give the user what he or she needs to make a decision. Our focus in this discussion is on the formal report, which tends to have the following type of organization.[32]

1. **Front cover.** The front cover should provide (a) the title of the program and its location, (b) the name of the evaluator(s), (c) the period covered by the report, and (d) the date the report was submitted. The front cover should be neat and attractively formatted.

2. **Summary.** Sometimes called the executive summary, this section of the report is a brief overview of the evaluation, explaining why and how it was done and listing its major conclusions and recommendations. Typically, this section is prepared for the individual who does not have time to read the full report. Therefore, it should not be longer than one or two pages. And even though the summary appears first in the report, it should be written *last.*

3. **Background information.** This section places the program in context, describing what the program was designed to do and how it began. The amount of detail provided in this section will depend on the needs and knowledge of the users. If most readers are unfamiliar with the program, it should be described in some detail; if most readers are involved with the program, this section can be kept short. A typical outline for this section might include the following:

- Origin of the program
- Goals of the program
- The program's target population
- Characteristics of the program materials, activities, and administrative procedures
- Staff involved with the program

4. **Description of the evaluation.** This section states the purpose of the evaluation, including why it was conducted and what it was intended to accomplish. Here you define the scope of the evaluation and describe how it was carried out. This section establishes the credibility of the evaluator and the evaluation findings (much like the Methods section of a research paper). Technical information about the evaluation design and analysis is presented here. Technical language should be kept to a minimum, however. Refer readers to appendices for specific technical information and for copies of any instruments used in the evaluation study. A general outline for this section might include the following headings:

- Purposes of the evaluation
- Evaluation design
- Outcome measures
  - Instruments used
  - Data collection and analysis procedures
- Process measures
  - Instruments used
  - Data collection and analysis procedures

5. **Results.** This section presents the results of the outcome or process evaluation. It is appropriate to present data or summarize findings in tables, figures, graphs, or charts. Before you begin writing this section, you should have already analyzed the data, tested for statistical significance (if appropriate), and prepared the tables, figures, and other illustrations.

6. **Discussion of results.** The results of the evaluation study are interpreted in this section, which should address two key issues: How certain is it that the program caused the results? How good were the results? The Results section explores some of the reasons why a certain outcome was reached and how the program compares to similar programs. Any strengths and weaknesses of the program are described here.

7. **Conclusions, recommendations, and options.** This section is an influential part of the report, because it outlines the major conclusions that can be drawn from the evaluation and suggests a course of action for enhancing the program's strengths and dealing with its flaws. The recommendations should address specific aspects of the program and should follow logically from your interpretation of the evaluation findings. Preparing a list of recommendations about the program's delivery or impact is especially important when the actual results differ from the predetermined objectives.

Once your report is written, you must decide how best to distribute it. Several options are available. You may send the full report to your immediate supervisor, division director, and board of directors and provide a copy of the executive summary to interested

FIGURE 14-3 Forms of Communicating Evaluation Findings to Various Audiences

Source: Adapted from L. L. Morris and coauthors, *How to Communicate Evaluation Findings,* pp. 9–10, copyright © 1987 by Sage Publications. Reprinted by permission of Sage Publications, Inc.

	Possible Communication Form											
Audience/Users	Technical Report	Executive Summary	Technical Professional Paper	Popular Article	News Release, Press Conference	Public Meeting	Media Appearance	Staff Workshop	Brochure	Memorandum	Personal Presentation	Internet Website/E-mail
Program Administrators	✔	✔	✔	✔	✔			✔		✔	✔	✔
Board Members, Trustees, Other Management Staff		✔		✔								✔
Advisory Committees	✔	✔	✔									✔
Funding Agencies	✔	✔									✔	✔
Community Groups		✔		✔		✔						✔
Current Clients				✔		✔	✔					✔
Potential Clients							✔					✔
Political Bodies (e.g., City Councils, Legislatures)		✔		✔								✔
Program Service Providers (e.g., Nutritionists, Health Educators)		✔		✔				✔	✔	✔	✔	✔
Organizations Interested in the Program Content			✔	✔								✔
Media					✔	✔						✔

community groups. You may inform the media and general public by distributing a press release and posting key findings on the organization's website. In some cases, the strategies for publishing the evaluation findings will have been specified upfront by the primary client; if not, you may suggest the formats that best communicate the findings to various audiences. Figure 14-3 shows how the findings of the evaluation study might be distributed.

THE CHALLENGE OF MULTICULTURAL EVALUATION

Multiculturalism poses some unique and difficult problems for program evaluation. Conducting a fair and democratic evaluation in a multicultural environment requires striking a balance between the rights of minority culture groups and the rights of the larger culture—a complex policy issue.

What does multiculturalism mean for the evaluator? First, the evaluator must strive to remain neutral in the face of competing minority interests. This is especially true when stakeholders have strong views about the evaluation outcome, try to influence the outcome, or downplay the possible contribution of the evaluation process.[33] Second, the evaluator must search out and define the views and interests of the minority groups to ensure that their needs are being met. When the minority group is defined as "poor" or "powerless," the evaluator has a compelling obligation to recognize the views and interests of this group.

Learning to affirm differences rather than deny them is the essence of what multiculturalism is all about.

– As cited in Kappa Omicron Nu's *Dialogue*, February 1997, p. 5

See Chapter 16 for more about cross-cultural communication and developing culturally appropriate intervention strategies.

Finally, the evaluator must be sensitive to the cultural differences that make implementing the evaluation difficult. The manner in which questions are asked, or the questions themselves, may be barriers to obtaining reliable data about the program's impact. Muslims, for example, are not comfortable answering personal questions about health and diet from a member of the opposite sex.[34] The Inuit of Nunavik, Quebec, are reluctant to answer questions about diet because they believe that thinking or talking too much about beluga whales and geese—traditional foods harvested by the Inuit—may result in their disappearance.[35] Perhaps the best message about multiculturalism was given by E. R. House at the University of Colorado at Boulder: "Treat minority cultures as you would be treated. Sooner or later, everyone may be a minority."[36]

Spreading the Word about the Program's Success

In her book *The Popcorn Report,* Faith Popcorn describes the importance of developing a vision of the future. One of her book's chapters bears the title "You Have to See the Future to Deal with the Present."[37] These are apt words for community nutritionists who are responsible for identifying a community's nutritional needs and developing programs to meet those needs. A good, effective nutrition program is achieved not by accident but by planning today to meet the needs of tomorrow.

The state of Arizona was following Popcorn's advice when it determined that existing training programs could not help child care workers acquire the skills needed to provide proper nutrition for the state's children with special health care needs. Its response to this needs assessment was to develop a program—Project CHANCE—to provide the necessary training. But the department's work didn't end when the course manual was published. The final step in the program planning process was to let stakeholders, community leaders, community nutritionists, child care program directors, and other interested parties know that the course was available. To do this, the department listed information about the course in national newsletters. It helped community colleges incorporate the course into their curricula. It had the course materials translated into Spanish to increase their dissemination. The course was listed as a resource for schools in the USDA's Healthy School Meals Program on the Internet and was described in a presentation at an annual meeting of the American Dietetic Association.[38] All of these activities served to get the word out to community nutritionists and other experts across the country that the Project CHANCE program had been designed, implemented, and tested and was ready for use in other states.

Entrepreneurship in Program Planning

Program planning is one of the most exciting aspects of the community nutritionist's job. It requires a great deal of creativity and offers opportunities to learn new skills and work with people in public relations, marketing, design, and communications. Go back to Chapter 1 and review the entrepreneurial activities listed in the margin on page 23. Three-quarters of the activities listed there are essential aspects of the planning process. You might be the first community nutritionist in your area to link clients in your weight-management program with support groups and sound nutrition information on the Internet. Or maybe you convince a popular local television personality to help raise awareness of the risk factors for osteoporosis. Perhaps you forge a partnership with a local company that never supported health promotion activities in the past. Perhaps your team introduces inner-city schoolchildren to the new "Apple Jane and the Cucumber King" board game—a program designed to encourage children to eat several fruits and vegetables daily. The possibilities for being an entrepreneur in this area are boundless.

Internet Resources

American Dietetic Association

The American Dietetic Association (ADA) supports several e-mail lists as a member benefit to promote networking and sharing of expertise: **www.eatright.org/Member/index_13294.cfm.**

Daily News **www.eatright.org/Member/index_16119.cfm**
Daily newsletter of news affecting food, nutrition, and health.

Dietetics-L **www.eatright.org/Member/index_16117.cfm**
General questions and answers regarding food, nutrition, and dietetics practice issues.

Foodservice-L **www.eatright.org/Member/index_16179.cfm**
An e-mail list for ADA members who specialize in commercial food service.

On the Pulse **www.eatright.org/Member/index_pulse.cfm**
Weekly online newsletter announcing policy developments affecting food, nutrition, and health.

ADA Dietetic Practice Groups

Many of the ADA Dietetic Practice Groups (DPGs) have their own EMLs; contact **practice@eatright.org.**

- CDHCF (Consultant Dietitians in Health Care Facilities)
- DBC (Dietitians in Business and Communications)
- DCE (Diabetes Care & Education)
- GN (Gerontological Nutritionists)
- HEN (Hunger and Environmental Nutrition)
- NCC (Nutrition in Complementary Care)
- NE (Nutrition Entrepreneurs)
- PNPG (Pediatric Nutrition)
- RDPG (Research)
- SCAN (Sports, Cardiovascular and Wellness Nutritionists)
- SNS (School Nutrition Services)

Dietetics Online
A discussion list for nutrition professionals around the world.
Send e-mail to: majordomo@empnet.com
Subject line: leave blank
In message area type: subscribe dietetics-online

Evaluation Working Group at CDC **www.cdc.gov/eval/resources.htm**
Lists resources about evaluation or conducting an evaluation; includes journals, online publications, and step-by-step manuals.

FDA Listservs Subscribe on-line at **www.fda.gov/emaillist.html.**
FDA has a variety of mailing lists to keep you up-to-date with news about the agency's activities and the products it regulates.

Food Safe Subscribe at **www.nal.usda.gov/foodborne/index.html.**
An interactive discussion group intended to link professionals interested in food safety issues.

FOOD TALK Subscribe at **http://vm.cfsan.fda.gov/~dms/nutrsub.html.**
A monthly e-mail newsletter for health professionals, educators, and consumers.

General Information about Listservs **www.lsoft.com/scripts/wl.exe?qL=nutrition&F=L&F=T**
Site contains a comprehensive list of nutrition-related listservs; several have a research focus.

Healthy People 2010 **http://hin.nhlbi.nih.gov**
Healthy People 2010 news and announcements from the National Institutes of Health (NIH).

Center for Food Safety and Applied Nutrition **http://vm.cfsan.fda.gov/~dms/nutrsub.html**
Site for subscribing to several food and nutrition listservs.

Meal Talk **http://schoolmeals.nal.usda.gov/Discussion/subscribemealtalk.html**
For those interested in healthful school meals. It is part of the Healthy School Meals Resource System developed by the Food and Nutrition Information Center (FNIC).

Success Talk **http://schoolmeals.nal.usda.gov/Discussion/subscribesuccesstalk.html**
Links school health professionals, child nutrition educators, and others interested in creating a healthful school nutrition environment.

NHLBI Information Center Notification List
An electronic newsletter that sends out news releases and announcements about new NHLBI materials.
Send e-mail to: listserv@list.nih.gov
In message area type: Subscribe NHLBIINFO-L

Public Health Nutrition Discussion and Information
A discussion list of professionals providing public health nutrition news.
Send e-mail to: http://mailman1.u.washington.edu/mailman/subscribe/phnutr-l
In message area type: Subscribe PHNUTR-L [your name]

USDA's Economic Research Service (ERS)
Announcements of new food and nutrition assistance program items at USDA.
To subscribe, send e-mail to LISTSERV@LISTSERV.ERS.USDA.GOV with the command:
SUBSCRIBE FOODASSISTANCE-AT-ERS

WIC-Talk Subscribe online at **www.nal.usda.gov/wicworks/Talk/subscribewictalk.html.**
A tool for professionals and paraprofessionals involved in providing nutrition services or educators interested in maternal and child health issues, nutrition, and breastfeeding.

Professional Focus

Leading for Success

Supreme Court Justice Potter Stewart once said about obscenity, "I cannot define it for you, but I know it when I see it."[1] The same might be said about leadership, although the wealth of research in this area has helped define the behaviors and attitudes of good leaders.

Leading is not the same as managing. The difference between the two has been described by Warren Bennis, an internationally recognized consultant, writer, and researcher on leadership: "Leaders are people who do the right thing; managers are people who do things right. Both roles are crucial, but they differ profoundly."[2] In general, managers crunch numbers, orchestrate activities, and control supplies, projects, and data. Leaders fertilize and catalyze. They enhance people, allowing them to stretch and grow. Where managers push and direct, leaders pull and expect.[3]

What exactly is leadership? Leadership is "the process whereby one person influences others to work toward a goal."[4] Leaders understand how an organization works (or doesn't work, as the case may be) and how people work—that is, what motivates them to be top performers. Leaders have the power to project onto other people their vision, their inspiration, their ideas.[5]

The traits of successful leaders number more than one hundred, according to recent studies, though not all experts agree on the importance of every trait.[6] Some of the attributes commonly ascribed to leaders are intelligence, credibility, energy, sociability, discipline, courage, and generosity. Integrity is essential. Integrity has been described as "the most important leadership principle that you will demonstrate to your leadership team and your followers."[7] Integrity means adhering to a high standard of honesty. Leaders with integrity strive to present their values and character honestly in their dealings with other people.

Accountability is also important. Accountability means doing what you say you are going to do. No matter whether the issue is large or small, leaders follow through on their promises and commitments. They know that *not* following through on a promise or commitment reduces their credibility and diminishes the trust other people have placed in them.[8] Leaders hold themselves accountable.

Leaders are forward thinking. They see the big picture and have a vision about where they are going—and where they want their teams to go. Leaders can describe what a team, department, or organization will look and feel like in six months, one year, or five years. Make no mistake, though, leaders are not ones to complain about the way things are or dwell on the way things might have been. They are optimistic about the future, seeing many possibilities and positive outcomes. Their enthusiasm is contagious. Leaders come in all shapes, sizes, and temperaments. They are found in the executive suite and on the factory floor, among support staff and across middle management. In their seminar programs on leadership, James Kouzes and Barry Posner used to say, "Leaders go places. The difference between managing and leading is the difference between what you can do with your hands and what you can do with your feet . . . . You can't lead from behind a desk. You can't lead from a seated position. The only way you can go anyplace is to get up from behind the desk and use your feet."[9] Then, they got a letter from the president of a computer software services company who indicated that although about one-third of his employees had disabilities and several were in wheelchairs, they led quite effectively. Kouzes and Posner reformulated their example to reflect the diversity of leaders. You can have a disability and be a leader; you can be a manager and be a leader; you can be a newly minted community nutritionist in your first job and be a leader.

Warren Bennis believes that leaders all have a guiding vision, passion, integrity, curiosity, and daring.[10] They are self-confident and instill self-confidence in others. They are willing to take risks and responsibility.[11] Following is a description of the principles of leadership outlined by General H. Norman Schwarzkopf, who guided U.S. troops to victory in the Gulf War.*

- **You must have clear goals.** Having specific goals and articulating them clearly makes it easy for everyone involved to understand the mission.
- **Give yourself a clear agenda.** First thing in the morning, write down the five most important things you need to accomplish that day. Whatever else you do, get those five things done first.
- **Let people know where they stand.** You do a great disservice to an employee or student when you give high marks for mediocre work. The grades you give the people who report to you must reflect reality.
- **What's broken, fix now.** If it's a problem, fix it now. Problems that aren't dealt with lead to other problems, and in the meantime, something else breaks down and needs fixing.
- **No repainting the flagpole.** Make sure that all the work your people are doing is essential to the organization.
- **Set high standards.** Too often we don't ask enough of people. People generally won't perform above your expectations, so it is important to expect a lot.
- **Lead and then get out of the way.** Yes, you must put the right people in the right place to get the job done, but then step back. Allow them to own their work.
- **People come to work to succeed.** Nobody comes to work to fail. Why do so many organizations operate on the principle that if people aren't watched and supervised, they'll bungle the job?
- **Never lie. Ever.** Lying undermines your credibility. Be straightforward in your thinking and actions.
- **When in charge, take command.** Leaders are often called upon to make decisions without adequate information. It is usually a mistake to put off making a decision until all the data are in. The best policy is to decide, monitor the results, and change course if necessary.
- **Do what's right.** "The truth of the matter," said Schwarzkopf, "is that you *always* know the right thing to do. The hard part is doing it."

References

1. As cited in M. Brown and B. P. McCool, High-performing managers: Leadership attributes for the 1990s, in *Human Resource Management in Health Care,* ed. M. Brown (Gaithersburg, MD: Aspen, 1992), p. 22.
2. W. Bennis, Learning some basic truisms about leadership, *National Forum* 71 (1991): 13.
3. J. D. Batten, *Tough-Minded Leadership* (New York: AMACOM, 1989), p. 2.
4. As cited in D. Hellriegel, J. W. Slocum, Jr., and R. W. Woodman, *Organizational Behavior,* 7th ed. (St. Paul, MN: West, 1995), p. 342.
5. P. J. Palmer, Leading from within, in *Insights on Leadership,* ed. L. C. Spears (New York: Wiley, 1998), pp. 200–1.
6. B. Czarniawska-Joerges and R. Wolff, Leaders, managers, entrepreneurs on and off the organizational stage, *Organization Studies* 12 (1991): 529–46.
7. N. L. Frigon, Sr., and H. K. Jackson, Jr., *The Leader: Developing the Skills and Personal Qualities You Need to Lead Effectively* (New York: AMACOM, 1996), pp. 15–36.
8. R. Carlson, *Don't Worry, Make Money* (New York: Hyperion, 1997), 151–52.
9. As cited in J. M. Kouzes and B. Z. Posner, *Credibility: How Leaders Gain and Lose It, Why People Demand It* (San Francisco: Jossey-Bass, 1993), p. 88.
10. W. Bennis, *On Becoming a Leader* (Reading, MA: Addison-Wesley, 1989), pp. 39–41.
11. R. A. Heifetz and D. L. Laurie, The work of leadership, *Harvard Business Review* 75 (1997): 124–34.

*From *Inc.* magazine, Goldhirsh Group, Inc., 38 Commercial Wharf, Boston, MA 02110 (www.inc.com). *Schwarzkopf on Leadership, Inc.* 14 (January 1992): 11.

Chapter *Fifteen*

Designing Community Nutrition Interventions

Learning *Objectives*

After you have read and studied this chapter, you will be able to:

- Describe five factors to consider when designing a community nutrition intervention.
- Describe three levels of intervention.
- Discuss five theories and models of consumer health behavior.

This chapter addresses such issues as health promotion theories and guidelines, public speaking, oral communications in presenting an educational session for a group, and current information technologies, which are Commission on Accreditation for Dietetics Education (CADE) *Foundation Knowledge and Skills* requirements for dietetics education.

Chapter *Outline*

Something To Think About...

We need to have visions, visions of healthy and safe lives in healthy and safe communities—even visions that seem impossible. Robert Kennedy used to say often, "Some people see things as they are and ask, 'Why?' I dream things that never were and ask, 'Why not?' " We need to dream things that never were and ask, "Why not?"

– **Barry S. Levy, *American Journal of Public Health,* February 1998, p. 191**

Introduction

In 1937 an inventor introduced a new product to the grocery store: the shopping cart. Until that time, people had shopped for their groceries using a small bag or basket. The inventor perceived the convenience and ease of using a cart on wheels for this activity and advertised his product with the question "Can you imagine winding your way through a spacious food market without having to carry a cumbersome shopping basket on your arm?"[1] Unfortunately, most people answered, no, they could not imagine doing this! They refused to accept the innovation. When queried about their behavior, customers claimed that the shopping cart looked like a baby carriage and that it made them feel weak and dependent. To get around this perception, the inventor hired women and men of various ages to come into his supermarkets and use the shopping carts to buy their groceries. This simple approach had the desired effect. Other customers saw the carts being used and elected to use them, too.[2]

This story illustrates two aspects of designing interventions. First, you must have information about your target population and why they do what they do. Understanding the behavior of your target population is an important step in developing strategies to influence—and eventually change—their **behavior.** Second, you must have an arsenal of tools for influencing behavior. Your toolbox might contain posters and table tents, cooking demonstrations, newspaper articles, and a health fair, or, as in the case of the shopping cart inventor, you might choose people from the target population to "model" the desirable behavior.

Recall from Chapter 1 that an intervention is a health promotion activity aimed at changing the behavior of a target population. This chapter describes the factors to consider when designing an intervention: program elements such as goals and objectives; the levels and types of interventions; the dietary habits, values, attitudes, and beliefs of the target population; and the theories of consumer behavior. Study each of these topics before reading how community nutritionists working on Case Study 1, on page 490, put all of the elements together.

Choose an Intervention Strategy

The first step in designing an intervention is to review the program's goals and objectives, which specify the program outcomes. At this point in the design process, your overall goal is to have a rough outline of what the intervention might look like. The details will come later.

Behavior The response of an individual to his or her environment.

Remember, the intervention or **intervention strategy** is the approach for achieving the program's goals and objectives. It addresses the question of *how* the program will be implemented to meet the target population's nutritional need. The intervention strategy can be directed toward one or more target groups: individuals, communities, and/or systems. Systems are the large, integrated environments in which all of us live and work. Targeting a system for intervention usually involves changing a public, corporate, or school policy, although it can include reorganizing a department to improve the manner in which a program is delivered.

The intervention strategy can also encompass one or more levels of intervention. That is, it can be designed to (1) build awareness, (2) change lifestyles, and/or (3) create a supportive environment.[3] At any one level, interventions may target individuals in small groups such as families, schools, worksites, and health clinics; people in social networks such as worksites, churches, and bridge clubs; entire organizations; or the community at large, which can be a city, province, state, or nation. Level I interventions focus on increasing awareness of a health or nutritional topic or problem. Awareness programs can be very successful in helping change attitudes and beliefs and increasing knowledge of risk factors, but they seldom result in actual behavior changes. Level II interventions are designed to help participants make lifestyle changes such as quitting smoking, being physically active, eating more fruits and vegetables and less saturated fat, and managing stress. These interventions can be successful when they call for small changes over time and when they use a combination of behavior modification and education. Level III interventions work toward creating environments that support the behavior changes made by individuals. Thus, a company's policy to promote heart-healthy and high-fiber foods in the company cafeteria makes it easier for employees who are trying to lose weight or lower their blood cholesterol concentration to make healthful food choices at work.

An intervention may be as simple as providing a brochure describing the benefits of breastfeeding to clinic clients or distributing a fact sheet listing the fat content of snack foods to grocery store shoppers. An intervention may be fairly complex, such as a mass media campaign that targets health writers across the nation or adolescents with low calcium intakes. Some interventions are full-service programs, complete with training manual, lesson plans, reproducible handouts, videos, and interactive websites.

Examples of intervention strategies are shown in Table 15-1. Intervention activities that increase awareness among individuals include health fairs, screenings, flyers, posters, table tents, newsletters, and Internet websites. Special events, websites, radio advertising, and television public service announcements promote awareness across the entire community. An example of a special event that increased awareness about heart disease risk among women is the Mother/Daughter Walk sponsored by the Heart and Stroke Foundation of Manitoba. Food labeling is an example of a system intervention that can increase awareness. When the Food and Drug Administration authorized a health claim for folate on food product and dietary supplement labels, it recognized the ability of product labels to inform women of childbearing age about the relationship between adequate folate intake and reduced risk of neural-tube defects.[4]

Level II interventions reach individuals through one-on-one counseling and small-group meetings. These interventions usually involve a formal program of assessing the individual's current attitudes, beliefs, and behaviors; setting goals for behavior change; developing the skills needed to change behavior; providing support for change; and evaluating progress. Examples of Level II interventions for communities are fitness programs in primary and secondary schools and health promotion programs for all city employees—activities that cut across broad sectors of the community. Systems interventions at this level include company incentives for employees to join local fitness clubs and the formation of a wellness committee composed of community and business leaders.

Intervention strategy An approach for achieving a program's goals and objectives.

TABLE 15-1 Examples of Intervention Strategies

TARGET GROUP	LEVEL OF INTERVENTION Level I: Build Awareness	Level II: Change Lifestyles	Level III: Create a Supportive Environment
Individuals	• Health fairs • Health screenings • Flyers, posters, table tents • Internet websites • Special events	• One-on-one counseling • Small-group sessions	• Worksite cafeteria programs • Peer leadership
Communities	• Media announcements • Internet websites • Special events	• Fitness programs in schools • Health promotion programs for city employees	• Municipal policy that supports food gleaning • Point-of-purchase labeling • Tax incentives for companies with health promotion programs
Systems	• Health claims on food labels • Legislation	• Company incentives for employees to join local fitness clubs • Formation of a community-based wellness committee	• Medicare coverage of medical nutrition therapy • School policy that restricts access to candy and soft drink machines • Legislation

Examples of Level III interventions that target individuals include worksite health promotion and cafeteria programs. Identifying peer leaders who can model behavior change and talk about how they changed their lifestyles is another way of creating a supportive environment. In the community at large, supportive environments are created through policies that encourage gleaning (a food recovery program), "point-of-purchase" labeling, and tax incentives for companies with health promotion programs. At the system level, supportive environments occur as a result of Medicare coverage for medical nutrition therapy, when school policy restricts access to candy and soft drink machines, and when eligibility requirements for food assistance programs are broadened.

Study the Target Population

When designing an intervention, study the target population's eating patterns and beliefs, values, and attitudes about foods and health. Some information about the target population was collected during the community needs assessment. Additional information can be obtained by conducting a library search of the literature, reviewing existing programs that deal with the target population, networking with colleagues who work with the group, and posting queries about the target population on Internet listservs. Consult the Internet Resources in this chapter, page 492, for sites that provide useful information on food and nutrition.

Nutrition and health-related listservs can be found among the Internet Resources on page 471.

See Chapter 3 for methods of assessing the target population's nutritional status.

The purpose of studying the target population is to understand their values, why they believe what they believe, and why they eat what they eat. For example, if your target population is overweight, sedentary, middle-class adults aged 30 to 50 years living in an urban area, you might ask these questions: Why do they choose pizza as their favorite lunch item within the food service market?[5] Why does a third of their calories come from food prepared away from home—whether eaten in restaurants, as takeout, or as home-delivered meals? Why do they prefer soft drinks rather than milk at meals? Why do they buy organic foods, which are often more expensive than comparable foods not labeled organic? Why do one-third of the members of this group use some form of alternative medicine, including dietary supplements such as chromium picolinate and shark cartilage, even when there is no scientific evidence that some of these products are effective? Why do more than two-thirds of this group never tell their physician about their use of these products?[6] Why do more than one-third of these consumers believe that cutting carbohydrates is a good strategy for improving long-term health?[7] Answering these questions provides clues to the target population's beliefs, values, and food patterns and to the interventions that may succeed in changing their behavior.

The target population's food-related behavior is important. For humans, foods are more than simply a source of nutrients and nourishment. They are also used to express friendliness and hospitality, maintain and strengthen personal relationships, enhance social status, relieve stress, and express religious and cultural beliefs.[8] Foods have symbolic meanings for humans, and the symbolic values we attach to foods, together with other factors, influence the decisions we make about them. Many environmental, sociocultural, and personal factors, some depicted in Figure 3-1 on page 62 and others described later, affect the food intake and nutritional status of the target population.

FOOD SUPPLY AND FOOD AVAILABILITY

Food choices are influenced by the types and amounts of foods available in the food supply and by the economies within a family. Food availability is affected by the food distribution system, the types of imported foods, facilities for food processing and production, and the regulatory environment. The target population's penchant for pizza is certainly due in part to pizza's widespread availability in supermarkets, convenience stores, and fast-food and take-out restaurants; it can even be ordered on the Internet. Their beliefs that "natural" is good and freshness is important steer them toward organic foods, producing a boom in sales in the total organic market from $3.7 billion in 1997 to $15 billion in 2004.[9]

INCOME AND FOOD PRICES

Income and food prices are two economic factors that affect food consumption. The relationship between income and food consumption is expressed by the Engel function, which is named for Ernst Engel, a Prussian mining engineer who was interested in sociological issues. In 1857 Engel published a study showing that the poorer a family is, the greater the proportion of its income it must spend on food purchases. The modern equivalent of this function states that as a consumer's income increases, the proportion of income spent on food decreases. A recent survey suggested that households with incomes over $50,000 spend 10 percent of their income on food, whereas the poorest households, with incomes of $5,000 to $9,999, spend about one-third of their disposable income on food.[10]

See Chapter 5 for more about the effects of food insecurity on low-income households.

Food prices affect consumption patterns. Households with higher incomes have more money to spend on food and choose whatever foods they want, regardless of price. They tend to buy convenience cuts of meat, poultry, and fish, which are more expensive than large roasts and whole chickens and fish.[11] Low-income households are more likely to have limited food budgets and to be concerned with price and value.

SOCIOCULTURAL FACTORS

Food choices are strongly influenced by **social groups.** Primary social groups such as families, friends, and work groups are more likely to affect behavior directly, and within this category, the family exerts the most influence.[12] This is not surprising, because the family is the first social group to which an individual belongs, and under most circumstances, it is the group to which an individual belongs for the longest time. The family is a paramount source of values for its members, and its values, attitudes, and traditions can have lasting effects on their food choices. This is especially true for children and teenagers. The calcium intakes of teenagers, for example, are higher in families in which teenagers perceive their parents' attention, care, support, and understanding than in families with low family connectedness.[13] Likewise, children whose parents did not regularly drink soft drinks were much less likely to consume soft drinks than children whose parents drank soft drinks on a regular basis.[14]

The culture in which we live affects our food behavior, and many of our food habits arise from its traditions, customs, belief systems, technologies, values, and norms. Culture dictates how foods are stored, processed, consumed, and disposed of—and even which foods are considered edible. North of the U.S.–Mexican border, for example, insects are seldom eaten, but in Mexico, the appetizer *los gusanos fritos* (fried caterpillars) may grace the menu in the finest restaurants. In the arctic region of Nunavik, near the Hudson Strait, the Inuit eat *niqituinnaq,* or real, natural, "country" food. Eating niqituinnaq is important, especially for older people, who prefer *ignuak* (fermented meat), seal, and other country food because it protects against disease and restores health. The Inuit rely on information handed down from generation to generation to know which animals are healthy and can be hunted for food.[15]

Religiosity—defined as a person's devoutness to his or her religious beliefs—may also influence food choices.[16] Religious beliefs affect the food choices of millions of people worldwide. Many religions, including Islam, Hinduism, Buddhism, Judaism, and Seventh-Day Adventism, specify the foods that may be eaten and how they should be prepared. Hinduism, for example, is the principal religion of India, and its food laws are steeped in ritual and meaning. Because the caste system is an integral part of Indian society, there are strict guidelines on how and with whom foods should be consumed. For Hindus, the focus of mealtime is eating, not conversation or socializing.[17] In North America, the daily food choices of Christians are not generally dictated by the basic doctrines of the Roman Catholic and Protestant churches, although some Christian church sacraments such as Holy Communion use food (bread and wine) symbolically. The principal dietary practices of several religions are summarized in Table 16-10 on page 510.

FOOD PREFERENCES, COGNITIONS, AND ATTITUDES

The environmental and sociocultural factors just described shape many of our personal attributes such as food preferences, cognitions, and attitudes. These attributes in turn affect our food choices. Preferences or likings for certain tastes and foods appear to develop quite early in humans. Not surprisingly, parents and their children tend to have similar food preferences.[18] Food preferences are shaped not only by family eating patterns but also by regional tastes. Mexican food is popular in western regions of the United States, whereas the Northeast prefers Italian foods.[19]

Food choices are affected by our **cognitions,** or what we think. It seems logical that consumers who have learned about food composition and healthful eating practices have the knowledge base needed to select foods for good health, but consumers do not always practice what they have learned. The most recent American Dietetic Association's

Social group A group of people who are interdependent and share a set of norms, beliefs, values, or behaviors.

Cognitions The knowledge and awareness we have of our environment and the judgments we make related to it.

Simply knowing which food choices consumers make does not tell us how they make their decisions.

David Young-Wolff/PhotoEdit, Inc.

Nutrition Trends Survey found that 85 percent of consumers believe that nutrition is important to health, but only 4 in 10 say they are doing all they can to eat healthfully. And although 82 percent say exercise is important to good health, only 66 percent indicate that they make a conscious effort to be physically active on a regular basis.[20] Thus, although knowledge can influence food-related behavior, the effect does not appear to be large. Having access to information about food, nutrition, and health does not guarantee that consumers will adopt healthful behaviors.[21] Nevertheless, experts predict that in the next few years the aging baby boomers, many of whom are in their fifties, may display more positive attitudes about nutrition and lifestyle.[22]

One of the most complex areas of consumer behavior is the relationship of **attitude** to behavior. Early attitude research, conducted at the turn of the century, suggested that an individual's behavior was determined to a large extent by her or his attitude toward that behavior. Beginning in the 1930s, however, some researchers began to suspect that there was not a predictable relationship between attitude and any given behavior. By the 1970s, some investigators concluded that attitudes could not be used to predict behavior; they maintained that the inconsistencies between attitude and behavior could be explained by any number of other variables, such as competing motives and individual differences.[23] Today, attitudes are believed to influence behavior indirectly.

HEALTH BELIEFS AND PRACTICES

Beliefs about foods, diet, and health influence our food choices. Two examples of the power of health beliefs can be drawn from different cultures. In Senegal, pregnant women avoid spicy condiments and foods, but they indulge their cravings for curdled milk, palm oil, meat, butter, and the traditional millet porridge. The Senegalese believe that if pregnant women are not allowed to satisfy their food cravings, their babies may be born with birthmarks.[24]

According to traditional Chinese beliefs, illnesses are caused by an excess of either yin (the dark, cold, feminine aspect) or yang (the bright, hot, masculine aspect) energy.

Attitude An individual's positive or negative evaluation of performing a behavior.

Foods and herbs, which themselves may be either yin or yang, are prescribed to treat certain symptoms of disease. Yang foods such as hot soups made from chicken, pork liver, or oxtail are prescribed to treat the clinical manifestations of excess yin, such as dry cough, muscle cramps, and dizziness. Yin foods include fish, most vegetables, and some fruits, such as bananas; they are used to treat the symptoms of hives and dry throat, which are believed to be caused by excess yang energy.[25] Thus health beliefs influence many consumer health practices, such as choosing leisure-time activities, eating comfort foods during illness, and using dietary supplements.

Draw from Current Research on Consumer Behavior

Simply knowing the factors that affect consumers' food choices does not tell us *how* consumers make their decisions. Many theories have been proposed to explain the decision-making process as it is related to changes in health behavior. Such theories are important because they suggest the questions that community nutritionists should ask to understand why consumers do what they do. Theories are sometimes presented in the form of *models*, which are simple images of the decision-making process. It is not possible to describe every health behavior theory in this chapter, but five deserve mention: the Stages of Change Model, the Health Belief Model, the Theory of Planned Behavior, the Social Cognitive Theory, and the Diffusion of Innovation Model. In the following sections, each theory is briefly described, and an example is given of how the theory can be applied to real-life situations. A summary of key concepts for several health behavior theories is provided in Table 15-2.

The Stages of Change Model

THE THEORY

The most widely used Stages of Change Model is the Prochaska and DiClemente transtheoretical model, originally developed to understand smoking cessation. The model is founded on three assumptions: (1) behavior change involves a series of different steps or stages, (2) there are common stages and processes of change across a variety of health behaviors, and (3) tailoring an intervention to the stage of change people are in at the moment is more effective than not considering the stage people are in. The transtheoretical model identifies five stages through which people move, although they don't always pass through these stages in a linear fashion, and relapses are common (see Figure 15-1).[26]

- **Precontemplation**—The individual is either unaware of or not interested in making a change.
- **Contemplation**—The individual is thinking about making a change, usually within the next 6 months. The individual may be weighing the risks and benefits of changing a behavior.
- **Preparation**—The individual actively decides to change and plans a change, usually within 1 month. The individual may have already tried changing in the recent past.
- **Action**—The individual is trying to make the desired change and has been working at making the change for less than 6 months. The individual has started making changes in his or her environment to support the changed behavior.
- **Maintenance**—The individual has sustained the change for 6 months or longer; the changed behavior has become a part of his or her daily routine.

TABLE 15-2 Summary of Key Concepts of Selected Models of Behavior Change

BEHAVIOR CHANGE MODEL/APPROACH	FOCUS	KEY CONCEPTS
Stages of Change Model (transtheoretical model)	Behavior change is explained as a readiness to change.	• Behavior change is described as a series of changes. • Specific behavior change strategies are identified for each stage.
Health Belief Model	Perception of the health problem and appraisal of proposed behavioral changes are central to a decision to change.	• Perceived susceptibility • Perceived impact • Perceived advantages of change • Appraisal of barriers • Self-efficacy
Theory of Planned Behavior or Theory of Reasoned Action	Behaviors are determined by a person's intentions to behave in certain ways.	• Intentions • Attitudes • Subjective norms • Perceived power and behavioral control
Theory of Trying	Behaviors are determined by a person's intentions to behave in certain ways but are influenced by past experiences with behavior.	• Intention to try • Frequency of past trying • Self-efficacy • Knowledge and skills required
Social Cognitive Theory	People and their environment interact continuously, each influencing the other.	• Learning occurs through taking action, observations of others taking action, and evaluation of the results of those actions.
Diffusion of Innovation	A process by which an innovation, idea, or behavior spreads and involves an ever-increasing number of individuals in a population.	• Innovators • Early adopters • Early majority • Late majority • Laggards
Self-efficacy	A component of numerous behavior change models; belief in ability to make a behavior change.	• Self-efficacy increases the probability of making a behavior change.
Motivational Interviewing	Client-centered approach to education and behavior change. Interviewer elicits client's intrinsic motivation to change; reinforces desire to change; seeks to diminish resistance to change; respects client's autonomy; interviewer-client relationship is a collaborative partnership.	• Motivation is fundamental to change. • Express empathy (acceptance facilitates change). • Articulate, explore, and resolve ambivalence to change. • Self-defined behavior change goals. • Support self-efficacy.

Source: Adapted from K. Bauer and C. Sokolik, *Basic Nutrition Counseling Skill Development* (Belmont, CA: Wadsworth/Thomson Learning, 2002).

The model resembles a spiral in some respects, with people moving around the spiral until they eventually achieve maintenance and termination. People in the contemplation stage are typically seeking information about a behavior change, whereas people in the maintenance stage are less likely to be looking for information and more likely to be searching for methods of strengthening the behavior and avoiding slipping back into old habits.

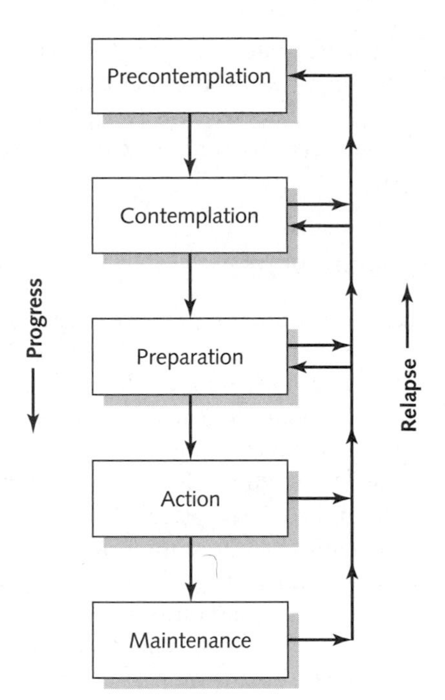

FIGURE 15-1 The Stages of Change Model

Source: S. Rollnick, P. Mason, and C. Butler, *Health Behavior Change: A Guide for Practitioners* (New York, NY: Churchill Livingstone, 2000), p. 19. Adapted from W. M. Sandoval et al., "Transtheoretical Model," in *Clinical Nutrition* 9: 64–69, © 1994, Aspen Publishers, Inc.

The key here is to develop an intervention strategy that will meet people's needs in a manner appropriate to the stage they are in. Both individuals and communities can be assessed for their state of readiness to change.[27]

THE APPLICATION—INDIVIDUAL

Fitness centers are interested in developing and implementing programs that promote fitness and health. A community nutritionist working at the Smithfield Fitness Club had read about a study that found an association between the presence of chronic disease such as hypertension, diabetes, heart disease, and dyslipidemia and readiness to change behavior in the areas of physical activity, fat intake, fruit and vegetable intake, and smoking. Study subjects were HMO members over the age of 40 years, a population much like the community nutritionist's own club members. The study found that members who were at highest risk of adverse health outcomes had the greatest readiness to change. A surprising finding was that members with heart disease were more ready to change their behavior to reduce disease risk than were members with diabetes.[28] Another study found that people in the precontemplation, contemplation, and preparation stages consumed fewer fruits and vegetables than people in the action or maintenance stages.[29]

The community nutritionist used these research findings and the Stages of Change Model to alter certain aspects of her health promotion programming. To increase awareness among clients in the precontemplation stage, she added special lectures on heart disease and hypertension to the roster of special events and developed brochures describing the risk factors for hypertension, diabetes, and heart disease. For clients in the contemplation and preparation stages, she offered four cooking demonstrations featuring low-fat recipes prepared with local fresh fruits and vegetables; the cooking demos were given in the fitness center's food center on Friday nights, when club attendance was high. These activities were designed to show clients who were mainly thinking about making changes that cooking the low-fat way is fun and easy. For clients in the action stage—that is, those who had participated in one of the special events or attended a cooking demonstration—she offered a 20-minute individual counseling session to answer questions about reading food labels, identifying food sources of fat, and calculating saturated fat intake. These actions were all designed to facilitate and support change among club members.

THE APPLICATION—COMMUNITIES

A recent university-sponsored study found that the prevalence of eating disorders among high school students in the city of Scottsville was nearly double the state average. A group of university researchers, community nutritionists, the city's chief medical officer, and a school nurse met to discuss what could be done to address the problem. The group determined that the city was in the precontemplation stage of readiness to change: No discussion of the problem among stakeholders (for example, parents, teachers, students, administrators, health authorities) had taken place, no plan for addressing the problem had been developed, and no activities had been undertaken to reduce the prevalence of eating disorders. The group believed that the problem was urgent and that some action should be taken. As a first step in moving the community to the contemplation stage, the group decided to hold a citywide meeting of stakeholders and key community leaders to assess their perception of the problem, identify resources for addressing the problem, and discuss desirable actions.

The Health Belief Model

THE THEORY

The Health Belief Model was developed in the 1950s by social psychologists with the U.S. Public Health Service as a means of explaining why people, especially people in high-risk groups, failed to participate in programs designed to detect or prevent disease.[30] The study of a tuberculosis screening program led G. M. Hochbaum to propose that participation in the program stemmed from an individual's perception of both his or her susceptibility to tuberculosis and the benefits of screening. Furthermore, an individual's "readiness" to participate in the program could be triggered by any number of environmental events, such as media advertising. Since Hochbaum's analysis, the Health Belief Model has been expanded to include all preventive and health behaviors, from smoking cessation to complying with diet and drug regimens.

The Health Belief Model has three components.[31] The first is the perception of a threat to health, which has two dimensions. An individual perceives that he or she is at risk of contracting a disease and is concerned that having the disease carries serious consequences, some of which may be physical or clinical (for example, death or pain), whereas others may be social (such as infecting family members or missing time at work). The second component is the expectation of certain outcomes related to a behavior. In other words, the individual perceives that a certain behavior (for example, choosing high-fiber foods to facilitate weight loss) will have benefits. Bound up in the perception of benefits is the recognition that there are barriers to adopting a behavior (for instance, choosing high-fiber foods requires skill in label reading and knowledge of food composition). The third component is **self-efficacy** or "the conviction that one can successfully execute the behavior required to produce the outcomes."[32] A key tenet of losing weight, for example, is the belief that one *can* lose weight. Other variables, such as education, income level, sex, age, and ethnic background, influence health behaviors in this model, but they are believed to act indirectly.

Self-efficacy The belief that one can make a behavior change.

THE APPLICATION

The American Cancer Society (ACS) recommends increasing fruit and vegetable intake to five to nine servings a day, limiting consumption of red meats high in fat, eating more whole grains, and adopting a physically active lifestyle as a means of lowering cancer risk.[33] Since these recommendations were first published, in 2001, Americans have increased their fruit and vegetable intake slightly, but their intake of dietary fat is still above the recommended level, and obesity is a widespread public health problem. The key question for community nutritionists and other practitioners in health promotion remains: Which factors promote dietary change in the general population?

Researchers in Washington state sought an answer to this question. They surveyed adults aged 18 years and older about their beliefs and health practices, using the Cancer Risk Behavior Survey, which consists of questions on risk factors for cancer, including dietary habits, alcohol consumption, sun exposure, smoking behavior, and preventive cancer screening. The researchers found that adults who believed strongly in a connection between diet and cancer, and who were knowledgeable about health recommendations, made a greater number of positive dietary changes than those who had little belief. Having knowledge of the fat and fiber content of foods and perceiving social pressure to eat a healthful diet did not predict who made behavioral changes in dietary patterns or weight.[34]

Knowing that beliefs play a key role in motivating people to make lifestyle changes, the community nutritionists with the Mississippi affiliate of the American Cancer Society increased the funding for their public awareness campaign. They designed three posters and a public service television announcement that reinforced the message that smoking, sun exposure, and certain dietary patterns are linked with increased cancer risk. This strategy was designed to influence the beliefs of people in high-risk groups.

The Theory of Planned Behavior

THE THEORY

The Theory of Planned Behavior, sometimes called the Theory of Reasoned Action, was developed by Icek Ajzen and Martin Fishbein. It "predicts a person's intention to perform a behavior in a well-defined setting."[35] The theory is a fundamental model for explaining social action and can be used to explain virtually any health behavior over which the individual has control. According to the model, behavior is determined directly by a person's intention to perform the behavior. **Intentions** are the "instructions people give to themselves to behave in certain ways."[36] They are the scripts that people use for their future behavior. In forming intentions, people tend to consider the outcome of their behavior and the opinion of significant others before committing themselves to a particular action. In other words, intentions are influenced by attitudes and **subjective norms.** Attitudes are determined by the individual's belief that a certain behavior will have a given outcome, by an evaluation of the actual outcome of the behavior, and by a perception of his or her ability to control the behavior. Subjective norms are determined by the individual's normative beliefs. In forming a subjective norm, the individual considers the expectations of various other people.

A modification of the theory, called the Theory of Trying, was proposed by Richard P. Bagozzi, who argued that more is needed to produce a behavior change than an expression of intention.[37] In the new model, shown in Figure 15-2, such factors as past experience (success or failure) with the behavior, the existence of mechanisms for coping with the behavior outcome (for instance, having a strategy for dealing with not meeting a weight-loss goal), and emotional responses to the process all influence the intention to try a behavior. Bagozzi and his colleagues hypothesized that when intentions are well

Intention A determination to act in a certain way.

Subjective norm The perceived social pressure to perform or not to perform a behavior.

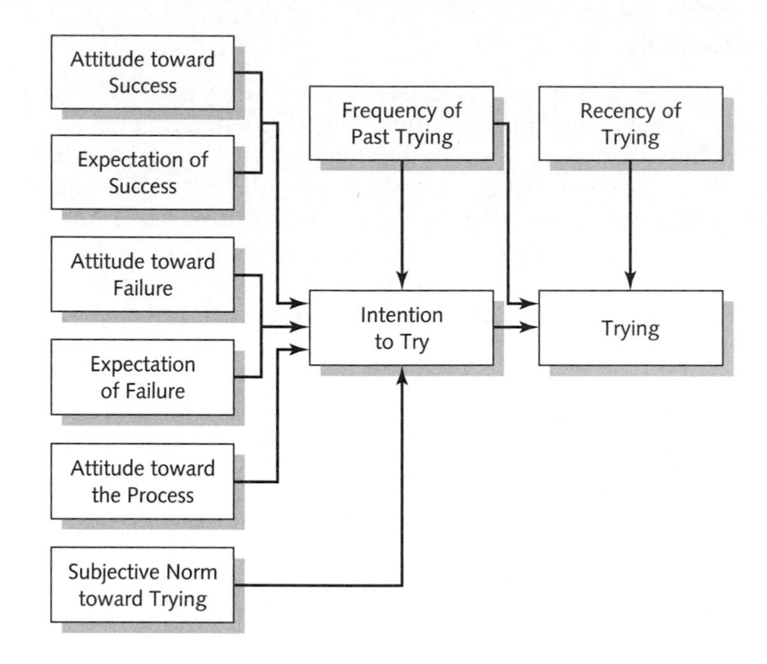

FIGURE 15-2 The Theory of Trying

Source: R. P. Bagozzi, The self-regulation of attitudes, intentions, and behavior, *Social Psychology Quarterly* 55 (1992): 179. Used with permission of R. P. Bagozzi and the American Sociological Association.

formed, they are strong mediators of behavior; when intentions are poorly formed, however, their influence on behavior is diminished, and that of attitudes grows stronger.[38]

THE APPLICATION

Dieting is a common method of trying to achieve an acceptably slim body shape. Young boys and girls often express the same dissatisfaction with their body shapes as do older adolescents and adults. Girls, in particular, express intentions to diet more frequently than boys express such intentions.[39] Research has shown, however, that even when the intention to diet is high, people have difficulty sticking with a weight-loss program.[40]

The community nutritionist at the Fairlawn Weight Management Center was experiencing a high dropout rate among adolescents participating in the center's "Get Fit Now" program, which included sessions on the principles of balanced eating, controlling eating impulses, and physical activity. Drawing on the principles of the Theory of Planned Behavior, he decided to survey the program participants about their intentions to lose weight, their attitudes about their body shapes, their level of self-esteem, their expectations related to success, their support from family and friends, and their perception of

their ability to control eating and lose weight. He used the survey results to add certain components to his "Get Fit Now" program. For example, he added two sessions: one to help participants clarify whether the time was right to lose weight and one to boost their coping skills so they could handle lapses. And he paired some participants, who had not been successful losing weight in the past, with others who had reached previous weight-loss goals. These actions were meant to improve the participants' intentions to master their eating habits and lose weight.

Social Cognitive Theory

THE THEORY

Social Cognitive Theory (SCT) explains behavior in terms of a model in which behavior, personal factors such as cognitions, and the environment interact constantly, such that a change in one area has implications for the others. For example, a change in the environment (say, the loss of a spouse's support for a weight-loss effort) produces a change in the individual (a decrease in the incentive to lose weight) and consequently a change in behavior (abandonment of a low-fat eating pattern). This theory, which is also known as social learning theory, was developed to explain how people acquire and maintain certain behaviors.

The major concepts in SCT, many of which were formulated by Albert Bandura, and their implications for interventions are given in Table 15-3. In this context, the environment

TABLE 15-3 Key Concepts in Social Cognitive Theory and Their Implications for Behavioral Intervention

CONCEPT	DEFINITION	IMPLICATIONS FOR BEHAVIORAL INTERVENTION
Environment	Factors that are physically external to the person	Provide opportunities and social support
Situation	Person's perception of the environment	Correct misperceptions and promote healthful norms
Behavioral capability	Knowledge and skill to perform a given behavior	Promote mastery learning through skills training
Expectations	A person's beliefs about the likely outcomes or results of a behavior	Model positive outcomes of health behavior
Expectancies	The values that the person places on a given outcome and incentives	Present outcomes of change that have a functional meaning
Self-control	Personal regulation of goal-directed behavior or performance	Provide opportunities for decision making, self-monitoring, goal setting, problem solving, and self-reward
Observational learning	Behavioral acquisition that occurs by watching the actions and outcomes of others' behavior	Include credible role models of the targeted behavior
Reinforcements	Responses to a person's behavior that increase or decrease the likelihood of its recurrence	Promote self-initiated rewards and incentives
Self-efficacy	The person's confidence in performing a particular behavior and in overcoming barriers to that behavior	Approach behavior change in small steps; seek specificity about the change sought
Emotional coping responses	Strategies or tactics that are used by a person to deal with emotional stimuli	Provide training and practice in problem solving and stress management skills
Reciprocal determinism	Dynamic interaction of the person, behavior, and the environment in which the behavior is performed	Consider multiple avenues to behavior change including environmental, skill, and personal change

Source: T. Baranowski, C. L. Perry, and G. S. Parcel, How individuals, environments, and health behavior interact: Social Learning Theory, in *Health Behavior and Health Education—Theory, Research, and Practice,* 3rd ed., ed. K. Glanz, F. M. Lewis, and B. K. Rimer (San Francisco: Jossey-Bass, 2002).

Programs in Action

Eat Healthy: Your Kids Are Watching

Five levels of influence for health education have been identified: intrapersonal/individual, interpersonal/group, institutional/organizational, community, and public policy. Programs can be targeted at just one level or at two or more levels. It is thought that a multilevel approach employing a combination of different strategies can be most effective.

The Theory of Planned Behavior explains behavior at the intrapersonal level by examining the relationships among an individual's beliefs, attitudes, intentions, and behavior. It assumes that the most important determinant of behavior is a person's intention regarding that behavior.[1] A person who believes that behavioral outcomes will be positive is likely to have a positive outlook on a change in behavior. Furthermore, a person who believes that others think he or she should perform certain behaviors will probably have a more positive attitude toward those behavioral changes. The Theory of Planned Behavior assumes that underlying reasons motivate people toward particular behaviors.[2] In planning education at the intrapersonal level, one can influence eating behavior by affecting personal attitudes and increasing awareness of subjective norms related to healthful eating. On a practical level, individuals can be given activities and suggestions for applying health messages.

Intervention at the interpersonal level is predicated on the assumption that the thoughts, advice, examples, assistance, and emotional support of others affect one's own feelings, behavior, and, therefore, health. People are influenced by and influential in their social environments.[3] Social Cognitive Theory (SCT) explains the interactions among behavior, personal factors, and environmental influences, including the opinions of others.[4] SCT states that people learn through their own experiences and by observing others. One aspect of SCT is observational learning or modeling; that is, one's beliefs are based in part on observing the behaviors of others and their behavioral outcomes. Observational learning is most effective when the role model is perceived to be powerful or respected—for example, when the role model is a parent. The case that follows exemplifies the use of both intrapersonal and interpersonal education to improve the nutritional health of low-income families.

The Michigan Nutrition Support Network is a public–private partnership to improve the nutritional health of Michigan's low-income families. The network's pilot partnership in Kent County included more than 40 active representatives from local business, health care, private practice, nonprofit agencies, and schools. *"Eat Healthy, Your Kids Are Watching"* was the network's focus group–tested message designed to prompt awareness in parents that they are role models for their children.

Goals and Objectives

The primary goal of the campaign was to improve the nutritional health of Kent County's low-income families through collaborative efforts among partners. The objectives were to develop and implement "awareness-building" activities promoting healthful eating to the target audience and to the public in general, and to construct a public–private partnership with businesses and agencies to assist with specific programs for the campaign.

includes both the social realm (family, friends, peers, coworkers) and the physical realm (the workplace, the layout of a kitchen). A strength of SCT is that it focuses on certain target behaviors rather than on knowledge and attitudes.[41]

THE APPLICATION

The prevalence of pica—a craving for and ingestion of nonnutritive substances such as clay, dirt, ice, baby powder, or laundry starch—among pregnant African-American women has been reported to range from 8 percent to 65 percent. Some reports have linked pica with low maternal hemoglobin levels and anemia.[42] A community nutritionist with the South Carolina Special Supplemental Nutrition Program for Women, Infants, and Children (WIC) had evidence that the prevalence of pica was quite high among the state's African-American WIC clients. She also recalled reading the results of a study in which peer counseling and motivational tapes were used to enhance breastfeeding among WIC clients.[43] Drawing on these research findings and on the principles of SCT, the community nutritionist designed a course that used peer counseling to

Methodology

Potential partners were located through personal contacts and written invitations to public agencies, commodity groups, food retailers, and others who work with community food and nutrition programs. Individual partners, once they became interested in the project, suggested others who they believed would benefit from the collaborative effort. The activities of the four-week campaign were categorized into two groups: awareness building and partnership programming. Awareness-building activities included 30-second cable spots, campaign newsletters in English and Spanish, signs on and in transit buses, a logo and slogan program with grocery stores and school districts, and a toll-free telephone number with messages in English and Spanish. Information on grocery store tours, cooking demonstrations, and a WIC module for nutrition education were among the materials and activities available for partners. The extensive partner kit included an events schedule, lesson plans, activity sheets, and recipes.

Results

The program reached an estimated 49,000 residents, including close to 7,000 low-income households. A random sample of 800 adults in households with children were surveyed to test awareness of the campaign and acceptance of its core message. Campaign awareness was 52 percent in households with children and 67 percent in the target population of low-income households. Approximately two-thirds of respondents indicated that they understood and agreed with the message when they heard it. An additional 20 percent indicated that they would adopt the message. School lunch menus, billboards, and television commercials were seen as most effective for reaching the target population.

Lessons Learned

An enthusiastic collaboration among businesses, community agencies, and community leaders was the key to the success of this venture. Partnerships forged between public and private organizations can grow strong as a result of working together on such campaigns.

References

1. D. E. Montano, D. Kasprzyk, and S. H. Taplin, The theory of reasoned action and the theory of planned behavior, in *Health Behavior and Health Education,* ed. K. Glanz, F. M. Lewis, and B. K. Rimer (San Francisco: Jossey-Bass, 1997).
2. I. Ajzen and M. Fishbein, *Understanding Attitudes and Predicting Social Behavior* (Englewood Cliffs, NJ: Prentice-Hall, 1980).
3. K. Glanz and B. K. Rimer, *Theory at a Glance: A Guide for Health Promotion Practice* (Washington D.C.: U.S. Department of Health and Human Services, Public Health Service, National Institutes of Health, 1997).
4. T. Baranowski, C. L. Perry, and G. S. Parcel, How individuals, environments, and health behavior interact, in *Health Behavior and Health Education.*

Source: Adapted from *Community Nutritionary* (White Plains, NY: Dannon Institute, Spring 2000). Used with permission. For more information about the *Awards for Excellence in Community Nutrition,* go to www.dannon-institute.org.

model healthful eating patterns and provide support for women who were trying not to practice pica. Although her budget did not allow for the cost of producing a videotape on the topic, she developed simple educational materials that explained what pica was and how to substitute other behaviors for it. Her idea drew on the SCT concepts of environment, situation, expectations, and observation.

The Diffusion of Innovation Model

THE THEORY

People often cannot or will not change their behavior, and many do not adopt innovations easily (recall the story about shopping carts at the beginning of this chapter). Even so, some people are more daring than others. Such people are the vanguard in the diffusion of innovation, the process by which an innovation spreads and involves an ever-increasing number of individuals within a population.[44] The Diffusion of Innovation

Model was developed by E. M. Rogers and F. F. Shoemaker in the 1970s to explain how a product or idea becomes accepted by a majority of consumers. The model consists of four stages:[45]

- **Knowledge**—The individual is aware of the innovation and has acquired some information about it.
- **Persuasion**—The individual forms an attitude either in favor of or against the innovation.
- **Decision**—The individual performs activities that lead to either adopting or rejecting the innovation.
- **Confirmation**—The individual looks for reinforcement for his or her decision and may change it if exposed to counter-reinforcing messages.

Innovations spread throughout a population largely by word of mouth. The speed of diffusion is a function, in part, of the number of people who adopt the innovation. Consumers can be classified according to how readily they adopt new ideas or products. *Innovators* adopt an innovation quite readily, usually without input from significant others. Innovators perceive themselves as popular and are financially privileged. This group is small. Like innovators, *early adopters* are integrated into the community and are well respected by their families and peers. Opinion leaders are often found in this group. Members of the *early majority* tend to be cautious in adopting a new idea or product, and persons in the *late majority* are skeptical and usually adopt an innovation only through peer pressure. Finally, the *laggards* are the last to adopt an idea or product, although they usually adopt it eventually. Members of this group tend to come from small families, to be single and older, and to be traditional.[46]

THE APPLICATION

A community nutritionist with Nutrition in Action, a company owned and operated by three registered dietitians, was concerned about several participants in her "Heart-Healthy Living" program. She perceived their lack of interest in making the kinds of dietary changes that would help lower their risk of having another heart attack. To boost their interest and enthusiasm for heart-healthy eating and cooking, she hit on the idea of contacting a popular local chef who had recently been interviewed on local television about the challenges he faced after surviving a heart attack. During the interview, the chef had indicated that he was just learning how to prepare low-fat foods and that he expected his new skills to make their way to his restaurant's menu. His comments agreed with those made by chefs who were surveyed about their food science knowledge and practice; that is, many chefs want to provide good nutrition to their customers but often lack the necessary knowledge and skills.[47] Believing the chef was a good early adopter, the community nutritionist convinced him to join the group and expand his heart-healthy cooking repertoire. She believed his enthusiasm would be catching and would influence participants who resisted adopting innovations related to cooking.

Put It All Together: Case Study 1

Recall from Chapter 14 that the results of formative, process, and outcome evaluations are critical for developing, implementing, and refining new programs.

Putting together all of the elements described in this chapter begins with reviewing the results of the needs assessment and the program goals and objectives. At this point in the program planning process, we use broad brush strokes to paint a picture of what an intervention strategy for Case Study 1 might look like. We chose health promotion activities and completed the grid shown in Table 15-4. It is not necessary or even practical to

TABLE 15-4 Case Study 1: Intervention Strategies for a Program Designed to Reduce Coronary Heart Disease Risk among Women

	LEVEL OF INTERVENTION		
TARGET GROUP	Level I: Build Awareness	Level II: Change Lifestyles	Level III: Create a Supportive Environment
Individuals	• Internet website • Posters, brochures	• "Heart Smart and Satisfying" cooking course • Smoking cessation course • Smoking reduction course	
Communities	• Special media events • Television and radio public service announcements • Antismoking campaigns in schools and worksites		• Work to secure legislation that prohibits smoking in city restaurants
Systems			• Work to secure legislation that prohibits tobacco-related advertising at all sporting events

fill in every box in the grid, because budgets, the number of staff available, and other factors will limit the scope of the intervention. Note that we chose to increase awareness at the individual and community levels (to meet program goal 2) and to build skills at the individual level (to meet program goal 1). We elected to promote supportive environments at the community and system levels.

The goals for Case Study 1 are listed on page 456.

We decided to conduct *formative evaluation* research to obtain information about the target population's skills. In focus group sessions, we asked women whether they had participated in cooking and smoking cessation courses, whether these programs worked, and what they found valuable about the course materials. We asked about their expectations related to reducing the risk of coronary heart disease (CHD). The results of these focus group discussions led us to consider adding an innovative element to our intervention strategy: a smoking reduction course, in which the goal would not be to get people to quit smoking but rather to help them cut down on the number of cigarettes smoked daily and adopt positive lifestyle behaviors (for example, eating less saturated fat and getting moderate exercise). Before finalizing the intervention strategy, we would plan to conduct a review of the literature related to smoking and health behaviors, develop a course outline, and then evaluate this element of the intervention strategy.

As we noted in Chapter 14, *formative evaluation* is the process of testing and assessing certain elements of a program before it is implemented fully.

As the overall intervention strategy began to take shape, we could see how our health promotion activities had been influenced by the theories of consumer behavior. The Internet website, posters, and brochures were aimed at reaching people who were in the contemplation and preparation stages of change. Our proposed activities in the policy arena—namely, working to secure legislation calling for smoke-free restaurants and a ban on advertising sponsored by tobacco companies at athletic events—were aimed at people in the maintenance stage. Some aspects of the "Heart Smart and Satisfying" cooking course, particularly those related to helping participants make simple dietary changes, reinforce healthful eating habits, and cope with setbacks, drew on the principles of the Health Belief Model and Social Cognitive Theory. And we could envision forming partnerships with other organizations, such as the local affiliates of the lung and cancer associations, to achieve some of our goals. The next step would be to develop the nutrition education component (described in Chapter 17).

Use Entrepreneurship to Steer in a New Direction

We already have a wealth of information about risk factors, epidemiology, healthful lifestyles, motivations for behavior change, and appropriate educational messages. The challenge comes in thinking of new ways of delivering health messages and services to vulnerable populations. One example of innovative thinking is The Well, a community-based drop-in wellness center for black women in a low-income housing complex in Los Angeles. The vision for this project emerged from a planning retreat attended by black women who were community leaders and activists. Their discussions focused on how to improve the poor health status of black women living in the district and led to the founding of The Well in 1994. The Well operates in partnership with the James Irvine Foundation, the UCLA Psychology Department, the John Wesley Community Health Institute, Inc., Fitness Funatics, and other groups to deliver services and programming in nutrition, fitness, HIV/AIDS education and intervention, avoidance of substance abuse, family planning, pregnancy, and parenting. Support groups called sister circles work to empower black women to take charge of their health. The Well's affiliation with UCLA, with the landlord of the housing project, and with the National Black Women's Health Project ensures its continued success in meeting the health and nutritional needs of its clients.[48]

Successful community interventions such as The Well have been made, but many people in the communities are not getting the message. What is needed to ensure the success of community interventions? Manning Feinleib, an associate editor of the *American Journal of Public Health,* writes that we need a better understanding of the community factors that influence change and of the reasons why consumers resist change.[49] When you plan community interventions, think of new ways to reach your target audience. Plan strategies for finding out why your clients are resisting a behavior change. Apply your creativity to influencing people to achieve behavior change. And when you are successful, consider helping other community nutritionists by publishing a carefully documented account of your own efforts and outcomes.

Internet Resources

Applications of Behavior Change

Guide to Community Preventive Services	**www.thecommunityguide.org**
HealthMedia	**www.healthmedia.com**
National Center for Chronic Disease Prevention and Health Promotion	**www.cdc.gov/nccdphp**
National Institute of Health, National Cancer Institute	**www.cancer.gov**

Other Nutrition and Health-Related Sites

Find timely information from the following sites, which serve as major directories of links:

American Dietetic Association	**www.eatright.org**

Arbor Nutrition Guide	**http://arborcom.com**
Extension Human Nutrition, Kansas State University	**www.oznet.ksu.edu/dp_fnut/freshfruitsandvegetables.htm**
Food and Nutrition Information Center (FNIC)	**www.nal.usda.gov/fnic**
Links to government sites related to nutrition and databases:	
Child Care Nutrition Resources	
Food Stamp Program Nutrition Resources	
Healthy School Meals Training Materials	
International Bibliographic Information on Dietary Supplements (IBIDS) Database	
USDA/FDA Foodborne Illness Educational Materials	
USDA/FDA HACCP* Training Programs	
WIC Works Database	
Healthfinder® or healthfinder® espanol	**www.healthfinder.gov or healthfinder.gov/espanol**
National Institutes of Health (NIH)	**http://health.nih.gov**
National Library of Medicine's MEDLINEplus	**www.nlm.nih.gov/medlineplus**
Tufts University's Nutrition Navigator *A website rating guide for categories including Educators, Health Professionals, General Nutrition, Journalists, Kids, Women, and Special Dietary Needs.*	**http://navigator.tufts.edu**

*HACCP = Hazard Analysis and Critical Control Point.

Professional Focus

Being an Effective Speaker

Public speaking ranks number one on most people's list of most dreaded activities. It causes churning stomachs, sweaty palms, dry mouths, and outright fear among many competent professional women and men. Many people would sooner have a root canal than give a 30-minute speech at a convention! If you feel this way about public speaking, take heart. You are not alone. More important, you *can* master the art of public speaking.

Tips for making an effective presentation are presented here. These basic principles apply to many situations, from a formal presentation at a scientific meeting to an informal update for your colleagues at a staff meeting. They also apply to many teaching situations.

Things to Do Before Your Presentation

Public speaking is a skill just like any other skill. Even so, you may find that you expect more of yourself when it comes to public speaking than you do in other settings. If you are learning to downhill ski, you don't expect to be skiing the double–black diamond trails and moguls at the end of your first season. If you are learning to speak Spanish, you don't expect to converse fluently after only a few lessons. So why should you expect to be a first-rate public speaker after a handful of presentations? Just like any other skill, public speaking requires practice, evaluation, and more practice and evaluation. It is an ongoing process. Even when you become skilled at public speaking, there will be room for improvement. In the beginning, try to remove a little pressure by remembering that you are working toward acquiring a skill that can be gained only by doing. You will improve over time. To accelerate your competency curve, follow these rules:

1. Organize your presentation around this basic principle of effective speaking: First, tell your audience what you are going to tell them; then tell them what you have to tell them; and finally, tell them what you told them! This strategy lets your audience know precisely what your presentation will cover and helps them remember the main points.

2. Prepare your visual aids so that they present your ideas effectively. Here are a few suggestions on preparing PowerPoint (or other presentation software) slides or other visual aids:[1]

- **Clear purpose.** An effective slide should have a main point or central theme.
- **Readily understood.** The slide's main point should be readily understood by the audience. If it is not, the audience will be trying to figure out what the slide has to say and will not be listening to the speaker.
- **Simple format.** The slide should be simple and uncluttered. Avoid slides that present large amounts of data, such as columns of numbers or many lines of text. Avoid long sentences: Use generally no more than 6 words per line and no more than 6 lines per slide. Limit bullets to 6 per slide; bulleted items should be 1 or 2 lines in length. A slide's text should contrast with its background. Design templates can help you achieve standardization in your choice of colors, styles, and positioning of text and graphics. Strive for consistency in your use of special effects.
- **Free of nonessential information.** Information not directly related to the slide's main point should be omitted.
- **Digestible.** The audience is capable of assimilating only so much information from a slide. It is better to have only a small amount of information (even just one sentence) than to cram numerous points onto a single slide.
- **Graphical format.** Some information is best presented graphically. In addition, the use of graphs and charts provides a visual change from slides containing only text.[2] Graphics and clip art can enhance and complement slide text. Balance their placement on the slide and use no more than two graphics per slide.
- **Visible.** Because most meeting rooms were not designed for projection, some people sitting in the back of the room may not be able to view the slides over the heads of those in front. For this reason, horizontal slides are more appropriate than vertical slides.
- **Legible.** Studies of projected image size and legibility show that the best slide template is about 42 spaces wide (9 centimeters) by 14 single-space lines (6 centimeters). The best type for slides is at *least* 5 millimeters, or 14 points. Larger font indicates more important information. Font size generally ranges from 18 to 48 point. Remember that decorative fonts can be hard to read.
- **Integrated with verbal text.** Slides should support and reinforce the verbal text. Conversely, the verbal text should lay a proper foundation for the slide.

Similar principles apply to preparing overheads. Keep the information simple, and limit the amount of information provided. To improve legibility, type the master copy in large type. Like slides, overheads should be integrated with your text.

3. Rehearse your presentation several times—you would do no less for a piano recital! If the presentation is formal, use a table or desk as a podium. Time the presentation from start to finish, including your opening and closing remarks, and adjust your presentation as needed. A general rule of thumb is that you should plan to spend about 1 minute per slide or overhead. If the presentation is informal, write down the key points you want to make and practice saying them. Rehearsing your presentation ensures that you will know your material and how you want to present it.

4. Use mental imaging to boost your self-confidence. What is mental imaging? It's a technique used by many successful businesspeople, politicians, actors, and athletes to develop and strengthen a positive mental picture of their performance. It is a way to relieve stress and reduce anxiety about speaking in public. It sounds hokey, but it works.

Picture Nikki Stone, a freestyle ski aerialist. She is preparing to do her routine at the 1998 Nagano Olympics, a performance watched by millions of spectators worldwide. She stands ready to begin. What is going through her head during those few seconds before she launches herself downhill into her routine? Is she thinking, "Oh, no, there's one spot where I always mess up. I'll never do it right. My coach, family, friends, and country will be humiliated." It isn't remotely possible that Stone's thought processes took this tack. You can be sure that when she stood poised to begin her routine in the Olympics, she pictured every move from beginning to end, all done flawlessly. It is an image she worked to cultivate both mentally and physically. The result was a gold medal performance.

You can use mental imaging to boost your performance and quell those butterflies in your stomach. On several occasions before your presentation, walk yourself mentally through the speech from beginning to end. Picture being introduced, standing up and walking to the podium, adjusting the microphone, smiling, giving your opening remarks, asking for the first slide, and so on, right down to the very end of your presentation. Picture giving your presentation and handling questions at the end with complete confidence. The key to using mental imaging successfully is to use it *whenever* a negative thought about your presentation intrudes. If a mental picture of you passing out behind the podium surfaces suddenly, use mental imaging to squash it. Force yourself instead to picture a confident, in-control YOU. Allow no negative thoughts about your presentation to take form. Encourage only positive thoughts. Mental imaging takes a little practice, but it is worth the effort. You will find that because you *think* you are more confident, you *are* more confident.

Things to Do During Your Presentation

Use the following techniques to ensure that you give a first-rate presentation:

- **Smile.** A smile will go a long way toward helping you relax and making you appear accessible to your audience. This is especially important when dealing with the general public.
- **Use eye contact.** Regardless of the size of the audience, select one person with whom to establish eye contact. Let your eyes dwell on this one individual a few moments, and then move on to another person. This gives the appearance of a one-on-one interactive discussion, which engages the audience in your presentation and helps ensure that they are listening to you.
- **Use gestures.** Gestures give energy to your presentation and provide additional emphasis for key points. Practice making them during your rehearsals. Exercise a little common sense here—wild arm movements and pirouettes will detract from your presentation.
- **Control the pace.** Although maintaining a steady pace will ensure that you complete your speech on time, it may make your audience sleepy. Vary the pace to keep the audience interested in what you are saying.
- **Use pauses.** Pauses, like gestures, can be used for emphasis. A well-timed pause keeps your audience engaged and allows them a moment to process what you've just said.
- **Vary the volume and pitch.** Changing the volume and pitch of your voice has more auditory appeal for the audience than speaking in a monotone.

Finally, two other points deserve mention. First, remember that the purpose of your presentation is to share information with your audience. Your listeners will generally be much less critical of your performance than you are. Being an effective speaker simply means that your audience is listening to your messages and absorbing the material you present. Second, despite all the tips and techniques listed here, you will want to develop your own style. Learn to be a relaxed, confident speaker *and* to be yourself.

References

1. The suggestions for preparing good slides were taken from the FASEB Call for Papers for the 1993 Annual Meeting, pp. 30–31; and V. Montecino, Creating an Effective PowerPoint Presentation, available at http://mason.gmu.edu/~montecin/powerpoint.html.
2. J. W. King and J. Rupnow, A primer on using visuals in technical presentations, *Food Technology* 46 (1992): 157–70.

Chapter *Sixteen*

Gaining Cultural Competence in Community Nutrition

Kathleen D. Bauer, PhD, RD

Learning *Objectives*

After you have read and studied this chapter, you will be able to:

- Define cultural competence as exhibited by community nutrition professionals.
- Identify and explain two cultural competence models.
- Describe the influence of culture on beliefs, values, and behaviors.
- Explain the importance of recognizing one's own cultural values and biases.
- Describe the basics of developing cross-cultural communication skills.
- Explain strategies for providing culturally competent nutrition interventions.

This chapter addresses such topics as sociocultural and ethnic food consumption patterns and the influence of cultural factors on food behavior, interpersonal communication skills, health behaviors and educational needs of diverse populations, and diversity issues, which are Commission on Accreditation for Dietetics Education (CADE) *Foundation Knowledge and Skills* requirements for dietetics education.

Chapter *Outline*

Something To Think About...

Father, Mother, and Me

Sister and Auntie say

All the people like us are We,

And every one else is They

And They live over the sea,

While We live over the way,

But—would you believe it?

—They look upon We

As only a sort of They!

We eat pork and beef

With cow-horn-handled knives.

They who gobble Their rice off a leaf,

Are horrified out of Their lives;

And They who live up a tree,

Feast on grubs and clay,

(Isn't it scandalous?) look upon We

As a simply disgusting They!...

– RUDYARD KIPLING, *We and They*

Introduction

How receptive would you be to a community education program featuring heart-healthy ways to prepare dog meat? How would you respond to a mandate that lobster could no longer be consumed because such practices are morally repugnant to the majority population? How would you like to be treated in a hospital where most of the décor was black and everyone wore black? If you were told to eat insects to improve the quality of your bones, would you readily change your eating habits? Does imagining such situations evoke feelings of confusion, shock, and anger? These examples may help you imagine how Hindus, who regard cows as sacred animals, may feel about community programs educating people on healthful ways to prepare beef; how Hmong may have felt about laws restricting healing ceremonies that include animal sacrifices; and how recent Asian immigrants may feel about going to a health center where everyone wears white, the color of mourning.

These scenarios highlight gaps in understanding food practices among various ethnic groups; however, differences between cultures occur on many levels—communication, sense of time, family practices, beliefs about the cause of illness, and healing beliefs, to name a few. Because North American society is composed of a large variety of groups, community health professionals and organizations need to have strategies to bridge cultural gaps. These strategies can be learned through gaining cultural competence.

Gaining Cultural Competence

Gaining cultural competence in community health care means developing **attitudes,** skills, and levels of awareness that enable one to provide culturally appropriate, respectful, and relevant interventions. The foundation of cultural competence is development of an awareness of one's own cultural matrix and an understanding that cultural beliefs influence our behavior and our conscious and unconscious thoughts. A nutrition

Attitude A collection of beliefs that includes an evaluative aspect.

professional needs to approach cross-cultural interactions with a nonjudgmental attitude and a willingness to explore and understand different values, beliefs, and behaviors. To work successfully with individuals from substantially different cultures or to develop cross-cultural programming, culturally sensitive communication, negotiation, and education skills are required. Consideration must also be given to the overall organizational cultural competence that community service agencies need in order to provide appropriate services and to support individual community health professionals in working effectively.

Terms Related to Cultural Competence

An understanding of the relevant concepts and terminology is an important first step toward gaining cultural competence.

CULTURE

Culture is shared history, consisting of "the thoughts, communication, actions, customs, beliefs, values, and institutions of racial, ethnic, religious or societal groups."[1] The societal groups can include gender, age, sexual orientation, physical or mental ability, health, occupation, and socioeconomic status. We develop cultural characteristics through life experiences and education. Culture directly and indirectly influences how we view the world and interact with others. Because each of us is a member of a number of societal groups and we have unique life experiences, no two people acquire exactly the same cultural attributes.

CULTURAL VALUES

Luckman[2] describes cultural values as "principles or standards that members of a cultural group share in common." Because cultural values are the grounding forces that provide meaning, structure, and organization in our lives, we hold on to them in the face of numerous obstacles. There are many examples in history of people practicing an outlawed religion in secrecy, despite the certainty of severe consequences if they were caught. See Table 16-1 for a list of functions of cultural values.

DIVERSITY

In the cultural context, diversity consists of differences among groups of people. Some forms of diversity are visible, such as physical differences, abilities and disabilities, and language differences. Other forms of diversity that may not be visible or obvious include sexual orientation, gender identification, socioeconomic status, and age.

TABLE 16-1 Functions of Cultural Values

- Provide a set of rules by which to govern lives.
- Serve as a basis for attitudes, beliefs, and behaviors.
- Guide actions and decisions.
- Give direction to lives and help solve common problems.
- Influence how to perceive and react to others.
- Help determine basic attitudes regarding personal, social, and philosophical issues.
- Reflect a person's identity and provide a basis for self-evaluation.

Source: Adapted from J. Luckmann, *Transcultural Communication in Health Care* (Albany, NY: Delmar/Thomson Learning, 2000), p. 23.

CROSS-CULTURAL

The term *cross-cultural* denotes interaction between or among individuals who represent distinctly different cultures. Because individuals develop behavior patterns and views of the world on the basis of unique life experiences and membership in several cultural groups, no two people exhibit identical behavior patterns. Therefore, all encounters between two people can be viewed as linking cultures. However, encounters between individuals or groups are not labeled as cross-cultural unless the attributes of the cultures they represent are substantially different.

ETHNOCENTRIC

People tend to be ethnocentric—that is, to consider the beliefs, values, customs, and viewpoints of their own group superior to those of every other group. Every culture teaches its members to regard its beliefs and views of reality as the best, and some cultures even teach that their beliefs are the *only* acceptable ones.[3]

Need for Cultural Competence

There is a compelling need for community health professionals to gain cultural and linguistic expertise. The reasons include demographic diversity and projected population shifts, increased utilization of traditional therapies, disparities in the health status of various racial and ethnic groups, under-representation of health care providers from culturally and linguistically diverse groups, and legislative, regulatory, and accreditation mandates.

DEMOGRAPHICS—POPULATION TRENDS

The United States has always had a rich mix of ethnic, racial, and societal groups, but the challenge of meeting the needs of a **multicultural** and dynamic population seems greater than ever. Since the 1970s, the United States has been moving toward a cultural plurality, where no single ethnic group is a majority. Census data indicate that diverse racial and ethnic groups in the United States have increased from approximately one-fourth to one-third of the population.[4] This trend is expected to continue, with minority groups climbing to 40 percent of the total population by the year 2030.[5] Hispanics, Asians, and Pacific Islanders have been increasing more rapidly than the rest of the U.S. population. By 2005, Hispanics accounted for nearly 13.7 percent of the U.S. population, exceeding African Americans as the largest minority group.

These changes have been brought about by alterations in immigration laws (the foreign-born population has more than doubled in the past 20 years), by corporate expansions into the global market, and by the tendency for minorities and immigrants to have higher birth rates.[6] In addition, the population mosaic is shifting in response to internal migration and the greater percentage of senior citizens.[7]

Linguistic diversity accompanies population shifts. Over 300 languages are spoken in the United States, and during the last decade the number of people who cannot speak English has increased substantially.[8] Census results revealed that approximately 14 percent of the U.S. population spoke a language other than English at home. In some states the percentages were substantially higher than the national average, such as New Mexico (36 percent) and California (31 percent).[9]

As the population of the country diversifies, so will the health care workforce. Although a number of the health professions lack diversity at this time, the picture is expected to change because the government and many professional organizations have instituted several programs to encourage minorities to train for health professional careers. (See the next section on under-representation of health care providers from culturally and linguistically

Multicultural In this chapter, a property of groups wherein several cultures are represented.

The United States has always had a rich mix of ethnic, racial, and societal groups; the challenge for community nutritionists is in meeting the needs of this diverse and dynamic audience.

Bob Thomas/GettyImages

diverse groups). Also, Census Bureau predictions for the 2000–2020 time period indicate that in general, a more diverse population will be available in the future to fill openings in the health care field. The number of women and aging baby boomers (46 to 64 years old) in the labor force will increase and will be accompanied by considerable increases in the numbers of Asians and Hispanics.[10] The need for effective communication skills will be universal. For example, recent immigrants will need cultural competence skills to work effectively within the North American society, and health care professionals will need cross-cultural communication skills to work effectively with clients and coworkers.

INCREASED UTILIZATION OF TRADITIONAL THERAPIES

Community health professionals cannot ignore the substantial increase in the utilization of complementary and traditional therapies, such as meditation, acupuncture, and herbal medicine. Diana Dyer, popular lecturer and author of *A Dietitian's Cancer Story: Information & Inspiration for Recovery & Healing from a 3-Time Cancer Survivor,* has been an advocate of learning about and combining conventional and complementary approaches to healing. She made many changes in her lifestyle to enhance healing but states in her book that "although I am a nutritionist to the core, learning to meditate, and doing it faithfully, has been the most important change I have made."[11] Kanjana also found a positive effect of using complementary therapies in her study of adolescent girls.[12] She found an improvement in various nutritional parameters after the girls participated in 80 minutes of yoga exercises each morning for six months. Developing an understanding and an appreciation of the health practices of various cultures can help health practitioners plan and implement meaningful interventions.

Health disparities Exist when a segment of the population bears a disproportionate incidence of a health condition or illness.

Ethnicity A property of a group that consists of its sharing cultural traditions, having a common linguistic heritage, and originating from the same land.

HEALTH DISPARITIES

Not all cultural groups have the same health status. There are substantial **health disparities** in segments of the population—disparities based on gender, race or **ethnicity,** education or income, disability, living in rural localities, or sexual orientation. The incidence

TABLE 16-2 Specific Examples of Health Disparities

African Americans	Infant mortality rate is double that of European Americans, heart disease death rates are more than 40 percent higher, and the death rate for all cancers is 30 percent higher. Life expectancy at birth is 77.1 years for European Americans but only 71.1 for African Americans.[13]
Hispanics	Prevalence of tuberculosis is twice the rate for the total population. Hispanics are twice as likely as non-Hispanic whites to develop diabetes and almost twice as likely to die from the disease.[14] In New York, the Puerto Rican infant mortality rate is 70 percent higher than that of the total population.[15]
Asian American and Pacific Islanders	Tuberculosis rates are five times higher than the total population, and 36 percent of those under age 65 do not have health insurance. Women of Vietnamese origin contract cervical cancer at almost five times the rate for European American women.[16]
Native Americans	Diabetes prevalence is 70 per 1,000, whereas it is only 30 per 1,000 for the total population. Cirrhosis deaths are 21.6 per 1,000 compared to 8 per 1,000 for the total population.[17] Pima Native Americans have the highest incidence of diabetes in the world.

of illness, disability, and death is higher among African Americans, Hispanic Americans, Native Americans, Asian Americans, Alaska Natives, and Pacific Islanders than in the general population.[18] (See Table 16-2 for specific examples of health disparities.)

The mission of the National Center on Minority Health and Health Disparities (NCMHD) is to promote minority health and to lead, coordinate, support, and assess the NIH effort to reduce and ultimately eliminate health disparities. For more information, see http://ncmhd.nih.gov/about_ncmhd/mission.asp.

CAUSES OF HEALTH DISPARITIES

A number of interrelated factors are thought to contribute to health disparities:

- **Socioeconomic status. Minorities** often have lower levels of income and education, reside in poorer housing, live in unsafe neighborhoods, and have fewer opportunities to engage in health-promoting behaviors.[19]
- **Lack of insurance.** Minorities are more likely to be uninsured than white Americans. Nearly 20 percent of blacks and Asians, 33 percent of Hispanics, and 27.5 percent of American Indian and Alaska natives did not have health insurance in 2003.[20]
- **Culture.** Some beliefs and health practices of minorities may contribute to health risks. In a study of African-American health attitudes, beliefs, and behaviors, only about 50 percent of respondents identified health as a high priority in their life.[21] An evaluation of Hispanic health beliefs and practices indicate a greater emphasis on the power of God and less on preventive health care.[22]
- **Access to and utilization of quality health care services.** Research has shown that many minority populations are less likely to receive routine medical checkups, obtain immunizations, undergo examinations for cancer, and receive treatment for hypertension.[23] Many of the factors found to limit utilization of available services can be attributed to a health care system that lacks cultural sensitivity.[24] They include inconvenient location, unawareness of services, feelings of discomfort with providers, health provider attitudes, lack of translators, and waiting in long lines.[25]
- **Discrimination/racism/stereotyping.** Individuals who perceive that they have been treated in a racist manner are more likely to exhibit psychological distress, depressive symptoms, substance use, and physical health problems.[26] The majority of health care professionals find prejudice morally repugnant, but several studies indicate that even well-meaning caregivers often demonstrate unconscious negative racial attitudes and make decisions based on **stereotypes.**[27]

Minorities Individuals designated by the 2000 U.S. Census Bureau as American Indians and Alaska Natives, Asian Americans, Native Hawaiians and Other Pacific Islanders, Blacks or African Americans, or Hispanics or Latinos.

Stereotypes Assumptions that information about a cultural group applies to all individuals who appear to represent that group.

- **Environment.** Minorities and the poor are more likely to live in polluted environments and to work in hazardous occupations that increase the likelihood of exposure to toxins.[28]

UNDER-REPRESENTATION OF HEALTH CARE PROVIDERS FROM CULTURALLY AND LINGUISTICALLY DIVERSE GROUPS

Community nutrition professionals are frequently challenged to provide services for cultural groups they have never encountered. Ideally, the health care workforce should be as diverse as the population it serves. Unfortunately, national diversity trends have not extended into many of the health professions. The majority of registered dietitians (91 percent) are non-Hispanic whites.[29] A survey of 7,550 public health nutrition personnel found that 91.6 percent of the respondents reported English as their primary language and only 6 percent stated that Spanish was a second language.[30] A more ethnically and linguistically diverse population of health care providers could help break down cultural barriers to health care access. In addition, minority health care professionals are more likely to work in medically underserved communities.[31] Recognizing the benefits of diversity, the American Dietetic Association has instituted several initiatives to promote the participation of under-represented groups in the professional organization.[32] The challenge of serving a diverse and rapidly changing public underscores the need for diversity in the health professions and also highlights the importance of cultural competence skills, because the mix of professionals will never be identical to the population it serves.[33]

LEGISLATIVE, REGULATORY, AND ACCREDITATION MANDATES

Recognizing the need to develop cultural competence skills, private and public organizations have taken action.[34] Health care accrediting agencies and professional organizations have set educational standards and core curriculum guidelines for developing cultural competence and for providing culturally competent services.[35] The American Dietetic Association has established a diversity-mentoring program. In 1998, a $400 million government initiative was instituted to eliminate health disparities—*The Initiative to Eliminate Racial and Ethnic Disparities in Health.*[36] The United States Department of Health and Human Services (DHHS) has created national standards for culturally and linguistically appropriate services in health care.[37] (See the section on organizational cultural competence later in this chapter).

In addition, Title VI of the Civil Rights Act of 1964 reads, in part, "No person in the United States shall, on ground of race, color or national origin, be excluded from participation in, be denied the benefits of, or be subjected to discrimination under any program or activity receiving federal financial assistance." Guidance issued by the Office of Civil Rights further clarifies Title VI as it relates to persons with limited English proficiency, stating that providers should include "reasonable steps to provide services and information in appropriate languages other than English to ensure that persons with limited English proficiency are effectively informed and can effectively benefit."[38]

Healthy People 2010 stresses the need to provide culturally competent, community-based health care systems in order to address health disparities among different segments of the population.[39] A culturally competent health care system may not completely eradicate health disparities, but it does provide the means to "respond to the needs of individuals, families, and communities in an acceptable, meaningful, and equitable manner."[40]

TABLE 16-3 Cultural Competence Continuum

STAGE	DESCRIPTION
Cultural destructiveness	Attitudes, practices, and policies that are destructive to other cultures.
Cultural incapacity	Paternalistic attitude toward the "unfortunates." No capacity to help.
Cultural blindness	Belief that culture makes no difference. Everyone is treated the same. Approaches of the dominant culture are applicable for everyone.
Cultural precompetence	Weaknesses in serving culturally diverse populations are realized, and there are some attempts to make accommodations.
Cultural competence	Differences are accepted and respected, self-evaluations are continuous, cultural skills are acquired, and a variety of adaptations are made to better serve culturally diverse populations.
Cultural proficiency	Engages in activities that add to the knowledge base, conducts research, develops new approaches, publishes, encourages organizational cultural competence, and works in society to improve cultural relations.

Source: Adapted from T. Cross, B. Bazron, K. Dennis, and M. Isaac, Toward a culturally competent system of care, Volume I (Washington, D.C.: Georgetown University Child Development Center, 1989).

Cultural Competence Models

Achieving cultural competence is a "developmental learning process that requires time, effort, active awareness, practice, and introspection."[41] To provide a framework for the process, a number of models have been developed. Two of them have particular significance for community nutrition professionals: the Cultural Competence Continuum and the Campinha-Bacote Cultural Competence Model.

CULTURAL COMPETENCE CONTINUUM MODEL

In the Cultural Competence Continuum Model, the process of gaining cultural competence is envisioned as a succession of stages (see Table 16-3). The Continuum provides a visual guide to assess individual or agency progress. However, movement through the stages cannot be expected to occur in unison for all cultural groups. For example, a person could be at a high level of proficiency for interacting with individuals who have disabilities but might be at a lower stage for working with lesbians and gays.

THE CAMPINHA-BACOTE CULTURAL COMPETENCE MODEL

A conceptual model of cultural competence developed by Campinha-Bacote[42] for health care professionals views cultural competence as a process rather than an end result: "the process in which the health care provider continuously strives to achieve the ability and availability to effectively work within the cultural context of a client (individual, family or community)." This model views cultural awareness, cultural skill, cultural knowledge, cultural encounters, and cultural desire as the five constructs of cultural competence (see Table 16-4). Cultural desire is the pivotal construct and begins the process of acquiring cultural competence (see Figure 16-1). Health care professionals can work on any one of the constructs to improve balance of the others with each construct influencing all others.[43]

CULTURAL AWARENESS

The foundation of cultural competence is an awareness of your own beliefs, values, and attitudes and an understanding that these attributes reflect your own **biases** and are really just one point of view among many. Without cultural self-awareness there is a tendency

Bias "A mental slant or leaning to one side; a highly personal and unreasoned distortion of judgment."[44]

TABLE 16-4 Constructs of the Campinha-Bacote Model of Cultural Competency

CULTURAL CONSTRUCT*	DESCRIPTION
Awareness	Health care providers become appreciative of the influence of culture on the development of values, beliefs, lifeways, practices, and problem-solving strategies. A basic requirement for cultural awareness is an in-depth exploration of one's own cultural background, including biases and prejudices toward other cultural groups.
Skill	Health care providers learn to perform culturally sensitive assessments and interventions.
Knowledge	Health care professionals develop a sound educational foundation concerning various worldviews in order to understand behaviors including food practices, health customs, and attitudes toward seeking help from health care providers. They also acquire knowledge of physical needs, such as common health problems and nutrition issues of different cultures.
Encounters	Providers seek and engage in cross-cultural encounters.
Desire	To appear **genuine** and to be effective cross-culturally, the health care provider must have a true inner feeling of wanting to engage in the process of becoming culturally competent.

*The mnemonic **ASKED** can assist health care professionals in assessing their level of cultural competence.

Source: Adapted for community nutrition professionals from J. Campinha-Bacote, A model and instrument for addressing cultural competence in health care, *Journal of Nursing Education* 38 (1999): 203–7.

Genuine Behavior and words are congruent, appear open and spontaneous; opposite is phony.

Worldviews The way individuals or groups view the universe to form values about their lives and the world.[45]

to be ethnocentric, devalue alternative cultural practices, blindly impose your own cultural procedures, and miss seeing opportunities for successful interventions.

Developing cultural self-awareness takes a concerted effort, because our views are part of our essence and feel so natural. They are the basic components of how we believe the world should function. Our **worldviews** have been repeatedly reinforced over our lifetime, because most of our life experiences occur within the same cultural context. We experience culture shock when we realize that our view of the world is not universally accepted. For example, many Americans are surprised to discover that some people from

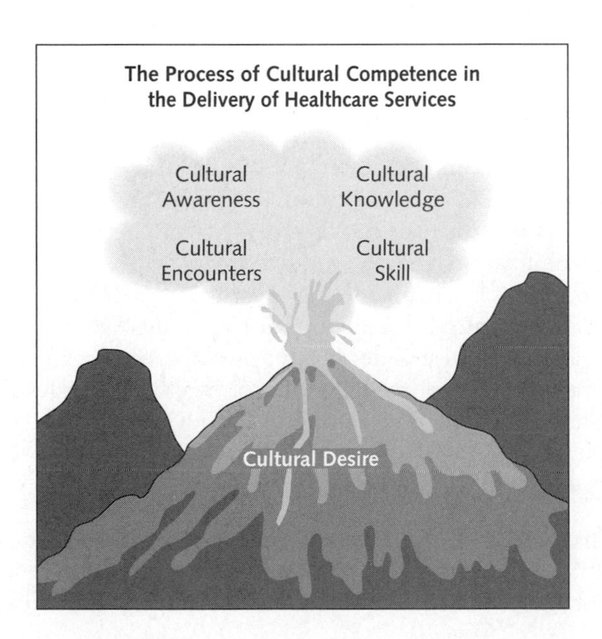

FIGURE 16-1 The Process of Cultural Competence in the Delivery of Health Care Services

According to the Campinha-Bacote model, cultural competence can be depicted as a volcano. The volcano symbolizes the fact that it is *cultural desire* that stimulates the process of cultural competence. When cultural desire erupts, it evokes the genuine desire to enter into the *process* of becoming culturally competent by actively seeking cultural awareness and cultural encounters, obtaining cultural knowledge, and conducting culturally sensitive assessments.

Source: J. Campinha-Bacote, *The Process of Cultural Competence in the Delivery of Healthcare Services,* 4th ed. (Cincinnati, OH: Transcultural C.A.R.E. Associates, 2002). Reprinted with permission of the author.

TABLE 16-5 Comparison of Common Values and Beliefs

MAJORITY AMERICAN CULTURE	OTHER CULTURES
Values/Beliefs	**Values/Beliefs**
Mastery over nature	Harmony with nature
Action, task oriented	Being
Time dominates, punctual	Personal interaction dominates
Human equality	Hierarchy/rank/status/authority
Individualism/privacy	Group welfare
Youth/thin/fit	Elders
Self-help/earned	Birthright/inheritance
Competition/free enterprise	Cooperation
Future orientation	Past or present orientation
Informality	Formality
Directness/openness/honesty	Indirectness/ritual/"face"
Practicality/efficiency	Idealism
Materialism	Spiritualism/detachment
Mind, body, and soul separate	Mind, body, and soul integrated
Disease is preventable	Humans cannot control disease
Confidentiality	Family decision making
Provider–client partnership	Provider-directed health care
Individuality/control over fate	Fate

Source: Adapted from readings of two authorities: K. P. Sucher and P. G. Kittler, *Food and Culture,* 4th ed. (Belmont, CA: Wadsworth/Thomson, 2004), and Debra P. Keenan, In the face of diversity: Modifying nutrition education delivery to meet the needs of an increasingly multicultural consumer base. *Journal of Nutrition Education* 28 (1996): 86–91.

other parts of the world view the direct communication style of Americans as impolite or even rude.

An awareness of the high degree of importance you place on your own particular beliefs, **values,** and cultural practices can help you appreciate individuals from a culture different from yours; they too hold dear certain specific beliefs, values, and cultural practices. You can then empathize with individuals from non-Western cultures who are experiencing confusion and problems as they try to participate in the North American health care system.

A method of becoming aware of your beliefs, values, and biases is to compare some of them to those of cultures different from your own. Table 16-5 provides a comparison of majority American values to those of various other cultures. One of the best ways to become aware of differences is to immerse yourself in the perceptual world or culture of others, as can be done by traveling or working in other countries.

CULTURAL KNOWLEDGE

Valuing diversity and viewing the world through multiple cultural lenses are at the heart of cultural competence. Both of these attributes are developed through the process of exploring unfamiliar cultures. Understanding various cultures provides a vehicle for developing attitudes congruent with cultural competence and discourages reliance on stereotypes.

Value "Any belief or quality that is important, desirable, or prized."[46]

TABLE 16-6 Values and Attitudes Congruent with Cultural Competence

For you to behave in a culturally competent manner, your attitudes need to convey an understanding and acceptance of diverse values and behaviors such as the following:

- Family is defined differently by different cultures (e.g., extended family members, fictive kin, godparents).
- Individuals from culturally diverse backgrounds may desire varying degrees of acculturation into the dominant culture.
- Male–female roles in families may vary significantly among different cultures (e.g., who makes major decisions for the family, play and social interactions expected of male and female children).
- Age and life-cycle factors must be considered in interactions with individuals and families (e.g., high value placed on the decisions of elders or the role of the eldest male or female in families).
- Meaning or value of medical treatment, health education, and wellness may vary greatly among cultures.
- Religion and other beliefs may influence how individuals and families respond to illnesses, disease, and death.
- Folk and religious beliefs may influence a family's reaction and approach to a child born with a disability or later diagnosed with a disability or special health care needs.
- Customs and beliefs about food, its value, preparation, and use differ from culture to culture.

Source: Adapted from material developed by T. D. Goode, National Center for Cultural Competence, Georgetown University Child Development Center.

(Table 16-6 lists values and attitudes congruent with cultural competence.) Cultural knowledge also provides the tools needed to develop culturally effective and relevant programs.

By exploring different cultures, you learn about new ways of interpreting reality and can develop alternative lenses through which to view your interactions with those who appear different from you. Keep in mind that it will be natural to experience some discomfort during your investigations as you learn about values and beliefs that conflict with yours. However, the process helps you develop attitudes congruent with cultural competence, such as appreciation, respect, and understanding of people who have cultural beliefs and behaviors different from your own.

An understanding of the generalities of cultural groups enables community nutrition professionals to develop relevant programming that builds on strengths and respects cultural differences. Without cultural understanding, there is a risk that the program you develop could conflict with common beliefs, values, and customs of the group. (See Table 16-7 for specific examples.)

Learning useful generalizations about a cultural group is only the starting point for developing relevant community interventions. Your programs need to be community based and must address the specific needs of the people you serve. That means as you

TABLE 16-7 Specific Examples of Value Conflicts That May Arise in Program Planning

- Messages that stress eating certain foods to prevent specific diseases may not have much of an effect if ill health is viewed as "God's will."
- Prevention may be looked at as a useless attempt to control fate. Doing good deeds and requesting forgiveness from a spiritual leader may appear to be the best courses of action for those who believe that illness is a curse for sins.
- Food programs that require that only particular family members eat donated foods may not be well received in cultures where the welfare of the group is placed before the individual.

Source: B. Schilling & E. Brannon, *Cross-Cultural Counseling: A Guide for Nutrition and Health Counselors* (Washington, D.C.: U.S. Government Printing Office, 1990).

assess a particular person or specific group, you must keep general characteristics in mind but make no assumptions. For example, even though you know that many Hindus are vegetarians, you would want to explore that behavior with the individuals involved, rather than just assuming that any particular Hindu is following a vegetarian lifestyle.

CULTURAL ENCOUNTERS

The driving force behind delivering culturally competent services is knowing and understanding the people you serve.[47] Be alert to economic, communication, religious, and familial factors (including eating rituals) when developing an appreciation for a different culture. Pay special attention to socioeconomic issues and environmental risks, because they can affect health and treatment. Also, you will develop a deeper appreciation for your client's attitudes, beliefs, and values by understanding the sociopolitical factors that influenced them. For example, lead is a common contaminant of inner-city environments because of the lead-based paint found on the interiors and exteriors of older buildings. Children who are chronically exposed to lead and who eat a high-fat diet that is also low in calcium and iron absorb more lead than those who eat a more nutritious diet.

There are many areas of cultural food practices that can be explored. Cultural food behaviors can be observed among descendants of immigrants even after many generations have passed. People of various cultures consume foods for religious, nutritional, and health reasons, as well as for hedonistic reasons. See Table 16-8 for examples of common cultural foods of various ethnic groups; see Table 16-9 for examples of traditional food practices used to influence health.

European Americans often value scientific reasoning, look for biological explanations (such as bacteria, viruses, or environmental toxins) for illness, and expect to find cures by using technology. This approach can sometimes conflict with the health care beliefs, customs, and traditions of other cultures, where emphasis may be placed on natural, supernatural, or religious/spiritual causes of a problem (yin and yang being out of balance, for instance, or the breaking of a taboo).[48]

Bob Daemmrich/StockBoston

At many American-style restaurants, you can experience other cultures by sampling from the various ethnic cuisines represented on the menu.

TABLE 16-8 Common Cultural Foods of Various Ethnic Groups

CULTURAL GROUP	COMMONLY CONSUMED FOODS	DIETARY CONCERNS/ISSUES
European Americans	Beef, chicken, pork, pasta, rice, bread, dairy foods, potatoes, bananas, apples, citrus juices, lettuce	High intake of fat, salt, sugar, and fast foods
Southern African Americans	Pork, organ meats, corn bread, rice, black-eyed peas, okra, greens, lard, hot sauce	Lactase deficiency is common; fried foods; low intake of fresh fruit and whole grains; pica common in rural South; breastfeeding rates are low
Asian/Pacific Island Americans	Pork, chicken, eggs, rice, wheat, bok choy, Chinese eggplant, mushrooms, water chestnuts, ginger root, soymilk, soy sauce	High salt intake; lactase deficiency is common; milk use is rare
Mexican Americans	Chicken, eggs, beans, flour or corn tortillas, rice, tomatoes, squash, lard, chili peppers, onions, tropical fruits, pine nuts	High intake of carbonated beverages; limited dental care among migrant workers
Native Americans	Game, fish, berries, roots, wild greens, commodity foods, fried bread	Broad differences exist among the subgroups; lack of refrigeration; high intake of refined sugar, cholesterol, fat, and energy; lactase deficiency and obesity are common
Puerto Ricans	Beans, various meats, rice, cornmeal, yams, sweet potatoes, onions, green peppers, tomatoes, lard, pineapple, bananas, sugar	Overweight and obesity are common; breastfeeding is not common; low intake of green leafy vegetables; dairy intake is low
Middle Eastern Americans	Fermented dairy products, feta cheese, lamb, legumes, pita bread, rice, olive oil, figs, dates, pomegranates, lemons, eggplants, phyllo pastries, honey	High incidence of lactose intolerance

Source: P. G. Kittler and K. P. Sucher, *Food and Culture,* 4th ed. (Belmont, CA: Wadsworth/Thomson, 2004); and Association for the Advancement of Health Education, *Cultural Awareness and Sensitivity: Guidelines for Health Educators* (Reston, VA: Association for the Advancement of Health Education, 1994).

In some cultures,there may be reluctance to seek help from health care providers. For example, members of some Hispanic groups may not seek medical attention because the act may signify a lack of strength and control over their lives.[49] Some consumers may seek traditional healers and have more confidence in their services than in those offered by Western biomedicine. Others may mistrust organizations that represent authority because they have experienced severe oppression or were victims of atrocities in their homelands. Positive experiences over time can help build trust and respect.[50] In many cultures, religion and health are not separate. Spirituality is viewed as a vital element in health, illness, and healing.[51] (See Table 16-10 for examples of common dietary practices of selected religious groups.)

Whenever possible, community nutrition professionals should invest time in learning about unfamiliar cultures that they are likely to encounter. There are a variety of strategies that health care professionals can employ to learn about other cultures. Eat at ethnic restaurants, explore stories about other cultures in the media, establish focus groups to gain insight into a target population's culture, read about cultural customs and etiquette, travel, take language lessons, familiarize yourself with diverse neighborhoods, and attend professional development and training classes.

TABLE 16-9 Traditional Health Beliefs Related to Food

CULTURAL GROUP	TRADITIONAL BELIEFS*
Chinese	The body is kept in harmony through a balance of yin and yang. Yin foods include those that are raw or cooked at low temperatures and are white or light in color. Yang foods usually are high-calorie foods, are cooked at high temperatures, and have red-orange-yellow colors. Some foods, such as rice, are considered neutral. Ginseng is used as a general health-promoting tonic and is thought to help cure a variety of ailments. Sometimes the "like cures like" concept is used to treat specific illnesses, such as attempting to cure impotence by eating male genital organs from sea otters.
Italians	Foods may be categorized as heavy or light, wet or dry, and acid or nonacid. Light foods (gelatin and soups) are thought to be easy to digest and are fed to those who are ill. Wet meals may be served once a week to cleanse the system. They include escarole, spinach, and cabbage cooked in fluid. Citrus foods, tomatoes, and peaches are considered acid foods and are believed to aggravate skin conditions. Too many dairy products are thought to cause kidney stones.
Korean	A balance of yin and yang maintains health. Eating too much or too little food can disrupt the balance. Too much food, even of good quality, can block ki (energy), resulting in cold hands and feet, cold sweats, or fainting.
Mexican	A balance of strengthening hot foods and weakness-promoting cold foods are needed to maintain health. If someone has a "hot" condition such as menstruation or pregnancy, then hot foods are avoided. Examples of hot foods include alcohol, beef, pork, chilies, cornhusks, oils, and onions; cold foods include citrus fruits, dairy products, most fresh vegetables, and goat.

*Note that many factors affect food intake and health practices. These are examples of traditional beliefs and behaviors that may or may not be practiced by individuals who represent the specific cultural group.

Source: P. G. Kittler and K. P. Sucher, *Food and Culture,* 4th ed. (Belmont, CA: Wadsworth/Thomson, 2004).

Cross-Cultural Communication

Cultural orientation has a major impact on the process of communication. When individuals share a common culture, or at least are familiar with each other's cultural background, it is much more likely that differences in perceptions will be minimal and communication will flow smoothly. Each society has a conscious and an unconscious set of expected reciprocal responses. For example, an Iranian woman may politely refuse an offer of coffee or tea and expect a second request, accompanied by insistence that she have something to drink. When such an expected interchange does not happen, a feeling of discomfort ensues that can set the stage for a breakdown in communication.

Communication Styles

Table 16-11 on page 511 provides a summary of key differences in communication styles among cultures. While reviewing this list, keep in mind that considerable variation exists within any particular cultural group. Community nutrition professionals are increasingly required to provide services for those who have communication styles that are unfamiliar. Professionals need to learn about these styles and find ways to communicate so that clients can be confident that their voices have been heard and that their values, beliefs, and behaviors are respected. The next section reviews some common cultural barriers to communication and offers practical suggestions for communicating in cross-cultural encounters.

TABLE 16-10 Dietary Practices of Selected Religious Groups

RELIGIOUS GROUP	DIETARY PRACTICES*
Buddhists	Dietary customs vary depending on sect. Many are lacto-ovo-vegetarians, because there are restrictions on taking a life. Some eat fish, and some eat no beef. Monks fast at certain times of the month and avoid eating solid food after the noon hour.
Hindus	All foods thought to interfere with physical and spiritual development are avoided. Many are lacto-vegetarians and/or avoid alcohol. The cow is considered sacred—an animal dear to the Lord Krishna. Beef is never consumed, and often pork is avoided.
Jewish	Kashrut is the body of Jewish law dealing with foods. The purpose of following the complex dietary laws is to conform to the Divine Will as expressed in the Torah. The term *kosher* denotes all foods that are permitted for consumption. (Many Jews also eat non-kosher foods.) To "keep kosher" means that the dietary laws are followed in the home. There is a lengthy list of prohibited foods, called treyf, which include pork and shellfish. The laws define how birds and mammals must be slaughtered, how foods must be prepared, and when they may be consumed. For example, dairy foods and meat products cannot be eaten at the same meal. During Passover, special laws are observed, such as the elimination of any foods that can be leavened.
Mormons	Alcoholic drinks and hot drinks (coffee and tea) are avoided. Many also avoid beverages containing caffeine. Mormons are encouraged to limit meat intake and to emphasize grains in the diet. Many store a year's supply of food and clothing for each member of the household.
Muslim	Overeating is discouraged, and consuming only two-thirds of capacity is suggested. Dietary laws are called halal. Prohibited foods are called haram, and they include pork and birds of prey. Laws define how animals must be slaughtered. Alcoholic drinks are not allowed. Fasting is required from sunup to sundown during the month of Ramadan.
Roman Catholics	Meat is not consumed on Fridays during Lent (40 days before Easter). No food or beverages (except water) are to be consumed 1 hour before taking communion.
Seventh Day Adventists	Most are lacto-ovo-vegetarians. If meat is consumed, pork is avoided. Tea, coffee, and alcoholic beverages are not allowed. Water is not consumed with meals but is drunk before and after meals. Followers refrain from using seasonings and condiments. Overeating and snacking are discouraged.

*Many of the religious guidelines regarding food have practical applications for the society. For example, the Hindu prohibition against killing cattle respects the need of Indian farmers to use cattle for power and their dung for fuel. Cows also supply milk to make dairy products, and the skin of dead cows is used to make leather goods.

Source: Adapted from Religious Food Practices, accessed at www.eatethnic.com.

BARRIERS TO CROSS-CULTURAL COMMUNICATION

In your career as a community nutrition professional, you and the client might not speak the same language. Even similar words that the two cultures share may not have the same meaning. Two individuals who speak the same language but are from different countries or even different areas of the same country may not give the same meaning to a specific word. For example, the word *bad* generally means something negative, but among some subgroups the word actually means something good.

Nonverbal behavior may not be interpreted correctly either. Studies indicate that no single aspect of nonverbal communication can be universally translated across all cultural groups.[52] More than 7,000 different gestures have been identified,[53] so there are

TABLE 16-11 Communication Styles of Various Cultural Groups

CATEGORY	COMMUNICATION STYLE
Emotional expressiveness	The dynamic and expressive body language of African Americans may be considered excessive and too intense to European Americans. Vigorous handshaking can be considered a sign of aggression for Native Americans but a gesture of good will for European Americans. Some Asian cultures value stoicism and may use a smile or laugh to mask other emotions.
Volume of speech	Asians tend to speak quietly, whereas African and European Americans generally speak loudly.
Touching	Friendly behavior for Hispanics often involves touching that Native Americans and Asians may find uncomfortable. Asians and Hispanics are not likely to appreciate a slap on the back. Many Asians totally avoid physical contact with strangers, even in transactions, such as giving change.
Vocal style	Latinos often use expressive language and engage in lengthy pleasant talk before getting down to business. European Americans prefer a quiet, controlled style that other groups may consider manipulative and cold.
Verbal following	Asians and Native Americans are more likely to use indirect and subtle forms of communication by avoiding direct questions and answers. The direct styles of African and European Americans are considered too confrontational. Native Americans find direct personal questions particularly offensive.
Eye contact	For European Americans making eye contact is a sign of respect, but among Hispanics, Native Americans, and Asians avoidance of eye contact is often considered proper behavior. Hispanics and Filipinos use sustained eye contact to challenge authority. Many African Americans use more eye contact during talking than when listening, whereas the opposite is true of most European Americans.
Physical space	Conversational space in Arab and Middle-Eastern cultures is commonly 6 to 12 inches, whereas among European Americans a comfortable distance is ordinarily "arm's length." Hispanics prefer closer proximity, and Asians a greater distance, than European Americans.
Silence	Duration of silence considered acceptable differs among cultures. Native Americans may take 90 seconds to formulate a response to a question, but that amount of silence can seem intolerable to others. European Americans are particularly uncomfortable with silence and will quickly fill the void with small talk. Some European Americans feel it is appropriate to start speaking before the other person has finished.
Question authority	Questioning authority comes naturally to Native and African Americans, but Asians are not likely to disagree with an elder or a person in a position of authority. Some clients will appear to agree by giving answers they believe are desired and then disregard information that does not make sense to them.
Aggression	Some cultural groups may have learned that aggressive behavior or bribery is required to get what they want from bureaucracy.
Gender roles	Different cultures prescribe who will talk during interviews. Even during female dietary and medical assessments, a husband may answer all questions, no matter how personal.

Sources: P. G. Kittler and K. P. Sucher, *Food and Culture,* 4th ed. (Belmont, CA: Wadsworth/Thomson, 2004); A. E. Ivey, N. Gluckstern, and M. B. Ivey, *Basic Attending Skills,* 3rd ed. (North Amherst, MA: Microtraining Associates, 1997); and C. Elliott, R. J. Adams, and S. Sockalingam, Multicultural Toolkit (Toolkit for Cross-Cultural Collaboration) (1999), www.awesomelibrary.org.

many opportunities for misunderstanding. For example, the same hand gesture that means "come here" in Nigeria signals "hello" in the United States. Nonverbal cues need to be interpreted very cautiously because meanings vary from one culture to the next. Your interpretation may be quite different from the speaker's intent.

Stereotyping means assuming that individuals will behave a certain way because they are from a particular culture or appear to be from the group. Kittler and Sucher[54] report that only about one-third of the individuals in a particular group of people actually behave in ways considered typical of that group.

PRACTICAL GUIDELINES FOR CROSS-CULTURAL COMMUNICATION

Learning culturally sensitive communication skills facilitates more favorable community intervention outcomes and increases the likelihood of more rewarding interpersonal experiences for health care providers. The following suggestions can enhance cross-cultural communication.[55,56]

1. Smile, show warmth, and be friendly.
2. Attempt to learn and use key words, especially greetings and titles of respect, in languages spoken by populations serviced by your organization.
3. Thank clients for trying to communicate in English.
4. Suggest that clients choose their own seat (to make comfortable personal space and eye contact possible).
5. Articulate clearly; speak in a normal volume. Often people mistakenly raise the volume of their voice when they feel someone is having difficulty understanding them.
6. Paying attention to children appeals to women of most cultures; however, some believe that accepting a compliment about a child is not appropriate, especially in front of the child.
7. When interacting with individuals who have limited English proficiency, always keep in mind that limitations in English proficiency do not reflect their level of intellectual functioning. Limited ability to speak the language of the dominant culture has no bearing on ability to communicate effectively in their language of origin. Clientele may or may not be literate in their language of origin or in English.
8. Explain to clients that you have some questions to ask and that there is no intention to offend. Request that they let you know if they prefer not to answer any of the questions.
9. If you are not sure how to interpret a particular behavior, Magnus[57] suggests that you should ask for clarification. For example, you could ask, "I notice that you are mostly looking down. Would you tell me what that means for you?"
10. Follow your intuition if you believe something you are doing is causing a problem. Magnus[58] suggests that you ask, "There seems to be a problem. Is something I am doing offending you?" Once informed of a difficulty, immediately apologize and admit, "I am sorry. I didn't mean to offend you."
11. Ask clients to identify their ethnicity. The specific terms used to identify an individual's ethnicity can be a touchy issue. For example, *Asian*, *Oriental*, *Chinese*, and *Chinese-American* have been used to describe individuals of a similar background, but not all of these terms are acceptable to all individuals. Magnus[59] suggests that to avoid alienation, a counselor should directly inquire about heritage with phrases such as "How do you describe your ethnicity?"

Suggestions for Communicating Information

1. Consider using a less direct approach than is common among Americans. Gardenswartz and Rowe[60] suggest some communication approaches that may lower the risk of misunderstanding and hurt feelings:

- Make observations rather than judgments about behaviors. For example, do not say, "Your dairy intake is low." Instead say, "You eat one dairy food a day and the authorities tell us to eat three."
- Refrain from using "you." For example, say "People who have a low intake of calcium are at an increased risk for osteoporosis" rather than "You are at an increased risk for osteoporosis."
- Be positive, saying what you want rather than what you do not want. For example, say, "Use a pencil to fill out the form" rather than "Don't use a pen to complete the form."

2. Use visual aids, food models, gestures, and physical prompts during interactions with those who have limited English proficiency.
3. If answers are unclear, ask the same question a different way.
4. Consider using alternatives to written communications, because word of mouth may be a preferred method of receiving information.
5. Write numbers down, just as they would appear in recipes, because spoken numbers are easily confused by those with limited skills in a language.

Ways in Which Discussions about Food Can Open Dialogue

1. Most people are pleased to educate others about their food ways, but some may feel that questions are probing.
2. Ask about foods used for celebrations and special occasions.
3. Ask about favorite foods and discuss how they can be incorporated into a diet plan.
4. Tell your own food stories. Letting clients know that you do not always make the best food choices can help them feel more comfortable being truthful about their behavior patterns.

Working with Interpreters

Professional interpreters or translators may be required in order for the community nutritionist to provide effective services. A **translator** works with written information, and an **interpreter** explains spoken words.

Community nutritionists should insist on using professional interpreters who have expertise in the language and are familiar with the culture of their clients. Too often, health care providers resort to using nonprofessional interpreters, such as friends or relatives of clients or housekeeping staff.[61] This has been shown to present numerous problems.[62] Sometimes clients are reluctant or embarrassed to discuss certain problems in front of close relations, or the nonprofessional interpreter may decide that certain information is irrelevant or unnecessary and may therefore not do a complete interpretation. Other nonprofessional interpreters may be unfamiliar with medical terminology and unknowingly make mistakes. All of these problems are compounded when a child is used as an interpreter. One study of untrained interpreters showed that 23 to 52 percent of phrases were misinterpreted; for example, the term *laxative* was used to describe diarrhea.[63] The difficulty of communication across cultures is illustrated in the story of a very sick epileptic Hmong child, Lia, who was moved from a community hospital to a children's hospital with an intensive care unit. With the help of an interpreter, the situation was explained to Lia's non-English-speaking parents. Later investigation revealed that the parents thought their child had to go to another hospital because the doctors at the community hospital were going on vacation.[64]

There are several phone-based interpreter services available, such as the AT&T Language Line, which provides 24-hour service in 140 languages. Phone interpreters are useful for emergencies but cannot take the place of in-person professionals; some clients

Translator A person who works in converting written words into another language.

Interpreter A person who works in converting spoken words into another language.

TABLE 16-12 Guidelines for Using an Interpreter

- Request an interpreter of the same gender and similar age. (Be sensitive and flexible in your selections since interpreters who are considerably older than a client may receive greater respect.)
- Decide before the meeting what questions will be asked.
- If possible, go over the questions with the interpreter before the meeting. A professional interpreter should be able to assist you in formulating new questions if certain ones are deemed offensive.
- Try to learn a few phrases of the client's language to use at the beginning and/or the end of the interview.
- Remember that sessions will take extra time. Schedule adequate time.
- Look at and speak directly to the client, not the interpreter.
- Speak clearly in short units of speech. Do not ask more than one question at a time.
- Avoid using slang, similes, metaphors, and idiomatic expressions. For example, do not say, "Do you have your ups and downs?"
- Listen carefully and watch body language for any changes in expression.
- Do not just follow prepared questions, but ask clients to expand upon new issues.
- To avoid misunderstandings, begin some of your sentences with "Did I understand you correctly that . . . " or "Tell me about . . . "
- To check on the client's understanding and the accuracy of the interpretation, ask the client to back-translate important dietary instructions or guidelines. This technique may also open the conversation to questions by the client.
- Be aware that interpreters come to sessions with their own cultural biases and may not completely convey everything that has been said.
- If your client appears tired, you and the interpreter may need to schedule another session.

Sources: Adapted from J. Luckman, *Transcultural Communication in Health Care* (Belmont, CA: Delmar/Thomson Learning, 2000); and K. Bauer and C. Sokolik, *Basic Nutrition Counseling Skill Development* (Belmont, CA: Wadsworth/Thomson Learning, 2002).

may have difficulty communicating personal issues with a faceless voice. Guidelines for using an interpreter are given in Table 16-12.

Culturally Appropriate Intervention Strategies

Learning about the food habits, health beliefs, and behaviors of specific cultural groups can be extremely valuable. Deep exploration of target populations helps community nutrition professionals identify resources and find ways to build on community strengths to find solutions to problems related to health and nutrition. However, given the time constraints of busy health professionals and the great variety of cultures, community health professionals cannot be expected to have intimate knowledge of many cultural groups. Therefore, what the community nutrition professional needs is universal skills in cultural competence that can be utilized with clients from any cultural group. Fundamentally, what providers must have in order to use intervention strategies effectively is an inherent caring, appreciation, and respect for their clients and the ability to display warmth, empathy, and genuineness. Such community nutrition professionals can "exemplify cultural competence in a manner that recognizes, values and affirms cultural differences among their clients."[65]

Explanatory Models

Medical anthropologists have developed explanatory models as a culturally sensitive tool for investigating a client's perception of why an illness developed and how the illness should be treated. Health care professionals have used them successfully as a way to gather information for assessments. Explanatory models explore five major concerns about an illness episode:[66]

1. Etiology
2. Time and mode of onset of symptoms
3. Pathophysiology
4. Course of sickness, including both degree and type of sick role—acute, chronic, impaired, and so on
5. Treatment

Table 16-13 lists examples of open-ended questions that can be used to create an explanatory model. In this respondent-driven interview approach, nutrition professionals ask simple, open-ended questions to initiate conversations, prompt clients for a better understanding when necessary, but for the most part exert little control over responses. By showing an unbiased and sincere desire to understand and accept traditional views and practices, you increase the likelihood that your clients will not fear criticism or ridicule, will feel comfortable telling their stories, will have a sense of control over their condition, and will be open to accepting suggestions for treatment.

The questions in Table 16-13 and in Table 3-7 on page 75 aid in understanding illness and food issues from a client's perspective. However, not every question is appropriate for every cross-cultural encounter, so health care professionals must use their judgment to select suitable ones. If you are going to use a lot of the questions, consider changing some of them into statements. A series of questions can feel like an interrogation, and in some

TABLE 16-13 Culturally Sensitive Respondent-Driven Interview Questions

Questions to understand view and treatment of health problems:

- What do you call this problem you are having? (*Note:* Use this term instead of *it* in the following questions.)
- What do you think caused your problem?
- When did it start and why did it start when it did?
- What does your sickness do to your body?
- Will you get better soon, or will it take a long time, in your opinion?
- What do you fear about your sickness?
- What problems has your sickness caused for you personally? for your family? at work?
- What kind of treatment will work for your sickness? What results do you expect from treatment?
- What home remedies are common for this sickness? Have you used them?
- Are there benefits to having this illness?
- Is there anyone else in your family that I should talk to?

Question to understand about traditional healers:

- How would a healer treat your sickness? Are you using that treatment?

Source: Modified from P. G. Kittler and K. P. Sucher, *Food and Culture in America: A Nutrition Handbook,* 3rd ed. (Belmont, CA: Wadsworth/Thomson, 1998), pp. 72–73; and A. Kleinman, L. Eisenberg, and B. Good, Culture, illness, and care: Clinical lessons from anthropologic and cross-cultural research. *Annals of Internal Medicine* 88 (1978): 251–58.

cultures questions may evoke defensiveness.[67] Variety can be accomplished by starting sentences with "Tell me . . ." or "Please describe . . . ."

LEARN Intervention Guidelines

Explanatory models are useful for assessment, and the LEARN guidelines given in Table 16-14 provide a framework for negotiating a culturally sensitive treatment plan to address a given illness episode.

1. **Listen.** Active listening is the foundation of successful cross-cultural communication. Your demeanor should come across as curious and nonjudgmental, and the speaker should be recognized as the expert regarding information about his or her experience.[68] Not only are you learning, but you are demonstrating that what your client has to say is very important to you—a key relationship-building skill. Make sure you come to a common understanding of the issues and problems. Request clarification when necessary by saying, "I didn't quite understand that."[69] Probe to understand who does the food preparation and shopping, and determine whether there are others in the extended family who are responsible for decision making.
2. **Explain.** Make sure that you have understood correctly by explaining back to the client your perception of what was related. For example: "You feel that diarrhea is a hot ailment, and your baby should not be given a hot food like infant formula but should drink barley water, a cool food. Did I understand you correctly?" Your explanation creates an opportunity to clarify any misunderstandings.
3. **Acknowledge.** Acknowledge the similarities and differences in your perspectives regarding the cause and/or treatment of the problem. For example: "Both you and your doctor feel that what your baby drinks will help her feel better. You feel your baby needs a cool food like barley water, and the health care providers at this clinic feel that your baby needs a drink with minerals to get better."
4. **Recommend.** The client should be given several options that are culturally relevant, concise, and practical. For example, an Indian woman who is a vegetarian and wishes to lose weight might be given the following recommendations: "You could start a walking program, reduce the amount of oil used when making rice or bean dishes, use skim milk to make yogurt, or eat smaller portions of fried bread."
5. **Negotiate.** After reviewing the options, negotiate a culturally sensitive plan of action with your client and any significant family members who are part of the decision-making process. Begin by asking, "Which of these options do you think would be a good place to start?" If there appears to be a conflict between the biomedical approach to healing and the client's cultural practices, then look for ways to "neutralize" the biomedical treatment.[70] Ask, "You feel that to drink water with minerals is a 'hot' remedy and inappropriate for a hot ailment. Is there a way to take this treatment but reduce the

TABLE 16-14 LEARN Communication Guidelines for Health Practitioners

L	Listen with sympathy and understanding to a client's perception of a problem.
E	Explain your perceptions of the problem.
A	Acknowledge and discuss differences and similarities.
R	Recommend treatment that is relevant, concise, and practical.
N	Negotiate agreement.

Source: Adapted from E. A. Berlin and W. C. Fowkes, A teaching framework for cross-cultural health care, *The Western Journal of Medicine* 139 (1983): 934–38.

hot effect?" Possibly the client will make a suggestion, such as combining treatments, or administering the treatment with a spiritual blessing, or giving the treatment at a certain time of day. If the condition is life-threatening and the cultural differences are great, consider including a respected member of the community in the negotiations. If the counselor appreciates the powerful influence of the client's culture as well as the equally powerful culture of biomedicine, then the need for compromise and mediation becomes obvious.[71]

Practical Considerations for Interventions

When dealing with people of various cultures, be sure that the items available in the agency (such as pictures, food guides, magazines, media resources, snack foods, and toys) reflect sensitivity to cultural backgrounds, literacy levels, and linguistic preferences. In order to respond to the needs and preferences of a particular community, encourage community members to participate at all levels of intervention—program development, implementation, and evaluation. Here are some suggestions to consider:

- Conduct focus groups to get opinions and suggestions.
- Have a representative of the population groups you are targeting review any publications you intend to use to ensure that they are meaningful and do not contain offensive material.[72]
- Community representatives can help identify places that your target audience frequents so that you can take your literature and programs to high-traffic areas.
- Ask volunteers to make the initial contacts with community organizations for outreach presentations.
- Recruit volunteers to distribute program literature to hair salons, barbershops, laundromats, dry cleaners, video stores, libraries, restaurants, and the like.
- Report outcomes to any individuals or groups that provided assistance. This will make them feel vested in the projects, and the likelihood of participation in the future will be greater.
- Show appreciation to volunteers—distribute certificates, provide appreciation lunches, list their names in publications.

Take time to learn about the community being served before designing a program. Find out who the principal people in the minority communities are (examples include clergy, funeral planners, politicians, marriage brokers, and healers) and enlist their support. Individuals in these roles can incorporate your message into their work, play, or prayer. It is also a good idea to utilize the local media outlets, such as minority radio or cable television programs. Radio and television stations are often required to provide airtime for community messages, which can publicize teaming up local programs with the American Diabetes Association or the American Heart Association.

When dealing with an immigrant population,[73] create a resource list of grocery stores and speciality markets and offer a chart that demonstrates lower-cost substitutions for high-cost familiar foods. On this chart, give the English words equivalent to a list of foreign words for traditional foods and spices. Keep on hand up-to-date public transportation schedules that will facilitate traveling to and from the markets.

If familiarizing an immigrant population with American styles of food preparation and ingredients is a goal, invite them to take part in a few informal cooking classes, particularly for those who are using food bank donations, WIC vouchers, or commodity foods. In these demonstrations, it is also a good idea to show how to use common kitchen appliances, such as refrigerators, freezers, stoves, ovens, or garbage disposals. The cooking classes might offer handouts on the basics of food preparation, as well as a cookbook that

Programs in Action

Encouraging Breastfeeding among African-American Women

Both the rate and the duration of breastfeeding are low in the United States, despite efforts to communicate its numerous health benefits. In 2002, only 33 percent of all mothers were breastfeeding 6 months after delivery, and the rates among African-American women were significantly lower.[1] Among the factors associated with choosing bottle-feeding over breastfeeding in African-American populations are that breastfeeding is too complicated, that it is not supported by family members, and that it takes too much time.[2,3] In contrast, a significant predictor is having a friend or relative who breastfed her infant, a correlation that emphasizes the importance of role models and peer support.[4] The Northside Breastfeeding Media Campaign is a grassroots, community-based breastfeeding promotion project.

Goals and Objectives

The goals of the campaign were to raise awareness and increase knowledge of breastfeeding in an African-American population and to create a supportive environment for breastfeeding through culturally specific images, messages, and materials.

Target Audience

The target audience was African-American women in the Near North Community of Minneapolis. The campaign aimed to reach 30 percent of African-American women in the community.

Rationale for the Intervention

In November 2000, the U.S. Department of Health and Human Services released the "Blueprint for Action on Breastfeeding," a comprehensive plan outlining the critical need to promote breastfeeding in minority communities as a way to reduce health disparities.[5] This campaign recognized the importance of community norms in influencing breastfeeding practices and involved extensive research to ensure that the right messages would reach the target audience.

Methodology

The campaign developed culturally specific materials and tested them prior to publication and distribution. Community advisors and African-American women who had breastfed were integrally involved in every step of the design and testing. A media advisory committee was formed to provide guidance to the media specialists and graphic designers to identify appropriate media channels, messages, and images. Media promotion materials from other campaigns, along with a survey of infant feeding practices, were used to develop message concepts. The concepts were pretested with the target audience.

Media strategies included bus stop posters, newspaper articles, public service announcements, radio and television newsrooms, and pamphlets to distribute to media audiences. The target audience was reached directly through pamphlets

incorporates unfamiliar "American" ingredients readily available for low cost, such as potatoes, carrots, and apples. For example, the cookbook might include recipes that modify some of the clients' traditional foods to help bring them in line with nutritional objectives, such as reducing or eliminating oil when cooking rice.

Finally, take the time to coordinate food education lessons with English as a Second Language (ESL) programs. Talking about food is fun. There is a great deal that you can do to help immigrants learn about American foods that their children may be eating at school or that can be used as substitutes for hard-to-find ethnic foods.

Organizational Cultural Competence

Health care delivery organizations have a responsibility to provide appropriate care that is sensitive to cultural norms, values, and beliefs of individuals; linguistically accessible; and physically available. Incentives to address these needs include state and federal guidelines, *Healthy People 2010* goals, and a desire to attract a growing share of business among racial and ethnic minorities.[74,75] To be effective, health care agencies must scrutinize all aspects of their organizational structure to infuse cultural competence at every level. These levels include:[76]

and promotional gifts that displayed the campaign themes—"Healthier Babies," "Faster and Easier," and "Get Back in Shape." The campaign educational and media materials were featured on the government's WIC site at www.nal.usda.gov/wicworks/Sharing_Center/statedev.html and are available on CD electronic files through www.ncemch.org.

The League of Catholic Women sponsored the campaign, with evaluation funding provided by a grant from the Allina Foundation. Participants in the overall Northside Breastfeeding Campaign included members of the community, along with representatives from WIC, two hospitals, several health clinics, and a Way to Grow program.

Results

Interviews were utilized to collect quantitative and qualitative data on the effectiveness of campaign intervention strategies in reaching the target audience. Thirty-one percent of females and 15 percent of males who were surveyed reported that they saw or heard the campaign messages. Bus stop posters, newspaper articles, and posters in health clinics were most effective. Acceptance of breastfeeding increased with age in both males and females, with females being more accepting.

Lessons Learned

The Northside Breastfeeding Campaign demonstrated the positive impact of involving community members and organizations in the development of nutrition materials and messages. Social marketing campaigns such as this can potentially increase the rate of breastfeeding initiation and the duration of breastfeeding in African-American communities.

References

1. U.S. Department of Health and Human Services, *Healthy People 2010: Conference Edition—Volumes 1 and II* (Washington, D.C.: Department of Health and Human Services, Public Health Service, Office of the Assistant Secretary for Health, 2000).
2. M. C. Mahoney and D. M. James, Predictors of anticipated breastfeeding in an urban, low-income setting, *Journal of Family Practice* 49 (2000): 529–33.
3. K. S. Corbett, Explaining infant feeding style of low-income black women, *Journal of Pediatric Nursing* 15 (2000): 73–81.
4. P. Kum-Nji and coauthors, Breast-feeding initiation: Predictors, attitudes, and practices among blacks and whites in rural Mississippi, *South Medical Journal* 92 (1999): 1183–8.
5. U.S. Department of Health and Human Services, *HHS Blueprint for Action on Breastfeeding* (Washington, D.C.: U.S. Department of Health and Human Services, Office of Women's Health, 2000).

Source: Adapted from *Community Nutritionary* (White Plains, NY: Dannon Institute, Fall 2001). Used with permission. For more information about the *Awards for Excellence in Community Nutrition*, go to www.dannon-institute.org.

- Culturally sensitive mission statements
- Structures to ensure consumer and community participation (the involvement of traditional healers, for example) in the planning, delivery, and evaluation of services
- Policies and procedures for the recruitment, hiring, and retention of a diverse and culturally competent workforce
- Policies and resources for staff development in cultural competence
- Adequate fiscal resources to support translation and interpretation services

Essential Organizational Elements of Cultural Competence

Five essential elements have been identified as necessary for an organization to provide culturally competent programming.[77]

1. **Valuing diversity.** Unless staff members and administrators view cultural competence as a plus for themselves and the populations they serve, there will not be a wholehearted attempt to develop culturally sensitive policies and procedures.
2. **Having the capacity for cultural self-assessment.** Organizations cannot develop goals without understanding their organizational needs. Self-assessment should be

Standard 1	Health care organizations should ensure that patients/consumers receive from all staff members effective, understandable, and respectful care that is provided in a manner compatible with their cultural health beliefs and practices and preferred language.
Standard 2	Health care organizations should implement strategies to recruit, retain, and promote at all levels of the organization a diverse staff and leadership that are representative of the demographic characteristics of the service area.
Standard 3	Health care organizations should ensure that staff at all levels and across all disciplines receive ongoing education and training in culturally and linguistically appropriate service delivery.
Standard 4	Health care organizations must offer and provide language assistance services, including bilingual staff and interpreter services, at no cost to each patient/consumer with limited English proficiency at all points of contact, in a timely manner during all hours of operation.
Standard 5	Health care organizations must provide to patients/consumers in their preferred language both verbal offers and written notices informing them of their right to receive language assistance services.
Standard 6	Health care organizations must ensure the competence of language assistance provided to limited English proficient patients/consumers by interpreters and bilingual staff. Family and friends should not be used to provide interpretation services (except on request by the patient/consumer).
Standard 7	Health care organizations must make available easily understood patient-related materials and post signage in the languages of the commonly encountered groups and/or groups represented in the service area.
Standard 8	Health care organizations should develop, implement, and promote a written strategic plan that outlines clear goals, policies, operational plans, and management accountability/oversight mechanisms to provide culturally and linguistically appropriate services.
Standard 9	Health care organizations should conduct initial and ongoing organizational self-assessments of CLAS-related activities and are encouraged to integrate cultural and linguistic competence-related measures into their internal audits, performance improvement programs, patient satisfaction assessments, and outcomes-based evaluations.
Standard 10	Health care organizations should ensure that data on the individual patient's/consumer's race, ethnicity, and spoken and written language are collected in health records, integrated into the organization's management information systems, and periodically updated.
Standard 11	Health care organizations should maintain a current demographic, cultural, and epidemiologic profile of the community as well as a needs assessment to accurately plan for and implement services that respond to the cultural and linguistic characteristics of the service area.
Standard 12	Health care organizations should develop participatory, collaborative partnerships with communities and utilize a variety of formal and informal mechanisms to facilitate community and patient/consumer involvement in designing and implementing CLAS-related activities.
Standard 13	Health care organizations should ensure that conflict and grievance resolution processes are culturally and linguistically sensitive and capable of identifying, preventing, and resolving cross-cultural conflicts or complaints by patients/consumers.
Standard 14	Health care organizations are encouraged to regularly make available to the public information about their progress and successful innovations in implementing the CLAS standards and to provide public notice in their communities about the availability of this information.

Source: USDHHS, Office of Minority Health, National Standards for Culturally and Linguistically Appropriate Services in Health Care (Washington, D.C.: U.S. Department of Health and Human Services, 2001).

TABLE 16-15 National Standards for Culturally and Linguistically Appropriate Services (CLAS) in Health Care

an ongoing process that continuously seeks to identify structural and cultural barriers to serving a fluid population base.

3. **Being conscious of the dynamics of the interaction of cultures.** Understanding interactions of minority cultures and the biomedical culture is necessary for making services culturally acceptable and usable to traditionally underserved populations. Community health professionals are in a strategic position for helping the two cultures interface so that minorities can fully participate in the biomedical health care system.

4. **Having institutionalized cultural knowledge.** Health care agencies need to provide training programs to increase cultural awareness, knowledge, and skills for all levels of the organizational hierarchy.
5. **Adapting service delivery on the basis of an understanding of cultural diversity.** Programming procedures and policies must be geared to interacting smoothly with the needs and traditions of the people served.

To provide guidance for instituting the above elements of organizational cultural competence, the Office of Minority Health of the USDHHS has issued National Standards for Culturally and Linguistically Appropriate Services (CLAS) in Health Care (see Table 16-15).[78]

Organizations can influence the cultural competence of health care delivery at three levels of intervention—the macro, mezzo, and micro levels.[79] At the macro level, health care providers can exert their influence in the development of culturally sensitive policies, laws, and regulations. Examples of macro-level interventions include *Healthy People 2010* and **Title VI of the Civil Rights Act.** The mezzo level addresses the design and delivery of culturally appropriate and effective community programs. Examples of mezzo-level interventions include WIC, Meals on Wheels, and the Head Start program. Organizations function on the micro level by providing resources and training programs to help professionals develop and utilize cultural competence skills.

Title VI of the Civil Rights Act The provision that a recipient of federal money may not discriminate on the basis of race, color, or national origin.

Internet Resources

Diversity Rx — **www.diversityrx.org**
Addresses state and federal laws, multicultural health best practices overview, and interpreter services.

Awesome Library — **www.awesomelibrary.org**
Multicultural Toolkit details communication patterns for various racial/ethnic groups; Bennett's stages of intercultural sensitivity.

Administration on Aging — **www.aoa.gov**
A manuscript developed by the Administration on Aging: Achieving Cultural Competence: A Guidebook for Providers of Services to Older Americans and Their Families. Outlines the principles of cultural competence and offers guidance on creating programs for racially and ethnically diverse older populations.

State University of the New York Institute of Technology — **www.sunyit.edu/library/html/culturedmed/index.html**
This site provides extensive bibliographies.

EthnoMed — **http://ethnomed.org**
This site contains information about cultural beliefs, medical issues, and other issues.

Office of Minority Health — **www.omhrc.gov**

Pacer Center — **www.pacer.org/about.htm**
Provides information for increasing opportunities and enhancing the quality of life of children and young adults with disabilities and of their families.

Orthodox Union — **www.ou.org/kosher/kosherqa**
Questions and answers about kosher food and kosher supervision of food production.

Islamic Food and Nutrition Council of America (IFANCA) — **www.ifanca.org**
Describes the foods appropriate under Muslim law for halal consumers.

Food and Nutrition Information Center — **www.nal.usda.gov/fnic/etext/000023.html**
Cultural food guide pyramids from the Food and Nutrition Information Center.

NIH's National Center on Minority Health and Health Disparities — **http://ncmhd.nih.gov/about_ncmhd/mission.asp**

Gaining Cultural Competence in a Muslim Community

By Alice Fornari, EdD, RD, and Alessandra Sarcona, MS, RD

Scenario

An Indian physician has contacted you for a nutrition intervention with a community group where he is also a participant. This group consists of Indian Muslims who have emigrated to the United States. The group meets one evening per week as a social support event, as well as to share a meal. The physician, as a member of the group, has noted a growing interest in nutrition among members but cannot field many of the questions that are directed to him. He has invited you to come and share the meal with them and to hold a question-and-answer informal discussion two times a month, as well as to set up nutrition counseling sessions with individuals who have health risks. He says that they are particularly interested in nutrition and in prevention of cardiovascular disease. You have done a lot of consulting for this physician and would enjoy the challenge of working with this Indian Muslim community; however, you do not know anything about their culture or their dietary habits. You decide you need to research the Muslim culture before you can provide any nutrition intervention with this group. You noted an article in the May 2002 issue of *Today's Dietitian* on "Understanding Muslims and Islamic Dietary Laws." You found information about the Islamic dietary laws, including foods considered haram and halal. The article also noted that women dress conservatively and cover their hair with a scarf, that men and women tend to avoid mixed groups, and that many Muslims avoid contact between sexes, such as shaking hands and hugging. The Indian physician stated that his group is traditional and consists mostly of married couples.

Learning Outcomes

- Identify the National Standards for Culturally and Linguistically Appropriate Services, and understand how these standards are related to health care.
- Cite various ways to explore unfamiliar cultures in an effort to become culturally competent.
- Interpret generalizations about the tastes and values of a particular cultural group in order to provide culturally competent nutrition interventions.

Foundation: Acquisition of Knowledge and Skills

1. Review the National Standards for Culturally and Linguistically Appropriate Services in Health Care from the chapter text (see Table 16-15).
2. Access the websites www.ifanca.org/ (go to top menu—what is halal) and www.iica.org/invitation/the_musu.html and outline the basic principles of Muslim Dietary Laws; identify foods considered haram and halal.

Step 1: *Identify Relevant Information and Uncertainties*

1. Using the section of this chapter called "Cultural Encounters," identify and discuss the cultural food practices and any other components that may be relevant to Muslims. Go to www.todaysdietitian.com and click on article archive to retrieve the May 2002 article *Understanding Muslims and Islamic Dietary Laws.*
2. Outline cultural beliefs, attitudes, or practices that you discovered in your research on the Muslim way of life that would be different from those of your ethnicity. How might this affect your interaction with the Muslim immigrants when you hold a group session, share a meal, and conduct individual counseling sessions?

Step 2: *Interpret Information*

1. Using Table 16-10, indicate what questions you may need to ask the physician before coming to a meeting and what questions you may need to ask the group when you meet with them, to gain a greater understanding of Muslim food habits and behaviors than you have acquired from your Web searching?
2. Review the sections titled "Cultural Encounters" and "Practical Considerations for Interventions" in the chapter text. From your appraisal, indicate which suggested strategies will help you to provide culturally sensitive interventions for your target group.

Step 3: *Draw and Implement Conclusions*

1. On the basis of your research on Muslim dietary practices, create a Muslim pictorial food guide. For ideas on

international pictorial food guides, locate the *Journal of the American Dietetic Association* in your library, or go to the ADA website www.eatright.org to search for the article in the April 2002 issue titled "Comparison of International Food Guide Pictorial Representations." This can be accessed with your ADA member number; then go to JADA. You can also search pub med at www.nlm.nih.gov for the article by Cronin, Reflections on food guides and guidance systems, *Nutrition Today* 33(5) (1998): 186–88.

2. Describe how you would use this pictorial food guide with your target group.

Step 4: *Engage in Continuous Improvement*

1. Develop additional interventions (resources, community participation, activities, shopping, food preparation, and the like) for presenting information about nutrition and prevention of heart disease as you become more familiar with the Muslim population.

Professional Focus

Cross-Cultural Nutrition Counseling

Community nutrition professionals are often involved in counseling clients who come from cultures substantially different from their own. Bauer and Sokolik's Cross-Cultural Nutrition Counseling Algorithm[1] provides a visual representation of a counseling session and illustrates how to incorporate basic cross-cultural communication tools, as covered in this chapter, into an intervention. See Figure 16-2. The components of this cross-cultural nutrition counseling algorithm are relationship-building skills and the four phases of a counseling session: involving, exploration–education, resolving, and closing. The following paragraphs offer an overview of each of these components. For a more in-depth analysis of them, refer to Basic Nutrition Counseling Skill Development.[2]

Relationship-Building Skills

Relationship-building skills lay the foundation for making connections, developing and maintaining rapport, and creating trust. A trusting relationship is essential for clients to feel free to discuss personal issues and to be open to hearing the messages you want to convey. In fact, a productive relationship can in itself be an instrument of change and should be continually nurtured during a counseling session. The following list details the relationship-building skills emphasized in the algorithm, with special consideration given to cross-cultural encounters.

1. **Attending.** You need to use attentive behavior and listen actively in order to understand your clients' needs clearly. Also, you want to show that you are genuinely concerned and to indicate to your clients that what they have to say is very important to you.
2. **Reflection.** Reflection statements provide a vehicle for expressing empathy—a way to express, accurately and with sensitivity, that you understand someone's feelings and the meaning of those feelings. Empathy is not the same as sympathy. Sympathy is "I feel sorry for you." Empathy is "If I put myself in your shoes, I see where you're coming from."
3. **Legitimation.** Reflection statements acknowledge a person's feelings; legitimation responses affirm that it is normal to have such feelings and reactions. For example, "You have a right to feel upset. Anyone would."
4. **Show respect.** Counselors can show respect through words and body language that convey unconditional positive regard for their clients and a sincere interest in their welfare. Clients are more likely to feel free to express their thoughts and explain their actions if providers are not judging their actions. Do not criticize cultural differences. Be open to understanding divergent ideas and perspectives. Respect can also be shown through compliments. For example, "I really admire how your family was able to escape . . . and make a life for yourselves in California."
5. **Personal support.** Your clients should know you want to help. Your words and body language should convey the message "I look forward to working with you."
6. **Partnership.** You should make it clear to your clients that a number of options and strategies for solving their problems are available, and that you look forward to working with your clients to find strategies that work for them.

FIGURE 16-2 Cross-Cultural Nutrition Counseling Algorithm

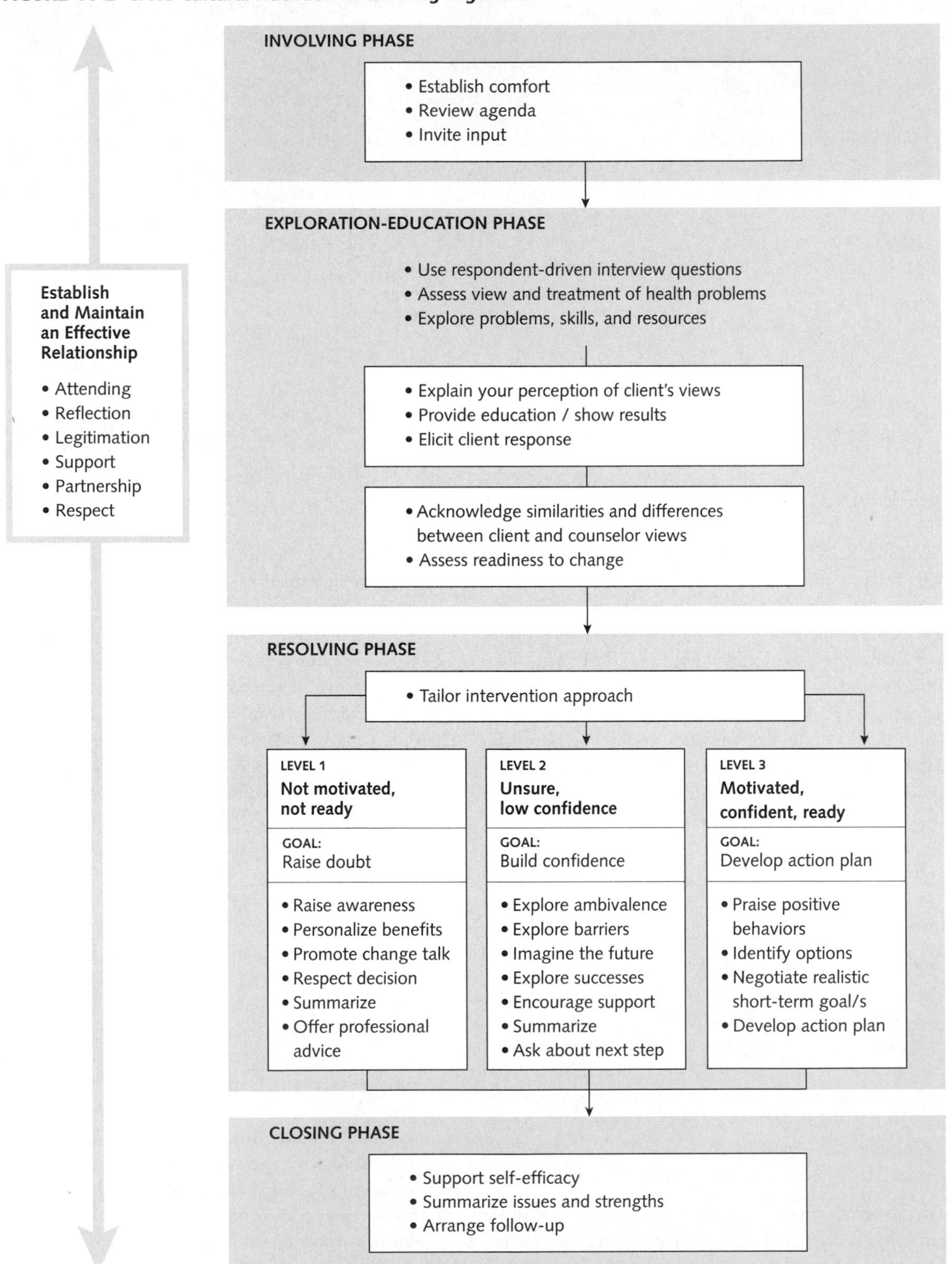

Source: K. Bauer and C. Sokolik, *Basic Nutrition Counseling Skill Development* (Belmont, CA: Wadsworth/Thomson Learning, 2002).

The Involving Phase

Since first impressions tend to be lasting impressions, you want your greeting to convey a sense of warmth and caring. Begin interactions in a formal manner and refrain from using first names or nicknames, since this could be considered disrespectful. Some small talk can aid in the development of a comfortable atmosphere and would be the expected course of action for certain cultural groups. Topics for small talk could include finding the office, the country of origin, or about adapting to U.S. living. A common opening after the greeting and small talk is asking an open-ended question in a curious manner, such as "What brings you here today?" During this phase, you should also explain something about your program and/or the counseling process, indicating what you can and cannot do. Set a short agenda for the session so your client knows what to expect.

The Exploration–Education Phase

During this phase of the algorithm, the counselor provides educational interventions and uses respondent-driven interview questions (refer to Table 16-13 and to Table 3-7 on page 75) to understand nutritional concerns, while focusing on identifying skills and resources that can be used to find solutions. The first four components of the LEARN guidelines in Table 16-14 (listen, explain your perceptions of workable strategies and your client's beliefs about treatment, acknowledge and discuss differences and similarities, and recommend options and strategies) provide a framework for discussing treatment strategies. After coming to the recommended options and strategies of the LEARN guidelines, the counselor needs to assess the client's motivational level for implementing any of the strategies. There are a number of ways to assess readiness to make changes,[3] but for people with limited English-speaking skills, the simplest method is simply to ask how they feel about working on a way to implement any of the strategies—in other words, whether they are not ready, are not sure, or are ready.

The Resolving Phase

The assessment of readiness is important because the algorithm takes into consideration the motivational level of your client. If your client is clearly not ready to make changes, your major goal will be to raise doubt about his or her present dietary behavior. Your major tasks are to raise awareness of the health/diet problems related to his or her dietary pattern, to personalize the benefits of change, to ask questions about the importance of changing (for example, "What do you believe will happen if you do not change what foods you eat?" or "What would have to be different for you to believe that it is important to change your diet?"), to summarize what has been discussed, to offer professional advice, and to express support if your client decides he or she is ready to make changes. Respect your client's decision. If his or her condition is life-threatening, consider seeking the aid of a respected community elder to help negotiate changes.

Clients who are unsure of their readiness to take action need something to shift the balance in favor of making a change. Your goal will be to build confidence by exploring their ambivalence. Some strategies for exploring ambivalence are examining pros and cons and asking questions such as "What are your barriers to making the recommended dietary changes?" or "What would need to be different for you to feel you are able to make changes in your diet?" If the barriers are cultural, use the last component of the LEARN guidelines to negotiate making the recommended dietary changes more acceptable, such as eating the foods at a certain time of the day or mixing them with the client's acceptable foods. At the end of your discussion, summarize and ask, "What's next?" Set some modest goals, if appropriate.

Clients who clearly indicate that they are ready to make dietary changes are ready to set goals and develop action plans for implementing the goals. Goals should be worked out jointly between client and counselor.

The Closing Phase

Review what occurred during the session, summarize issues, identify strengths, support self-efficacy, and restate goals. Plan for the next counseling encounter, which could be a phone call, an e-mail, a fax, or a counseling session.

References

1. K. Bauer and C. Sokolik, *Basic Nutrition Counseling Skill Development* (Belmont, CA: Wadsworth/Thomson Learning, 2002), p. 79.
2. Ibid, p. 64.
3. Ibid, p. 60.

Chapter *Seventeen*

Principles of Nutrition Education

Learning *Objectives*

After you have read and studied this chapter, you will be able to:

- Develop a nutrition education plan for a program intervention.
- Design nutrition messages.
- Describe four strategies for increasing program participation.
- Describe three basic principles of effective writing.

This chapter addresses such issues as educational theories and techniques, educational materials development, educational needs of diverse populations, presenting an educational session for a group, and lay and technical writing, which are Commission on Accreditation for Dietetics Education (CADE) *Foundation Knowledge and Skills* requirements for dietetics education.

Chapter *Outline*

Something To Think About...

To cease smoking is the easiest thing I ever did. I ought to know because I've done it a thousand times.

– MARK TWAIN

Introduction

Consumers are bombarded daily with dozens of health messages. Some are short and sweet: "Just Do It!" "Just Say No!" Others are complex and require time to process and understand. For example, an article in a consumer health magazine titled "The B Vitamin Breakthrough" describes research related to homocysteine and the risk of coronary heart disease (CHD) and takes 20 minutes to read and study.[1] Regardless of the format in which the health message is delivered—television and radio commercials, Internet and billboard advertising, print (magazines, newspapers, brochures, books) and electronic media (television, CD-ROM, the Internet), even sports clothing and T-shirts—the "successful" health or nutrition message has a favorable impact on the target audience. It gets them to examine their belief system, evaluate the consequences of a certain behavior, or *change* their behavior.

This chapter describes how the community nutritionist develops the nutrition education component of an intervention. **Nutrition education** is an instructional method that promotes healthful behaviors by imparting information that individuals can use to make informed decisions about food, dietary habits, and health.[2] This chapter discusses the principles of nutrition education. Two topics are as important in this as in previous chapters: the target population—what they think, feel, believe, want, and do—and entrepreneurship.

Applying Educational Principles to Program Design

Nutrition and health behaviors are complex issues. Individuals and groups can possess very different combinations of background, culture, health risks, health beliefs, motivators, learning style, environment, goals, and expectations.[3] An effective nutrition intervention program integrates good instructional design and learning principles and uses media that facilitate a high degree of individualization. Educational research shows that the effect of an intervention on the target population's knowledge and behavior depends on the intervention's application of six basic educational principles: consonance, relevance, individualization, feedback, reinforcement, and facilitation.[4] Table 17-1 offers examples of methods for applying these principles to the development of effective nutrition education interventions.

Learning across the Life Span

People of any age learn best if they have the prerequisite knowledge, if content is broken into small pieces, if what they learned is used and reinforced, if they have an opportunity to practice what they have learned, and if the content seems relevant.[5] It is also important to consider that people's learning needs, preferences, and abilities change as they age.[6]

Nutrition education An instructional method for promoting healthful food-related behavior.

TABLE 17-1 Applications of Educational Principles to Program Design

CRITERIA	DESCRIPTION	EXAMPLE
Consonance	Degree of fit between program and its objectives, or degree to which communication is directed toward accomplishing the intended outcome	In intervention to improve proper self-monitoring of blood glucose, teaching benefits of monitoring without showing how to monitor would be insufficient.
Relevance	Degree to which intervention is geared to clients, including reading level and visual acuity	Program should be tailored to clients' knowledge, beliefs, circumstances, and prior experience, determined by pre-tests, baseline questionnaires, or interviews.
Individualization	Allows clients to have personal questions answered or instructions paced according to individual learning progress	Because clients learn in different ways and at different rates, a program is more likely to be effective if the education is tailored to individual needs.
Feedback	Helps clients learn by providing a measuring stick to determine how much progress they are making	Feedback can be based on achieved learning objectives (such as increased knowledge about a given subject) or outcomes (such as increased adherence to a prescribed diet).
Reinforcement	Components of the program (other than feedback) designed to reward the desired behavior	Praise and congratulations are very effective in rewarding changed behavior.
Facilitation	Measures taken to accomplish desired actions or eliminate obstacles	For example, a weekly food diary sheet facilitates a client's ability to record actual food intake.

Source: Adapted from R. Patterson, ed. *Changing Patient Behavior: Improving Outcomes in Health and Disease Management* (San Francisco, CA: Jossey-Bass, 2001), pp. 16–17.

Consider the tips listed in Table 17-2 for teaching across the life span. Although children and adults learn in many of the same ways, there are some aspects of adult learning that are important to consider, as discussed in the next section. **Adult education** is a generic term that refers to formal, informal, vocational, and continuing education for the purpose of learning.[7] For adults, learning is an intentional, purposeful activity. Adult learners approach learning differently than children do, and they have different motivations for learning.

ADULT LEARNERS

Nutrition professionals who provide education for adults need to know how adults learn best and to be familiar with ways to enhance instructional effectiveness in this age group. The basic principles guiding adult education are that (1) adult roles, responsibilities, and previous experiences influence learning, (2) adult learning is constantly occurring, and (3) the purpose of the adult educator is to facilitate this continuous learning process.

Adult education The process whereby adults learn and achieve changes in knowledge, attitudes, values, or skills.

Adult learners learn best when the subject matter is tied directly to their own realm of experience, and their learning is facilitated when they can make connections between their past experiences (a parent died of cancer) and their current concerns

TABLE 17-2 Tips to Consider when Teaching across the Life Span

Children

- Keep your message short, clear, and simple.
- Emphasize positive points; avoid negative or judgmental statements.
- Relate the message to the child's interests.
- Make practical, concrete suggestions.
- Involve the child (ask questions, relate to his or her experiences and activities).
- Show the child how to, not why.

Adolescents

- Relate to the adolescent's interests.
- Consider the impact of peer pressure.
- Consider the client's rebelliousness and attitudes toward authority.
- Address his or her insecurities about physical changes.
- Discuss mood changes and impulsiveness.
- Tie teaching concepts to adolescent concerns, such as appearance and athletic performance.

Pregnant Women

- Relate present needs to the expected performance during labor and delivery and a healthy outcome.
- Address her anxieties and concerns.
- Consider the impact of the physiological, psychological, and emotional changes taking place.
- Acknowledge her needs.

Adults

- Acknowledge and relate to the client's needs and concerns.
- Consider his or her experiences with and knowledge of the subjects discussed.
- Personalize your interaction to the client's health profile.

Elderly

- Address the anxieties and concerns associated with isolation, chronic disease, and economic constraints.
- Consider the impact of chronic conditions and diseases on the client's ability to communicate and his or her attention span.
- Relate to his or her needs with empathy.
- Consider your recommendations in the context of the client's quality of life.

Source: Beyond Nutrition Counseling: Achieving Positive Outcomes through Nutrition Therapy, American Dietetic Association, 1996.

(whether eating a low-fat, high-fiber diet will protect them from cancer).[8] Adults are motivated to learn by the relevance of the topic to their lives, and they stick with a program or activity because they believe it will help them meet their learning goals.[9] Adults retain new information best when they are actively involved in problem-solving exercises and hands-on learning that enable them to assimilate, practice, and use the information in meaningful ways.[10] In short, an effective program takes into account the learning style and motivations of the target population.[11] Table 17-3 provides a summary of points to consider when providing educational experiences for adult learners.[12]

What we know about adult learners suggests the following recommendations for nutrition educators:

- Make learning problem-centered and meaningful to the learner's life situations.
- Make information concrete; adults prefer concrete knowledge to abstract information. Define all abstract terms.

TABLE 17-3 Providing Learner-Centered Educational Experiences for Adult Learners

Many of the following are considerations for all types and ages of learners but should be carefully addressed when working with adults.

Create an Environment That Supports Learning

Physical Dimensions

- Geographically convenient location
- Easy-to-locate classroom or meeting room
- Furniture of appropriate size
- Adequate heat and light

Psychological, Sociological Dimensions

- Appropriate location for topic being presented to audience
- Friendly, open manner of the instructor
- Instructor respects learners
- Opportunities for learners to get to know each other are provided
- Guidelines and procedures allow all learners equal opportunity to participate in learning

Preparing the Learning Episode

Assess learners' needs.
Assess learners' interests.
Assess learners' abilities/previous knowledge.
Encourage learner participation in establishing goals and content of the learning experience.

Providing Instruction

Encourage high learner interactive participation.
Emphasize material that learners perceive as relevant.
Use repetition and plan time for learner "practice."
Use a variety of teaching methods.
Use instructional resources that "fit" learners' needs.
Plan audiovisuals that all can see and easily interpret.
Provide frequent feedback.

Source: Adapted from E. J. Hitch and J. P. Youatt, *Communicating Family and Consumer Sciences,* (Tinley Park, IL: Goodheart-Willcox Company, 1995), p. 59.

- Make learning a collaboration between the educator and the learner. Use the adult learners' experiences as educational resources. Adults do not learn well with a lecture style of teaching; instead, invite learners to share their personal experiences and knowledge.
- Encourage participatory approaches to learning, such as small-group discussion and responding to questions. Engage adults in a dialogue with their peers, and allow them to share their experiences.
- Ask open-ended questions to draw out what the adults already know about the topic you are teaching.
- Seize the "teachable moments." Adults who are in a transition phase in their lives, such as pregnancy, mid-life, or older adult years, are generally more open to learning. Identify these specific life transitions and utilize them as opportunities for learning.
- Increase the adult learners' sense of self-worth by validating their experiences. Provide plenty of evaluation information and use positive feedback.
- Establish a positive learning environment. Ensure that the physical environment is attractive, comfortable, and well ventilated. Minimize distracting sounds.
- Recognize individual and cultural differences because they affect learning styles.[13]

To ensure successful educational interventions, it is imperative to study your potential audience, learning what motivates them, what they want, and what they need.[14]

Research your target population by (1) reviewing the literature, (2) conducting formative research (for example, one-on-one interviews or focus groups), and (3) asking representatives from the audience to help you with the planning and development of the program or material.[15] This will ensure that your efforts to educate adults will result in a successful experience for the learner as well as the educator.

Developing a Nutrition Education Plan

The nutrition education plan outlines the strategy for disseminating the intervention's key messages to the target population. The key nutrition messages may be designed to change consumer behavior, as in the "5 a Day—for Better Health" message to "Eat five to stay alive." (The "5 a Day—for Better Health" program is a national media program designed to get Americans to eat five or more servings of fruits and vegetables daily. The program is a joint venture of the National Cancer Institute [NCI] and the Produce for Better Health Foundation, which has more than 800 members.[16]) Nutrition messages may also inform or educate. When community nutritionists lobby legislators for more funding to support a statewide network of food banks and soup kitchens, they use nutrition messages to educate legislators about food insecurity and inform them of the consequences to young children of being malnourished.

The nutrition education plan is a written document that describes the needs of the target population, the goals and objectives for intervention activities, the program format, the lesson plans (including instructional materials such as handouts and videos), the nutrition messages to be imparted to the target population, the marketing plan, any partnerships that will support program development or delivery, and the evaluation instruments.[17] These aspects of the plan are often organized into a manual that can be used by staff who work with the program. Having a detailed nutrition education plan can prevent confusion, especially when new staff members join the team or a substitute instructor must be found on short notice.

A nutrition education plan is developed for each intervention target group. Table 17-4 illustrates the similarities among nutrition education plans developed for individuals, communities, and systems. Nutrition education plans for individuals and communities are identical, the same activities being appropriate to both types of interventions. At the systems level, the nutrition education plan might properly be called a strategy. System-level strategies do not require formal lesson plans or program identifiers such as logos or

TABLE 17-4 Activities Related to Developing a Nutrition Education Plan

	TARGET GROUP		
ACTIVITY	INDIVIDUALS	COMMUNITIES	SYSTEMS
Assess needs.	✓	✓	✓
Set goals and objectives.	✓	✓	✓
Specify the format.	✓	✓	✓
Develop a lesson plan.	✓	✓	
Specify nutrition messages.	✓	✓	✓
Choose program identifiers.	✓	✓	
Develop a marketing plan.	✓	✓	✓
Specify partnerships.	✓	✓	✓
Conduct evaluation research.	✓	✓	✓

Programs in Action

Behaviorally Focused Nutrition Education Programs for Children

Nutrition programs for school-age children traditionally have two sets of goals.[1] The first set strives to enhance the child's basic nutrition and food knowledge and to help children select a healthful diet. The second set aims to reduce disease risk by encouraging health-promoting eating behaviors. Behaviorally focused programs have been somewhat successful in meeting these goals.

Behaviorally focused nutrition education addresses three domains of learning: cognitive, affective, and behavioral.[2] Cognitive teaching presents children with the "how" of eating more healthfully: which foods should be part of a healthful diet and which should be eaten only on occasion. Affective teaching addresses factors that motivate children to change the way they eat. The behavioral component of nutrition education helps children build new eating skills and behaviors.

Successful education programs rely heavily on the behavioral component of learning.[3] Incentives, reinforcement, and rewards can be classified as behavioral elements that keep children interested in learning. Other factors that have been found to increase the nutrition knowledge of children include adequate time and intensity devoted to nutrition education and family involvement. Parents and family members serve as role models and influencers in the child's environment. Their involvement is particularly important when educating younger children, who still model parent behavior rather than the behavior of their peers. Parents often are receptive to participating in the education process, particularly at home. In one program, a short video for children and other nutrition education materials were sent home to interested parents. Nearly all parents reported preferring to receive information at home rather than attending a workshop.[4]

The following are among the strategies recommended to promote healthful eating among students in the lower elementary school grades.[5]

- Involve parents in nutrition education through homework or take-home videos.
- Provide role models (parents and teachers) for healthful eating.
- Use incentives to reinforce healthful eating.
- Identify easy-to-prepare, tasty, and healthful snacks such as fruits and vegetables.
- Increase students' confidence in their ability to make healthful eating choices.

The "5 a Day—for Better Health" program has inspired schools nationwide to create multilevel, integrated programs that teach and motivate children to eat more fruits and vegetables daily. A behavioral curriculum of classroom education, parental involvement, changes in school food service, and industry support can increase fruit and vegetable consumption.[6] The "Five for Kids, Too!" program is a successful example of this type of curriculum.

Children living in lower-income homes are at higher risk for health problems. This may be attributed, in part, to poor diet. According to the National Cancer Institute, American children average only 3.4 servings of fruits and vegetables per day, compared with a national goal of 5 daily servings. "Five for Kids, Too!" was developed to teach young children simple dietary changes that would keep them healthy and also delay the onset of diseases such as cancer and heart disease.

Goals and Objectives

The primary goal of the "Five for Kids, Too!" program was to increase daily consumption of fruits and vegetables among the target audience. Other goals included increasing awareness of the importance of fruits and vegetables, promoting willingness to try fruits and vegetables, and decreasing resistance to eating fruits and vegetables.

action figures, but they may draw on these program elements to reinforce key messages. A marketing plan is just as crucial to system-level activities as to those aimed at individuals and communities.

DEVELOPING LESSON PLANS

Nutrition educators have opportunities to provide nutrition education in many different settings. However, nutrition professionals are usually not trained as teachers, and they rarely have all the skills needed to develop effective, interesting, and creative lesson plans.[18] An effective lesson plan is straightforward and should be easy for whoever is delivering the lesson to follow.

Methodology

The "Five for Kids, Too!" program was created by a Maryland elementary school for kindergarten through third-grade students. Year-round activities were designed to reach children, as well as parents and caregivers, who have the greatest influence on their dietary habits. Activities used free materials from industry and from the American Cancer Society and were supplemented with letters home to parents, incentive gifts for the children, and a classroom visit from a community nutritionist.

A new media center made possible the creation of monthly 5-minute instructional videos. The videos were designed to be entertaining and featured the school nutritionist. The program is currently offered to and conducted by first-grade teachers.

Results

The "Five for Kids, Too!" program has been evaluated through presurveys and postsurveys of parents and children and through teacher evaluations. The children's survey contains just one question: "How many servings of fruits and vegetables do you think you should eat each day?" The parents' survey includes several questions related to the child's eating behaviors and attitudes about fruits and vegetables.

The results of the children's survey showed an increase from 17 percent to 87 percent of children answering that they need to eat five servings daily. Parent postsurvey results demonstrated that a higher percentage of children were consuming three to four or five to six daily servings of fruits and vegetables, and fewer were consuming only one to two servings. More than two-thirds of parents stated that their child was more willing to taste fruits and vegetables and that the program helped their child and the whole family to eat more fruits and vegetables. Parents expressed gratitude and appreciation for the efforts of the teacher. Teachers generally rated the program very highly. Approximately 77 percent described its effectiveness as excellent or very good, and more than 90 percent stated that they would be willing to repeat the program.

Lessons Learned

- Television is a powerful medium for this age group. The five-minute instructional videos were well received by the children. Although producing the videos took a lot of time, they could be used repeatedly without additional work.
- The program is now designed as a "copy and collect" project for the teachers—forms are copied, distributed, and later collected—because teachers were more likely to participate if their perceived workload was not increased.

References

1. L. Lytle, Nutrition education for school-aged children, Chapter 4 in I. Contento et al., eds., Nutrition education and implications, *Journal of Nutrition Education* (1995) 27: 298–311.
2. Ibid.
3. Ibid.
4. P. C. Dunn and coauthors, At-home nutrition education for parents and 5- to 8-year-old children: The HomePlate pilot study, *Journal of the American Dietetic Association* (1998) 98: 807–9.
5. Guidelines for school health programs to promote lifelong healthy eating, *Journal of School Health* (1997) 67: 9–26.
6. C. L. Perry and coauthors, Changing fruit and vegetable consumption among children: The 5-a-Day Power Plus Program in St. Paul, Minnesota, *American Journal of Public Health* (1998) 88: 603–9.

Source: Adapted from *Community Nutritionary* (White Plains, NY: Dannon Institute, Fall 1999). Used with permission. For more information about the *Awards for Excellence in Community Nutrition,* go to www.dannon-institute.org.

The first step in developing a lesson plan is to know your target audience, the setting, and the content. Consider the following principles when you are developing your lesson plan.[19]

- Your lesson plan should be centered on the learner's interests, needs, and motivation. This information should be available to you if you have done a complete assessment of your target audience.
- When learning is related to real-life situations, learners can readily identify with the situation. The examples that you provide should be directly related to the learners' lives and experiences.
- Adults learn best when they are involved in the process. Ask them at the beginning of class what they expect to learn. People learn by doing and by being involved in solving the problem at hand.

Structuring Your Knowledge

The first component of lesson writing is to identify the major concept that you are communicating. To attempt to design a lesson plan without organizing the main and related concepts within the subject matter is similar to putting the cart before the horse. It won't go anywhere. To help educators structure their knowledge, researchers have identified several questions that educators should ask themselves before creating their lesson plan.[20] Three of these questions are

- What is the telling question? Ask yourself: What am I trying to teach? Have I identified the major concept to be taught, and what do I need to know about it? Here are two examples of telling questions in nutrition: What are calcium rich foods? What are *trans* fatty acids?
- What are the key concepts? Once you have defined the telling questions, list the concepts you plan to define and teach about. For example, if your telling question is "What are calcium rich foods?" some of the key concepts you will want to define are dairy sources of calcium, plant sources of calcium, and other alternative sources of calcium, such as calcium-fortified soymilk.
- What methods of inquiry are used? What teaching method would you like to use to teach this lesson? Is the lesson more conducive to a group discussion with related activities? Or is a lecture more appropriate for the target audience? Would demonstrations be more useful? Choosing the most effective teaching method is important in the transfer of knowledge. Consider using some of the active learning experiences cited in Figure 17-1.

Writing Instructional Objectives

Developing useful lesson plans is based on writing effective objectives. Ask yourself, "What will the learner be able to do as a result of the learning experience?" Effective specific instructional objectives should (1) concentrate on the learner and not the teacher,

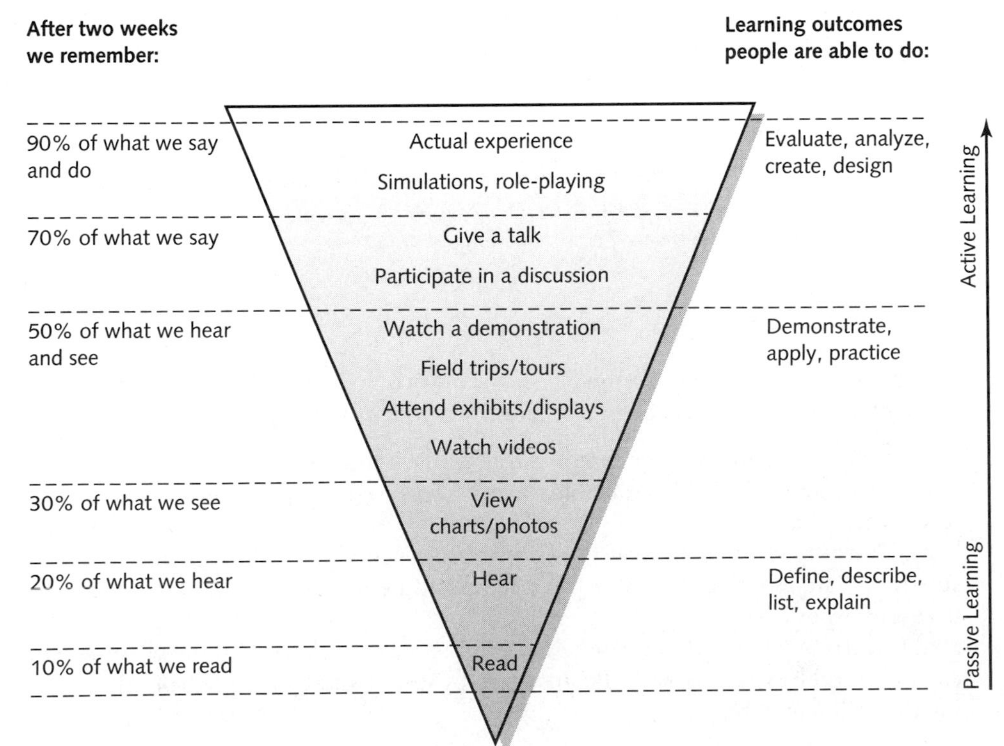

FIGURE 17-1 The Cone of Experience: From Passive Learning to Active Learning

Source: Adapted from Edgar Dale's Cone of Experience, *Audio-Visual Methods in Teaching,* 3rd ed. (New York: Holt, Rinehart, and Winston, 1969).

CHARACTERISTICS OF EFFECTIVE OBJECTIVES	COMMON MISTAKES TO AVOID	EXAMPLE OF INCORRECTLY STATED OBJECTIVES	EXAMPLES OF WELL-STATED OBJECTIVES
Concentrate on the learner.	Avoid describing teacher's performance.	To demonstrate to students how to read the label.	At the end of the demonstration, students are expected to: Identify the serving size on the food label.
Clearly communicate your intent using verbs such as *write, define, list, identify, compare, draw, differentiate.**	Avoid using broad terms such as *grasp, believe, have faith, internalize, enjoy, see, love, realize.*	Sees the value of breastfeeding.	At the end of the session, students will be able to: Identify three benefits of breastfeeding.
State objectives in terms of the expected learning outcomes rather than in terms of procedure.	Avoid terms such as *increase, gain, improve, add, develop.*	Gains knowledge of fiber.	At the end of the session, students will be able to: Differentiate between low- and high-fiber foods.
Include one learning outcome "specific" in describing the behavior of the learner as a result of the learning experience.	Avoid including *two active verbs* in one objective.	*Understands* how to count carbohydrate grams and *applies* the knowledge effectively.	Upon completion of the class, students will be able to: Apply the knowledge of counting carbohydrates effectively.

*Verbs used with general objectives: understand, know, learn, comprehend, apply, use, interpret, evaluate, demonstrate.
Verbs used with specific objectives: write, identify, compare, describe, list, state, differentiate, distinguish, explain.

Source: Adapted from R. AbuSabha, *Effective Nutrition Education for Behavior Change* (Clarksville, MD: Wolf Rinke Associates, 1998), pp. 81–82.

TABLE 17-5 Characteristics of Effective Objectives That Are Measurable and Attainable and Common Mistakes to Avoid

(2) clearly communicate a specific instructional intent, (3) be stated in terms of the *end product* (the learning outcomes) and not in terms of the process of learning, and (4) describe one type of learning outcome per objective that is specific in describing learners' performance (see Table 17-5).[21]

Components of a Lesson Plan

Planning of lessons should not be done haphazardly. Certain essential elements should always be part of your plan: objectives, body of the lesson, activities, and evaluation. A common format used to structure lesson plans follows.[22]

1. **Lesson Title.** Every lesson should have a title. Choose a catchy title when you can, but be sure it is not ambiguous. A title is a one-sentence (limit it to ten words) summary of your topic.
2. **Target Audience.** In order to maximize the usefulness of your lesson plan, clearly identify the target audience and their grade or educational level.
3. **Duration.** Identify the duration of the lesson (for example, 2 hours).
4. **General Objectives.** This is the goal for the class: a statement of what will be accomplished or learned. A general objective for the lesson plan on "5 a Day" might be "Learn the importance of eating five fruits and vegetables a day."
5. **Specific Objectives.** Identify the expected learning outcomes using measurable statements, as discussed earlier.
6. **Procedure.** Describe in detail your plan for the lesson, such as how the goals and objectives will be accomplished. Your lesson plan should include an introduction, a body, and a conclusion.

    - *Introduction.* Describe how the instructor will introduce the class. What activity, story, or segment will be used to start the session? Be sure that the introduction is interesting enough to grab the attention of the audience. Help learners make a connection with their own experiences.

- *Body of the Lesson.* The body of the lesson should contain two pieces of information: background and the way the lesson is organized. It is helpful to provide background on the subject matter, along with references to familiarize future instructors with the major concepts, and give them the opportunity to obtain further information when they need it. In addition, the instructor needs to know the organization of the lesson: what to do first, when to introduce an activity, and so on.
- *Closure.* In one or two sentences, summarize the lesson. For example, if the lesson is about the bread/starch group and is targeted to children in the third grade, you may close your lesson with the following sentence: "Remind learners that breads, cereals, rice, and pasta are important in our diet because they contain many vitamins and minerals that are essential to energy and good health."

7. **Learning Experiences or Activities.** List the different activities you expect the instructor and/or the learners to be involved with during the educational session. Activities may include showing slides or a video, group discussion, or a virtual learning experience using the Internet. Be creative with the activity, and aim to involve learners in applying the principles learned.
8. **Method of Evaluation.** Describe how the instructor will assess whether the expected outcomes have been achieved. Provide learners with experiences and/or tests designed to apply and evaluate what has been learned. Creative assignments, knowledge tests, exams, and homework are all methods of evaluating learners' performance. In addition, be sure to include an evaluation of the educator's performance to assess how well the information was presented.
9. **Materials Needed.** List all of the necessary materials for use by the educator and those needed by the learner. Plan to include as many of these materials in the lesson plan as possible (for example, paper copies of all PowerPoint slides, overheads, and handouts).

Nutrition Education to Reduce CHD Risk: Case Study 1

This section describes the development of a nutrition education plan for the "Heartworks for Women" program, a health promotion activity designed to help individuals (women) reduce their CHD risk. Consult Table 17-4 as you read this section.

The senior manager responsible for developing, implementing, and evaluating the intervention for reducing CHD risk among women (Case Study 1) reviews the proposed intervention levels and activities outlined in Table 15-4 on page 491. The senior manager's goal is to develop a coordinated plan for carrying out the intervention and evaluating its effectiveness. She begins by reviewing the proposed intervention activities and the expertise and time commitments of her staff. She decides to organize intervention activities into two areas: smoking, which she assigns to team 1, and nutrition, which she assigns to team 2. Team 1 members have expertise in health promotion, medicine, and epidemiology; team 2 consists mainly of community nutritionists and a health educator. Table 17-6 shows how the manager divided the intervention activities. The manager expects the teams to collaborate on some activities, such as developing content for the website, and to take leadership on others.[23] Some activities, such as creating education materials (flyers and brochures, for example) will be developed by each team for its respective programs. The senior manager designates a leader for each team.

The community nutritionist assigned to lead team 2 is given responsibility for developing the nutrition component of three intervention activities: the "Heartworks for

TABLE 17-6 Case Study 1: Assignment of Responsibility for Carrying Out Intervention Activities

TARGET GROUP	TEAM 1 (Smoking)	TEAM 2 (Nutrition)
Individuals	Smoking cessation course Smoking reduction course	"Heartworks for Women" course Smoking reduction course
Community	Special media events TV and radio public service announcements Antismoking campaigns in schools and worksites	Special media events
Systems	Legislative activities	Legislative activities

Women" program, the nutrition content of the website, and the "smoking reduction" program. Her team will work with team 1 to develop the smoking reduction program, which will probably include messages and activities in the areas of nutrition and physical activity. The teams will work together to secure antismoking legislation, because this is a departmental activity. The team leader maps a strategy for the activities she's been assigned, breaking each major activity into smaller ones. For the "Heartworks for Women" program, for example, she indicates that goals and objectives must be developed, a format chosen, nutrition messages specified, and evaluation research conducted. Next, she reviews the interests, skills, and current assignments of her staff and assigns a staff member to take responsibility for each of the major activities. The next sections describe how the "Heartworks for Women" program was developed.

"Heartworks for Women" Program: Assessing Participants' Needs

The community nutritionist responsible for developing the "Heartworks for Women" program first identifies the target population's educational needs. She asks these questions: What learning style is best suited for the potential program participants? What kinds of instructional tools (for example, videos or printed handouts) will have the greatest impact with this group? Can Internet activities be incorporated into lesson plans? Will participants be comfortable in group settings? Should some individual nutrition counseling be provided? Where should group sessions be taught? The answers to these questions can be found by reviewing the data obtained during the community needs assessment and by conducting formative evaluation research. For instance, focus groups can be organized to gather information not obtained previously from members of the target population.[24] Focus group participants might be asked about suitable locations for group activities, whether they have access to the Internet, and what they would like to learn about diet and CHD risk.

SET GOALS AND OBJECTIVES

See Table 15-1 on page 477 for examples of three levels of intervention.

The next step is to develop goals and objectives for the program. Recall that the "Heartworks for Women" program is a Level II intervention, which means that it is a skills-building program. It will be designed to address three of the outcome measures specified in Chapter 14: (1) Decrease women's mean saturated fat intake to 11 percent from their current intake of 14 percent of total energy within 1 year, (2) increase the percent of women who can specify CHD as the leading cause of death in the city of Jeffers by 50 percent within 1 year, and (3) increase the percent of women who can specify two major CHD risk factors by 25 percent within 1 year. The "Heartworks for

Women" program is not the only means of achieving these objectives; some Level I activities are designed to address these objectives at the individual and community levels.

After reviewing the larger goals and objectives (described in Chapter 14), the community nutritionist determines that the "Heartworks for Women" program has two goals: (1) to educate individuals about the contributions of diet to CHD risk and (2) to build skills related to heart-healthy cooking and eating. Specific objectives are as follows:

- Increase awareness of the relationship of diet to CHD risk so that by the end of the course, the percent of participants who can name two dietary factors that raise blood total cholesterol will increase to 75 percent from 25 percent.
- Increase knowledge of dietary sources of fat so that by the end of the course, the percent of participants who can name three major sources of dietary fat that contribute to heart disease will increase to 75 percent from 30 percent.
- Increase knowledge of low-fat cooking methods so that by the end of the course, the percent of participants who can describe and use five low-fat cooking methods will increase to 75 percent from 60 percent.
- Increase label-reading skills so that by the end of the course, the percent of participants who can specify accurately the fat content of foods using the nutrition information provided on food labels will increase to 75 percent from 20 percent.

Using these objectives as a guide, the community nutritionist sketches a rough outline of the program sessions, as shown in Table 17-7. The outline shows the link between the program objectives and the individual sessions. Session 1, for example, will provide general information about the major CHD risk factors and the contribution of diet to CHD risk. In this manner, the community nutritionist can be certain that any information that must be imparted to participants to meet the program objectives has been included in the program outline.

SPECIFY THE PROGRAM FORMAT

Program formats vary, depending on what the program is intended to accomplish and what resources are available to implement it. Choosing a format is much like choosing an intervention strategy: Begin with the big brush strokes. The format might consist of only

TABLE 17-7 Case Study 1: Rough Outline Showing the Link between the Objectives and Proposed Sessions for the "Heartworks for Women" Program

PROGRAM OBJECTIVE	PROPOSED SESSION TO MEET OBJECTIVE
Increase awareness of the relationship of diet to CHD risk so that by the end of the course, 75 percent of participants can name two dietary factors that increase blood total cholesterol.	• Introduction to course
Increase knowledge of dietary sources of fat so that by the end of the course, 75 percent of participants can name three major sources of dietary fat that contribute to heart disease.	• Major food sources of various types of fats • Low-fat meats • Low-fat dairy products • Shopping for heart-healthy fats
Increase knowledge of low-fat cooking methods so that by the end of the course, 75 percent of participants can describe and use five low-fat cooking methods.	• Low-fat meats • Low-fat dairy products • Fruits and vegetables • Reading restaurant menus
Increase label reading skills so that by the end of the course, 75 percent of participants can specify accurately the fat content of foods using the nutrition information provided on food labels.	• How to read food labels • Shopping for heart-healthy foods

three didactic lectures, or it might require six lectures and two cooking demonstrations, or it might involve three individual counseling sessions and ten group sessions. The community nutritionist chooses a format that suits the topic and the amount of information that must be presented. She anticipates making some changes to the program format after analyzing the results of evaluation research and estimating projected program costs.

The results of the community needs assessment and focus group sessions showed that most potential participants for the "Heartworks for Women" program have at least a high school education. Many have participated in group classes (for example, weight management classes). About 15 percent of those surveyed in focus groups have access to the Internet. The community nutritionist considers these and other factors when choosing a format. She decides on an 8-week program designed to fulfill the goals and objectives outlined previously. The program will consist of 90-minute sessions in which participants will have an opportunity to set target dietary goals, try new behaviors, and assess their success. The key strategy will be to seek small behavioral changes. The participants' skill level at entry and readiness to learn will be evaluated at the beginning of the program. The sessions will be organized as follows:

Session 1: Getting Started
Session 2: Looking for Fat in All the Right Places
Session 3: Cooking Meat the Low-Fat Way
Session 4: Dairy Goes Low-Fat
Session 5: Focus on Fruits and Vegetables
Session 6: Reading Food Labels
Session 7: Grocery Shopping Made Easy
Session 8: Reading Restaurant Menus

When choosing a format, the community nutritionist considers many details related to implementing the program. If the format calls for individual counseling, the facility must have private rooms for this activity. Likewise, if the format calls for small-group sessions, there should be conference rooms or classrooms for teaching groups. If cooking demonstrations are included in the format, the facility should have counters, sinks, electrical outlets, and other equipment. Her decision about the final program format is influenced by the availability of facilities, equipment, and staff.

DEVELOP LESSON PLANS

The community nutritionist considers the instructional method (for instance, group sessions or one-on-one counseling) best suited for teaching heart-healthy cooking skills in the "Heartworks for Women" program. She chooses to present the material in group sessions, knowing that the participants will learn from one another.[25] Moreover, a program consisting mainly of group sessions is more likely to fit within the budget, because group sessions tend to be less costly than individual counseling. She develops objectives, selects instructional materials, and specifies other materials (such as goal sheets) required for each lesson. Table 17-8 shows this information for the first two sessions.

She must decide whether to develop nutrition education materials herself, use existing materials, or do both. To save time, she reviews existing programs and their nutrition education materials to determine whether they can be used with or adapted for this population. For example, in Session 1, which describes the major risk factors for CHD, the community nutritionist elects to develop her own handout on homocysteine because she cannot locate one, but she plans to use an existing brochure that describes other leading CHD risk factors. In Session 2, she plans to use a dietary fats chart developed by a leading food company and used widely in nutrition counseling. For Session 3, which will

SESSION	TITLE	SESSION OBJECTIVES	INSTRUCTIONAL MATERIALS	LEARNING ACTIVITIES	NUTRITION MESSAGES
1	Getting Started	At the end of the session, participants will be able to: • Describe the program's two goals and four objectives. • Describe five major risk factors for CHD.	• Participant information form • Description of course goals and objectives • Handout: "Is Homocysteine a Risk Factor for Heart Disease?" • Handout: "Recipe for Summer Salsa and Baked Pita Chips" • Brochure: "Get the Facts about Heart Health"	• CHD and nutrition knowledge pre-test • Handout: "Am I Ready for Change?" • Taste test = Summer Salsa and Baked Pita Chips (a low-fat recipe)	• Diets high in fat and saturated fat raise blood total cholesterol. • High blood cholesterol levels are a risk factor for CHD. • Low-fat cooking is easy.
2	Looking for Fat in All the Right Places	• Define four types of dietary fats. • Describe the major food sources of dietary fat. • Describe the major sources of fat in the typical U.S. diet.	• Handout: "Definitions of Fats" • Handout: "Dietary Fats Chart" • Handout: Goal sheet	• Dietary Fats Quiz • Completion of goal sheet: Reducing saturated fat intake	• Choose low-fat foods more often than high-fat foods.

TABLE 17-8 Case Study 1: Lesson Plans for the First Two Sessions of the "Heartworks for Women" Program

demonstrate low-fat cooking methods for meat, she decides to use reproducible masters and other materials from the *Lean 'N Easy* educational kit developed by the National Cattlemen's Beef Association.[26] In the selection of program materials, factors such as the cost of purchasing existing educational programs and materials must be weighed against the time required to produce educational materials in-house and the cost of duplicating materials for participants. Check the Internet Resources on page 546 for a list of websites that provide nutrition handouts and other useful educational tools.

SPECIFY THE NUTRITION MESSAGES

Nutrition messages should be specified for each lesson plan. The messages should convey a simple, easy-to-understand concept related to heart-healthy eating and cooking: "Choose low-fat dairy products," "Choose lean cuts of meat," and so forth. One or two of these messages may be used in the nutrition education plan developed for the community- or systems-level interventions. The nutrition message for Session 5—"Eat five or more servings of fruits and vegetables every day"—does double-duty: It is used in the individual session for the "Heartworks for Women" program, and it is a key nutrition message used in the media campaign to build awareness among women. More detailed information about nutrition messages is given later in this chapter.

CHOOSE PROGRAM IDENTIFIERS

Tag line A simple, short message that conveys a key intervention message and is used on promotional materials.

The community nutritionist and her teammates choose program identifiers such as the program name, a logo, an action figure, or a **tag line.** Tag lines are short, simple messages that convey a key theme of the program. They are typically used in promotional materials such as flyers and brochures. The tag line "Good Food for Good Health" might appear on departmental stationery, along with the program name and logo. These elements give the program its own identity and foster a sense of ownership among participants. The program name is important. It is usually selected after consultation with colleagues and members of the target population.

DEVELOP A MARKETING PLAN

"If you don't exist in the media, for all practical purposes you don't exist," remarked National Public Radio's Daniel Schorr.[27] The community nutritionist develops a marketing plan to promote the "Heartworks for Women" program to the target population. Details about this plan are presented in Chapter 18.

SPECIFY PARTNERSHIPS

Forming partnerships with grocery stores, local farmers, retail establishments such as Target stores, government agencies, nonprofit organizations, and other groups is one way of controlling the cost and increasing the reach of programs.[28] The community nutritionist with the "Heartworks for Women" program established a partnership with a local grocery store chain that enabled her to use one of its stores as the setting for one session on shopping for heart-healthy foods and reading labels. (The benefits to the store are obvious.) She also networked with a national food company and the local affiliate of the American Heart Association to obtain complimentary nutrition education materials for the course.

Consider creative ways to build partnerships with:

- Businesses
- Cooperative extension offices
- Fitness and recreational facilities
- Food and pharmaceutical companies
- Hospitals, HMOs, and health departments
- Food and nutrition assistance programs
- Nonprofit and civic organizations
- Parent, church, and community groups
- Schools and universities

CONDUCT FORMATIVE EVALUATION

Formative evaluation should be conducted throughout the program design process. Examples of formative evaluation research include the focus group sessions held early in the design phase and additional focus group testing of dietary messages and program instructional materials, such as the handout on homocysteine for Session 1 and the

Mary Kate Denny

Focus group sessions can be used to obtain the target population's opinions and impressions about program elements. Here, high school students offer their views on a program's name, logo, and tag line.

dietary fats chart for Session 2. Here, the community nutritionist invites members of the target population to review these materials for reading level and ease of understanding. In other words, the target population helps determine whether the materials are appropriate and useful. Educational materials such as brochures, fact sheets, and handouts should be checked for reading grade level using a word processor's readability formula or a formula such as the Fry Index or SMOG grading test (see Appendix C and the list of Internet Resources near the end of this chapter).

For an annotated bibliography of "Health Materials for Low-Literate Audiences," visit www.worlded.org/us/health/docs/comp.

At least one trial run of the program should be completed to ensure that the lectures fit within the designated time frame, that the dietary messages are understandable and appropriate for the target population, and that the nutrition education plan is sufficiently detailed to curtail glitches. The results of formative evaluation are used to change and improve program delivery.

Designing Nutrition and Health Messages

Consumers process nutrition and health messages at different levels, depending on their interest in the topic and their past experience with similar messages. Their responses to nutrition and health messages can be viewed as a continuum. On one end is a state of mindless passivity, wherein consumers pay little attention to the message; at the other end is the state of active attention, where consumers respond to the message by thinking about it and considering what it might mean for their own health and well-being.[29] The important question is, How can nutrition messages be formulated to influence consumer behavior? In this section, several general ideas for designing nutrition and health messages are described.

General Ideas for Designing Messages

Studies of consumer behavior suggest several ways of designing nutrition and health messages to grab consumers' attention. For example, present information in a novel or unusual fashion. The "5 a Day—for Better Health" campaign logo, shown in Figure 17-2, appears on aprons, bookmarks, buttons, clothing (caps, T-shirts, golf shirts, and jackets), pens, magnets, coloring books, key chains, watches, patches, mugs, tote bags, and magnetic memo boards.[30] No matter what the medium, the message to eat more fruits and vegetables reaches consumers in unexpected ways.

FIGURE 17-2 Logo for the "5 a Day—for Better Health" Campaign

Source: This logo appears on the National Cancer Institute's website at http://cancercontrol.cancer.gov/5aday/ and on the Produce for Better Health Foundation's website at www.5aday.com. Used with permission of the Produce for Better Health Foundation.

Use language that says to the consumer, "Listen to this. It's important." For example, if you are giving a radio interview to promote the "Heartworks for Women" program, you might say something like "How many listeners know the number of women who die from heart attacks every year?" This approach is a cue for listeners to pay attention to the message; it engages their thought processes.

Use language that is immediate. A good example is available from the antidrug program: "This is your brain. . . . This is your brain on drugs. . . . Any questions?" Messages that use verbs in the present tense and demonstratives such as *this, these,* and *here* make consumers pay attention. Whenever possible, avoid using qualifiers such as *perhaps, may,* and *maybe* that express uncertainty. Consumers prefer straightforward statements such as "High-fat diets raise blood cholesterol levels" rather than tentative statements such as "A diet high in fat may increase blood cholesterol levels." Of course, the challenge for community nutritionists is deciding when strong, clear statements about research findings can be made and when more conservative statements are appropriate.

Think about the target population and consider the style and format of messages that will get their attention. Children, for example, are not small adults, and they process information differently than adults do. A common mistake when presenting information to children has been to assume that children will reject a behavior portrayed as unhealthful or bad (for instance, skipping breakfast or snacking on junk foods all day).[31] Sometimes a behavior is appealing to children and teenagers precisely *because* it is portrayed as bad or unhealthful! When designing messages for children and teenagers, go directly to the source: Ask them which messages they respond to and what type of messages they prefer. Also, start early. Messages directed to children and adolescents have the potential for a lasting effect. The "Gimme 5" program, for example, is a "5 a Day" program that targets high school students. Nutrition messages are delivered through "Fresh Choices" (a school meal and snack program), "Raisin' Teens" (a program for parents), and other program activities.[32]

A consortium of professional organizations such as the American Dietetic Association; trade organizations such as the National Cattlemen's Beef Association, the National Dairy Council, and the Food Marketing Institute; and federal government agencies such as the U.S. Department of Agriculture and the U.S. Department of Health and Human Services have formed a partnership to promote positive, simple, consistent messages to help consumers achieve healthful, active lifestyles.[33] The messages developed by the partnership were derived from focus group discussions with consumers who were asked, "Why aren't Americans doing more to eat right and get fit?" The answer was perhaps not too surprising: Consumers cite many reasons for not eating better and not being more active—lack of time and energy, no "will power," bad habits, lack of motivation, the perception that "healthy" foods cost more and are hard to find, and the belief that being fit requires expensive equipment or club membership. Moreover, the focus group participants had clear ideas about how nutritionists and other health professionals should communicate with them. Here is a summary of their opinions on effective communications:

- "Give it to me straight." Use simple, straightforward language. Don't use technical, scientific jargon or complicated instructions.
- "Make it simple and fun." Make it clear that eating healthful diets and being physically active are not time-consuming, complicated chores. Place the emphasis on improving habits, not on trying to achieve perfection. Provide practical, easy-to-implement strategies.
- "Explain what's in it for me." Make the benefits of healthful lifestyles clear. Motivate consumers by citing outcomes that are meaningful to the audience, such as being happier and having more energy.
- "Stop changing your minds." Be consistent in making recommendations. The wealth of dietary recommendations and scientific findings is overwhelming and confusing for most consumers.
- "Offer choices." Consumers are empowered when they can make their own choices for behavior changes.[34] For example: "You can enjoy sweet foods in moderation. Split a dessert with a friend *or* take half home to enjoy the next day."

The ideas and opinions expressed by focus group participants led to the creation of the "It's All About You" campaign. The campaign's nutrition messages—designed to be simple, appealing, and "action-able"—aim to help consumers make healthful choices that fit their lifestyles:[35]

Be Realistic. Make small changes over time in what you eat and the level of activity you do. Small steps work better than giant leaps.

Be Adventurous. Expand your tastes to enjoy a variety of foods.
Be Flexible. Go ahead and balance what you eat over several days. No need to worry about just one meal or one day.
Be Sensible. Enjoy all foods, just don't overdo it.
Be Active. Walk the dog, don't just watch the dog walk.

In tests of the nutrition messages, consumers reported that the messages were positive, motivating statements that got their attention. The messages can be paired with supporting tips that provide ideas for specific changes in behavior. For example, the message "Be Realistic" can be paired with supporting tips such as "Park your car in the farthest spot" or "To save calories and fat, use a cooking spray instead of oil to sauté foods." These messages can be incorporated into a variety of programs as a way of helping consumers do more to achieve good health.

IMPLEMENTING THE PROGRAM

After the program has been designed and tested, it is ready for implementation. The goal of this phase of the planning process is to deliver as faithfully as possible the program laid out in the nutrition education plan. There will be glitches, of course. Perhaps an ingredient or cooking utensil was omitted from the list of materials needed for a cooking demonstration, or no overhead projector was available at the facility, or a program flyer featured the wrong time for the session. Anything can happen. Make a record of any unexpected problems so that you can devise a strategy for preventing them in future programs. Once the program is under way, the community nutritionist and her team work to keep it running smoothly. Two questions arise during this phase: How can program participation be increased? And how can the program be improved?

ENHANCING PROGRAM PARTICIPATION

Let's state the obvious: The higher the level of participation in a program, the better. High participation increases the likelihood that a program will be effective and that people will be involved with it long enough to make a behavior change. But what is participation? And how do program planners maximize participation in their programs? **Participation** is the number of people who take part in a health promotion activity. If the health promotion activity is a group education program, participation consists of the number of people involved at the end of each educational session. If the activity is a newsletter, participation is the number of people who receive the newsletter times the number of newsletter editions or mailings per year. Participation rates vary, depending on how new the activity is and whether an incentive for participating is offered. For group education sessions offered on a voluntary basis and without incentives, the participation rate may range from 5 percent to 35 percent of the target population. A newsletter may have a much higher participation rate of 85 to 95 percent.[36]

What can be done to improve participation rates? First, understand the target population and their needs and interests. Second, use evaluation research to improve the program design. Make the activity enjoyable and relevant to the target population's needs. Remove barriers to participation. Remember, people participate in health promotion activities for different reasons: to have fun, be with friends or family, learn something new, be challenged, fulfill a goal, or seek support. Find ways to help them see the immediate benefits of participating. Make it easy for people to sign up for or attend the activity. Schedule the activity at a convenient time. Third, use incentives for participating.

Participation The number of people who take part in a health promotion activity.

Incentives range from formal recognition for achieving goals to raffle prizes and treats such as T-shirts, magnets, and cookbooks. Fourth, build "ownership" of the program among participants by using slogans, action figures, and logos to enhance the program's identity. Finally, promote, promote, promote—in other words, make the program highly visible for the target population.

CONDUCTING SUMMATIVE EVALUATION

Summative evaluation, which is designed in the planning stage but conducted at the end of the program, provides information about the effectiveness of the program.[37] Summative evaluation is designed to obtain data about the participants' reaction to all aspects of the program, including the topics covered, the instructors or presenters, any instructional materials, the program activities (for example, cooking demonstrations and taste tests), the physical arrangements for the program (including the location, room temperature, and availability of parking), registration procedures, advertising and promotion, and any other aspect of the program. Participants are asked to rate these program elements, perhaps scoring their assessment on a five-point rating scale. They may be asked to explain, in their own words, what aspects of the program they liked most or least and to suggest ways in which the program can be improved.[38] As with formative evaluation, the data obtained from summative evaluation are used to improve the program's delivery and effectiveness and to make the program an inviting place for learning.[39]

Summative evaluation Research conducted at the end of a program that helps determine whether the program was effective and how it might be improved.

See Chapter 14 for more discussion of various types of program evaluation, including formative, process, impact, outcome, structure, and fiscal evaluation.

Entrepreneurship in Nutrition Education

Creativity and innovation—the twin elements of entrepreneurship—can be applied to many aspects of nutrition education, from the development of action figures and logos to the use of communications media such as the Internet, CD-ROM, and DVD.[40] Program planning is definitely a good venue for practicing entrepreneurship. Consider the smoking reduction program mentioned earlier in this chapter. This program idea represents a new approach to reducing CHD risk. It focuses on smokers—a target population that contributes the lion's share toward the costs of treating heart attacks and stroke—but in a new way. Rather than trying to get smokers to quit smoking altogether, the program is designed to build positive health behaviors in other areas of their lives (for example, eating heart-healthy foods, taking a walk five days a week, and learning to manage stress). The program's goals are to help smokers to cut down on the number of cigarettes smoked and to reduce their CHD risk by adopting some of the positive health behaviors seen typically in nonsmokers. Before this program idea can be implemented, however, many questions must be asked: Has this approach been tried before? Does evidence from the literature indicate that it will work? Are there sufficient data to support making dietary and physical activity recommendations to smokers? What do the experts in this area think about this approach? What is the theoretical framework that supports this approach? If the approach has never been tried, should it be undertaken as a formal, scientific study? Some of these questions fall into the policy arena, but all must be addressed before program development can move forward.

See also Chapter 15 for a review of popular behavior change theories.

Examine Mark Twain's comment at the beginning of this chapter. It highlights a perpetual problem in the field of health promotion: How can we help consumers change their behavior? One approach to motivating consumers is to design effective nutrition messages and programs.

Internet Resources

Food and Nutrition Information Center's Resources for Educators	**www.nal.usda.gov/fnic/pubs_and_db.html**
Facilitated Dialogue Basics: Let's Dance	**www.unce.unr.edu/publications/SP04/SP0421.pdf**

Self-study guide for planning facilitated discussions using learner-centered education techniques.

Washington Dairy Council's Nutrition Education Resources	**www.eatsmart.org**

Resources for Assessing Readability of Print Materials

Readability Testing	**www.healthsystem.virginia.edu/internet/health-education/read.cfm**
National Cancer Institute's Clear and Simple Method	**www.cancer.gov/aboutnci/oc/clear-and-simple/page1**

International Food Guides

Asian Diet Pyramid	**www.news.cornell.edu/science/Dec95/st.asian.pyramid.html**
Latin American Diet Pyramid*	**www.oldwayspt.org/pyramids/latin/p_latin.html**
Mediterranean Diet Pyramid	**www.oldwayspt.org/pyramids/med/p_med.html**
Ohio State University; scroll down to Cultural Diversity—Eating in America	**http://ohioline.osu.edu/lines/food.html#FOODN**
Puerto Rican Food Guide Pyramid	**www.hispanichealth.com/pyramid.htm**
USDA MyPyramid and Special Audience Food Guides	**www.nal.usda.gov/fnic/Fpyr/pyramid.html**
Vegetarian Food Guide Pyramid	**www.eatright.org**

*Asian and vegetarian food guides are also available on this site, which is sponsored by Harvard University and Oldways Preservation and Exchange Trust.

Professional Focus

Being an Effective Writer

The written word, Rudyard Kipling once observed, is "the most powerful drug used by mankind." It inspires, educates, and engages us. Because it is so powerful, we want to be sure that we can express ourselves well, regardless of the audience for whom we are writing or the topic being discussed.

But how do you become a good writer? Any number of strategies, some of which are described here, can help you improve your writing, but one of the most important steps you can take is to practice. To learn to express yourself clearly and concisely, you must practice, practice, practice. Fortunately, most of us have ample opportunity to practice our craft. Our job or school requires us to write many types of documents: business letters, informal memos, reports, study proposals, scientific articles, fact sheets for the general public, and project updates, to name just a few. Mastering the basic principles of grammar and syntax is also important. Knowing how to use these tools properly will serve you well in all writing situations.

Three Basic Rules of Writing

Although there are different types of writing—for example, professional and business writing versus copy writing for an advertisement—several basic principles apply:

1. *Know what you want to say.* A well-known fiction writer once remarked, "I don't find writing particularly difficult; it's figuring out what I want to say that's so hard!" If you don't know what you want to tell the reader, your writing will meander around and leave the reader bewildered and dissatisfied. Your first step, then, is to decide what point or points you want to make. Jot them down before you begin to write your article

or report. Once you clarify in your own mind precisely what it is that you want to say, the writing itself will flow more smoothly, and the reader will follow your thinking more easily.

2. *Eliminate clutter.* "The secret of good writing is to strip every sentence to its clearest components."[1] Every word in every sentence must serve a useful purpose. If it doesn't, mark it out. Consider the approach used by Franklin D. Roosevelt to convert a federal government memo into plain English. The original blackout memo read as follows.[2]

> Such preparations shall be made as will completely obscure all Federal buildings and non-Federal buildings occupied by the Federal government during an air raid for any period of time from visibility by reason of internal or external illumination.

"Tell them," Roosevelt said, "that in buildings where they have to keep the work going to put something across the windows." By changing gobbledygook into plain English, Roosevelt made the memo simple and direct—and comprehensible. This strategy is important in all types of writing. Consider the following statement from one computer company's "customer bulletin": "Management is given enhanced decision participation in key areas of information system resources." What on earth does that mean? It might mean "The more you know about your system, the better it will work," or it could mean something else.[3] The wording is so jumbled that the customer can't be sure what the company is trying to tell her and will probably take her business elsewhere as a result.

Whenever possible, eliminate clutter and jargon. To free your writing from clutter, clear your head of clutter. Clear writing comes from clear thinking.

3. *Edit, edit, edit.* A well-written piece does not occur by accident. It is crafted through diligent editing. Writers often edit their manuscripts eight, nine, ten, or more times. Editing pares the piece to the bare bones. It makes the writing stronger, tighter, and more precise. The paragraph shown in Figure 17-3 is an example of an edited manuscript. This particular paragraph was taken from William Zinsser's book *On Writing Well.* What you see in the figure is a draft that had already been edited four or five times. Zinsser says that he is "always amazed at how much clutter can still be profitably cut."[4]

Reading and Writing

Russell Baker, who won the Pulitzer Prize for his book *Growing Up,* was asked to write a piece about punctuation as part of a series on writing published by the International Paper Company. Baker began his piece by saying, "When you write, you make a sound in the reader's head. It can be a dull

> This carelessness can take any number of ~~different~~ forms. Perhaps a sentence is so excessively ~~long and~~ cluttered that the reader, hacking his way through ~~all~~ the verbiage, simply doesn't know what ^it ~~the writer~~ means. Perhaps a sentence has been so shoddily constructed that the reader could read it in any of ^several ~~two or three different~~ ways. ~~He thinks he knows what the writer is trying to say, but he's not sure.~~ Perhaps the writer has switched pronouns in mid-sentence, or ~~perhaps he~~ switched tenses, so the reader loses track of who is talking ~~to whom~~ or ~~exactly~~ when the action took place. Perhaps Sentence B is not a logical sequel to Sentence A -- the writer, in whose head the connection is ~~perfectly~~ clear, has not ^bothered to provide ~~given enough thought to providing~~ the missing link. Perhaps the writer has used an important word incorrectly by not taking the trouble to look it up ~~and make sure~~. He may think that "sanguine" and "sanguinary" mean the same thing, but ~~I can assure you that~~ the difference is a bloody big one ~~to the reader~~. ^The reader ~~He~~ can only ~~try to~~ infer ~~as to~~ (speaking of big differences) what the writer is trying to imply.

FIGURE 17-3 An Example of an Edited Manuscript

Source: Copyright © 1976, 1980, 1985, 1988, 1990, 1994, 1998 by William K. Zinsser. From *On Writing Well,* 6th ed., published by HarperCollins. Reprinted by permission of the author.

mumble—that's why so much government prose makes you sleepy—or it can be a joyful noise, a sly whisper, a throb of passion." He went on to speak of the importance of punctuation in letting your voice speak to the reader. "Punctuation," he wrote, "plays the role of body language. It helps readers hear you the way you want to be heard."[5]

How can you learn to master the rules of punctuation and the principles of writing? Read. The more you read, the better you write. The better you write, the better you can communicate. The better you communicate, the better you inform, inspire, and educate. To improve your writing skills, consider adding one or more of the following resources to your professional library:

- *On Writing Well* by W. Zinsser
- *The Elements of Style* by W. Strunk, Jr., and E. B. White
- *A Manual of Style* by The University of Chicago Press
- *The Elements of Grammar* by M. Shertzer
- *Fowler's Modern English Usage* by H. W. Fowler
- *A Writer's Reference* by D. Hacker
- *E-Writing: 21st Century Tools for Effective Communication* by D. Booher

Different Strokes for Different Folks

Not all writing assignments are the same. Some require the formal language of the scientific method, whereas others are meant to entertain (*and* inform). Choose a style, format, and tone of voice appropriate for the piece you are writing.

Writing for Professional Audiences

Materials written for professional groups must conform to a more rigorous, traditional format and style than those aimed at consumers. Scientific articles, for example, have specific subheadings and a formal tone (refer to the Professional Focus feature in Chapter 4). The best way to learn how to write these types of documents is to study published articles. Once again, editing is important. When possible, ask your colleagues to review your manuscripts. Their comments will help you identify places where your meaning is unclear.

Writing for the General Public

Writing for the general public is an important part of the community nutritionist's job. Newspaper and magazine articles, fact sheets, brochures, pamphlets, posters, and websites can all be used to teach and inform consumers. In addition to the basic writing principles outlined in the previous section, two things are important to bear in mind when writing for the general public:

1. The most important sentence in any piece of writing is the first one. If it doesn't engage readers and induce them to read further, your article or brochure is dead. Therefore, the lead sentence must capture the reader immediately. Consider this lead to an article about designer tomatoes: "Strap on your goggles, consumers, this one's getting messy."[6] The reader wonders instantly why tomatoes should be stirring up trouble. Or this lead from an article about dieting: "Dieters are a diverse group, but they share one common goal: to make their current diet their last one. Unfortunately, lasting results aren't what most get. In fact, the odds are overwhelming—9 to 1—that people who've lost weight will gain it back."[7] Anyone who has tried to lose weight will find this lead enticing. In addition to capturing the reader's attention, the lead must do some work. It must provide a few details that tell the reader why the article was written and why she or he should read it.
2. Know when to close. Choosing an end point is as important as choosing a lead. A closing sentence works well when it surprises readers or makes them think about the article's topic. An article about the challenge of change concluded with this comment, made by a man who participated in Dean Ornish's lifestyle program for high-risk heart disease patients: "What Ornish has given me is that opening and an understanding of what members of the group have said many times: A longer life may be important, but a better life is of the essence."[8] This closing remark works because it is personal and thought-provoking.

References

1. W. Zinsser, *On Writing Well* (New York: Harper & Row, 1988), p. 7.
2. Ibid., p. 8.
3. Ibid., p. 153.
4. Ibid., p. 10.
5. The quotation by Russell Baker was taken from an advertisement that appeared in *Discover,* April 1987, pp. 30–31.
6. B. Carey, Tasty tomatoes: Now there's a concept, *Health* 7 (1993): 24.
7. C. Simon, The triumphant dieter, *Psychology Today* (June 1989): 48.
8. G. Leonard, A change of heart, *In Health* 5 (1992): 51.

Chapter *Eighteen*

Marketing Nutrition and Health Promotion

Learning *Objectives*

After you have read and studied this chapter, you will be able to:

- Develop a marketing plan.
- Conduct a situational analysis.
- Describe how to apply the four P's of marketing to the development of a marketing strategy.
- Explain how social marketing is used to promote community interventions.

This chapter addresses such issues as needs assessment, marketing theories and techniques, applying marketing principles, negotiation techniques, and current information technologies, which are Commission on Accreditation for Dietetics Education (CADE) *Foundation Knowledge and Skills* requirements for dietetics education.

Chapter *Outline*

Something To Think About...

I can give you a six-word formula for success: "Think things through—then follow through."

– Edward Rickenbacker

Introduction

On her way home from work, a New York woman hears a familiar jingle on the car radio reminding her which soap to use for beautiful skin. Halfway around the world, in a remote Sri Lankan village, another woman hears a message on the community radio that teaches her about oral rehydration therapy for her child's diarrhea. Both of these women are part of a target audience for a well-planned marketing campaign. But whereas the New York woman is listening to traditional Madison Avenue marketing, the woman in Sri Lanka is a "consumer" receiving a message grounded in social marketing.[1] The same basic principles underlie both the commercial and the social approaches to marketing. The aim of both approaches is to strengthen the fit between the products, services, and programs offered and the needs of the population. As you will see, marketing is for everyone, regardless of job description. Whether you are a dietitian in private practice seeking referrals from physicians for new clients, a public health nutritionist developing nutrition education materials for pregnant teens, or a community nutritionist coordinating citywide screenings for hypertension, you can use marketing strategies. This chapter provides an overview of basic marketing principles and strategies.

What Is Marketing?

Buying and selling have a long history, but comprehensive and systematic marketing research evolved only in recent decades.[2] Peter Drucker, a management consultant, is credited with demonstrating the benefits of marketing to business. In the 1950s, he suggested that the primary focus of any business should be the consumer, not the product.[3]

Most people think that marketing means selling and promotion. Perhaps that is not too surprising, given that every day someone is trying to sell us something. Selling and promotion are only part of marketing. What, then, is marketing? **Marketing** is the process by which individuals and groups get what they need and want by creating and exchanging products and values with others.[4] Informally, marketing is the process of finding and keeping customers.[5] Many companies are very successful at commercial marketing, which employs powerful techniques for selecting, producing, distributing, promoting, and selling an enormous array of goods and services to a wide variety of people in every possible political, social, and economic context.[6]

Marketing The process by which individuals and groups get what they need and want by creating and exchanging products and values with others.

Social marketing A method for changing consumer behavior; the design, implementation, and management of programs that seek to increase the acceptability of a social idea or practice among a target group.

Social marketing draws on many of the techniques and technologies of commercial marketing, but it seeks to increase the acceptability of an idea, a practice, a product, or all three among a certain group of people—the target population.[7] Social marketing, described in more detail later in this chapter, is a strategy for changing consumer behavior. It combines the best elements of consumer behavior theory with marketing tools and skills to help consumers change their beliefs, attitudes, values, actions, or behaviors.[8] Social marketing implies that individuals do new things or give up old things in exchange for benefits they hope to receive. Social marketing promotes ideas and behaviors as "products." In the nutrition arena, a social practice to be marketed, for example, is summed up in each of the following slogans: "Got Milk?" "Eat Fresh, Buy Local" and "5 a Day for Better Health." A social idea to be marketed is the theme "Low-fat foods taste good."

Marketing, whether commercial or social, is a tool for managing change. It helps companies, government agencies, nonprofit organizations, dietitians in private practice, and others recognize, define, interpret, and cope with changing consumer values, interests, lifestyles, and purchasing behaviors.[9] The purpose of marketing is to find a problem, need, or want (through marketing research) and to fashion a solution to it. The solution to the problem, need, or want is outlined in the marketing plan, which is described in the next section.

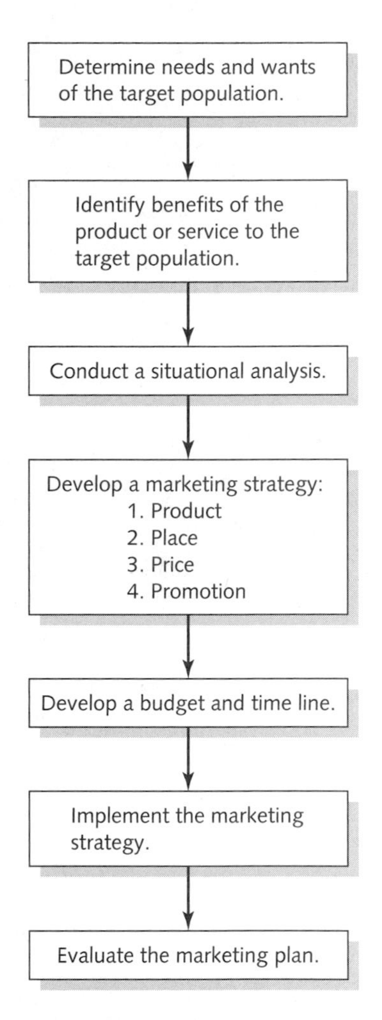

FIGURE 18-1 A Marketing Plan for Health Promotion

Source: Adapted from S. C. Parks and D. L. Moody, A marketing model: Applications for dietetic professionals, *Journal of the American Dietetic Association* 86 (1986): 40; and J. C. Levinson and S. Godin, *The Guerrilla Marketing Handbook* (Boston: Houghton Mifflin, 1994), pp. 5–7.

Develop a Marketing Plan

When completed, the marketing plan will outline the steps for achieving the goals and objectives of the overall intervention strategy and the program plan. It describes precisely how and in what form the nutrition and health messages will be delivered to the target population. Figure 18-1 shows the steps the community nutritionist takes in developing a marketing plan.

First, determine the needs and wants of the target population, because marketing starts with the customer.[10] Some ideas about their needs and wants can be gleaned from the community needs assessment and from focus group sessions held earlier in the program planning process. Additional information can be collected by asking questions of the target population: What are your perceptions about this nutritional problem or need? About this product or service? About this agency or organization? What products or services are you buying? What health benefits do you desire? The goal of this step is to build a knowledge base from which to develop a marketing strategy.[11]

Second, specify the benefits of the product or service to the target population. Remember, people generally want intangible things when they buy a product or service: safety, security, happiness, attractiveness, fun. Women who sign up for the "Heartworks for Women" program (Case Study 1 on page 540) may seek benefits such as reducing their risk of having a heart attack or stroke, learning new cooking skills, enjoying new recipes, and trying something new with a friend. It's the benefit that sells the product or service.[12]

Third, conduct a situational analysis. Analyze your potential **market** (those customers who share some common life characteristics), the environment in which your product or service will be positioned, and the competition.[13] Select a **target market,** which will be the primary, distinct customer group for your product, program, or service. In this step, you may be required to split your target population into smaller groups, each of which will respond differently to a given marketing strategy. For example, the target population for Case Study 2, which consists of independent elderly people over the age of 75 years, may be very diverse, some having high incomes and personal computers with Internet access and others living near the poverty threshold. A marketing strategy that includes the Internet may reach the former but will probably not reach the latter.

Next, develop a marketing strategy for ensuring a good fit between the goals and resources of the organization and the needs and wants of the target population.[14] The marketing strategy specifies a target market and four distinct elements traditionally known as the four P's: product, place, price, and promotion.[15] For example, every time a consumer buys a box of Shredded Wheat rather than a box of Cheerios, he or she makes the purchase at a competitive price in a grocery store, possibly because of a television advertisement. Such a sale was the result of a marketing strategy devised months before

Market Potential customers for a product or service; a group of unique customers who share some characteristics.

Target market One particular market segment pinpointed as a primary customer group.

by Kraft Foods regarding the issues of product, place, price, and promotion. (These elements are described in a later section.) This phase requires setting goals and objectives to indicate what the marketing strategy is expected to accomplish.

Before the marketing plan can move forward, a budget and timetable must be developed. The budget accounts for all expenditures related to implementing the marketing strategy, such as the cost of designing logos, Web pages, action figures, brochures, videos, and so forth, and printing all educational and promotional materials. The timetable specifies the marketing activities to be done each month both before the launch of the product, program, or service and after the launch, when the goal is to keep awareness high among the target market.

After all aspects of the marketing plan have been decided, implement the plan according to the original design and then evaluate its effectiveness. Did the marketing strategy reach the right audience at the right time with the right messages? Did the target market's beliefs, attitudes, actions, or behavior change? Use the results of the evaluation to make alterations to the marketing strategy and improve the positioning of the product or service in the marketplace.

CONDUCT A SITUATIONAL ANALYSIS

A situational analysis is a detailed assessment of the environment, including an evaluation of the consumer, the competition, and any other factors that may affect the program or business.[16] This step, which is critical to the ultimate success of the entire marketing plan, is sometimes referred to as a **SWOT analysis.** Conducting such an analysis (SWOT is an acronym for **S**trengths, **W**eaknesses, **O**pportunities, and **T**hreats) entails describing the present state of the business or agency, including its strengths and weaknesses, programs, and services, as well as the threats and opportunities present in the external environment (competition, pending legislation, and the like).[17]

Getting to Know Your Market

The first step in the marketing process is the identification and analysis of all "consumers" of your product or service—your current and potential markets. Consumers can be categorized as one of three types: users of services, referral sources, and other decision makers.[18] The users of services are the clients themselves or potential clients. For example, the Special Supplemental Nutrition Program for Women, Infants, and Children (WIC) program identifies its users as low-income pregnant or breastfeeding women and mothers of children under 5 years of age. The National Dairy Council targets a different market—"leader groups," such as educators, dietitians, health teachers, and dental hygienists.[19] Users of the "Heartworks for Women" program are women aged 18 years and older.

Referral sources include anyone who refers clients or customers to you. They may include physicians, social workers, teachers, and former clients. Other decision makers are those who influence the client's decision to use a service or join a program. Such people include family members (spouses, parents) and third-party payers, among others. It is important to identify these three types of users for your particular setting. Table 18-1 illustrates how consumers of the "Heartworks for Women" program are classified.

Market Research: Target Markets

An inherent part of the situational analysis is to determine and target your clients or audiences. Each target market should be viewed as a separate and different audience. For example, one survey of dietitians in private practice found they listed the following as their target populations.[20]

- Overweight women between 20 and 40 years of age

SWOT An acronym that stands for strengths, weaknesses, opportunities, and threats. A situational analysis technique often used in market research.

TABLE 18-1 Worksheet to Identify Consumers: Case Study 1

TYPE	CONSUMERS	CHARACTERISTICS	BENEFITS
Users	Women living in the city of Jeffers	Age 18+ years	Improved quality of life; decreased risk of CHD; better health
Referral sources	Cardiologists Other physicians Social workers Former clients	Health care providers, coworkers	Delayed CHD development
Decision makers	Spouse/significant other, coworkers	Age 18+ years	Want spouse, significant other, or coworker to be "healthy"

Source: Adapted from C. B. Matthews, Marketing your services: Strategies that work, *ASHA Magazine* 30 (1988): 23.

- Athletic men
- Middle-class women for weight reduction
- People interested in sports nutrition
- People seeking healthful lifestyles
- Individuals wanting basic nutrition information
- Physicians for referrals
- Corporations to provide employee wellness workshops

Ideally, if resources are available, you should develop a specific marketing strategy for each target audience. Once a population group has been identified for targeting, it is important to determine its prevailing patterns of lifestyle, eating, drinking, working conditions, attitudes toward nutrition and health, and current and past state of health.[21]

Most programs and organizations find it unrealistic to serve the total target market effectively. For this reason, actual and potential markets should be divided further into distinct and homogeneous subgroups—a process called **market segmentation.** Market segmentation offers the following benefits:[22]

- A more precise definition of consumer needs and behavior patterns
- Improved identification of ways to provide services to population groups
- More efficient utilization of nutrition and health education resources through a better fit among products, programs, services, and consumers

As an example of market segmentation, consider as your potential market the population of adults aged 45 years or older—sometimes referred to as 45+ consumers. This total market can be divided into several more homogeneous parts, some of which are shown in Figure 18-2. Each of these segments can then be reached with a distinct marketing strategy (to be discussed shortly).

Market Research: Market Segmentation

Market research enables community nutritionists to target specific groups for health promotion and disease prevention in terms of their geography, demography, and psychography. This type of analysis is helpful in many ways: (1) targeting those at risk, (2) carrying out strategic marketing planning, (3) developing marketing media strategies, (4) examining the feasibility of various promotional tools (for example, direct mailing of nutrition education materials), and (5) determining the appropriate mix of nutrition programs and services to offer on the basis of demographics (such as concentration of women, infants, children, and the elderly).

Market segmentation The separation of large groups of potential clients into smaller, distinct groups with similar characteristics and/or needs. Advantages include simpler, more accurate analysis of each group's needs and more customized delivery of service.

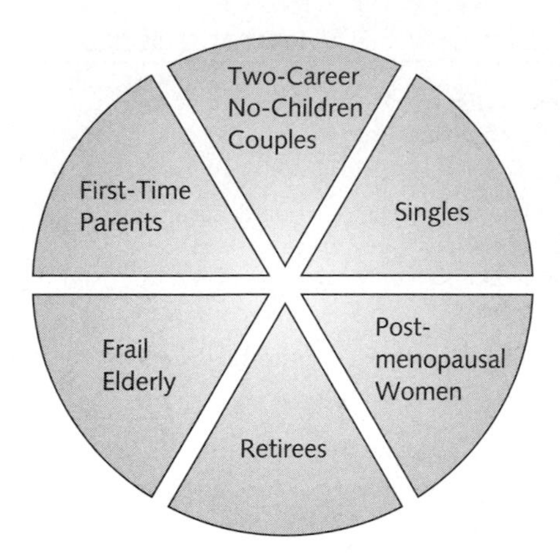

FIGURE 18-2 Market Segmentation: The 45+ Consumers

Four classes of variables are typically used for market segmentation:[23]

1. *Geographical segmentation* is the grouping of people according to the location of their residence or work (region, county, census tract). This can be done on a simple geographical basis or according to other variables, such as population density or climate.
2. *Demographic segmentation* is the grouping of individuals on the basis of such variables as age, sex, income, occupation, education, family size, religion, race, marital status, and life cycle stage.
3. *Psychographic segmentation* is based on such criteria as personal values, attitudes, opinions, personality, behavior, lifestyle, and level of readiness for change.
4. *Behavioristic segmentation* is based on such criteria as purchase frequency and occasion, benefits sought, and attitude toward product.

An understanding of demographics is essential to the development and targeting of nutrition and public health programs. Consider the public health agency or wellness center with programs that promote such "products" as smoking cessation, heart-healthy diets, fitness programs, maternal and infant care, cancer prevention, hypertension screening, diabetes education, infant mortality reduction, and prevention of AIDS.[24] The community agency or wellness center must first analyze the demographics of the areas served by these programs in order to identify its clients. The information to categorize and examine includes

- Total population of the area that the program is intended to serve
- Rate of change of the population
- Age and sex distribution
- Racial, ethnic, and religious composition
- Socioeconomic status
- Housing information
- Fertility patterns

Of these, age distribution of the population, trends over time, and fertility rates are particularly important to public health nutrition programs. Major segmentation variables for consumer markets are given in Table 18-2. Figure 18-3 summarizes the steps required for market segmentation and target marketing.

Obviously, the situational analysis demands a significant amount of market research. This research includes the use of both primary (direct) and secondary (indirect) data. Primary data are new data collected for the first time through random-sampling surveys,

TABLE 18-2 Major Segmentation Variables for Consumer Markets

VARIABLE	TYPICAL BREAKDOWNS
Geographical	
Region	Pacific, Mountain, West North Central, West South Central, East North Central, East South Central, South Atlantic, Middle Atlantic, New England
County size	A, B, C, D
City size	Under 5,000, 5,000–20,000, 20,000–50,000, 50,000–100,000, 100,000–250,000, 250,000–500,000, 500,000–1,000,000, 1,000,000–4,000,000, over 4,000,000
Density	Urban, suburban, rural
Climate	Northern, southern
Demographic	
Age	Under 6, 6–11, 12–19, 20–34, 35–49, 50–64, 65+
Gender	Male, female
Family size	1–2, 3–4, 5+
Family life cycle	Young, single; young, married, no children; young, married, youngest child under 6; young, married, youngest child 6 or over; older, married, with children; older, married, no children under 18; older, single; other
Income	Under $2,500, $2,500–$5,000,$5,000–$7,500, $7,500–$10,000, $10,000–$15,000, $15,000–$20,000, $20,000–$30,000, $30,000–$50,000, over $50,000
Occupation	Professional and technical; managers, officials, and proprietors; clerical, sales; artisans, forepersons; operatives; farmers; retired; students; homemakers; unemployed
Education	Grade school or less, some high school, graduated from high school, some college, graduated from college
Religion	Catholic, Protestant, Jewish, Hindu, Muslim, other
Race	Asian, black, Hispanic, American Indian, white
Nationality	American, British, French, German, Italian, Japanese, Latin American, Middle Eastern, Scandinavian, etc.
Social class	Lower-lower, upper-lower, lower-middle, upper-middle, lower-upper, upper-upper
Psychographic	
Lifestyle	Straight, swinger, longhair, yuppie, conservative, liberal
Personality	Compulsive, gregarious, authoritarian, ambitious, leader, follower, independent, dependent
Generational identity	Older Americans, baby boomers, generation X, generation Y
Behavioristic	
Purchase occasion	Regular occasion, special occasion
Benefits sought	Quality, service, economy, convenience, health
User status	Nonuser, ex-user, potential user, first-time user, regular user
Usage rate	Light user, medium user, heavy user
Loyalty status	None, medium, strong, absolute
Readiness stage	Unaware, aware, informed, interested, desirous, intending to buy
Attitude toward product	Enthusiastic, positive, indifferent, negative, hostile

Source: Adapted from P. Kotler and A. Andreasen, *Strategic Marketing for Health Care Organizations* (Upper Saddle River, NJ: Prentice-Hall, 1995). Used with permission.

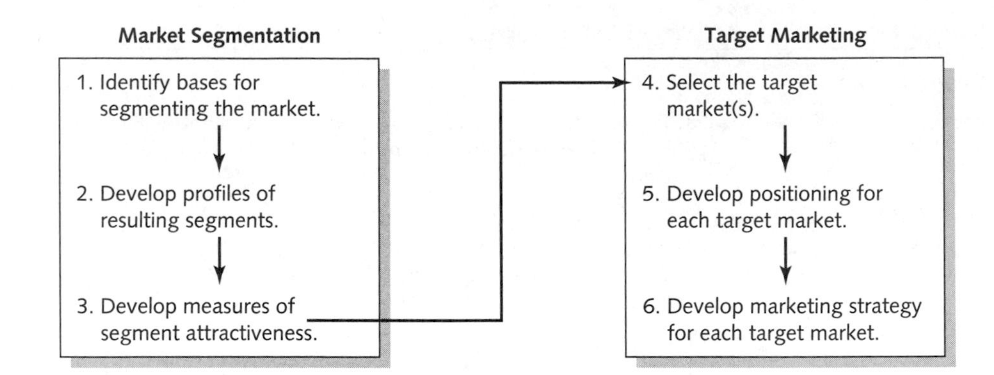

FIGURE 18-3 Steps in Market Segmentation and Target Marketing

Source: Adapted with permission from P. Kotler and R. N. Clarke, *Marketing for Health Care Organizations* (Upper Saddle River, NJ: Prentice-Hall, 1987), p. 234.

questionnaires, and qualitative methods such as personal interviews and focus groups. Table 18-3 describes six methods frequently used to collect primary data about a market.

Secondary data are those gathered by government agencies, private market research companies, and nonprofit organizations. Federal government sources of secondary data include the Bureau of Economic Analysis, Bureau of Justice Statistics, Bureau of Labor Statistics, Census Bureau, National Center for Health Statistics, National Technical Information Service, Social Security Administration, and U.S. Department of Agriculture. (Many of these were described in Chapter 2.) The *American Statistics Index, Statistical Abstract of the United States,* and *Survey of Current Business* are excellent sources of business and general economic statistics.

Private market research companies include A. C. Nielsen and America's Research Group, both of which conduct consumer behavior surveys. Some nonprofit groups, such

TABLE 18-3 Methods Used to Collect Primary Data

METHOD	DESCRIPTION
Mail survey	The most frequently used technique in market research; often misused. Tips to improve reliability: Use homogeneous sample, keep length reasonable, pre-test and rewrite as necessary.
Telephone survey	Can involve errors as in mail surveys. Helpful to ask, is the true meaning of the question being reflected by the interviewer, or is it distorted? Is the wording of the question likely to elicit a biased response?
Internet survey	Gathers data from people who have e-mail accounts or visit a particular website; can obtain highly specific information about people who use the Internet.
Personal interview	A recommended supplement to mail or telephone surveys; helpful in observing subtle feedback that would otherwise be unavailable.
Consumer panel	Often used to test new products using the same persons in several tests; a problem is that panel members do not always represent the buying population and may not always give honest answers.
Focus group	Frequently used to gather information from small homogeneous groups; allows the researcher to see how people view a product, intervention, or other issue; serves as a communications bridge between the researcher and the people the researcher is trying to reach; not to be used to persuade, convince, or reach a consensus.

Source: Adapted from S. C. Parks, Research techniques used to support marketing management decisions, in *Research: Successful Approaches,* 2nd ed., E. R. Monsen (Chicago: American Dietetic Association, 2003), pp. 364–81. © 2003 The American Dietetic Association. Reprinted with permission. Also, J. Sterne, *World Wide Web Marketing* (New York: Wiley, 1995).

as the American Dietetic Association and other professional associations, are sources of secondary data. Directories such as *Dun & Bradstreet, Moody's Manuals,* and *Thomas Register* provide data on businesses in a given market area. The *Sales Management Annual Survey of Buying Power* provides local information on population, income, and retail establishments. If you can't locate the secondary data you require, consider working with a specialist or information broker who can provide research services for a fee.[25]

A number of vendors also provide information about various segments of a population based on demographics, geography, and psychographics. Each vendor has a clustering system that groups individuals on the basis of like characteristics. This technique uses the statistical method called cluster analysis to classify neighborhoods by their residents' demographics, attitudes, media habits, and buying patterns.

ANALYZING THE ENVIRONMENT AND WATCHING FOR TRENDS

The next step in the situational analysis is to identify any external environmental factors or social trends that may influence the needs of the program or organization. Such issues include health care reform, legislative and regulatory changes, shifting demographics, and behavioral changes such as eating on the run and sedentary, indoor lifestyles. For example, a declining birth rate and a growing population of older people are affecting the composition of many communities. Changes in the typical family, including older first-time parents, the high divorce rate, and increasing numbers of working mothers, also affect the needs of a given community.

The general age of the target area is influential in determining needed programs. In an area largely inhabited by people of retirement age, classes on prenatal nutrition counseling will have far less impact than, for instance, a class on heart-healthy cooking for one or two. Other significant trends include the aging of the baby boom generation and the increasing cultural and ethnic diversity of the population, which must be considered when developing or expanding any public health program.

Analyzing the Competition

Once you have a thorough understanding of your consumers, you must determine how your existing competitors are positioned in the marketplace. What are their strengths and weaknesses? What is the attitude of your target market toward the competition? Table 18-4 shows the results of such an analysis completed by a community nutritionist who specializes in cardiovascular disease.

Your aim is to find a **market niche** for your program or service in which your strengths can be matched with the needs of your particular target market. To satisfy those

Market niche The particular area of service or the particular product suited to the specific clients to be reached. The underlying philosophy is that you cannot be all things to all people, so you must find the spot that fits your objectives and goals and enables you to meet a particular unmet need.

TABLE 18-4 Worksheet for Analyzing the Competition

COMPETITOR	STRENGTHS	WEAKNESSES	MY COMPETITIVE ADVANTAGE
Wellness center	Great location; personable staff	Large and diverse; no known specialty; little follow-up after program completion	I specialize in cardiovascular nutrition counseling; I provide 6 months of follow-up (phone calls, e-mail, e-newsletter).
Sports medicine clinic	Personable staff; good programming in sports nutrition; good name recognition	Outdated brochures and videos; nearly impossible to find parking	I maintain up-to-date educational resources; I travel to the client in my "Nutritionist-on-Wheels" mobile.

Source: Adapted from C. B. Matthews, Marketing your services: Strategies that work, *ASHA Magazine* 30 (1988): 23.

FIGURE 18-4 The Four P's of the Marketing Mix

Source: Adapted with permission from M. Ward, *Marketing Strategies: A Resource for Registered Dietitians* (Binghamton, NY: Niles & Phipps, 1984), p. 111.

Marketing Mix

Product
- Quality
- Features
- Style
- Packaging
- Services

Place
- Channels
- Coverage
- Locations
- Inventory
- Transport

Price
- List price
- Discounts
- Credit terms

Promotion
- Advertising
- Personal promotion
- Publicity
- Branding

Target Market

needs, you must know and understand your target audience so well that your service provides the perfect fit and "sells" itself, setting you apart from other providers in the same market and improving your **competitive edge.** Examples of competitive edges include your area of expertise, professional image, size, location, and customer service. You will want your target market to perceive that it can benefit from using your services rather than those of your competition.

Competitive edge An advantage over others who are in the business, gained through use of business strategies, market research, expert management, new product development, or other sound business techniques.

Develop a Marketing Strategy

The four elements of the marketing strategy—product, place, price, and promotion—are usually referred to as the **marketing mix.** The development of the appropriate marketing mix should result directly from the previous step of the marketing process—analysis of the consumer, environment, and competition. Once the needs of the target audience have been identified and analyzed, the set of four P's can be constructed. The primary focus of the four P's is the identified target market, as shown in Figure 18-4. Successful marketers get the right product, service, or program to the right place at the right time for the right price.[26]

Marketing mix Four universal elements of marketing that are often called the "four P's"—product, price, place, and promotion. Product encompasses the range of services offered; price encompasses the monetary and intangible value of the product; place is where the product is available; promotion is persuasive communication aimed at targeted users.

PRODUCT

The term **product** refers to all the characteristics of the product or service that are to be exchanged with the target market. Characteristics such as style, special features, packaging, quality, brand names, and options must be designed to fit the needs and preferences of the target market. From a marketing standpoint, the product or service—whether it's a new automobile, a diet soft drink, or a nutrition class for a congregate meal site—is viewed as a collection of tangible and intangible attributes that may be offered to a market to satisfy a want or need.[27]

Product Anything offered in the marketplace to be exchanged for money or something of value. It may be either a tangible good or a service.

In community nutrition, the product is often a service to be delivered. These services should be of high quality, tailored to fit the needs of the target market, and adapted to be congruent with the consumers' social characteristics (for example, culture, ethnicity, or language skills). A group of dietitians in private practice delineated their services as including individual counseling for modified diets, weight reduction, sports nutrition, normal nutrition, prenatal diets, and eating disorders; group programs and workshops on nutrition topics; consulting services to schools, health care facilities, supermarkets, health spas, and restaurants; and the teaching of courses at community colleges and universities.[28]

PLACE

Place is the actual location where the exchange takes place. Accessibility, convenience, and comfort for the client are the criteria to consider. Are your hours of operation convenient and flexible? Is parking available? Place also includes the channels of distribution required to deliver the service or product to the consumer. The distribution channels are the intermediaries—individuals, facilities, or agencies—that control or influence the consumer's choice of service or product.[29] Health providers, employers, school boards, voluntary health organizations, shopping malls, and commercial retailers are a few of the important intermediaries for community nutritionists. They are viewed as channels for reaching identified target markets.

In other instances, distribution channels are more like "gatekeepers"—you cannot reach your target except by going through them. Physicians, parents, media program directors, corporate executives, members of Congress, and insurance company decision makers can all be gatekeepers, depending on the service being offered and the target audience. The overall marketing strategy may include one approach for the client and a different approach to reach the intermediary or gatekeeper.

Distribution channels vary depending on the target market and the service provided. Because third-party reimbursement by insurance companies for nutrition services usually requires a physician referral, a dietitian in private practice may want to target obstetricians, cardiologists, or other medical practice groups identified as distribution channels. Another approach the dietitian might take is to target insurance companies or state legislators in an effort to enhance the current third-party reimbursement system. A wellness program dietitian might identify corporate executives or employers as distribution channels, because approval for a wellness program usually rests with a company's management.

PRICE

Price includes both tangible costs (fee for service) and intangible commodities (time, effort, and inconvenience) that the consumer must bear in the marketing exchange. Once you understand what the consumer perceives as the costs involved in adopting a health behavior or participating in a given program, you have a better chance of influencing the exchange. You may do this by persuading the consumer that the benefits to be received outweigh the perceived costs. Alternatively, incentives (money, groceries, gifts, or personal recognition) can be offered to increase motivation and facilitate consumer participation. Likewise, costs can be reduced (less waiting time) or prices discounted to certain groups (senior citizens, students). Any of these tactics can considerably reduce the "price" consumers perceive themselves as paying for the program or service.

PROMOTION

The last P in the marketing mix—**promotion**—consists of the agency's or organization's informative or persuasive communication with the target market. What do you want to say? To whom do you want to say it? When do you want to say it?[30] The communication

TABLE 18-5 Promotional Tools

PERSONAL PROMOTION AIDS	MEDIA AND PUBLIC EVENTS	GRAPHIC/PRINT MATERIALS
One-to-one communication	Public relations	Logos
Networking	Publicity	Brochures
Business cards	News releases	Flyers
Letters	Press kits	Portfolios
Résumés	Media interviews	Proposals
Letters of reference	News conferences	Posters
Use of a name	Press briefings	Banners
Seminars	Special events	Audiovisual aids
Workshops	Celebrities	Computer graphics
Consulting	Advertising	Giveaways (T-shirts, mugs, tote bags, product samples)
Public speaking	Direct mail	Internet websites
Writing	Contests	Statement stuffers
	Trade shows	Catalogs
		Screen savers

Source: Adapted from K. K. Helm, *The Competitive Edge: Advanced Marketing Strategies for Dietetics Professionals,* 2nd ed. (Chicago: American Dietetic Association, 1995), p. 79. © The American Dietetic Association. Reprinted with permission. Also, J. Kremer and J. D. McComas, *High-Impact Marketing on a Low-Impact Budget* (Rocklin, CA: Prima, 1997), pp. 237–92.

messages are designed to have a measurable effect on the knowledge, attitude, and/or behavior of the target market.[31] The medium, message content, and message format are chosen to complement the target market's communication needs.[32]

Promotion has four general objectives:[33]

- To inform and educate consumers about the existence of a product or service and its capabilities (what the community program has to offer)
- To remind present and former users of the product's continuing existence (for example, prenatal nutrition counseling)
- To persuade prospective purchasers that the product is worth buying (improved health status, other benefits)
- To inform consumers about where and how to obtain and use the product (accessibility, location, and time)

Although people often assume that promotion is limited to advertising, promotion actually includes much more than advertising, as shown in Table 18-5. As one author has noted:

> Marketers generally agree that although mass media approaches are appropriate for developing consumer awareness in the short term, face-to-face programs such as workplace encounters are more effective (though not always cost-effective) in changing behavior in the long term. In designing any communication policy, the marketer commonly considers the effects that may be achievable through the use of a mix of approaches, capitalizing on the strengths of each.[34]

Advertising Any paid form of nonpersonal presentation and promotion of ideas, goods, or services by an identified sponsor.

Chapter 7's Professional Focus feature on page 233 offers tips for working with the media.

The four most common promotional tools are advertising, sales promotion, personal promotion, and public relations.

Advertising is standardized communication in print or electronic media that are purchased.[35] Examples of advertising media include telephone directory yellow pages, billboards, newspapers, radio, trade and professional journals, magazines, and the Internet. The role of advertising is to communicate a concise and targeted message that ultimately stimulates action by the carefully defined audience. Advertising can reach large numbers of people in many locations and can help build image. In advertising, you control the

TABLE 18-6 Advantages and Limitations of Major Media Categories

MEDIUM	ADVANTAGES	LIMITATIONS
Newspapers	Flexibility; timeliness; good local market coverage; broad acceptance; high believability	Short life; poor reproduction quality; small "pass-along" audience
Television	Combines sight, sound, and motion; appealing to the senses; high attention; high reach	High absolute cost; high clutter; fleeting exposure; less audience selectivity
Direct mail	Audience selectivity; flexibility; no ad competition within the same medium; personalization	Relatively high cost; "junk mail" image
Radio	Mass use; high geographical and demographic selectivity; low cost	Audio presentation only; lower attention than television; nonstandardized rate structures; fleeting exposure
Magazines	High geographical and demographic selectivity; credibility and prestige; high-quality reproduction; long life; good pass-along readership	Long ad purchase lead time; some waste circulation; no guarantee of position
Outdoor	Flexibility; high repeat exposure; low cost; low competition	No audience selectivity; creative limitations
Internet	Low cost; high selectivity; personal; is interactive one-on-one marketing	Not always statistically representative; can't verify user identity

Source: Adapted from P. Kotler and R. N. Clarke, *Marketing for Health Care Organizations* (Upper Saddle River, NJ: Prentice-Hall, 1987), p. 451; and J. Sterne, *World Wide Web Marketing* (New York: Wiley, 1995), pp. 239–68.

nature and timing of the message because you are buying the time or space. Which media you choose will depend on the characteristics of your target market, the size of your budget, and the goals of your advertising campaign. The advantages and limitations of the major media categories are listed in Table 18-6.

Sales promotion consists of such things as coupons, free samples, point-of-purchase materials, and trade catalogs. The use of these activities as part of the promotion strategy encourages potential consumers to purchase or use a particular product or service.

Personal promotion or communication can be done through small group meetings, counseling sessions and nutrition classes, formal presentations to organizations or community groups, displays and booths at health fairs and related conferences, and telephone conversations or personal meetings with the public and other professionals (such as your referral sources). Unlike the standardized message presented in advertising, the message presented in personal promotion can be tailored to fit the needs of the particular individual or group. Personal promotion also offers other advantages:[36]

- Direct contact provides positive feedback to the listener through both the verbal communications and the nonverbal gestures of the communicator.
- The interpersonal contact facilitates the transfer of the message better than other methods.
- The communicator can ensure comprehension by asking questions and monitoring responses.
- The communicator has an opportunity to probe for resistance to change and then is in a position to address each issue.
- The listener may believe that someone is now in a position to monitor his behavior and thus hold him accountable for it.

The Professional Focus at the end of Chapter 15 offers numerous tips for effective public speaking.

Sales promotion Short-term incentives to encourage purchases or sales of a product or service.

Personal promotion/ communication Oral presentation in a conversation with one or more prospective purchasers for the purpose of making sales or building goodwill.

Public relations An organized effort to promote a favorable image of a person or product through news coverage or goodwill (free services, charitable work, etc.).

Public relations, or publicity, is used to create a positive image of an individual or organization in the mind of the consumer. The American Dietetic Association receives publicity every March during National Nutrition Month activities. Personal branding is the process of "creating a world of meaning and relevancy for others to know what is genuinely unique about you." [37] See the accompanying box for examples of ADA members who have had success in branding their services.

Brand Name Dietetics

Nancy Clark, RD, sports nutritionist, is showcased on the Wheaties box advertising the Summer 2004 Olympics.

Nancy Clark, and Courtesy of Wheaties® and General Mills

For most grocery shoppers, choosing a store is a relatively straightforward process, with cost and convenience being paramount; however, once inside, shoppers have a vast array of choices to make between many similar products by different manufacturers. Cost can be the primary factor in deciding which product to purchase, but many shoppers make purchases based on the brands themselves. The same process is used when selecting a provider of health-related services, such as a community nutritionist. *Branding*—what a product or service stands for and is designed to do—is established and understood among consumers to help them make decisions as to what goods and services they will purchase. Much like the branding of consumer products, dietetics professionals must establish a strong **brand image,** defined by the American Marketing Association as "a mirror reflection of the brand personality or product being; it is what people believe about a brand: their thoughts, feelings, expectations."

Branding Works

Brand image A mirror reflection of the brand personality or product being marketed; it is what people believe about a brand: their thoughts, feelings, and expectations.

Several members of the American Dietetic Association have had success in branding their services, and many have taken different approaches in beginning the brand-creation process. Getting one's name out there is one way to establish a brand. Nancy Clark, MS, RD, sports nutritionist, and author of Nancy Clark's *Sports Nutrition Guidebook*, adds her name to all her products, including her books and website. Similarly, Becky Dorner, RD, president of Becky Dorner and Associates, says that her company uses its logo on all marketing pieces, as well as on its website, letterhead, note cards, envelopes, publication flyers, quarterly newsletter, consulting services brochure, ads, manuals, and books—and on the company's e-zine, which reaches more than 3,000 long-term-care health professionals across the United States. According to Dorner, an eye-catching logo and a memorable name are of the essence.

Other members have found that discovering a niche market is an effective means for brand creation. Sylvia E. Meléndez-Klinger, MS, RD, of Hispanic Foods Communication, discovered that the Hispanic market was virtually untapped. "There are so few dietitians with my food industry marketing experience," Meléndez-Klinger explains. "It helped to have a great education, be bilingual, and have a great knowledge of many of the Hispanic cultures, having lived in Mexico, Puerto Rico, Spain, and Central America." Cathy Leman, RD, owner of NutriFit, found her combination of being a personal trainer and a dietitian, as well as marketing specifically to women, to be an effective means of differentiating herself among dietitians and personal trainers.

Accepting speaking engagements is also recommended as a means of establishing oneself as a brand. "My name has become associated with reliable, helpful sports nutrition information. This helps me to get asked to be a speaker at professional and lay organizations, as well as an author of articles for magazines," says Clark. Similarly, Meléndez-Klinger rarely turns down an opportunity to speak in front of an audience, but she also recommends networking, becoming a media spokesperson, and volunteering as additional ways to become known.

There are many branding and marketing communications specialists—several focused specifically on health care services—that can help you create your brand and market your services. However, even without using such services, with the right combination of patience, enthusiasm, and savvy, success is within your reach.

Source: Excerpted from K. Stein, Brand name dietetics, *Journal of the American Dietetic Association* 104 (2004): 1530–33.

Publicity Tools

Publicity tools include articles in newsletters or local newspapers, informational brochures and newsletters, radio and television interviews, other forms of public speaking, displays, posters, audiovisual materials, thank-you notes, Internet websites, and other tools that present a favorable image of the organization or professional to the target market.

Public service announcements (PSAs) can also be used as a form of publicity. They offer the advantage of being free of charge and have the potential of reaching a large audience. PSAs are brief messages—often only 30 to 60 seconds in length—used to promote programs, activities, and services of federal, state, or local governments and the activities of nonprofit organizations when they are regarded as serving a community interest. A nutrition message can be a PSA with credit given to the organization submitting it. Here are five tips for creating effective PSAs:[38]

PSA Research offers an on-line information library of public service advertising with access to reviews of PSA campaigns to affect health behaviors; go to www.psaresearch.com.

- Produce announcements of professional quality. One professional PSA is better than several low-quality ones.
- Get the audience involved. Sound effects, questions, repetition, and humor are sometimes more effective at grabbing the attention of the audience than the factual approach.
- Market the service offered, not just the PSA topic. Offer a brochure, a toll-free information hotline, or a screening service.
- Simplify the response action needed. Advertise a local phone number or an easy-to-remember Post Office box number or Web address.
- Develop and nurture a good rapport with the television and radio public service directors. They may be able to improve the PSA's production quality and increase scheduled air time.

Two additional promotional tools deserve mention: direct mail and word-of-mouth referrals. Many marketers develop brochures, newsletters, fliers, and other promotional items that are mailed to specific targeted groups or geographical areas. Others claim that the word-of-mouth referral—having associates and clients do some of the promoting for you—is one of the most important promotional strategies. A successful program or service will generate its own word-of-mouth publicity, because satisfied consumers will tell others about the program and may encourage their participation. The promotional strategies used by individuals or organizations will vary with the populations they are trying to reach, the goals of the program, and the resources available for promoting it.

Monitor and Evaluate

Evaluation is the key to the success of any marketing program. Evaluation methods include tracking changes in volume or net profit, referral sources, and customer satisfaction.[39] In this step of the marketing process, you need to ask the following questions: Are you accomplishing your goals? Who benefited from the service? What changes in knowledge, attitudes, and practices occurred? What are the actual costs of providing the service? What changes are warranted to make the service more effective in the future? For example, once you understand the profile of clients who were successful with a given program, your promotional efforts can be targeted more effectively.

If your marketing plan included a thorough situational analysis and you carefully translated these results into a marketing mix for your targeted audience, a periodic assessment may be all that is needed. Marketing is an ongoing process, however, and situations change—sometimes affecting your marketing strategy. If so, you may need to reevaluate your objectives and goals. Are they still achievable? Do you need to take an alternative direction? Should you redirect your strategy?

In *Guerilla Marketing,* J. C. Levinson cautions against abandoning your marketing plan once it begins generating favorable results. Instead, he advises an ongoing commitment to your marketing orientation and continued investment in your marketing strategy. He offers "ten truths you must never forget":[40]

1. The market is constantly changing.
2. People forget fast.
3. Your competition isn't quitting.
4. Marketing strengthens your identity.
5. Marketing is essential to survival and growth.
6. Marketing enables you to hold on to your old customers.
7. Marketing maintains morale.
8. Marketing gives you an advantage over competitors who have ceased marketing.
9. Marketing allows your business to continue operating.
10. You have invested money that you stand to lose.

Social Marketing: Community Campaigns for Change

Social marketing makes a comprehensive effort to influence the acceptability of social ideas in a population, usually for the purpose of changing behavior.[41] Health ideas such as cardiovascular fitness, heart-healthy eating, and hypertension screening can be "sold" in the same way as presidential candidates or toothpaste.[42] Examples of social marketing include the public service messages produced by electronic and print media, such as messages intended to change behavior related to smoking, hypertension, use of seat belts, teenage pregnancy, driving after drinking, drug use, safe sex, suicide, and similar concerns.

Whereas traditional marketing seeks to satisfy the needs and wants of targeted consumers, social marketing aims to change their attitudes and/or behavior. To do so, the marketing process must be followed in its entirety.[43] Marketers identify four types of behavior change, listed here in order of increasing difficulty:

- *Cognitive change.* A change in knowledge is the easiest to market, but there appears to be little connection between knowledge change and behavior change. Examples include campaigns to explain the nutritional value of different foods and campaigns to expand awareness of government programs such as the Food Stamp Program or WIC.
- *Action change.* This kind of change is more difficult to achieve than cognitive change, because the individual must first understand the reason for change and then invest something of value (time, money, or energy) to make the change. Screening programs for hypertension, hypercholesterolemia, or breast cancer involve this type of change.
- *Behavior change.* This type of change is more difficult to achieve than either cognitive change or action change because it costs the consumer more in terms of personal involvement on a *continuing* basis—for example, the adoption of a low-fat, high-fiber daily diet or the addition of regular physical exercise to one's lifestyle.
- *Value change.* This type of change is the most difficult to market. An example is population control strategies to persuade families to have fewer children.

Social marketing goes beyond advertising. It seeks to bring about changes in the behavior of the target audience as well as in its attitudes and knowledge.[44] Social marketing can be applied to a wide variety of social problems but is particularly appropriate in three situations:[45]

1. When new research data and information on practices need to be disseminated to improve people's lives. Examples include campaigns for cancer prevention, breastfeeding promotion, and childhood immunization.
2. When countermarketing is needed to offset the negative effects of a practice or the promotional efforts of companies for products that are potentially harmful—for example, cigarettes, alcoholic beverages, or (in developing countries) infant formula.
3. When activation is needed to move people from intention to action—for example, motivating people to lose weight, exercise, or floss their teeth. Without movement, there is no marketing.[46]

Social Marketing at the Community Level

The Pawtucket Heart Health Program (PHHP) is presented here as an example of how marketing principles can be used in the planning, implementation, and evaluation of a community-wide social marketing campaign. The Pawtucket "Know Your Cholesterol" campaign was one of the earliest cholesterol awareness and screening efforts.[47] The following objectives for the campaign were formulated on the basis of national random sampling data for both the general population and physicians.

- Increased physician education on blood cholesterol and heart disease
- Increased awareness among the general population of blood cholesterol as a risk factor for coronary heart disease
- Increased numbers of people knowing their cholesterol level as a result of attending PHHP-sponsored screening, counseling, and referral events (SCOREs)
- Large numbers of people showing reductions in their blood cholesterol level at two-month follow-up measurements

In segmenting the community of Pawtucket, Rhode Island, adults were a primary focus because demographics (gender and age) showed that awareness levels were equivalent for

Programs in Action

Motivating Children to Change Their Eating and Activity Habits

Many children do not eat a healthful daily diet or get the exercise they need to combat nutrition-related chronic diseases and to promote lifelong health. Children eat less than the recommended number of daily servings of fruits and vegetables.[1] A majority of children in this country eat too much fat, especially saturated fat.[2] Only about half of young people regularly participate in vigorous physical activity.[3] One in four children in the United States is overweight.[4] Children do not always have the chance to reap the benefits of a lifestyle marked by a good diet and adequate physical activity. Children spend more time watching television in a year than they spend in school,[5] many school systems are limiting or eliminating physical education,[6] and food industry advertising may influence children to choose foods high in fat and/or sugar.[7,8]

As we noted earlier in the chapter, social marketing seeks to bring about changes in the behavior of a target audience, as well as in its attitudes and knowledge. Social marketing is highly applicable when countermarketing is needed to offset the negative effects of a practice (for example, lack of physical activity) or the advertising campaigns of companies for certain products (for example, high-fat or high-sugar snacks for children). "Eat Well & Keep Moving" applied the principles of social marketing in the development of a program to improve the nutrition status and health of children.

Goals and Objectives

The goals of the research phase of Eat Well & Keep Moving were to decrease students' consumption of total and saturated fat, to increase their intake of fruits and vegetables, to reduce television viewing, and to increase moderate and vigorous physical activity.

Target Audience

The target audience was 479 fourth- and fifth-grade students in six intervention and eight matched control elementary schools in Baltimore. Eighty-five percent of participating students received free/reduced-price school lunch; over 90 percent were African American.

Rationale for the Intervention

Unlike traditional health curricula, Eat Well & Keep Moving was a multifaceted program encompassing all aspects of the learning environment from the classroom, cafeteria, and gymnasium to the school hallways, the home, and even community centers. This varied approach helped reinforce important messages about nutrition and physical activity, and it increased the chance that students would eat well and be physically active throughout their lives.

Methodology

The research phase of Eat Well & Keep Moving was conducted in collaboration with the Baltimore City Department of Education. It was taught by classroom teachers over the course of 2 years, beginning in the fall of 1995. Using the interdisciplinary approach, the curriculum was integrated into the core subjects of math, science, language arts, and social studies. Thirteen lessons on nutrition and health-related fitness concepts were taught each year. Four of the classroom lessons involved children practicing a "safe workout" routine while learning concepts related to nutrition and physical activity. To further integrate nutrition and physical activity, five supplementary physical education lessons were taught, using nutrition and food as the themes of the activities. Modules that were developed as extensions of classroom lessons provided opportunities for

men and women in the national sample. Cardiologists, family practice physicians, and physicians in general practice were targeted for direct-mail educational packages and grand rounds presentations on blood cholesterol and heart disease at the community hospital. Middle-aged men who had had previous contact with the PHHP were also the focus of a direct-mail and telemarketing campaign to attend SCOREs during the campaign period.

Early program development steps included a pilot test of a self-help "nutrition kit" on lowering cholesterol levels and pre-testing of the SCORE protocol at the community hospital. Promotional tools selected to reach the general and targeted audiences included newspapers; print media distributed through worksites, churches, and schools; direct mail and telemarketing; and SCORE delivery at worksites, churches, and various other community locations. In order to avoid spillover into the control community, television and radio were not used.

students to participate in activities related to program goals. These included *Freeze My TV* to reduce television viewing, *3 At School* and *Five a Day* to promote consumption of fruits and vegetables, and *Walking Clubs* to promote physical activity and fitness. Educational materials established links to school food service, using the cafeteria as a learning laboratory for nutrition. To reinforce concepts at home, families received nutrition and fitness information through newsletters and other vehicles. Teachers were motivated through a wellness session that was part of their teacher training.

Results

Fourteen Baltimore elementary schools successfully participated in the 4-year demonstration program. Students rated the lessons and activities highly, and 100 percent of responding teachers said that they would utilize the program again. Diet was evaluated with 24-hour recall measures. Longitudinal data collected from 479 students demonstrated significant decreases in percent of total calories from fat and saturated fat, a significant increase in fruit and vegetable consumption, and a marginal reduction in television viewing. Student knowledge on nutrition and healthful activity also increased significantly.

Since September 1997, schools involved in the study and the Baltimore Department of Education have sustained Eat Well & Keep Moving on their own. To date, personnel in 65 Baltimore elementary schools have been trained, and 40 schools have been implementing the program.

Lessons Learned

For a program to be successful, those who develop its components must consider the school constituents' needs, constraints, and motivations. Obtain inputs not only from teachers, food service staff, and principals, but also from students and parents. Keeping the program inexpensive to implement helps, too.

References

1. S. M. Krebs-Smith and coauthors, Fruit and vegetable intakes of children and adolescents in the United Sates, *Archives of Pediatrics and Adolescent Medicine* 150 (1996): 81–86.
2. E. Kennedy and J. Goldberg, What are American children eating? Implications for public policy, *Nutrition Reviews* 53 (1995): 111–26.
3. National Center for Chronic Disease Prevention and Health Promotion, *Physical Activity and Health: A Report of the Surgeon General* (Atlanta, GA: Centers for Disease Control and Prevention, 1996).
4. R. P. Troiano and coauthors, Overweight prevalence and trends for children and adolescents: The National Health and Nutrition Examination Surveys, 1963–1991, *Archives of Pediatrics and Adolescent Medicine* 149 (1995): 1085–91.
5. W. H. Dietz and S. L. Gortmaker, Do we fatten our children at the TV set? Obesity and television viewing in children and adolescents, *Pediatrics* 75 (1985): 807–12.
6. U.S. Department of Health and Human Services, Public Health Service, *Healthy People 2010* (Washington, D.C.: U.S. Government Printing Office, 2000).
7. H. L. Taras and coauthors, Television's influence on children's diet and physical activity, *Journal of Developmental Behavior in Pediatrics* 10 (1989): 176–80.
8. T. N. Robinson and J. D. Killen, Ethnic and gender differences in the relationships between television viewing and obesity, physical activity, and dietary fat intake, *Journal of Health Education* 26 (1995): S91–S98.

Source: Adapted from *Community Nutritionary* (White Plains, NY: Dannon Institute, Spring 2001). Used with permission. For more information about the *Awards for Excellence in Community Nutrition,* go to www.dannon-institute.org.

In the marketing mix, SCOREs were initially priced at $5 per person for both an initial and a follow-up measurement. The researchers reasoned that people who had already paid for a second measurement would be more likely to have a follow-up test than if they had to pay for it separately. Price reductions and specials were also offered. Promotional publicity strategies included the "kick-off" SCORE at a St. Patrick's Day parade and six weekly advice columns in the local newspaper.

Results showed that 1,439 adults attended 39 SCOREs. Sixty percent were identified as having elevated serum cholesterol levels. Two months after the campaign, 72.3 percent of these persons had returned for a second measurement. Nearly 60 percent of this group had reduced their serum cholesterol levels.

The essential components of the campaign's marketing strategy were integrated into the ongoing activities of the PHHP. During the first 2 years following the campaign, over

Think about novel ways to reach your target population. Here, an eye-catching city bus promotes the "Got Milk?" message.

Terry Heffernan

10,000 persons had their blood cholesterol level measured, were given information on dietary management of high serum cholesterol, and were referred to physicians when necessary. A later survey of local physicians showed that they were more aggressive in initiating either diet or drug therapy than physicians in the neighboring community or those who had participated in the national survey. The local physicians cited, as the major reason for changing their practice, increased patient requests for blood cholesterol measurements and/or nutrition information. The researchers concluded that informed consumers had influenced changes in their physicians' treatment of high serum cholesterol levels. Their overall conclusion was that "a well-functioning marketing operation can lead to more effective and efficient use of resources and improved consumer satisfaction. . . . **Health marketing** has the potential of reaching the largest possible group of people at the least cost with the most effective, consumer-satisfying program."[48] Check this chapter's Internet Resources related to social marketing on page 571.

A Marketing Plan for "Heartworks for Women": Case Study 1

Health marketing Health promotion programs that are developed to satisfy consumer needs, are strategically planned to reach as broad an audience as is in need of the program, and thereby enhance the organization's ability to effect population-wide changes in targeted risk behaviors.

Team 2—the team responsible for the nutrition intervention activities in the "Heartworks for Women" program—reviewed the results of the community needs assessment (see Table 14-1 on page 454), the goals and objectives for the intervention strategy (see page 456), and the program goals (see page 538). The following needs and wants of the target population were identified in the community needs assessment and in focus group discussions: want to stop smoking (one participant's view: "I want to get this monkey off my back"); want to feel better; want to look better; don't want to have a heart attack (one participant remarked, "My mother died of a heart attack at the age of 44. She was shopping at the mall and just went like that. I don't want that to happen to me"). Some women wanted better health and more information about low-fat cooking (one participant said, "I just buy whatever is convenient to cook. Junk food is quick and easy").

The benefits of the "Heartworks for Women" program, shown in Table 18-1, addressed these needs and wants. The results of the situational analysis revealed one major target segment: working women (64 percent of the total market), who were mostly aged 25 to 58 years. Two smaller segments were university/college students (20 percent), aged 18 to 24 years; and an "other" category (16 percent), which included stay-at-home mothers, retirees, and other women who did not work outside the home or go to school. The latter group had two main age segments: 18 to 30 years and 55+ years. Drawing on an analysis of the broad environment in which these women live and work, team 2 found both positive and negative elements in the environment. On the positive side, smoking was not allowed in government buildings, and many women were aware of low-fat eating messages in the media. On the negative side, the media feature smoking as attractive and seldom present women, especially middle-aged women, as fit and active. Competition for the "Heartworks for Women" program included weight-management programs offered by all private health/fitness clubs, and counseling provided by 11 dietitians in private practice in the areas of weight management and cholesterol reduction.

The objectives for the marketing strategy were designed to meet the broad goals outlined for the "Heartworks for Women" program: (1) Increase women's awareness of the relationship of diet to CHD risk, and (2) build skills related to heart-healthy eating and cooking. Specific objectives of the marketing strategy for the program are as follows:

- The "Heartworks for Women" program will be offered in 60 companies within the city by the end of one year.
- The "Heartworks for Women" program will be offered in all five universities and colleges within the city by the end of one year.
- At least 100,000 women living in the city of Jeffers will be exposed to "Heartworks for Women" messages through the following channels: advertising (city bus), flyers, promotional brochures, posters, radio interviews, newspaper articles, and the Internet website.

The marketing mix for the "Heartworks for Women" program is shown in Figure 18-5. It focuses on worksites and universities/colleges as the primary gates for delivering program

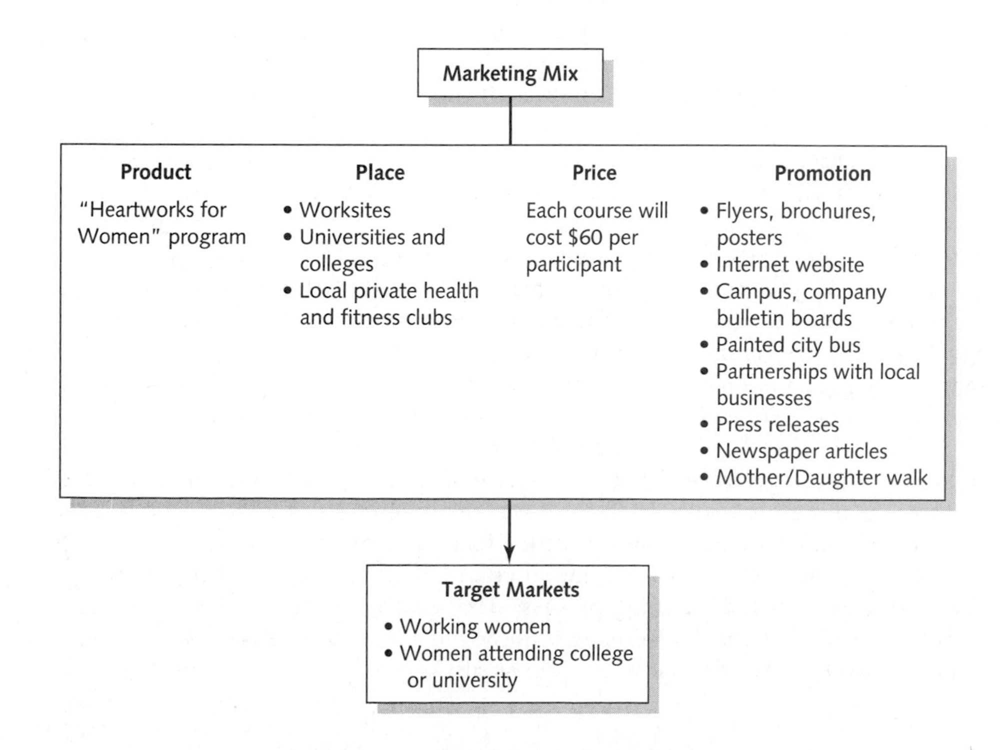

FIGURE 18-5 Marketing Mix for the "Heartworks for Women" Program

Launch (Apr.)

Marketing Tool	Jan.	Feb.	Mar.	Apr.	May	Jun.	Jul.	Aug.	Sept.	Oct.	Nov.	Dec.
Bulletin boards												
Campus			✔	✔	✔			✔	✔	✔		
Worksite			✔	✔	→	→	→	→	→	→	→	→
Flyers			✔	✔				✔	✔			
Brochures			✔	→	→	→	→	→	→	→	→	→
City bus				✔	→	→	→	→	→	→	→	→
Internet website			✔	→	→	→	→	→	→	→	→	→
Mother/Daughter walk									✔			
Press releases				✔					✔			
Radio announcement				✔					✔			
Newspaper articles				✔					✔			

FIGURE 18-6 Marketing Timetable for the "Heartworks for Women" Program

messages and promoting the program. Partnerships with local private health/fitness clubs provide another opportunity for boosting the program's visibility in the community. The program will be promoted through flyers, brochures, posters in company cafeterias, press releases related to special events, the painted city bus, the Internet website, and the Mother/Daughter Walk. The program will be priced competitively at $60 per participant.

The projected budget for the marketing strategy must take into account both staff time for designing materials and the cost of printing and duplicating promotional materials such as flyers and posters. (Additional information about the projected budget for the marketing strategy is presented in the next chapter.) The timeline allows team 2 to schedule all marketing activities prior to the program launch date. An example of a timetable is shown in Figure 18-6. Of course, in real life, the timetable is considerably larger than this, because it shows all activities for the program, including those designed to sustain awareness and increase participation over the life of the program.

As with all other aspects of program planning, the marketing strategy must be evaluated. In this example, the summative evaluation conducted 6 months after the launch determined that only 35 companies had signed on for the program, and most of these were large companies with a human resources department and more than 200 employees. Team 2 realized that two changes should be made in the marketing strategy. First, an important segment of the target group—women who worked in light-manufacturing companies with no human resources department and fewer than 100 employees—were not aware of the program. Thus the marketing strategy and the program format had to be redesigned to be more attractive to this group. For example, rather than offering a full eight-session course, small companies were offered only two or three 45-minute sessions, each held over the lunch hour. Each individual session cost only $5 per participant. The marketing mix was changed accordingly. Second, the decision to paint a city bus with the program logo and one nutrition message had been made early in the planning process, after team 2 learned of the success of this approach in another city. However, the chosen bus route was one that included a small mall and two large residential sections in an affluent part of town. Team 2 decided to choose another bus route, one that circulated through two industrial parks that featured light-manufacturing businesses. This bus route would provide more program visibility for women in the target market and for small companies, which were more difficult to recruit to the program than large companies. Adjustments in the marketing strategy would be made again in another 6 months when another evaluation was to occur.

Entrepreneurship Leads the Way

The challenge for the next decade will be to use the marketing strategies described in this chapter to remind consumers of the benefits of good health and to motivate them to make behavior changes. The important thing is to stay focused on the needs and wants of the "consumer" of health promotion activities. Remember, too, that consumers are a diverse lot, and broad target groups such as men aged 18 to 49 years or teenagers or postmenopausal women no longer describe today's consumer market. Consumers can and should be grouped into smaller, better-defined categories. In today's marketplace, the Internet makes one-to-one marketing feasible. Even mass marketing, a successful marketing tool of the past, is giving way to selective, target marketing. Tomorrow's consumer promises to be an independent thinker, highly educated and sophisticated, demanding, a seeker of innovation, and a pursuer of wellness.[49] Marketing will help you capture this changing profile, and entrepreneurship will help you plan for it.

Internet Resources

Social Marketing Resources

Hispanic Health Council, Inc.	**www.hispanichealth.com/pana.htm**
Center for Chronic Disease Prevention	**www.cdc.gov/nccdphp/dnpa/social_marketing_resources.htm**
Novartis Foundation for Sustainable Development	**www.foundation.novartis.com/social_marketing.htm**
Social Marketing Institute	**www.social-marketing.org**
Canadian Social Marketing Network	**www.hc-sc.gc.ca/hppb/socialmarketing**
Tools of Change	**www.toolsofchange.com**
Turning Point Social Marketing Collaborative	**www.turningpointprogram.org**
USDA Social Marketing Training Center	**www.nal.usda.gov/foodstamp/Training/social_marketing.html**

Social Marketing Campaigns

America on the Move	**www.americaonthemove.org**
Breastfeeding Awareness	**www.4women.gov**
Calcium: Select to Protect Campaign	**http://rutgers.njfsnep.org/social/default.asp**
California Project Lean	**www.californiaprojectlean.org**
5-a-Day Campaign	**www.5aday.gov**
VERB	**www.cdc.gov/youthcampaign/index.htm**
BAM! Body and Mind	**www.bam.gov**

Professional Focus

The Art of Negotiating

Whether we like it or not, we are all negotiators. Every day we negotiate with family, friends, and coworkers to get something we want. Although each of us negotiates something every day, most of us don't negotiate particularly well. We tend to find ourselves in situations where the negotiations leave us feeling frustrated, taken advantage of, dissatisfied, or just plain worn out. This is unfortunate, because negotiation is the heart of all business deals.

The reason why people sometimes emerge from negotiations feeling this way is that they tend to see only two ways to negotiate: hard or soft.[1] Hard negotiators view the other participants as adversaries. Their goal is victory—their side wins and the other side loses. Hard negotiators tend to distrust the other party and to make threats. They perceive the negotiation process as a contest of wills. Soft negotiators, on the other hand, think of the participants as friends. Their goal is agreement among the parties. They tend to be trusting, avoid a battle of wills, change positions easily, and make concessions to maintain the relationship. Many of us use soft negotiation tactics in our dealings with our parents, siblings, friends, or other people who are important to us.

Are these the only ways of negotiating? There is a better way to negotiate than either the soft or the hard approach, according to Roger Fisher and William Ury of the Harvard Negotiation Project. The method they developed is called *principled negotiation.* Its main precept is that a decision about an issue should be based on its merits, not on what each side says it will or will not do. The method can be boiled down to four basic elements:

- People—Separate the people from the problem.
- Interests—Focus on interests, not positions.
- Opinions—Generate a variety of possibilities before deciding what to do.
- Criteria—Insist that the result be based on some objective standard.

Separate the People from the Problem

When people sit down at the bargaining table, they bring with them certain perceptions about the relationships among the participants and about the problem itself. Whether these perceptions are accurate or false, they pervade the proceedings. There is a tendency to confuse the participants' relationships with the "issue" or "problem." One of the first steps to take in negotiating is to separate the problem from the people and to deal with relationship goals and problem goals separately. This means thinking about how to get good results from the negotiation and what kind of relationship is likely to produce those results.

Another challenge in negotiating is having to deal with a problem when emotions are running high. People sometimes come to the bargaining table with strong feelings. They may be angry or frightened, may feel threatened or misunderstood, or may be worried about the outcome. A good way to handle the emotional aspects of the negotiation process is to recognize such emotions and give them legitimacy. The bottom line, write Fisher and Brown, is to "do only those things that are both good for the relationship and good for us."[2]

Focus on Interests, Not Positions

The purpose of negotiating is to serve our interests. Interests motivate people to reach certain decisions. The primary problem in most negotiations is not the difference in positions but the conflicts between the two sides' needs, fears, desires, and concerns—in other words, their interests. In addition, most of us tend to think that the other party's interests are similar to our own. This is almost never the case. When negotiating, begin by defining, as precisely as possible, your own interests, and then allow the other participants to define theirs. Work through the discussion until mutual interests are identified.[3]

Consider a Variety of Options

We sometimes approach a negotiating session with only one outcome in mind. We operate with blinders on and fail to see other dimensions to the problem that may be a source of possible solutions. To get around this barrier, bring the participants together to brainstorm about potential solutions and options. In a good brainstorming session, judgments about possible solutions are suspended, and everyone involved in the negotiation is allowed to contribute ideas. At the end of the session, the parties discuss the various options, picking several that offer the most promise. The parties give themselves time to evaluate each of these "best and brightest" ideas and consider which of them, if any, would best suit their purpose. Once again, when exploring options, consider the interests of both parties.

Use Objective Criteria

Suppose your roommate wants to buy your car, but you cannot agree on a price. She thinks that your asking price is too high; you think her offer is too low. Where do you go from here? One option is to consult the Blue Book price for your car's make and model. Another is to examine the newspaper listings of used cars for sale to determine the asking price for cars like yours. These options are the criteria or standards that help you reach an agreement. The type of criteria you use will depend on the nature of the issue being negotiated. In this case, the criterion was the fair market value of your car. In other situations, a court decision, tradition, precedent, scientific judgment, cost, or moral standard might serve as the criterion. The important thing is to choose an objective standard that all parties are comfortable with.

Build Good Relationships

The negotiating process is much like a tango—a little give and a little take on both sides. Regardless of the issue, a "good" negotiation is fueled by a good relationship. When next you enter into a negotiation, take a few minutes to evaluate your relationship with the other person or party. The questions below will help you determine how good your working relationship is and where improvements can be made.[4] With practice, you can help build good working relationships.

Do we want to work together? In a good relationship, people want to work together. They respect each other and actively pursue strategies for sorting out differences. They work to keep problems to a minimum.

Are we reliable? Good relationships are built on trust and constancy. All parties have confidence that verbal and written commitments will be kept. The parties work to allay any concerns about trustworthiness.

Do we understand each other? Even in the best of working relationships, there will be differences of opinion, values, perceptions, and motives. In good relationships, the parties strive to accept each other and work toward understanding their differences.

Do we use our powers of persuasion effectively? In good relationships, persuasion is used to influence and inform the other party about an issue or proposed action. The parties refrain from using coercive tactics and rely instead on rational, logical discussions of the merits of a particular position or action.

Do we communicate well? Good communication is based on sound and compassionate reasoning. In good relationships, sensitive issues can be discussed in a supportive environment where candor is valued. The parties in a good relationship communicate often, consult each other before making decisions, and practice "active listening," where the parties work to hear each other with an open, flexible mind.

Work toward Success

Good negotiating means that all parties leave the bargaining table feeling they have won. The successful negotiation is one in which the outcome advances both parties' interests. It is seen as fair. The solution was arrived at with an efficient use of everyone's time. Neither party feels that he or she is at a disadvantage. And the solution will be implemented according to plan. A successful negotiation leaves the parties feeling respect for their counterparts and a desire to work together again.[5]

References

1. The discussion of hard and soft negotiation was adapted from R. Fisher and W. Ury, *Getting to Yes—Negotiating Agreement without Giving In* (New York: Penguin, 1981), pp. 8–9.
2. The quotation is taken from R. Fisher and S. Brown, *Getting Together—Building a Relationship That Gets You to Yes* (Boston: Houghton Mifflin, 1988), p. 38.
3. K. Albrecht and S. Albrecht, Added value negotiating, *Training* 39 (1993): 26–29.
4. J. D. Batten, *Tough-Minded Leadership* (New York: AMACOM, 1989), pp. 122–32; M. DePree, *Leadership Is an Art* (New York: Doubleday, 1989), pp. 89–96; and R. Fisher and S. Brown, *Getting Together*, pp. 178–79.
5. J. Allan, Talking your way to success, *Accountancy* 111 (1993): 62–63.

Chapter *Nineteen*

Managing Community Nutrition Programs

Learning *Objectives*

After you have read and studied this chapter, you will be able to:

- Differentiate between strategic and operational planning.
- Describe the four functions of management.
- Describe methods to coordinate an organization's activities.
- Outline methods for obtaining peak performance from employees.

This chapter addresses such issues as strategic management, program planning and documentation of appropriate activities, human resource management, preparation of a budget, and current information technologies, which are Commission on Accreditation for Dietetics Education (CADE) *Foundation Knowledge and Skills* requirements for dietetics education.

Chapter *Outline*

Something To Think About...

Do you want to be a positive influence in the world? First, get your own life in order. Ground yourself in the single principle so that your behavior is wholesome and effective. If you do that, you will earn respect and be a powerful influence. Your behavior influences others through a ripple effect. A ripple effect works because everyone influences everyone else. Powerful people are powerful influences.

If your life works, you influence your family.

If your family works, your family influences the community.

If your community works, your community influences the nation.

If your nation works, your nation influences the world.

If your world works, the ripple effect spreads throughout the cosmos.

– LAO TZU, ***Tao Te Ching (The Tao of Leadership)***

Introduction

Community nutritionists must be good planners and managers. One of the past presidents of the American Dietetic Association (ADA), Judith L. Dodd, said this about dietitians' need for management expertise: "It's not enough to have technical knowledge. Other skills are necessary, whether you want to call them leadership skills or communication skills or simply survival skills. We must know how to communicate, how to negotiate, how to persuade, and how to work with various groups within any of our market environments."[1] In other words, community nutritionists must have good management skills. **Management** is the process of achieving organizational goals through planning, organizing, leading, and controlling.[2] In this chapter, we examine the functions of management and consider how they are used by community nutritionists. In the following chapter, we review the principles of grant writing, because community nutritionists often seek extramural funding for program activities.

The Four Functions of Management

The main activities and responsibilities of managers can be grouped into four areas, or functions, as shown in Figure 19-1. *Planning* is the forward-looking aspect of a manager's job; it involves setting goals and objectives and deciding how best to achieve them. *Organizing* focuses on distributing and arranging human and nonhuman resources so that plans can be carried out successfully. *Leading* involves influencing others to carry out the work required to reach the organization's goals. *Controlling* consists of regulating certain organizational activities to ensure that they meet established standards and goals. This section describes each management function.

Management The process of achieving organizational goals through engaging in the four major functions of planning, organizing, leading, and controlling.

FIGURE 19-1 The Four Functions of Management

Source: K. M. Bartol and D. C. Martin, *Management,* p. 7. Copyright 1991 by McGraw-Hill, Inc. Used with permission of the McGraw-Hill Companies.

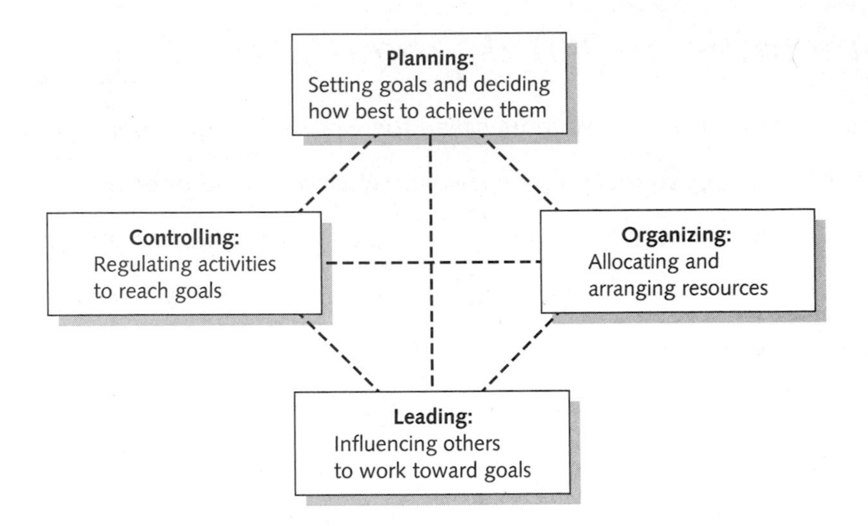

Planning

A colossal marketing blunder made headlines around the world in the spring of 1993. Hoover, a subsidiary of the Maytag Corporation, launched a promotional campaign in Ireland and Britain in which consumers who purchased any of its household appliances (worth at least $150 U.S.) were offered two free return plane tickets to Europe or the United States (valued at about $500 U.S.). Hoover's aim was to stimulate interest in its products, a feat it readily achieved. For reasons not entirely clear, Hoover's management failed to anticipate the number of customers who would accept the promotion's restrictions (such as inconvenient departure times) and demand the promised air tickets. The result was a $30 million loss, the firing of three top executives, and a place in management textbooks as "one of the great marketing gaffes of all time."[3]

What is the lesson in this story for community nutritionists? In two words: plan ahead. To those of us standing outside the Hoover snafu, it seems incredible that no one foresaw the shortfall between anticipated revenues and actual expenses. What happened on the inside may never be fully known. Regardless, the magnitude of the event suggests a glitch in the planning process—failure of managers to calculate accurately the direct and indirect costs associated with the promotional campaign.

For the individual manager or managerial team, **planning** involves deciding what to do and when, where, and how to do it. It focuses on future events and on finding solutions to problems. Planning is ongoing. It involves performing a number of activities in a generally logical, predetermined sequence and considering a variety of solutions before a plan of action is chosen.[4]

TYPES OF PLANNING

Planning Deciding what to do and when, where, and how to do it. Planning facilitates finding solutions to problems and is the basis for good management.

Strategic planning Long-term planning that addresses an organization's overall goals.

There are different types of planning. **Strategic planning** is broad in scope and addresses the organization's overall goals. Strategic planning occurs over a period of several years and is usually undertaken by the organization's senior managers. It focuses on formulating objectives; assessing past, current, and future conditions, trends, and events; evaluating the organization's strengths and weaknesses; and making decisions about the appropriate course of action. The American Dietetic Association's 2004–2008 strategic plan for "Leading the Future of Dietetics" includes the six strategic goals listed in the accompanying box. The plan allows for flexibility and innovation while providing direction for the organization in its mission to advance the profession of dietetics and to improve the nutritional health of all people.[5]

American Dietetic Association Strategic Plan, 2004–2008

Mission

Leading the future of dietetics

Vision

American Dietetic Association members are the most valued source of food and nutrition services.

Values

- *Customer Focus*—operates with consideration for the needs and expectations of internal and external customers
- *Integrity*—acts ethically, with accountability and attention to excellence
- *Innovation*—fosters an environment of positive change through creativity and continuous improvement
- *Life-Long Learning*—takes personal accountability for own competence and seeks opportunities for continued learning
- *Collaboration*—promotes open dialogue, cooperation and the sharing of knowledge
- *Inclusivity*—demonstrates respect and sensitivity toward and appreciation for, the backgrounds, differences, and points of view of others
- *Social Responsibility*—guides decisions and actions by considering economic, environmental and social implications

Strategic Goals

1. Build an aligned, engaged, and diverse membership.
2. Influence key food, nutrition, and health initiatives.
3. Impact the research agenda and facilitate research supporting the dietetics profession.
4. Increase demand for and utilization of services provided by members.
5. Empower members to compete successfully in a rapidly changing environment.
6. Proactively focus on emerging areas of food and nutrition.*

*ADA has identified several selected priority food and nutrition issues in which ADA can make a significant impact. They include: Healthy aging; Integrative medicine; Food supply; Genetics; and Obesity.

Source: The American Dietetic Association, 2004; available at www.eatright.org.

The strategic plan guides the development of operational plans. **Operational planning** is short-term planning typically done by mid-level managers. It deals with specific actions, expenditures, and controls and with the timing of these activities in a formal, structured process.[6] After setting policy as a major strategic initiative in its 1996–1999 strategic plan, for example, the ADA set into motion an operational plan to obtain data it could use in meeting its policy goal. It commissioned the Lewin Group to

Operational planning Short-term planning that focuses on the activities and actions required to meet the organization's goals.

determine the cost savings associated with medical nutrition therapy services and then launched its first public relations campaign to educate policy makers about the study findings.[7]

Another type of planning is **project management.** Community nutritionists who implement diabetes education programs, conduct citywide hypertension screenings, and assess the iron status of inner-city school children are all involved in project management. You have participated in project management through the case study activities in this book. Project management involves coordinating a set of activities that are typically limited to one program or intervention. It requires setting goals and objectives and outlining the project's **critical path,** the series of tasks and activities that will take the longest time to complete. Any delay in the activities on this path will delay the project's completion.[8] Managers work to spot and then remove bottlenecks, thus improving work flow along the critical path.[9] Refer to the case study later in this chapter for a discussion of the critical path for the "Heartworks for Women" program.

Organizing

Organizing is the process by which carefully formulated plans are carried out. Managers must arrange and group human and nonhuman resources into workable units to achieve organizational goals. Thus certain structures must be in place to guide employees in their activities and decision making. Imagine the confusion that would result if you had a problem and didn't know who your supervisor was! In this section, we describe organization structures and how people are managed as part of the process of organizing.

ORGANIZATION STRUCTURES

Just as a building's layout—the arrangement of offices, windows, hallways, restrooms, and stairwells—affects the way people work, so does an organization's structure.[10] The organization structure is the formal pattern of interactions and activities designed by management to link the tasks that employees perform to achieve the organization's goals. The word *formal* is used intentionally here to distinguish management's official operating structure from the informal patterns of interaction that exist in all organizations.[11] In developing the organization structure, managers consider how to assign tasks and responsibilities, how to define jobs, how to group individual employees to carry out certain tasks, and how to institute mechanisms for reporting on progress.

In organizing their employees, managers may find an organization chart helpful. Organization charts give employees information about the major functions of departments, relationships among departments, channels of supervision, lines of authority, and certain position titles within units.[12] According to the organization chart for a typical public health department (shown in Figure 19-2), the community nutritionist who coordinates the maternal, infant, and child care programs reports to the department manager, the director of nutrition, who in turn reports to the medical health officer. Nontraditional organization charts, in which employees have roughly the same rank, also exist. A "web" format, such as the one shown in Figure 19-3, exhibits no strict hierarchy of authority. This format has the advantage of promoting teamwork and consensual decision making, but it can create confusion about authority and responsibility. This design seems to work well in small companies and organizations.

Organization charts help establish lines of communication and procedures, but they do not depict rigid systems. An organization's informal structure, depicted humorously in Figure 19-4 on page 580, is often quite powerful and sometimes paints a more realistic picture of how the organization actually works. The informal structure arises spontaneously

Project management A plan that coordinates a limited set of activities around a single program or intervention.

Critical path In a complex project, the series of tasks that takes the longest amount of time to complete.

Organization chart A line diagram that depicts the broad outlines of an organization's structure and suggests the reporting relationships among employees.

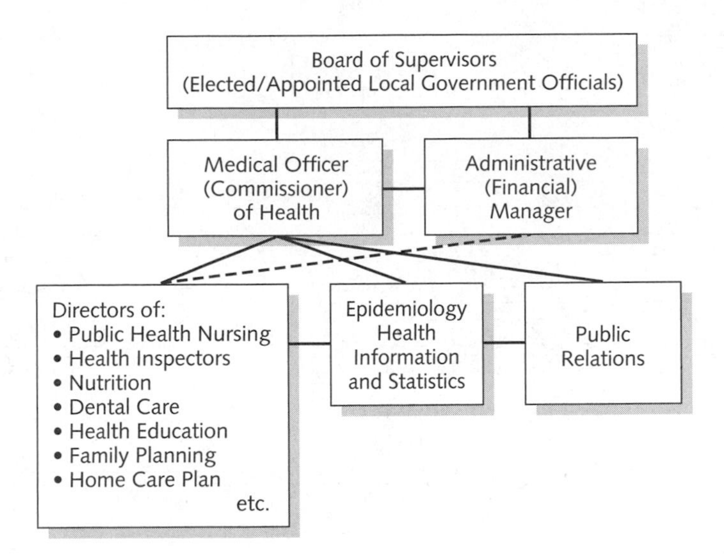

FIGURE 19-2 Organization Chart for a Hypothetical Local Health Department

Source: J. M. Last, *Public Health and Human Ecology* (Stamford, CT: Appleton & Lange, 1998), p. 323. Used with permission. Reprinted by permission of the McGraw-Hill Companies.

from employees' interactions, brief alliances, and friendships with coworkers throughout the organization.[13]

Another essential dimension of organization structure is departmentalization, or the manner in which employees are clustered into units, units into departments, and departments into divisions or other larger categories. Departmentalization directly affects how managers carry out their duties, supervise their employees, and monitor group dynamics and perspectives. An important aspect of departmentalization is the **span of management,** or **span of control**—that is, the number of subordinates who report directly to a specific manager.[14] Deciding how many employees should report to a single manager

Span of management or **span of control** The number of subordinates who report directly to a specific manager.

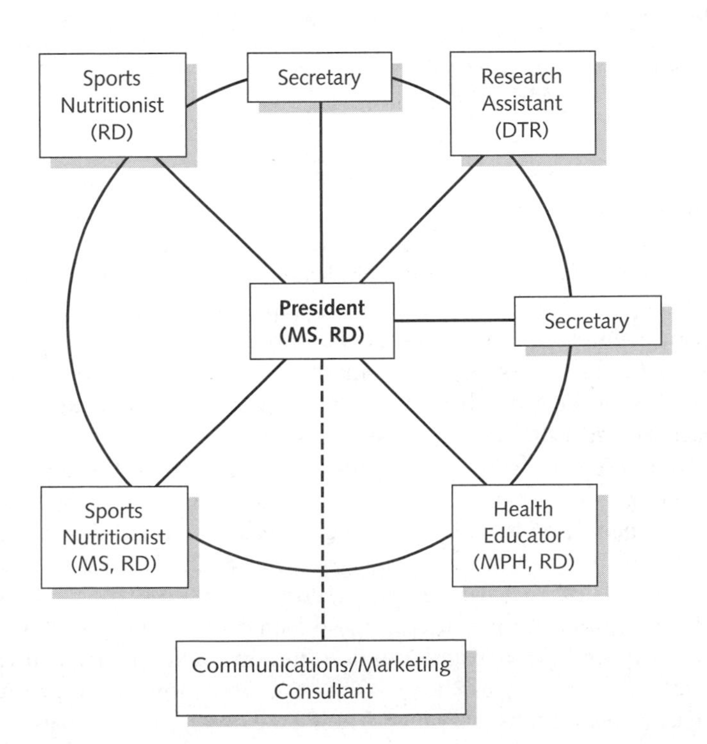

FIGURE 19-3 Nontraditional Organization Chart for Nutrition in Action

In this nontraditional organization structure, employees have roughly the same rank, and there is no strict hierarchy of authority.

FIGURE 19-4 The Typical Organization Chart on Paper and in Practice

Source: Organizational Behavior: Concepts and Applications, 4th ed. by Gray/Starke, © 1988. Reprinted by permission of Prentice-Hall, Inc., Upper Saddle River, New Jersey.

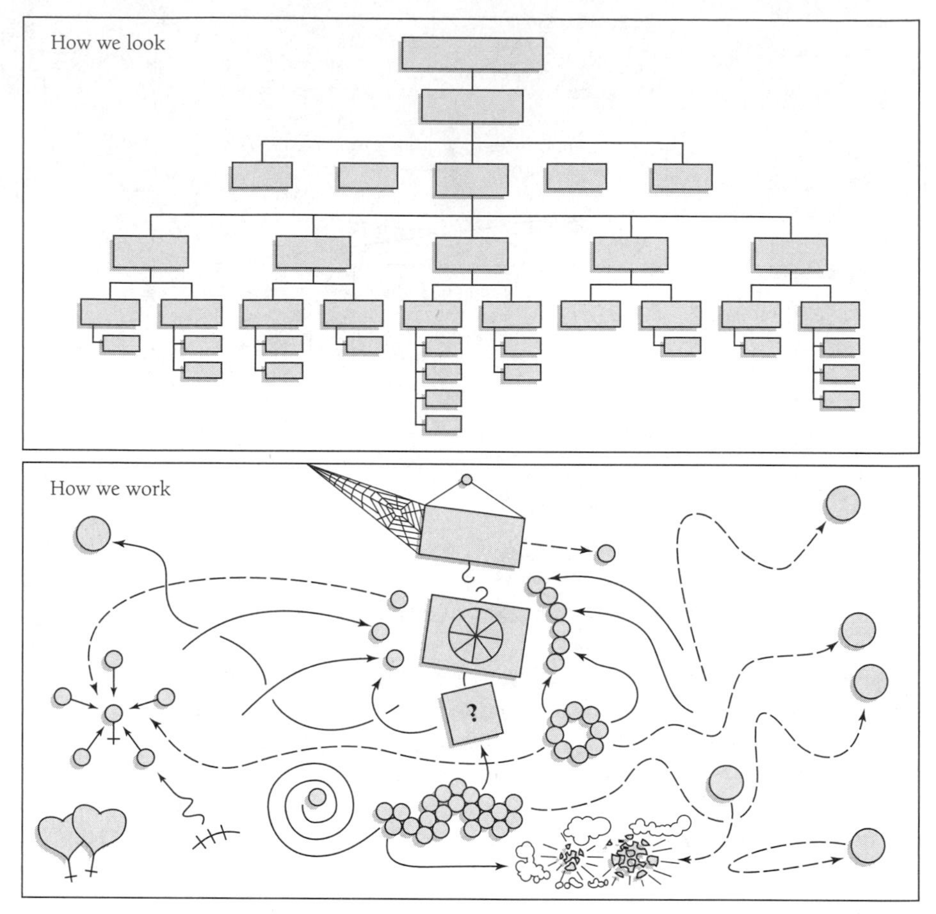

How an organization looks on paper differs from how it really works. Whereas the formal organization structure shows the lines of authority and how communications should travel, the informal organization reveals the actual patterns of social interaction between employees in different units of the organization.

is not a trivial decision. When a manager must supervise a large number of employees directly, she may feel overwhelmed and have difficulty coordinating tasks and keeping on schedule; with too few employees to supervise, she may feel underutilized and disaffected. The ideal span of management has not been identified precisely. Some researchers argue that the range is about 5 to 25 employees, depending on the level of organization; theoretically, lower-level managers can supervise more employees directly than managers higher in the hierarchy. Napoleon once remarked, "No man can command more than five distinct bodies in the same theater of war." Most management experts now recommend that each manager directly supervise only 3 to 7 subordinates.[15]

Delegation The assignment to others of part of a manager's work, along with both the responsibility and the authority necessary to achieve the expected results.

Line position A position with the authority and responsibility to achieve the organization's main goals and objectives.

Another method of coordinating an organization's activities is through **delegation,** or the assignment of part of a manager's work to others. "Everyone knows about delegating, but not too many people do it, and fewer still do it well."[16] The inability to delegate work to others has felled many a fast-track manager.[17] Some people do the work themselves because they don't realize they can delegate it, don't know how to get the most out of a computer, or don't appreciate how delegating work enables them to work more efficiently. Table 19-1 gives tips on how managers can delegate some activities to their employees.

Line and staff relationships also help clarify an organization's structure. A person in a **line position** has direct responsibility for achieving the organization's goals and objectives.

TABLE 19-1 Tips for Delegating Your Work to Others

- The first step to take when delegating your work to subordinates is to evaluate their skill levels and the difficulty of the task. The trick is to select the employee with the appropriate skills who will find the task challenging but not frustrating.
- Once you have chosen the best employee, give him or her all the information needed to do the job well. Be honest about the work; if it's drudgery, say so.
- Do these three things to ensure success:
  1. Give the employee *responsibility* for completing the project or task.
  2. Hold the employee *accountable* for the results.
  3. Provide the employee with the *authority* to make needed decisions, direct others to help with the project, and carry out actions as required.
- Make sure the employee has the necessary materials, equipment, time, and funds to complete the project.
- Evaluate the employee's progress periodically and be prepared to accept a less-than-perfect result.

Source: Adapted from L. Baum, Delegating your way to job survival, *Business Week,* November 2, 1987, p. 206.

The term *staff* is commonly used to refer to the groups of employees who work in a particular unit or department—for example, the director of nutrition in a public health department is said to supervise the nutrition staff. In management practice, however, *staff* has a particular meaning. An employee in a **staff position** assists those in line positions. To some extent, the concept of line and staff positions is a holdover from eighteenth- and nineteenth-century management theory, but the terms are still used today in many organizations to help employees identify individuals who have the authority and responsibility for fulfilling the organization's mandate.

JOB DESIGN AND ANALYSIS

Managers in community nutrition are responsible for **job design**—that is, for determining the various duties associated with each job in their area. They sometimes need to conduct a **job analysis** to determine the purpose of a job, the skill set and educational background required to carry it out, and the manner in which the employee holding that job interacts with others. The formal outcome of a job analysis is the preparation of a job description.[18]

A job description helps employees understand what is expected of them and to whom they report. Although there is no standard format, most job descriptions include the items shown in Table 19-2.[19] The job description serves as a basis for rating and classifying jobs, setting wages and salaries, and conducting performance appraisals (described later). It helps organizations comply with the accrediting, contractual, legal, and regulatory directives of the government or other institutions, and it provides a basis for decisions about promotions and training. In addition, it can be adapted for use in recruiting and hiring prospective employees. For example, the text of a position announcement to be published in a newspaper or journal can be taken from the job description. A sample job description for a nutritionist with the FirstRate Spa and Health Resort is given in Figure 19-5.

Staff position A position whose primary purpose is to assist those in line positions.

Job design The specification of tasks and activities associated with a particular job.

Job analysis The systematic collection and recording of information about a job's purpose, its major duties, the conditions under which it is performed, and the knowledge, skills, and abilities needed to perform the job effectively.

HUMAN RESOURCE MANAGEMENT

Paying attention to the people who produce a product or service is what human resource management is all about. Community nutritionists report being involved in several aspects of managing people, including recruiting people for the organization and evaluating the performance of their employees.[20] Both of these activities are described in the next paragraphs.

TABLE 19-2 Components of a Job Description

- **Job title.** The job title clarifies the position and its skill level (for example, *director, supervisor, assistant,* or *secretary*).
- **Immediate supervisor.** The job description states the position and title of the immediate supervisor.
- **Job summary.** The job summary is a short statement of the purpose of the job and its major tasks and activities.
- **Job duties.** This section of the job description is a detailed statement of the specific duties of the job and how these duties are carried out.
- **Job specifications.** This is a list of minimum hiring requirements, usually derived from the job analysis; it includes the skill requirements for the job, such as educational level, licensure requirements, and experience (usually expressed in years), and any specific knowledge, advanced training, manual skills, and communication skills (both oral and written) required to do the job. The physical demands of the job, if any, including working conditions and job hazards, are included here.

Source: Adapted from J. G. Liebler and C. R. McConnell, *Management Principles for Health Care Professionals,* 4th ed. (Sudbury, MA: Jones and Bartlett Publishers, 2004).

FIGURE 19-5 Sample Job Description

FirstRate Spa and Health Resort
Department of Human Resources

Position Title: Nutritionist

Responsible to: Director of health promotion

Responsibilities: Under the general direction of the director of health promotion, the nutritionist is responsible for developing, implementing, and evaluating the spa's risk reduction programs.

Job Duties: The nutritionist has the following duties and responsibilities:
- Develops, implements, and evaluates the spa's risk reduction programs in the areas of smoking cessation, blood cholesterol reduction, stress management, and diabetes control.
- Assists the director in developing marketing strategies for the risk reduction program.
- Assists the director in developing budgets.
- Tracks program revenue and expenses.
- Works with other spa nutritionists and staff to develop marketing tools and teaching/instructional aids.
- Participates as required in coordinating research projects.

Qualifications and Experience: The nutritionist should have the following qualifications and experience:
- A university degree in nutrition
- Certification as a registered dietitian
- Two years of experience teaching community-based courses or programs

Skills: The nutritionist should have the following skills:
- Demonstrated ability to write and give presentations
- Experience using word processing and spreadsheet software is desirable
- Experience using the Internet is desirable
- Ability to work with people of all ages and backgrounds

Date: June 2005

Staffing

Staffing is the set of human resource activities designed to recruit individuals to help meet the organization's goals and objectives. Recruitment has two distinct phases: attracting applicants and hiring candidates.[21] Both direct and indirect recruiting strategies are often used to attract applicants. Direct methods include placing "Help Wanted" ads in newspapers and journals, mailing personalized letters to potential applicants, and participating in job recruitment fairs. Indirect recruiting strategies include activities that keep the organization's name in the public's eye, such as collaborative projects with other institutions and training sessions for professionals.

An important aspect of recruiting is **affirmative action,** which includes all activities designed to ensure and increase equal employment opportunities for groups protected by federal laws and regulations. A significant law in this area is Title VII of the Civil Rights Act of 1964, which forbids employment discrimination on the basis of gender, race, color, religion, or national origin. Most organizations have some type of affirmative action plan that details its goals and policies related to hiring, training, and promoting protected groups. In fact, organizations with federal contracts exceeding $50,000 and with 50 or more employees must file an affirmative action plan with the Department of Labor's Office of Federal Contract Compliance Programs; others may develop such plans on a voluntary basis.[22]

The hiring process involves conducting job interviews with potential candidates, screening applicants, selecting the best candidate, checking references, and, finally, offering the position. In some cases, two or three suitable candidates will be invited back for interviews with key managers in other departments or divisions before a final candidate is chosen.

Evaluating Job Performance

Providing feedback to employees about their performance is essential to maintaining good working relationships and can occur informally at any time. In the daily course of solving unexpected problems and reviewing the progress on a project, managers in community nutrition have ample opportunity to update their employees about their performance. In fact, the best managers spend up to 40 percent of their time in face-to-face interactions with their employees that allow time for performance feedback.[23]

The **performance appraisal** is a formal method of providing feedback to an employee. It is designed to influence the employee in a positive, constructive manner and involves defining the organization's expectations for employee performance and discussing how the employee's recent performance compares with those expectations. The performance appraisal is used to help define future performance goals, determine training needs and merit pay increases, and assess the employee's potential for advancement. Because employees are often uneasy about the review process, managers must take care to be objective in their rating of performance. One way to reduce employees' discomfort is to let them know in advance how the process works and what to expect during the performance appraisal interview. Employees are most likely to have a favorable impression of the process when they are aware of the evaluation criteria used, have an opportunity to participate in the process, and are able to discuss their career development.[24]

Employees want to hear from management, and they want management to listen to them.

– D. Keith Denton

The performance appraisal interview is a good time for reviewing and changing both organizational and personal goals. Managers who spend time helping subordinates develop their personal goals are more likely to achieve their organization's goals, because setting challenging personal goals is linked strongly with performance.[25]

The key to conducting a good performance appraisal interview is to start with clear objectives, focus on observable behavior, and avoid vague, subjective statements of a

Affirmative action Any activity undertaken by employers to increase equal employment opportunities for groups protected by federal equal employment opportunity (EEO) laws and regulations.

Performance appraisal A manager's formal review of an employee's performance on the job.

TABLE 19-3 How to Provide Constructive Criticism

Use the process outlined here to offer constructive criticism to your employees without making them defensive. These questions can be used in a performance appraisal interview, but they lend themselves to many other job-related discussions. Most people questioned in this manner make specific suggestions to improve their performance.

- **Initiate.** "How would you rate your performance during the last six months?" or "How do you feel about your performance during the last quarter?"
- **Listen.** If you get a general response like "fine," or "pretty good," or "8 on a 10-point scale," follow up with a more focused question: "What in particular comes to mind?" or "What have you been particularly pleased with?" Your objective is to discuss a positive topic raised by the employee.
- **Focus.** "You mentioned that you were particularly satisfied with. . . . Let's talk further about that aspect of your job."
- **"How" probe.** "How did you approach . . . ?" or "What method did you use?"
- **"Why" probe.** "How did you happen to choose that approach?" or "What was your rationale for that method?"
- **"Results" probe.** "How has it worked out for you?" or "What results have you achieved?"
- **Plan.** "Knowing what you know now, what would you have done differently?" or "What changes would you make if you worked on this again?"

Source: J. G. Goodale, Seven ways to improve performance appraisals, *HRMagazine* 38 (1993): 79. Reprinted with the permission of *HRMagazine,* published by the Society for Human Resource Management, Alexandria, VA.

personal nature. Consult Table 19-3 for a list of questions that can be used in talking about performance and giving constructive criticism.[26]

Leading

Leading is the management function that involves influencing others to achieve the organization's goals and objectives. It focuses on the horizon, on asking what and why, whereas managing focuses on the bottom line, on asking how and when.[27] (Reread the Professional Focus feature on leadership that appears in Chapter 14.) Organizations need strong leaders *and* good managers.

MOTIVATING EMPLOYEES

An important function of managers is to motivate their employees to "get the job done." Barcy Fox, a vice president with Maritz Performance Improvement Company, comments, "Your employees aren't you. What works for you won't work for them. What motivates you to pursue a goal won't motivate them. Until you grasp that, you won't set attainable goals for your workers."[28] There is no single strategy for motivating employees to perform, and opinions differ on exactly what managers need to do to stimulate motivation. Peter Drucker, a management consultant, argues that managers can do four things to obtain peak performance from their employees:[29]

- Set high standards and stick to them. Employees become demoralized when managers do not hold themselves to the same high standards set for other workers in the organization. Employees are stimulated to work hard with managers who expect a lot of themselves, believe in excellence, and set a good example.
- Put the right person in the right job. Employees take pride in meeting a challenge and doing a job well. Their sense of accomplishment enhances their performance. Underutilizing employees' skills destroys their motivation.

- Keep employees informed about their performance. Employees should be able to control, measure, and evaluate their own performance. To do this, they need to know and understand the standard against which their performance is being compared. The job description and performance appraisal interview help keep employees informed about what managers expect of them. Keep in mind the unwritten rule of managing people: Put praise in writing. But when you must reprimand an employee, do so verbally and in private.
- Allow employees to be a part of the process. One means of motivating employees is to give them a managerial vision, a sense of how their work contributes to the success of the project, program, or organization. In a sense, a manager cannot "give" his own vision to his employees, but he can allow them to help mold and shape the vision, thereby making it theirs as well as his.

COMMUNICATING WITH EMPLOYEES

Communicating is a critical managerial activity that can take many forms. Verbal communication is the written or oral use of words to communicate messages. Written communications can take the form of reports, résumés, telephone messages, memoranda, procedure manuals, policy manuals, and letters. Oral communications, or the spoken word, usually take place in forums such as telephone conversations, committee meetings, and formal presentations. Good managers are skilled in both of these areas (review the Professional Focus features in Chapter 15 ("Being an Effective Speaker") and Chapter 17 ("Being an Effective Writer"). Nonverbal communication in the form of gestures, facial expressions, and so-called body language is also important. Tone of voice—the *way* something is said, as opposed to *what* is said—sometimes relays information more effectively than the actual words. Even the placement of office furniture communicates certain elements of attitude and style. Reportedly, nonverbal communication accounts for between 65 and 93 percent of what actually gets communicated.[30]

Becoming a good communicator means paying attention to people and events, observing the nuances of nonverbal and verbal communication, and becoming a good listener. Open communication in an organization does not occur by accident. It results from the daily use of certain techniques and skills that promote communication. Merck, the pharmaceutical company, uses an approach called face-to-face communication. The four rules on which this communication is built are listed in Table 19-4.[31] Communicating effectively with coworkers is a skill that, like other skills described in this book, can be acquired with practice and determination.

The team leader for the "Heartworks for Women" course in Case Study 1, for example, learns of a conflict between two staff members—each believes that she is the main liaison with the design company hired to prepare the marketing materials. The liaison has a key role in keeping the project on schedule and also has the opportunity to become involved in the organization's other marketing campaigns. The team leader determines, first, that the design company will want to deal with only one staff member and, second, that the rivalry between these two staff members has been friendly but has the potential to be disruptive. She schedules a meeting with the two of them, asking them to think about the problem and how it might be resolved. During the meeting, the three of them talk about the liaison's role, the responsibilities and activities of each staff member, and certain projects "in the works" within the department. The leader aims to keep the discussion open and amiable. She succeeds in getting the more senior staff person to agree to allow the junior staff person, who has worked previously with the design company, to serve as liaison; she gives the senior staff person the responsibility for developing the website content and some design elements. The leader's role in this communication

TABLE 19-4 Four Rules of Face-to-Face Communication

- **Be candid.** Any questions or comments that arise during face-to-face interactions (be they performance appraisal interviews or committee meetings) should be addressed truthfully. If appropriate, indicate that you cannot answer the question because of its confidential nature. If you do not know the answer to a question, say so.
- **Be prepared.** Ask participants to submit questions or topics for discussion several weeks or days before the meeting is scheduled to occur. Organize an agenda, preferably a written one. This strategy has the advantage of ensuring that topics of interest to you and your employees are covered. If you develop a written agenda, distribute it to the participants before the meeting so that they can arrive prepared.
- **Encourage participation.** Conduct the face-to-face meeting in a relaxed and friendly atmosphere where people feel comfortable expressing their ideas.
- **Focus on listening, not judging.** When sensitive topics arise, try not to appear upset or judgmental. Listen carefully to questions and the discussion they prompt. Pay attention to nonverbal cues and group dynamics. The essence of good communications is simple: Treat others with respect.

Source: Adapted from Merck's face-to-face communication, in D. K. Denton, *Recruitment, Retention, and Employee Relations* (Westport, CT: Quorum Books, 1992), pp. 158–59.

process is to help the staff members appreciate each other's skills and contribution to the organization.

Controlling

Controlling is the management function concerned with regulating organizational activities so that actual performance meets accepted organizational standards and goals.[32] The controlling function involves determining which activities need control, establishing standards, measuring performance, and correcting deviations. Managers set up control systems—mechanisms for ensuring that resources, quality of products and services, client satisfaction, and other activities are regulated properly. Such controls help managers cope with the uncertainties that arise when plans go awry. An unforeseen reduction in the public health department's budget, for example, might force the manager of the nutrition division to reallocate personnel and impose tighter spending controls.

Community nutritionists often have responsibility for program and departmental budgets and for managing information. The ADA's role delineation study determined that managers in community or public health programs were responsible for preparing financial analyses and reports, controlling program costs, monitoring a program's financial performance, documenting a program's operations, and making decisions about capital expenditures.[33] Even entry-level community nutritionists are expected to be able to define basic financial terms, read and understand financial reports, maintain cost control of materials and projects, and, in some settings, develop plans to generate revenue.[34] Strong management skills enable community nutritionists to detect irregularities in client service or cost overruns and to handle complex projects and programs. Two aspects of the control process are described in the following paragraphs.

Balance sheet A financial statement that depicts, at given intervals (such as monthly), an organization's assets and the claims against those assets (liabilities) at that time.

FINANCIAL AND BUDGETARY CONTROL

The primary tools of financial control are the two types of financial statements: balance sheets and income statements. The **balance sheet** lists the organization's assets and liabilities. Assets include cash, accounts receivable (sales of a product or service for which payment has not yet been received), inventories, and items such as buildings, machinery, and equipment. Liabilities include accounts payable (bills that must be paid to outside suppliers and the like), short- and long-term loans, and shareholders' equity such as common stock. (Government agencies and most small, independent companies do not have shareholders' equity.) The **income statement** summarizes the organization's operations over a specific time period, such as a quarter or a year. It generally lists revenues (the assets derived from selling goods and services) and expenses (the costs incurred in producing the revenues). The difference between revenues and expenses is the organization's profit or loss—sometimes called the bottom line.

Financial control is typically managed through an operating budget, which is "a plan for the accomplishment of programs related to objectives and goals within a definite time period, including an estimate of resources required, together with an estimate of the resources available."[35] Budgets can be planned for the organization as a whole or for subunits such as divisions, departments, and programs. **Budgeting** is closely linked to planning. It is the process of stating in quantitative terms (usually dollars) the planned organizational activities for a given period of time. Although the terms *budgeting* and *accounting* are sometimes used interchangeably, they are not the same thing. Accounting is a purely financial activity, whereas budgeting is both a financial and a program planning activity.

Because all expenses of an organization—everything from pencils and paper clips to salaries and benefits—must be accounted for, the budgeting process requires considerable planning and managing. It usually begins with a review of the previous budget period and proceeds to the development of a new budget. Most programs and projects receive a monthly budgetary review. The primary areas to be examined include total project funding, expenditures to date, current estimated cost to complete the project, anticipated profit or loss, and an explanation of any deviations from the planned budget expenditure. Periodic status summaries help managers control resources and adhere to the planning schedule.[36] The internal approval process for a new budget can involve bargaining, compromise, and outright competition for scarce resources. Managers who can justify their budget requests are more likely to be successful in getting funds appropriated for their program's projects and activities.

INFORMATION CONTROL

In the course of your community work, you will need to collect, organize, retrieve, and analyze many types of data and information. At the very least, you will need to know about the health and nutritional status of the population in your district, the demographic profile of your client base, how people in your community use your organization's programs and services, how cost-effective your programs are, and what expenses your staff incurred in carrying out their program activities. At this point, a distinction must be made between data and information. **Data** are unanalyzed facts and figures. For example, a survey of infants in your community shows that over the past 12 months, 36 out of 478 infants had birthweights less than 2,500 grams. These data don't mean much on their own; to be meaningful, they must be transformed into **information.** For example, by processing the data further (for example, converting the data percentages or calculating the mean weights), the manager can compare these figures with those from previous reporting years. A change in the figures may signal the need to reassess program goals and reallocate resources to help prevent low birthweight.

The primary purpose of information is to help managers make decisions. Information is a decision support tool in that it helps managers answer these questions: What do we

Income statement A financial statement that summarizes the organization's financial position over a specified period, such as a quarter or a year.

Budgeting The process of stating in quantitative terms, usually dollars, the planned organizational activities for a given period of time.

Data Unanalyzed facts and figures.

Information Data that have been analyzed or processed into a form that is meaningful for decision makers.

want to do? How do we do it? Did we do what we said we were going to do? The first question addresses the planning function of management, the second the organizing function, and the third the controlling function. Managers begin the decision-making process by organizing the data into usable information. Then they specify the means by which the data are to be analyzed. This analysis usually involves sorting, formatting, extracting, and transforming the data via certain statistical methods. Finally, managers interpret the output and prepare reports that summarize the information. Decisions are then based on the figures in the reports.[37] Of course, managers' decisions are only as good as their information. Constant vigilance is required to prevent the GIGO (garbage in/garbage out) syndrome.

Computers are essential to community nutrition practice. They are used to prepare letters, memos, grant proposals, reports, manuals, fact sheets, and brochures; to track revenue and expenses; to exchange e-mail with colleagues; and to access the Internet for timely information.

With the advent of computers, processing information has become both simpler and more complex. It is simpler because many aspects of data analysis are carried out on computers and not by hand. It is more complex because one must choose from hundreds of computer systems and software programs. Because computers can produce reams of data very quickly, managers are apt to become overwhelmed by the magnitude and diversity of the data available to them. To be useful, information must be[38]

1. *Relevant.* Although it is sometimes difficult to know precisely what information is needed, managers should try to identify the information most likely to help in making decisions.
2. *Accurate.* If information is inaccurate or incorrect, the quality of the decision will certainly be questionable.
3. *Timely.* Information must be available when it is needed.
4. *Complete.* Information must cover all of the areas important to the decision-making process.
5. *Concise.* Information should be concise, providing a summary of the items central to the decision.

Management Issues for "Heartworks for Women": Case Study 1

The development and design of the "Heartworks for Women" program has spanned several chapters in this text, beginning with Chapter 2. We have discussed how it fit into a larger intervention strategy, the choice of program format and nutrition messages, and the elements of the marketing plan. In this section, we describe three management issues related to the program: the critical path, the budget, and grantsmanship.

The Critical Path

The community nutritionist who leads team 2—the team responsible for the nutrition intervention activities for the "Heartworks for Women" program, as discussed in Chapter 17—develops a timeline early in the planning stages, when the program plan (described in Chapter 14) is outlined. A variety of formats can be used for this activity. Figure 18-6 on page 570 is one example of a timetable and shows key marketing activities for the "Heartworks for Women" program. (In reality, the marketing timetable would be developed after the time line showing the critical path.) Because the program will be launched the following April, the team 2 leader lays out the paths that must be completed to ensure that all program elements have been finalized before the launch date. She plots a time

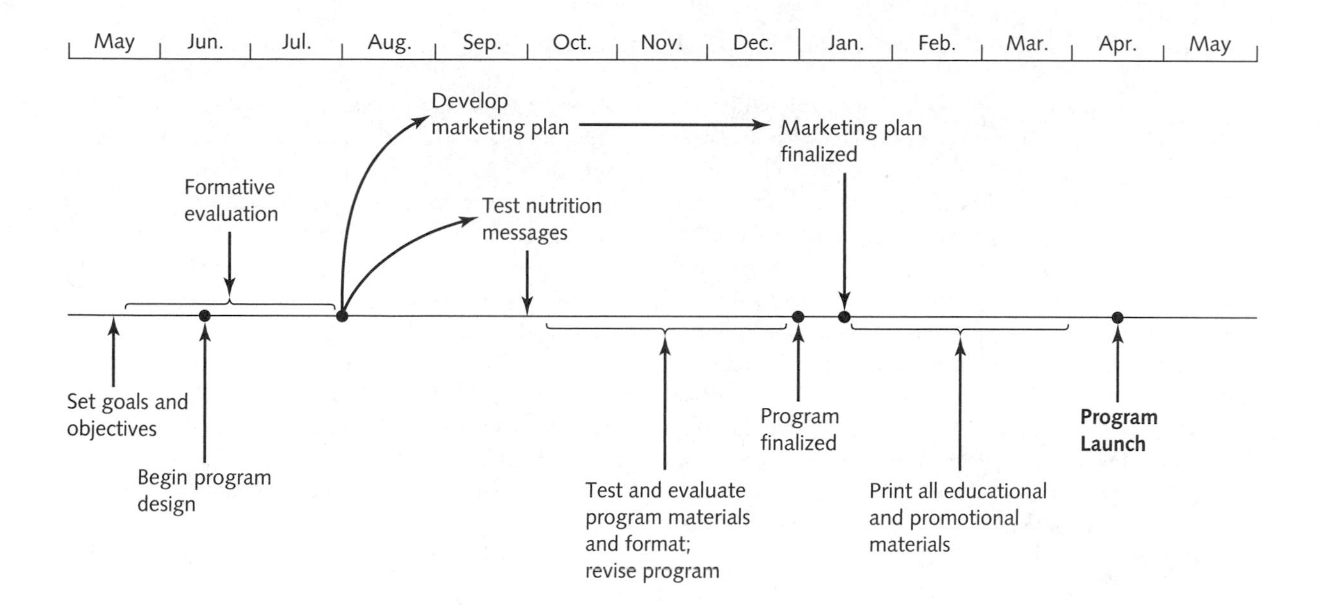

FIGURE 19-6 Time Line for Determining the Critical Path for the "Heartworks for Women" Program: Case Study 1

line, shown in Figure 19-6. The critical path consists of the steps involved in program design and evaluation—a process estimated to take 7 months. Two minor paths—developing the marketing plan and testing the appropriateness of the nutrition messages for the target population—are estimated to require several months as well. Any glitches in finalizing the program will produce delays in finalizing the marketing plan. For example, some elements of the marketing plan will be incomplete if the nutrition messages aren't finalized on time.

Several milestones appear on the timetable: early June, when program planning begins; August 1, when the formative evaluation has been completed; December 31, when the program must be finalized; January 15, when the marketing plan is finalized; and April 18, the date of the launch. These are the points at which major tasks must be completed on time or the program launch date will be delayed.

Once the time line has been developed, what is the next step for the community nutritionist? First, she shares it with her team members to confirm that it is reasonable and that all activities have been accounted for. Next, she breaks down each path into a group of activities that can be monitored. For example, the program launch requires that the following tasks be completed: choose and confirm the date, location, and time for the launch; prepare a list of invitees; develop, print, and mail invitations; prepare and mail press releases; prepare and finalize all advertising about the launch; choose a menu; arrange for delivery of food and beverages; outline the launch program (for instance, opening remarks); contact guest speakers and mail a confirmation letter to each of them; arrange for any decorating of the facility; and arrange interviews with the media. Completion dates for these activities are also determined.

THE OPERATING BUDGET

The operating budget is a statement of the financial plan for the program. It outlines the revenues and expenses related to the program's operation. Revenues for the "Heartworks for Women" program will be derived from the enrollment fees for individual participants

Programs in Action

The Better Health Restaurant Challenge

Public and private partnerships can advance nutrition education efforts by helping to ensure the delivery of consistent messages. Perhaps most important, such partnerships can combine resources to achieve an effective nutrition education program—more than any one party may be able to accomplish alone.[1]

A variety of partners can become involved and provide leadership for the design and development of health promotion and chronic disease prevention programs. Coalition building can take place among community-based nutrition and health professionals, opinion leaders, and local businesses. It is important to involve . . . the food industry in the implementation of prevention programs. Voluntary community and health organizations involved in disease prevention work include the American Cancer Society and numerous others. Such volunteer organizations are often effective in promoting community-based nutrition and dietary interventions and risk-reduction programs.[2]

An increasing proportion of food dollars is spent on foods eaten or purchased away from home. Portion sizes of foods prepared away from home often exceed those recommended in the Food Pyramid—three-cup portions of pasta (equal to six Pyramid servings) and nine-ounce cooked portions of meat (equal to three Pyramid servings) are not unusual. Furthermore, estimating portion sizes when eating out is difficult—even for nutrition students. Balanced meals—with appropriate portions from at least three food groups—can be difficult to find. Foods tend to be prepared with generous amounts of fat. Additionally, healthier items are limited or may not be on the menu.

Restaurants making nutrition claims about dishes are required to list nutrition information. With the exception of fast-food chains and national restaurants, few provide nutrition information. Many restaurants are resistant because they have too many dishes and menu variations; the menu would be too cluttered and harder to change.

Consumers want healthful menu options. The availability of healthful foods in restaurants helps individuals eat more responsibly. Sales of menu items that are promoted as healthful have been shown to increase during special programs and campaigns.

Partnerships and coalitions can be a highly effective way to coordinate healthful-eating programs in the community. They extend and enhance individual resources as partners work together to deliver consistent nutrition messages. A partnership between health providers and restaurants is win–win for the partners and for the community: The health provider's message is reinforced in local restaurants, the restaurant's name and reputation are enhanced, and consumers have access to healthful food when dining out.

The Better Health Restaurant Challenge demonstrates how health professionals, HMOs, restaurants, consumers, and health organizations can benefit from partnership. The month-long Better Health Restaurant Challenge (BHRC) was an annual contest to determine the best-tasting low-fat restaurant menu items in the Minneapolis/St. Paul area. The program was sponsored by HealthPartners, a large Minnesota-based managed-care organization. Health organizations such as the American Heart Association and the American Cancer Society received a financial contribution from HealthPartners on behalf of winning restaurants.

Goals and Objectives

The goals of the BHRC were to increase the availability of tasty, low-fat menu items in restaurants and to increase restaurant patrons' selection of low-fat menu items, even among diners who would not ordinarily choose low-fat foods. To achieve these goals, program objectives included promoting

and the sales of merchandise such as coffee mugs and T-shirts, as shown in Table 19-5. Some expenses, such as the design of the program logo and printing of educational materials, are directly related to the program; others, such as office rent and electricity, contribute indirectly to the program's costs. At the beginning of the fiscal year, the operating budget has a zero balance (net revenue minus total expenses). Expenses will be offset by anticipated revenue. Any discrepancies in either revenue or expenses will be reflected in the operating budget's monthly net balance.

EXTRAMURAL FUNDING

The art of seeking and being awarded extramural funding is the subject of Chapter 20.

Early in the planning process, the team 2 leader identifies several activities for which extramural funding might be obtained. She perceives that some expenses related to the participating restaurants, menu items, HealthPartners, and low-fat eating to the community; encouraging the long-term availability of low-fat items; providing low-fat dining options in a variety of eating establishments; and increasing consumer and restaurant participation in the BHRC.

Methodology

Restaurants in the Minneapolis area were contacted by mail and telephone to acquaint them with the program and interest them in participating. Dietitians from HealthPartners also attended trade shows and restaurant association meetings to solicit participants. (In the first years of the program, dietitians cold-called area restaurants. Many restaurants now call on their own to sign up.)

Participating restaurants agreed to work with HealthPartners' registered dietitians to create at least two low-fat menu items, to train restaurant staff to promote the healthful items, to keep at least one low-fat item on the menu for one month, and to extend a 20 percent discount on low-fat food items to members of HealthPartners. Participating restaurants and their menu items were widely promoted by HealthPartners through the media and extensive advertising.

HealthPartners and its dietitians developed and funded numerous materials, including point-of-purchase promotions of the low-fat menu items. They also created ballots so that diners could rate the low-fat items based on taste; the ballots were passed out by restaurant staff to diners who ordered the healthful items. The 1997 slogan to encourage diners to participate in the BHRC was "Enjoy Being a Food Critic? We Savor Your Opinion." (Slogans using the term *low-fat* did not score well during testing.)

Restaurants with the highest overall taste ratings were declared winners. Winners in the nine categories in which diners voted were announced to the media.

Results

The number of restaurants and locations participating in the BHRC has more than tripled since the program's inception. Over that same period, the number of different menu items has increased from 60 to 280.

Approximately 14,000 diners completed ballots. Over half reported that items tasted better than expected. The average taste rating was 4.13 out of 5. An overwhelming majority of diners said that they would order the low-fat item again and that the BHRC program increased the likelihood of their ordering low-fat items in the future. Over 90 percent of participating restaurants kept their low-fat menu items on the menu after the program ended.

Lessons Learned

Treating diners as food critics increased participation, as did the challenge's focus on foods that taste good. Big advertising budgets are not necessary. Partners such as the local chapter of the American Heart Association, area hospitals, and even community leaders can help get the word out.

References

1. Position of the American Dietetic Association: Nutrition Education for the Public, 1996.
2. Position of the American Dietetic Association: The Role of Nutrition in Health Promotion and Disease Prevention Programs, 1998.

Source: Adapted from *Community Nutritionary* (White Plains, NY: Dannon Institute, Fall 1998). Used with permission. For more information about the *Awards for Excellence in Community Nutrition,* go to www.dannon-institute.org.

launch, such as the costs of printing banners and balloons and providing food for invitees, could be underwritten, as might the costs of printing program manuals and training materials. She sets a target figure of about $8,000 and begins networking in her community. She contacts two colleagues whose organizations have sponsored health promotion activities in the past—the manager of a leading supermarket and a registered dietitian who is the district sales representative for a national drug company—who agree to provide financial support for these activities on behalf of their organizations. She expands her network to include the owner of a popular ladies' dress shop, who is interested in "doing some community work." This contact proves profitable—the owner of the dress shop is willing to contribute $1,000 toward the launch. Other commitments are confirmed over a period of several months. The team 2 leader assures all grantors that their

TABLE 19-5 Sample Operating Budget for the "Heartworks for Women" Program

Revenue		
Program income ($60/participant × 75 participants/worksite × 60 worksites)		$270,000
Sales of merchandise		4,500
	Total revenue:	$274,500
Expenses		
Direct expenses		
Personnel—salaries		$163,500
Telephone/Internet		4,250
Equipment		2,000
Travel		1,575
Postage		3,500
Office supplies		8,000
Instructional materials		5,500
Advertising and promotion		10,000
Design		3,750
Printing		3,000
	Total direct expenses:	$205,075
Indirect expenses		
Personnel—benefits (28.7%)		$ 46,925
Rent		7,800
Utilities		1,200
Equipment depreciation		7,200
Maintenance and repairs		2,000
Janitorial services		4,300
	Total indirect expenses:	$69,425
	Total expenses:	$274,500
	Net revenue (expenses):	$0

contributions will be acknowledged publicly and negotiates such details as the amount of the grant and the timing of the payment.

The Business of Community Nutrition

Whether you work in the public or the private sector, you need strong management skills. You must be able to set a direction for your business or program; define goals and objectives; organize the delivery of your product or service; motivate people to help your organization reach its goals; allocate materials, equipment, personnel, and funds to your operations; control data systems; and provide leadership. More than ever before, management and leadership skills are needed to gain a competitive edge in what has become an increasingly competitive health care environment. As you move into the community with your ideas, products, and nutrition services, keep these four strategies for success in mind:[39]

- Continually assess the competitive environment.
- Continually assess your strengths.
- Build organizational skills.
- Build managerial (people and process) skills.

Internet Resources

American Society for Healthcare Foodservice Administrators	**www.ashfsa.org**
Business Tools	**www.score.org/business_toolbox.html**
Consumer Information Center	**www.pueblo.gsa.gov**
National Food Service Management Institute	**www.olemiss.edu/depts/nfsmi/index.html**
School Nutrition Association	**www.schoolnutrition.org**
Links to Business Journals and Management Magazines	**www.brint.com/magazine.htm#ISJbus**
Leadership: Getting It Done by R. Campbell	**www.ssu.missouri.edu/faculty/rcampbell/leadership/default.htm**
U.S. Department of Labor	**www.dol.gov**
Americans with Disabilities Act Home Page	**www.usdoj.gov/crt/ada/adahom1.htm**
U.S. Department of Labor, Employment and Training Administration	**www.doleta.gov**
Free Management Library	**www.mapnp.org/library**
U.S. Small Business Administration (SBA)	**www.sba.gov**
The Service Corps of Retired Executives	**www.score.org**
Official Business Link to U.S. Government	**http://business.gov**
The Peter Drucker Archives	**www.druckerarchives.net**

Professional Focus

Time Management

Time is a nonrenewable resource, a precious commodity. "Time is life," writes Alan Lakein in his best-selling book *How to Get Control of Your Time and Your Life.*[1] "It is irreversible and irreplaceable. To waste your time is to waste your life, but to master your time is to master your life and make the most of it."

Many of us have difficulty deciding how to spend our time. Every day we must choose among dozens of activities and duties, all competing for our time and attention: family, friends, work, school, shopping, sports, television, movies, hobbies. The list is endless. Given that we all have a limited amount of time available each day, how can we choose from the multitude of options that surround us? The key to effective time management, argues Lakein, is *control.*

Control Is Essential

Control means recognizing that it is easy to become overwhelmed by the number of decisions we face about how we spend our time. The secret to controlling time effectively is making conscious decisions about which activities deserve our attention and effort—and which don't. Taking control of your time means *planning* how you will spend it. Planning starts with deciding what your priorities are.

Quadrant II Is Where the Action Should Be

Stephen Covey, author of *The 7 Habits of Highly Effective People,* writes that the essence of good time management can be summarized in the following phrase: Organize and execute (in other words, *plan*) around priorities. Covey developed a time management matrix, shown in Figure 19-7, to demonstrate how most people spend their time.[2]

Covey defines activities as either "urgent" or "important." Urgent activities demand immediate attention. Important activities are those that will yield long-term results; they require a more proactive approach than urgent activities.

Look at the time management matrix. Quadrant I activities are both urgent and important. These activities are called "crises" or "problems." All of us have to deal with Quadrant I activities at times. The problem with focusing constantly on Quadrant I activities, though, is that doing so tends to lead to stress, burnout, and a sense that we are always putting out brushfires.

Some people spend quite a bit of their time in Quadrant III, thinking that they are in Quadrant I. They are reacting to things that are urgent and appear important but that actually yield short-term results. People who spend their time in Quadrant III are likely to feel out of control. Covey believes that people who devote most of their time to Quadrant III and IV activities lead irresponsible lives, because they do not manage themselves and their actions.

Highly effective people spend their time in Quadrant II activities, such as those listed in Figure 19-7. It is through these activities that we accomplish the truly important things in life.

It's as Easy as ABC

The first step in taking control of your time is setting priorities—determining those things that are most important to you. Lakein calls this process the ABC Priority System. "A" activities have a high value, "B" items a medium value, and "C" items a low value.

Take 3 minutes to write down all of the things you should accomplish, or would like to accomplish, in the next 3 months. These activities probably fall into several categories: personal, family, community, social, school, career, financial, and spiritual goals. Don't hesitate to include some things that are off-the-wall. Now, for each item, assign a priority of A, B, or C. Remember, the A activities are those with the highest value in your life. In addition, rank each activity within each category: A-1, A-2, and so on; B-1, B-2, and so on.

Your A activities are top priorities, the items you want to find time for. The key to finding time for them lies not in prioritizing the activities on your schedule but in scheduling your priorities. In other words, if your A-1 activity is learning Japanese, then schedule time for it, even if you can spend only 10 or 15 minutes a day.

Keep three things in mind as you carry out this task. First, only you can decide what your priorities are. No one else can do this for you. Second, your A list will probably change over time and should be reviewed periodically. What is important to you now may not be important to you in 6 months. Finally, because planning in advance is the best thing you can do for yourself, write down your priorities. "A daily plan, in writing, is the single most effective time management strategy, yet not one person in ten does it. The other nine will always go home muttering to themselves, 'Where did the day go?'"[3]

The Top-10 Time Wasters

People from different jobs and disciplines have similar problems managing time. A survey of 40 sales representatives and 50 engineering managers from 14 countries identified the following activities as the top time wasters:[4]

- Telephone interruptions
- Drop-in visitors
- Meetings (scheduled and unscheduled)
- Crises
- Lack of objectives, priorities, and deadlines
- Cluttered desk and personal disorganization
- Ineffective delegation
- Attempting too much at once

FIGURE 19-7 Time Management Matrix™

Source: Excerpted from *The 7 Habits of Highly Effective People,* p. 151, by Stephen R. Covey. Reprinted by permission of the Free Press, a division of Simon & Schuster Adult Publishing.

	Urgent	Not Urgent
Important	**Quadrant I** Activities: Crises Pressing problems Deadline-driven projects	**Quadrant II** Activities: Prevention Relationship building Recognizing new opportunities Planning, recreation
Not Important	**Quadrant III** Activities: Interruptions, some calls Some mail, some reports Some meetings Proximate, pressing matters Popular activities	**Quadrant IV** Activities: Trivia, busy work Some mail Some phone calls Time wasters Pleasant activities

- Indecision and procrastination
- Lack of self-discipline

These are not the only time bandits by any means. Paper work, inadequate staffing, too much socializing, and travel all undermine our ability to manage time well. The key to managing these time wasters is to take control of them. Schedule your telephone calls for certain periods of the day, and do not accept interruptions during "quiet" times unless an emergency arises. Minimize interruptions from colleagues and visitors. Attend only meetings that are directly related to your work.

Learn to Say No

When you set your priorities—that is, when you say yes to one thing—you must say no to something else. You will always be saying no to something. You can never do everything that you want to do or everything that everyone else wants you to do! The key to managing your time effectively is to learn to say no, firmly and courteously. Sometimes this can be difficult, for a task that is a B or C priority for you may be an A priority for someone important to you. When this happens, take a moment to consider the consequences for you, the other person, and your mutual relationship if you say no. In many cases, you can reach a compromise and maintain goodwill by deciding together how to rank an activity.

Work Smarter, Not Harder

It is a myth that the harder you work, the more you accomplish. Many workaholics fall into this trap. They are more attuned to time spent working than to time spent working effectively. Avoid the trap of thinking that time spent working is automatically time spent valuably. "If it is not quality time and quality work, you are wasting time."[5]

Mastering the principles of good time management can take several months or even years. If you experience difficulty achieving control of your time, don't give up. Keep working at it. Although your progress may seem slow, you are further along than if you had never begun this challenging process at all.

References

1. The description of the ABC Priority System was adapted from A. Lakein, *How to Get Control of Your Time and Your Life* (New York: Penguin, 1973).
2. S. R. Covey, *The 7 Habits of Highly Effective People.* New York: Simon & Schuster. © 1989 Stephen R. Covey. Used with permission. All rights reserved.
3. The quotation is from A. Mackenzie, *The Time Trap* (New York: AMACOM, 1990), p. 41.
4. M. LeBoeuf, Managing time means managing yourself, in *The Management of Time,* ed. A. D. Timpe (New York: Facts on File Publications, 1987), pp. 31–36.
5. The quotation is from J. C. Levinson, *The Ninety-Minute Hour* (New York: Dutton, 1990), p. 168.

Chapter *Twenty*

Building Grantsmanship Skills

Carol Byrd-Bredbenner, PhD, RD, FADA

Learning *Objectives*

After you have read and studied this chapter, you will be able to:

- Outline the general process for preparing a grant proposal.
- Conduct the foundational work needed to generate a grant proposal.
- Develop a grant proposal.
- Analyze grant proposals and identify their strengths and weaknesses.

This chapter addresses such issues as written communications and technical writing, documenting appropriately a variety of activities, preparation of a budget, and current information technologies, which are Commission on Accreditation for Dietetics Education (CADE) *Foundation Knowledge and Skills* requirements for dietetics education.

Chapter *Outline*

Something To Think About...

Determine that the thing can and shall be done and then find the way.

– Abraham Lincoln

Introduction

Community nutritionists increasingly find themselves seeking extramural funding for program activities and interventions. Perhaps extra funds are needed for printing banners and T-shirts, publishing a newsletter, underwriting the advertising costs for a program launch, buying food supplies for a cooking demonstration, marketing nutrition messages on a local television station, developing and evaluating new program materials, or expanding services. Annually, grants totaling over $150 billion are distributed by public and private groups. These grants are vital in helping today's organizations fund many of the important projects that often exceed the resources they have available. To capture grant dollars for their organizations, grant seekers will find that well-honed grantsmanship skills (proposal seeking, writing, and management abilities) are essential.

Although you may not realize it, you have been creating proposals that resulted in awards of services, goods, and money throughout your life. Babies present a compelling argument for attention every time they cry. As children grow and mature, they may use their rapidly developing verbal and reasoning skills to explain why they require an increased allowance or how their communication needs warrant the purchase of a cell phone. In fact, both a child's informal request and a formal grant proposal submitted to a public or private grant sponsor are designed to achieve the same goal: convince a decision maker to provide the services, goods, and/or money the recipient needs to achieve the objectives stated in the proposal.

Proposals range in length from a single page to more than a hundred pages. Regardless of the length, the development of proposals can be divided into three main steps: laying the foundation, building the grant proposal, and assembling the final product. The fourth step, reviewing the grant proposal, is completed by the grant reviewers selected by the grant sponsor. The final step is putting the proposal into action if it is funded or revising it in hopes of future success.

Laying the Foundation for a Grant

Writing a grant is a lot like building a house. To build a house that will be sound and pass the reviews of building inspectors, builders must first lay a solid foundation. The grant proposals deemed most worthy of funding by grant review panels also grow out of a solid foundation. The foundation work for grant writing includes generating ideas, describing goals, and identifying funding sources and potential collaborators.

Generate Ideas

Proposal preparation begins with an idea. Ideas can be generated in a variety of ways, including noting a legislative initiative, reading the implications for research included at the end of many journal articles, observing societal trends or needs in the community, brainstorming with colleagues, and reviewing statistical data such as census figures or morbidity and mortality rates and causes. Ideas also may come from a grant sponsor (see the section, later in this chapter, on identifying funding sources).

In community nutrition, the ideas that lead to grant proposals often are related to a desire to better understand the parameters affecting dietary behaviors and/or a desire to improve dietary behaviors. As ideas gel, they become goals such as the following.

- Gather baseline data. Examples: Survey adult women to determine their ability to use the nutrition facts given on food labels to make dietary planning decisions. Examine the effect that television food advertisements have on children's food preferences.
- Create a nutrition education intervention based on a new learning theory or innovative teaching method, and assess its effectiveness. Examples: Assess the effectiveness of a series of nutrition classes based on the transtheoretical model on lowering the dietary fat intake of cardiac bypass patients. Compare the impact of nutrition education comic books and nutrition website activities on teens' vitamin C and vitamin A intake. Evaluate the impact of a social marketing campaign to increase calcium and iron intake on blood lead levels in urban children. Evaluate the impact of a Web-based, self-paced instructional program focusing on common problems in feeding children and on parents' comfort and satisfaction in dealing with these problems.
- Provide a service and evaluate its impact. Examples: Evaluate the impact of participation in the School Breakfast Program on school tardiness and absence rates. Distribute nutritious foods to pregnant women and assess the impact of these foods on maternal weight gain, infant birthweight, and length of gestation. Compare bone density changes in women who received vouchers for calcium-rich foods with those in women who continued eating their regular diet and received calcium supplements. Determine the relationship between the availability of nutrition peer counselors in college dorms and the rate of problems with weight gain and weight loss among residents.

Once an idea is formulated, the next step is to review the literature for *at least* the last 5 years and preferably more. In your search of the literature, you should address all facets of the idea, keeping in mind that many facets may not be directly related to nutrition. For example, let's say your idea is to design a school-based nutrition education class series and determine its impact on teens' calcium intake. A literature search for this project should include investigating current studies on teen calcium needs and intake levels; factors affecting teens' intake of calcium-rich foods; preferences of teens and teachers with regard to learning activities; physical and emotional developmental characteristics of teens; and learning theory applied to teen education programs, especially those aimed at changing health and nutrition behavior.

Surveying the literature is a worthwhile investment of time. It enables grant seekers to assess the value of an idea and adjust its focus to describe their goals more clearly.

- Grant seekers may find that someone else thought of the idea first. If so, the grant seeker should think about how the idea differs from or complements existing resources or activities. For example, several well-designed, effective curricula addressing calcium intake may already be available. In that case, the grant seeker may decide to narrow the focus of the idea to implement only one of these existing curricula and evaluate its impact rather than also developing a curriculum.
- Grant seekers may discover that the idea is too narrow. For instance, they may learn that the main barrier to getting teen girls to increase their milk intake is that the girls believe they'll gain weight. If so, the grant seeker might choose to widen the focus of the idea to include healthful, low-fat choices and fat trade-offs from all the food groups.
- Grant seekers may find that they should alter their focus if they discover that the need for programs that help teens control their portion sizes hasn't been addressed effectively. Alternatively, they might discover that the face-to-face instruction they were planning isn't how today's technology-savvy teens want to learn about healthful eating.

Not only will the literature search help grant seekers to more fully develop and refine an idea and to describe their goals; it will also help identify other professionals who could be sources of valuable advice and/or potential collaborators. And the literature search will provide the background data they will require to write a compelling needs statement for the proposal. When the literature search nears completion, grant seekers should be able to describe the following, clearly and concisely, in two or three minutes: their idea, the uniqueness of the proposed project, and why it is likely to succeed. This preparedness will help them when the time comes to contact potential collaborators and grant sponsors.

KEY TO SUCCESS

The most successful grant writers are up-to-date in their subject matter area. They read key journals regularly and are well informed about current trends and activities. The time needed to complete a literature review can be substantial if the idea is in a new area. In recent years, however, online abstracts and journals and Internet search engines have greatly facilitated the task of locating and retrieving pertinent resources. Also, keep in mind that grant seekers can reap long-term benefits from the literature review by focusing their work on a particular subject matter area (for example, diabetes or weight control) and/or audience (for example, children or women) over the course of several years. Once the initial time investment has been made in developing an expertise in a subject matter area and/or audience, only a small time investment is needed weekly or monthly to remain abreast of new developments. Moreover, focusing one's work helps to develop a track record of working in a particular topic, which, in turn, establishes one's credibility as an authority on the topic.

Describe Goals

Goals are broad statements describing desired long-range improvements. The two broad goal statements in *Healthy People 2010* are good examples of how to state goals:

- Increase the quality and years of healthy life.
- Eliminate health disparities.

Note that the goals grant seekers set are likely to be less expansive than the *Healthy People 2010* goals because grant seekers are setting goals for their organizations, not for the entire nation.

In community nutrition, there are so many worthwhile goals and vital needs that it is sometimes difficult to know which should be addressed first. However, perceptive grant seekers know that they improve their chances of receiving funding if they focus on addressing important goals that can be achieved with the combined finite resources available from their organization and the grant sponsor. *Healthy People 2010* goals and objectives, along with the following questions, can help grant seekers narrow the list of possible goals. If a question can be accurately answered yes, proceed to the next question. Otherwise, seriously consider adjusting the goal.

1. Are the goal and the need(s) it will fulfill congruent with the grant seeker's mission? Most employers have a mission statement that describes the broad scope of their work. A review of the organization's mission statement can quickly help grant seekers determine whether a goal is appropriate to pursue. For example, the mission statement in Figure 20-1 clearly indicates that the ABC Institute's mission is to help people improve their lives through educational processes.* Thus, grants designed to deliver prenatal care

**Note:* Figures 20-1 through 20-3 and 20-5 through 20-14 were developed for illustrative purposes only. Figures 20-5 through 20-14 are a composite of several actual grant proposals; these figures are not from an actual proposal.

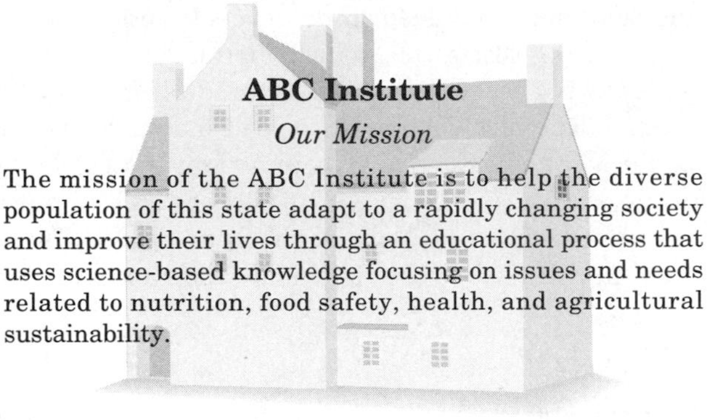

FIGURE 20-1 Example of an Organization Mission Statement

to teenage girls would not be an appropriate pursuit for a grant seeker employed by the ABC Institute. However, the grant seeker might collaborate with other organizations, such as a hospital that could provide the obstetrical care, and propose to provide only nutrition counseling and education to the pregnant teens. If the answer to question 1 is clearly no, then the goal is not an appropriate one for the grant seeker's organization to address.

2. Is working to meet this goal a priority of the grant seeker's organization? If not, think carefully before proceeding to work on this goal and the need(s) it will fulfill. For success (and sometimes to gain permission to work on the goal), it is important to have the organization's support.

3. Is it possible to achieve the goal in a timely manner? Some goals and needs are extremely difficult to address. Before deciding to tackle such a goal, think carefully about the commitment, time frame, funding, and other resources that will be required. Weigh the required inputs against the possibility of meeting several other needs, instead, with the same time, money, and energy. It may be more worthwhile to focus the organization's efforts in a way that generates the greatest community impact and likelihood for success than on needs arising from virtually intractable problems.

4. Is working to achieve the goal an objective stated in *Healthy People 2010*? *Healthy People 2010* is our nation's health agenda. It can serve as the foundation for all work in community nutrition related to health promotion and disease prevention. If the goal is not part of *Healthy People 2010* objectives, community nutritionists should consider whether other, more critical or more widespread needs should be addressed.

5. Would working to achieve the goal meet the needs of a substantial number of people? Is their need severe? If this need is not met in the near future, are the long-term consequences likely to be severe? If the answer to all of these questions is no, there are probably other goals and needs that it is more important for the grant seeker's organization to address. However, if the need is severe in the short term or long term for even a small proportion of the population, grant seekers should consider addressing it.

6. Have those in need expressed an interest in having this need attended to? If not, consider why. Perhaps they are not aware of their need, distrust those who have offered help in the past, or do not see this need as a problem. The reason for their lack of interest may indicate that a different need must be met before the proposed need is addressed. If those in need are interested in having the need attended to, grant seekers should think about whether their organization is willing to work collaboratively with

them. If not, the likelihood of successfully achieving the goal and meeting their need is limited.

Once the goals have been selected, the grant seeker is ready to write clearly defined, specific goals that will meet important needs and answer important questions. When a project's goals are too broad or poorly defined, it is difficult to establish measurable objectives, create an evaluation plan, or develop a proposal budget and time schedule.

In today's world, it is still possible to find grants specifically designed to establish or enhance a community service, such as distributing food packages or providing nutrition counseling. However, more and more grant sponsors also want to know what impact their funds had on the community. For example, a grant sponsor is likely to want to know more than the fact that the $50,000 it gave was used to buy fresh produce that was distributed throughout the year to 29 limited-resource families. The sponsor might also want to know that participation in the food packages program was associated with improved growth patterns among toddlers or that participants had fewer school or work absences due to infections. Because of this increasing need to demonstrate the impact of grant dollars, it is often helpful to begin by stating grant goals in the form of precisely worded questions, as shown in Table 20-1. In this table, the question on the right is just one example of how the question on the left can be made more specific. For practice, try to write a different, more precise question for each of the questions that are too broad.

Note that the questions in the right column of the table can easily be converted to a goal statement. For instance, the first could be rephrased as "Determine whether a weight-loss clinic that incorporates behavior modification techniques helps obese women decrease their intake of simple carbohydrates." Try to write a goal statement for each item in Table 20-1.

TABLE 20-1 Framing Broad versus Precise Goal-Related Questions

TOO BROAD	MORE PRECISE
Does nutrition education change people's food patterns?	Can a weight-loss clinic that incorporates behavior modification techniques help obese women to decrease their intake of simple carbohydrates?
Can counseling sessions improve blood lipid levels?	Do counseling sessions geared to a teen's stage of change for fat intake lead to lower LDL values?
Does poor personal hygiene increase the rate of foodborne illness?	Can the rate of school absences due to gastrointestinal distress be reduced by having middle school children wash their hands with warm soapy water before eating lunch?
Does poor diet reduce the effectiveness of education?	Is regular participation in the School Breakfast Program associated with increased reading scores in third-grade children?
Are women who participate in WIC better nourished?	Do pregnant women who participate in WIC for at least 5 months before delivery give birth to fewer low-birthweight and/or preterm infants?
Can social marketing campaigns reduce health care costs?	Can a social marketing campaign promoting increased physical activity levels reverse the trend toward increased body weight and related disease among residents in Green County?
Do Nutrition Facts labels help people eat better diets?	Can an interactive on-line presentation that teaches women with Type 1 diabetes mellitus how to use nutrition labels improve their hemoglobin A_{1c} (HbA_{1c}) levels?

KEY TO SUCCESS

Successful grant seekers choose worthwhile goals that match their organization's mission and interests, and address important needs that can be met in a meaningful and timely manner. They also sharply focus their goals. They know that most grants have a finite dollar amount and time frame, and they are well aware of the resources their own employer has available for the proposed project. Therefore, to stay within the parameters of the grant and the grant seeker's resources (such as available personnel or office space), successful grant seekers divide multifaceted projects into manageable pieces and then decide which pieces to work on first.

Identify Funding Sources

Finding a funding source generally takes two forms: generating an idea in response to a grant sponsor's request or finding a grant sponsor to fund the grant seeker's idea.

- The grant seeker generates an idea in response to a grant sponsor's Request for Proposals (RFP) or Request for Quotation (RFQ). Both an RFP and an RFQ invite grant seekers to submit proposals. An RFP (also called a Request for Applications [RFA]) tends to be much less specific in regard to the activities that can be proposed (see Figure 20-2), whereas an RFQ tends to be very specific about the activities that the grant recipient must engage in. That is, the issuers of RFQs are looking to award

FIGURE 20-2 Two Sample Requests for Proposals (RFPs)
Note the general nature of an RFP.

Good Health Foundation

The Good Health Foundation operates under the laws of the state. Only nonprofit organizations are eligible to receive a Good Health Foundation grant. Grant recipients must submit an annual progress report and audited financial statement of the expenditures of the foundation grant. The Good Health Foundation invites eligible applicants to submit grant proposals that:

- Support community nutrition education programs for children and young adults that improve their health and development.
- Provide nutrition counseling for pregnant women and mothers of infants so that they can assist their children in developing good nutritional practices at an early age.
- Train persons to work as educators and demonstrators of good nutritional practices.
- Promote dissemination of information regarding healthful nutritional practices.

Federal Government Agency—Brand New Research Initiative

99.0 Improving Nutrition and Food Handling for Optimal Health

Investigators are encouraged to contact the Program Director at (000) 000-0000 regarding questions about the suitability of research topics (or at programdirector.gov to arrange a telephone consultation).

- Standard Research project awards for this program are not likely to exceed a total budget (including indirect costs) of $300,000 for 3 to 4 years of support.
- Program Deadline: November 15

The consumption of a nutritious diet is important for maintaining health and decreasing the risk for chronic diseases. There is a need to improve understanding of the role of foods and their components (e.g., phytochemicals) in promoting health. The primary objective of this program is to support research that contributes to our understanding of appropriate dietary practices throughout the life cycle and factors that affect these requirements such as gender. Studies to determine the effects of dietary components on immune function are encouraged. In addition, new insights are needed about factors that affect the attitudes and behavior of consumers toward food.

The following areas of research will be emphasized: (a) nutritional requirements including metabolism and utilization for all age groups; (b) effect of community projects on lowering the incidence of obesity through increased physical activity; (c) identification of obstacles to adopting healthful food habits, with particular emphasis on factors affecting consumer attitudes and behavior; (d) development of recommendations for interventions to improve nutritional status; and (e) improvement of safe food handling behaviors among those at increased risk for foodborne illness. Support will not be provided for research on dietary requirements as related to therapies for infectious diseases, cancer, and alcohol-related disorders.

The XYZ Government agency seeks a researcher with established expertise in the evaluation of classroom nutrition education materials to conduct a study to assess the use and effectiveness of the Healthy Eating Kit as an instructional resource in high schools. Information from the study will be used to revise the Kit; develop future nutrition education classroom materials; and document the nutrition-related subject matter, types of lessons, and hands-on activities that are of greatest interest to teachers and have the most impact on students. The Kit will be assessed by brief interviews of 150 teachers that focus primarily on questions from the evaluation form included in the Kit and in-depth qualitative interviews of 30 teachers. In-depth interviews will focus on the effectiveness of the Kit in communicating with culturally diverse youth, methods for adapting the Kit for use with younger children, and suggestions for communicating nutrition topics in the future. Study participants may be recruited through the evaluation forms returned to the XYZ Government Agency, State Departments of Education, and advertisements in professional journals. Efforts must be made to represent all geographic regions in the United States, culturally diverse student segments, and urban, suburban, and rural areas.

FIGURE 20-3 A Sample Request for Quotation (RFQ)
Note that this RFQ is much more specific than the sample RFPs in Figure 20-2.

a contract for work they specified and want grant seekers to provide a cost estimate for that work (see Figure 20-3).

- The grant seeker generates an idea for which he or she seeks an appropriate grant sponsor. Regardless of the goal and need, there is probably a government or community agency, industry trade group, food or drug company, or local business willing to support the program or intervention activities. According to some, "If you can find an effective solution to a pressing community problem, you can find funders to help you implement it."[1] The challenge is to find the grant sponsor who is willing to fund the proposal.

Many types of grants are available (see Figure 20-4). How do community nutritionists find out about these grants? They network with colleagues to learn about upcoming funding opportunities; they contact granting agencies for information about RFPs; they call local businesses to seek small grants to support community projects. In the United States, grant funding usually comes from government agencies, foundations and community trusts, and business and industrial organizations (see Table 20-2).The Internet Resources section near the end of this chapter lists the websites of potential funding sources.

When grant seekers begin to explore potential grant sponsors, they should think about each facet of the project.[2] For example, say the goal is to determine the effect of early intervention on the incidence of Type 2 diabetes mellitus in overweight teenage boys. Grant seekers should ask themselves:

- Will the project provide nutrition education to the boys or training to physical education teachers who work with the boys? If so, the grant seekers could try a sponsor that funds after school projects or in-service training for teachers.
- Will the project use exercise equipment? If so, the grant seekers could try sporting goods manufacturers or sneaker makers.
- Will the project involve the use of monitoring equipment? If so, they might try a pedometer maker or pharmaceutical company that markets glucose monitoring supplies.
- Will the project include a community awareness component? If so, it would make sense to try a sponsor that funds social marketing campaigns or a nonprofit organization that works to raise awareness about health issues.
- Will the project involve providing more low-calorie, fiber-rich foods such as fresh produce and whole grains? If so, they might try a supermarket chain, produce association, or grain trade group.

FIGURE 20-4 Types of Grants

Block grant: A grant from the federal government to states or local communities for broad purposes as authorized by legislation. Recipients have great flexibility in distributing such funds as long as the basic purposes are fulfilled. The five federal block grant areas are maternal and child health, community services, social services, preventive health and health services, and primary care.

Capitation grant: A grant made to an institution to provide a dependable support base, usually for training purposes. The amount of the grant is dependent on the size of enrollment or number of people served.

Categorical grant: A grant similar to a block grant, except funds must be expended within specific categories, such as maternal and child care. Funds available may be based on a specific formula (see *Formula grant* below).

Challenge grant: A grant that serves as a magnet to attract additional funding.

Conference grant: A grant awarded to support the costs of meetings, symposia, or special seminars.

Demonstration grant: A grant, usually of limited duration, made to establish or demonstrate the feasibility of a theory or an approach.

Equipment grant: A grant that provides money to purchase equipment.

Formula grant: A grant in which funds are provided to specified grantees based on a specific formula, prescribed in legislation or regulation, rather than based on an individual project review. The formula is usually based on such factors as population, enrollment, per capita income, morbidity and mortality, or a specific need. Capitation grants are one type of formula grant, and block grants usually are awarded based on a formula.

Matching grant: A grant that requires the grant recipient to match the money provided with cash or in-kind gifts (such as donated products, supplies, equipment, services) from another source.

Planning grant: A grant made to support planning, developing, designing, and establishing the means for performing research or accomplishing other approved objectives.

Project grant: The most common form of grant, made to support a discrete, specified project to be performed by the named investigator(s) in an area representing his or her specific interest and competencies.

Research grant: A grant made to support investigation or experimentation aimed at the discovery and interpretation of facts, the revision of accepted theories in light of new facts, or the application of new or revised theories.

Service grant: A grant that supports cost of organizing, establishing, providing, or expanding the delivery of health or other essential services to a specified community or area. This also may be done through a block grant.

Training grant: A grant awarded to an organization to support costs of training students, personnel, or prospective employees in research, or in the techniques or practices pertinent to the delivery of health services in the particular area of concern.

- Is this a demonstration project? If so, grant seekers could try a local foundation or municipal government.

Once grant seekers have identified a potential grant sponsor, they must decide whether to compete for it. According to the old adage, "It takes two to tango." Nowhere is this truer than with procuring grant funds. For success, the needs of the grant seeker must be in tune with the needs and interests of the funding source—that is, the proposal must clearly demonstrate how the grant seeker's goals complement those of the grant sponsor.[3] The following questions can help grant seekers decide whether to compete for a grant from a particular grant sponsor.

TABLE 20-2 Types of Funding Agencies

TYPE	EXAMPLES
Federal	National Institutes of Health; National Cancer Institute; National Institute on Aging; Department of Health and Human Services; U.S. Agency for International Development; National Heart, Lung, & Blood Institute; U.S. Department of Agriculture; Food & Drug Administration
State and local	Department of education, department of human services, health department, arts council, local school district
Foundations	Ford Foundation, Spencer Foundation, Rockefeller Foundation, International Life Sciences Institute—Nutrition Foundation, March of Dimes
Nonprofit organizations	American Diabetes Association, American Heart Association, American Dietetic Association, American Cancer Society, local civic groups
Industry	Pharmaceutical companies, trade associations, commodity groups, local supermarkets
Institutional	Local universities and hospitals

Source: Adapted from M. R. Schiller and J. C. Burge, How to write proposals and obtain funding, in *Research: Successful Approaches*, 2nd ed., ed. E. R. Monsen (Chicago: American Dietetic Association, 2003), pp. 49–68.

1. Do the grant sponsor's funding priorities include the project's goals? If not, can the project's goals be easily revised to be congruent? Even the most eloquently written proposal is unlikely to be funded if its goals do not match those of the grant sponsor.[4]

2. Are sufficient grant funds available for the grant seeker's organization to achieve the goals in a successful and timely manner? If not, ask whether your organization is willing and/or able to provide sufficient resources to augment grant funds? Is there another organization that could serve as a partner and furnish some funds? If grant funds combined with resources from the grant seeker's organization and partners are insufficient, the likelihood of achieving the goals is very low.

3. What will this grant sponsor fund (for example, personnel, travel, or office supplies)? Is that what is needed? Every proposal has specific needs. If the grant sponsor is not able to support those needs, the grant objectives cannot be achieved. Let's say the project is one that would require travel to several states to train dietitians to use a teaching kit, but the sponsor won't fund travel. If face-to-face instruction is essential, this sponsor is not a good match unless travel funds are available elsewhere. Alternatively, if face-to-face instruction is not essential, the proposal could be revised to use other methods of training, such as Web conferencing or a self-instructional manual, which would be eligible for funding.

4. Who is eligible to apply for this grant? Some grants are limited to specific types of organizations such as nonprofit agencies, accredited universities, or outpatient clinics. If the grant seeker's organization is not eligible, an alternative may be to collaborate with another organization that is.

5. Can the grant seeker's organization meet any special requirements of the grant sponsor? For example, some grants may require audited quarterly financial reports. Some require the grant recipient to match grant funds dollar for dollar. If the organization cannot meet the special requirements, other funding sources will need to be considered.

6. When is the grant application due? Can the grant seeker's organization respond in time? If not, when will this grant opportunity be available again? Writing a grant proposal, depending on the sponsor's requirements and the applicant's prior experience and readiness, can take as long as 6 months. Thus grant seekers must honestly assess prior time

commitments and how much time will be required to generate a thoughtful, well-written proposal. Proposals that arrive after the due date are almost always discarded. And hastily written proposals may establish a negative reputation for the grant seeker and his or her organization.

7. What are the guidelines for writing this proposal? What forms and format are required? Every grant sponsor publishes some type of proposal-writing guidelines; some are extremely specific, others more general. For success, it is essential that grant seekers follow such guidelines, forms, and format exactly. In addition, grant seekers need to be certain to use the forms provided by the sponsor; trying to revise the forms to better fit one's proposal will create extra work (and frustration) for grant reviewers, who are expecting to find information presented in a specific way. Many grant sponsors will discard proposals that are not prepared according to their instructions. Even though proposal guidelines, forms, and formats vary from sponsor to sponsor, you will see below that the elements in proposals tend to be quite similar across grant sponsors.

8. Is information available on the types of projects the sponsor has funded in the past? Many grant sponsors publish abstracts of projects they have funded. Increasingly, the abstracts are available on their websites. Reviewing these can help grant seekers determine how well their idea is likely to be received. They may find that their project is too similar to grants already funded to reveal anything new. On the other hand, they may learn that the sponsor says it has six funding priorities but seems to fund only two priorities. Grant seekers also may want to contact a few grant recipients and ask them to share their experiences and insights related to procuring a grant from this sponsor and how they would approach this activity differently next time. Some may be willing to share a copy of their winning proposals.

9. What are the credentials of the grant reviewers? Some sponsors reveal the names of past reviewers, but many do not. However, nearly all sponsors will indicate the general background of reviewers. Knowing the reviewers' level of expertise can help grant seekers determine the level of detail and documentation to use in their proposal. For example, if dietitians and physicians are to review the proposal, grant seekers will be able to use precise medical terminology, which might not be appropriate if the reviewers have a nonmedical background. If the sponsor does reveal the names of past reviewers, grant seekers might ask them to share their insights on how the review process worked, pitfalls to avoid, and general advice.

10. Is it possible to contact the sponsor before preparing the proposal? Most government sponsors and some private sponsors welcome inquiries (see Figure 20-2 on page 602). Contacting the sponsor before preparing a proposal can save grant seekers and the sponsor time. Moreover, it helps to verify the accuracy of the information in the call for proposals—sometimes changes occur after the RFPs or RFQs are printed. After carefully reading the call for proposals, reviewing past funded projects, and speaking with reviewers, grant seekers may still have questions or need more information on how to fine-tune project goals to be fully congruent with the grant sponsor's priorities. For example, if the sponsor seems to be funding only two of its six priorities—and the two are not the priority area the proposal will be in—grant seekers might ask why. It could be that the sponsor did not receive any worthy proposals in other areas or that the two priorities funded are its top priorities.

KEY TO SUCCESS

The most successful grant seekers select a grant sponsor before writing a proposal. They understand the value of doing their homework so that they do not waste time writing a proposal for a sponsor that is unlikely to fund work in the area proposed. They also know that it is essential to follow the grant guidelines, forms, and formats exactly and to observe all deadlines. In addition, they write the proposal using terminology familiar to the grant reviewers—after all, it is the reviewers who will make funding recommendations to the sponsor. Even successful grant writers experience failure; but they capitalize on the time invested in writing a proposal by preparing it using word processing software. This way, they can easily cut, paste, and reformat the proposal for other grant sponsors or revise it for the same sponsor for the next funding cycle.

ANOTHER KEY TO SUCCESS

The most successful grant seekers avoid "chasing" grants. That is, they keep their work focused and do not jump from topic to topic just to capture the latest stream of grant funding. Unfocused grant chasers usually feel frustrated because they are constantly changing gears and trying to develop expertise in many, often unrelated areas. In addition, without a history of successful related projects, grant chasers will have difficulty demonstrating to grant reviewers that they can deliver the work proposed. Keeping work focused does not mean that one is forever locked into a single topic or audience. Successful grant seekers can and do change gears, but the progression is usually an evolutionary one. That is, their work naturally branches into new areas as science progresses, community demographics change, or new problems arise. For instance, a professional working with Type 2 diabetes mellitus (DM) treatment in adults may branch into prevention of this disease in children. Or, because this disease is a risk factor for cardiovascular disease (CVD), this professional may branch into CVD prevention work. Alternatively, because obesity is a risk factor for this type of DM, this professional's work may branch into weight-control programming for teens. Grant seekers, however, who *do* want to make an abrupt or radical change in their focus can increase their chances of success by working with collaborators to generate a history of successful work in the new focus area.

Identify Potential Collaborators

It is possible to complete the work in some grant proposals alone, but one person usually does not have all the expertise or time needed to complete all the proposed tasks. For instance, depending on the project, a community nutritionist may need the expertise of a statistician, epidemiologist, or physician to complement his or her nutrition expertise. Collaboration is the rule rather than the exception.

Collaborators may come from within the grant seeker's organization or from outside. For example, if the organization is not eligible to apply for a particular grant, it will need to partner with another organization that is eligible. Collaboration with other organizations is important when one organization lacks expertise in a particular area vital to the proposal. Collaborating may be critical to success for novice grant seekers or those who are changing their focus. Increasingly, government grant sponsors are requiring—or giving bonus points to—proposals that are multidisciplinary and/or multi-institutional, thus making collaboration essential. Often, proposals that reflect a multidisciplinary approach to a problem are highly rated.[5]

Collaboration can greatly improve a proposed project. A well-conceived and well-managed collaboration plan gives grant seekers the opportunity to assemble a team that can effectively and efficiently achieve the project goals. The best collaborators are those who know what is expected of them and when, are excited about and committed to the project, and are willing to follow through on all their responsibilities related to the project. A clear system of frequent communication with all collaborators will help to keep everyone on track and moving toward achieving the project goals in a timely manner.

See the Professional Focus feature on page 631 for more about the benefits of teamwork.

Collaborative efforts that are poorly managed or lack teamwork and open communication will hinder success.

Community nutritionists can find collaborators by networking with colleagues and corresponding with individuals identified via the literature search who are working on similar projects. Before inviting someone to be a collaborator, explain the project's objectives, time frame, and budget and gauge his or her interest and ability to participate. You may also want to interview a few references to learn more about the potential collaborator's prior performance record.

KEY TO SUCCESS

Select collaborators carefully, and try to get them involved as early in the grant writing process as possible. Make certain every collaborator knows and agrees to the duties involved, time lines, budget allocation, and sequence of names on publications before the project begins. Keep all collaborators informed of the progress of the grant application.

Building the Proposal

Throughout the grant writing process, keep in mind that the goal is to write a clear, complete, concise, and compelling document that will persuade the grant sponsor to award the grant seeker a portion of its finite resources. The most successful grant proposals demonstrate that the proposed project matches the sponsor's priorities; that it is among the most worthwhile, well planned, sound, and likely to succeed of all the proposal alternatives; and that it is in the hands of qualified professionals who are committed to seeing the project through to completion.

Components of a Proposal

The components of a proposal vary from one grant sponsor to another (carefully check the sponsor's instructions!), but they generally include the items discussed below. Note that the sequence of these items is the order in which they usually appear in a proposal; however, they are not necessarily written in this sequence. For example, the transmittal letter frequently is written last, and the abstract should not be written until after the proposal is in final form.

LETTERS OF INTENT

Increasingly, grant sponsors are encouraging or requiring grant seekers to submit a letter of intent prior to submitting a full proposal. In some cases, the grant sponsor wishes to receive only a very brief letter of intent stating that the grant seeker intends to submit a proposal in response to a specific grant program. This type of letter of intent helps the grant sponsor determine how many proposals it will receive, estimate the time it will need to spend processing the proposals, and determine the number of grant reviewers it is likely to need. Often, grant sponsors do not send a reply to this type of letter of intent.

In other cases, grant sponsors indicate that the letter of intent is a preproposal that includes a synopsis of most or all of the parts of a full proposal. This type of letter of intent should give the grant sponsor a clear picture of the type of proposal a grant seeker intends to submit. Using this type of letter of intent, the grant sponsor can get an idea whether the proposal is suitable for its grant program. Most grant sponsors respond to this type of letter of intent and indicate to the grant seeker whether or not it should submit a full proposal.

January 11, 20XX

Federal Government Agency
Brand New Research Initiative
Proposal Services Unit
Room 000 Canal Place
000 Main Street, N.W.
Washington, DC 00000

Dear Sir or Madam,

My colleagues and I are pleased to submit the enclosed proposal, *Improving Safe Food Handling Behaviors of Immunocompromised Adults*, in response to the Federal Government Agency's Brand New Research Initiative call for proposals, category 99.0, Improving Nutrition and Food Handling for Optimal Health. The proposal requests $185,289 for a two-year project. Enclosed please find the original and 19 copies of this proposal, as specified in the Request for Proposals.

If you have questions or need additional information, please contact me. I am looking forward to your reply.

Sincerely,

Greta Grantwriter

Greta Grantwriter, Ph.D., R.D., F.A.D.A.
Nutrition Services Director
Super State University
000 Elm Street
University City, USA 00000
Phone: 222-222-0000
Fax: 222-222-1111
E-mail: grantwriter@superuniversity.edu

FIGURE 20-5 Proposal Transmittal Letter

The letter of intent should be sent, by the due date, to the individual specified on the call for proposals. This letter should include all of the information requested and should indicate how the grant sponsor can contact the grant seeker.

TRANSMITTAL LETTER

The transmittal letter is a brief, friendly communication addressed to the individual designated on the call for proposals. As shown in Figure 20-5, this letter should indicate the reason for submitting the proposal. That is, is the proposal in response to a call for proposals, is it an invitation to submit a full proposal after submitting a preproposal or letter of intent, or is it an unsolicited proposal? Some grant sponsors have several grant programs under way simultaneously, so grant seekers must clearly indicate the grant program to which they are submitting the proposal. The transmittal letter also should state the name of the proposal and the number of copies enclosed. And grant seekers must be sure to indicate how the grant sponsor can contact them.

TITLE PAGE

Many grant sponsors, especially government agencies, have a specific form that must serve as the title page. If the grant sponsor does not have a specific form or provide information on how to set up the title page, include the following information on the title page: project title; the grant program the proposal is being submitted to; proposed start and end dates of the project; funds requested; project director's name, address, phone number, fax number, and e-mail address; legal name of the organization to which the award should be made; and authorized organizational representative's name, title, address,

FIGURE 20-6 Sample Proposal Title Page When No Form or Format Is Specified

Project Title: Improving Safe Food Handling Behaviors of Immunocompromised Adults
Grant Program: Federal Government Agency's Brand New Research Initiative, category 99.0, Improving Nutrition and Food Handling for Optimal Health
Proposed Start Date: September 15, 20XX
Proposed End Date: September 14, 20XX
Funds Requested: $185,289

Project Director:
Greta Grantwriter, Ph.D., R.D., F.A.D.A.
Nutrition Services Director
Super State University
000 Elm Street
University City, USA 00000
Phone: 222-222-0000
Fax: 222-222-1111
E-mail: grantwriter@superuniversity.edu

The award should be made to:
Super State University

Authorized Organizational Representative:
Joseph Jones, Dean of Research
Office of Research & Sponsored Programs
00 University Plaza
University City, USA 00000
Phone: 222-222-2222
Fax: 222-222-3333
E-mail: authorizedorganizationalrep@superuniversity.edu

telephone number, fax number, and e-mail address. See Figure 20-6. Note that some of the information from the transmittal letter is repeated on the title page—it is important that this information appear in *both* places. Keep in mind that although the title page is the first page of every copy of the submitted grant proposal, only the individual designated to receive the proposal submission is likely to see the transmittal letter.

The proposal title should be brief, specific, and informative. It should set the stage for the proposal and quickly and accurately inform the grant reviewers about what they will be reading. Note that some sponsors impose a limit on the number of characters (and spaces) permitted.

ABSTRACT

The proposal abstract or summary outlines the proposed project and appears at the beginning of the proposal. The abstract is usually limited to one page or a specified number of words. Some sponsors provide a structure that must be used. Regardless of the limitations, the abstract should contain a needs statement (why the project is important and needs to be done) and should describe the main goals of the project. A sample abstract is shown in Figure 20-7. Note that some grant sponsors may instruct grant seekers also to describe the project methods, time frame, and budget in the abstract.

It is best to write the abstract after the final proposal draft is complete to ensure that no key information is omitted. Nevertheless, do not let the abstract be an afterthought. Be sure to allow sufficient time to write it carefully, because the abstract is usually the first part of the proposal that reviewers read. Grant reviewers form their initial impression of a project from the abstract, which means the abstract can be critical to success. If the abstract does not convince reviewers that the project is well matched to the sponsor's priorities or does not address an important need, the grant reviewer may be inclined to spend little time on the proposal and to rate it poorly.

FIGURE 20-7 Sample Abstract

Improving Safe Food Handling Behaviors of Immunocompromised Adults

Foodborne disease caused by microbial pathogens remains a significant public health problem in the 21st century. Each year upwards of 5,000 deaths and 325,000 hospitalizations in the United States are food related. Hospital costs for treating victims of foodborne illnesses likely top $3 billion and lost productivity costs amount to as much as $9 billion—these costs do not include the total societal burden of chronic, long-term consequences of some foodborne illnesses. A primary contributor to the high rate of foodborne illness in this country is that many consumers are simply unaware of the vital role they play in preventing foodborne illness. In reality, food mishandling in home kitchens likely causes a significant amount of illness. Further exacerbating the rate of foodborne illness is that recent generations have not been taught about safe food handling at school or at home. As a result, a large proportion of adults have limited food preparation experience, have never learned basic principles of food safety, and thus lack the critical knowledge needed to help them proactively protect themselves and their families. For most people, a bout with foodborne illness is a relatively short-lived, minor inconvenience. But, for the one quarter of the U.S. population "at increased risk" (e.g., persons with weakened immune systems due to disease [e.g., HIV/AIDS] or pharmaceutical or radiological treatments; pregnant women and their fetuses; lactating mothers; infants and young children; preschool children in day care; and elderly persons), foodborne illness can be debilitating, even deadly. The high rate and exorbitant cost of foodborne disease, coupled with the devastating effects it can have on the nutritional status and overall health of individuals with weakened immune systems, clearly indicates a need for an effective, behaviorally focused, and audience-specific food safety education program. Previous work by the applicant indicates that immunocompromised adults and their caregivers prefer food safety information that addresses key foodborne illness prevention measures and issues in the context of their value to them and delivers this education via interactive, online instruction. Thus, the objectives of this project are to: (a) develop a behaviorally focused food safety intervention program that is responsive to the documented needs and interests related to food safety education of immunocompromised adults and their caregivers; (b) evaluate the effectiveness of the intervention program in effecting positive food safety behavior change and reducing foodborne illness risk; (c) refine the intervention program; and (d) disseminate the program to health professionals. This project will span 24 months. The budget request is: $185,289.

After writing the abstract, grant seekers should ask these questions:

1. Does the abstract succinctly summarize all the important elements of the project?
2. Is it prepared according to the format specified in the call for proposals?
3. Will it create in the grant reviewer's mind an accurate, complete picture of what is being proposed and why?

When a grant seeker is able to answer yes to all of these questions, he or she should have a colleague read the abstract and then ask the colleague to describe the project. If the colleague misses any major points, the grant seeker would be wise to check to be certain that the abstract is complete and clear.

GRANT NARRATIVE

The grant narrative includes these sections: needs statement, goals and objectives, and methods. Some grant sponsors indicate that the time and activity charts from the methods section, along with the capability statements, should be included in the grant narrative; others indicate that they should appear in another section.

Needs Statement

This section includes a clear, concise, and well-supported problem statement (or needs assessment), with a review of the current literature related to the problem. The literature review should establish the grant seeker's familiarity with and expertise in every facet of the study. Pilot test data and/or needs assessments data can be included, too. The introduction should clearly demonstrate that the problem described represents a gap between what currently exists or is known and what could (or should) exist or be known.[6] It should build a convincing case that the proposed project will help bridge that gap and extend the current knowledge base. The most successful grant proposals convey a sense of urgency regarding the problem and its resolution (that is, the goals of the project).

Figure 20-8 shows a sample needs statement. The introduction and literature review found in many journal articles provide other helpful examples.

FIGURE 20-8 Grant Narrative: Improving Safe Food Handling Behaviors of Immunocompromised Adults

Improving Safe Food Handling Behaviors of Immunocompromised Adults

NEEDS STATEMENT

Foodborne disease caused by microbial pathogens remains a significant public health problem in the 21st century, so much so that it is one of the priorities in the Healthy People 2010 initiative (1).* In addition, a new guideline, "keeping food safe to eat," was added to the Dietary Guidelines for Americans (2). Headline news stories focusing on widespread outbreaks of foodborne illness caused by lapses in personal hygiene or emerging pathogens are vivid reminders that the food that nourishes and sustains us also can be debilitating, and in some cases deadly (3). Recently, the Centers for Disease Control and Prevention estimated that 5,000 deaths and 325,000 hospitalizations each year in the United States are food related (4). Hospital costs for treating victims of foodborne illnesses likely top three billion dollars and lost productivity costs amount to as much as nine billion dollars—these costs do not include the total societal burden of chronic, long-term consequences of some foodborne illnesses (5). In addition, these costs do not account for the vast majority of foodborne disease cases caused by mishandling of food at home that go unreported (4–7) because consumers believe they have the "flu" (8).

A primary contributor to the high rate of foodborne illness in this country is that very few consumers believe foodborne illness originates in their homes—almost two-thirds never questioned whether someone in their household with "flu-like" symptoms (i.e., fever, chills, and nausea) could actually have had a foodborne disease caused by foods prepared at home (9–10). Concomitantly, many consumers do not understand the magnitude of the control they have in their own kitchen to reduce the risk of acquiring microbial foodborne illness (3).

Further exacerbating the rate of foodborne illness is that recent generations have not learned about safe food handling and food preparation in school (11–12). Increasing numbers of working mothers and growing reliance on frozen meals, restaurant dining, and take-out foods have decreased opportunities for children to learn safe food handling via observation. As a result, a large proportion of adults have limited food preparation experience, have never learned basic principles of food safety, and thus lack the critical knowledge needed to help them proactively protect themselves and their families (3, 13–18).

The importance of foodborne illness as a current public health concern is underscored by the increasing numbers of "at risk" or highly susceptible populations (19). One-quarter of the United States population is considered at increased risk for foodborne illness (1). Those considered "at risk" include persons with weakened immune systems due to disease (e.g., HIV/AIDS) or pharmaceutical or radiological treatments; pregnant women and their fetuses; lactating mothers; infants and young children; preschool children in day care; and elderly persons (8). Others who may be at a disproportionately greater risk include those living in institutional settings and financially disadvantaged individuals such as homeless persons and migrant farmworkers (9).

The high rate and exorbitant cost of foodborne disease, coupled with the devastating effects it can have on the nutritional status and overall health of individuals with weakened immune systems, clearly indicates a need for an effective, behaviorally focused food safety education program for these individuals and their caregivers. Although a wide variety of food safety educational efforts initiated by the FDA and USDA have been launched in recent years as a direct result of the National Food Safety Initiative, the target audience for these efforts has been very broadly defined (e.g., American consumers). That is, the same message and delivery mode is used to reach everyone regardless of personal needs and/or interests. Food safety education is most likely to be most effective if the messages are tailored to the needs and interests of a specific audience (20–22).

*The numbers in parentheses in this sample narrative refer the grant reviewer to the proposal's bibliography. The style used to note references in the text and the bibliographic referencing style may be specified in the call for proposals. If it is not, citations should be presented in an accepted journal format, including the title and complete reference. Authors should be listed in the same order as they appeared on the published paper.

1

When evaluating a needs statement, ask the following questions. If any are answered no, the needs statement must be revised until every question can be answered yes.

1. Does the needs statement demonstrate a clear understanding of the problem or need and why it is significant to address at this time?
2. Does the needs statement explicitly describe the focus (goals) of the proposed project and provide a conceptually sound rationale for the focus?
3. Is it clear that the proposed project is feasible, that it is different from prior work, and that it will resolve a significant, timely problem or meet a pressing need?
4. Does the needs statement's organization logically and methodically build a case for the proposed solution (objectives)?
5. Does the needs statement indicate how this project will build on previous work completed by the grant seeker or others?

FIGURE 20-8 Grant Narrative: Improving Safe Food Handling Behaviors of Immunocompromised Adults—*continued*

Findings of 17 focus group interviews ($n = 97$) conducted by the grant applicant to document the needs and interests related to food safety education of immunocompromised adults and their caregivers revealed that these individuals wanted food safety information that specifically addressed key foodborne illness prevention measures and issues in the context of their value to immunocompromised individuals. They also indicated that the preferred delivery mode for food safety education is via interactive, online instructional modules, dedicated chat rooms, and monthly e-mail or voice mail food safety tips and news (23). Currently, food safety education targeted to this population and delivered in the manner they prefer does not exist.

GOALS AND OBJECTIVES

The goal of this project is, like that of the grant sponsor, to improve the safe food handling behaviors of those at increased risk for foodborne illness, specifically immunocompromised adults and their caregivers. The project objectives are:

1. To develop a behaviorally focused food safety intervention program that is responsive to the needs, interests, and delivery mode preferences of immunocompromised adults and their caregivers that will enable them to increase their food safety knowledge, develop more positive attitudes toward practicing safe food handling, and improve their self-reported food safety practices.
2. To evaluate the effectiveness of the food safety intervention program on improving food safety knowledge, attitudes, and self-reported food safety practices.
3. To revise the food safety education intervention program based on the findings of the evaluation study.
4. To make health professionals throughout the region aware of the food safety intervention program and encourage them to promote the program to their patients.

METHODS

Project Design. This project has two primary phases: intervention lesson development and intervention lesson summative and impact evaluation.

Intervention Lesson Development. The intervention program to be developed will include six on-line, interactive, 15-minute lessons, focusing on food safety for immunocompromised adults. The lessons will be on-line because research indicates that computer-based learning can heighten learner interest and motivation, meet the needs of learners with varying learning styles, increase achievement levels, and shorten learning time (24–29). In addition, focus group data collected by the applicant revealed that immunocompromised adults and their caregivers preferred computer-based learning because it is convenient, provides constant interaction, and is under their direct control. Another important feature of computer-based learning is that it permits features that are not possible in printed materials (e.g., hot links to other Web sites, animation, video, narration), allows the use of full color graphics at no extra production costs, and can be updated more cost-effectively and easily than printed matter.

Each lesson will include a brief introduction, the lesson itself, and a brief conclusion. The introduction will be designed to "grab" the attention of the participants and will include interactive quizzes, games, and short video clips. Each lesson will focus on one aspect of food safety. (Figure 0 summarizes the primary generalization addressed and change process(es) emphasized in each lesson.) All lessons will have a strong visual component (e.g., simulation of bacteria growth rate in the danger zone; demonstration of how cross contamination occurs). Lesson conclusions will reiterate important points and provide participants with concrete methods for immediately applying the lesson's content. At the end of each lesson, participants will be able to print out fact sheets that summarize the lesson and post questions to the project staff.

2

6. Are current, reputable references such as refereed journal articles, statistical data, and reference books cited?
7. Does the needs statement explain why this particular grant sponsor should be interested in this proposal?
8. Does the needs statement convey a sense of urgency regarding the problem (need) and its resolution?
9. Is the needs statement interesting enough to compel the grant reviewer to continue reading?
10. Does the length of the needs statement conform to the sponsor's instructions?

For practice, read the needs statement in Figure 20-8 carefully and answer the questions above. For any question answered no, grant seekers should think about how to improve the needs statement.

FIGURE 20-8 Grant Narrative: Improving Safe Food Handling Behaviors of Immunocompromised Adults—*continued*

Lesson development procedures are described below in a stepwise fashion.

1. Develop plan for the lesson series by creating (a) conceptual outline of the specific subject matter to be included; (b) generalizations to be addressed; and (c) objectives that reflect expected learner outcomes.
2. Partition outline into lessons (i.e., single topics in the conceptual outline). Create map to show the links or interrelationships between lessons.
3. Create storyboard for each lesson. Transform storyboard into a multimedia format as each lesson storyboard is complete (i.e., text is written and accompanying images and sounds assembled).
4. Conduct formative evaluation of each lesson as soon as the lesson is transformed into a multimedia format. The cognitive response method will gauge learner reactions and satisfaction (i.e., learners will verbalize the thoughts, feelings, and ideas that come to mind immediately after exposure to each screen). This means of inquiry was selected because it has proved useful in evaluating food and nutrition education materials (30). The formative evaluation of each lesson will involve three to five dietetic interns and three to five immunocompromised adults and/or caregivers of these adults. Formative evaluation data will be analyzed and needed refinements identified.
5. Link lessons in the series using the map created in Step 2 above.

Intervention Lesson Summative and Impact Evaluation. When the intervention program is complete, summative and impact evaluation will occur as follows.

Sample. Study participants will be recruited from adults (age 18 and above) with weakened immune systems due to disease (e.g., HIV/AIDS) or pharmaceutical or radiological treatments who are receiving outpatient treatment from the Super State University Health Care System during spring 20XX and caregivers of immunocompromised adults. The University Health Care System includes 11 hospitals (2 specialize in cancer treatment; 1 is recognized as the premier treatment center for HIV/AIDS in the tri-state metropolitan area) and has nearly 1,700 physicians on staff. The attending physicians of eligible adults will receive information about the study and will be asked to encourage their patients to enroll. Enrollment invitation posters and flyers will be distributed to all physician offices and placed in key areas of each hospital in the system. Interested patients and caregivers will have the option of contacting project staff via telephone, e-mail, fax, or letter. One hundred individuals will be recruited to participate in each of the four study groups. Immunocompromised adults and their caregivers often experience greater than normal time pressures due to increased need for medical interventions. Thus, to increase the likelihood of their participation in this project, each treatment group participant who completes all phases of the study will receive $50 and an instant-read thermometer. Control group participants who complete all study phases will receive $10 and an instant-read thermometer.

Study Design. A Solomon 4-Group experimental design will be used (31). This research design includes a pre/post treatment group, post-only treatment group, pre/post control group, and post-only control group. Participants will be randomly assigned to the study groups. The pre/post treatment group will complete the pre-test and post-test and participate in the six intervention lessons. The post-only treatment group will complete the posttest and participate in the six intervention lessons. The pre/post control group will complete the pre-test and post-test while the post-only control group will complete only the post-test. Those in the control groups will have the opportunity to complete the lessons after the post-test data are collected. The Solomon 4-Group design was selected because it maintains a great deal of internal and external validity (31). To assess the long-term effectiveness of the intervention lessons, a follow-up post-test will be administered approximately 6 months after the post-test to all study groups.

Participants in the treatment groups will be instructed that one intervention lesson will be posted at one-week intervals, for a total of six weeks. They will receive an e-mail and/or voice mail notification when each lesson is posted. Each lesson will be available for a one-week period during which participants will complete it at a time convenient to their schedules. One week after each lesson is posted, the project staff will host a

3

Goals and Objectives

The project's goals and objectives describe what the grant seeker plans to achieve. Goals tend to be broad, general-purpose statements or ideals, such as "to improve the food handling practices of immunocompromised adults and their caregivers." The project objectives are much more specific and measurable than goals. They specify what will be accomplished when the project is completed. See the goals and objectives section in Figure 20-8. Also, refer to the section on program goals and objectives in Chapter 14 on pages 455–456.

Once all the objectives are written, the grant seeker should place them in a logical sequence. For example, if the project objectives will be addressed in phases, the grant seeker should place them in chronological order. If some objectives are more important than others, they should be presented from most important to least important.

FIGURE 20-8 Grant Narrative: Improving Safe Food Handling Behaviors of Immunocompromised Adults—*continued*

one-hour chat room discussion that will address questions participants posted to researchers about the lesson's content as well as offer an opportunity to discuss related issues. Each chat room session will be scheduled at three different times (one weekday, one weekend day, and one weekday evening). A six-week intervention period was chosen because the applicant's previous research revealed that this is a sufficient intervention time for individuals to significantly improve their food safety knowledge, attitudes, and reported practices (32). Once the participants complete the lesson series and chat rooms that accompany each lesson, the participants will have the opportunity to post questions to the researchers and/or engage in monthly chat rooms for the next six months. They also will receive monthly e-mail and/or voice mail food safety tips and news. The food safety lessons will remain available for them to consult if they desire.

Questionnaires. Two self-report questionnaires will be used in this study. The first will collect demographic data. The second, the Food Safety Questionnaire–Model 2 (FSQ) (33), developed and validated previously by the applicant, is a highly reliable 55-item objective instrument that assesses food safety knowledge, attitudes, and reported food handling behaviors and consumption of foods at high risk for bacterial contamination.

The pre-test will be administered via mail or e-mail one week before the first intervention lesson is posted. The pre/post treatment and pre/post control groups will complete both study questionnaires during the pre-test. The post-test will be administered one week after the last intervention lesson. All study groups will complete the FSQ during the post-test. The post-only treatment and post-only control groups also will complete the demographics questionnaires during the post-test. Approximately six months after the post-test, the study groups again will complete the FSQ. To minimize potential learning effects of the questionnaires, the purpose of the questionnaires will not be discussed with study participants and feedback on participant performance on the questionnaires will not be provided. To protect participant confidentiality, only key project staff will have access to data collected, and upon completion of the study all identifying characteristics will be stripped from the data files.

Data Analysis. Analysis of variance on the post-test scores will be conducted to determine whether pre-testing affected post-test performance. Analysis of covariance, with pre-test scores serving as the covariate, will be conducted to determine the impact of the interventions on knowledge, attitudes, and reported behaviors.

Intervention Program Revisions and Dissemination. The summative and impact evaluation data will be reviewed to determine needed lesson refinements. The lessons will be revised and posted permanently on the university's website.

To extend the impact of this project and enhance its value, the findings of the study and availability of the free online program will be disseminated via the media; professional organization publications and conferences; university publications distributed to alumni, faculty, and students; appropriate electronic listservs, discussion groups, and electronic bulletin boards; and the university's website and food- and nutrition-related websites such as the National Agricultural Library. All health care professionals employed by the applicant institution will be made aware of these products through e-mail advisories sent to the faculty directors of these programs. The availability of this program also will be noted in the research presentations and journal articles that result from this project. The results of this project will be further disseminated by presentations at health and nutrition professional meetings (e.g., American Dietetic Association, American Public Health Association) and in journal articles and related educational publications (e.g., *Journal of Nutrition Education and Behavior, Journal of Food Protection*).

4

Although the goals and objectives section tends to be short (half a page or less), the importance of thoughtfully and precisely constructed goals and objectives cannot be overemphasized. The goals and objectives are the heart of the proposal—*they are the reason why one seeks the grant sponsor's help.* "When sponsors fund your projects, they are literally 'buying' your objectives."[7]

A grant seeker should ask the following questions to assess the quality of the project goals and objectives. It will be important to revise project goals and objectives until all of these questions can be answered yes.

1. Is each objective clearly and logically related to the sponsor's goal(s) and priorities?
2. Is each objective clearly and logically related to the goal and mission of the grant seeker's organization?

3. Does the grant seeker's organization have the skill, personnel, and resources to achieve each goal in a timely and cost-efficient manner?
4. Does each objective contribute to meeting the project goals?
5. Does each objective provide a realistic, feasible, attainable, and manageable solution to the problem or need?
6. Is the list of objectives complete? That is, do the objectives address all of the project's major activities?
7. Do the objectives clearly describe what will be different when the project is completed? For example, will people gain new skills that will improve their health? Will a more effective nutrition intervention program be available? Will we discover how to deliver community nutrition services more effectively?
8. Do the objectives clearly indicate what must be measured to determine progress toward meeting each objective? Is it clear that the grant seeker will know when each objective has been reached?
9. Are the objectives presented in a logical sequence?
10. Does each objective use an action verb?
11. Are the objectives written so clearly and precisely that writing the evaluation section will be a simple task?

Methods

The methods section (sometimes called project description) describes in detail the procedures for achieving the study's objectives, justifies those procedures, and explains why the plan is likely to work. A full account of how the objectives will be achieved usually includes describing the project design, how success will be measured (for example, through an evaluation plan), the participants (or sample) who will be involved, the sequence and time frame of activities (time and activity charts), and the duties and capabilities of the project staff.

Many grant sponsors place more emphasis on this part of the proposal than on other parts—this is where the grant seeker must convince the sponsor that he or she knows how to achieve the project's objectives. Explanations are bolstered by citing pertinent past work that the grant seeker or others have completed and by highlighting innovative or unique features of the present project.

It is important that the methods section clearly describe all the resources needed to complete the project activities. This description will make it easier to construct the budget and justify every proposed expense.

Project Design

This section describes the overall organization of the proposed project. The sample proposal in Figure 20-8 indicates that this project has two primary phases: intervention lesson development and intervention lesson summative and impact evaluation. Both are described in detail.

Participants (Sample)

If people will be involved in the project (for example, answering questions or participating in classes), grant seekers need to provide a complete description of their characteristics (for example, women with osteoporosis or overweight preschoolers), how many will be involved, how they will be involved and/or what they will be asked to do, and how participants will be recruited. If participants will be assigned to various groups (such as experimental vs. control), grant seekers should explain how the assignments will be made (for example, randomly or as whole classrooms).

To strengthen a proposal, grant seekers can indicate how accessible participants with the desired characteristics are to them. For instance, if the grant seeker is a weight-loss counselor and proposes a weight-control intervention, he or she could indicate the number of clients coming to the organization's clinic each month who might agree to participate. If participants will be offered an incentive to secure their participation, grant seekers should be sure to explain why the incentive is needed, what the incentive will be, and when it will be awarded. Many grant sponsors realize that recruiting participants can be a challenge and incentives are sometimes necessary. However, it is a good idea to verify that the sponsor will fund an incentive before including it in a proposal. Note that anytime grant seekers propose to work with people, they will need to comply with their institution's Institutional Review Board (IRB) policy on dealing with human subjects (see the accompanying box, "Basic Principles of the Protection of Human Subjects).

Basic Principles of the Protection of Human Subjects

Scientific research has produced substantial social benefits, but it has raised some troubling ethical questions, too. Public attention focused on these questions after reported abuses of human subjects in biomedical experiments, especially during World War II. During the Nuremberg War Crimes Tribunal, the Nuremberg code was drafted as a set of standards by which to judge physicians and scientists who had conducted biomedical experiments on concentration camp prisoners. This code became the prototype of many later codes designed to ensure that research involving human subjects would be conducted ethically. (*Note:* Human subjects are living individuals about whom a researcher obtains data by intervention or interaction with the individual or identifiable private data).

The code consists of rules that guide researchers or the reviewers of research in their work. However, sometimes these rules are inadequate, conflicting, and/or difficult to interpret or apply. Thus, in 1978, the National Commission for the Protection of Human Subjects met to articulate the ethical principles that could be used to formulate, criticize, and interpret rules. This commission's report, *The Belmont Report: Ethical Principles and Guidelines for the Protection of Human Subjects of Research,* describes three fundamental ethical principles for acceptable conduct of research with human subjects: respect for persons, beneficence, and justice.

- Respect for persons recognizes the personal dignity and autonomy of individuals. It also requires special protection of vulnerable groups, such as children and prisoners. Researchers must get full consent from individuals before including them as research subjects. Before giving full consent, subjects must be informed about the research purpose and procedures, as well as its risks and benefits. Subjects must have a chance to ask questions and must be able to withdraw from the research at any time.
- Beneficence is the obligation to protect persons from harm by maximizing benefits and minimizing possible risks. The appropriateness of involving vulnerable populations must be clearly demonstrated, and the consent process must thoroughly disclose risks and benefits.
- Justice requires that the benefits and burdens of research be distributed fairly. Subjects should not be selected simply because they are readily available or because they are vulnerable as a consequence of illness, age, or socioeconomic condition. Research should not overburden individuals already burdened by their environments or conditions.

In community nutrition, our work nearly always involves human subjects. Sometimes community nutrition work is limited to practice, other times it involves research, still other times it is a combination of research and practice. The term *practice* tends to refer to interventions designed solely to enhance the patient or client well-being and to have a reasonable expectation of success. The purpose of medical or behavioral practice is to provide diagnosis, preventive treatment, or therapy to individuals. On the other hand, the term *research* refers to any systematic gathering and analysis of information designed to develop or contribute to generalizable knowledge (expressed, for example, in theories, principles, and statements of relationships, or shared at public meetings or in publications). Research includes

- Interviews, surveys, tests, or observations that are designed to gather nonpublic information about individuals or groups.
- Studies of existing data, either public or private, where the identity of individuals is known.
- Studies designed to change subjects' physical or psychological states or environments.

Whenever human subjects are involved in research of any type, the Belmont Report's principles should guide the researchers' actions. Keep in mind that the protection of human subjects is important in all types of research, from behavioral to biomedical. Most colleges, universities, and hospitals, as well as many other organizations, have an Institutional Review Board (IRB) that is responsible for educating researchers about the protection of human subjects and for approving research involving humans.

Source: Adapted from Department of Health, Education, and Welfare, Office of the Secretary, the National Commission for the Protection of Human Subjects of Biomedical and Behavioral Research, *The Belmont Report: Ethical Principles and Guidelines for the Protection of Human Subjects of Research* (April 18, 1979).

Evaluation Plan (Study Design)

An evaluation plan explains how the grant seeker proposes to measure the outcomes or impact of the project. That is, how will the grant seeker determine that the objectives have been met?

Many grant sponsors have long required an evaluation plan—they want to know whether the granted funds are having the desired impact. In other words, they want evidence that they are getting their money's worth! Even if the sponsor does not require evaluation as a condition of funding, it is a good idea to include one. That's because programs without an evaluation plan cannot demonstrate their impact and, thus, often have difficulty sustaining their funding, especially during an economic downturn. Furthermore, if grant seekers genuinely want to achieve the goals set, a well-designed and well-executed evaluation plan will help them track the project's progress and its strengths as well as pinpoint weaknesses and point up new directions for future work.

When planning an evaluation, grant seekers need to review the call for proposals very carefully to see whether the sponsor requires the plan to include special methods. For example, the sponsor may require that a specific test be used to measure blood glucose levels or may specify the use of a particular software package to report data. For large, complex projects that involve numerous staff working in more than one organization, a management plan also may be a required part of the evaluation plan. For ideas on how to evaluate projects, review journal articles summarizing similar projects.

Recall that there are three types of evaluation: formative, summative, and impact. *Formative evaluation* is the process of testing and assessing program elements before the program is fully implemented. *Summative evaluation* occurs at the end of a program to document its success and the extent to which project objectives were achieved. Finally, *impact evaluation* measures the overall worth and value of the program. In community nutrition, it tells us to what extent participants improved their knowledge, attitudes, values, skills, and/or health behaviors and physiological parameters. Grant seekers need to consider carefully how each of these evaluation types will be addressed in the evaluation plan.

To create the evaluation plan, grant seekers begin by asking, "What questions do we want to answer as a result of the evaluation?" For instance, during formative evaluation they may want to learn how people respond to a fact sheet design or whether the contents are accurate. As a part of the summative and impact evaluation, they may want to know how much program participants learned about a topic, the amount their bone density increased, or how their intake of fruits and vegetables changed. This question is easy to answer when the project objectives are written in measurable terms. If answering this question is difficult, the grant seeker should revisit the project objectives and fine-tune them until they are expressed in clear, precise, measurable terms and clearly state the questions he or she wants to answer as a result of the evaluation. Then the grant seeker can use answers to the questions below to write the evaluation plan in three parts: measurements, data analysis, and dissemination.

1. What specific information is needed to generate those answers?
2. What is the source of the information?
3. How will the information be gathered? Specifically, what data instruments are needed to generate the data?
4. What evaluation design will be used?
5. What analyses will be conducted?

Measurements

There are many ways to measure intervention outcomes, including tests, questionnaires, surveys, clinical examinations, food recalls, personal logs or memos, checklists, observational rating cards used by program staff or experts, and program records (such as daily activity reports and e-mail files). In any one study, there are often several methods for gathering the same information, and depending on the circumstances, some methods are better than others. Let's say you want to determine dietitians' understanding of genetics. To find out, you could create a questionnaire or use an existing one. Then you could interview each person face-to-face-or by phone, or you could survey them by mail or e-mail. Dietary intake could be assessed via a food frequency form, 24-hour recall, or 3-day food record or in several other ways.

The measures selected to generate needed information will depend on several factors, including the purpose and scope of the evaluation, resources available for evaluation (such as personnel time, money, and skill), instrument choices available, measurement precision needed, and burden to participants (such as the time they must spend and the commitment they must make). The best measurement choices are feasible, justifiable, valid, and reliable. Feasible measures fit resource constraints, yield measurements that are suitably precise, are manageable, and are not overly burdensome to study participants. (For example, a 14-day weighed food record probably isn't feasible in a study of preschool eating patterns.) Justifiable measures are among the best choices available to measure a particular characteristic, are as safe and nonintrusive as possible, and are truly necessary and within the project's scope. Valid measures actually measure what they are intended to measure. Reliable measures are just that; they give similar results each time they are used. Many excellent research design resources are available that can provide more detail on choosing feasible, justifiable, valid, and reliable measures for community nutrition programs, as shown in the accompanying feature, "Research Design Resources for the Community Nutritionist."

The measurements section will flow most logically if it is written in the same sequence as that in which the objectives appear in the grant narrative. The grant seeker can further increase clarity by describing the interrelationships among the project activities. This section of the proposal should have the following characteristics.

- Provide sufficient detail about how the grant seeker will measure progress toward meeting each objective to demonstrate the technical soundness of the measurements selected, his or her in-depth understanding of the method, and his or her credibility as an evaluator. A critical discussion of the feasibility, justifiability, validity, and reliability of the instrument, along with the advantages associated with the method and how any disadvantages will be overcome, helps demonstrate the technical soundness of the proposed measures.

Research Design Resources for the Community Nutritionist

Research: Successful Approaches, 2nd ed., edited by E. R. Monsen (Chicago: American Dietetic Association, 2003)

A comprehensive guide to planning, executing, presenting, and evaluating various types of research. Includes topics such as evidence-based practice, behavioral theory–based research, and research methods in complementary and alternative medicine.

The Practice of Social Research, 10th ed., by E. Babbie (Belmont, CA: Wadsworth, 2003)

An authoritative research methods text with a broad set of topics, including illustrative examples such as welfare and poverty, gender issues, the AIDS epidemic, and more.

Handbook in Research and Evaluation: A Collection of Principles, Methods, and Strategies Useful in the Planning, Design, and Evaluation of Studies in Education and the Behavioral Sciences, 3rd ed., by S. Isaac and W. B. Michael (San Diego, CA: Edits Publishing, 1997)

An authoritative handbook on research and evaluation methods designed for researchers in education or behavioral sciences.

Research in Education, 9th ed., by J. W. Best and J. V. Kahn (Boston, MA: Allyn and Bacon, 2003)

A reference book covering topics such as identifying problems, forming hypotheses, constructing and using data-gathering tools, designing research studies, and employing statistical procedures to analyze data.

Health Behavior and Health Education: Theory, Research, and Practice, 3rd ed., edited by K. Glanz, B. K. Rimer, and F. M. Lewis (San Francisco, CA: Jossey-Bass, 2002)

A textbook that provides information on health behavior theories and research practice, drawing on fields such as cognitive psychology, marketing, and communications to explain diverse factors affecting health behavior in diverse populations.

Planning, Implementing, and Evaluating Health Promotion Programs: A Primer, 4th ed., by J. F. McKenzie, B. L. Neiger, and J. L. Smeltzer (Redwood City, CA: Benjamin Cummings, 2004)

A comprehensive text providing information needed to plan, implement, and evaluate health promotion programs in a variety of settings, with discussion of needs assessment, how to acquire needs assessment data, and how to apply theory to practice.

▸ Describe the procedure for administering the measurement in a stepwise fashion. If it is a complex or confusing procedure, consider including a diagram. Example: The participants will be placed in a private room, instructed to change into a paper robe and slippers, and weighed on a calibrated balance-beam scale.

▸ Describe how the measurements have been used previously. Indicate the grant seeker's experience with these measures. Example: The project director developed and validated the Nutrition Attitude Scale with individuals having characteristics similar to those of participants in the proposed study sample. The scale yielded highly reliable results (Cronbach Alpha reliability coefficient = 0.89).

▸ Describe who will collect the data and how they will be trained. Example: Three dietetics graduate research assistants will be trained to be observers who can uniformly complete the observation checklist. This training will include a detailed review of the recipe and observation instrument, preparation of the recipe several times, role-playing practice sessions, analysis of the practice sessions highlighting inconsistencies, and further practice sessions to achieve uniformity. Each observer also will gain field practice by completing practice observations with a minimum of five individuals who have characteristics similar to those of the study participants. During the field practice, the observers and project director will simultaneously observe the same individuals. An observer will be considered ready to collect data when the interrater reliability rate for the observer and project director is a minimum of 90 percent. Dr. Harris, a senior project associate, is nationally known for his evaluation expertise and will conduct all phases of the observer training.

- Indicate where the data will be collected. Example: Each person will individually participate in a face-to-face interview with a trained interviewer in a room relatively free of distractions.

- Indicate how the data will be recorded. Example: The interviewer will take notes during the interviews, and the interviews will be audiotaped for later transcription. Example: The questionnaire is a pencil-and-paper, self-report, objective instrument designed to measure attitudes toward nutrition.

- Indicate how participant confidentiality will be protected. Example: All study questionnaires will be kept in a locked file cabinet in the project director's office. Once data are transcribed into computer files, all identifying data will be removed and the paper questionnaires will be shredded.

Data Analysis

The data analysis section describes how the data collected will be analyzed to determine the success of the project. It is a good idea to consult a statistician during preparation of the proposal, and periodically throughout the project, to ensure that the data collected will be valid, usable, and generalizable and can be interpreted to provide answers to study questions. Example: Frequencies, percentages, and means will be calculated to determine the participants' scores on knowledge about nutrition label reading, their attitudes toward diet and health, and their label usage behaviors.

Dissemination

The purpose of this section is to describe how interested audiences will learn about the project and its outcomes. The dissemination section should explain why dissemination activities are important for the project; concisely describe a feasible, effective, and suitable dissemination plan; and indicate who will be responsible for this task.

Dissemination offers many benefits to both the sponsor and the grant recipient. For instance, members of the public learn more about the sponsor and its priorities. They become aware of the priorities, mission, and activities of the organization receiving the funds. This increased awareness can lead to other funding opportunities, help locate new clients and collaborators, and enhance information sharing among colleagues. It also helps build the reputation of the grant recipient.

Unfortunately, many worthwhile projects are funded and completed without anyone except those involved knowing about their outcomes. Increasingly, grant sponsors are requiring grant seekers to declare, either in the methods section or in a special dissemination section, specifically how they will disseminate information generated from the project.

The methods section in Figure 20-8 is an example of how this section might be written for the objectives stated earlier in the figure. Note how the example details the manner in which each project objective will be achieved and indicates why the grant seeker selected that method. Also note how the flow of this section parallels the sequence in which the objectives are presented earlier in the figure and addresses each of the items

FIGURE 20-9 Time and Activity Chart

This incomplete sample chart for the proposal in Figure 20-8 shows what activities will occur, when they will occur, and who is responsible. For practice, finish this chart using the proposal in Figure 20-8.

Task	Time to Complete	Start Date	End Date	Personnel Responsible*
Lesson Development and Evaluation				
1. Develop plan for lesson series, conceptual outline, generalizations, and objectives.	8 weeks	9/15/10	11/15/10	PD, SAs, RA
2. Partition outline into lessons. Create map of lesson links.	2 weeks	11/15/10	11/30/10	PD, SAs, RA
3a. Create storyboard for each lesson (complete research, writing, image selection/creation, and preparation of all materials for Step 3b).	30 weeks	12/1/11	6/15/11	PD, SAs, RA
3b. Transform storyboards into multimedia as each lesson is completed.	30 weeks	3/1/11	8/15/11	PD, RA, CPT
4a. Conduct formative evaluation of each lesson as it is completed.	22 weeks	3/15/11	8/31/11	PD, RA
4b. Refine lessons.	22 weeks	4/1/11	9/15/11	PD, SAs, RA, CPT, CE
5. Link lessons as the formative evaluation concludes for the lesson series.	2 weeks	9/1/11	9/15/11	PD, SAs, RA, CPT
6. Recruit study participants.	8 weeks	9/15/11	11/15/11	RA

*PD = Project Director; SAs = Senior Associates; RA = Research Assistant; CPT = Computer Programming Team; CE = Copy Editor.

in the bulleted list on pages 620–621. Many journal articles provide other excellent examples of how to develop the methods section.

Time and Activity Chart

A time and activity chart breaks the entire project into manageable steps that clearly show grant reviewers how the project will proceed. This chart can take several forms. Two common formats appear in Figures 20-9 and 20-10. All formats of these charts should chronologically list all steps that must be taken to complete the project and should indicate when work on each step will begin and end. Note in Figure 20-9 how the chart also shows the duration of the activity and who is responsible for the work. Figure 20-10 reveals how project activities overlap. If proposal page limitations permit, consider submitting charts formatted like both Figures 20-9 and 20-10. If there is room for only one, choose a format like Figure 20-9 because it also reveals who is responsible for completing each task, which can greatly facilitate writing the budget narrative. Time and activity charts often appear in the methods section; however, some grant sponsors instruct applicants to place them in a separate section or in the appendix.

Accurate, neatly constructed time and activity charts are essential to all but the most simplistic grant proposals. These charts not only guide the grant seeker's work if the grant is funded but also help convince grant reviewers that the grant seeker understands the process involved in reaching the proposed goals and objectives.

Task	Year: 2010				Year: 2011												Year: 2012								
	9	10	11	12	1	2	3	4	5	6	7	8	9	10	11	12	1	2	3	4	5	6	7	8	9
Lesson Development and Evaluation																									
1. Develop plan for lesson series, conceptual outline, generalizations, and objectives.																									
2. Partition outline into lessons. Create map of lesson links.																									
3a. Create storyboard for each lesson.																									
3b. Transform storyboards into multimedia as each lesson is completed.																									
4a. Conduct formative evaluation of each lesson as it is completed.																									
4b. Refine lessons.																									
5. Link lessons as the formative evaluation concludes for the lesson series.																									
6. Recruit study participants.																									

FIGURE 20-10 Time and Activity Chart
This incomplete sample chart for the proposal in Figure 20-8 shows the overlap among activities. For practice, finish this chart using the proposal in Figure 20-8.

Capability

The capability section establishes the credibility of the grant seeker, including project staff, and the organization's ability to complete the proposed project with expertise, on time, and within the budget. It is essential for the grant seeker to convince reviewers of his or her capabilities in this section *and* throughout the entire proposal. The quality of the proposal itself will tell reviewers a great deal about the grant seeker—it will indicate whether he or she thinks clearly and logically, expresses thoughts cogently and succinctly, has technical skill, analyzes data objectively and accurately, interprets results correctly and conservatively, and discriminates between significant and inconsequential information.[8]

The capability section should describe the grant-seeking organization's goals and philosophy (mission), activities, success stories, and distinguishing characteristics that make it uniquely qualified to carry out the proposed project. It also describes the facilities, equipment, reference materials, support personnel, and any other resources that are key to the success of the proposed project. In some cases, the organization's capability statement appears in the proposal introduction to establish why the organization is proposing a particular program and how it is uniquely qualified to deliver the program with excellence.

The credibility and capability of project staff can be established by including curricula vitae (résumés) for all key project personnel and/or a brief (one page or less) personal statement describing in narrative form each person's training and expertise, previous experience with projects and grants like the one proposed, and role in the proposed project.

Many grant sponsors set limits on the length, content, and/or style of curricula vitae. In addition, they may instruct grant seekers to place curricula vitae in a different location, such as in the appendix. Whenever possible, it is helpful to weave the project staff's expertise (or that of the organization) into the proposal's needs statement or methods section. This technique builds credibility and indicates to grant reviewers that the applicant has a proven record of accomplishment. Note in Figure 20-8 how the writer took the opportunity to begin establishing her expertise by citing prior work ("focus group data collected by the applicant revealed that . . ."; "the applicant's previous research revealed that . . ."; and "the questionnaire developed and validated previously by the applicant").

For success, the project director (or principal investigator) must clearly have the expertise needed to direct all the activities of the grant. If the project director lacks expertise, a codirector can provide complementary expertise. For multiphase, complex projects, an interdisciplinary team usually is needed. Teams that have a successful track record of working together are much more credible than teams hastily assembled to compete for a newly announced grant.

Budget

The budget is one of the most important parts of the proposal. It not only describes the expected cost of the project but also demonstrates the grant seeker's planning and management expertise. Incomplete, inflated, or inadequate budgets signal inexperience and/or incompetence.

To begin, grant seekers should carefully review the grant guidelines to determine exactly which expenses can be charged and which are excluded. In addition, they need to determine exactly how the grant sponsor wants the budget presented. Some grant sponsors provide forms that must be used. Others list the budget categories that must be included. Still others offer little direction regarding the budget. Next, grant seekers need to scrutinize the grant narrative and list all anticipated expenses, being sure to build in some flexibility to cover unanticipated events, cost increases, and salary raises. Grant seekers also need to remember that all project costs must be incurred during the proposed period of time. Paying for expenses that occurred prior to the grant period (including costs associated with proposal preparation) is rarely allowed. In addition, prepaying people for work that will occur after the grant ends or buying supplies, materials, postage, and the like to be used after the grant ends is rarely permitted. In nearly all cases, all costs must be auditable (that is, receipts are needed).

The most successful grant proposals have budgets that are totally consistent with their grant narrative—that is, every budgeted expense is clearly related to the project goals, objectives, and methods. No budget item should make grant reviewers wonder why it is included. For instance, if participants will be offered incentives, grant seekers need to be sure to describe them in the grant narrative. If the incentives are not described, the reviewers will wonder about their purpose and expense—and about the grant seeker's skill as a grant manager. To be certain that the budget and grant narrative are clearly linked, grant seekers should analyze the narrative carefully and list all anticipated expenses. Then, once the budget is constructed, they need to examine each expense and verify that the need for the expense is clearly justified in the grant narrative.

The importance of an accurate budget cannot be overstated. Grant seekers who overextend their resources, propose to do too much in a funding period, or underestimate their costs will appear unqualified and inexperienced to seasoned grant reviewers. If somehow the grant seeker does receive the grant in spite of such underestimates, the insufficient resources, time, and money will cause a great deal of frustration and anxiety and may result in a failure to achieve the proposed goals.

YEAR 1	
DIRECT COSTS	
A. Salaries and Wages	
1. Senior Personnel	
a. 1 Project Director (PD): Greta Grantwriter	$17,271
b. 2 Senior Associates (SAs): John Smith and Margaret Anderson	$23,832
2. Other Personnel	
a. 1 Research Assistant	$4,500
b. 1 Copy Editor	$1,400
c. 1 Computer Programming Team	$25,000
B. Fringe Benefits	$13,321
C. Total Salaries, Wages, and Fringe Benefits	**$85,324**
D. Equipment	$3,000
E. Materials and Supplies	$1,000
F. Travel	$263
G. All Other Direct Costs	$530
H. Total Direct Costs (Items C to G)	**$90,117**
INDIRECT COSTS (23% of direct costs)	$22,530
TOTAL COST	**$112,647**
YEAR 2	
DIRECT COSTS	
A. Salaries and Wages	
1. Senior Personnel	
a. 1 Project Director (PD): Greta Grantwriter	$9,673
b. ...	

FIGURE 20-11 Budget
This incomplete sample budget for the proposal in Figure 20-8 lists each anticipated financial cost. For practice, finish this budget using the proposal in Figure 20-8.

BUDGET CATEGORIES

Budgets are usually divided into two main categories: direct costs and indirect costs. Some also require cost sharing, which includes cash or in-kind gifts (for example, donated products, supplies, equipment, and services).

Direct Costs

Direct costs are concrete project expenditures that are directly listed line by line (that is, they are line items). Direct costs usually include personnel (that is, salaries, wages, consultant fees, and fringe benefits), equipment, supplies, and travel (see Figure 20-11).

Indirect Costs

Indirect costs are also called overhead, administrative costs, or facilities and administrative costs. Indirect costs are real costs associated with a grant that usually are not specifically listed on budgets, primarily because they tend to be difficult to price individually. They may include the costs of administering the grant, facility and equipment usage, and utilities. Because it is hard to estimate the extra costs the grant recipient will incur in administering the grant, indirect costs are usually calculated as a percentage of the direct costs. As you can see in Figure 20-11, indirect costs were calculated as 23 percent of the direct costs.

The percentage charged for indirect costs varies with the grant-seeking organization and the grant sponsor. Organizations that frequently receive federal grants usually have an approved federal indirect costs percentage rate. The percentage of indirect costs that grant sponsors will pay ranges from zero to the full federal indirect cost rate. Grant seekers need to be sure to check the grant guidelines or call the grant sponsor to determine how much they can charge for indirect costs. If the grant sponsor does not pay indirect costs, grant seekers can estimate the administrative costs of the project and list them as direct costs. For example, a grant seeker could estimate the cost of renting workspace and the percentage of a payroll clerk's time that will be devoted to this project and list them as direct cost line items.

Some people erroneously view indirect costs as an unwarranted windfall to the grant recipient's organization. However, indirect costs are indeed actual costs that must be incurred by the grant-seeking organization if the grant is to be completed successfully. For example, the grant recipient's organization must administer the grant; these administrative duties are likely to increase the work of nongrant staff (for example, the payroll department will need to pay and keep records for new grant-paid employees; the accounting department will have the added responsibility of processing purchasing requests for items used in the grant and of auditing grant expenditures). Sometimes the increase is so great that administrative departments need to hire extra help or pay overtime. Also, to do the work of the grant, staff will need to use facilities, equipment, and utilities, thereby making them unavailable or less accessible to employees doing nongrant work. Remember that if indirect costs are not covered by the grant sponsor, these costs nearly always must be drawn from the grant-seeking organization's other resources. Thus, if a grant sponsor will not pay indirect costs, the grant-seeking organization must assess whether it is able to bear the burden of the additional indirect costs it will incur in the course of completing the project funded by the grant.

Cost Sharing

Cost sharing, which is also called cost matching or matching, occurs when the grant-seeking organization agrees to pay certain costs (by contributing cash or in-kind goods or services) of the project. For instance, an organization may agree to contribute the cost of office space, incentives, photocopying, a percentage of a worker's time, and/or indirect costs. Grant seekers should check the grant guidelines or call the sponsor to determine whether cost sharing is required and, if so, how much of the grant costs must be matched. If cost sharing is required, the grant-seeking organization usually is required to contribute (or match) a percentage of the funds requested. For example, some grant sponsors require that the grant seeker contribute one dollar for every dollar they award (a 100 percent match). Others may require the grant seeker to contribute more or less than a dollar for every dollar they award. Figure 20-12 shows a sample budget for an RFP that required a minimum of a 100 percent match.

Budget Narrative

The budget narrative follows the budget in the proposal. Its purpose is to explain or justify all expenditures. It is wise to include a budget narrative even if the grant sponsor does not request one. The narrative may be as extensive as the one shown in Figure 20-13, which explains all the expenses in Figure 20-11. Alternatively, it could be quite brief, as shown in Figure 20-14. The most useful budget narratives show the basis for all calculations. For example, instead of just saying that Fringe Benefits equal $13,321, give explicit calculations like those shown in Figure 20-13.

Appendixes

Appendixes are placed at the end of the proposal. They should contain carefully selected materials that directly support the proposal. Appendixes may include strong letters of support and endorsements from individuals critical to the project's success, assurances of cooperation between institutions and subcontractors, reprints of articles published by the project staff, data collection questionnaires, organization charts, financial statements, and recent annual reports. In some proposals, curricula vitae from all key project personnel are included in the appendixes; in others they are placed elsewhere.

YEAR 1			
DIRECT COSTS			
A. Salaries and Wages	**Amount Requested**	**Cost Sharing**	**Total Amount**
1. Senior Personnel			
a. 1 Project Director (PD): Greta Grantwriter	$7,271	$10,000	$17,271
b. 2 Senior Associates (SAs): John Smith and Margaret Anderson	$11,000	$12,832	$23,832
2. Other Personnel			
a. 1 Research Assistant	$4,500	$0	$4,500
b. 1 Copy Editor	$0	$1,400	$1,400
c. 1 Computer Programming Team	$25,000	$0	$25,000
B. Fringe Benefits	$3,000	10,321	$13,321
C. Total Salaries, Wages, and Fringe Benefits	**$50,771**	**$34,553**	**$85,324**
D. Equipment	$3,000	$0	$3,000
E. Materials and Supplies	$1,000	$0	$1,000
F. Travel	$263	$0	$263
G. All Other Direct Costs	$530	$0	$530
H. Total Direct Costs (Items C to G)	**$55,564**	**$34,553**	**$90,117**
INDIRECT COSTS (23%)	$0	$22,530	$22,530
TOTAL COST	**$55,564**	**$57,083**	**$112,647**
YEAR 2			
DIRECT COSTS			
A. **Salaries and Wages**			
1. Senior Personnel			
a. 1 Project Director (PD): Greta Grantwriter	$5,000	$4,673	$9,673
b. ...			

FIGURE 20-12 Cost-Sharing Budget

This incomplete sample cost-sharing budget for the proposal in Figure 20-8 lists each anticipated financial cost. For practice, finish this budget using the proposal in Figure 20-8.

Appendixes should not be used to get around the page limits set by the grant sponsor. Keep in mind that grant reviewers are very busy and may skip over the appendixes. Some grant sponsors do not forward appendixes to reviewers. Therefore, the grant seeker should be certain that all the information reviewers need to make an informed decision about the proposal is included in the grant narrative.

Assembling the Final Product

When all parts of the proposal parts are finished, the next step is to assemble them into a complete package. Then use a word processor's spelling and grammar checker. Next, the grant seeker should carefully proofread the proposal to find obvious errors as well as to identify ways to refine the proposal further. Next, the grant seeker needs to format the proposal in an inviting, professional style, following all instructions provided by the grant sponsor, and give it to conscientious colleagues who will supply thorough, honest feedback. Using this feedback to refine the proposal can lead to important and valuable improvements. Once the proposal is finished, it is time to make the specified number of copies and send it to the designated address before the deadline passes.

KEYS TO SUCCESS

The most inviting and easy-to-read proposals tend to have these characteristics.[9]

- A clear, cohesive, highly readable writing style (that is, active voice, sentences of varying length but none more than 30 words long).

FIGURE 20-13 Budget Narrative

This incomplete sample budget narrative for the proposal in Figure 20-8 describes each anticipated financial cost and how it was derived. For practice, finish this narrative using the proposal in Figure 20-8 and the budget completed in Figure 20-11.

YEAR 1

DIRECT COSTS

A. Salaries and Wages

1. Senior Personnel

a. 1 Project Director (PD): Greta Grantwriter: 20% time for a total of $17,271. Responsibilities include: serving as instructional design expert for the design of all lessons; overseeing all lessons including development of outline, generalizations, and objectives; direction of partitioning of outline into lessons; creation of map of all lessons; creation of storyboards for each lesson; directing transformation of storyboards for all lessons into multimedia as each is completed; working with programming team to develop on-line lesson structure; overseeing formative evaluation of each lesson as it is completed; identifying refinements needed in all lessons based on formative evaluation results; overseeing refinement of all lessons.

b. 2 Senior Associates (SAs): John Smith and Margaret Anderson: 15% time each for a grand total of $23,832. Responsibilities of each include: developing outline, generalizations, and objectives for three lessons; partitioning developed outline into lessons; creating map of the lessons; creating storyboards for the lessons (completing research, writing, image identification); identifying refinements needed in lessons based on formative evaluation results.

2. Other Personnel

a. 1 Research Assistant: 10% time for a total of $4,500; Responsibilities include: assisting PD and SAs to create storyboards for each lesson (gathering research materials, locating and/or creating images); working with programming team to transform storyboards for all lessons into multimedia as each is completed; conducting formative evaluation; identifying refinements needed in all lessons based on formative evaluation results.

b. 1 Copy Editor: The copy editor will edit all written portions of the lessons under the direction of PD: 40 hours @ average rate of $35 per hour = $1,400.

c. 1 Computer Programming Team: The computer programming team will transform storyboards into multimedia as each lesson is completed; refine lessons; link lessons; make CD-ROM masters. $2,500/month x 10 months = $25,000.

B. Fringe Benefits: Fringe Benefits are 24% for 1 Project Director, 2 Senior Associates, and 1 Research Assistant @ 0.24 x ($17,271 + $23,832 + $4,500) for a total of $10,945. Fringe Benefits are 9% for 1 Copy Editor and 1 Computer Programming Team 0.09 x ($1,400 + $25,000) for a total of $2,376. Total Fringe Benefits equal $13,321 ($10,945 + $2,376).

C. Total Salaries, Wages, and Fringe Benefits: Total Salaries, Wages, and Fringe Benefits = $85,324 [$72,003 (Total Salaries and Wages) + $13,321 (Total Fringe Benefits)].

D. Equipment

1. Windows-based multimedia-capable computer hardware with scanner: Equipment to be used by Research Assistant for storyboard work @ $3,000.

E. Materials and Supplies

1. **Research Articles:** (i.e., cost for article reprints from on-line sources for research for each lesson) Estimated 20 articles x 6 lessons @ $2.50 each = $250.
2. **Storyboard Creation Materials:** (i.e., cost of clip art, office supplies) Estimated at $125 per lesson x 6 lessons = $750.

F. Travel

Mileage to meetings with Computer Programming Team ($0.35/mile x 10 trips x 75 mile average round trip). Total mileage cost equals $263.

G. All Other Direct Costs

1. Telephone: Telephone costs are estimated to be $15 per month ($15 x 12 months) for a total cost of $180.
2. Printing & Copying: Miscellaneous copying: $50.
3. Duplicated CD-ROMs: 100 copies for formative evaluation @ $3.00/CD = $300.

H. Total Direct Costs

Total costs for C through G above equal $90,117.

INDIRECT COSTS

Indirect Costs equal $22,530 (23% of total direct costs)

TOTAL COST

Total Direct and Indirect costs equal $112,647 ($90,117 + 22,530).

YEAR 2

A. Salaries and Wages

1. Senior Personnel

a. 1 Project Director (PD): Greta Grantwriter: 10% time for a total of $9,673. Responsibilities include: ...

- Appropriate vocabulary is used, any terms likely to be unknown to reviewers are defined, and jargon is avoided.
- Paper is white, 20-pound bond, of standard letter size.
- Margins are 1-inch (top, bottom, and sides).
- Typeface is 12-point serif (such as Times Roman).
- Pages are numbered.
- Bold headings and subheadings are used to guide the reader to each section.
- The right margin is ragged (not justified).
- Upper- and lowercase letters are used (not all uppercase).

YEAR 1

DIRECT COSTS	
A. Salaries and Wages	
1. Senior Personnel	
a. 1 Project Director (PD): Greta Grantwriter	$17,271[a]
b. 2 Senior Associates (SAs): John Smith and Margaret Anderson	$23,832[b]
2. Other Personnel	
a. 1 Research Assistant	$4,500[c]
b. 1 Copy Editor	$1,400[d]
c. 1 Computer Programming Team	$25,000[e]
B. Fringe Benefits	$13,321[f]
C. Total Salaries, Wages, and Fringe Benefits	**$85,324**
D. Equipment	$3,000[g]

Budget Narrative

[a]20% time for a total of $17,271.
[b]15% time each for a grand total of $23,832.
[c]10% time for a total of $4,500.
[d]40 hours @ average rate of $35 per hour = $1,400.
[e]$2,500/month x 10 months = $25,000.
[f]Fringe Benefits are 24% for Project Director, Senior Associates, and Research Assistant and 9% for Copy Editor and Computer Programming Team.
[g]Windows-based multimedia-capable computer hardware with scanner.

FIGURE 20-14 Budget This incomplete sample budget with brief budget narrative for the proposal in Figure 20-8 lists each anticipated financial cost. For practice, finish this budget using the proposal in Figure 20-8.

- Italics and underlining are used sparingly.
- Paragraphs are indented five spaces.
- White space is used to break up the pages. For example, double spacing between paragraphs helps increase white space and makes the document appear inviting and readable.
- The document is bound with only a staple in the upper lefthand corner.
- Illustrations, graphs, and charts are clear and enhance understanding.

Review of the Grant Proposal

After receiving a grant proposal, grant sponsors usually send grant seekers a note confirming that they received the proposal and indicating approximately when they plan to announce what proposals will be funded. If the grant seeker does not receive a letter within a few weeks of submitting the proposal, he or she should contact the individual designated on the call for proposals by phone, e-mail, or letter to confirm that the proposal was received and ask when a funding decision is likely to be made. Keep in mind that making funding decisions sometimes takes longer than the grant sponsor originally planned. However, if the grant seeker does not receive a funding decision (or a notice that the decisions have been delayed) within a few weeks of the date provided by the grant sponsor, the grant seeker should contact the individual designated on the call for proposals and politely ask when a funding decision is likely to be made.

After receiving a grant proposal, the grant sponsor organizes the proposals submitted and distributes them to grant reviewers. The job of grant reviewers is to review each proposal, compare it to the sponsor's priorities and criteria, judge the relative merit of the project and its likelihood for success, assess whether the project staff and organization can deliver what is proposed within the time frame specified, and evaluate the adequacy of the budget. In many ways, the reviewers are stewards of the grant sponsor's funds, and most take their responsibilities very seriously. Grant seekers are most likely to get the best review possible when they prepare their proposals carefully and thoughtfully and observe all the instructions provided by the grant sponsor.

Everyone hopes that his or her grant proposals are fully funded the first time around. Sometimes, however, grant sponsors will contact an applicant and ask him or her to modify the proposal (such as by reducing the budget, trimming away some objectives, or increasing the number of participants, for example). When this happens, the grant seeker will need to analyze the proposal carefully and decide whether the changes are feasible and how they will be made. It is important for both grant sponsor and grant seeker to understand clearly the work that will be completed before the grant is awarded.

Successful grant seekers will need to apply their management and budgeting skills carefully so that they will achieve grant goals expertly, on budget, and on schedule. Remember, the proposal is what the sponsor is buying. Therefore, it is important for the grant recipient to deliver the product promised. If the grant recipient needs or wants to diverge from the proposal, he or she will probably need to renegotiate with the grant sponsor. Grant recipients should carefully review all the rules governing the grant before accepting it.

Learning that a grant wasn't funded isn't good news; it's not necessarily bad news either. That's because many grant proposals can be revised and resubmitted for consideration in the next go-round. In fact, many funded grants are resubmissions. Remember that in grant writing, "failure often precedes success."[10]

The most common reasons why proposals are not funded are that there were problems with planning and time management, the proposal did not follow the guidelines, the deadline for submitting the application was missed, the proposal lacked a well-conceived plan of action or idea, and/or the proposal had errors in the budget estimates. In many cases, a proposal is rejected because it does not match the grant sponsor's funding priorities.[11] A "generic" proposal sent out to several grant sponsors is *not* recommended—every proposal should be tailored to the specific sponsor's priorities and guidelines.

Most government grant sponsors and some private grant sponsors provide grant seekers with a written summary of the reviewers' comments. These summaries can help all applicants—successful and unsuccessful—learn how to improve future proposals. Grant seekers use these comments to rework a rejected proposal and put it in a more competitive position next time. Another great way to improve grant-writing skills and better understand the review process is to become a reviewer. Grant sponsors do recruit reviewers, so grant seekers should consider sending them a letter along with a copy of their curriculum vitae.

There is no doubt that grant writing takes considerable time and effort, but the rewards of success are worth it! The most successful grant seekers demonstrate four critical qualities: (1) diligence in researching and identifying grant sponsors; (2) creativity in matching project goals with those of sponsors; (3) attentiveness to detail in proposal preparation; and (4) persistence in revising proposals and resubmitting them to potential grant sponsors.

Internet Resources

Agency for Healthcare Research and Quality	**www.ahrq.gov**
American Federation for Aging Research	**www.afar.org**
Centers for Disease Control and Prevention/Funding	**www.cdc.gov**
Canadian Foundation for Dietetic Research	**www.dietitians.ca/cfdr/**
Cooperative State Research, Education, and Extension Service Grant and Funding Opportunities	**www.csrees.usda.gov**
Department of Agriculture	**www.usda.gov**
Department of Education	**www.ed.gov**

Department of Health and Human Services Access to CRISP and Tools for Grantwriting	**www.hhs.gov/grantsnet/**
Economic Research Service Research on Food Assistance and Nutrition	**www.ers.usda.gov**
The Foundation Center	**www.fdncenter.org**
Grantsmanship Center	**www.tgci.com**
Henry J. Kaiser Family Foundation	**www.kff.org**
Life Sciences Research Foundation	**www.lsrf.org**
National Center for Research Resources	**www.ncrr.nih.gov**
National Dairy Council	**www.nationaldairycouncil.org**
National Institutes of Health	**www.nih.gov**
National Science Foundation	**www.nsf.gov**
Notices of Funding Availability	**http://ocd1.usda.gov/nofa.htm**
Office of Research on Women's Health	**http://www4.od.nih.gov/orwh/**
Robert Wood Johnson Foundation	**www.rwjf.org**

Professional Focus

Teamwork Gets Results

At the end of the day, you bet on people, not on strategies.
– Larry Bossidy, CEO, AlliedSignal, as cited in N. M. Tichy's *The Leadership Engine*

Few things are accomplished in community nutrition without teamwork. Community nutritionists often work in teams to lobby state legislators, plan mass media campaigns, identify perceived needs in the community, design programs, and raise awareness about a nutritional problem. When the Arizona Department of Health Services designed a program to help the staff at child care facilities learn how to meet the needs of children with special health care needs, they used a team approach. The course instructors for Project CHANCE included a parent of a child with special health care needs, a nutritionist, an occupational therapist, and a director of a child care facility.[1] Each member of the teaching team brought personal insights and experiences to the discussion of how to meet the nutritional needs of children with cerebral palsy, cystic fibrosis, epilepsy, cleft palate, and other challenges.

A team is a *small number of people* with *complementary skills* who are committed to a *common purpose* for which they hold themselves *mutually accountable.*[2] Each aspect of this definition is important. The number of people on a team can range from 2 members to a maximum of about 16 members. Twelve members is a size that allows team members to interact easily with one another. Large groups can be unwieldy to manage and direct.[3]

The people chosen to be on the team should be selected carefully so that the team has balance, variety, and essential expertise.[4] When pulling a team together, prepare a list of potential team members. Think about their similarities and differences. A strong team consists of people with different problem-solving styles and methods of organizing information. Consider diversity of age, sex, educational training, and the like among team members. Diversity ensures a wealth of ideas and views.

Select team members who offer the expertise you believe will be needed to help achieve a goal or solve a problem. If you are developing a restaurant program, for example, you might want team members with expertise in nutrition, food service, marketing, and communications. If you are planning an intervention to address a nutrition problem such as hypertension, include someone from the target population—someone who can speak personally about hypertension—on the team.

"The heart of any team is a shared commitment by the members to their collective performance."[5] Achieving the team's goals requires the cooperation of every team member. In other words, a team probably won't be successful in meeting

Achieving a team's goals requires the cooperation of every team member. Here, children participate in *the Healthy Kids Challenge (HKC)*—a nonprofit school-based health initiative now active in over 700 schools in 35 states. HKC seeks to raise awareness and encourage healthy changes in the eating and activity habits of schoolchildren, maintains a partnership with *Cooking Light* magazine, and seeks outside funding for its programs. For more information, visit www.healthykidschallenge.com.

its goals if one or more members don't participate fully. Successful teams are teams whose members decide how they will work together and know what goal they are working toward.

Team members are mutually accountable for results. They are willing as individuals to assume responsibility for a portion of the work assigned to the team. They are active participants, not passive bench sitters. They prepare in advance for meetings, contribute their ideas, and evaluate themselves.

Most teams have a single leader or captain, but some teams have a "rotating leader"—that is, the team leadership shifts among team members. The team leader is responsible for preparing the meeting agenda, raising important issues, asking questions, keeping the discussion on friendly terms, maintaining order during the meeting, steering the team toward its goal, and recording the team's accomplishments. Diplomacy and humor are two tools the team leader can use to keep the team on track. Much like the football coach or symphony conductor, the team leader's main role is to help team members work together effectively.

Because teams are composed of individuals with different backgrounds, perspectives, ideas, and opinions, conflicts can arise. Thus team leaders must be able to deal with conflict, a by-product of how people interact in groups. Conflict often has a negative connotation and typically is viewed as a signal that something is wrong. Most people, in fact, try to avoid conflict at all costs. Yet conflict is inevitable in all relationships. It can arise from competition among workers for scarce resources, from questions about authority relationships, from pressures arising outside the organization, and even from solutions developed for dealing with previous conflicts. The problem in most organizations and work groups is that conflicts are avoided and left unresolved. When this occurs, team members are likely to be defensive and hostile and to suspect that they can't trust one another.[6]

To handle conflict positively, team leaders should remember that conflict usually emerges because of differences in attitudes, values, beliefs, and feelings. Appreciating and being sensitive to such differences helps establish a climate of cooperation. Next, team leaders work to pinpoint the underlying cause of the conflict. Finally, they resolve the differences through open negotiation. If necessary, rules and procedures are changed to address the underlying problem. At all times, leaders work to ensure that team members feel their complaints and concerns are perceived as legitimate. The trick to managing a team on which there are strong differences of opinion is to work toward finding common ground.[7]

References

1. L. Rider, MS, RD, Arizona Department of Education, personal communication, April 6, 1998.
2. J. R. Katzenbach, The myth of the top management team, *Harvard Business Review* 75 (1997): 83–91.
3. D. Hellriegel and J. W. Slocum, Jr., *Organizational Behavior,* 10th ed. (Mason, OH: South-Western/Thomson Publishers, 2003).
4. N. L. Frigon, Sr., and H. K. Jackson, Jr., *The Leader: Developing the Skills and Personal Qualities You Need to Lead Effectively* (New York: AMACOM, 1996), pp. 63–64.
5. Hellriegel and Slocum, *Organizational Behavior,* 2003.
6. M. S. Corey and G. Corey, *Groups—Process and Practice* (Belmont, CA: Wadsworth, 2001).
7. J. G. Liebler and C. R. McConnell, *Management Principles for Health Care Professionals,* 4th ed. (Sudbury, MA: Jones and Bartlett Publishers, 2004).

Appendix A

Nutrition Assessment and Screening

A-1 How to Plot Measures on a Growth Chart

FIGURE A1-1 How to Plot Measures on a Growth Chart

You can assess the growth of infants and children by plotting their measurements on a percentile graph. Percentile graphs divide the measures of a population into 100 equal divisions so that half of the population falls at or above the 50th percentile, and half falls below. Using percentiles allows for comparisons among people of the same age and gender.

To plot measures on a growth chart, follow these steps:

- Select the appropriate chart based on age and gender. For this example, use the accompanying chart, which gives percentiles for weight for girls from birth to 36 months.
- Locate the infant's age along the horizontal axis at the bottom of the chart (in this example, 6 months).
- Locate the infant's weight in pounds or kilograms along the vertical axis of the chart (in this example, 17 pounds or 7.7 kilograms).
- Mark the chart where the age and weight lines intersect (shown here with a black dot), and read off the percentile.

Source: E. Whitney and S. Rolfes, *Understanding Nutrition*, 9th ed. (Belmont, CA: Wadsworth Publishing Co., 2002), p. 536.

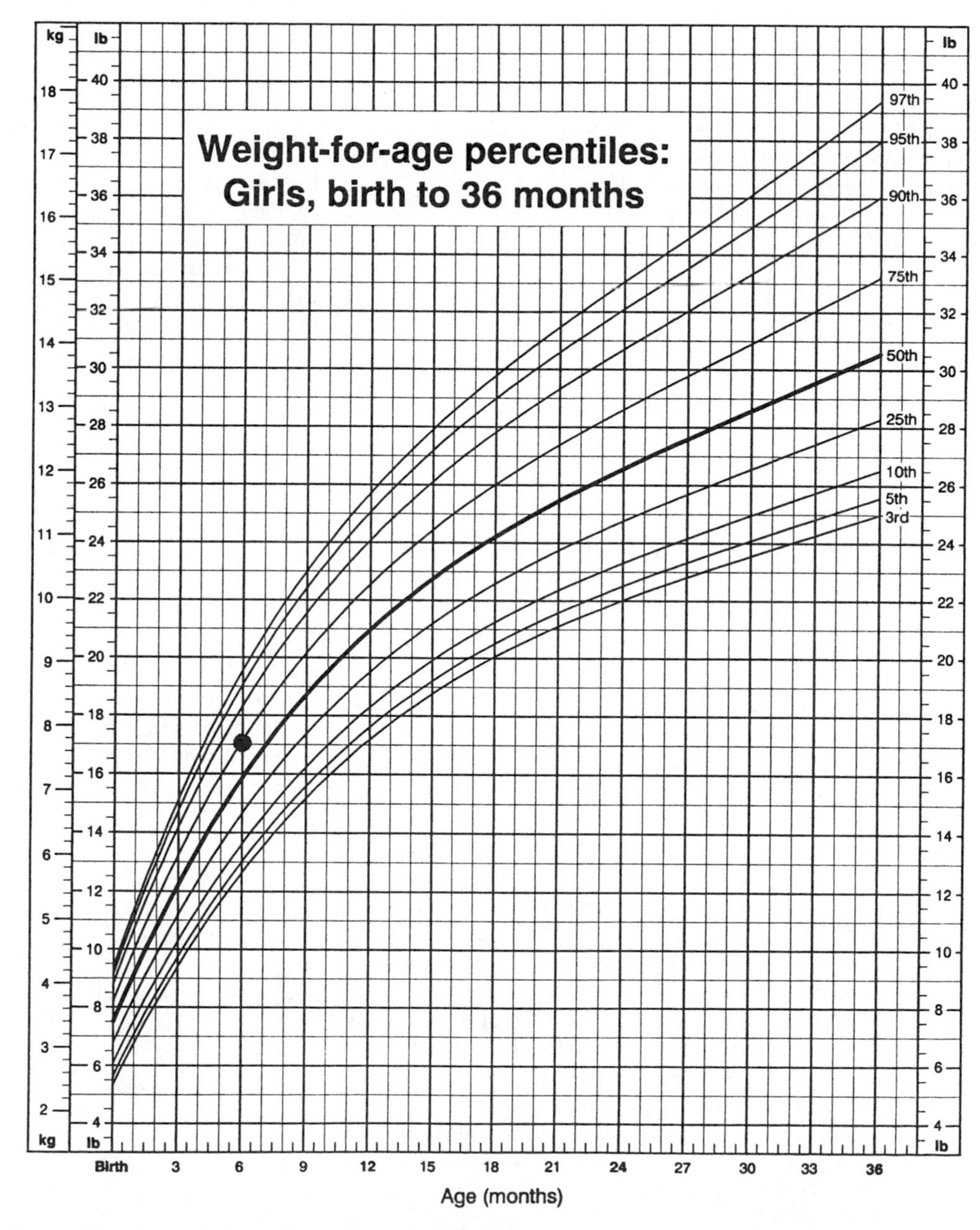

This 6-month-old infant is at the 75th percentile. Her pediatrician will weigh her again over the next few months and expect the growth curve to follow the same percentile throughout the first year. In general, dramatic changes or measures much above the 80th percentile or much below the 10th percentile may be cause for concern.

A-2 Assessment of Children's Growth Status

Several indices for assessing growth are derived from a child's height, weight, and age. The actual values for height, weight, and age are plotted on a nomogram or growth chart as shown in Figure A1-1, which makes it possible to compare a particular child's height and weight for age with a national standard. The anthropometric reference data used in the United States for assessing physical growth are based on a large, nationally representative sample of children from birth to 20 years of age. The percentile curves (for example, 5th, 10th, 25th, 50th, 75th, 90th, and 95th) for this reference population are displayed on growth charts developed by the National Center for Health Statistics (NCHS). The physical growth charts for girls and boys are available at www.cdc.gov/growthcharts. A discussion of indices derived from growth measurements is given here.

Stature/Length-for-Age

Low stature-for-age is defined as a stature-for-age value below the 5th percentile. In other words, low stature-for-age reflects a child's failure to achieve for a given age group distribution a height that conforms to standards established for a well-nourished, healthy population of children. Low stature-for-age is sometimes referred to as growth stunting, which may be the consequence of poor nutrition, a high frequency of infections, or both.

Although shortness in an individual child may be a normal reflection of the child's genetic heritage, a high prevalence rate of growth stunting reflects poor socioeconomic conditions. In some developing countries, the prevalence rate is as high as 60 to 70 percent. In developed countries, the rate is about 2 to 5 percent.[1] The prevalence of stunting is highest during the second or third year of life.[2]

Weight-for-Length/Stature

Low weight-for-stature, or thinness, is defined as a weight-for-stature value below the 5th percentile. This indicator is a sensitive index of current nutritional status and is often associated with recent severe disease. Between the ages of 1 and 10 years, the indicator is relatively independent of age. It is relatively independent of ethnic groups, particularly among children aged 1 to 5 years.[3]

Weight-for-stature differentiates between nutritional stunting, when a child's weight may be appropriate for her height, and wasting, when weight is very low for height owing to reductions in both tissue and fat mass. The prevalence of low weight-for-stature is usually less than 5 percent except during periods of famine, war, or other extreme conditions. A prevalence rate greater than 5 percent indicates the presence of serious nutritional problems.

At the other end of the spectrum, a high weight-for-stature (a value greater than the 95th percentile) correlates well with overweight. It typically indicates excess food consumption, low activity levels, or both.

Weight-for-Age

Weight-for-age is a composite index of stature-for-age and weight-for-stature. In children aged 6 months to about 7 years, weight-for-age is an indicator of acute malnutrition. It is widely used to assess protein-energy malnutrition and overnutrition. However, one limitation of using weight-for-age as an indicator of protein-energy malnutrition is that it does not consider height differences. For example, a child who has low weight-for-age may be genetically short with proportionally low height and low weight, rather than too thin with a low weight for height. As a result, the prevalence of protein-energy malnutrition in small children may be overstated if only this indicator is used.[4] Weight-for-age is most useful in clinical settings where repeated measurements of the indicator are used to evaluate children who are not gaining weight.[5]

Body Mass Index (BMI)-for-Age

BMI-for-age is an anthropometric index of weight and height combined with age. BMI-for-age charts for boys and girls aged 2 to 20 years are a major addition to the new CDC Pediatric Growth Charts and are used to classify children and adolescents as underweight, overweight, or at risk of overweight. CDC recommends the use of BMI-for-age for children aged 2 years and older in place of the weight-for-stature charts developed in 1977. For children, BMI is gender specific and age specific. However, weight-for-stature performs equally well in preschool aged children (primarily between 2 and 5 years of age) and can be used in this age group.

The recommendations are to classify BMI-for-age at or above the 95th percentile as overweight and between the 85th and 95th percentile as at risk of overweight.[6] Overweight, rather than obesity, is the term preferred for describing children and adolescents with a BMI-for-age equal to or greater than the 95th percentile of BMI-for-age or weight-for-length/stature. The 85th percentile is included on the BMI-for-age and the weight-for-stature charts to identify those at risk of overweight. The cutoff for underweight of less than the 5th percentile is based on recommendations by the World Health Organization Expert Committee on Physical Status.[7] Figure A2-1 plots a child's measurements on the BMI-for-age, weight-for-age, and stature-for-age charts.

Example

Angelica is a 13-year-old girl with intermittent weight and height measurements since age 2. Her measurements have been plotted on the BMI-for-age, weight-for-age, and stature-for-age charts.

AGE	WEIGHT	STATURE	BMI
2	$23\frac{1}{4}$	33	15
3	$28\frac{1}{2}$	$36\frac{1}{4}$	15.2
4	$36\frac{1}{2}$	$39\frac{1}{4}$	16.7
7	60	46	19.9
11	$94\frac{1}{2}$	$52\frac{1}{2}$	24.1
13	143	$60\frac{1}{4}$	27.7

The BMI-for-age chart shows a steady increase in BMI-for-age from age 2 to 4. By age 7, Angelica's BMI-for age was slightly above the 95th percentile and it remains in this channel or above to the present indicating that she is overweight. Her stature-for-age has consistently been below the 50th percentile while after age 3 her weight has been consistently above the 50th percentile and at age 13 it is above the 90th percentile. We see that Angelica is slightly shorter than other girls her age while her weight is higher.

FIGURE A2-1 BMI-for-Age, Weight-for-Age, and Stature-for-Age

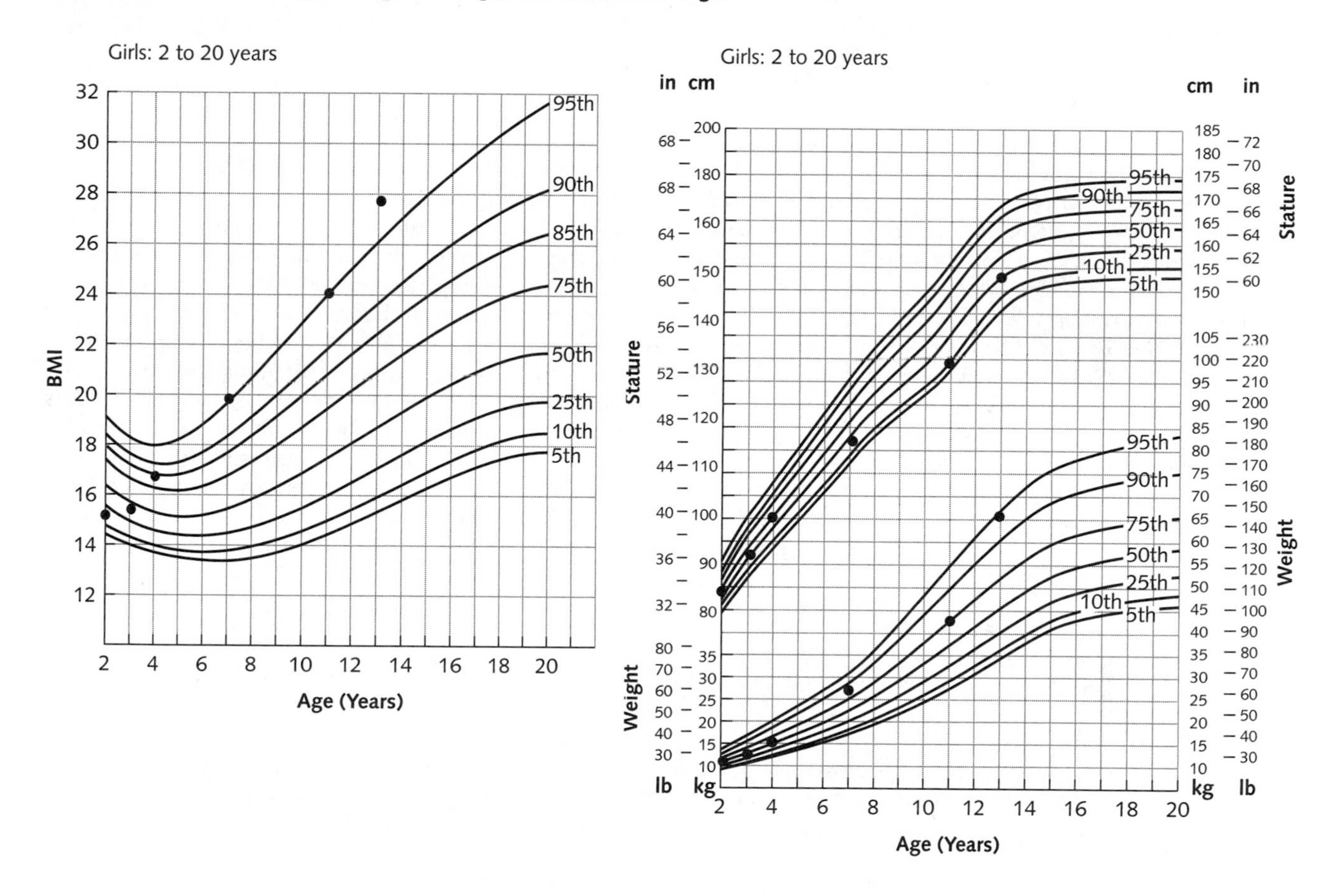

Appendix B

Complementary Nutrition and Health Therapies

Paralleling a paradigm shift from "sickness" to "wellness" is an expansive interest in and use of *complementary* and *alternative medicine (CAM)*—whether it be acupuncture, herbs, or chiropractic. Table B-1 lists the seven major categories of complementary and alternative medicine. Many of the CAM therapies defined in the glossary that follows are called "holistic," which means that the health care provider considers the whole person, including physical, mental, emotional, and spiritual health. In the past decade, a National Center for Complementary and Alternative Medicine was established at the National Institutes of Health, courses on complementary medicine were introduced into medical schools, and herbs and supplements became best-sellers for national pharmaceutical and supermarket chains. See Table B-2 for a list of herbs thought to be effective and those that should be avoided.

A number of important factors have led to the current proliferation of alternative health therapies, such as an interest in returning to a more natural lifestyle, dissatisfaction with the current state of Western health care, the unwanted side effects of prescription drugs, the spiraling cost and disarray of managed health care, and aging baby boomers who want a better quality of health. As members of the ADA, dietetics professionals serve the public through the promotion of optimal nutrition, health, and well-being.[1] Therefore, the implications for the dietetics profession are noteworthy. Clients and others may ask for your opinion on various alternative therapies, and you will want to inquire about the use of CAM therapies by clients and educate them about the state of scientific knowledge with regard to alternative nutrition and health therapies.

TABLE B-1 Classification and Examples of Complementary and Alternative Medical (CAM) Practices

I. Mind-Body Medicine

Involves behavioral, psychological, social, and spiritual approaches to health. Examples include: yoga, psychotherapy, art therapy, tai chi, hypnosis, dance therapy, meditation, biofeedback, music therapy, imagery, support groups, humor therapy, prayer.

II. Alternative Medical Systems

Involves complete systems of theory and practice that have been developed outside the Western biomedical approach. It is divided into four subcategories:

A. Acupuncture and Oriental medicine. Examples include: acupuncture, herbal formulas, tai chi.

B. Traditional indigenous systems. Examples include: American Indian medicine, ayurvedic medicine, traditional African medicine, traditional Aboriginal medicine, Central and South American practices.

C. Unconventional Western systems. Examples include: homeopathy, Cayce-based systems, orthomolecular medicine.

D. Naturopathy

III. Lifestyle and Disease Prevention

This category involves theories and practices designed to prevent the development of illness, identify and treat risk factors, or support the healing and recovery process. To be classified as CAM, the lifestyle therapies for behavior change, dietary change, exercise, stress management, and addiction control must be based on a nonorthodox system of medicine or be applied in unconventional ways.

IV. Biologically Based Therapies

Includes natural and biologically based practices, interventions, and products. Many overlap with conventional medicine's use of dietary supplements. There are four subcategories as shown below.

A. *Individual herbs* (e.g., ginkgo biloba, garlic, ginseng, echinacea, saw palmetto, ginger, green tea, psyllium)

B. *Special diet therapies* (e.g., Pritikin, Omish, Gerson, vegetarian, fasting, macrobiotic, Mediterranean, Paleolithic, Atkins, natural hygiene)

TABLE B-1 Classification and Examples of Complementary and Alternative Medical (CAM) Practices—*continued*

C. *Orthomolecular medicine.* This subcategory refers to products used as nutritional supplements for preventive or therapeutic purposes. They are usually used in combinations and at high doses (e.g., ascorbic acid, carotenes, tocopherols, niacinamide, choline, lysine, boron, melatonin, dehydroepiandrosterone [DHEA], amino acids, carnitine, probiotics, glucosamine sulfate, chondroitin sulfate)

D. *Pharmacological, biological, and instrumental interventions.* Includes products and procedures applied in an unconventional way (e.g., cartilage, enzyme therapies, hyperbaric oxygen, bee pollen, cell therapy, iridology)

V. Manipulative and Body-Based Systems

Refers to systems that are based on manipulation and/or movement of the body. It has three categories:

A. *Chiropractic medicine*

B. *Massage and body work* (e.g., Swedish massage, applied kinesiology, reflexology, rolfing, polarity)

C. *Unconventional physical therapies* (e.g., colonics, hydrotherapy, heat and electrotherapies)

VI. Biofield

Involves systems that use subtle energy fields in and around the body for medical purposes. Examples include: therapeutic touch, healing touch, reiki.

VII. Bioelectromagnetics

Refers to the unconventional use of electromagnetic fields for medical purposes.

Source: Adapted from National Institutes of Health, National Center for Complementary and Alternative Medicine, *Classification of Complementary and Alternative Medical Practices: General Information Package* (Silver Spring, MD: Office of Alternative Medicine Clearinghouse, 1998).

Glossary of Complementary and Alternative Medicine Terminology

Additional information about the following therapies can be found on the National Center for Complementary and Alternative Medicine's website at www.nccam.nih.gov.

Acupuncture: a technique that involves piercing the skin with long thin needles at specific anatomical sites to stimulate, disperse, and regulate the flow of *chi,* or vital energy, and relieve pain or illness. Acupuncture sometimes uses heat, pressure, friction, suction, or electromagnetic energy to stimulate the points.

Aromatherapy: a technique that uses essential oils (the volatile oils distilled from plants) to enhance physical, psychological, and spiritual health. The oils are inhaled, massaged into the skin in diluted form, or used in baths.

Ayurveda (EYE-your-VAY-dah): a traditional Hindu system of improving health practiced in India for more than 5,000 years. In this approach, illness is a state of imbalance among the body's systems and can be detected through procedures involving reading the pulse and checking the tongue. It uses herbs, diet, meditation, massage, and yoga to stimulate the body to make its own natural drugs.

Bioelectromagnetic applications: the use of electrical energy, magnetic energy, or both, to stimulate bone repair, wound healing, and tissue regeneration.

Biofeedback: the use of special devices to convey information about heart rate, blood pressure, skin temperature, muscle relaxation, and the like, to enable a person to learn how to consciously control these medically important functions.

Chelation therapy: the use of ethylenediamine tetraacetic acid (EDTA) to bind with metallic ions, thus healing the body by removing toxic metals.

Chinese (Oriental) medicine: practitioners are trained to use a variety of ancient and modern therapeutic methods, including acupuncture, herbal medicine, massage, moxibustion (heat therapy), and nutrition and lifestyle counseling, to treat a broad range of acute and chronic diseases.

Chiropractic: a manual healing method of manipulating or adjusting vertebrae to relieve musculoskeletal pain suspected of causing problems with internal organs.

Guided imagery: a technique that guides clients to achieve a desired physical, emotional, or spiritual state by visualizing themselves in that state.

Herbal medicine: the use of plants to treat disease or improve health; also known as *botanical medicine* or *phytotherapy.*

Homeopathic medicine: a practice based on the theory that "like cures like"—that is, that substances that cause symptoms in healthy people can cure those symptoms when given in very dilute amounts.

Hypnotherapy: a technique that uses hypnosis and the power of suggestion to improve health behaviors, relieve pain, and heal.

Iridology: the study of changes in the iris of the eye and their relationships to disease.

Massage therapy: a healing method in which the therapist manually kneads muscles to reduce tension, increase blood circulation, improve joint mobility, and promote healing of injuries.

Meditation: a self-directed technique of relaxing the body and calming the mind.

Naturopathic medicine: a system that integrates traditional medicine with botanical medicine, clinical nutrition, homeopathy, acupuncture, hydrotherapy, and manipulative therapy.

Orthomolecular medicine: the use of large doses of vitamins to treat chronic disease.

Reflexology: manipulation of specific areas on the feet and hands that correspond to a particular organ or zone in the upper body.

Therapeutic touch: a technique that uses graceful, sweeping movements of the hands, a few inches from the body, to scan the patient's energy flow, replenish it when necessary, release congestion or obstruction, and generally restore order and balance.

TABLE B-2 Selected Herbs, Their Common Uses, and Risks

COMMON NAME	CLAIMS AND USES	RISKS*
Aloe (gel)	Promote wound healing	Generally considered safe
Black Cohosh (roots)	Ease menopause symptoms	May cause clotting in blood vessels of eye, change curvature of cornea
Chamomile (flowers)	Relieve indigestion	Generally considered safe
Chaparral (leaves and twigs)	Slow aging, "cleanse" blood, heal wounds	Acute, toxic hepatitis
Comfrey (leafy plant)	Soothe nerves	Liver disease
Echinacea (roots)	Alleviate symptoms of colds, flus, and infections; promote wound healing; boost immunity	Generally considered safe
Ephedra (stems)	Promote weight loss	Rapid heart rate, tremors, seizures, insomnia, headaches
Feverfew (leaves)	Prevent migraine headaches	Generally considered safe; may cause mouth irritation, swelling, ulcers, and GI distress
Garlic (bulbs)	Lower blood lipids and blood pressure	Generally considered safe; may cause garlic breath, body odor, gas and GI distress; inhibits blood clotting
Ginger (roots)	Prevent motion sickness, nausea	Generally considered safe
Ginkgo biloba (tree leaves)	Improve memory, relieve vertigo	Generally considered safe; may cause headache, GI distress, and dizziness
Ginseng (roots)	Boost immunity, increase endurance	Generally considered safe; may cause insomnia and high blood pressure
Goldenseal (roots)	Relieve indigestion, treat urinary infections	Generally considered safe
Kava (roots)	Relieve anxiety, promote relaxation	Liver failure
Saw palmetto (ripe fruits)	Relieve symptoms of enlarged prostate; diuretic; enhance sexual vigor; enlarge mammary glands	Generally considered safe
St. John's wort (leaves and tops)	Relieve depression and anxiety	Generally considered safe; may cause fatigue and GI distress
Valerian (roots)	Calm nerves, improve sleep	Generally considered safe
Yohimbe (tree bark)	Enhance "male performance"	High blood pressure, fatigue, kidney failure, seizures

*Allergies are always a possible risk. Call your physician or the FDA Med Alert hot line at 800-332-1088 if you experience adverse effects.

Sources: S. Foster and V. E. Tyler, *Tyler's Honest Herbal: A Sensible Guide to the Use of Herbs and Related Remedies* (New York: Haworth Press, 1999); New safety measures are proposed for dietary supplements containing ephedrine alkaloids, *Journal of the American Medical Association* 278 (1997): 15; C. L. Bartels and S. J. Miller, Herbal and related remedies, *Nutrition in Clinical Practice* 13 (1998): 5–19; N. Spaulding-Albright, A review of some herbal and related products commonly used in cancer patients, *Journal of the American Dietetic Association* 97 (1997): S208–S15; adapted from E. Whitney and S. Rolfes, *Understanding Nutrition,* 10th ed. (Belmont, CA: Wadsworth/Thomson Learning, 2005).

Appendix **C**

The SMOG Readability Formula*

To calculate the SMOG reading grade level, begin with the entire written work that is being assessed, and follow these four steps:

1. Count off 10 consecutive sentences near the beginning, in the middle, and near the end of the text.
2. From this sample of 30 sentences, circle all of the words containing three or more syllables (polysyllabic), including repetitions of the same word, and total the number of words circled.
3. Estimate the square root of the total number of polysyllabic words counted. This is done by finding the nearest perfect square, and taking its square root.
4. Finally, add a constant of three to the square root. This number gives the SMOG grade, or the reading grade level that a person must have reached if he or she is to fully understand the text being assessed.

A few additional guidelines will help to clarify these directions:

- A sentence is defined as a string of words punctuated with a period (.), an exclamation point (!), or a question mark (?).
- Hyphenated words are considered as one word.
- Numbers that are written out should also be considered, and if in numeric form in the text, they should be pronounced to determine if they are polysyllabic.
- Proper nouns, if polysyllabic, should be counted, too.
- Abbreviations should be read as unabbreviated to determine if they are polysyllabic.

Not all pamphlets, fact sheets, or other printed materials contain 30 sentences. To test a text that has fewer than 30 sentences:

1. Count all of the polysyllabic words in the text.
2. Count the number of sentences.
3. Find the average number of polysyllabic words per sentence as follows:

$$\text{Average} = \frac{\text{Total number of polysyllabic words}}{\text{Total number of sentences}}$$

4. Multiply that average by the number of sentences *short of 30.*
5. Add that figure to the total number of polysyllabic words.
6. Find the square root and add the constant of 3.

Source: U.S. Department of Health and Human Services, Public Health Service, *Pretesting in Health Communications,* NIH Pub. No. 84-1493, (Bethesda, MD: National Cancer Institute, 1984), pp. 43–45.

Perhaps the quickest way to administer the SMOG grading test is by using the SMOG conversion table. Simply count the number of polysyllabic words in your chain of 30 sentences and look up the approximate grade level on the chart shown on page 640.

An example of how to use the SMOG Readability Formula is provided below. The SMOG Conversion Table is shown on page 640.

In Controlling Cancer—You Make a Difference

The key is ACTION. You can help protect yourself against cancer. Act promptly to:

Prevent some cancers through simple changes in lifestyle.
Find out about early detection tests in your home.
Gain peace of mind through regular medical checkups.

Cancers You Should Know About

Lung Cancer is the number one cancer among men, both in the number of new cases each year (79,000) and deaths (70,500). Rapidly increasing rates are due mainly to cigarette smoking. By not smoking, you can largely prevent lung cancer. The risk is reduced by smoking less, and by using lower tar and nicotine brands. But quitting altogether is by far the most effective safeguard. The American Cancer Society offers Quit Smoking Clinics and self-help materials.

Colorectal Cancer is second in cancer deaths (25,100) and third in new cases (49,000). When it is found early, chances of cure are good. A regular general physical usually includes a digital examination of the rectum and a guaiac slide test of a stool specimen to check for invisible blood. Now there are also Do-It-Yourself Guaiac Slides for home use. Ask your doctor about them. After you reach the age of 40, your regular check-up may include a "Procto," in which the rectum and part of the colon are inspected through a hollow, lighted tube.

Prostate Cancer is second in the number of new cases each year (57,000) and third in deaths (20,600). It occurs mainly in men over 60. A regular rectal exam of the prostate by your doctor is the best protection.

A Check-Up Pays Off

Be sure to have a regular general physical including an oral exam. It is your best guarantee of good health.

How Cancer Works

If we know something about how cancer works, we can act more effectively to protect ourselves against the disease. Here are the basics:

1. Cancer spreads; time counts—Cancer is uncontrolled growth of abnormal cells. It begins small and if unchecked, spreads. If detected in an early, local stage, the chances for cure are best.
2. Risk increases with age—This is not a reason to worry, but a signal to have more regular thorough physical check-ups. Your doctor or clinic can advise you on what tests to get and how often they should be performed.
3. What you can do—Don't smoke and you will sharply reduce your chances of getting lung cancer. Avoid too much sun, a major cause of skin cancer. Learn cancer's Seven Warning Signals, listed on the back of this leaflet, and see your doctor promptly if they persist. Pain usually is a late symptom of cancer; don't wait for it.

Unproven Remedies

Beware of unproven cancer remedies. They may sound appealing, but they are usually worthless. Relying on them can delay good treatment until it is too late. Check with your doctor or the American Cancer Society.

More Information

For more information of any kind about cancer—free of cost—contact your local Unit of the American Cancer Society.

Know Cancer's Seven Warning Signals

1. Change in bowel or bladder habits.
2. A sore that does not heal.
3. Unusual bleeding or discharge.
4. Thickening or lump in breast or elsewhere.
5. Indigestion or difficulty in swallowing.
6. Obvious change in wart or mole.
7. Nagging cough or hoarseness.

If you have a warning signal, see your doctor.

(This pamphlet is from the American Cancer Society.)

We have calculated the reading grade level for this example. Compare your results to ours, then check both with the SMOG Conversion Table:

Readability Test Calculations

Total number of polysyllabic words	= 38
Nearest perfect square	= 36
Square root	= 6
Constant	= 3
SMOG reading grade level	= 9

SMOG Conversion Table*

Total Polysyllabic Word Counts	Approximate Grade Level (+1.5 Grades)
0–2	4
3–6	5
7–12	6
13–20	7
21–30	8
31–42	9
43–56	10
57–72	11
73–90	12
91–110	13
111–132	14
133–156	15
157–182	16
183–210	17
211–240	18

*Developed by Harold C. McGraw, Office of Educational Research, Baltimore County Schools, Towson, Maryland.

References

Chapter 1

1. W. B. Johnston, Global work force 2000: The new world labor market, *Harvard Business Review* 69 (1991): 115–27.
2. P. H. Mirvis, Human resource management: Leaders, laggards, and followers, *Academy of Management Executive* 11 (1997): 43–56.
3. G. I. Balch, Employers' perceptions of the roles of dietetics practitioners: Challenges to survive and opportunities to thrive, *Journal of the American Dietetic Association* 96 (1996): 1301–5.
4. G. A. Hillery, Jr., Definitions of community: Areas of agreement, *Rural Sociology* 20 (1955): 111–23.
5. The discussion of the concept of community was adapted from T. N. Clark, *Community Structure and Decision-Making: Comparative Analyses* (San Francisco: Chandler, 1968), pp. 83–9; A. D. Edwards and D. G. Jones, *Community and Community Development* (The Hague: Mouton, 1976), pp. 11–39; R. M. MacIver, *On Community, Society and Power* (Chicago: University of Chicago Press, 1970), pp. 29–34; and D. E. Poplin, *Communities* (New York: Macmillan, 1979), pp. 3–25.
6. L. A. Pal, *Beyond Policy Analysis: Public Issue Management in Turbulent Times* (Scarborough, Ontario: ITP, 1997), p. 1.
7. Community Nutrition Institute, Looking for solutions: Food recovery, recycling, and education, *Nutrition Week* 27 (July 11, 1997): 4–6.
8. Community Nutrition Institute, USDA to support "human gleaning" with new fund, *Nutrition Week* 27 (March 21, 1997): 3.
9. T. A. Nicklas and coauthors, Development of a school-based nutrition intervention for high school students: *Gimme 5, American Journal of Health Promotion* 11 (1997): 315–22.
10. Institute of Medicine, *The Future of Public Health* (Washington, D.C.: National Academy Press, 1988).
11. American Heart Association. Heart Disease and Stroke Statistics—2005 Update. Dallas, TX: American Heart Association, 2005.
12. *The Burden of Chronic Diseases as Causes of Death,* 2002; accessed at www.cdc.gov/nccdphp.
13. Joint United Nations Programme on HIV/AIDS, *Report of the Global HIV/AIDS Epidemic* (Geneva: UNAIDS, 2004).
14. National Center for Health Statistics, *National Vital Statistics Report,* 2004.
15. National Institute of Allergy and Infectious Diseases, Tuberculosis, *Health Matters,* March 2002 (National Institutes of Health, Bethesda, MD).
16. Centers for Disease Control and Prevention, *Reported Tuberculosis in the United States, 2003* (Atlanta, GA: U.S. Department of Health and Human Services, CDC), September 2004.
17. Statistics Canada; available at www.statcan.ca/english/Pgdb/health36.htm.
18. National Cancer Institute of Canada, Canadian Cancer Statistics 2004; available at www.cancer.ca.
19. P. Lee and D. Paxman, Reinventing public health, *Annual Review of Public Health* 18 (1997): 1–35; Institute of Medicine, *The Future of the Public's Health in the 21st Century* (Washington, D.C.: National Academy Press, 2003).
20. The discussion of the concepts of health and public health was adapted from J. M. Last, *Public Health and Human Ecology* (East Norwalk, CT: Appleton & Lange, 1987), pp. 1–26; and G. Pickett and J. J. Hanlon, *Public Health: Administration and Practice* (St. Louis: Times Mirror/Mosby College Publishing, 1990), pp. 3–20.
21. M. P. O'Donnell, Definition of health promotion, *American Journal of Health Promotion* 1 (1986): 4–5.
22. J. Stokes III and coauthors, Definition of terms and concepts applicable to clinical preventive medicine, *Journal of Community Health* 8 (1982): 33–41.
23. M. Minkler, Health education, health promotion, and the open society: An historical perspective, *Health Education Quarterly* 16 (1989): 17–30.
24. The quote is from O'Donnell, Definition of health promotion, p. 4.
25. J. P. Elder and coauthors, *Motivating Health Behavior* (Albany, NY: Delmar, 1994), pp. 1–7.
26. The discussion of the types of prevention efforts was adapted from American Dietetic Association, Position of the American Dietetic Association: The role of nutrition in health promotion and disease prevention programs, *Journal of the American Dietetic Association* 98 (1998): 205–8; and R. H. Fletcher, S. W. Fletcher, and E. H. Wagner, *Clinical Epidemiology: The Essentials,* 4th ed. (Baltimore, MD: Williams & Wilkins, 2005).
27. As cited in Last, *Public Health and Human Ecology,* p. 16.
28. World Health Organization, Fifty-sixth World Health Assembly, International Conference on Primary Health Care, Alma-Ata: Twenty-fifth Anniversary, April 20, 2003, Document A56/27.
29. World Summit on Sustainable Development, Johannesburg, South Africa, 2002; available at www.who.int/wssd/en.
30. World Health Organization, *Targets for Health for All* (Copenhagen: World Health Organization Regional Office for Europe, 1985).
31. H. Nielsen, *Achieving Health for All:* A framework for nutrition in health promotion, *Journal of the Canadian Dietetic Association* 50 (1989): 77–80; *Toward a Healthy Future: Second Report on the Health of Canadians,* 1999; available at www.hc-sc.gc.ca.
32. U.S. Department of Health and Human Services, Public Health Service, *Healthy People 2000: National Health Promotion and Disease Prevention Objectives* (Washington, D.C.: U.S. Government Printing Office, 1990).
33. U.S. Department of Health and Human Services, *Healthy People 2010: Understanding and Improving Health,* 2nd ed. Washington, D.C.: U.S. Government Printing Office, November 2000.
34. U.S. Department of Health and Human Services, Public Health Service, *Healthy People: The Surgeon General's Report on Health Promotion and Disease Prevention,* PHS Pub. No. 79-55071 (Washington, D.C.: U.S. Government Printing Office, 1979); U.S. Department of Health and Human Services, Public Health Service, *Promoting Health/Preventing Disease: Objectives for the Nation* (Washington, D.C.: U.S. Government Printing Office, 1980).
35. U.S. Department of Health and Human Services, *Healthy People 2010.*
36. U.S. Department of Health and Human Services, Public Health Service, *The Surgeon General's Report on Nutrition and Health,* DHHS Pub. No. 88-50210 (Washington, D.C.: U.S. Government Printing Office, 1988), pp. 1–20.
37. U.S. Department of Health and Human Services, *Healthy People 2010.*

38. The definition of surveillance was adapted from Federation of American Societies for Experimental Biology, Life Sciences Research Office, *Third Report on Nutrition Monitoring in the United States,* Vol. 1 (Washington, D.C.: U.S. Government Printing Office, 1995), pp. 1–8.
39. H. Blackburn, Research and demonstration projects in community cardiovascular disease prevention, *Journal of Public Health Policy* 4 (1983): 398–421.
40. Centers for Disease Control and Prevention, *National Vital Statistics Report, 2004.*
41. M. T. Molla and coauthors, *Summary Measures of Population Health: Report of Findings on Methodological and Data Issues* (Hyattsville, MD: National Center for Health Statistics, 2003); National Center for Health Statistics, *Health, United States, 2004 with Chartbook on Trends in the Health of Americans* (Hyattsville, MD: 2004).
42. Public Health Service, U.S. Department of Health and Human Services, HP 2010 Progress Review: Nutrition and Overweight, January 21, 2004; available at www.cdc.gov/nchs/hphome.htm.
43. M. Kaufman, Preparing public health nutritionists to meet the future, *Journal of the American Dietetic Association* 86 (1986): 511–14.
44. Update on state licensure laws; available at www.eatright.org.
45. Pan American Health Organization, *Food and Nutrition Issues in Latin America and the Caribbean* (Washington, D.C.: Pan American Health Organization/Inter-American Development Bank, 1990).
46. Food and Agriculture Organization of the United Nations white paper on the Food Policy and Nutrition Division, pp. 24–29.
47. L. M. Brown and M. F. Fruin, Management activities in community dietetics practice, *Journal of the American Dietetic Association* 89 (1989): 373–77.
48. C. J. Gilmore and coauthors, Determining educational preparation based on job competencies of entry-level dietetics practitioners, *Journal of the American Dietetic Association,* April 2002; and Commission on Accreditation for Dietetics Education, American Dietetic Association, Foundation Knowledge and Skills and Competency Requirements for Entry-Level Dietitians, 2002, available at www.eatright. org/cade.
49. Winnipeg Prenatal Nutrition Initiative, *Healthy Start for Mom & Me,* Winnipeg, Manitoba, Canada R2W 4J5.
50. The discussion of the role delineation study and the responsibilities of dietitians was taken in part from American Dietetic Association, *Role Delineation for Registered Dietitians and Entry-Level Dietetic Technicians;* ADA Reports, President's Page: Beyond the RD, *Journal of the American Dietetic Association* 90 (1990): 1117–21; Of Professional Interest, Commentary on the role delineation study, *Journal of the American Dietetic Association* 90 (1990): 1122–23; and M. T. Kane and coauthors, Role delineation for dietetic practitioners: Empirical results, *Journal of the American Dietetic Association* 90 (1990): 1124–33.
51. Brown and Fruin, Management activities, 373–77.
52. American Dietetic Association, *Role Delineation.*
53. G. Gates and W. Sandoval, Teaching multi-skilling in dietetics education, *Journal of the American Dietetic Association* 98 (1998): 278–84; M. R. Schiller and K. N. Wolf, Preparing for practice in the 21st century, *Topics in Clinical Nutrition,* 16 (2000): 1–12.
54. The discussion of entrepreneurship was adapted from B. J. Bird, *Entrepreneurial Behavior* (Glenview, IL: Scott, Foresman, 1989), pp. 1–33, 57–76, and 349–75; and J. G. Burch, *Entrepreneurship* (New York: Wiley, 1986), pp. 4–42.
55. Bird, *Entrepreneurial Behavior,* p. 3.
56. *Health Start,* Pawnee, OK 74058-0462.
57. *Dietetics Online®* can be found on the Internet at www.dietetics.com.
58. Burch, *Entrepreneurship,* pp. 28–29.
59. A. L. F. Foong and coauthors, Entrepreneurs and intrapreneurs: Common blood, different languages, *Entrepreneurship, Innovation, and Change* 6 (1997): 67–72.
60. Community Nutrition Institute, WIC on Wheels Winnebago serves a sprawling Vegas, *Nutrition Week* 27 (August 1, 1997): 6.
61. Gilmore and coauthors, *Journal of the American Dietetic Association.*
62. S. C. Parks, The fractured anthill: A new architecture for sustaining the future, *Journal of the American Dietetic Association* 101 (2001): 133.
63. V. R. Young, Good nutrition for all: Challenge for the nutritional sciences in the new millennium, *Nutrition Today* 36 (2001): 6–16.
64. The Institute for the Future, *Health and Health Care 2010* (San Francisco, CA: Jossey-Bass, 2000).
65. M. N. Haan and coauthors, The impact of aging and chronic disease on use of hospital and outpatient services in a large HMO: 1971–1991, *Journal of the American Geriatric Society* 45 (1997): 667–74.
66. The role of the ADA environment scan: Shaping the future of the profession, *Journal of the American Dietetic Association* 102 (2002): S1820–S1839.
67. M. Kennedy, Boomers versus busters, *Healthcare Executive,* 13 (1998): 6–10.
68. National Center for Health Statistics, *Health, United States, 2004 with Chartbook on Trends in the Health of Americans* (Hyattsville, MD: 2004).
69. J. L. Zaichkowsky, Consumer behavior: Yesterday, today, and tomorrow, *Business Horizons* 34 (1991): 51–58.
70. This list is adapted from M. Fierro, The obesity epidemic—how states can trim the fat, *Issue Brief,* National Governor's Association, Health Policy Studies Division, June 2002; available at www.nga.org/cda/files/OBESITYIB.pdf.
71. The role of the ADA environment scan: Shaping the future of the profession, *Journal of the American Dietetic Association* 102 (2002): S1820–S1839.

Chapter 2

1. U.S. Census Bureau, Public Information Office, Hispanic and Asian Americans increasing faster than overall population, *U.S. Census Bureau News,* June 14, 2004.
2. A. L. Martinez, *Hunger in Latino Communities* (Washington, D.C.: Congressional Hunger Center and Congressional Hispanic Caucus Institute, 1995).
3. G. Christakis, Community assessment of nutritional status, in H. S. Wright and L. S. Sims, *Community Nutrition—People, Policies, and Programs* (Belmont, CA: Wadsworth, 1981), pp. 83–97.
4. E. B. Perrin, Information systems for health outcome analysis, in *Oxford Textbook of Public Health,* 3rd ed., Vol. 2, ed. R. Detels, W. W. Holland, J. McEwen, and G. S. Omenn (New York: Oxford University Press, 1997), pp. 491–97.
5. Department of Health and Human Services and Department of Agriculture, *Nutrition Monitoring in the United States: The Directory of Federal and State Nutrition Monitoring and Related Research Activities* (Hyattsville, MD: U.S. Government Printing Office, 2000).
6. B. Haglund, R. R. Weisbrod, and N. Bracht, Assessing the community: Its services, needs, leadership, and readiness, in *Health Promotion at the Community Level,* ed. N. Bracht (Newbury Park, CA: Sage, 1999), pp. 91–292.
7. S. Dubois and coauthors, Ability of the Higgins Nutrition Intervention Program to improve adolescent pregnancy outcome, *Journal of the American Dietetic Association* 97 (1997): 871–78.
8. The description of the BRFSS is from the Centers for Disease Control and Prevention website at www.cdc.gov/brfss, accessed October 2004.
9. W. Carr, *Measuring Avoidable Deaths and Diseases in New York State,* Paper Series 8 (New York: United Hospital Fund, 1988), p. 48; and U.S. Department of Health and Human Services, Public Health Service, *Health United States and Prevention Profile, 1991* (Hyattsville, MD: Public Health Service, 1991), pp. 158–60.
10. National Center for Health Statistics, National Health and Nutrition Examination Survey (1999–2002 NHANES) Public-Use Data Files; available at the center's website at www.cdc.gov/nchs.
11. T. A. Hammad and coauthors, Withdrawal rates for infants and children participating in WIC in Maryland, *Journal of the American Dietetic Association* 97 (1997): 893–95.
12. A. L. Owen and G. M. Owen, Twenty years of WIC: A review of some effects of the

program, *Journal of the American Dietetic Association* 97 (1997): 777–82.

13. National Center for Chronic Disease Prevention and Health Promotion, *U.S. Obesity Trends 1985 to 2000,* available at www.cdc.gov/nccdphp/dnpa/obesity/trend/maps/index.htm; and A. H. Mokhad and coauthors, The continuing epidemics of obesity and diabetes in the United States, *Journal of the American Medical Association* 286 (2001): 1195–1200.
14. J. Hill and R. Wing, The National Weight Control Registry, *The Permanente Journal* 7 (2003): 34–37.
15. National Center for Health Statistics, *Healthy People 2000 Final Review* (Hyattsville, MD: Public Health Service, 2001).
16. J. Pomerleau and coauthors, Food intake of immigrants and non-immigrants in Ontario: Food group comparison with the recommendations of Canada's Food Guide to Healthy Eating, *Journal of the Canadian Dietetic Association* 58 (1997): 68–76.
17. The reports on the evaluation of the FitzIn Program are available from the Heart & Stroke Foundation of Manitoba, Winnipeg, Manitoba, Canada R3B 2H8.
18. The description of the purposes of community needs assessment was adapted from Christakis, Community assessment of nutritional status; D. B. Jelliffe and E. F. P. Jelliffe, *Community Nutritional Assessment* (Oxford: Oxford University Press, 1990), pp. 142–55; and Office of Disease Prevention and Health Promotion, Public Health Service and American Dietetic Association, *Worksite Nutrition—A Guide to Planning, Implementation, and Evaluation* (Chicago: American Dietetic Association, 1993), pp. 14–15.
19. T. G. Rundall, Health planning and evaluation, in *Public Health and Preventive Medicine,* ed. J. M. Last et al. (Norwalk, CT: Appleton & Lange, 1998).
20. C. W. Tyler, Jr., and J. M. Last, Epidemiology, in *Public Health and Preventive Medicine,* ed. J. M. Last et al., pp. 11–39; C. S. Reichardt and T. D. Cook, Beyond qualitative versus quantitative methods, in *Qualitative and Quantitative Methods in Evaluation Research,* ed. T. D. Cook and C. S. Reichardt (Beverly Hills, CA: Sage, 1979), pp. 7–27; and R. Walker, An introduction to applied qualitative research, in *Applied Qualitative Research,* ed. R. Walker (Aldershot, England: Gower, 1985), pp. 1–24.
21. R. H. Fletcher, S. W. Fletcher, and E. H. Wagner, *Clinical Epidemiology—The Essentials,* 4th ed. (Baltimore, MD: Lippincott, Williams & Wilkins, 2005).
22. Infant mortality data were taken from "Eliminating Racial and Ethnic Disparities in Health," available on the U.S. Department of Health and Human Service's website at raceandhealth.hhs.gov.
23. Information about the United Way of America was obtained from its website at www.unitedway.org.
24. J. Harvey-Berino and coauthors, Food preferences predict eating behavior of very young Mohawk children, *Journal of the American Dietetic Association* 97 (1997): 750–53.
25. M. B. Dignan and P. A. Carr, *Program Planning for Health Education and Promotion,* 2nd ed. (Philadelphia, PA: Lea & Febiger, 1992), pp. 17–58.
26. P. R. Voss and coauthors, Role of secondary data, in *Needs Assessment: Theory and Methods,* ed. D. E. Johnson et al. (Ames: Iowa State University Press, 1987), pp. 156–70.
27. Jelliffe and Jelliffe, *Community Nutritional Assessment,* pp. 355–83.
28. Information about the University of Michigan data archive services was obtained from a booklet published by the Institute for Social Research, University of Michigan (Ann Arbor); Catalog of Data Collections (Ann Arbor, MI: Interuniversity Consortium for Political and Social Research, 1992); and C. Campbell, 1990 Census public use microdata samples (PUMS), *ICPSR Bulletin* 13 (1993): 1–7.
29. Information about the Institute for Research in Social Science was obtained from a booklet published by the institute at the University of North Carolina (Chapel Hill, 1991).
30. S. J. Kunitz, The history and politics of U.S. health care policy for American Indians and Alaska Natives, *American Journal of Public Health* 86 (1996): 1464–73.
31. The definition of culture was taken from *Merriam-Webster's Online Dictionary,* 2004; available at www.m-w.com/dictionary.htm, accessed October 2004.
32. Pan American Health Organization, *Health Conditions in the Americas* (Washington, D.C.: Pan American Health Organization, 1994).
33. Department of Health and Human Services, *Nutrition Monitoring in the United States,* 2000; and Federation of American Societies for Experimental Biology, Life Sciences Research Office, prepared for the Interagency Board for Nutrition Monitoring and Related Research, *Third Report on Nutrition Monitoring in the United States,* Vols. 1 and 2 (Washington, D.C.: U.S. Government Printing Office, 1995).
34. The Centers for Disease Control and Prevention's website is www.cdc.gov.
35. U.S. Department of Health and Human Services, *Healthy People 2010* (Washington, D.C.: U.S. Government Printing Office, January 2000).
36. Dignan and Carr, *Program Planning for Health Education and Promotion,* pp. 51–57.
37. Jelliffe and Jelliffe, *Community Nutritional Assessment,* pp. 452–64.
38. A. D. Spiegel and H. H. Hyman, *Strategic Health Planning: Methods and Techniques Applied to Marketing and Management* (Norwood, NJ: Ablex, 1991), p. 203.
39. Perrin, Information systems, p. 492.
40. M. Samuelson, Commentary: Changing unhealthy lifestyle: Who's ready . . . Who's not? An argument in support of the stages of change component of the transtheoretical model, *American Journal of Health Promotion* 12 (1997): 13–14.
41. N. Wellman, Power to propel your program, *Community Nutritionary* (Tarrytown, NY: Dannon Institute, Fall 2000), p. 2.
42. *Healthy People in Healthy Communities,* Stock No. 017-001-00546-1 (Washington, D.C.: U.S. Government Printing Office, 2001), pp. 1–40.
43. See the *Healthy People Toolkit,* which provides examples of state and national experiences in setting and using objectives, at: www.health.gov/healthypeople/state/toolkit.
44. B. J. Bird, *Entrepreneurial Behavior* (Glenview, IL: Scott, Foresman, 1989), p. 3.
45. The United Way of America Internet address is www.unitedway.org.
46. The Internet address for the Initiative on Race is raceandhealth.hhs.gov.

Chapter 3

1. A. R. Folsom and coauthors, Physical activity and incidence of coronary heart disease in middle-aged women and men, *Medicine & Science in Sports and Exercise* 29 (1997): 901–9.
2. T. D. Dye and coauthors, Unintended pregnancy and breast-feeding behavior, *American Journal of Public Health* 87 (1997): 1709–11.
3. J. Brug and coauthors, The relationship between self-efficacy, attitudes, intake compared to others, consumption, and stages of change related to fruit and vegetables, *American Journal of Health Promotion* 12 (1997): 25–30.
4. J. Lykkesfeldt and coauthors, Ascorbic acid and dehydro-ascorbic acid as biomarkers of oxidative stress caused by smoking, *American Journal of Clinical Nutrition* 65 (1997): 959–63.
5. Federation of American Societies for Experimental Biology, Life Sciences Research Office, prepared for the Interagency Board for Nutrition Monitoring and Related Research, *Third Report on Nutrition Monitoring in the United States,* Vol. 1 (Washington, D.C.: U.S. Government Printing Office, 1995), pp. 1–17.
6. E. Velempini and K. D. Travers, Accessibility of nutritious African foods for an adequate diet in Bulawayo, Zimbabwe, *Journal of Nutrition Education* 29 (1997): 120–27.
7. M. Story and J. Stang, eds., *Nutrition and the Pregnant Adolescent: A Practical Reference Guide* (Washington, D.C.: Maternal and Child Health Bureau, Health Resources and Services Administration, 2000).
8. D. Albanes and coauthors, Effects of supplemental beta-carotene, cigarette smoking, and alcohol consumption on serum carotenoids in the Alpha-Tocopherol, Beta-Carotene Cancer Prevention Study, *American Journal of Clinical Nutrition* 66 (1997): 366–72.

9. T. Goode, *Policy Brief 4: Engaging Communities to Realize the Vision of One Hundred Percent Access and Zero Health Disparities: A Culturally Competent Approach* (Washington, D.C.: National Center for Cultural Competence, Georgetown University Child Development Center, 2001).
10. S. Staveteig and A. Wigton, *Racial and Ethnic Disparities: Key Findings from the National Survey of America's Families* (The Urban Institute New Federalism Series B No B-5, February 2000); and B. D. Smedley, A. Y. Stith, and A. R. Nelson, eds., *Unequal Treatment: Confronting Racial and Ethnic Disparities in Health Care* (Washington, D.C.: National Academy Press, 2002).
11. Maternal and Child Health Bureau, *Child Health USA, 2001* (Washington, D.C.: U.S. Government Printing Office, 2002).
12. B. D. Weiss and coauthors, Health status of illiterate adults: Relation between literacy and health status among persons with low literacy skills, *Journal of the American Board of Family Practitioners* 5 (1992): 257–64.
13. R. Perez-Escamilla and coauthors, *Community Nutritional Problems Among Latino Children in Hartford, Connecticut* (Hartford: University of Connecticut and the Hispanic Health Council, 1997).
14. D. Cohen, *Consumer Behavior* (New York: Random House, 1991), pp. 76–127.
15. A. A. Davies-Adetugbo, Sociocultural factors and the promotion of exclusive breastfeeding in rural Yoruba communities of Osun State, Nigeria, *Social Science in Medicine* 45 (1997): 113–25.
16. J. W. Creswell, *Research Design: Qualitative, Quantitative, and Mixed Methods Approaches,* 2nd ed. (Thousand Oaks, CA: Sage, 2002); and J. E. Perkin, Design and use of questionnaires in research, in E. R. Monsen, ed., *Research: Successful Approaches,* 2nd ed. (Chicago: American Dietetic Association, 2003).
17. C. E. Woteki and coauthors, Selection of nutrition status indicators for field surveys: The NHANES III design, *Journal of Nutrition* 120 (1990): 1440–45.
18. J. H. Sabry, Purposes of food consumption studies, in M. E. Cameron and W. A. Van Staveren, *Manual on Methodology for Food Consumption Studies* (Oxford: Oxford University Press, 1992), pp. 25–31.
19. I. H. E. Rutishauser, Practical implementation, in Cameron and Van Staveren, *Manual on Methodology for Food Consumption Studies,* pp. 223–45.
20. V. J. Schoenbach, Appraising health risk appraisal [editorial], *American Journal of Public Health* 77 (1987): 409–11.
21. D. H. Gemson and R. P. Sloan, Efficacy of computerized health risk appraisal as part of a periodic health examination at the worksite, *American Journal of Health Promotion* 9 (1995): 462–66.
22. D. R. Anderson and M. J. Staufacker, The impact of worksite-based health risk appraisal on health-related outcomes: A review of the literature, *American Journal of Health Promotion* 10 (1996): 499–508.
23. L. Breslow and coauthors, Development of a health risk appraisal for the elderly (HRA-E), *American Journal of Health Promotion* 11 (1997): 337–43.
24. K. W. Smith and coauthors, The validity of health risk appraisal instruments for assessing coronary heart disease risk, *American Journal of Public Health* 77 (1987): 419–24.
25. J. J. Korelitz and coauthors, Health habits and risk factors among truck drivers visiting a health booth during a trucker trade show, *American Journal of Health Promotion* 8 (1993): 117–23.
26. J. M. Last, *Public Health and Human Ecology,* 2nd ed. (New York: McGraw-Hill Professional, 1998); and *Third Report of the NCEP Expert Panel on Detection, Evaluation, and Treatment of High Blood Cholesterol in Adults,* NIH Publication No. 01-3305 (Washington, D.C.: DHHS, May 2001).
27. Nutrition Screening Initiative, *A Physician's Guide to Nutrition in Chronic Disease Management in Older Adults* (Washington, D.C.: Nutrition Screening Initiative, 2002).
28. E. R. Monsen, ed., *Research: Successful Approaches,* 2nd ed. (Chicago: American Dietetic Association, 2003).
29. I. Seidman, *Interviewing as Qualitative Research: A Guide for Researchers in Education and the Social Sciences,* 2nd ed. (New York: Teachers College Press, 1998), pp. 63–78.
30. R. A. Krueger, *Focus Groups: A Practical Guide for Applied Research* (Thousand Oaks, CA: Sage, 1994).
31. C. E. Basch, Focus group interview: An underutilized research technique for improving theory and practice in health education, *Health Education Quarterly* 14 (1987): 411–48.
32. R. B. Masters and coauthors, The use of focus groups in the design of cholesterol education intervention programs, *American Journal of Health Promotion* 8 (1993): 95–97.
33. J. L. Kristeller and R. A. Hoerr, Physician attitudes toward managing obesity: Differences among six specialty groups, *Preventive Medicine* 26 (1997): 542–49.
34. C. S. Haignere and coauthors, One method for assessing HIV/AIDS peer-education programs, *Journal of Adolescent Health* 21 (1997): 76–79.
35. E. Helsing, On the clear need to meet and learn to speak clearly: Statement from the World Health Organization, *American Journal of Clinical Nutrition* 65 (suppl.) (1997): 1098S–99S.
36. R. S. Gibson, *Principles of Nutritional Assessment* (New York: Oxford University Press, 1990), pp. 3–20.
37. K. S. Kubena, Accuracy in dietary assessment: On the road to good science, *Journal of the American Dietetic Association* 100 (2000): 775–76; and W. A. van Staveren and M. C. Ocke, Estimation of dietary intake, in B. A. Bowman and R. M. Russell, eds., *Present Knowledge in Nutrition,* 8th ed. (Washington, D.C.: International Life Sciences Institute, 2001), pp. 605–16.
38. R. S. Burke and coauthors, Nutrition studies during pregnancy, *American Journal of Obstetrics and Gynecology* 46 (1943): 38–52; and B. S. Burke and H. C. Stuart, A method of diet analysis, *Journal of Pediatrics* 12 (1938): 493–503.
39. The discussion of diet assessment methods was adapted from G. Block, A review of validations of dietary assessment methods, *American Journal of Epidemiology* 115 (1982): 492–505; and R. K. Johnson and J. H. Hankin, Dietary assessment and validation, in E. R. Monsen, ed., *Research: Successful Approaches,* 2nd ed. (Chicago: American Dietetic Association, 2003).
40. G. H. Beaton and coauthors, Sources of variance in 24-hour dietary recall data: Implications for nutrition study design and interpretation, *American Journal of Clinical Nutrition* 32 (1979): 2456–2559.
41. The discussion of the problems with the 24-hour dietary recall method was adapted from J. L. Forster and coauthors, Hypertension prevention trial: Do 24-h food records capture usual eating behavior in a dietary change study? *American Journal of Clinical Nutrition* 51 (1990): 253–557.
42. W. C. Willett and coauthors, Validation of a semi-quantitative food frequency questionnaire: Comparison with a 1-year diet record, *Journal of the American Dietetic Association* 87 (1987): 43–47.
43. R. J. Coates and C. P. Monteilh, Assessments of food-frequency questionnaires in minority populations, *American Journal of Clinical Nutrition* 65 (suppl.) (1997): 1108S–15S.
44. R. Briefel and coauthors, Assessing the nation's diet: Limitations of the food frequency questionnaire, *Journal of the American Dietetic Association* 92 (1992): 959–62; and G. Block and A. F. Subar, Estimates of nutrient intake from a food frequency questionnaire: The 1987 National Health Interview Survey, *Journal of the American Dietetic Association* 92 (1992): 969–77.
45. G. P. Sevenhuysen and L. A. Wadsworth, Food image processing: A potential method for epidemiologic surveys, *Nutrition Reports International* 39 (1989): 439–50; and T. A. Fox and coauthors, Telephone surveys as a method for obtaining dietary information: A review, *Journal of the American Dietetic Association* 92 (1992): 729–32.
46. S. K. Kumanyika and coauthors, Dietary assessment using a picture-sort approach, *American Journal of Clinical Nutrition* 65 (suppl.) (1997): 1123S–29S.
47. L. Kohlmeier and coauthors, Computer-assisted self-interviewing: A multimedia approach to dietary assessment, *American Journal of Clinical Nutrition* 65 (suppl.) (1997): 1275S–81S.

48. Gibson, *Principles of Nutritional Assessment,* pp. 292–97.
49. Ibid., pp. 178–79.
50. Ibid., pp. 155–62.
51. R. D. Lee and D. C. Nieman, *Nutritional Assessment,* 3rd ed. (New York: McGraw-Hill, 2002).
52. The discussion of sensitivity and specificity was taken from R. H. Fletcher and coauthors, *Clinical Epidemiology—The Essentials,* 4th ed. (Baltimore, MD: Williams & Wilkins, 2005).
53. G. Block and A. M. Hartman, Issues in reproducibility and validity of dietary studies, *American Journal of Clinical Nutrition* 50 (1989): 1133–38.
54. A. Fink and J. Kosecoff, *How to Conduct Surveys: A Step-by-Step Process,* 2nd ed. (Thousand Oaks, CA: Sage, 1998).
55. A. K. Rundle and coauthors, *Cultural Competence in Health Care* (San Francisco, CA: Jossey-Bass, 2002).
56. T. Barer-Stein, *You Eat What You Are: A Study of Ethnic Food Traditions* (Toronto: McClelland & Stewart, 1979), pp. 13–23.
57. K. D. Travers, Using qualitative research to understand the sociocultural origins of diabetes among Cape Breton Mi'kmaq, in *Chronic Diseases in Canada* (Ottawa: Health Canada, 1995), pp. 140–43.
58. J.-P. Habicht and D. L. Pelletier, The importance of context in choosing nutritional indicators, *Journal of Nutrition* 120 (1990): 1519–24.
59. Nutrition Screening Initiative, *Keeping Older Americans Healthy at Home* (Washington, D.C.: Greer, Margolis, Mitchell, Burns & Associates, 1996).
60. S. A. Anderson, Core indicators of nutritional state for difficult-to-sample populations, *Journal of Nutrition* 120 (1990): 1559–600.
61. M. Stouthamer-Loeber and W. B. van Kammen, *Data Collection and Management: A Practical Guide* (Thousand Oaks, CA: Sage, 1995), pp. 114–18.
62. The discussion of reference data was adapted from Gibson, *Principles of Nutritional Assessment,* pp. 209–46 and 349–76; and F. E. Johnston and Z. Ouyang, Choosing appropriate reference data for the anthropometric assessment of nutritional status, in J. H. Himes, ed., *Anthropometric Assessment of Nutritional Status* (New York: Wiley, 1991), pp. 337–46.
63. Nutrition Screening Initiative, *Keeping Older Americans Healthy at Home.*
64. Gibson, *Principles of Nutritional Assessment,* pp. 137–53.

Chapter 4

1. V. Heiser, *An American Doctor's Odyssey* (New York: Norton, 1936), pp. 100–3.
2. The discussion of the uses and basic concepts of epidemiology were adapted from C. W. Tyler, Jr., and J. M. Last, Epidemiology, in *Public Health & Preventive Medicine,* 14th ed., ed. J. M. Last and R. B. Wallace (Norwalk, CT: Appleton & Lange, 1998).
3. The discussion of folic acid is from L. Langseth, International Life Sciences Institute Monograph Series: *Nutritional Epidemiology: Possibilities and Limitations* (Brussels: ILSI Europe Press, 1996).
4. E. A. Jacobs and coauthors, Fetal alcohol syndrome and alcohol-related neurodevelopmental disorders, *Pediatrics* 106 (2000): 358–61; and *Summary of Findings from the 1999 National Household Survey on Drug Abuse* (Washington, D.C.: Department of Health and Human Services, 2000).
5. *Healthy People 2010* (Washington, D.C.: Department of Health and Human Services, 2000).
6. G. P. Holmes and coauthors, Chronic fatigue syndrome: A working case definition, *Annals of Internal Medicine* 108 (1988): 387–89.
7. A. D. Langmuir, The territory of epidemiology: Pentimento, *Journal of Infectious Diseases* 155 (1987): 349–58.
8. The discussion of risk and risk factors and the explanations for research results were adapted from R. H. Fletcher, S. W. Fletcher, and E. H. Wagner, *Clinical Epidemiology—The Essentials,* 4th ed. (Baltimore, MD: Williams & Wilkins, 2005).
9. The discussion of relative risk and and the examples of risk were adapted from L. Langseth, International Life Sciences Institute Monograph Series: *Nutritional Epidemiology: Possibilities and Limitations* (Brussels: ILSI Europe Press, 1996).
10. The definitions and discussion of relative risk are adapted from J. F. Jekel, J. G. Elmore, and D. L. Katz, *Epidemiology Biostatistics and Preventive Medicine* (Philadelphia, PA: W. B. Saunders, 1996), pp. 21, 76–77.
11. U.S. Department of Health and Human Services, *The Surgeon General's Report on Nutrition and Health* (Washington, D.C.: U.S. Government Printing Office, 1988), pp. 83–137.
12. A. Keys, Coronary heart disease in seven countries, *Circulation* 41 and 42 (Suppl. 1) (1970): 1–4.
13. S. A. Glantz, *Primer of Biostatistics,* 5th ed. (Norwalk, CT: Appleton & Lange, 2001).
14. H. Kato and coauthors, Epidemiologic studies of coronary heart disease and stroke in Japanese men living in Japan, Hawaii, and California: Serum lipids and diet, *American Journal of Epidemiology* 97 (1973): 372–85.
15. R. B. Shekelle and coauthors, Diet, serum cholesterol, and death from coronary heart disease—the Western Electric study. *New England Journal of Medicine* 304 (1981): 65–70.
16. M. R. Garcia-Palmieri and coauthors, Relationship of dietary intake to subsequent coronary heart disease incidence. The Puerto Rico Heart Health Program, *American Journal of Clinical Nutrition* 33 (1980): 1818–27.
17. L. H. Kushi and coauthors, Diet and 20-year mortality from coronary heart disease—the Ireland–Boston Diet–Heart Study, *New England Journal of Medicine* 312 (1985): 811–18.
18. *Report of the Expert Panel on Population Strategies for Blood Cholesterol Reduction* (Bethesda, MD: National Institutes of Health, 1990), pp. 33–61; see also Adult Treatment Panel III Report, 2001, available at www.nhlbi.nih.gov.
19. D. D. Gorder and coauthors, Dietary intake in the Multiple Risk Factor Intervention Trial (MRFIT): Nutrient and food group changes over 6 years, *Journal of the American Dietetic Association* 86 (1986): 744–51.
20. T. A. Dolecek and coauthors, A long-term nutrition intervention experience: Lipid responses and dietary adherence patterns in the Multiple Risk Factor Intervention Trial, *Journal of the American Dietetic Association* 86 (1986): 752–58.
21. S. M. Grundy and coauthors, Coronary risk factor statement for the American public: A statement of the nutrition committee of the American Heart Association, *Arteriosclerosis* 5 (1985): 678A–82A.
22. Council on Scientific Affairs, Dietary and pharmacologic therapy for the lipid risk factors, *Journal of the American Medical Association* 250 (1983): 1873–79.
23. The Expert Panel, Report of the National Cholesterol Education Program Expert Panel on detection, evaluation, and treatment of high blood cholesterol in adults, *Archives of Internal Medicine* 148 (1988): 36–69.
24. *Dietary Guidelines for Americans* (Washington, D.C.: U.S. Government Printing Office, 2005).
25. G. Sorensen and coauthors, Work-site nutrition intervention and employees' dietary habits: The Treatwell Program, *American Journal of Public Health* 82 (1992): 877–80.
26. A. R. Feinsten, Scientific standards in epidemiologic studies of the menace of daily life, *Science* 242 (1988): 1257–63.
27. The discussion of the types of epidemiologic studies was adapted from G. D. Friedman, *Primer of Epidemiology,* 5th ed. (New York: McGraw-Hill, 2002); and Tyler and Last, Epidemiology, in *Public Health & Preventive Medicine.*
28. L. Kaizer and coauthors, Fish consumption and breast cancer risk: An ecological study, *Nutrition and Cancer* 12 (1989): 61–68.
29. N. F. Butte and coauthors, Human milk intake and growth faltering of rural Mesoamerindian infants, *American Journal of Clinical Nutrition* 55 (1992): 1109–16.
30. T. R. Dawber, *The Framingham Study: The Epidemiology of Atherosclerotic Disease* (Cambridge, MA: Harvard University Press, 1980).
31. W. P. Castelli and coauthors, Incidence of coronary heart disease and lipoprotein cholesterol levels: The Framingham Study,

Journal of the American Medical Association 256 (1986): 2835–38.
32. E. Arnesen and coauthors, Serum total homocysteine and coronary heart disease, *International Journal of Epidemiology* 24 (1995): 704–9; and M. R. Malinow, Plasma homocysteine—a risk factor for arterial occlusive diseases, *Journal of Nutrition* 126 (1996): S1238–S1243.
33. A. B. Miller and coauthors, A study of diet and breast cancer, *American Journal of Epidemiology* 107 (1978): 499–509.
34. The discussion of the uses of epidemiology in the nutritional sciences was adapted from M. L. Burr, Epidemiology for nutritionists: 1. Some general principles, *Human Nutrition: Applied Nutrition* 37A (1983): 259–64.
35. C. S. Johnston, Vitamin C, in B. A. Bowman and R. M. Russell, eds., *Present Knowledge in Nutrition,* 8th ed.(Washington, D.C.: International Life Sciences Institute, 2001), pp. 175–83.
36. Centers for Disease Control and Prevention, Indicators for chronic disease surveillance, *Morbidity and Mortality Weekly Report* 53 (2004): 1–4.
37. World Health Organization, *The World Health Report 2002: Reducing Risks, Promoting Healthy Life* (Geneva: World Health Organization, 2002).
38. World Health Organization, *Global Strategy on Diet, Physical Activity, and Health* (Geneva: World Health Organization, 2004).
39. The discussion of the BRFSS and YRBSS is from the Centers for Disease Control and Prevention website at www.cdc.gov.
40. W. Willett, *Nutritional Epidemiology,* 2nd ed. (New York: Oxford University Press, 1998), pp. 33–49.
41. P. P. Basiotis and coauthors, Number of days of food intake records required to estimate individual and group nutrient intakes with defined confidence, *Journal of Nutrition* 117 (1987): 1638–41.
42. The discussion of food consumption was adapted from R. S. Gibson, *Principles of Nutritional Assessment* (New York: Oxford University Press, 1990), pp. 21–54.
43. K. S. Kubena, Accuracy in dietary assessment: On the road to good science, *Journal of the American Dietetic Association* 100 (2000): 775–76; and W. A. van Staveren and M. C. Ocke, Estimation of dietary intake, in B. A. Bowman and R. M. Russell, eds., *Present Knowledge in Nutrition,* 8th ed. (Washington, D.C.: International Life Sciences Institute, 2001), pp. 605–16.
44. This discussion of dietary assessment methods is adapted from L. Langseth, International Life Sciences Institute Monograph Series: *Nutritional Epidemiology: Possibilities and Limitations* (Brussels: ILSI Europe Press, 1996).

Chapter 5

1. C. M. Olson, Nutrition and health outcomes associated with food insecurity and hunger, *Journal of Nutrition* 129 (1999): 521S–524S.
2. D. H. Holben, An overview of food security and its measurement, *Nutrition Today* 37 (2002): 156–61; and G. Bickel and coauthors, *Guide to Measuring Household Food Security, Revised 2000* (Alexandria, VA: U.S. Department of Agriculture, Food and Nutrition Service, 2000).
3. G. Bickel, 2000.
4. P. L. Splett, Federal Food Assistance Programs: A step to food security for many, *Nutrition Today,* March/April 1994, pp. 6–13.
5. L. Schwartz-Nobel, *Starving in the Shadow of Plenty* (New York: Putnam, 1981), pp. 35–36.
6. Bread for the World Institute, *Hunger 2002: A Future with Hope* (Washington, D.C.: Bread for the World Institute, 2002).
7. C. DeNavas and coauthors, U.S. Census Bureau, Current Population Reports, P60-226, *Income, Poverty, and Health Insurance Coverage in the United States: 2003* (Washington, D.C.: U.S. Government Printing Office, 2004).
8. G. M. Fisher, Department of Health and Human Services, *The Development of the U.S. Poverty Thresholds—A Brief Overview,* Newsletter of Government Statistics Section and the Social Statistics Section of the American Statistical Association, Winter 1997, pp. 6–7, available at http://aspe.os.dhhs.gov/poverty/papers/hptgssiv.htm; G. M. Fisher, U.S. Census Bureau Poverty Measurement Workings Papers, *The Development of the Orshansky Poverty Thresholds and Their Subsequent History as the Official U.S. Poverty Measure,* available at http://www.census.gov/hhes/poverty/povmeas/papers/orshansky.html; and U.S. Census Bureau, *How the Census Bureau Measures Poverty,* available at http://www. census.gov/hhes/poverty/povdef.html.
9. G. M. Fisher, 1997.
10. U.S. Census Bureau, *How the Census Bureau Measures Poverty.*
11. M. Nord and coauthors, *Household Food Security in the United States, 2003* (Alexandria, VA: Economic Research Service, Food and Rural Economics Division, 2004), Report No. 42 (FANRR-42).
12. USDA Economic Research Service Briefing Room, Food security in the United States, available at www.ers.usda.gov/briefing/foodsecurity; and G. Bickel, 2000.
13. G. Bickel, 2000.
14. M. Nord and G. Bickel, *Measuring Children's Food Security in U.S. Households, 1995–1999* (FANRR-25) (Alexandria, VA: Food and Rural Economics Division, Economic Research Service, U.S. Department of Agriculture, 2002).
15. E. J. Adams, L.Grummer-Strawn, and G. Chavez, Food insecurity is associated with increased risk of obesity in California women, *Journal of Nutrition* 133 (2003): 1070–1074; K. Alaimo, C. Olson, and E. A. Frongillo, Jr. Low family income and food insufficiency in relation to overweight in children: Is there a paradox? *Archives of Pediatrics and Adolescent Medicine,* 155 (2001): 1161–1167; A. Drewnowski and S. E. Specter, Poverty and obesity: The role of energy density and energy costs, *American Journal of Clinical Nutrition* 79 (2004): 6–16; J. E. Stuff and coauthors, Household food insecurity is associated with adult health status, *Journal of Nutrition* 134 (2004): 2330–2335; and N. T. Vozoris and V. S. Tarasuk, Household food insufficiency is associated with poorer health. *Journal of Nutrition* 133 (2003): 120–126.
16. M. Nord and coauthors, 2004.
17. D. Rose, Economic determinants and dietary consequences of food security in the United States, *Journal of Nutrition* 129 (1999): 517S–520S.
18. M. Nord and coauthors, 2004.
19. U.S. Department of Labor, *Minimum Wage Laws in the States;* available at http://www.dol.gov/esa/minwage/america.htm.
20. Institute for Food and Development Policy, The true state of the nation: How is American really doing? *Food First Backgrounder,* 2001; available at www.foodfirst.org.
21. V. Peterson, *Homeless: Struggling to Survive* (Farmington Hills, MI: Gale Group, 2000), p. 31; T. P. Hall, *Balancing the Budget on the Backs of the Poor,* 1999 Report Card on Hunger, November 1999; available at www.housegov/tonyhall.
22. M. Kim, J. Ohls, and R. Cohen, *Hunger in America, 2001. National Report Prepared for America's Second Harvest* (Princeton, NJ: Mathematica Policy Research, Inc., 2001).
23. M. Nord and coauthors, 2004.
24. M. Nord and G. Bickel, 2002; available at www.ers.usda.gov/publications/fanrr25.
25. M. Nord and coauthors, 2004.
26. K. Alaimo, C. Olson, and E. A. Frongillo, Food insufficiency and American school-aged children's cognitive, academic, and psychosocial development, *Pediatrics* 108 (2001): 44–51.
27. K. Alaimo and coauthors, Food insufficiency, family income, and health in U.S. preschool and school-aged children, *American Journal of Public Health,* May 2001.
28. M. Nord and coauthors, 2004.
29. C. DeNavas, 2004.
30. K. Alaimo, May 2001.
31. C. DeNavas, 2004.
32. M. Nord, 2004.
33. K. Alaimo, 2001.
34. M. Nord, 2004.
35. C. M. Olson and coauthors, Factors protecting against and contributing to food insecurity among rural families,

Family Economics and Nutrition Review (2004), 16(1): 12–20.

36. D. H. Holben and coauthors, Food security status of households in Appalachian Ohio with children in Head Start, *Journal of the American Dietetic Association* 104 (2004): 238–241.
37. M. Nord and coauthors, 2004.
38. Ibid.
39. A. Mittal and M. Kawaai, Freedom to trade? Trading away American family farms, *Food First Backgrounder* 7 (2001): 1–6; and A. K. Mishra and coauthors, *Income, Wealth, and the Economic Well-being of Farm Households* (Washington, D.C.: Economic Research Service, 2002), Report No. AER812; available at www.ers.usda.gov/publications/aer812/.
40. Kim, Ohls, and Cohen, 2001.
41. The United States Conference of Mayors, *A Status Report on Hunger and Homelessness in America's Cities,* December 2001, available at http://usmayors.org; and The United States Conference of Mayors–Sodexho Hunger and Homelessness Survey, *A Status Report on Hunger and Homelessness in America's Cities, A 25-City Survey,* December, 2003, available at http://usmayors.org.
42. The United States Conference of Mayors–Sodexho Hunger and Homelessness Survey, 2003.
43. National low-income housing coalition, *Out of Reach 2003: America's Housing Wage Climbs;* available at www.nlihc.org.
44. J. L. Wiecha, J. T. Dwyer, and M. Dunn-Strohecker, Nutrition and health services needs among the homeless, *Public Health Reports* 106 (1991): 364–74; N. L. Oliveira and J. P. Goldberg, The nutrition status of children and women who are homeless, *Nutrition Today* 37 (2002): 70–77.
45. Bread for the World Institute, *Hunger 1993: Uprooted People* (Washington, D.C.: Bread for the World Institute, 1992), p. 107.
46. C. Olson and D. H. Holben, Position of the American Dietetic Association on Domestic Food and Nutrition Security, *Journal of the American Dietetic Association* 102 (2002): 1840–47.
47. Ibid.
48. M. Nord and coauthors, 2004.
49. D. Rose, Economic determinants and dietary consequences of food insecurity in the United States, *Journal of Nutrition* 129 (1999): 517S–520S.
50. Kim, Ohls, and Cohen, 2001.
51. The United States Conference of Mayors–Sodexho Hunger and Homelessness Survey, 2003.
52. N. Kotz, *Hunger in America: The Federal Response* (New York: Field Foundation, 1979), p. 17.
53. P. Wilde, Strong economy and welfare reforms contribute to drop in Food Stamp rolls, *ERS Food Review* (2001), 24(1): 2–7; and Caster L. Trends in FSP participation rates: Focus on 1994–1998 (Washington, D.C.: Mathematica Policy Research, Inc., 2000).
54. The U.S. Conference of Mayors, *A Status Report on Hunger and Homelessness in America's Cities,* December 2001, available at http://usmayors.org; and The United States Conference of Mayors–Sodexho Hunger and Homelessness Survey, 2003.
55. The Administration for Children and Families, *HHS Fact Sheet: Work Not Welfare* (Washington, D.C.: ACF Office of Public Affairs, November 1997), pp. 1–3.
56. P. Wilde, 2001; and Working toward Independence, available at www.whitehouse.gov/news/releases/2002/02/print/welfare-book-01.html.
57. Center on Hunger and Poverty, What comes after welfare reform? December 2001, available at www.centeronhunger.org; and P. Loprest, *How Families That Left Welfare Are Doing: A National Picture* (Washington, D.C.: The Urban Institute, 1999).
58. Children's Defense Fund, *Families Struggling to Make It in the Workforce: A Post Welfare Report* (Washington, D.C., Children's Defense Fund, 2000).
59. E. Kennedy, P. M. Morris, and R. Lucas, Welfare reform and nutrition programs: Contemporary budget and policy realities, *Journal of Nutrition Education* 28 (1996): 67–70; D. Rose and M. Nestle, Welfare reform and nutrition education: Alternative strategies to address the challenges of the future, *Journal of Nutrition Education* 29 (1996): 61–66.
60. Department of Health and Human Services, *Indicators of Welfare Dependence: Annual Report to Congress, 2003,* Executive Summary; available at aspe.hhs.gov/hsp/indicators03/execsum.htm.
61. Ibid.
62. L. Caster, *Trends in FSP participation rates: Focus on 1994–1998* (Washington, DC: Mathematica Policy Research, Inc., November 2000); and M. Andrews and coauthors, *Household Food Security in the United States, 1999* (FANRR-8) (Alexandria, VA: Food and Rural Economics Division, Economic Research Service, U. S. Department of Agriculture, 2000).
63. Olson and Holben, 2002.
64. V. Oliveira, *The Food Assistance Landscape* (Alexandria, VA: Economic Research Service, Food and Rural Economics Division, 2004), Food Assistance and Nutrition Research Report No. 28-4, available at www.ers.usda.gov/publications/fanrr28-4; Economic Research Service, USDA, Food and Nutrition Assistance Programs and the General Economy Briefing Room, available at http://www.ers.usda.gov/Briefing/GeneralEconomy/.
65. P. Casey and coauthors, Maternal depression, changing public assistance, food security, and child health status, *Pediatrics* 113 (2004): 298–304; J. T. Cook and coauthors, Welfare reform and the health of young children, *Archives of Pediatric and Adolescent Medicine* 156 (2002): 678–684; C. M. Devine and coauthors, Sandwiching it in: Spillover of work onto food choices and family roles in low- and moderate-income urban households, *Social Science & Medicine* 56 (2003): 617–630; and M. Nord, 2004.
66. V. Oliveira, 2004.
67. M. Nord, 2004; and V. Oliveira, 2004.
68. M. Nord, 2004; S. J. Jones and coauthors, Lower risk of overweight in school-aged food insecure girls who participate in food assistance, *Archives of Pediatrics & Adolescent Medicine* 157 (2003): 780–784.
69. V. Oliveira, 2004.
70. Community Nutrition Institute, *Nutrition Week,* January 22, 1993, p. 6; and Olson and Holben, 2002.
71. Food and Nutrition Service, *Food Stamp Program* website; available at www.fns.usda.gov/fsp.
72. Food and Nutrition Service, A short history of the Food Stamp Program; available at www.fns.usda.gov/fsp/rules/Legislation/history.htm.
73. Food and Nutrition Service, Food Stamp Program data, available at www.fns.usda.gov/pd/fspmain.htm; and V. Oliveira, 2004.
74. Economic Research Service, Food Stamp Program map machine; available at www.ers.usda.gov/data/foodstamps.
75. Food and Nutrition Service, Characteristics of food stamp households: Fiscal year 2002; available at www.fns.usda.gov/oane/MENU/Published/FSP/FILES/Participation/2002Characteristics.htm.
76. Food and Nutrition Service, Food Stamp Program applicants and recipients; available at www.fns.usda.gov/fsp/applicant_recipients/fs_Res_Ben_Elig.htm.
77. S. Crixell and B. J. Friedman, Food insecurity in the barrio: Availability of affordable and nutritious foods in local grocery stores, *Journal of the American Dietetic Association* 103 (2003): A45; Food Research and Action Center, *Community Childhood Identification Project* (Washington, D.C.: FRAC, 1995), pp. 33–34; P.R. Kaufman and coauthors, Do the Poor Pay More for Food? Item Selection and Price Differences Affect Low-Income Household Food Costs. Agricultural Economic Report No. 759, November 1997; E. S. Leibtag and coauthors, Exploring Food Purchase Behavior of Low-Income Households. How Do They Economize? ERS Agriculture Information Bulletin No. 747-07, June 2003; K. Morland and coauthors, Neighborhood characteristics associated with the location of food stores and food service places, *American Journal of Preventive Medicine* 22 (2002): 23–29; and D. Rose and R. Richards, Food store

access and household fruit and vegetable use among participants in the US Food Stamp program, *Public Health Nutrition* 7 (2004):1081–1088.
78. J. Allen and K. Gadson, Food consumption and nutritional status of low-income households, *National Food Review* 26 (1984): 27–31; Food Research and Action Center, *Community Childhood Identification Project* (Washington, D.C.: FRAC, 1995), pp. 33–34.
79. Economic Research Service, USDA, Food and Nutrition Assistance Programs and the General Economy Briefing Room; available at www.ers.usda.gov/Briefing/GeneralEconomy.
80. M. Nord, 2004.
81. Food Research and Action Center, *State of the States: A Profile of Food and Nutrition Programs across the Nation* (Washington, D.C.: FRAC), February 2002.
82. Food Research and Action Center, 1995.
83. Economic Research Service, Food and Nutrition Assistance Programs: Image Gallery, available at www.ers.usda.gov/Briefing/FoodNutritionAssistance/gallery/programs02.htm; The National Nutrition Safety Net, available at www.fns.usda.gov/fsec/toolkit/other_nutrition_programs.htm; and Social Welfare Programs in the Territories, available at aspe.os.dhhs.gov/98gb/12terri.htm.
84. Food and Nutrition Service, Food Distribution Programs website; available at www.fns.usda.gov/fdd.
85. Food and Nutrition Service, CSFP Homepage, available at www.fns.usda.gov/fdd/programs/csfp/; Food and Nutrition Service, CSFP Fact Sheet, available at www.fns.usda.gov/fdd/programs/csfp; Food and Nutrition Service, Annual Summary of Food and Nutrition Service Programs, available at www.fns.usda.gov/ pd/annual.htm; and Economic Research Service, Food and Nutrition Assistance Programs: Image Gallery, available at www.ers.usda.gov/Briefing/FoodNutritionAssistance/gallery/programs02.htm.
86. Food and Nutrition Service, FDPIR Homepage, available at http://www.fns.usda.gov/fdd/programs/fdpir/; Food and Nutrition Service, FDPIR Fact Sheet, available at http://www.fns.usda.gov/fdd/programs/fdpir/.
87. Food and Nutrition Service, TEFAP Homepage, available at www.fns.usda.gov/fdd/programs/tefap/; Food and Nutrition Service, TEFAP Fact Sheet, available at www.fns.usda.gov/fdd/programs/fdpir/.
88. Food and Nutrition Service, NSIP Homepage, available at /www.fns.usda.gov/fdd/programs/nsip; and Administration on Aging, Nutrition Services Incentive Program, available at www.aoa.gov/eldfam/ Nutrition/Nutrition_services_incentive.asp.
89. Food and Nutrition Service, Food Distribution Disaster Assistance Homepage; available at www.fns.usda.gov/fdd/programs/fd-disasters.
90. Food and Nutrition Service, National School Lunch Program, available at www.fns.usda.gov/cnd/lunch/; Food and Nutrition Service, School Breakfast Program, available at www.fns.usda.gov/cnd/breakfast; and V. Oliveira, 2004.
91. Food and Nutrition Service, Afterschool Snack in the NLSP, available at www.fns.usda.gov/cnd/Afterschool/default.htm; and V. Oliveira, 2003.
92. Food and Nutrition Service, Special Milk Program; available at www.fns.usda.gov/cnd/milk.
93. Food and Nutrition Service, Summer Food Service Program; available at www.fns.usda.gov/cnd/summer/.
94. Food and Nutrition Service, Child and Adult Care Food Program, available at www.fns.usda.gov/cnd/Care/CACFP/cacfphome.htm; and V. Oliveira, 2004.
95. Food and Nutrition Service, Women, Infants, and Children, available at www.fns.usda.gov/wic; and V. Oliveira, 2004.
96. Food and Nutrition Service, WIC Farmers' Market Nutrition Program; available at www.fns.usda.gov/wic/FMNP/FMNPfaqs.htm.
97. Department of Health and Human Services Administration on Aging, The Elderly Nutrition Program Fact Sheet, May 1998, pp. 1–4.
98. Administration on Aging, Elderly Nutrition Program; available at www.aoa.gov/press/fact/alpha/fact_elderly_nutrition.asp.
99. Food and Nutrition Service, Senior Farmers' Market Nutrition Program; available at www.fns.usda.gov/wic/SeniorFMNP/SFMNPmenu.htm.
100. The U.S. Conference of Mayors, *A Status Report on Hunger and Homelessness in America's Cities,* December 2001; available at usmayors.org.
101. M. Nord and coauthors, 2004.
102. America's Second Harvest; available at http://www.secondharvest.org.
103. B. O. Daponte, Food pantry use among low-income households in Allegheny County, Pennsylvania, *Journal of Nutrition Education* 30 (1998): 50–57; M. Edlefsen and C. M. Olson, Perspectives of volunteers in emergency feeding programs on hunger, its causes, and solutions, *Journal of Nutrition Education and Behavior* 34 (2002): 93–99.
104. The U.S. Conference of Mayors, 2001; and The U.S. Conference of Mayors–Sodexho Hunger and Homelessness Survey, 2003.
105. The discussion of community food security is taken in part from L. S. Kantor, Community food security programs improve food access, *Food Review* 24 (2001): 20–26.
106. The definition in the margin and the discussion of community food security are adapted in part from D. Bichler and coauthors, *Getting Food on the Table: An Action Guide to Local Food Policy* (Venice, CA: Community Food Security Coalition, March 1999).
107. USDA Community Food Security Initiative, Economic Research Service Briefing Room; available at ers.usda.gov/briefing/FoodSecurity/community/index.htm, accessed September 2002.
108. USDA, Food and Nutrition Service, *Gleaning and Food Recovery,* available at www.fns.usda.gov/fns/MENU/GLEANING; and A. Hoisington and coauthors, Field gleaning as a tool for addressing food security at the local level: Case study, *Journal of Nutrition Education* 33 (2001): 43–48.
109. D. Pelletier, Enhancing local food access by building community food security, *Community Nutritionary* (Tarrytown, NY: Dannon Institute, Fall 2000), p. 13.
110. J. L. Brown and D. Allen, Hunger in America, *Annual Reviews in Public Health* 9 (1988): 503–26.
111. L. Sims and J. Voichick, Our perspective: Nutrition education enhances food assistance programs, *Journal of Nutrition Education* 28 (1996): 83–85; A. B. Joy and C. Doisy, Food stamp nutrition education program: Assisting food stamp recipients to become self-sufficient, *Journal of Nutrition Education* 28 (1996): 123–126.
112. Food Research and Action Center, *Good Choices in Hard Times: Fifteen Ideas for States to Reduce Hunger and Stimulate the Economy* (Washington, D.C.: Food Research and Action Center, July 2002); available at www.frac.org.
113. A. Tagtow, Does food insecurity exist in America's heartland? *Hunger and Environmental Newsletter,* Summer 2002, p. 7.
114. R. Lobosco, A commentary on domestic hunger: A problem we can solve, *Topics in Clinical Nutrition* 9 (1994): 8–12.
115. The quotation is from Edward Everett Hale (1822–1909), *For the Lend-a-Hand Society.*

Chapter 6

1. Centers for Disease Control and Prevention, *Physical Activity and Good Nutrition: Essential Elements for Good Health* (Washington, D.C.: U.S. Department of Health and Human Services, 2000); available at www.cdc.gov/nccdphp.
2. Office of the Surgeon General, *The Surgeon General's Call to Action to Prevent and Decrease Overweight and Obesity,* 2001 (Rockville, MD: Department of Health and Human Services, 2001); and Centers for Disease Control and Prevention, Overweight and Obesity: Frequently Asked Questions (FAQs); available at http://www.cdc.gov/nccdphp/dnpa/obesity/faq.htm.
3. P. R. Lee, Nutrition policy: From neglect and uncertainty to debate and action, *Journal of the American Dietetic Association* 72 (1978): 581–88; American Dietetic Association,

Position paper on a national nutrition policy, *Journal of the American Dietetic Association* 76 (1980): 596–99; and S. Crutchfield and J. Weimer, Nutrition policy in the 1990s, *Food Review* (Washington, D.C.: Economic Research Service, U.S. Department of Agriculture, 2000).

4. As cited in J. E. Austin and C. Hitt, *Nutrition Intervention in the United States* (Cambridge, MA: Ballinger, 1979), pp. 357–85.
5. Quoted in ibid., p. 356.
6. D. J. Palumbo, *Public Policy in America—Government in Action* (San Diego: Harcourt Brace Jovanovich, 1988), p. 17.
7. R. R. Briefel, Nutrition Monitoring in the United States, in *Present Knowledge in Nutrition*, 8th ed. (Washington, D.C.: International Life Sciences Institute, 2001), pp. 617–35.
8. *Nutrition Monitoring in the United States: An Update Report on Nutrition Monitoring*, DHHS (PHS) Pub. No. 89-1255 (Washington, D.C.: U.S. Government Printing Office, 1989); J. B. Mason and coauthors, *Nutrition Surveillance* (Geneva: World Health Organization, 1984); the discussion of the five types of data collection and end-use activities was adapted from E. Yetley, A. Beloian, and C. Lewis, Dietary methodologies for food and nutrition monitoring, in *Vital and Health Statistics: Dietary Methodology Workshop for the Third National Health and Nutrition Examination Survey* (Washington, D.C.: U.S. Government Printing Office, 1992), pp. 58–67.
9. Mason and coauthors, *Nutrition Surveillance;* and R. R. Briefel and C. T. Sempos, Introduction, in U.S. Department of Health and Human Services, *Vital and Health Statistics*, pp. 1–2.
10. The definition of nutrition monitoring and related research was taken from Federation of American Societies for Experimental Biology, prepared for the Interagency Board for Nutrition Monitoring and Related Research, *Third Report on Nutrition Monitoring in the United States*, Vol. 1 (Washington, D.C.: U.S. Government Printing Office, 1995), p. xxiii.
11. G. Ostenso, National Nutrition Monitoring System: A historical perspective, *Journal of the American Dietetic Association* 84 (1984): 1181–85.
12. U.S. Department of Health and Human Services, U.S. Department of Agriculture, *Nutrition Monitoring in the United States* (Washington, D.C.: U.S. Government Printing Office, 1986); and *Nutrition Monitoring in the United States: An Update Report.*
13. Federation of American Societies for Experimental Biology, *Third Report on Nutrition Monitoring in the United States*, Vol. 1, pp. 1–17.
14. N. W. Jerome and J. A. Ricci, Food and nutrition surveillance: An international overview, *American Journal of Clinical Nutrition* 65 (suppl.) (1997): 1198S–1202S.
15. The margin definition of the NNMRRP was adapted from Briefel, *Present Knowledge in Nutrition*, pp. 617–635.
16. *Nutrition Monitoring in the United States: An Update Report*, pp. 2–3.
17. The section that follows was taken from Department of Health and Human Services, U.S. Department of Agriculture, *Nutrition Monitoring in the United States: The Directory of Federal and State Nutrition Monitoring and Related Research Activities* (Hyattsville, MD: National Center for Health Statistics, 2000); and *Third Report on Nutrition Monitoring in the United States*, Vols. 1 and 2.
18. Federation of American Societies for Experimental Biology, *Third Report on Nutrition Monitoring in the United States*, Vol. 1, p. 25.
19. G. M. McQuillan and coauthors, Update on the seroepidemiology of human immunodeficiency virus in the United States household population: NHANES III, 1988–1994, *Journal of Acquired Immune Deficiency Syndromes and Human Retrovirology* 14 (1997): 355–60; and National Health and Nutrition Examination Survey, Data accomplishments, accessed December 22, 2004, at www.cdc.gov/nchs/about/major/nhanes/DataAccomp.htm.
20. C. E. Woteki and coauthors, National Health and Nutrition Survey—NHANES: Plans for NHANES III, *Nutrition Today* (January/February 1988): 26.
21. Federation of American Societies for Experimental Biology, *Third Report on Nutrition Monitoring in the United States*, Vol. 1, pp. 24–25.
22. W. C. Chumlea and coauthors, Stature prediction equations for elderly non-Hispanic white, non-Hispanic black, and Mexican-American persons developed from NHANES III data, *Journal of the American Dietetic Association* 98 (1998): 137–42.
23. N. Dupree, NHANES—What's available from NHANES III and Plans for NHANES IV, available on the website for the Centers for Disease Control and Prevention at www.cdc.gov/brfss.
24. R. Briefel, *Present Knowledge in Nutrition*, 2001.
25. J. Dwyer and coauthors, Integration of the Continuing Survey of Food Intakes by Individuals and the National Health and Nutrition Examination Survey, *Journal of the American Dietetic Association* 101 (2001): 1142–43.
26. Information about the BRFSS was obtained from the website of the Centers for Disease Control and Prevention at www.cdc.gov/brfss.
27. The margin definition of food disappearance data was taken from *Third Report on Nutrition Monitoring in the United States*, Vol. 1, pp. 1–3.
28. Mason and coauthors, *Nutrition Surveillance*, p. 12.
29. The description of uses of the BRFSS was adapted from the CDC website at http://www.cdc.gov/brfss.
30. G. E. Brown, Jr., National Nutrition Monitoring System: A congressional perspective, *Journal of the American Dietetic Association* 84 (1984): 1185–89.
31. The description of the Food and Nutrition Board and Institute of Medicine is adapted from the website of the National Academy of Sciences at www.nas.edu.
32. Information about the DRIs was obtained from the Institute of Medicine at www.nas.edu/iom; and A. A. Yates and coauthors, Dietary Reference Intakes: The new basis for recommendations for calcium and related nutrients, B vitamins, and choline, *Journal of the American Dietetic Association* 98 (1998): 699–706.
33. Institute of Medicine, *Dietary Reference Intakes for Energy, Carbohydrate, Fiber, Fat, Fatty Acids, Cholesterol, Protein, and Amino Acids* (Washington, D.C.: National Academy Press, 2002); available at www.nap.edu.
34. Institute of Medicine, *Dietary Reference Intakes: Applications in Dietary Assessment* (Washington, D.C.: National Academy Press, 2001); and S. I. Barr, S. P. Murphy, and M. I. Poos, Interpreting and using the dietary reference intakes in dietary assessment of individuals and groups, *Journal of the American Dietetic Association* 102 (2002): 780–88.
35. Institute of Medicine, *Dietary Reference Intakes for Calcium, Phosphorus, Magnesium, Vitamin D, and Fluoride* (Washington, D.C.: National Academy Press, 1997), p. 186.
36. E. Black and T. J. Cole, Biased over- or under-reporting is characteristic of individuals whether over time or by different assessment methods, *Journal of the American Dietetic Association* 101 (2001): 70–80; and R. K. Johnson, What are people really eating and why does it matter? *Nutrition Today* 35 (2000): 40–46.
37. U.S. Department of Agriculture, Center for Nutrition Policy and Promotion. Healthy Eating Index, available at www.usda.gov/cnpp/healthyeating.html; and P. P. Basiotis, A. Carlson, S. A. Gerrior, W. Y. Juan, and M. Lino, *The Healthy Eating Index: 1999–2000.* U.S. Department of Agriculture, Center for Nutrition Policy and Promotion (2002). CNPP-12.
38. P. P. Basiotis and coauthors, *The Healthy Eating Index: 1999–2000.*
39. L. R. Young and M. Nestle, The contribution of expanding portion sizes to the U.S. obesity epidemic, *American Journal of Public Health* 92 (2002): 246–49.
40. J. B. Richmond, Forward, *American Journal of Clinical Nutrition* 32 (1979): 2621–22.
41. The Report of the U.S. Senate Select Committee on Nutrition and Human Needs, Dietary goals for the United States, *Nutrition Today* (September/October 1977): 20–30.
42. R. E. Patterson and coauthors, Is there a consumer backlash against the diet

and health message? *Journal of the American Dietetic Association* 101 (2001): 37–41.
43. K. W. McNutt, An analysis of the Dietary Goals for the United States, second edition, *Journal of Nutrition Education* 10 (1978): 61–62.
44. The Report of the U.S. Senate Select Committee on Nutrition and Human Needs, Dietary goals, p. 22.
45. United States Department of Agriculture and Department of Health and Human Services, *2005 Dietary Guidelines for Americans*; available at www.healthierus.gov/dietaryguidelines.
46. U.S. Department of Health and Human Services, Public Health Service, *The Surgeon General's Report;* Committee on Diet and Health, National Research Council, *Diet and Health: Implications for Reducing Chronic Disease Risk—Executive Summary* (Washington, D.C.: U.S. Government Printing Office, 1989); The Expert Panel, 2001; and United States Department of Agriculture and Department of Health and Human Services, *2005 Dietary Guidelines for Americans.*
47. United States Department of Agriculture and Department of Health and Human Services, *2005 Dietary Guidelines for Americans.*
48. T. Byers and coauthors, American Cancer Society Guidelines on Nutrition and Physical Activity for Cancer Prevention: Reducing the risk of cancer with healthy food choices and physical activity, *CA: A Cancer Journal for Clinicians* 52 (2002): 92–119; available at www.cancer.org.
49. P. M. Behlen and F. J. Cronin, Dietary recommendations for healthy Americans summarized, *Family Economics Review* 3 (1985): 17–24.
50. O. Hayes, M. F. Trulson, and F. J. Stare, Suggested revision of the basic 7, *Journal of the American Dietetic Association* 31 (1955): 1103–7.
51. U.S. Department of Agriculture, *The Food Guide Pyramid,* Home and Garden Bulletin No. 252 (Hyattsville, MD: Human Nutrition Information Service, 1992).
52. S. Borra and coauthors, Developing actionable dietary guidance messages: Dietary fat as a case study, *Journal of the American Dietetic Association* 101 (2001): 678–84; C. Geiger, Communicating Dietary Guidelines for Americans: Room for improvement, *Journal of the American Dietetic Association* 101 (2001): 793–97.
53. R. M. Mullis and coauthors, Developing nutrient criteria for food-specific dietary guidelines for the general public, *Journal of the American Dietetic Association* 90 (1990): 847–51.
54. C. Bittle and coauthors, ADA's historic commitment to improve nutrition policy, *Journal of the American Dietetic Association* 101 (2001): 406–7; and C. Bittle and K. J. Gorton, Nutrition policy: Your opportunity to be involved, *Journal of the American Dietetic Association* 101 (2001): 177.
55. Information about the U.S. Action Plan on Food Security was taken from two Internet documents: *Framework for the U.S. Action Plan on Food Security* and *Discussion Paper on Domestic Food Security*. Both are available at www.fas.usda.gov/icd/summit.

Chapter 7

1. *Winnipeg Free Press,* December 17, 1997, p. B-1.
2. *Time,* December 29, 1997/January 5, 1998, pp. 6–7.
3. M. Y. Kubik, L. A. Lytle, and M. Story, Soft drinks, candy, and fast food: What parents and teachers think about the middle school food environment, *Journal of the American Dietetic Association* 105 (2005): 233–39.
4. M. Mezzetti and coauthors, Population attributable risk for breast cancer: Diet, nutrition, and physical exercise, *Journal of the National Cancer Institute* 90 (1998): 389–94; and S. A. Smith-Warner and coauthors, Alcohol and breast cancer in women: A pooled analysis of cohort studies, *Journal of the American Medical Association* 279 (1998): 535–40.
5. As cited in L. A. Pal, *Beyond Policy Analysis* (Scarborough, Ontario: ITP, 1997), p. 72.
6. Ibid., pp. 1–12.
7. The discussion of the policy cycle was adapted from W. Lyons, J. M. Scheb II, and L. E. Richardson, Jr., *American Government: Politics and Political Culture* (St. Paul, MN: West, 1995), pp. 467–75; and D. J. Palumbo, *Public Policy in America—Government in Action* (San Diego: Harcourt Brace Jovanovich, 1988), pp. 1–155.
8. U.S. Census Bureau, *Health Insurance Coverage, 2004.*
9. Pal, *Beyond Policy Analysis,* pp. 1–12; and R. B. Denhardt, *Public Administration: An Action Orientation* (Belmont, CA: Wadsworth, 1995), pp. 245–48.
10. K. Schlossberg, Nutrition and government policy in the United States, in *Nutrition and National Policy,* ed. B. Winikoff (Cambridge, MA: MIT Press, 1978), pp. 334–39.
11. R. Carson, *Silent Spring* (Boston: Houghton Mifflin, 1962).
12. As cited in Palumbo, *Public Policy in America,* p. 23.
13. H. W. Schultz, *Food Law Handbook* (Westport, CT: Avi, 1981), pp. 3–21.
14. Provisions of the Nutrition Labeling and Education Act of 1990, as cited in the *Congressional Record*—House, July 30, 1990, pp. II 5836–40, and Legislative Highlights, Wrap-up of the ADA's issues in the 101st Congress, *Journal of the American Dietetic Association* 90 (1990): 1653–55.
15. Pal, *Beyond Policy Analysis,* pp. 8–9.
16. Information about the Department of Health and Human Services was obtained from www.hhs.gov. Information about the U.S. Department of Agriculture was obtained at www.ars.usda.gov.
17. Community Nutrition Institute, Gleaning and greening in the Puget Sound region, *Nutrition Week* 27 (November 21, 1997): 3.
18. Center on Hunger and Poverty, What comes after welfare reform, December 2001, available at www.centeronhunger.org; P. Loprest, *How families that left welfare are doing: A national picture* (Washington, D.C.: The Urban Institute, 1999); Children's Defense Fund, *Families Struggling to Make It in the Workforce: A Post Welfare Report* (Washington, D.C., Children's Defense Fund, 2000).
19. C. O. Jones, *An Introduction to the Study of Public Policy* (Belmont, CA: Wadsworth, 1970), pp. 1–15.
20. B. G. Peters, *American Public Policy—Process and Performance* (New York: Franklin Watts, 1982), p. 69.
21. H. H. Schauffler and J. Wilkerson, National health care reform and the 103rd Congress: The activities and influence of public health advocates, *American Journal of Public Health* 87 (1997): 1107–12.
22. J. E. Austin and C. Hitt, *Nutrition Intervention in the United States* (Cambridge, MA: Ballinger, 1979), pp. 37–41.
23. L. Carroll, *Alice in Wonderland* (New York: Washington Square Press, 1951), pp. 74–75.
24. The discussion of how a bill is introduced into Congress was adapted from T. R. Dye, *Politics in States and Communities,* 7th ed. (Upper Saddle River, NJ: Prentice-Hall, 1991), pp. 156–75.
25. *How Congress Works,* 2nd ed. (Washington, D.C.: Congressional Quarterly, 1991), p. 140.
26. Society for Nutrition Education, *Influencing Food and Nutrition Policy—A Public Policy Handbook* (Oakland, CA: Society for Nutrition Education, 1987), p. 5.
27. The discussion of the federal budget process was adapted from Lyons and coauthors, *American Government,* pp. 507–20; and G. J. Gordon and M. E. Milakovich, *Public Administration in America,* 5th ed. (New York: St. Martin's, 1995), pp. 314–55.
28. Denhardt, *Public Administration,* pp. 148–60.
29. Gordon and Milakovich, *Public Administration in America,* pp. 315–16.
30. Denhardt, *Public Administration,* pp. 148–60.
31. Ibid.
32. Schultz, *Food Law Handbook.*
33. American Dietetic Association, Position of the American Dietetic Association: Affordable and accessible health care services, *Journal of the American Dietetic Association* 92 (1992): 746–48.
34. American Dietetic Association, White paper on health care reform, *Journal of the American Dietetic Association* 92 (1992): 749.
35. ADA's lobbying efforts focus on health care reform, *Journal of the American Dietetic Association* 93 (1993): 754.
36. American Dietetic Association, Health care reform initiatives stress grass-roots lobbying and coalition building, *Journal of the American Dietetic Association* 93 (1993): 528.

37. ADA continues push for medical nutrition therapy, improved child nutrition programs, and labeling of dietary supplements, *Journal of the American Dietetic Association* 94 (1994): 721.
38. American Dietetic Association, ADA urges Congress to expand Medicare coverage for medical nutrition therapy, *Journal of the American Dietetic Association* 95 (1995): 974.
39. American Dietetic Association, ADA mobilizes grassroots action to secure Medicare coverage for medical nutrition therapy, *Journal of the American Dietetic Association* 96 (1996): 1241.
40. Medicare Medical Nutrition Therapy Act of 1997 introduced in Congress, *ADA Courier* 36 (1997): 1.
41. New ADA campaign seeks more cosponsors for Medicare Medical Nutrition Therapy Act; update on child/elderly bills, *Journal of the American Dietetic Association* 97 (1997): 1372.
42. Remarks made by the Hon. John E. Ensign on introducing the Medical Nutrition Therapy Act of 1997 in the House of Representatives, from the *Congressional Record,* April 17, 1997, p. E696.
43. American Dietetic Association's Public Relations Team, *Improving health and saving health-care dollars: Medical nutrition therapy works* (Chicago, IL: American Dietetic Association, April 9, 2002); R. E. Smith and coauthors, Medical nutrition therapy: The core of ADA's advocacy efforts (Part I), *Journal of the American Dietetic Association* 105 (2005): 825–34.
44. The legislative priority areas of the American Dietetic Association are from Government Affairs, Legislation: Priority Areas; available at www.eatright.org/Public/Government Affairs/98_9015.cfm.
45. American Dietetic Association, White paper: Pubic Policy Strategies for Nutrition and Aging; available at www.eatright.org/Public/GovernmentAffairs/98_11128.cfm.
46. American Dietetic Association, Improved Nutrition and Physical Activity Act to be introduced, *ADA Policy Initiatives and Advocacy Report: On the Pulse,* May 17, 2002.
47. The discussion of state licensure is from: Update on state licensure laws; available at www.eatright.org.
48. Ibid.
49. As presented in General Session: "Bioterrorism and Food: Are We Ready?" at the 2002 National Food Policy Conference, Washington, D.C., April 22–23, 2002; and M. M. Cody, Bioterrorism: What does it mean for dietetics professionals and the American public? In Public Health/Community Nutrition Practice Group, *The Digest,* Spring 2002, pp. 8–11.
50. CSPI Newsroom, Bill to Establish Single Food-Safety Agency Would Help Prevent Food-borne Illnesses, October 10, 2001, available at www.cspinet.org; for current status of the Safe Food Act of 2004 (S. 2910), go to thomas.loc.gov.
51. D. Knorr and A. J. Sinskey, Biotechnology in food production and processing, *Science* 229 (1985): 1224–29.
52. J. Q. Wilkinson, Biotech plants: From lab bench to supermarket shelf, *Food Technology* 51 (1997): 37–42.
53. N. A. Higley and J. B. Hallagan, Safety and regulation of ingredients produced by plant cell and tissue culture, *Food Technology* 51 (1997): 72–74.
54. Wilkinson, Biotech plants, p. 41; B. C. Babcock and C. A. Francis, Solving global nutrition challenges requires more than new technologies, *Journal of the American Dietetic Association* 100 (2000): 1308–11; and C. McCullum, Food biotechnology in the new millennium: Promises, realities, and challenges, *Journal of the American Dietetic Association* 100 (2000): 1311–15.
55. Americans' Opinions About Genetically Modified Foods Remain Divided, But Majority Want a Strong Regulatory System, available at http://pewagbiotech.org/newsroom/releases/112404.php3; and A. Wilson, J. Latham, and R. Steinbrecher, Genome Scrambling—Myth or Reality? Transformation-induced mutations in transgenic crop plants, pp. 1–36; available at www.econexus.info.
56. D. Shattuck, Complementary medicine: Finding a balance, *Journal of the American Dietetic Association* 97 (1997): 1367–69.
57. A. Barrocas, Complementary and alternative medicine: Friend, foe, or OWA? *Journal of the American Dietetic Association* 97 (1997): 1373–76.
58. American Dietetic Association, *On the Pulse,* December 14, 2001; Position of the American Dietetic Association, Functional foods, *Journal of the American Dietetic Association* 99 (1999): 1278–85.
59. The definition of the term *nutraceuticals* was taken from D. E. Pszczola, Highlights of "The nutraceutical initiative: A proposal for economic and regulatory reform," *Food Technology* 46 (1992): 77–79.
60. The discussion of DSHEA and herbal supplements is from M. Boyle, *Personal Nutrition,* 5th ed. (Belmont, CA: Wadsworth Publishing Co., 2004), p. 187.
61. J. B. German, Genetic dietetics: Nutrigenomics and the future of dietetics practice, *Journal of the American Dietetic Association* 105 (2005): 530–31; and R. M. DeBusk and coauthors, Nutritional genomics in practice: Where do we begin, *Journal of the American Dietetic Association* 105 (2005): 589–98.
62. As cited in T. Peregrin, The new frontier of nutrition science: Nutrigenomics, *Journal of the American Dietetic Association* 101 (2001): 1306.
63. Linda Goodwin, RD, Alliance Program Director, American Dietetic Association, personal communication, July 1998.
64. Capitol Resources, Grassroots targeting, American Dietetic Association 80th Annual Meeting, October 28, 1997.
65. E. C. Ladd, *The American Polity—The People and Their Government* (New York: Norton, 1985), p. 351.
66. G. Starling, *Understanding American Politics* (Homewood, IL: Dorsey, 1982), pp. 184–85.
67. T. D. Bevels, Public interest groups and the public manager, *Bureaucrat* 25 (Winter 1996–97): 8–12.
68. J. M. Burns, J. W. Peltason, and T. E. Cronin, *Government by the People,* 12th ed. (Upper Saddle River, NJ: Prentice-Hall, 1984), pp. 167–68; and M. L. Watts, How dietitians can make a difference through political action, *Journal of the American Dietetic Association* 102 (2002): 1226–27.
69. Ibid., p. 169.
70. H. J. Rubin and I. S. Rubin, *Community Organizing and Development* (New York: Macmillan, 1992), pp. 274–95.
71. Ibid., p. 282.
72. Alliance for the Prevention of Chronic Disease (Winnipeg, Manitoba, Canada, 1997); see also M. J. Feeney and J. V. White, Corporate alliances: A strategy for success, *Journal of the American Dietetic Association* 100 (2000): 1124.
73. The description of how to communicate effectively with elected officials was taken from How to be heard on Capitol Hill, *Journal of the American Dietetic Association* 92 (1992): 296.
74. American Dietetic Association, *The ADA Advocacy Guide* (Chicago, IL: American Dietetic Association).

Chapter 8

1. A. H. Mokdad and coauthors, Prevalence of obesity, diabetes, and obesity-related health risk factors, *Journal of the American Medical Association* 289 (2001): 76–79; A. A. Hedley and coauthors, Prevalence of overweight and obesity among U.S. children, adolescents, and adults, 1999–2002, *Journal of the American Medical Association 291* (2004): 2847–50; National Heart, Lung, and Blood Institute Expert Panel on the Identification, Evaluation, and Treatment of Overweight and Obesity in Adults, Executive summary of the clinical guidelines on the identification, evaluation, and treatment of overweight and obesity in adults, *Journal of the American Dietetic Association* 98 (1998): 1178–91; and T. Lobstein, L. Baur, and R. Uauy for the IASO International Obesity Task Force, Obesity in children and young people: A crisis in public health, *Obesity Reviews* 5 (Suppl 1) (2004): 4–85.
2. T. Lobstein and coauthors, 2004; and A. A. Hedley and coauthors, 2004.
3. A. Must and coauthors, The disease burden associated with overweight and obesity, *Journal of the American Medical Association* 282 (1999): 1523–29; NHLBI, 1998; G. Wang and W. H. Dietz, Economic burden of obesity in youths aged 6 to 17

years: 1979–1999, *Pediatrics* 109 (2002): E81. [erratum appears in *Pediatrics* 109 (2002): 1195]; and E. A. Finkelstein, I. C. Fiebelkorn, and G. Wang, State-level estimates of annual medical expenditures attributable to obesity, *Obesity Research* 12 (2004): 18–24.
4. K. Clement and P. Ferre, Genetics and the pathophysiology of obesity, *Pediatric Research* 53 (2003): 721–25. Epub 2003 Mar 05.
5. Ibid; M. P. Galvez, T. R. Frieden, and P. J. Landrigan, Obesity in the 21st century, *Environmental Health Perspectives* 111 (2003): A684–85; and C. B. Ebbeling, D. B. Pawlak, and D. S. Ludwig, Childhood obesity: Public-health crisis, common sense cure. *Lancet* 360 (2002): 473–82.
6. Surgeon General's Call to Action to Prevent and Decrease Overweight and Obesity, 2001; E. E. Calle and coauthors, Body-mass index and mortality in a prospective cohort of U.S. adults, *New England Journal of Medicine* 341 (1999): 1097–1105; and T. L. S. Visscher and coauthors, Obesity and unhealthy life-years in adult Finns, *Archives of Internal Medicine* 164 (2004): 1413–20.
7. NHLBI, 1998.
8. Executive Summary of the Third Report of the National Cholesterol Education Program (NCEP) Expert Panel on Detection, Evaluation, and Treatment of High Blood Cholesterol in Adults (Adult Treatment Panel III), *Journal of the American Medical Association* 285 (2001): 2486–97.
9. Centers for Disease Control and Prevention Growth Charts; available at www.cdc.gov/growthcharts.
10. T. H. Cole and coauthors, Establishing a standard definition for child overweight and obesity worldwide: International survey, *British Medical Journal* 320 (2000): 1240–43.
11. National Health and Nutrition Examination Survey (NHANES); available at www.cdc.gov/nchs/nhanes.htm.
12. A. A. Hedley and coauthors, 2004.
13. Ibid.
14. Ibid.
15. R. S. Strauss and H. A. Pollack, Epidemic increase in childhood overweight, 1986–1998, *Journal of the American Medical Association* 286 (2001): 2845–48; and D. M. Hoelscher and coauthors, Measuring the prevalence of overweight in Texas school children, *American Journal of Public Health* 94 (2004): 1002–8.
16. A. Must and R. S. Strauss, Risks and consequences of childhood and adolescent obesity, *International Journal of Obesity & Related Metabolic Disorders,* 23 Suppl 2 (1999): S2–S11; and G. Wang and W. H. Dietz, 2002.
17. Centers for Disease Control and Prevention, "U.S. Obesity Trends, 1985 to 2003"; available at www.cdc.gov/nccdphp/dnpa/obesity/trend/maps/.
18. Centers for Disease Control and Prevention, "Prevalence Data for Weight Classifications Based on BMI"; available at http://apps.nccd.cdc.gov/brfss/.
19. Centers for Disease Control and Prevention, "Selected Metropolitan/Micropolitan Area Risk Trends (SMART). Risk Factors and Calculated Variables: Weight Classifications Based on BMI"; available at http://apps.nccd.cdc.gov/brfss-smart/.
20. J. A. Grunbaum and coauthors, Youth Risk Behavior Surveillance—United States, 2003 (abridged), *Journal of School Health* 74 (2004): 307–324.
21. Trust for America's Health, Issue Report: F as in fat: How obesity policies are failing in America. *Trust for America's Health,* December 2004.
22. J. A. Grunbaum and coauthors, 2004; and Trust for America's Health (TFAH), 2004.
23. D. M. Hoelscher and coauthors, 2004.
24. Arkansas Center for Health Improvement, "The Arkansas Assessment of Childhood and Adolescent Obesity," 2004; available at www.achi.net.
25. B. Sherry and coauthors, Trends in state-specific prevalence of overweight and underweight in 2- through 4-year-old children from low-income families from 1989 through 2000, *Archives of Pediatric and Adolescent Medicine* 158 (2004): 1116–24.
26. Centers for Disease Control and Prevention (CDC). Nutritional status of children participating in the Special Supplemental Nutrition Program for women, infants, and children—United States, 1991–1998, *MMWR Morbidity and Mortality Weekly Report* 45 (1996): 65–69.
27. TFAH, 2004.
28. U.S. Department of Health and Human Services, *The Surgeon General's Call to Action to Prevent and Decrease Overweight and Obesity* (Rockville, MD: Office of the Surgeon General, Public Health Service, US Department of Health and Human Services, 2001).
29. E. A. Finkelstein, I. C. Fiebelkorn, and G. Wang, 2004.
30. Ibid.
31. K. E. Thorpe and coauthors, Trends: The impact of obesity on rising medical spending, *Health Affairs,* October 2004: W4:480–W4:486 (Web version).
32. G. Wang and W. H. Dietz, 2002.
33. R. Sturm and K. B. Wells, Does obesity contribute as much to morbidity as poverty or smoking? *Public Health* 115 (2001): 229–35; and A. Must and coauthors, 1999.
34. E. S. Ford, W. H. Giles, and W. H. Dietz, Prevalence of the metabolic syndrome among U.S. adults: Findings from the third National Health and Nutrition Examination Survey, *Journal of the American Medical Association,* 287 (2002): 356–59; and NCEP ATP III, 2001.
35. E. S. Ford, W. H. Giles, and W. H. Dietz, 2002.
36. S. Cook and coauthors, Prevalence of a metabolic syndrome phenotype in adolescents: Findings from the third National Health and Nutrition Examination Survey, 1988–1994, *Archives of Pediatrics and Adolescent Medicine* 157 (2003): 821–27.
37. A. Jain, What words for obesity? A summary of the research behind obesity interventions, BMJ Publishing Group, 30 April 2004; available at http://www.unitedhealthfoundation.org/obesity.pdf.
38. C. B. Ebbeling, D. B. Pawlak, and D. S. Ludwig, 2002.
39. R. S. Strauss and H. A Pollak, Social marginalization of overweight children, *Archives of Pediatrics and Adolescent Medicine* 157 (2003): 746–52; M.E. Eisenberg, D. Neumark-Sztainer, and M. Story, Associations of weight-based teasing and emotional well-being among adolescents, *Archives of Pediatrics and Adolescent Medicine* 157 (2003): 733–38.
40. T. Lobstein, L. Baur, and R. Uauy, 2004; C. B. Ebbeling, D. B. Pawlak, and D. S. Ludwig, 2002; TFAH, 2004; and A. Jain, 2004.
41. J. O. Hill and coauthors, Obesity and the environment: Where do we go from here? *Science* 299 (2003): 853–55; and K. D. Brownell and K. B. Horgen, *Food Fight: The Inside Story of the Food Industry, America's Obesity Crisis, and What We Can Do About It* (New York: McGraw-Hill/Contemporary Books, 2003).
42. B. Swinburn and G. Egger, Preventive strategies against weight gain and obesity, *Obesity Reviews* 3 (2002): 289–301.
43. M. I. Goran, Energy metabolism and obesity, *Medical Clinics of North America* 84 (2000): 347–62.
44. J. O. Hill, H. R. Wyatt, E. L. Melanson, Genetic and environmental contributions to obesity, *Medical Clinics of North America* 84 (2000): 333–46.
45. T. Rankinen and coauthors, The human obesity gene map: The 2001 update, *Obesity Research* 10 (2002): 196–243.
46. L. Perusse and C. Bouchard, Gene–diet interactions in obesity, *American Journal of Clinical Nutrition* 72, 5 Suppl (2000): 1285S–1290S.
47. Ibid; and S. B. Roberts and A. S. Greenberg, The new obesity genes, *Nutrition Reviews* 54 (1999): 41–49.
48. D. S. Pine and coauthors, The association between childhood depression and obesity, *Pediatrics* 107 (2001): 1049–56; E. Goodman and R. C. Whitaker, A prospective study of the role of depression in the development and persistence of adolescent obesity, *Pediatrics* 109 (2003); 497–504; and M. F. Dallman and coauthors, Chronic stress and obesity: A new view of "comfort food," *Proceedings of the National Academy of Science U.S.A.* 100 (2003): 11696–701.
49. D. S. Pine and coauthors, 2001.
50. E. Goodman and R. C. Whitaker, 2003.

51. M. F. Dallman and coauthors, 2003.
52. Ibid.
53. Centers for Disease Control and Prevention, Trends in intake of energy and macronutrients—United States, 1971–2000, *Morbidity and Mortality Weekly Report (MMWR)* 53 (2004): 80–82; and J. Putnam, J. Allshouse, and L. S. Kantor, U.S. per capita food supply trends: More calories, refined carbohydrates, and fats, *Economic Research Service Food Review* 25-3 (2002): 2–15.
54. J. Putnam, J. Allshouse and L. S. Kantor, 2002.
55. Centers for Disease Control and Prevention, 2004.
56. G. Block, Food contributing to energy intake in the U.S.: Data from NHANES III and NHANES 1999–2000, *Journal of Food Composition and Analysis* 17 (2004): 439–47.
57. S. J. Nielsen and B. M. Popkin, Changes in beverage intake between 1977 and 2001, *American Journal of Preventive Medicine* 27 (2004): 205–10.
58. G. Mrdjenovic and D. A. Levitsky, Nutritional and energetic consequences of sweetened drink consumption in 6- to 13-year-old children, *Journal of Pediatrics* 142 (2003): 604–10; D. S. Ludwig, K. E. Peterson, and S. L. Gortmaker, Relationship between consumption of sugar-sweetened drinks and childhood obesity: A prospective, observational analysis, *Lancet* 357 (2001): 505–508; C. Zizza, A. M. Siga-Riz, and B. M. Popkin, Significant increase in young adults' snacking between 1977–1978 and 1994–1996 represents a cause for concern! *Preventive Medicine* 32 (2001): 203–10; L. J. Gillis and O. Bar-Or, Food away from home, sugar-sweetened drink consumption, and juvenile obesity, *Journal of the American College of Nutrition* 6 (2003): 539–45; and S. A. French, M. Story, and R. W. Jeffrey, Environmental influences on eating and physical activity, *Annual Review of Public Health* 22 (2001): 309–35.
59. Institute of Medicine, *Dietary Reference Intakes for Energy, Carbohydrate, Fiber, Fat, Fatty Acids, Cholesterol, Protein and Amino Acids* (Washington, D.C.: The National Academy Press, 2002); D. A. Schoeller, K. Shay, and R. F. Kushner, How much physical activity is needed to minimize weight gain in previously obese women? *American Journal of Clinical Nutrition* 66 (1997): 551–56; J. P. DeLany and coauthors, Energy expenditure in African American and white boys and girls in a 2-y follow-up of the Baton Rouge Children's Study, *American Journal of Clinical Nutrition* 79 (2004): 268–73; and K. Patrick and coauthors, Diet, physical activity, and sedentary behaviors as risk factors for overweight in adolescence, *Archives in Pediatric and Adolescent Medicine* 158 (2004): 385–90.
60. K. Patrick and coauthors, 2004.
61. L. Dong, G. Block, and S. Mandel, Activities contributing to total energy expenditure in the United States: Results from the NHAPS Study, *International Journal of Nutrition and Physical Activity* 1 (2004): 4.
62. U.S. Department of Health and Human Services, *Data2010: The Healthy People 2010 Database;* available at http://wonder.cdc.gov/data2010.
63. Kaiser Family Foundation, 1999. "Kids and media at the new millennium: A comprehensive national analysis of children's media use. A report of the Kaiser Family Foundation"; available at http://www.kff.org/.
64. R. J. Hancox, B. J. Milne, and R. Poulton. Association between child and adolescent television viewing and adult health: A longitudinal birth cohort study, *Lancet* 364 (2004): 257–62.
65. R. G. McMurray and coauthors, The influence of physical activity, socioeconomic status, and ethnicity on the weight status of adolescents, *Obesity Research* 8 (2000): 130–39.
66. D. M. Cultler, E. K. Glaeser, and J. M. Shapiro. Why have Americans become more obese? In: *The Economics of Obesity/E-FAN-04-004* (Washington, D.C.: Economic Research Service, U.S. Department of Agriculture, May 2004) pp. 2–5.
67. D. Lakdawalla, and T. Philipson, The growth of obesity and technological change. In: *The Economics of Obesity/E-FAN-04-004* (Washington, D.C.: Economic Research Service, U.S. Department of Agriculture, May 2004), pp. 6–9.
68. Ibid.
69. L. R. Young and M. Nestle, The contribution of expanding portion sizes to the U.S. obesity epidemic, *American Journal of Public Health* 92 (2002): 246–49; and S. J. Nielsen and B. M. Popkin, Patterns and trends in food portion sizes, 1977–1998, *Journal of the American Medical Association* 289 (2003): 450–53.
70. S. J. Nielsen and B. M. Popkin, 2003.
71. B. J. Rolls, E. L. Morris, and L. S. Roe, Portion size of food affects energy intake in normal-weight and overweight men and women, *American Journal of Clinical Nutrition* 76 (2002): 1207–13.
72. M. Nestle, Increasing portion sizes in the American diets: More calories, more obesity, *Journal of the American Dietetic Association* 103 (2003), 39–40.
73. S-Y Chou, M. Grossman, and H. Saffer, An economic analysis of adult obesity: Results from the Behavioral Risk Factor Surveillance System. In: *The Economics of Obesity/E-FAN-04-004* (Washington, D.C.: Economic Research Service, U.S. Department of Agriculture, May 2004) pp. 10–13.
74 J. Maddock, The relationship between obesity and the prevalence of fast-food restaurants: State-level analysis, *American Journal of Health Promotion* 19 (2004): 137–43.
75. C. B. Ebbeling, D. B. Pawlak, and D. S. Ludwig, 2002; M. A. Pereira and coauthors, Fast-food habits, weight gain, and insulin resistance (the CARDIA study): 15-year prospective analysis, *Lancet* 365 (2005).
76. M. A. Pereira and coauthors, 2005.
77. J. P. Block, R. A. Scribner, and K. DeSalvo, Fast food, race/ethnicity, and income: A geographical analysis, *American Journal of Preventive Medicine* 27 (2004): 211–37.
78. Ibid.
79. J. Levine, Food industry marketing in elementary schools: Implications for school health professionals, *Journal of School Health* 69 (1999): 290–91; and P. Cram and coauthors, Fast food franchises in hospitals, *Journal of the American Medical Association* 287 (2002): 2945–46.
80. J. F. Guthrie, B-H Lin, and E. Frazao, Role of food prepared away from home in the American diet, 1977–78 versus 1994–96: Changes and consequences, *Journal of Nutrition Education and Behavior* 34(2002): 140–50.
81. B. Swinburn and G. Egger, 2002.
82. Institute of Medicine, *Preventing Childhood Obesity: Health in the Balance* (Washington, D.C.: The National Academies Press, 2005).
83. S.A. Bowman and coauthors, Effects of fast-food consumption on energy intake and diet quality among children in a national household survey, *Pediatrics* 113 (2004): 112–18.
84. P. M. Anderson, K. F. Butcher, and P. B. Levine, "Maternal Employment and Overweight Children," Dartmouth College Department of Economics, January 2002.
85. R. Lopez, Urban sprawl and risk for being overweight or obese, *American Journal of Public Health* 94 (2004): 1574–79.
86. Ibid; R. Ewing and coauthors, Relationship between urban sprawl and physical activity, obesity, and morbidity, *American Journal of Health Promotion* 18 (2003): 47–57; L. D. Frank, M. A. Andersen, and T. L. Schmid, Obesity relationships with community design, physical activity, and time spent in cars, *American Journal of Preventive Medicine* 27 (2004): 87–96; B. E. Saelens and coauthors, Neighborhood-based differences in physical activity: An environment scale evaluation, *American Journal of Public Health* 93 (2003): 1552–58; and B. E. Saelens, J. F. Sallis, and L. D. Frank, Environmental correlates of walking and cycling: Findings from the transportation, urban design, and planning literature, *Annals of Behavioral Medicine* 25 (2003): 80–91.
87. M. J. Aboelata and coauthors, *The Built Environment and Health: 11 Profiles of Neighborhood Transformation* (Oakland, CA: The Prevention Institute, 2004); available at http://www.preventioninstitute.org.

88. K. M. Booth, M. M. Pinkston, and W. S. C. Poston, Obesity and the built environment, *Journal of the American Dietetic Association* 105 (2005): S110–S117.
89. R. Ewing and coauthors, 2003.
90. B. E. Saelens, J. F. Sallis, and L. D. Frank, 2003.
91. L. D. Frank, M. A. Andersen, and T. L. Schmid, 2004.
92. U.S. Department of Health and Human Services, *Healthy People 2010* (Washington, D.C.: U.S. Government Printing Office, 2000).
93. L. Mancino, L. Biing-Hwan, and N. Ballenger. *The Role of Economics in Eating Choices and Weight Outcomes-AIB-792* (Washington, D.C.: Economic Research Service, U.S. Department of Agriculture, October 2004).
94 A. A. Hedley and coauthors, 2004.
95. A. Drewnowski and S. E. Specter, Poverty and obesity: The role of energy density and energy costs, *American Journal of Clinical Nutrition* 79 (2004): 6–16.
96. B. Rolls and R. A. Barnett, *The Volumetrics Weight-Control Plan: Feel Full on Fewer Calories* (New York: HarperCollins, 2000); A. Drewnoski, Palatability and satiety: Models and measures, *Annals Nestle [Fr]* 56 (1998): 32–42; J. E. Blundell and J. I. MacDiarmid, Passive overconsumption: Fat intake and short-term energy balance, *Annals of New York Academy of Sciences* 827 (1997): 392–407; and A. Drewnowski, Energy intake and sensory properties of food, *American Journal of Clinical Nutrition* 6 (1995): 1081S–1085S.
97. B. J. Rolls, F. H. Castellanos, and J.C. Halford, Volume of foods consumed affects satiety in men, *American Journal of Clinical Nutrition* 67 (1998): 1170–77.
98. A. Drewnowski, Energy density, palatability, and satiety: Implications for weight control, *Nutrition Reviews* 56 (1998): 347–55.
99. B-H. Lin and R. M. Morrison, Higher fruit consumption linked with lower body mass index. *Food Review* 25 (2003): 28–32.
100. P. R. Kaufman and coauthors, *Do the Poor Pay More for Food? Item Selection and Price Differences Affect Low-Income Household Food Costs* (Washington, D.C.: Agricultural Economic Report No. 759, Economic Research Service, U.S. Department of Agriculture); and K. Morland and coauthors, Neighborhood characteristics associated with the location of food stores and food service places, *American Journal of Preventive Medicine* 22 (2002), 23–29.
101. K. Morland and coauthors, 2002.
102. D. Rose and R. Richards, Food store access and household fruit and vegetable use among participants in the U.S. Food Stamp Program, *Public Health Nutrition* 7 (2004): 1081–88.
103. A. Cheadle and coauthors, Community-level comparisons between the grocery-store environment and individual dietary practices, *Preventive Medicine* 20 (1991): 250–61.
104. K. Morland, S. Wing, and A. Diez Roux, The contextual effect of the local food environment on residents' diets: The Atherosclerosis Risk in Communities Study, *American Journal of Public Health* 92 (2002): 1761–67.
105. NHLBI, 1998.
106. R. E. Glasgow, E. Lichtenstein, and A. C. Marcus, Why don't we see more translation of health promotion research to practice? Rethinking the efficacy-to-effectiveness transition, *American Journal of Public Health* 93 (2003): 1261–67.
107. K. D. Brownell and R. W. Jeffrey, Improving long-term weight loss: Pushing the limits of treatment, *Behavioral Therapy* 18 (1987): 353–74.
108. M. L. Klem and coauthors, A descriptive study of individuals successful at long-term maintenance of substantial weight loss, *American Journal of Clinical Nutrition* 66 (1997): 239–46.
109. Ibid; H. R. Wyatt and coauthors, Long-term weight loss and breakfast in subjects in the National Weight Control Registry. *Obesity Research* 10 (2002): 78–82; and A. A. Gorin and coauthors, Promoting long-term weight control: Does dieting consistency matter? *International Journal of Obesity and Related Metabolic Disorders* 28 (2004): 278–81.
110. Jain, 2004.
111. Ibid.
112. Active Living Research program, Robert Wood Johnson Foundation; available at http://www.activelivingresearch.org/, accessed January 2005.
113. D. J. Hennrikus and R. W. Jeffery, Work-site intervention for weight control: A review of the literature, *American Journal of Health Promotion* 10 (1996): 471–98.
114. Ibid.
115. U.S. Department of Health and Human Services, National Institutes of Health (NIH), *Strategic Plan for NIH Obesity Research: A Report of the NIH Obesity Research Task Force,* NIH Publication No. 04-5493 (Washington, D.C.: U.S. Department of Health and Human Services, 2004).
116. T. Baranowski and coauthors, School-based obesity prevention: A blueprint for taming the epidemic, *American Journal of Health Behavior* 26 (2002): 486–93.
117. S. L. Gortmaker and coauthors, Reducing obesity via a school-based interdisciplinary intervention among youth, *Archives in Pediatrics and Adolescent Medicine* 153 (1999): 409–18; and T. N. Robinson, Reducing children's television viewing to prevent obesity: A randomized controlled trial, *Journal of the American Medical Association* 282 (1999): 1561–67.
118. T. Dwyer and coauthors, An investigation of the effects of daily physical activity on the health of primary school students in South Australia, *International Journal of Epidemiology* 12 (1983): 308–13.
119. A. Datar and R. Sturm, Physical education in elementary school and body mass index: Evidence from the Early Childhood Longitudinal Study, *American Journal of Public Health* 94 (2004): 1501–506; and K. J. Coleman and coauthors, Impacting epidemic increases in child risk for overweight in low-income schools: The El Paso Coordinated Approach to Child Health (El Paso CATCH), *Archives of Pediatrics and Adolescent Medicine,* 159 (2005): 217–24.
120. J. James and coauthors, Preventing childhood obesity by reducing consumption of carbonated drinks: Cluster randomized controlled trial, *British Medical Journal* 328 (2004):1237.
121. Barnowski and coauthors, 2002.
122. D. M. Hoelscher and coauthors, Designing effective nutrition interventions for adolescents, *Journal of the American Dietetic Association* 102 (2002): S52–S63.
123. TFAH, 2004.
124. University of Baltimore study; available at www.ubalt.edu/experts/obesity/index.html.
125. K. D. Brownell and K. B. Horgen, 2003.
126. *Healthy People 2010, 2000.*
127. TFAH, 2004.
128. Ibid.
129. American Academy of Pediatrics, Prevention of pediatric overweight and obesity, Policy Statement, *Pediatrics* 112 (2003): 424–30.
130. Institute of Medicine, 2005.
131. Arkansas Center for Health Improvement, 2004.
132. TFAH, 2004.
133. U.S. Department of Health and Human Services, *Steps to a HealthierUS: A Community-Focused Initiative to Reduce the Burden of Asthma, Diabetes, and Obesity, Program Announcement;* available at http://edocket.acces s.gpo.gov/2004/04-10416.htm.
134. TFAH, 2004.
135. Centers for Disease Control and Prevention, *Youth Media Campaign,* www.cdc.gov/youthcampaign/ and VERB, www.verbnow.com/.
136. Institute of Medicine, 2005.
137. Ibid.
138. Centers for Disease Control, National Center for Chronic Disease Prevention and Health Promotion, Physical activity and good nutrition: Essential elements to prevent chronic diseases and obesity 2003, *Nutrition in Clinical Care* 6 (2003): 135–38.
139. L. M. Grummer-Strawn and M. Zuguo, Does breastfeeding protect against pediatric overweight? Analysis of longitudinal data from the Centers for Disease Control and Prevention Pediatric Nutrition Surveillance System, *Pediatrics* 113 (2004): e81–86; M. W. Gillman and coauthors, Risk of overweight among adolescents

who were breastfed as infants, *Journal of the American Medical Association* 285 (2001): 2461–67; and CDC, 2003.

140. Centers for Disease Control and Prevention, *Overweight and Obesity State Programs;* available at www.cdc.gov/nccdphp/dnpa/obesity/state_programs/index.htm.
141. Centers for Disease Control and Prevention (CDC), Division of Adolescent and School Health, *School Health Defined: Coordinated School Health Program;* available at www.cdc.gov/nccdphp/dash/about/school_health.htm.
142. Centers for Disease Control and Prevention (CDC), Division of Adolescent and School Health, *School Health Index;* available at http://apps.nccd.cdc.gov/shi/.
143. J. Gerberding and J. Marks, Making America fit and trim—Steps big and small, *American Journal of Public Health* 94 (2004): 1478–79.
144. Centers for Disease Control and Prevention, *Prevention Research Centers;* available at www.cdc.gov/prc/about/index.htm.
145. USDHHS, *Strategic Plan for NIH Obesity Research.*
146. NIH, National Heart, Lung, and Blood Institute, *NHLBI Obesity Education Initiative;* available at www.nhlbi.nih.gov/about/oei/oei_pd.htm.
147. Ibid.
148. NIH, National Cancer Institute, *NCI Organization;* available at www.cancer.gov/aboutnci/organization.
149. NIH, National Cancer Institute, *5 A Day for Better Health Evaluation Report;* available at www.cancer.gov.
150. NIH, National Cancer Institute, *Eat 5 to 9 A Day for Better Health: About the Program;* available at http://5aday.gov/index-about.shtml#background.
151. NIH, National Institute of Diabetes and Digestive and Kidney Diseases. *Weight-Control Information Network;* available at www.niddk.nih.gov/health/nutrit/win.htm.
152. NIEHS Press Release, 2003.
153. U.S. Department of Agriculture, *Nutrition Assistance Programs, Food and Nutrition Service Overview;* available at www.fns.usda.gov/fsp/faqs.htm.
154. U.S. Department of Agriculture, Team Nutrition; available at www.fns.usda.gov/tn.
155. U.S. Department of Agriculture (USDA) News Release No. 0433.04, "Secretary Veneman Announces the HEALTHIERUS School Challenge. Kicks Off National School Lunch Week, October 10–16" (Washington, D.C.: USDA, October 7, 2004).
156. U.S. Department of Agriculture, *Center for Nutrition Policy and Promotion Homepage;* available at www.usda.gov/cnpp.
157. X. Guo and coauthors, Healthy eating index and obesity, *European Journal of Clinical Nutrition* 58 (2004): 1580–86.
158. U.S. Food and Drug Administration, *Backgrounder Report of the Working Group on Obesity;* available at www.fda.gov/oc/initatives/obesity/backgrounder.html.
159. T. J. Muris, Chairman, Federal Trade Commission, *Statement on the Announcement of FDA Obesity Working Group Report, March 12, 2004;* available at www.ftc.gov/speeches/muris/040312obesity.htm.
160. Federal Trade Commission, *Consumer Information;* available at www.ftc.gov/bcp/conline/edcams/fitness/coninfo.html.
161. Federal Trade Commission, *FTC Releases Guidance to Media on False Weight-Loss Claims;* available at www.ftc.gov/opa/2003/12/weightlossrpt.htm.
162. C. L. Hayne, P. A. Moran, and M. M. Ford, Regulating environments to reduce obesity, *Journal of Public Health Policy* 25 (2004): 48–64.
163. B. Swinburn and G. Egger, 2002.
164. M. Nestle and M. F. Jacobson, Halting the obesity epidemic: A public health policy approach, *Public Health Reports* 115 (2000): 12–24.
165. FDA, *Backgrounder Report of the Working Group on Obesity.*
166. C. L. Hayne, P. A. Moran, and M. M. Ford, 2004.
167. Institute of Medicine, 2005.
168. Ibid.
169 C. Hawkes. *Marketing Food to Children: The Global Regulatory Environment.* (Geneva, Switzerland: World Health Organization, 2004).
170. S. E. Linn, Food marketing to children in the context of a marketing maelstrom. *Journal of Public Health Policy* 24 (2004): 24–35.
171. C. Hawkes, 2004.
172. Institute of Medicine, 2005.
173. TFAH, 2004.
174. U.S. Department of Agriculture, *Foods Sold in Competition with USDA School Meal Programs: A Report to Congress,* Washington, D.C.: U.S. Department of Agriculture, 2001.
175. U.S. Department of Agriculture, *Foods Sold in Competition with USDA School Meal Programs: A Report to Congress;* and M.Y. Kubik and coauthors, The association of the school food environment with dietary behaviors of young adolescents, *American Journal of Public Health* 93 (2003): 1168–73.
176. U.S. Department of Agriculture, *Foods Sold in Competition with USDA School Meal Programs: A Report to Congress.*
177. Food Research and Action Center, *Highlights of the Child Nutrition and WIC Reauthorization Act of 2004;* available at www.frac.org/.
178. United States General Accounting Office, *Report to Congressional Requesters. School Meal Programs. Competitive foods are available in many schools; actions taken to restrict them differ by state and locality* (Washington, D.C.: GAO-04-673, April 2004).
179. K. D. Brownell and K. B. Horgen, 2003.
180. National Association for Sport and Physical Education, *Shape of the Nation Report;* available at www.mcph.org/reports/ Shape%20of%20the%20Nation%20REport%2001.pdf.
181. Robert Wood Johnson Foundation, 2003.
182. National Center for Health Statistics, U.S. Department of Education, *Nutrition Education in Public Elementary and Secondary Schools Highlights;* available at http://nces.ed.gov/surveys/frss/publications/96852/.
183. C. L. Hayne, P. A. Moran, and M. M. Ford, 2004.
184. FRAC, Highlights of the Child Nutrition and WIC Reauthorization Act of 2004; NASBE Policy Update, October 14, 2004.
185. NASBE Policy Update, October 14, 2004.
186. M. J. Aboelata and coauthors, *The Built Environment and Health: 11 Profiles of Neighborhood Transformation.*
187. R. Ewing and coauthors, 2004; and R. Lopez, 2004.
188. Institute of Medicine, 2005.
189. C. L. Hayne, P. A. Moran, and M. M. Ford, 2004.
190. T. Pollard, 2003.
191. C. L. Hayne, P. A. Moran, and M. M. Ford, 2004.
192. D. Perry, Fighting fat at the supermarket. *Philadelphia Enquirer,* October 15, 2004.
193. TFAH, 2004.
194. R. I. Weiss and J. A. Smith, Legislative approaches to the obesity epidemic. *Journal of Public Health Policy* 25 (2004): 36–48.
195. Personal Responsibility in Food Consumption Act, H.R. 339, 108th Cong., 2nd Sess. 2004.
196. Commonsense Consumption Act of 2003, S. 1428, 108th Congress (2004); and R. I. Weiss and J. A. Smith, 2004.
197. B. Keller and J. A. Smith, Legal approaches to the obesity epidemic: An introduction, *Journal of Public Health Policy* 25 (2004): 3–9.
198. S. Roberts, *Policy options for promoting healthy environments to address the global obesity epidemic.* Presentation made at the 37th Annual Conference of the Society for Nutrition Education, Salt Lake City, Utah, July 21, 2004.
199. M. Nestle, *Food Politics: How the Food Industry Influences Nutrition and Health.* (Berkeley, CA: University of California Press, 2002); and K. D. Brownell and K. B. Horgen, 2003.
200. L. Cohen and coauthors, "Cultivating Common Ground: Linking Health and Sustainable Agriculture" (Prevention Institute, September 2004).
201. Ibid.
202. S. Fields, The fat of the land. Do agricultural subsidies foster poor health? *Environmental Health Perspectives* 112 (2004); available at http://ehp.niehs.nih.gov/members/2004/112-14/spheres.html.
203. TFAH, 2004.

204. National Governors Association's Center for Best Practices, 2002; and World Health Organization, *Global Strategy on Diet, Physical Activity, and Health* (Geneva, Switzerland: World Health Organization, May 2004).
205. F. Kuchler, T. Abebayehu, and M. Harris, *Taxing Snack Foods: What to Expect for Diet and Tax Revenues* (Washington, D.C.: Agriculture Information Bulletin No. 747-08, Economic Research Service, U.S. Department of Agriculture, August 2004); and M. F. Jacobson and K. D. Brownell, Small taxes on soft drinks and snack foods to promote health, *American Journal of Public Health* 90 (2000).
206. TFAH, 2004.
207. The National Review Online, *The American Diet: The Center for Science in the Public Interest Should Move to France;* available at www.nationalreview.com/comment/comment060500d.html.
208. World Health Organization (WHO), *Obesity: Preventing and Managing the Global Epidemic. Report of a WHO Consultation* (Geneva, Switzerland: WHO Technical Report Series No. 894. WHO, 2000).
209. World Health Organization (WHO), *World Health Report 2002: Reducing Risks, Promoting Healthy Life* (Geneva, Switzerland: World Health Organization, 2002).
210. N. J. Rigby, S. Kumanyika, and W. P. T. James, Confronting the epidemic: The need for global solutions, *Journal of Public Health Policy* 25 (2004): 75–91.
211. Ibid.
212. Ibid.
213. S. Kumanyika and coauthors, Public health approaches to the prevention of obesity. Working Group of the International Obesity Taskforce, Obesity prevention: The case for action, *International Journal of Obesity* 26 (2002): 425–36.
214. S. J. Blumenthal, J. Hendi, and L. Marsillo, A public health approach to decreasing obesity, *Journal of the American Medical Association* 288 (2002): 2178.
215. A. M. Prentice, and S. A. Jebb, Fast foods, energy density and obesity: A possible mechanistic link, *Obesity Reviews* 4 (2003):187–94.
216. Institute of Medicine, 2004; *Surgeon General's Call to Action to Prevent Overweight and Obesity,* 2001.

Chapter 9

1. Health Care Financing Administration, National Health Expenditures; J. Levi, Managed care and public health, *American Journal of Public Health* 90 (2000): 1823–24; M. McCarthy, Fragmented U.S. health care system needs major reform, *The Lancet* 357 (2001): 782; and Centers for Medicare and Medicaid Services. Health care spending reaches $1.6 trillion in 2002. Released January 8, 2004. Available at www.cms.hhs.gov/media/press/release.asp?Counter-935.
2. Ibid.
3. U.S. Department of Health and Human Services, National Health Expenditures, March 2001, as cited by Health Insurance Association of America at www.ahip.org.
4. U. E. Reinhardt, P. S. Hussey, and G. F. Anderson, U.S. health care spending in an international context, *Health Affairs* (2004), 23(3): 10–25.
5. President's Column, A new era for prevention, *The Nation's Health* (January 1990): 2.
6. R. H. Carmona, Remarks before the Joint Economic Committee, U.S. Congress, October 1, 2003; available at www.surgeongeneral.gov.
7. *Health People 2010;* available at www.healthypeople.gov.
8. J. S. Hampl, J. V. Anderson, and R. Mullis, Position of the American Dietetic Association: The role of dietetics professionals in health promotion and disease prevention. *Journal of the American Dietetic Association* 102 (2002): 1680–1687.
9. Ibid.
10. L. W. Sullivan, Healthy People 2000: Promoting health and building a culture of character, *American Journal of Health Promotion* 5 (1990): 5–6.
11. Ibid., p. 6.
12. G. E. A. Dever, *Community Health Analysis,* 2nd ed. (Gaithersburg, MD: Aspen, 1991), p. xiii.
13. National Center for Health Statistics, *Health, United States, 2003* (Hyattsville, MD, 2003).
14. Ibid.
15. Ibid.
16. U.S. Census Bureau, *2004 Health Insurance* (Table HI05).
17. The definitions provided in the margin throughout this section are from *Reimbursement and Insurance Coverage for Nutrition Services* (Chicago, IL: American Dietetic Association, 1991), pp. 121–25.
18. U.S. Census Bureau, *Health Insurance Coverage 2003.* Available at www. census.gov/hhes/hlthins/hlthin03/hlth03asc.html.
19. R. S. Stern, A comparison of length of stay and costs for health maintenance organization and fee-for-service patients, *Archives of Internal Medicine* 149 (1989): 1185–88.
20. National Center for Health Statistics, *Health, United States, 2003.*
21. J. Gabel and coauthors, Health benefits in 2004: Four years of double-digit premium increases take their toll on coverage, *Health Affairs* (2004), 23(5): 200–209.
22. American Dietetic Association, *Medical Nutrition Therapy across the Continuum of Care* (Chicago, IL: American Dietetic Association, 1998), pp. 1–2.
23. S. Wolfe, Handling your health care, *Buyer's Market* 1 (1985): 2.
24. Stern, A comparison of length of stay and costs.
25. HHS Fact Sheet: The State Children's Health Insurance Program (SCHIP), March 2002.
26. L. Shi and D. A. Singh, *Delivering Health Care in America* (Gaithersburg, MD: Aspen Publishers, 2001).
27. Centers for Medicare and Medicaid Services, *Why is CMS in Baltimore?,* available at www.cms.hhs.gov/about/history/whybalto.asp; and Centers for Medicare and Medicaid Services, Department of Health and Human Services, *Who is eligible for Medicare?,* available at medicare.custhelp.com.
28. Centers for Medicare and Medicaid Services and U.S. Department of Health and Human Services, *Medicare and You 2005.* Publication No. CMS–10050. September 2004.
29. Centers for Medicare and Medicaid, Department of Health and Human Services, *What are the Medicare premiums and coinsurance rates for 2005?,* available at medicare.custhelp.com.
30. Ibid.
31. P. Michael, Impact and components of the Medicare MNT benefit, *Journal of the American Dietetic Association* 101 (2001): 1140–41; and M. Ochs, New Medicare changes to expand MNT access in 2002, *Journal of the American Dietetic Association* 102 (2002): 30.
32. Centers for Medicare and Medicaid; available at www.cms.hhs.gov.
33. U.S. Department of Health and Human Services, *The Facts About Upcoming New Benefits in Medicare.* Publication No. CMS-11054. February 17, 2004.
34. Centers for Medicare and Medicaid Services, U.S. Department of Health and Human Services. Centers for Medicare and Medicaid Services, *Your Medicare Benefits.* Publication No. CMS – 10116. July 2004.
35. Centers for Medicare and Medicaid Services, *Your Medicare Benefits;* and Centers for Medicare and Medicaid Services and Department of Health and Human Services, *What Is a Medicare Advantage Plan?,* available at medicare.custhelp.com.
36. U.S. Census Bureau, *2004 Health Insurance* (Table HI05).
37. Centers for Medicare and Medicaid Services, *Medicaid: A Brief Summary;* available at www.cms.hhs.gov/publications/overview-medicare-medicaid/default4.asp.
38. L. Stollman, ed., *Nutrition Entrepreneur's Guide to Reimbursement Success* (Chicago, IL: American Dietetic Association, 1995), pp. 1–3.
39. Institute of Medicine, *America's Health Care Safety Net: Intact but Endangered* (Washington, D.C.: National Academy Press, 2000), pp. 1–20.
40. J. S. Ahluwalia, Health care in the United States: Our dynamic jigsaw puzzle, *Archives of Internal Medicine* 150 (1990): 256–58.

41. Centers for Medicare and Medicaid Services, State Children's Health Insurance Program (SCHIP), available at www.cms.hhs.gov/schip/consumers_default.asp; and H. Ross, State Children's Health Insurance Program, *Closing the Gap* (Washington, D.C.: Office of Minority Health), March 2000.
42. Centers for Medicare and Medicaid Services, State Children's Health Insurance Program (SCHIP), Site Portal; available at www.cms.hhs.gov/schip.
43. Centers for Medicare and Medicaid Services, State Children's Health Insurance Program (SCHIP).
44. Centers for Medicare and Medicaid Services, SCHIP Enrollment Reports; available at www.cms.hhs.gov/schip/enrollment.
45. Consumers Union, Health care today: Who's hurting, *Consumer Reports* 63 (1998): 7.
46. U.S. Census Bureau, *2004 Health Insurance* (Table H105); and L. Shi, The convergence of vulnerable characteristics and health insurance in the U.S., *Social Science & Medicine* 53 (2001): 519.
47. S. A. Schroeder, Prospects for expanding health insurance coverage, *New England Journal of Medicine* 344 (2001): 847–52.
48. H. Deets, Early retirees, others need health insurance, *Modern Maturity* 41 (1998): 82.
49. U.S. Census Bureau, *Children with Health Insurance: 2001* (Washington, D.C.: U.S. Department of Commerce, 2003).
50. B. C. Strunk and P. J. Cunningham, Trends in Americans' access to needed medical care, 2001–2003, *Track Rep.* 10 (August 2004): 1–4.
51. Gabel and coauthors, Health benefits in 2004.
52. U.S. Census Bureau, *Children with Health Insurance: 2001.*
53. U.S. Census Bureau, *Historical Health Insurance Tables;* available at www.census.gov/hhes/hlthins/historic/index.html.
54. Strunk and Cunningham, Trends in Americans' access to needed medical care.
55. Dever, *Community Health Analysis,* p. 13.
56. Legislative Highlights: Finn offers testimony on nutrition and the elderly, *Journal of the American Dietetic Association* 92 (1992): 1064–65.
57. Ibid.
58. Institute of Medicine, Committee on Quality of Health Care in America, *Crossing the Chasm: A New Health System for the 21st Century* (Washington, D.C.: National Academy Press, 2001).
59. Wolfe, Handling your health care, p. 2.
60. Mr. Galley, Editorial, *Boston Globe,* September 1991, as quoted by M. K. Fox, Reimbursement practices and trends, presented at the American Dietetic Association's annual meeting in Dallas.
61. Ibid.
62. The discussion of bureaucracy in government-run health care is from E. Brodsky, Government-run health care isn't worth the wait, *Health and You* (Spring 1992): 2.
63. As stated by J. Neel, Healthcare: U.S. looks to German model, *Nature* 351 (1992): 433.
64. U. E. Reinhardt, P. S. Hussey, and G. F. Anderson, U.S. health care spending in an international context, *Health Affairs* (2004), 23(3): 10–25.
65. Centers for Medicare and Medicaid Services, Health care spending reaches $1.6 trillion in 2002. Released January 8, 2004; available at www.cms.hhs.gov/media/press/release.asp?Counter-935.
66. *Delivering Health Care in America,* 2001.
67. High cost of malpractice insurance; available at www.cbsnews.com/stories/2004/04/02/eveningnews/consumer/main610201.shtml.
68. W. M. Sage, The forgotten third: Liability insurance and the medical malpractice crisis, *Health Affairs* (2004), 23(4): 10–21.
69. The definitions provided in this section are from *Reimbursement and Insurance Coverage for Nutrition Services* (Chicago, IL: American Dietetic Association, 1991), pp. 121–25.
70. R. G. Abramson and coauthors, Generic drug cost containment in Medicaid: Lessons from five state MAC programs, *Health Care Financing Review* (2004), 25: 25–34.
71. The discussion on trends is adapted from M. K. Fox, *Overview of Third-Party Reimbursement* (Chicago, IL: American Dietetic Association, 1991), pp. 3–5.
72. *Delivering Health Care in America,* 2001, pp. 189–90.
73. C. Cowan and coauthors, National health expenditures, 2002, *Health Care Financing Review* (2004), 25: 143–66.
74. *Delivering Health Care in America,* 2001, pp. 207–8.
75. Health care: People's chief concerns. Survey results available at www.publicagenda.org/issues/pcc_detail.cfm?issue_type=healthcare&list=2.
76. P. deVise, *Misuses and Misplaced Hospitals and Doctors: A Locational Analysis of the Urban Health Care Crisis,* Resource Paper no. 2 (Commission on College Geography, 1973), p. 1.
77. President's Commission for the Study of Ethical Problems in Medical and Biomedical and Behavioral Research, *Report: The Ethical Implications of Differences in the Availability of Health Services* (Washington, D.C.: U.S. Government Printing Office, 1983), p. 22.
78. *Healthy People 2010:* The role of family physicians in addressing health disparities, *American Family Physician* 62 (2000): 1971–75; and B. D. Smedley and coauthors, *Unequal Treatment: Confronting Racial and Ethnic Disparities in Health Care* (Washington, D.C.: National Academy Press, 2002).
79. C. Grogan, A comparison of Canada, Britain, Germany, and the United States, *Journal of Health Politics, Policy, and Law* 17 (1992): 213–32.
80. Ibid., p. 226.
81. S. Staveteig and A. Wigton, *Racial and ethnic disparities: Key findings from the National Survey of America's Families* (The Urban Institute New Federalism Series B No B-5, February 2000); and Public Health Service, *Healthy People 2000: Trends in Racial and Ethnic-Specific Rates for the Health Status Indicators, United States, 1990–1998,* Statistical Note No. 23, 2002, pp. 1–16.
82. U.S. Department of Health and Human Services, *Healthy People 2010* (Washington, D.C.: U.S. Government Printing Office, January 2000).
83. Institute of Medicine, *Unequal Treatment: Confronting Ethnic and Racial Disparities in Health Care* (Washington, D.C.: National Academy Press, 2002), available from www.nap.edu; and A. C. Monheit and V. P. Vistnes, Race/ethnicity and health insurance status: 1987 and 1996, *Medical Care Research and Review* 57 (Suppl 1) (2000): 11–35.
84. U.S. Department of Health and Human Services, *The Initiative to Eliminate Racial and Ethnic Disparities in Health* (Washington, D.C.: U.S. Government Printing Office, 1998).
85. The discussion of national health care coverage is adapted from American College of Physicians, Access to health care, *Annals of Internal Medicine* 112 (1990): 641–61.
86. The discussion of U.S. history is adapted from Ahluwalia, *Health Care in the United States,* pp. 256–58.
87. R. K. Johnson and A. M. Coulston, Medicare: Reimbursement rules, impediments, and opportunities for dietitians, *Journal of the American Dietetic Association* 95 (1995): 1378–80.
88. H. Holman, Chronic disease—the need for a new clinical education, *Journal of the American Medical Association* 292 (2004): 1057–59.
89. G. S. Omenn, Challenges facing public health policy, *Journal of the American Dietetic Association* 93 (1993): 643.
90. Division of Government Affairs, Health care reform, *Legislative Newsletter* (October 1991): 2.
91. C. S. Chima and H. A. Pollack, Position of the American Dietetic Association: Nutrition services in managed care, *Journal of the American Dietetic Association* 102 (2002): 1471–78.
92. Ibid.
93. G. H. Brundtland, Nutrition, health, and human rights; available at www.who.int/nut/nutrition1.htm.
94. Nutrition Screening Initiative, *Managing Nutrition Care in Health Plans* (Washington, D.C.: Greer, Margolis, Mitchell, Burns, & Associates, 1996), pp. 1–5.
95. Position of the American Dietetic Association: Cost-effectiveness of medical

nutrition therapy, *Journal of the American Dietetic Association* 95 (1995): 88–91.

96. Position of the American Dietetic Association, Nutrition services in managed care, *Journal of the American Dietetic Association* 96 (1996): 391–95.
97. Position of the American Dietetic Association: Cost-effectiveness of medical nutrition therapy, *Journal of the American Dietetic Association* 95 (1995): 88–91; Position of the American Dietetic Association: Nutrition services in managed care; and J. G. Pastors and coauthors, How effective is medical nutrition therapy in diabetes care? *Journal of the American Dietetic Association* 103 (2003): 827–831.
98. M. D. Simko and M. T. Conklin, Focusing on the effectiveness side of the cost-effectiveness equation, *Journal of the American Dietetic Association* 89 (1989): 485–87.
99. Ibid., p. 486.
100. American College of Physicians, Access to health care, *Annals of Internal Medicine* 112 (1990): 641–61.
101. A. Inman-Felton, K. G. Smith, and E. Q. Johnson, eds., *Medical Nutrition Therapy across the Continuum of Care* (Chicago, IL: American Dietetic Association, 1997).
102. R. Gould, The next rung on the ladder: Achieving and expanding reimbursement for nutrition services, *Journal of the American Dietetic Association* 91 (1991): 1383–84.
103. P. Splett, Effectiveness and cost effectiveness of nutrition care: A critical analysis with recommendations, *Journal of the American Dietetic Association* 91 (1991): S1–S50.
104. M. R. Schiller and C. Moore, Practical approaches to outcomes evaluation, *Topics in Clinical Nutrition* 14 (1999): 1–12; and C. W. Biesemeier, Demonstrating the effectiveness of medical nutrition therapy, *Topics in Clinical Nutrition* 14 (1999): 13–24.
105. G. E. Gray and L. K. Gray, Evidence-based medicine: Applications in dietetic practice, *Journal of the American Dietetic Association* 102 (2002): 1263–72; and Position of the American Dietetic Association, Cost-effectiveness of medical nutrition therapy, *Journal of the American Dietetic Association* 95 (1995): 88.
106. Ibid.
107. Legislative Highlights, Finn offers testimony on nutrition and the elderly, *Journal of the American Dietetic Association* 92 (1992): 1064–65; Institute of Medicine, Committee on Nutrition Services for Medicare Beneficiaries, *The Role of Nutrition in Maintaining Health in the Nation's Elderly: Evaluating Coverage of Nutrition Services for the Medicare Population* (Washington, D.C.: National Academy Press, 2000).
108. P. Michael, Advocacy for coverage of nutrition services, *Journal of the American Dietetic Association* 105 (2005): 701–02.
109. American Dietetic Association's Public Relations Team, *Improving Health and Saving Health-Care Dollars: Medical Nutrition Therapy Works* (Chicago, IL: American Dietetic Association, April 9, 2002); and M. Ochs, New Medicare changes to expand MNT access in 2002, *Journal of the American Dietetic Association* 102 (2002): 30.
110. Community Nutrition Institute, Dietitians pushing a bill for medical nutrition, *Nutrition Week* 27 (May 2, 1997): 2.
111. E. R. Monsen, From the environment to MNT: Dietitians face key issues, *Journal of the American Dietetic Association* 97 (1997): 360.
112. H. Holman, Chronic disease—the need for a new clinical education, *Journal of the American Medical Association* 292 (2004): 1057–1059.
113. L. W. Sullivan, Partners in prevention: A mobilization plan for implementing *Healthy People 2000, American Journal of Health Promotion* 5 (1991): 291–97.
114. A. Hawkins, The future of health care: Well care, *Personal Health Management 97 News* (Summer 1997): 3–55.
115. Public Policy News: Medicare reform offers ADA opportunity to promote MNT, *Journal of the American Dietetic Association* 97 (1997): 378.
116. The list of benefits is from K. Smith, How to argue for nutrition services, *DBC Dimensions* (Fall 1992): 7.

Chapter 10

1. E. Hackman and coauthors, Maternal birthweight and subsequent pregnancy outcome, *Journal of the American Medical Association* 250 (1983): 2016–19.
2. L. H. Allen, Pregnancy and lactation, in B. A. Bowman and R. M. Russell, *Present Knowledge in Nutrition* (Washington, D.C.: International Life Sciences Institute, 2001), pp. 403–25.
3. R. Lewis, New light on fetal origins of adult disease, *The Scientist* 14 (2000): 1, 16.
4. K. M. Godfrey and D. J. Barker, Fetal nutrition and adult disease, *American Journal of Clinical Nutrition* 71 (2000): 1344S–1352S.
5. National Center for Health Statistics, Department of Health and Human Services, *Healthy People 2010 Progress Review: Maternal, Infant, and Child Health* (Hyattsville, MD: Public Health Service, 2003).
6. *National Vital Statistics Reports,* Volume 52, No. 10 (Hyattsville, MD: National Center for Health Statistics, 2002).
7. *Unequal Treatment: Confronting Racial and Ethnic Disparities in Health Care* (Washington, D.C.: National Academy Press, 2002); *Percentage of Low-Birthweight and Very-Low-Birthweight Infants by Race and Hispanic Origin of Mother in the United States, 2002,* National Center for Health Statistics National Vital Statistics Report, Vol. 52, No. 10, available at www.cdc.gov/nchs.
8. U.S. Department of Health and Human Services, *Healthy People 2010* (Washington, D.C.: U.S. Government Printing Office, 2000); and Centers for Disease Control and Prevention, Surveillance Summaries, February 21, 2003, *MMWR* 52 (2003): (No. SS-2).
9. U.S. Department of Health and Human Services, Public Health Service, *Healthy People 2000: National Health Promotion and Disease Prevention Objectives,* DHHS Pub. No. 91-50212 (Washington, D.C.: U.S. Department of Health and Human Services, 1990), pp. 366–90; and *Healthy People 2010,* 2000.
10. Position of the American Dietetic Association, Nutrition and lifestyle for a healthy pregnancy outcome, *Journal of the American Dietetic Association* 102 (2002): 1479–90.
11. National Academy of Sciences, *Nutrition During Pregnancy* (Washington, D.C.: National Academy Press, 1990); and Institute of Medicine, National Academy of Sciences, *Nutrition during Lactation* (Washington, D.C.: National Academy Press, 1991).
12. *Healthy People 2000 Final Review,* 2001.
13. *Healthy People 2010,* 2000, p. 16–9.
14. *Healthy People 2010 Progress Review.*
15. Ibid.
16. *Healthy People 2010,* pp. 16–1 through 16–62.
17. U.S. Department of Health and Human Services, Public Health Service, *Progress Review: Maternal, Infant and Child Health,* October 22, 2003.
18. Position of the American Dietetic Association, Nutrition and lifestyle for a healthy pregnancy outcome, 2002.
19. *Healthy People 2010,* 2000, pp. 16–36.
20. J. E. Brown and coauthors, Development of a prenatal weight gain intervention program using social marketing methods, *Journal of Nutrition Education* 24 (1992): 21–29.
21. Centers for Disease Control and Prevention, *Births, Final Data, 2002* (Washington, D.C.: National Center for Health Statistics, 2002); and National and state specific pregnancy rates among adolescents—United States, 1995–1997, *Morbidity and Mortality Weekly Report* 49 (2000): 605–611.
22. R. J. Trissler, The child within: A guide to nutrition counseling for pregnant teens, *Journal of the American Dietetic Association* 99 (1999): 916–18; and M. Story and J. Stang (eds.), *Nutrition and the Pregnant Adolescent: A Practical Reference Guide* (Washington, D.C.: Maternal and Child Health Bureau, 2000).
23. Ibid.
24. M. S. Bergman, Improving birth outcomes with nutrition intervention, *Topics in Clinical Nutrition* 13 (1997): 74–79.
25. B. Worthington-Roberts and L. Klerman, Maternal nutrition, in *New Perspectives in Prenatal Care,* ed. I. R. Merkatz, J. E. Thompson, and R. Goldenberg (New York: Elsevier Science, 1990).

26. Ibid.
27. The description of head circumference measurement was taken from Whitney, Cataldo, and Rolfes, *Understanding Normal and Clinical Nutrition,* p. 616.
28. Office of Women's Health, *Breastfeeding: HHS Blueprint for Action on Breastfeeding* (Washington, D.C.: Department of Health and Human Services, 2000), pp. 1–21; available at www.4woman.gov.
29. D. C. S. James and B. Dobson, Position of the American Dietetic Association: Promoting and supporting breastfeeding, *Journal of the American Dietetic Association* 105 (2005): 810–18; American Academy of Pediatrics, Policy Statement: Breastfeeding and the use of human milk, *Pediatrics* 115 (2005): 496–506; and L. F. DiGiorgio, Promoting breastfeeding to mothers in the Special Supplemental Nutrition Program for Women, Infants, and Children, *Journal of the American Dietetic Association* 105 (2005): 716–17.
30. *Healthy People 2010.*
31. *Mothers' Survey: Breastfeeding Trends through 2002* (Cleveland, OH: Ross Products Division, Abbott Laboratories, 2002).
32. Position of the American Dietetic Association, Breaking the barriers to breastfeeding, *Journal of the American Dietetic Association* 101 (2001): 1213–20; and M. E. Bentley and coauthors, Sources of influence on intention to breastfeed among African-American women at entry into WIC, *Journal of Human Lactation* 15 (1999): 27–34.
33. World Health Organization, *Evidence for the Ten Steps to Successful Breastfeeding,* Pub. No. WHO/CHD 98.9 (Geneva: WHO, 1998), pp. 1–5; and UNICEF/WHO, *The UNICEF Baby Friendly Hospital Initiative: Ten Steps to Successful Breastfeeding* (New York, NY: UNICEF, 1992).
34. U.S. Department of Agriculture, *Women, Infants, and Children National Breastfeeding Promotion Campaign;* available at www.nal.usda.gov.
35. La Leche League International, Breastfeeding Peer Counselor Program, personal communication, 2002; available at www.lalecheleague.org.
36. B. Dobson, Community-based coalition building for breastfeeding promotion and support, *Nutrition Education for the Public Networking News,* Autumn 1996, 3, 5, 12.
37. The discussion of focus group findings was taken from C. Bryant and D. Bailey, The use of focus group research in program development. Unpublished manuscript, Lexington, Kentucky, 1989.
38. R. E. Kleinman, ed., *Pediatric Nutrition Handbook,* 4th ed. (Elk Grove Village, IL: American Academy of Pediatrics, 1998); and S. J. Fomon, Feeding normal infants: Rationale for recommendations, *Journal of the American Dietetic Association,* 101 (2001): 1002–5.
39. N. Butte and coauthors, The Start Healthy Feeding Guidelines for Infants and Toddlers, *Journal of the American Dietetic Association,* 104 (2004): 442–454.
40. Ibid.
41. M. C. Holst, Developmental and behavioral effects of iron deficiency anemia in infants, *Nutrition Today* 33 (1998): 27–36.
42. J. S. Forsyth and coauthors, Relation of infant diet to childhood health: Seven year followup of cohort of children in Dundee infant feeding study, *British Medical Journal,* January 3, 1998, pp. 27–32.
43. Food Allergy Network, *Information about Food Allergies,* 1999; and H. Sampson, Food allergy, *Journal of the American Medical Association* 278 (1997): 22–26.
44. The Start Healthy Feeding Guidelines for Infants and Toddlers, 2004.
45. Select Panel for the Promotion of Child Health, WIC: The Special Supplemental Food Program for Women, Infants, and Children, in *Better Health for Our Children: A National Strategy,* 2 vols. (Washington, D.C.: U.S. Department of Health and Human Services, 1981), Vol. 2, pp. 57–68.
46. Institute of Medicine, *Dietary Risk Assessment in the WIC Program* (Washington, D.C.: National Academy Press, 2002), pp. 1–167; and A. R. Swensen, L. J. Harnack, and J. A. Ross, Nutritional assessment of women enrolled in the Special Supplemental Program for Women, Infants, and Children (WIC), *Journal of the American Dietetic Association* 101 (2001): 903–8.
47. Institute of Medicine, *Proposed Criteria for Selecting the WIC Food Packages: A Preliminary Report of the Committee to Review the WIC Food Packages* (Washington, D.C.: National Academy Press, 2004), available at www.nap.edu.
48. A. L. Owen and G. M. Owen, Twenty years of WIC: A review of some effects of the program, *Journal of the American Dietetic Association* 97 (1997): 777–82.
49. *Early Intervention: Federal Investments Like WIC Can Produce Savings,* Publication No. 92-18 (Washington, D.C.: US General Accounting Office, 1992); S. Avruch and A. P. Cackley, Savings achieved by giving WIC benefits to women prenatally, *Public Health Reports* 110 (1995): 27–34; and M. Kotelchuck and coauthors, WIC participation and pregnancy outcomes: Massachusetts statewide evaluation project, *American Journal of Public Health* 74 (1994): 1086–89.
50. Food Research and Action Center, *WIC in the States: Twenty-Five Years of Building a Healthier America* (Washington, D.C.: Food Research and Action Center, 1999).
51. D. Rush and coauthors, The national WIC evaluation: Evaluation of the Special Supplemental Food Program for Women, Infants, and Children, *American Journal of Clinical Nutrition* 48 (1988): 412.
52. Mathematica Policy Research, *The Savings in Medicaid Costs for Newborns and Their Mothers from Prenatal Participation in the WIC Program,* Report prepared for the U.S. Department of Agriculture, Washington, D.C., 1990.
53. B. Devaney and A. Schirm, *Infant Mortality among Medicaid Newborns in Five States: The Effects of Prenatal WIC Participation* (Princeton, N.J.: Mathematica Policy Research, 1993).
54. WIC reform means more mothers and children served, *Nutrition Forum Newsletter* (February 1988): 9–11; 77. Food Research and Action Center, *WIC Works: Let's Make It Work for Everyone* (Washington, D.C.: Food Research and Action Center, 1993), pp. 1–19.
55. Community Nutrition Institute, Breastfeeding rates rise significantly among at-risk groups, WIC moms, *Nutrition Week* 27 (21) (1997): 4–6; D. L. Montgomery and P. L. Splett, Economic benefit of breastfeeding infants enrolled in WIC, *Journal of the American Dietetic Association* 97 (1997): 379–85; and Community Nutrition Institute, Food assistance: A variety of practices may lower the costs of WIC, *Nutrition Week* 27 (39) (1997): 4–5.
56. D. Herman and coauthors, The effect of the WIC Program on food security status of pregnant, first-time participants, *Family Economic and Nutrition Review* 16 (2004): 21–29.
57. M. Ponza and coauthors, Nutrient intakes and food choices of infants and toddlers participating in WIC, *Journal of the American Dietetic Association* 104 (2004): S71–S79.
58. Community Nutrition Institute, Working women's access to WIC benefits, *Nutrition Week* 27 (41) (1997): 4–6.
59. P. A. Buescher and coauthors, Prenatal WIC participation can reduce low birthweight and newborn medical costs: A cost–benefit analysis of WIC participation in North Carolina, *Journal of the American Dietetic Association* 93 (1993): 163–66.
60. Community Nutrition Institute, WIC on wheels Winnebago serves a sprawling Vegas, *Nutrition Week* 27 (29) (1997): 6.
61. Select Panel for the Promotion of Child Health, *Better Health for Our Children: A National Strategy,* 2 vols. (Washington, D.C.: U.S. Department of Health and Human Services, 1981), Vol. 1, p. 168.
62. Community Nutrition Institute, Farmers' market program to grow with new funds, *Nutrition Week* 28(2) (1998): 2–3.
63. Ibid., Vol. 1, p. 156.
64. The discussion regarding the Title V Maternal and Child Health Programs is from the Maternal and Child Health Bureau, www.mchb.hrsa.gov.
65. Select Panel for the Promotion of Child Health, *Better Health,* 2:17–22.
66. Ibid., Vol. 2, p. 20.
67. Ibid., Vol. 2, p. 21.
68. Position of the American Dietetic Association, Nutrition services for children with special health care needs, *Journal of the American Dietetic Association* 95 (1995): 809–14.

69. H. H. Cloud, Nutrition services for children with developmental disabilities and special health care needs, *Topics in Clinical Nutrition* 16 (2001): 28–40.
70. Select Panel for the Promotion of Child Health, *Better Health,* Vol. 2, p. 44.
71. Ibid., Vol. 1, p. 165.
72. *Telling the Healthy Start Story: A Report on the Impact of the 22 Demonstration Projects* (Washington, D.C.: National Center for Education in Maternal and Child Health, 1999).
73. R. E. Brennan and M. N. Traylor, Components of nutrition services, in *Call to Action: Better Nutrition for Mothers, Children, and Families,* ed. C. Sharbaugh (Washington, D.C.: National Center for Education in Maternal and Child Health, 1991), pp. 243–55.
74. B. Barber-Madden and coauthors, Nutrition for pregnant and lactating women: Implications for worksite health promotion, *Journal of Nutrition Education* 18 (1986): S72–S75.
75. Worthington-Roberts and Williams, *Nutrition in Pregnancy and Lactation,* pp. 239–40.
76. The list of recommendations was adapted from Worthington-Roberts and Pitkin, Women's nutrition, p. 133.

Chapter 11

1. Federal Interagency Forum on Child and Family Statistics, *America's Children: Key National Indicators of Well-Being, 2002* (Vienna, VA: Health Resources and Services Administration, 2002).
2. U.S. Department of Health and Human Services, *Healthy People 2010* (Washington, D.C.: U.S. Government Printing Office, January 2000).
3. *Healthy People 2000 Final Review* (Hyattsville, MD: Public Health Service, 2001).
4. Centers for Disease Control and Prevention, *Prevalence of Overweight among Children and Adolescents, United States, 1999,* accessed at www.cdc.gov/nchs; and D. Neumark-Sztainer and coauthors, Overweight status and eating patterns among adolescents: Where do youths stand in comparison with the *Healthy People 2010* objectives? *American Journal of Public Health* 92 (2002): 844–51.
5. *HP 2010 Progress Review: Nutrition and Overweight,* January 21, 2004.
6. Ibid.
7. Ibid.
8. U.S. Department of Agriculture, Changes in children's diets: 1989–1991 to 1994–1996, www.fns.usda.gov/oane/MENU/Published/CNP/FILES/changessum.htm; and P. Gleason and C. Suitor, *Food for Thought: Children's Diets in the 1990s* (Princeton, NJ: Mathematica Policy Research, Inc., 2001); R. Rajeshwari and coauthors, Secular trends in children's sweetened-beverage consumption (1973–1994): The Bogalusa Heart Study, *Journal of the American Dietetic Association* 105 (2005): 208–14.
9. P. Basiotis and coauthors, *The healthy eating index: 1999–2000,* U.S. Department of Agriculture, Center for Nutrition Policy and Promotion, CNPP-12.
10. Center for Nutrition Policy and Promotion, Report card on the diet quality of children ages 2 to 9, *Nutrition Insights,* September 2001.
11. U.S. Department of Agriculture, Changes in children's diets.
12. Ibid.
13. Ibid.
14. Ibid.
15. Ibid.
16. Position of the American Dietetic Association: Dietary guidance for healthy children ages 2 to 11 years, *Journal of the American Dietetic Association,* 104 (2004).
17. Ibid.
18. M. Story, D. Neumark-Sztainer, and F. Simone, Individual and environmental influences on adolescent eating behaviors, *Journal of the American Dietetic Association* 102 (2002): S40–S51; L. D. Ritchie and coauthors, Family environment and pediatric overweight: What is a parent to do? *Journal of the American Dietetic Association* 105 (2005): S70–S79.
19. Ibid.
20. Ibid.
21. A. J. Rainville, K. Choi, and D. Brown, Healthy school nutrition environment: A nationwide survey of school personnel, *National Food Service Management Institute Insight*, R-122-03, March 2004.
22. Story, 2002.
23. Ibid.
24. Ibid.
25. Ibid.
26. G. C. Rampersaud and coauthors, Breakfast habits, nutritional status, body weight, and academic performance in children and adolescents, *Journal of the American Dietetic Association* 105 (2005): 743–60.
27. Prospect Associates, Ltd., Environmental scan and audience analysis for phase II of *Eat Smart. Play Hard,* U.S. Department of Agriculture, Food and Nutrition Service (March 2003).
28. Ibid.
29. Ibid.
30. Story, 2002.
31. Environmental scan and audience analysis for phase II of *Eat Smart. Play Hard.* 2003.
32. Ibid.
33. Story, 2002.
34. Teenagers with eating disorders, July 2004.
35. *Prevalence of Overweight among Children and Adolescents*; C. L. Ogden and coauthors, Mean body weight, height, and body mass index, United States 1960–2002, Centers for Disease Control and Prevention, Advance Data.347), October 2004.
36. W. H. Dietz, Health consequences of obesity in youth: Childhood predictors of adult disease, *Pediatrics* 101 (1998): 518–24.
37. Ibid.
38. G. Wang and W. H. Dietz, Economic burden of obesity in youths aged 6 to 17 years: 1979–1999, *Pediatrics* 109 (2002): E81.
39. S. Kirk, B. J. Scott, and S. R. Daniels, Pediatric obesity epidemic: Treatment options, *Journal of the American Dietetic Association* 105 (2005): S44–S51.
40. Dietz, 1998.
41. Ibid.
42. American Diabetes Association, Obesity speeds onset of both types of diabetes in kids, Statistics about Youth & Diabetes; available at www.diabetes.org/diabetes-statistics/children.jsp.
43. Ibid.
44. Ibid.
45. Dietz, 1998.
46. Ibid.
47. Ibid.
48. J. T. Dwyer and coauthors, Predictors of overweight and overfatness in a multiethnic pediatric population, *American Journal of Clinical Nutrition* 67 (1998): 602–10.
49. K. A. Coon and coauthors, Watching television at meals is related to food consumption patterns in children, *Pediatrics* 107 (2001): E7; and C. Byrd-Bredbenner and coauthors, Nutrition messages on prime-time television programs, *Topics in Clinical Nutrition* 16 (2001): 61–72.
50. Office of the Surgeon General, *The Surgeon General's Call to Action to Prevent and Decrease Overweight and Obesity, 2001* (Rockville, MD: Department of Health and Human Services, 2001).
51. Ibid.
52. The general descriptions of assessment methods were adapted from E. N. Whitney and S. R. Rolfes, *Understanding Nutrition,* 9th ed. (Belmont, CA: Wadsworth/Thomson Learning, 2002), p. 535.
53. Position of the American Dietetic Association: Dietary guidance for healthy children ages 2 to 11 years, 2004.
54. Maternal and Child Health Bureau, *Child Health USA, 2001* (Washington, D.C.: U.S. Government Printing Office, 2001).
55. P. C. DuRousseau and coauthors, Children in foster care: Are they at nutritional risk? *Journal of the American Dietetic Association* 91 (1991): 83–85; and M. L. Taylor and S. A. Koblinsky, Dietary intake and growth status of young homeless children, *Journal of the American Dietetic Association* 93 (1993): 464–66.
56. D. J. Sherman, The neglected health care needs of street youth, *Public Health Reports* 107 (1992): 433–40.
57. U.S. Department of Agriculture, Economic Research Service, *Household Food*

Security in the United States, 2003, Publication FANRR-42.
58. Ibid.
59. Ibid.
60. Sherman, 1992.
61. D. Neumark-Sztainer and coauthors, Primary prevention of disordered eating among preadolescent girls: Feasibility and short-term effect of a community-based intervention, *Journal of the American Dietetic Association* 100 (2000): 1466–73; see also P. Packard and K. Stanek Krogstrand, Half of rural girls aged 8 to 17 years report weight concerns and dietary changes, with both more prevalent with increased age, *Journal of the American Dietetic Association* 102 (2002): 672–77; and L. L. Birch and J. P. Fischer, Development of eating behaviors among children and adolescents, *Pediatrics* 101 (1998): 539–49.
62. Youth Risk Survey.
63. American Academy of Child & Adolescent Psychiatry, Teenagers with eating disorders (July 2004), accessed December 11, 2004, from www.aacap.org/publications/factsfam/eating/htm.
64. D. L. Bogen and R. C. Whitaker, Anemia screening in the special supplemental nutrition program for women, infants, and children: Time for change? *Archives of Pediatrics and Adolescent Medicine,* 156 (2002): 969–70; S. Saloojee and J. M. Pettifor, Iron deficiency and impaired child development, *British Medical Journal,* 323 (2001): 1377–78.
65. *Healthy People 2010.*
66. Centers for Disease Control and Prevention, Recommendations to prevent and control iron deficiency in the United States, *Morbidity and Mortality Weekly Report* 47, No. RR-3 (1998): 1–29.
67. *Healthy People 2000 Final Review,* 2001.
68. Bogan, 2002.
69. L. Brabin and B. J. Brabin, The cost of successful adolescent growth and development in girls in relation to iron and vitamin A status, *American Journal of Clinical Nutrition* 55 (1992): 955–58.
70. National Cattlemen's Beef Association, *Iron in Human Nutrition.* (Chicago, IL: National Cattlemen's Beef Association, 1998), p. 24; and L. Hallberg and L. Rossander-Hultén, Iron requirements in menstruating women, *American Journal of Clinical Nutrition* 54 (1991): 1047–58.
71. Ibid.
72. The discussion of high blood lead levels was adapted from Centers for Disease Control and Prevention, *Health, United States, 2001,* and Position of the American Dietetic Association, Dietary guidance for healthy children aged 2 to 11 years, 2004.
73. Center for Disease Control, Fact Sheet, *Preventing Dental Caries,* October 2002; available at www.cdc.gov/OralHealth/factsheets/dental_caries.htm.
74. Ibid.
75. C. Knittel, Prevention task force backs fluoridation, sealants, *Journal of the California Dental Association,* 30 (2002): 813–14.
76. Ibid.
77. Ibid.
78. *Healthy People 2000 Final Review,* 2001; and Centers for Disease Control and Prevention, *Recommendations for Using Fluoride to Prevent and Control Dental Caries in the United States,* August 2001; available at www.cdc.gov/mmwr.
79. Knittel, 2002.
80. Ibid.
81. W. Bao and coauthors, Usefulness of childhood low-density lipoprotein cholesterol level in predicting adult dyslipidemia and other cardiovascular risks: The Bogalusa Heart Study, *Archives of Internal Medicine* 156 (1996): 1315–320.
82. National Cholesterol Education Program, *Report of the Expert Panel on Blood Cholesterol Levels in Children and Adolescents,* NIH Pub. No. 91-2732 (Washington, D.C.: U.S. Department of Health and Human Services, 1991), pp. 1–22; and American Academy of Pediatrics, Committee on Nutrition, Cholesterol in childhood, *Pediatrics* 101 (1998): 141–47.
83. R. E. Kreipe and C. P. Dukarm, Eating disorders in adolescents and older children, *Pediatric Reviews* 20 (1999): 410–20; Position of the American Dietetic Association, Nutrition intervention in the treatment of anorexia nervosa, bulimia nervosa, and eating disorders not otherwise specified, *Journal of the American Dietetic Association* 101 (2001): 810; and L. A. Lytle, Nutritional issues for adolescents, *Journal of the American Dietetic Association* 102 (2002): S8–S12.
84. American Academy of Child & Adolescent Psychiatry, Teenagers with eating disorders, July 2004.
85. Ibid.
86. Position of the American Dietetic Association, Providing nutrition services for infants, children, and adults with developmental disabilities and special health care needs, *Journal of the American Dietetic Association* 104 (2004).
87. U.S. Department of Agriculture, Accommodating Children with Special Dietary Needs in the School Nutrition Programs.
88. Position of the American Dietetic Association, Providing nutrition services for infants, children, and adults with developmental disabilities and special health care needs, 2004.
89. Ibid.
90. H. H. Cloud, Nutrition services for children with developmental disabilities and special health care needs, *Topics in Clinical Nutrition* 16 (2001): 28–40.
91. L. A. Wodarski, An interdisciplinary nutrition assessment and intervention protocol for children with disabilities, *Journal of the American Dietetic Association* 90 (1990): 1563–68.
92. Position of the American Dietetic Association, Providing nutrition services for infants, children, and adults with developmental disabilities and special health care needs, 2004.
93. *Healthy People 2010.*
94. The quotation was cited in J. E. Austin and C. Hitt, *Nutrition Intervention in the United States* (Cambridge, MA: Ballinger, 1979), p. 93.
95. U.S. Department of Agriculture, Food and Nutrition Service, Fact Sheet, *School Lunch Program,* accessed November 26, 2004, from www.fns.usda.gov/cnd/lunch/aboutlunch/NSLPfactSheet.htm.
96. U.S. Department of Agriculture, Food and Nutrition Service, Fact Sheet, *School Breakfast Program,* accessed November 26, 2004, from www.fns.usda.gov/cnd/breakfast/aboutBFast/bfastfacts.htm.
97. Minnesota Department of Children, Families & Learning, Fast break to learning; School breakfast program: A report of the third year results, 2001–2002.
98. Maryland State Department of Education, School and Community Nutrition Program, Findings from year III of the Maryland meals for achievement classroom breakfast pilot program.
99. U.S. Department of Agriculture, Food and Nutrition Service, Fact Sheet, *After School Snack Programs.*
100. U.S. Department of Agriculture, Food and Nutrition Service, Fact Sheet, *Summer Food Service Programs.*
101. U.S. Department of Agriculture, Food and Nutrition Service, Evaluation of the 14-state summer food service program pilot project; available at www.fns.usda.gov/oane/MENU/Published/CNP/FILES/SFSPPilot.htm.
102. Department of Agriculture, Food and Nutrition Service, Fact Sheet, Child nutrition commodity programs from www.fns.usda.gov/fdd/programs.
103. Information about the DHHS's Head Start Program was adapted from www.acf.dhhs.gov/programs/hsb.
104. Information about the DHHS's Early Head Start Program was adapted from www.acf.dhhs.gov/programs/hsb.
105. U.S. Department of Agriculture, The school meals initiative implementation study third year report; available at www.fns.usda.gov/oane/MENU/Published/CNP/FILES/SMIYear3.htm.
106. Changes in children's diets.
107. Ibid.
108. The school meals initiative implementation study third-year report.
109. Ibid.
110. S. B. Templeton, M. A. Marlette, and M. Panemangalore, Competitive foods increase the intake of energy and decrease the intake of certain nutrients by adolescents consuming school lunch, *Journal of*

the *American Dietetic Association,* 105 (2005): 215–20.

111. K. Anderson and coauthors, Eat smart: North Carolina's recommended standards for all foods available in school. North Carolina Department of Health and Human Services, North Carolina Division of Public Health, 2004.
112. Texas Department of Agriculture. Texas public school nutrition policy, revised, August 1, 2004, accessed November 29, 2004, from Healthykids@agr.state.tx.us.
113. *The Surgeon General's Call to Action to Prevent and Decrease Overweight and Obesity,* 2001.
114. S. M. Gross and B. Cinelli, Coordinated school health program and dietetics professionals: Partners in promoting healthful eating, *Journal of the American Dietetic Association* 104 (2004): 793–98.
115. U.S. Department of Agriculture, Evaluation of the USDA fruit and vegetable pilot program: Report to congress, Economic Research Service Publication, E-FAN 03-006, April 2003.
116. U.S. Department of Agriculture, Fact Sheet. Join the team—becoming a team nutrition school; available at www.fns.usda.gov/tn/join/index.htm.
117. Information about USDA's *YourSELF* program was obtained from the USDA's website at www.usda.gov.
118. Anderson, 2004.
119. Produce for Better Health, 5 a Day Newsletter, Fruit/Veggie Snacks in school are huge success; available at www.5aday.com/html/press/pressrelease.php?recordid=67.
120. Centers for Disease Control, Youth Media Campaign. VERB, It's What You Do.
121. National Osteoporosis Foundation, National Bone Campaign, Powerful girls have powerful bones; available at www.nof.org/powerfulbones.
122. American Heart Association Schoolsite Program (Dallas: National Center, American Heart Association).
123. American Cancer Society "Changing the Course" Program (Minneapolis, MN: Minnesota Division, American Cancer Society).
124. Information about the Kids Café program was adapted from www.secondharvest.org and Feeding Children Better Organization, Kids Café, available at www.feedingchildrenbetter.org/pages/ourmission/kidscafe/index.jsp.
125. J. Brooke, Indians revive a running tradition, *New York Times,* August 2, 1998, p. 16.
126. National Clearinghouse on Families and Youth, *Reconnecting Youth and Community: A Youth Development Approach,* July 1996; S. Hamdan and coauthors, Perceptions of adolescents involved in promoting lower-fat foods in schools: Associations with level of involvement, *Journal of the American Dietetic Association* 105 (2005): 247–51.
127. Select Panel for the Promotion of Child Health, Vol. 1, p. 2.

Chapter 12

1. National Center for Health Statistics, *Health United States, 2004* (Hyattsville, MD: U.S. Department of Health and Human Services, December 2004).
2. A. E. Harper, Nutrition, aging, and longevity, *American Journal of Clinical Nutrition* (Suppl.) 36 (October 1982): 737–49; the margin entry is from Steve Dickenson, Copley News. Reprinted with permission.
3. Administration on Aging, *Profile of Older Americans: 2003* (Hyattsville, MD: U.S. Department of Health and Human Services, 2004).
4. Federal Interagency Forum on Aging-Related Statistics, *Older Americans 2004: Key Indicators of Well-Being* (Washington, D.C.: U.S. Government Printing Office, 2004); K. Davis and S. Raetzman, Meeting future health and long-term care needs of an aging population, *The Commonwealth Fund,* December 1999; and American Dietetic Association, *Nutrition Care of the Older Adult,* 2nd ed. (Chicago: American Dietetic Association, 2004).
5. Administration on Aging, *Profile of Older Americans, 1997* (Hyattsville, MD: U.S. Department of Health and Human Services, 1997).
6. AHA Conference Proceedings, Unified Dietary Recommendations, Summary of a Scientific Conference on Preventive Nutrition: Pediatrics to Geriatrics, *Circulation* 100 (1999): 450.
7. National Cholesterol Education Program, *Report of the Expert Panel on Detection, Evaluation, and Treatment of High Blood Cholesterol in Adults* (Bethesda, MD: National Institutes of Health, 2001); NHLBI, Facts about the *DASH Eating Plan,* May 2003; and *Dietary Guidelines for Americans,* 2005.
8. U.S. Department of Health and Human Services. *Healthy People 2010* (Washington, D.C.: U.S. Government Printing Office, 2000).
9. The following discussion is adapted from *Healthy People 2010 Progress Review: Nutrition and Overweight,* January 21, 2004, and *Progress Review: Physical Activity and Fitness,* April 14, 2004.
10. The Boomer Report, Vol. II, Nos. 4, 5, 6, and 7, 1990; and S. T. Borra, Food and nutrition education for baby boomers: Challenges and opportunities, *Nutrition News,* 54(2) 1991: 5–6; and American Society on Aging, *America the Wise: Boomers, Elders, and the Longevity Revolution.*
11. L. J. Hughes and coauthors, Expanding nutrition opportunities from hospital into the community: The role of cooperative extension, *Topics in Clinical Nutrition* 15 (2000): 10–18.
12. *Report of the Expert Panel on Detection, Evaluation, and Treatment of High Blood Cholesterol in Adults,* 2001.
13. The National 5 a Day for Better Health Program; available at www.5aday.gov.
14. K. Lancaster and coauthors, Evaluation of a nutrition newsletter by older adults, *Journal of Nutrition Education* 29 (1997): 145–51.
15. Finding a Weight Loss Program that Works for You, *Food Insight,* November/December 2004; available at www.ific.org.
16. A. Eldridge and coauthors, Development and evaluation of a labeling program for low-fat foods in a discount department store foodservice area, *Journal of Nutrition Education* 29 (1997): 159–61.
17. Food Marketing Institute, Supermarket Consumer Affairs, *Directory of Nutrition and Health Programs* (Washington, D.C.: Food Marketing Institute, 1997).
18. R. R. Wing and J. O. Hill, Successful weight loss maintenance, *Annual Review of Nutrition* 21 (2001): 323; J. O. Hill, H. Thompson, and H. Wyatt, Weight maintenance: What's missing? *Journal of the American Dietetic Association* 105 (2005): S63–S66.
19. Public Health Service, *Making a Difference in Worksites* (Atlanta, GA: Centers for Disease Control and Prevention, 1995); and M. P. O'Donnell, Health Promotion in the Workplace, 3rd ed. (Albany, NY: Delmar/Thomson Publishing, 2002).
20. R. W. Jeffery, The healthy worker project, *American Journal of Public Health* 83 (1993): 395–401.
21. E. Larson, MNT: An innovative employee-friendly benefit that saves, *Journal of the American Dietetic Association* 101 (2001): 24–26.
22. The American Dietetic Association and U.S. Public Health Service, *Worksite Nutrition: A Guide to Planning, Implementation, and Evaluation* (Chicago: American Dietetic Association, 1993).
23. M. F. Kuczmarski and D. O. Weddle, Position paper of the American Dietetic Association: Nutrition across the spectrum of aging, *Journal of the American Dietetic Association* 105 (2005): 616–33.
24. Administration on Aging, *A Profile of Older Americans, 2003* (Hyattsville, MD: U.S. Department of Health and Human Services, 2003).
25. E. L. Smith, P. E. Smith, and C. Gilligan, Diet, exercise, and chronic disease patterns in older adults, *Nutrition Reviews* 46 (1998): 52–61.
26. J. F. Guthrie and B. H. Lin, Overview of the diets of lower- and higher-income elderly and their food assistance options, *Journal of Nutrition Education and Behavior* 34 (2002): S31–S41; and L. T. Lee, W. M. Drake, and D. L. Kendler, Intake of calcium and vitamin D in three Canadian long-term care facilities, *Journal of the American Dietetic Association* 102 (2002): 244–247.
27. Center for Nutrition Policy and Promotion, A focus on nutrition for the elderly: It's

time to take a closer look, *Food Insights*, July 1999; and L. McBean and coauthors, Healthy eating in later years, *Nutrition Today* 36 (2001): 192–201.
28. American Society for Nutritional Science, Modified food guide pyramid for people over seventy years of age, *Journal of Nutrition* 129 (1999): 751–53.
29. *Mortality & Morbidity Weekly Report*, 2003.
30. K. M. Mehta, K. Yaffe, and K. E. Covinsky, Cognitive impairment, depressive symptoms, and functional decline in older people, *Journal American Geriatric Society* 50 (2002):1045–50; and H. A. Wygaard and G. Albreksten, Risk factors for admission to a nursing home: A study of elderly people receiving home nursing, *Scandinavian Journal of Primary Health Care* 10 (1992):128–33.
31. M. Bernard, V. Lampley-Dallas, and L. Smith, Common health problems among minority elders, *Journal of the American Dietetic Association* 97 (1997): 771–76; and M. L. Lopez and coauthors, Building educational partnerships to serve Latinos in central California, *Journal of Family and Consumer Sciences,* Summer 1997.
32. *Older Americans 2004: Key Indicators of Well-Being,* 2004.
33. Ibid. *Nutrition Care of the Older Adult*, 2004; and R. S. Wold and coauthors, Increasing trends in elderly persons' use of nonvitamin, nonmineral dietary supplements and concurrent use of medications, *Journal of the American Dietetic Association* 105 (2005): 54–64.
34. American Dietetic Association, White paper: Public Policy Strategies for Nutrition and Aging; available at www.eatright.org/Public/GovernmentAffairs/98_12031.cfm.
35. M. Nestle and J. A. Gilbride, Nutrition policies for health promotion in older adults: Education priorities for the 1990s, *Journal of Nutrition Education* 22 (1990): 316.
36. Position paper of the American Dietetic Association: Nutrition across the spectrum of aging, 2005.
37. *A Profile of Older Americans, 2003;* Institute of Medicine, Committee on Nutrition Services for Medicare Beneficiaries, *The Role of Nutrition in Maintaining Health in the Nation's Elderly: Evaluating Coverage of Nutrition Services for the Medicare Population* (Washington, D.C.: National Academy Press, 1999).
38. K. Gray-Donald, The frail elderly: Meeting the nutritional challenges, *Journal of the American Dietetic Association* 95 (1995): 538–40.
39. Nutrition Screening Initiative, *A Physician's Guide to Nutrition in Chronic Disease Management for Older Adults* (Washington, D.C.: Nutrition Screening Initiative, 2002).
40. M. A. Hess, President's page: ADA as an advocate for older Americans, *Journal of the American Dietetic Association* 91 (1991): 847–49.
41. J. V. White and coauthors, Nutrition Screening Initiative: Development and implementation of the public awareness checklist and screening tools, *Journal of the American Dietetic Association* 92 (1992): 163–67.
42. The discussion of tools used by the Nutrition Screening Initiative was adapted from White and coauthors, Nutrition Screening Initiative, pp. 163–67.
43. *Nutrition Care of the Older Adult*, 2004; see also Mini nutritional assessment: A practical assessment tool for grading the nutritional status of elderly patients, available at www.mna-elderly.com/index.htm.
44. W. C. Chumlea, A. F. Roche, and M. L. Steinbaugh, Estimating stature from knee height for persons 60 to 90 years of age, *Journal of the American Geriatric Society* 33 (1985): 116–20.
45. M. M. Henshaw and J. M. Calabrese, Oral health and nutrition in the elderly, *Nutrition in Clinical Care* 4 (2001): 34–42.
46. J. Weinberg, Psychologic implications of the nutritional needs of the elderly, *Journal of the American Dietetic Association* 60 (1972): 293–96.
47. N. Wellman, L. Y. Rosenzweig, and J. L. Lloyd, Thirty years of the Older American Nutrition Program, *Journal of the American Dietetic Association* 102 (2002): 348–50; and *Food Assistance: Options for Improving Nutrition for Older Americans* (Washington, D.C.: U.S. General Accounting Office, August 2000), GAO/RCED-00-238.
48. S. Saffel-Shrier and B. M. Athas, Effective provision of comprehensive nutrition case management for the elderly, *Journal of the American Dietetic Association* 93 (1993): 439–44; and L. Rhee, N. Wellman, V. Castellanos, and S. Himburg, Continued need for increased emphasis on aging in dietetics education, *Journal of the American Dietetic Association* 104 (2004): 645–649.
49. *Nutrition Care of the Older Adult*, 2004.
50. R. M. Morrison, The Graying of America, *Food Review,* Vol. 25, No. 2 (Washington, D.C.: Economic Research Service, 2002).
51. Administration on Aging, Department of Health and Human Services, Older Americans Act; available at www.aoa.gov/about/legbudg/oaa/legbudg_oaa.asp, accessed February 2005.
51. R. Voelker, Federal program nourishes poor elderly, *Journal of the American Medical Association* 278 (1997): 1301.
52. D. A. Roe, Development and current status of home-delivered meals programs in the United States: Who is served? *Nutrition Reviews* 48 (1990): 181–85.
53. N. S. Wellman and coauthors, Elder insecurities: Poverty, hunger, and malnutrition, *Journal of the American Dietetic Association* 97 (1997): S120–S122.
54. Mathematica Policy Research, Inc., *Serving Elders at Risk, The Older Americans Act Nutrition Programs: National Evaluation of the Elderly Nutrition Program 1993–1995, Volume 1: Title III, Evaluation Findings* (Washington, D.C.: U.S. Department of Health and Human Services, 1996); and B. E. Millen and coauthors, The Elderly Nutrition Program: An effective national framework for preventive nutrition interventions, *Journal of the American Dietetic Association* 102 (2002): 234–40.
55. D. L. Edwards and coauthors, Home-delivered meals benefit the diabetic elderly, *Journal of the American Dietetic Association* 93 (1993); 585–87; D. L. MacLellan, Contribution of home-delivered meals to the dietary intake of the elderly, *Journal of Nutrition for the Elderly* 16 (1997): 23–27; and E. Fogler-Levitt and coauthors, Utilization of home-delivered meals by recipients 75 years of age or older, *Journal of the American Dietetic Association* 95 (1995): 552.
56. L. Y. Yamaguchi and coauthors, Improvement in nutrient intake by elderly meals-on-wheels participants receiving a liquid nutrition supplement, *Nutrition Today* 33 (1998): 37–44.
57. *Serving Elders at Risk, The Older Americans Act Nutrition Programs: National Evaluation of the Elderly Nutrition Program,* 1996.
58. D. L. McKay and coauthors, The impact of home-delivered meals on elderly patients, *Care Management* 6 (2000): 32–36; N. Fey-Yensan and coauthors, Food safety risk identified in a population of elderly home-delivered meal participants, *Journal of the American Dietetic Association* 101 (2001): 1055–57; and N. Wellman, Nutrition 2030: The Elderly Nutrition Program: Contributing to the health and independence of older adults, *A White Paper on Lessons from Federal Nutrition Assistance Programs,* December 1999, available from National Policy and Resource Center on Nutrition and Aging at www.fiu.edu.
59. E. A. Gollub and D. O. Weddle, Improvements in nutritional intake and quality of life among frail homebound older adults receiving home-delivered breakfast and lunch, *Journal of the American Dietetic Association* 104 (2004): 1227–35; and M. B. Moran, Challenges in the meals on wheels program, *Journal of the American Dietetic Association* 104 (2004): 1219–21.
60. N. Fey-Yensan, C. English, and H. R. Museler, *Nutrition to Go:* A nutrition newsletter for older, limited-resource, home-delivered-meal participants, *Journal of Nutrition Education and Behavior* 34 (2002): S69–S70.
61. M. L. Watts, Improving nutrition for American seniors: A new look at the Older Americans Act, *Journal of the American Dietetic Association* 105 (2005): 527–29.
62. Thirty years of the Older American Nutrition Program, 2002.
63. The information regarding the Meals On Wheels Association of America is from its website at www.mowaa.org.
64. C. C. Rogers, Changes in the Older Population and Implications for Rural Areas,

Rural Development Research Report No. 190 (Washington, D.C.: Economics Research Service, 1999); and D. Bower, *Rural America*, Vol. 17, No. 2 (Washington, D.C.: Economic Research Service, Summer 2002).
65. E. D'urso-Fischer and coauthors, Reaching out to the elderly, *Journal of Nutrition Education* 23 (1991): 20–25.
66. Ibid., pp. 20–21.
67. The information regarding the Food Marketing Institute's *To Your Health* Program is available at www.fmi.org.
68. J. W. McClelland and coauthors, Extending the reach of nutrition education for older adults: Feasibility of a train-the-trainer approach in congregate nutrition sites, *Journal of Nutrition Education and Behavior* 34 (2002): S48–S52; and C. J. Rainey and K. L. Cason, Nutrition interventions for low-income, elderly women, *American Journal of Health Behavior* 25 (2001): 245–51.
69. C. M. Wellington and B. A. Piet, Living younger: A lifestyle and nutrition program for seniors 54 and better, *Journal of the American Dietetic Association* 101 (2001): A-79; and N. Sahyoun, Nutrition education for the healthy elderly population: Isn't it time? *Journal of Nutrition Education and Behavior* 34 (2002): S42–S47.
70. The list of advantages is adapted from D. A Roe, *Geriatric Nutrition,* 3rd ed. (Upper Saddle River, NJ: Prentice-Hall, 1992), pp. 1–9.

Chapter 13

1. United Nations Children's Fund, *The State of the World's Children 2005* (New York: UNICEF, 2005).
2. Economic and Social Department, *The State of Food Insecurity in the World 2004* (Rome: Food and Agriculture Organization, 2004).
3. P. A. Sanchez and M. S. Swaminathan, Millenium Project: Hunger in Africa: The link between unhealthy people and unhealthy soils, *Lancet*, 365 (2005): 442–444.
4. United Nations Administrative Committee on Coordination Sub-Committee on Nutrition (ACC/SCN) *Fifth Report on the World Nutrition Situation,* Geneva: ACC/SCN in collaboration with the International Food Policy Research Institute (IFPRI), 2004; and J. R. Lupien, Hunger after the millennium: Perspectives and demands, *Nutrition Today* 37 (2002): 96–102.
5. L. R. Brown, *Who Will Feed China? Wake Up Call for a Small Planet* (New York: Norton, 1995); and United Nations Population Fund, *The State of World Population, 2001* (New York: United Nations Population Fund, 2001).
6. A. Durning, Life on the brink, *World Watch* 3 (1990): 22–30.
7. *State of Food Insecurity in the World 2004.*
8. Bread for the World Institute, *Hunger 2002* (Washington, D.C.: Bread for the World Institute, 2002).
9. *State of Food Insecurity in the World 2004.*
10. B. Hartman and J. Boyce, *Quiet Violence: View from a Bangladesh Village* (San Francisco: Institute for Food and Development Policy, 1983).
11. P. Uvin, The state of world hunger, *Nutrition Reviews* 52 (1994): 151–61.
12. Bread for the World Institute, *Hunger 1998: Hunger in a Global Economy* (Silver Spring, MD: Bread for the World Institute, 1997).
13. *State of Food Insecurity in the World 2004.*
14. UNICEF, 2005.
15. Ibid.
16. K. A. Annan, *We the Children: Meeting the Promises of the World Summit for Children* (New York: United Nations, 2001); and UNICEF, 2005.
17. *State of Food Insecurity in the World 2004.*
18. C. D. Williams, N. Baumslag, and D. B. Jelliffe, *Mother and Child Health: Delivering the Services,* 3rd ed. (New York: Oxford University Press, 1994).
19. *State of Food Insecurity in the World 2004.*
20. Ibid.
21. UNICEF, 2005.
22. Ibid.
23. UNICEF, *State of the World's Children* (London: Oxford University Press, 1989), p. 82.
24. *State of Food Insecurity in the World 2004.*
25. Ibid.
26. Ibid.
27. Ibid.
28. UNICEF, 2005.
29. Independent Commission on International Issues, *North-South: A Program for Survival* (Cambridge, MA: MIT Press, 1980), pp. 49–50.
30. U.S. National Committee for World Food Day, *World Food Summit: Promises and Prospects,* Fourteenth Annual World Food Day Teleconference materials, 1997.
31. HIPC, 2004.
32. E. R. Shaffer and coauthors, Global trade and public health, *American Journal of Public Health* 95 (2005): 23–34.
33. G. Kent, Food trade: The poor feed the rich, *Food and Nutrition Bulletin* 4 (1982): 25–33.
34. UN Millenium Project, Task Force on Hunger, Investing in development: A practical guide to achieve the Millenium Development Goals (London, Sterling, VA: Earthscan, 2005).
35. F. M. Lappe and J. Collins, *Food First: Beyond the Myth of Scarcity* (Boston: Houghton Mifflin, 1978), p. 15; and Bread for the World Institute, *Hunger 1996: Countries in Crisis* (Silver Spring, MD: Bread for the World Institute, 1995).
36. Interreligious Taskforce on U.S. Food Policy, *Identifying a Food Policy Agenda for the 1990s: A Working Paper* (Washington, D.C.: Interreligious Taskforce on U.S. Food Policy, 1989), pp. 1–30.
37. Brown, *Who Will Feed China?* The list in the margin is from Hunger after the millennium: Perspectives and demands, 2002.
38. J. Kocher, Not too many but too little, in J. D. Gussow, *The Feeding Web: Issues in Nutritional Ecology* (Palo Alto, CA: Bull, 1978), pp. 81–83.
39. Bread for the World Institute, *Hunger 1997: What Governments Can Do* (Silver Spring, MD: Bread for the World Institute, 1996).
40. International Food Policy Research Institute, *Feeding the World, Preventing Poverty, and Protecting the Earth: A 2020 Vision* (Washington, D.C.: IFPRI, 1996).
41. Bread for the World Institute, *Hunger 1998.*
42. G. Gardner and B. Halweil, *Underfed and Overfed: The Global Epidemic of Malnutrition* (Washington, D.C.: Worldwatch Institute, 2000); and *Fourth Report on the World Nutrition Situation,* 2000, as cited in Position of the American Dietetic Association, 2003.
43. E. O'Kelly, Appropriate technology for women, *Development Forum* (June 1984): 2.
44. T. Gura, New genes boost rice nutrients, *Science* 285 (1999): 994–95.
45. B. C. Babcock and C. A. Francis, Solving global nutrition challenges requires more than new biotechnologies, *Journal of the American Dietetic Association* 100 (2000): 1308–11.
46. *State of Food Insecurity in the World 2004.*
47. G. Gardner, *Shrinking Fields: Cropland Loss in a World of Eight Billion* (Washington, D.C.: Worldwatch Institute, 1996).
48. L. Brown, *State of the World Report 1990* (Washington, D.C.: Worldwatch Institute, 1990).
49. M. Boyle and L. Aomari, Position of the American Dietetic Association, Addressing world hunger, malnutrition, and food insecurity, *Journal of the American Dietetic Association* 103 (2003): 1046–57.
50. *Oxfam America News,* Fall 1992, p. 8.
51. Mobilization for nutrition: Results from Iringa, *Mothers and Children: Bulletin on Infant Feeding and Maternal Nutrition* 8 (1989): 1–3; Improving child survival and nutrition, *Evaluation Report: Joint WHO/UNICEF Nutrition Support Program in Iringa, Tanzania* (United Republic of Tanzania: WHO/UNICEF, 1989).
52. *Evaluation Report: Joint WHO/UNICEF Nutrition Support Program.*
53. Durning, Life on the brink, p. 29.
54. World Declaration on Nutrition, *Nutrition Reviews* 51 (1993): 41–43.
55. L. Miring and C. R. Mumaw, Needs assessment for in-service training for community nutrition educators in the Kiambu district in Kenya, *Journal of Nutrition Education* 25 (1993): 70–73.
56. Food and Agriculture Organization, *The Sixth World Food Survey* (Rome: FAO

Statistical Analysis Service, 1996); and FAO, *World Food Summit Papers* (Rome: FAO, 1996).

57. V. R. Young, Good nutrition for all: Challenge for the nutritional sciences in the new millennium, *Nutrition Today* 36 (2001): 6–16.
58. Bread for the World Institute, *Hunger* 2002 (Washington, D.C.: Bread for the World Institute, 2002).
59. *State of Food Insecurity in the World 2004.*
60. Ibid.
61. Ibid.
62. S. Lewis, Food security, environment, poverty, and the world's children, *Journal of Nutrition Education* 24 (1992): 3S–5S; and C. L. Detweiler, Baby weighings and village folklore groups, *Journal of the American Dietetic Association* 103 (2003): 959–61.
63. UNICEF, *The State of the World's Children 1998* (New York: Oxford University Press, 1998).
64. Williams et al., *Mother and Child Health.*
65. K. A. Annan, *We the Children: Meeting the Promises of the World Summit for Children.*
66. World Health Organization, *Feeding and Nutrition of Infants and Young Children* (Copenhagen: World Health Organization, 2000).
67. L. Robertson, Breastfeeding practices in maternity wards in Swaziland, *Journal of Nutrition Education* 23 (1991): 284–87.
68. UNICEF, 2005.
69. World Health Organization, *Evidence for the Ten Steps to Successful Breastfeeding,* Pub. No. WHO/CHD 98.9, (Geneva: WHO, 1998), pp. 1–5; and UNICEF/WHO, *The UNICEF Baby-Friendly Hospital Initiative: Ten Steps to Successful Breastfeeding,* (New York: UNICEF, 1992).
70. The discussion of HIV and breastfeeding is adapted from Position of the American Dietetic Association, Addressing world hunger, malnutrition, and food insecurity.
71. American Academy of Pediatrics, Policy Statement: Breastfeeding and the use of human milk, *Pediatrics,* 115 (2005): 496–506.
72. Joint United Nations Programme on HIV/AIDS, *Report of the Global HIV/AIDS Epidemic* (Geneva: UNAIDS, June 2000); and World Health Organization, HIV and Infant Feeding (Geneva: WHO, 1998).
73. United Nations Development Programme, *Human Development Report 2003* (New York: Oxford University Press, 2003); and J. R. Lupien, Hunger after the millennium: Perspectives and demands, *Nutrition Today* 37 (2002): 96–102.
74. UNICEF, 2005; and J. P. Habicht, J. DaVanzo, and W. P. Butz, Mother's milk and sewage: Their interactive effects on infant mortality, *Pediatrics* 81 (1988): 456–61.
75. Dr. Carol Dyer's findings related to social and cultural beliefs about food in India are from A. Berg, *The Nutrition Factor* (Washington, D.C.: Brookings Institute, 1973), p. 46.
76. G. H. Pelto, Improving complementary feeding practices and responsive parenting as a primary component of interventions to prevent malnutrition in infancy and early childhood, *Pediatrics* 106 (2000): 1300–06.
77. UNICEF, 2005.
78. Ibid.
79. Lewis, Food security.
80. Ibid., p. 5S.
81. UNICEF, 2005.
82. K. A. Annan, *We the Children.*
83. A. R. Quisumbing and coauthors, *Women: The Key to Food Security* (Washington, D.C.: The International Food Policy Research Institute Food Policy Report, 1995); and *Fifth Report on the World Nutrition Situation,* 2004.
84. *Women: The Key to Food Security,* 1995; and D. Hinrichsen, Winning the food race, *Population Reports,* Vol. XXV, December 1997.
85. Trade and Development Program, *Exploring the Linkages: Trade Policies, Third World Development, and U.S. Agriculture* (Washington, D.C.: Bread for the World Institute, 1989), p. 23, as adapted from M. Carr, *Blacksmith, Baker, Roofing Sheet Maker* (London: Intermediate Technology Publications, 1984).
86. T. van den Briel and P. Webb, Fighting world hunger through micronutrient fortification programs, *Food Technology* 57 (2003): 44–47; S. Dalton, An education and research opportunity in Nepal: Dietetics in a developing country, *Topics in Clinical Nutrition* 11 (1996): 39–46; and W. Fawzi and coauthors, A prospective study of malnutrition in relation to child mortality in the Sudan, *American Journal of Clinical Nutrition* (1997): 1062.
87. The list of types of programs was adapted from G. M. Wardlaw, Hunger and undernutrition in the world, *Nutri-News* (St. Louis, MO: Mosby–Year Book, 1990), p. 14.
88. V. R. Young, Good nutrition for all: Challenge for the nutritional sciences in the new millennium, *Nutrition Today* 36 (2001): 6–16.
89. *National Conference on Primary Health Care* (Kathmandu: Ministry of Health, Health Services Coordination Committee, World Health Organization, and UNICEF, 1977), pp. 9, 25, as cited by M. E. Frantz, Nutrition problems and programs in Nepal, *Hunger Notes* 2 (1980): 5–8.
90. C. Aroney-Sine, Health care crisis: The global challenge, *Seeds* 15 (1993): 9–11.
91. Joint United Nations Programme on HIV/AIDS, *Report of the Global HIV/AIDS Epidemic* (Geneva: UNAIDS, 2004).
92. Ibid.
93. Ibid.
94. International Food Policy Research Institute, *2020 Vision News & Views* (Washington, D.C.: IFPRI, December 2000).
95. M. W. Rosegrant and M. A. Sombilia, Critical issues suggested by trends in food, population, and the environment for the year 2020, *American Journal of Agricultural Economics* 79 (1997): 1467–71; and L. R. Brown, G. B. Gardner, and B. Halweil, 16 impacts of population growth, *The Futurist* 33 (1999): 36–41.
96. *State of Food Insecurity in the World 2004.*
97. B. Scott and coauthors, The dietitian's role in ending world hunger: As citizen and health professional, *Topics in Clinical Nutrition* 13 (1998): 31–45.
98. The case for optimism, in E. Cornish, *The Study of the Future: An Introduction to the Art and Science of Understanding and Shaping Tomorrow's World* (Washington, D.C.: World Future Society, 1977), pp. 34–37.
99. L. Brown, *State of the World Report* (Washington, D.C.: Worldwatch Institute, 1990).
100. D. Elgin, *Voluntary Simplicity: Toward a Way of Life That Is Outwardly Simple, Inwardly Rich* (New York: Morrow, 1981), p. 25.
101. G. McGovern, *The Third Freedom: Ending Hunger in Our Time* (New York: Simon and Schuster, 2001).

Chapter 14

1. Arizona Department of Health Services, Office of Nutrition Services, *Project CHANCE—A Guide to Feeding Young Children with Special Needs* (Phoenix: Arizona Department of Health Services, 1995).
2. L. Rider, MS, RD, and L. C. Patty, RD, CPM, personal communication, May 1998.
3. National Cholesterol Education Program, Report of the National Cholesterol Education Program Expert Panel on detection, evaluation, and treatment of high blood cholesterol in adults, *Archives of Internal Medicine* 148 (1988): 36–69; and National Cholesterol Education Program, *Report of the Expert Panel on Population Strategies for Blood Cholesterol Reduction,* NIH Publication No. 90-3046 (Bethesda, MD: National Cholesterol Education Program, National Institutes of Health, U.S. Department of Health and Human Services, 1990); and The Expert Panel, National Cholesterol Education Program, *Report of the Expert Panel on Detection, Evaluation, and Treatment of High Blood Cholesterol in Adults* (Bethesda, MD: National Institutes of Health, 2001).
4. A. S. Pickoff, G. S. Berenson, and R. C. Schlant, Introduction to the symposium celebrating the Bogalusa Heart Study, *The American Journal of the Medical Sciences* 310 (Suppl. 1) (1995): S1–S2; C. C. Johnson and T. A. Nicklas, Health ahead—The Heart Smart Family approach to the prevention of cardiovascular disease, *The American Journal of the Medical Sciences* 310 (Suppl. 1) (1995): S127–S132.

5. New states will participate in WIC Farmers' Market, *Nutrition Week* 28 (1998): 6; and United States Department of Agriculture, Farmers' markets continue to increase nationwide, *USDA News Release* No. 0413.02, September 30, 2002.
6. K. M. Bartol and D. C. Martin, *Management,* 3rd ed. (New York: McGraw-Hill, 1998).
7. A. L. Martinez, *Hunger in Latino Communities* (Washington, D.C.: Congressional Hunger Center and the Congressional Hispanic Caucus Institute, 1995).
8. R. Perez-Escamilla, D. A. Himmelgreen, and A. Ferris, *Community Nutritional Problems among Latino Children in Hartford, Connecticut* (Storrs and Hartford: University of Connecticut and the Hispanic Health Council, 1997).
9. Information about developing goals and objectives was adapted from K. L. Probert, ed., *Moving to the Future: Developing Community-Based Nutrition Services* (Washington, D.C.: Association of State and Territorial Public Health Nutrition Directors, 1996), pp. 21–24.
10. U.S. Department of Health and Human Services, *Healthy People 2010: Understanding and Improving Health,* 2nd ed. (Washington, D.C.: U.S. Government Printing Office, November 2000).
11. A. Fink, *Evaluation Fundamentals: Guiding Health Programs, Research, and Policy* (Newbury Park, CA: Sage, 1993), pp. 1–17.
12. N. Estabrook, *Teach Yourself the Internet in 24 Hours* (Indianapolis: Sams, 1997), pp. 91–99.
13. G. J. Gordon and M. E. Milakovich, *Public Administration in America,* 5th ed. (New York: St. Martin's Press, 1995), p. 369.
14. P. H. Rossi, H. E. Freeman, and M. W. Lipsey, *Evaluation—A Systematic Approach,* 6th ed. (Beverly Hills, CA: Sage, 1999).
15. A. D. Spiegel and H. H. Hyman, *Strategic Health Planning: Methods and Techniques Applied to Marketing and Management* (Norwood, NJ: Ablex, 1991), pp. 324–25.
16. H. J. Rubin and I. S. Rubin, *Community Organizing and Development,* 3rd ed. (Boston: Allyn & Bacon, 2000).
17. A. D. Grotelueschen and coauthors, *An Evaluation Planner* (Urbana: Office for the Study of Continuing Professional Education, University of Illinois at Urbana-Champaign, 1974).
18. American Dietetic Association's Public Relations Team, *Improving Health and Saving Health Care Dollars: Medical Nutrition Therapy Works* (Chicago, IL: American Dietetic Association, April 2002); and Position of the American Dietetic Association, Cost-effectiveness of medical nutrition therapy, *Journal of the American Dietetic Association* 95 (1995): 88.
19. D. J. Caron, Knowledge required to perform the duties of an evaluator, *Canadian Journal of Program Evaluation* 8 (1993): 59–78.
20. Rossi, Freeman, and Lipsey, *Evaluation–A Systematic Approach.*
21. Fink, *Evaluation Fundamentals,* p. 167.
22. J. L. Breault and R. Gould, Formative evaluation of a video and training manual on feeding children with special needs, *Journal of Nutrition Education* 30 (1998): 58–61.
23. G. E. A. Dever, *Community Health Analysis—Global Awareness at the Local Level,* 2nd ed. (Gaithersburg, MD: Aspen, 1991), pp. 45–73.
24. J. B. McKinlay, The promotion of health through planned sociopolitical change: Challenges for research and policy, *Social Science and Medicine* 36 (1993): 109–17; and J. B. McKinlay, More appropriate evaluation methods for community-level health interventions, *Evaluation Review* 20 (1996): 237–43.
25. S. A. McGraw and coauthors, Using process data to explain outcomes: An illustration from the Child and Adolescent Trial for Cardiovascular Health (CATCH), *Evaluation Review* 20 (1996): 291–312.
26. Ibid.
27. The discussion of impact evaluation was adapted from T. C. Timmreck, *Planning, Program Development, and Evaluation: A Handbook for Health Promotion, Aging, and Health Services,* 2nd ed. (Sudbury, MA: Jones & Bartlett, 2002); M. Thorogood and Y. Coombes, eds., *Evaluating Health Promotion: Practice and Methods* (London: Oxford University Press, 2000); J. L. Smeltzer and J. F. McKenzie, *Planning, Implementing, and Evaluating Health Promotion Programs: A Primer,* 3rd ed. (San Francisco, CA: Benjamin Cummings, 2000); and D. J. Petersen and G. A. Alexander, *Needs Assessment in Public Health: A Practical Guide for Students and Professionals* (New York: Kluwer Academic, 2001).
28. A. Fink, *Evaluation Fundamentals: Guiding Health Programs, Research, and Policy* (Newbury Park, CA: Sage, 1993), pp. 1–4.
29. *Moving to the Future: Developing Community-Based Nutrition Services,* 1996, pp. 59–66.
30. Cost-effectiveness of medical nutrition therapy, 1995; and Food Research and Action Center, *WIC in the States: Twenty-Five Years of Building a Healthier America* (Washington, D.C.: Food Research and Action Center, 1999).
31. L. L. Morris and coauthors, *How to Communicate Evaluation Findings* (Newbury Park, CA: Sage, 1988), pp. 9–10.
32. Ibid., pp. 20–22.
33. M. O'Brecht, Stakeholder pressures and organizational structure for program evaluation, *Canadian Journal of Program Evaluation* 7 (1992): 139–47.
34. C. Kemp, Islamic cultures: Health-care beliefs and practices, *American Journal of Health Behavior* 20 (1996): 83–89.
35. J. D. O'Neil, B. Elias, and A. Yassi, Poisoned food: Cultural resistance to the contaminants discourse in Nunavik, *Arctic Anthropology* 34 (1997): 29–40.
36. E. R. House, Multicultural evaluation in Canada and the United States, *Canadian Journal of Program Evaluation* 7 (1992): 153.
37. F. Popcorn, *The Popcorn Report* (New York: HarperCollins, 1991), p. 12.
38. L. Rider and L. C. Patty, personal communication, May 1998, *A Guide to Feeding Young Children with Special Needs;* available at www.azdhs.gov/phs/oncdps/children.

Chapter 15

1. The man who put your groceries on wheels, *New York Post,* April 28, 1977, p. 37.
2. D. Cohen, *Consumer Behavior* (New York: Random House, 1981), p. 427.
3. M. P. O'Donnell, Definition of health promotion: Part II: Levels of programs, *American Journal of Health Promotion* 1 (1986): 6–9.
4. Food and Drug Administration, Food labeling: Health claims and label statements; folate and neural tube defects; final rule, *Federal Register* 61 (March 5, 1996): 8752–81.
5. A. E. Sloan, Food industry forecast: Consumer trends to 2020 and beyond, *Food Technology* 52 (1998): 37–38, 40, 42, 44; and J. Guthrie and coauthors, Understanding economic and behavioral influences on fruit and vegetable choices, *Amber Waves* (United States Department of Agriculture, Economic Research Service), April 2005.
6. D. M. Shaffer and coauthors, Nonvitamin, nonmineral supplement use over a 12-month period by adult members of a large health maintenance organization, *Journal of the American Dietetic Association* 103 (2003): 1500–505; and D. W. Kaufman and coauthors, Recent patterns of medication use in the ambulatory adult population in the United States: The Slone Survey, *Journal of the American Medical Association* 287 (2002): 337–44.
7. A. E. Sloan, Top ten trends to watch and work on in 2003, *Food Technology* 57 (2003): 30–50.
8. P. G. Kittler and K. P. Sucher, *Food and Culture,* 4th ed. (Belmont, CA: Wadsworth/Thomson, 2004).
9. E. Lipson, *The Organic Foods Sourcebook* (New York: McGraw-Hill Contemporary, 2001); and E. Lipson, Food and Culinary Dietetic Practice Group Teleforum: An Expert's Guide to the Organic Market, January 14, 2005.
10. B. Senauer, E. Asp, and J. Kinsey, *Food Trends and the Changing Consumer* (St. Paul, MN: Eagan, 1991), pp. 133–53; and Interagency Board for Nutrition Monitoring and Related Research, *Nutrition Monitoring in the United States,* Vol. 1 (Washington, D.C.: U.S. Government Printing Office, 1995), pp. 59–60.
11. Interagency Board for Nutrition Monitoring and Related Research, *Nutrition Monitoring in the United States, Chartbook I,* DHHS Pub. No. 93-1255-2 (Hyattsville,

MD: Department of Health and Human Services, 1993), p. 55.
12. Cohen, *Consumer Behavior,* pp. 76–127.
13. D. Neumark-Sztainer and coauthors, Correlates of inadequate consumption of dairy products among adolescents, *Journal of Nutrition Education* 29 (1997): 12–20.
14. G. C. Grimm, L. Harnack, and M. Story, Factors associated with soft drink consumption in school-aged children, *Journal of the American Dietetic Association* 104 (2004): 1244–49.
15. J. D. O'Neil, B. Elias, and A. Yassi, Poisoned food: Cultural resistance to the contaminants discourse in Nunavik, *Arctic Anthropology* 34 (1997): 29–40.
16. H. G. Koenig, M. E. McCullough, and D. B. Larson, *Handbook of Religion and Health* (New York: Oxford University Press, 2001); and A. Hart and coauthors, Is religious orientation associated with fat and fruit/vegetable intake? *Journal of the American Dietetic Association* 104 (2004): 1292–96.
17. A. Kilara and K. K. Iya, Food and dietary habits of the Hindu, *Food Technology* 46 (1992): 94–104.
18. J. O. Fisher and coauthors, Parental influences on young girls' fruit and vegetable, micronutrient, and fat intakes, *Journal of the American Dietetic Association* 102 (2002): 58–64.
19. Regional food preferences outlined in study of consumer eating trends, *Journal of the American Dietetic Association* 90 (1990): 1727.
20. American Dietetic Association, *Nutrition & You: Trends 2002* (Chicago, IL: American Dietetic Association, 2002).
21. I. M. Parraga, Determinants of food consumption, *Journal of the American Dietetic Association* 90 (1990): 661–63.
22. The role of the ADA environment scan: Shaping the future of the profession, *Journal of the American Dietetic Association* 102 (2002): S1820–S1839.
23. I. Ajzen and M. Fishbein, *Understanding Attitudes and Predicting Social Behavior* (Upper Saddle River, NJ: Prentice-Hall, 1997), pp. 12–28.
24. C. S. Wilson, Nutritionally beneficial cultural practices, *World Review of Nutrition and Dietetics* 45 (1985): 68–96.
25. American Dietetic Association and American Diabetes Association, *Ethnic and Regional Food Practices: Chinese American Food Practices, Customs, and Holidays* (Chicago: American Dietetic Association, 1990), pp. 2–3.
26. M. K. Campbell, Stages of change: The transtheoretical model, in *Charting the Course for Evaluation: How Do We Measure the Success of Nutrition Education and Promotion in Food Assistance Programs?* Summary of Proceedings (Alexandria, VA: U.S. Department of Agriculture, 1997), pp. 19–21.
27. The description of the Stages of Change Model was adapted from R. J. Budd and S. Rollnick, The structure of the readiness to change questionnaire: A test of Prochaska & DiClemente's transtheoretical model, *British Journal of Health Psychology* 1 (1996): 365–76; M. H. Read, Age, dietary behaviors and the stages of change model, *American Journal of Health Behavior* 20 (1996): 417–24; and D. J. Bowen, S. Kinne, and N. Urban, Analyzing communities for readiness to change, *American Journal of Health Behavior* 21 (1997): 289–98.
28. R. G. Boyle and coauthors, Stages of change for physical activity, diet, and smoking among HMO members with chronic conditions, *American Journal of Health Promotion* 12 (1998): 170–75.
29. M. K. Campbell and coauthors, Stages of change and psychosocial correlates of fruit and vegetable consumption among rural African-American church members, *American Journal of Health Promotion* 12 (1998): 185–91.
30. N. K. Janz and coauthors, The Health Belief Model, in *Health Behavior and Health Education—Theory, Research, and Practice,* 3rd ed., ed. K. Glanz, B. K. Rimer, and F. M. Lewis (San Francisco, CA: Jossey-Bass, 2002), pp. 45–66.
31. The discussion of the Health Belief Model was adapted from A. Caggiula, Health Belief Model, in *Charting the Course for Evaluation,* pp. 15–16; and Use of health belief model with older adults' food-handling practices, *Journal of Nutrition Education and Behavior* 34 (2002): S25–S30.
32. A. Bandura, *Social Learning Theory* (Upper Saddle River, NJ: Prentice-Hall, 1977), p. 79.
33. T. Byers and coauthors, American Cancer Society Guidelines on Nutrition and Physical Activity for Cancer Prevention: Reducing the risk of cancer with healthy food choices and physical activity, *CA: A Cancer Journal for Clinicians* 52 (2002): 92–119; available at www.cancer.org.
34. R. E. Patterson, A. R. Kristal, and E. White, Do beliefs, knowledge, and perceived norms about diet and cancer predict dietary change? *American Journal of Public Health* 86 (1996): 1394–1400.
35. D. E. Montano and D. Kasprzyk, The theory of reasoned action and the theory of planned behavior, in *Health Behavior and Health Education—Theory, Research, and Practice,* 2002, pp. 67–98.
36. R. P. Bagozzi and Y. Yi, The degree of intention formation as a moderator of the attitude–behavior relationship, *Social Psychology Quarterly* 52 (1989): 266–79.
37. R. P. Bagozzi, The self-regulation of attitudes, intentions, and behavior, *Social Psychology Quarterly* 55 (1992): 178.
38. Ibid., pp. 178–204.
39. M. Conner, E. Martin, and N. Silverdale, Dieting in adolescence: An application of the theory of planned behaviour, *British Journal of Health Psychology* 1 (1996): 315–25; and P. Packard and K. Stanek Krogstrand, Half of rural girls aged 8 to 17 years report weight concerns and dietary changes, with both more prevalent with increased age, *Journal of the American Dietetic Association* 102 (2002): 672–77.
40. National Heart, Lung, and Blood Institute, *The Practical Guide: Identification, Evaluation, and Treatment of Overweight and Obesity in Adults* (Washington, D.C.: National Institutes of Health, 2000); and R. R. Wing, A. Gorin, and D. Tate, Strategies for changing eating and exercise behavior, *Present Knowledge in Nutrition* (Washington, D.C.: International Life Sciences Institute, 2001, pp. 650–61.
41. D. B. Abrams and coauthors, Social learning principles for organizational health promotion: An integrated approach, in *Health and Industry—A Behavioral Medicine Perspective,* ed. M. F. Cataldo and T. J. Coates (New York: Wiley, 1986), pp. 28–51.
42. A. J. Rainville, Pica practices of pregnant women are associated with lower maternal hemoglobin level at delivery, *Journal of the American Dietetic Association* 98 (1998): 293–96.
43. S. M. Gross and coauthors, Counseling and motivational videotapes increase duration of breast-feeding in African-American WIC participants who initiate breast-feeding, *Journal of the American Dietetic Association* 98 (1998): 143–48.
44. E. M. Rogers, *Diffusion of Innovations* (Upper Saddle River, NJ: Prentice-Hall, 1997).
45. As cited in Cohen, *Consumer Behavior,* pp. 429–30.
46. S. Ram and J. N. Sheth, Consumer resistance to innovations: The marketing problem and its solutions, *Journal of Consumer Marketing* 6 (1989): 5–14.
47. G. Reichler and S. Dalton, Chefs' attitudes toward healthful food preparation are more positive than their food science knowledge and practices, *Journal of the American Dietetic Association* 98 (1998): 165–69.
48. K. A. E. Brown and coauthors, The Well: A neighborhood-based health promotion model for black women, *Health & Social Work* 23 (1998): 146–52.
49. M. Feinleib, editorial: New directions for community intervention studies, *American Journal of Public Health* 86 (1996): 1696–98.

Chapter 16

1. U.S. Department of Health and Human Services, OPHS Office of Minority Health, *National Standards for Culturally and Linguistically Appropriate Services in Health Care Final Report* (Washington, D.C.: U.S. Government Printing Office, March 2001).
2. J. Luckmann, *Transcultural Communication in Health Care* (Albany, NY: Delmar/Thomson Learning, 2000), p. 23.
3. R. Edge, One middle-age white male's perspective on racism and cultural competence: A view from the bunker where we

wait to have our privilege stripped away, *Mental Retardation* 40 (2002): 83–85.
4. U.S. Department of Health and Human Services, *Healthy People 2010* (Washington, D.C.: U.S. Government Printing Office, January 2000).
5. T. Goode, *Policy Brief 4: Engaging Communities to Realize the Vision of One Hundred Percent Access and Zero Health Disparities: A Culturally Competent Approach* (Washington, D.C.: National Center for Cultural Competence, Georgetown University Child Development Center, 2001).
6. S. Staveteig and A. Wigton, *Racial and ethnic disparities: Key finding from the National Survey of America's Families* (The Urban Institute New Federalism Series B No B-5, February 2000).
7. MHS Staff, The changing face of health care consumers, *Mark Health Services* 21(4) (2001): 4–10.
8. T. Goode, S. Sockalingam, M. Brown, and W. Jones, *Policy Brief 2: Linguistic Competence in Primary Health Care Delivery Systems: Implications for Policy Makers* (Washington, D.C.: National Center for Cultural Competence, Georgetown University Child Development Center, 2000).
9. Diversity Rx, *Promoting language and cultural competence to improve the quality of health care for minority, immigrant, and ethnically diverse communities;* available at www.diversityrx.org.
10. H. Fullerton and M. Toossi, Labor force projections to 2010: Steady growth and changing composition, *Monthly Labor Review* (Washington, D.C.: U.S. Government Printing Office, November 2001).
11. D. Dyer, *A Dietitian's Cancer Story. Information & Inspiration for Recovery and Healing from a 3-Time Cancer Survivor,* 4th ed. (Ann Arbor, MI: Swan Press, 1999), p. 218.
12. K. Kanjana, A study on the effects of yoga therapy and diet counseling on nutritional and personality profile of female adolescents, Ph.D. Dissertation, Presidency College, University of Madras, Chepauk, Chennai, March 2001.
13. U.S. Census Bureau. *Overview of Race and Hispanic Origin Census 2000 Brief* (Washington, D.C.: U.S. Government Printing Office, March 2001). U.S. Administration on Aging, *Cultural Competency—Fact Sheet, The Many Faces of Aging;* available at www.aoa.gov.
14. U.S. Census Bureau, *Overview of Race and Hispanic Origin Census 2000 Brief.*
15. P. G. Kittler and K. P. Sucher, *Food and Culture,* 3rd ed. (Belmont, CA: Wadsworth/Thomson, 2001), p. 241.
16. U.S. Census Bureau. *Overview of Race and Hispanic Origin Census 2000 Brief.*
17. C. Elliott, R. J. Adams, and S. Sockalingam, *Multicultural Toolkit (Toolkit for Cross-Cultural Collaboration);* available at www.awesomelibrary.org/multicultural-toolkit.html.
18. E. Cohen and T. Goode, *Policy Brief 1: Rationale for Cultural Competence in Primary Health Care* (Washington, D.C.: National Center for Cultural Competence, Georgetown University Child Development Center, 1999).
19. U.S. Census Bureau, *Overview of Race and Hispanic Origin Census 2000 Brief.*
20. U.S. Census Bureau, Historical Health Insurance Tables; available at www.census.gov/hhes/hlthins/historic/index.html.
21. R. K. Lewis and B. L. Green, Assessing the health attitudes, beliefs, and behaviors of African Americans attending church: A comparison from two communities, *Journal of Community Health* 25 (2000): 211–24.
22. Leadership, Education and Training (LET) Program in Maternal and Child Nutrition; available at: *Causes of Health Disparities,* www.epi.umn.edu/let/causes.html.
23. E. Cohen and T. Goode, *Policy Brief 1: Rationale for Cultural Competence in Primary Health Care.*
24. R. E. Zambrana and L. A. Logie, Latino child health: Need for inclusion in the U.S. national discourse, *American Journal of Public Health* 90 (2000): 1827–33.
25. R. K. Lewis and B. L. Green, Assessing the health attitudes, beliefs, and behaviors of African Americans attending church.
26. D. R. Williams, Race, socioeconomic status, and health, *Annals of New York Academy of Sciences* 806 (1999): 173–88.
27. B. D. Smedley, A. Y. Stith, and A. R. Nelson, eds., *Unequal Treatment: Confronting Racial and Ethnic Disparities in Health Care* (Washington, D.C.: National Academy Press, 2002).
28. National Institute of Environmental Health Sciences, *National Institutes of Health, Health (2001) Disparities Research Booklet;* available at www.niehs.nih.gov/oc/factsheets/disparity/phome.htm.
29. J. A. Bryk and T. Kornblum, Report on the 1997 membership database of the American Dietetic Association, *Journal of the American Dietetic Association* 99 (1999): 102–7.
30. B. Haughton, M. Story, and B. Keir, Profile of public health nutrition personnel: Challenges for population/system-focused roles and state-level monitoring, *Journal of the American Dietetic Association* 98 (1998): 664–70.
31. B. D. Smedley, A. Y. Stith, and A. R. Nelson, eds., *Unequal Treatment: Confronting Racial and Ethnic Disparities in Health Care.*
32. American Dietetic Association, *ADA Diversity Philosophy Statement;* available at www.eatright.com.
33. F. G. Donini-Lenhoff and H. L. Hedrick, Increasing awareness and implementation of cultural competence principles in health professions education, *Journal of Allied Health* 29 (2000): 241–45.
34. L. S. Robin, Cultural competence in diabetes education and care, in *A Core Curriculum for Diabetes Education,* 4th ed. (Chicago: American Association of Diabetes Educators, 2002), pp. 99–119.
35. U.S. Department of Health and Human Services, OPHS Office of Minority Health, *National Standards for Culturally and Linguistically Appropriate Services in Health Care, Final Report.*
36. U.S. Department of Health and Human Services, *The Initiative to Eliminate Racial and Ethnic Disparities in Health* (Washington, D.C.: U.S. Government Printing Office, 1998).
37. R. C. Like, R. P. Steiner , and A. J. Rubel, STFM core curriculum guidelines Recommended Core Curriculum Guidelines on Culturally Sensitive and Competent Health Care, *Family Medicine* 28 (1996): 291–97.
38. T. Goode, S. Sockalingam, M. Brown, and W. Jones, *Policy Brief 2: Linguistic Competence in Primary Health Care Delivery Systems: Implications for Policy Makers.*
39. U.S. Census Bureau, *Overview of Race and Hispanic Origin Census 2000 Brief.*
40. Recommended Core Curriculum Guidelines on Culturally Sensitive and Competent Health Care, 1996.
41. L. S. Robin, Cultural competence in diabetes education and care, in *A Core Curriculum for Diabetes Education,* 4th ed.
42. J. Campinha-Bacote, *The Process of Cultural Competence in the Delivery of Healthcare Services,* 4th ed. (Cincinnati, OH: Transcultural C.A.R.E. Associates, 2003), p. 6.
43. J. Campinha-Bacote, A model and instrument for addressing cultural competence in health care, *Journal of Nursing Education* 38 (1999): 203–7.
44. Association for the Advancement of Health Education, *Cultural Awareness and Sensitivity: Guidelines for Health Educators* (Reston, VA: Association for the Advancement of Health Education, 1994).
45. L. Purnell, Purnell's model for cultural competence, in L. Purnell and B. Paulanka, eds., *Transcultural Health Care: A Culturally Competent Approach* (Philadelphia: Davis, 1998), pp. 7–51.
46. Association for the Advancement of Health Education, *Cultural Awareness and Sensitivity: Guidelines for Health Educators.*
47. Administration on Aging, *Achieving Cultural Competence: A Guidebook for Providers of Services to Older Americans and Their Families;* available at www.aoa.gov.
48. Ibid.
49. B. Schilling and E. Brannon, *Cross-Cultural Counseling: A Guide for Nutrition and Health Counselors* (Washington, D.C: U.S. Government Printing Office, 1990).
50. K. McCullough-Zander, ed., *Caring across Cultures: The Provider's Guide to Cross-Cultural Health* (Minneapolis, MN: The Center for Cross-Cultural Health, 2000).
51. Ibid.
52. D. Arthur, The importance of body language, *HR Focus* 72 (1995): 22–23.
53. L. Haffner, Translation is not enough: Interpreting in a medical setting, *Western Journal of Medicine* 157 (1992): 255–259.

54. P. G. Kittler and K. P. Sucher, *Food and Culture,* 3rd ed. (Belmont, CA: Wadsworth/Thomson, 2001), p. 46.
55. T. D. Goode, Georgetown University Child Development Center, *Promoting Cultural Diversity and Cultural Competence;* available at gucdc.georgetown.edu/nccc/nccc7.html.
56. D. E. Graves and C. W. Suitor, *Celebrating Diversity: Approaching Families through Their Food* (Arlington, VA: National Center for Education in Maternal and Child Health, 1998).
57. M. Magnus, What's your IQ on cross-cultural nutrition counseling? *The Diabetes Educator* 96 (1996): 57–62.
58. Ibid.
59. Ibid.
60. L. Gardenswartz and A. Rowe, *Managing Diversity in Health Care* (San Francisco, CA: Jossey-Bass, 1998), p. 108.
61. P. G. Kittler and K. P. Sucher, *Food and Culture,* 3rd ed., p. 57.
62. L. Haffner, Translation is not enough: Interpreting in a medical setting, *Western Journal of Medicine* 157 (1992): 255–259.
63. P. G. Kittler and K. P. Sucher, *Food and Culture,* 3rd ed., p. 57.
64. A. Fadiman, *The Spirit Catches You and You Fall Down* (New York: The Noonday Press, 1997), p. 145.
65. Administration on Aging, *Achieving Cultural Competence: A Guidebook for Providers of Services to Older Americans and Their Families.*
66. A. V. Blue, The Provision of Culturally Competent Health Care; available at www.musc.edu, accessed May 25, 2002.
67. L. Gardenswartz and A. Rowe, *Managing Diversity in Health Care.*
68. R. Edge, One middle-age white male's perspective on racism and cultural competence stripped away.
69. M. Magnus, What's your IQ on cross-cultural nutrition counseling?
70. E. B. Berlin and W. C. Fowkes, A teaching framework for cross-cultural health care—Application in family practice, *Western Journal of Medicine* 139 (1983): 934–38.
71. A. Kleinman, L. Eisenberg, and B. Good, Culture, illness, and care: Clinical lesions from anthropologic and cross-cultural research, *Annals of Internal Medicine* 88 (1978): 251–58.
72. Administration on Aging, *Achieving Cultural Competence: A Guidebook for Providers of Services to Older Americans and Their Families.*
73. D. E. Graves and C. W. Suitor, *Celebrating Diversity: Approaching Families through Their Food.*
74. U.S. Census Bureau, *Overview of Race and Hispanic Origin Census 2000 Brief.*
75. B. D. Smedley, A. Y. Stith, and A. R. Nelson, eds., *Unequal Treatment: Confronting Racial and Ethnic Disparities in Health Care.*
76. Administration on Aging, *Achieving Cultural Competence: A Guidebook for Providers of Services to Older Americans and Their Families.*
78. U.S. Department of Health and Human Services, OPHS Office of Minority Health, *National Standards for Culturally and Linguistically Appropriate Services in Health Care, Final Report.*
79. Administration on Aging, *Achieving Cultural Competence: A Guidebook for Providers of Services to Older Americans and Their Families.*

Chapter 17

1. M. Mason, The B vitamin breakthrough, *Health* 9 (1995): 69–73.
2. L. K. Guyer, Outcome-based nutrition education, in *Nutrition and Food Services for Integrated Health Care: A Handbook for Leaders* (Gaithersburg, MD: Aspen, 1997), pp. 206–41.
3. R. Patterson, ed., *Changing Patient Behavior: Improving Outcomes in Health and Disease Management* (San Francisco, CA: Jossey-Bass, 2001).
4. Ibid.
5. The discussion of how people learn best is from E. J. Hitch and J. P. Youatt, *Communicating Family and Consumer Sciences* (Tinley Park, IL: The Goodheart-Wilcox Company, 1995), p. 59.
6. H. Osborne, *Overcoming Communication Barriers in Patient Education* (Gaithersburg, MD: Aspen Publishers, 2000), pp. 26–31.
7. A. C. Tuijnman, Concepts, theories, and methods, in *International Encyclopedia of Adult Education and Training,* 2nd ed. (New York: Pergamon, 1996), pp. 3–8.
8. K. Wagschal, I became clueless teaching the genXers, *Adult Learning* 8 (1997): 21–25.
9. D. R. Garrison, Self-directed learning: Toward a comprehensive model, *Adult Education Quarterly* 48 (1997): 18–33.
10. M. S. Knowles, E. F. Holton, and R. A. Swanson, *The Adult Learner: The Definitive Classic in Adult Education and Human Resource Development,* 5th ed. (Houston, TX: Gulf Publishing Company, 1998).
11. K. L. Cason, J. F. Scholl, and C. Kassab, A comparison of program delivery methods for low income nutrition audiences, *Topics in Clinical Nutrition* 17 (2002): 63–73; and S. Brookfield, *Understanding and Facilitating Adult Learning* (San Francisco, CA: Jossey-Bass, 1991).
12. R. AbuSabha, *Effective Nutrition Education for Behavior Change* (Clarksville, MD: Wolfe Rinke Associates, 1998), pp. 39–52.
13. K. Resnicow, Applying theory to culturally diverse and unique populations, in *Health Behavior and Health Education—Theory, Research, and Practice,* 3rd ed., ed. K. Glanz, B. K. Rimer, and F. M. Lewis (San Francisco, CA: Jossey-Bass, 2002).
14. C. K. Miller, G. L. Jensen, and C. L. Achterberg, Evaluation of a food label nutrition intervention for women with type 2 diabetes mellitus, *Journal of the American Dietetic Association* 99 (1999): 39–45; and C. Burnett, *Welcome to WIC: Do We Really Mean It?* (Davis, CA: University of California, 1993).
15. J. Goldberg, R. E. Rudd, and W. Dietz, Using three data sources and methods to shape a nutrition campaign, *Journal of the American Dietetic Association* 99 (1999): 717–22; and M. Warrix and coauthors, Development of educational materials: Sensitizing nutrition educators to cultural diversity, *Multicultural Perspectives* 2 (2000): 14–18.
16. The "Eat five to stay alive" message was taken from P. Jaret, Only 5 a day, *Health* 12 (1998): 78.
17. K. L. Probert, ed., *Moving to the Future: Developing Community-Based Nutrition Services* (Washington, D.C.: Association of State and Territorial Public Health Nutrition Directors, 1996), pp. 25–34.
18. R. AbuSabha, *Effective Nutrition Education for Behavior Change,* 1998, pp. 39 52.
19. E. A. Martin and V. R. Beal, *Nutrition Work with Children* (Chicago, IL: University of Chicago, 1978), pp. 271–89; Position of the American Dietetic Association, Nutrition education for the public, *Journal of the American Dietetic Association* 96 (1996): 1183–87; and I. Contento and coauthors, The effectiveness of nutrition education and implications for nutrition education policy, programs and research: A review of research, *Journal of Nutrition Education* 27 (1995): 291–346.
20. D. B. Gorwin, *Educating* (Ithaca, NY: Cornell University Press, 1981), pp. 23–61.
21. N. E. Gronlund, *How to Write and Use Instructional Objectives,* 4th ed. (New York: Macmillan, 1991); and R. F. Mager, *Preparing Instructional Objectives,* 2nd ed. (Belmont, CA: Lake Publishing Co., 1984).
22. R. AbuSabha, *Effective Nutrition Education for Behavior Change,* 1998, pp. 77–87.
23. D. Michalczyk, Impact your practice: Communicate effectively online, *Journal of the American Dietetic Association* 102 (2002): 778–79.
24. R. A. Krueger and M. A. Casey, *Focus Groups: A Practical Guide for Applied Research* (Thousand Oaks, CA: Sage Publications, 2000).
25. P. Cranton, Types of group learning, *New Directions for Adult and Continuing Education* 71 (1996): 25–32.
26. *Lean 'N Easy* education kit (Chicago: National Cattlemen's Beef Association, 1996).
27. As cited in T. MacLaren, Messages for the masses: Food and nutrition issues on television, *Journal of the American Dietetic Association* 97 (1997): 733–34.
28. A. L. Eldridge and coauthors, Development and evaluation of a labeling program for low-fat foods in a discount department store foodservice area, *Journal of Nutrition Education* 29 (1997): 159–61; and K. J. Harris and coauthors, Community partnerships:

Review of selected models and evaluation of two case studies, *Journal of Nutrition Education* 29 (1997): 189–95.
29. R. L. Parrott, Motivation to attend to health messages, in E. Maibach and R. L. Parrott, eds., *Designing Health Messages: Approaches from Communication Theory and Public Health Practice,* ed. E. Maibach and R. L. Parrott (Thousand Oaks, CA: Sage, 1995), pp. 7–23.
30. These items are all available for purchase on the 5 a Day for Better Health website, dccps.nci.nih.gov/5aday.
31. E. W. Austin, Reaching young audiences, in *Designing Health Messages: Approaches from Communication Theory and Public Health Practice,* pp. 114–41.
32. T. A. Nicklas and coauthors, Development of a school-based nutrition intervention for high school students: Gimme 5, *American Journal of Health Promotion* 11 (1997): 315–22.
33. *Reaching Consumers with Meaningful Health Messages: A Handbook for Nutrition and Food Communicators* (Chicago: Dietary Guidelines Alliance, 1996).
34. L. S. Chapman, Maximizing program participation, *The Art of Health Promotion* 2 (1998): 1–8.
35. Nutrition Messages from the "It's All About You" Campaign of the Dietary Guidelines Alliance. Copyright © 1996 The Dietary Guidelines Alliance.
36. International Food Information Council, A new nutrition conversation with consumers, *Food Insight,* September/October 2000.
37. J. J. Moran, *Assessing Adult Learning: A Guide for Practitioners* (Malabar, FL: Krieger, 1997), pp. 1–20.
38. A. E. Goody and C. E. Kozoll, *Program Development in Continuing Education* (Malabar, FL: Krieger, 1995), pp. 62–68.
39. P. A. Sissel, Participation and learning in Head Start: A sociopolitical analysis, *Adult Education Quarterly* 47 (1997): 123–37.
40. L. J. Hughes, M. Rochford, and D. L. Minch, Expanding nutrition opportunities from hospital into the community: The role of cooperative extension, *Topics in Clinical Nutrition* 15 (2000): 10–18; and J. Burnie and B. Haughton, EFNEP: A nutrition education program that demonstrates cost–benefit, *Journal of the American Dietetic Association* 102 (2002): 39–45.

Chapter 18

1. The opening vignette was adapted from E. Clift, Social marketing and communication: Changing health behavior in the Third World, *American Journal of Health Promotion* 3 (1989): 17–24.
2. P. F. Basch, *International Health* (New York: Oxford University Press, 1990), p. 285.
3. P. F. Drucker, *Managing the Nonprofit Organization: Principles and Practices* (New York: Harper Business, 1992); and P. F. Drucker, *The Essential Drucker* (New York: Harper Business, 2001).
4. P. Kotler and K. L. Keller, *Marketing Management,* 12th ed. (Upper Saddle River, NJ: Prentice-Hall, 2005).
5. J. Trivers, *One Stop Marketing* (New York: Wiley, 1996), p. 31.
6. Clift, Social marketing, p. 17. The margin definition was adapted from A. Andreasen and P. Kotler, *Strategic Marketing for Nonprofit Organizations* (Upper Saddle River, NJ: Prentice-Hall, 2002).
7. P. Kotler, N. Roberto, and N. Lee, *Social Marketing: Improving the Quality of Life,* 2nd ed. (Thousand Oaks, CA: Sage Publications, 2002); and R. Alcalay and R. A. Bell, Strategies and practices in community-based campaigns promoting nutrition and physical activity, *Social Marketing Quarterly* 7 (2001): 3–15.
8. A. A. Andreasen, *Marketing Social Change: Changing Behavior to Promote Health, Social Development, and the Environment* (San Francisco, CA: Jossey-Bass, 1995); and M. Siegel and L. Donner, *Marketing Public Health: Strategies to Promote Social Change* (Gaithersburg, MD: Aspen Publishers, 1998).
9. E. W. Maibach and coauthors, Social marketing, in *Health Behavior and Health Education—Theory, Research, and Practice,* 3rd ed., ed. K. Glanz, B. K. Rimer, and F. M. Lewis (San Francisco, CA: Jossey-Bass, 2002), pp. 437–61.
10. Trivers, *One Stop Marketing,* p. 44.
11. P. Francese, *Marketing Know-How: Your Guide to the Best Marketing Tools and Sources* (Ithaca, NY: American Demographics Books, 1996), pp. 15–29.
12. J. C. Levinson and S. Godin, *The Guerrilla Marketing Handbook* (Boston: Houghton Mifflin, 1994), p. 7.
13. Trivers, *One Stop Marketing,* p. 135.
14. P. Kotler, *Marketing Management.*
15. W. D. Perreault and E. J. McCarthy, *Basic Marketing: A Global-Managerial Approach* (New York: McGraw-Hill, 1999).
16. The discussion of the marketing planning process was adapted from Matthews, Marketing your services: Strategies that work, *ASHA Magazine* 30 (1988): 21–25.
17. K. K. Helm, *The Competitive Edge: Advanced Marketing Strategies for Dietetics Professionals* (Chicago: American Dietetic Association, 1995).
18. Matthews, Marketing your services, p. 23.
19. M. Ward, *Marketing Strategies: A Resource for Registered Dietitians* (Binghamton, NY: Niles & Phipps, 1984), p. 63.
20. Ibid.
21. W. Lancaster, T. McIllwain, and J. Lancaster, Health marketing: Implications for health promotion, *Family and Community Health* (February 1983): 47.
22. Ibid., p. 45.
23. Dever, *Community Health Analysis,* pp. 253–64.
24. The section on demographics and psychographics was adapted from Dever, *Community Health Analysis,* pp. 255–77.
25. S. C. Parks, Research techniques used to support marketing management decisions, in *Research: Successful Approaches,* ed. E. R. Monsen (Chicago: American Dietetic Association, 1992), p. 306.
26. D. J. Breckon, *Managing Health Promotion Programs: Leadership Skills for the 21st Century* (Gaithersburg, MD: Aspen Publishers, 1997); pp. 217–31.
27. W. D. Novelli, Health care, politicians, toothpaste: All can be marketed the same way, *Marketing News* 14 (1981): 1, 7.
28. Ward, *Marketing Strategies,* p. 81.
29. Novelli, Health care, p. 7.
30. Matthews, Marketing your services, p. 25.
31. Kotler, *Marketing Management,* Chapter 3.
32. Kotler, *Marketing in Non-profit Organizations,* as cited by Parks and Moody, A marketing model, p. 40.
33. Dever, *Community Health Analysis,* p. 252.
34. J. A. Quelch, Marketing principles and the future of preventive health care, Millbank Memorial Fund Quarterly/Health and Society 58 (1980): 317, as cited by Dever, *Community Health Analysis,* p. 283.
35. The section on promotional tools was adapted from Ward, *Marketing Strategies,* pp. 17–22.
36. The list of advantages is from T. Golaszewski and P. Prabhaker, Applying marketing strategies to worksite health promotion efforts, *Occupational Health Nursing* 32 (1984): 188–92.
37. American Marketing Association, *Dictionary of Marketing Terms;* available at http://marketingpower.com/live/mg-dictionary.php; and K. Stein, Brand name dietetics, *Journal of the American Dietetic Association* 104 (2004): 1530–33.
38. The discussion about PSAs and the list of five tips were adapted from Ward, *Marketing Strategies,* pp. 19, 72–73.
39. Matthews, Marketing your services, p. 25.
40. J. C. Levinson, *Guerilla Marketing* (Boston: Houghton Mifflin, 1998), pp. 29–31.
41. P. Kotler and A. Andreasen, *Strategic Marketing for Nonprofit Organizations* (Upper Saddle River, NJ: Prentice-Hall, 1995).
42. Novelli, Health care, p. 7.
43. Dever, *Community Health Analysis,* p. 280.
44. This synopsis of the social marketing process was adapted from Clift, Social marketing, p. 18.
45. The three situations described here are from K. F. Fox and P. Kotler, The marketing of social causes: The first 10 years, *Journal of Marketing* 44 (1980) 24–33.
46. P. E. Smith, Cost–benefit analysis and the marketing of nutrition services, in *Benefits of Nutrition Services: A Costing and Marketing Approach* (Columbus, OH: Ross Laboratory, 1987), p. 30.
47. The discussion of the PHHP was adapted from Lefebvre and Flora, Social marketing, pp. 309–11.

48. Ibid., p. 302.
49. S. Roberts, The new consumer, *Marketing Magazine* 14 (1998): 12–13.

Chapter 19

1. How the new strategic plan works—for you, *Journal of the American Dietetic Association* 92 (1992): 1070.
2. The definition of management was taken from K. M. Bartol and D. C. Martin, *Management* (New York: McGraw-Hill, 1991), p. 6. Other definitions appearing in this chapter were also taken or adapted from this textbook.
3. Hoover: It sucks, *The Economist* 327 (1993): 66.
4. B. B. Longest, Jr., J. S. Rakich, and K. Darr, *Managing Health Services Organizations and Systems,* 4th ed. (Baltimore, MD: Health Professions Pr, 2000).
5. President's page: "Leading the future of dietetics," *Journal of the American Dietetic Association* 103 (2003): 420.
6. K. G. Hardy, The three crimes of strategic planning, *Business Quarterly* 57 (1992): 71–74.
7. President's page: Taking stock of ADA's strategic initiatives, *Journal of the American Dietetic Association* 97 (1997): 429; and President's page: Checking the strategic framework road map, *Journal of the American Dietetic Association* 98 (1998): 588.
8. Bartol and Martin, *Management,* pp. 174 and 308.
9. J. Elton and J. Roe, Bringing discipline to project management, *Harvard Business Review* 76 (1998): 153–59.
10. D. R. Dalton and coauthors, Organization structure and performance: A critical review, *Academy of Management Review* 5 (1980): 49–64.
11. S. P. Robbins, *Organizational Behavior: Concepts, Controversies, and Applications* (Upper Saddle River, NJ: Prentice-Hall, 2000).
12. J. G. Liebler and C. R. McConnell, *Management Principles for Health Care Professionals,* 4th ed. (Sudbury, MA: Jones and Bartlett Publishers, 2004).
13. *Organizational Behavior: Concepts, Controversies, and Applications,* 2000.
14. Bartol and Martin, *Management,* p. 348.
15. D. D. Van Fleet and A. G. Bedeian, A history of the span of management, *Academy of Management Review* 2 (1977): 356–72.
16. J. C. Levinson, *The Ninety-Minute Hour* (New York: Dutton, 1990), p. 51.
17. M. W. McCall, Jr., and M. M. Lombardo, What makes a top executive? *Psychology Today* 17 (1983): pp. 26–31.
18. Bartol and Martin, *Management,* p. 407.
19. Liebler et al., *Management Principles,* 2004.
20. American Dietetic Association, *Role Delineation for Registered Dietitians and Entry-Level Dietetic Technicians* (Chicago: American Dietetic Association, 1990), pp. 255–90.
21. The discussion of staffing was adapted from Liebler and McConnell, *Management Principles,* 2004.
22. Bartol and Martin, *Management,* pp. 410–11.
23. F. Rice, Champions of communication, *Fortune* 123 (1991): 111–112, 116, 120.
24. B. R. Nathan and coauthors, Interpersonal relations as a context for the effects of appraisal interviews on performance and satisfaction: A longitudinal study, *Academy of Management Journal* 34 (1991): 352–69.
25. A. A. Chesney and E. A. Locke, Relationships among goal difficulty, business strategies, and performance on a complex management simulation task, *Academy of Management Journal* 34 (1991): 400–24.
26. J. G. Goodale, Seven ways to improve performance appraisals, *HRMagazine* 38 (1993): 77–80.
27. W. Bennis, *On Becoming a Leader* (Reading, MA: Addison-Wesley, 1989), pp. 39–45.
28. R. McGarvey, Goal-getters, *Entrepreneur* 21 (1993): 144.
29. P. F. Drucker, *The Practice of Management* (New York: Harper & Row, 1986), pp. 302–11.
30. Bartol and Martin, *Management,* p. 520.
31. D. K. Denton, *Recruitment, Retention, and Employee Relations* (Westport, CT: Quorum, 1992), pp. 155–71.
32. Bartol and Martin, *Management,* p. 594.
33. American Dietetic Association, *Role Delineation Study,* pp. 273–78.
34. J. Sneed, E. C. Burwell, and M. Anderson, Development of financial management competencies for entry-level and advanced-level dietitians, *Journal of the American Dietetic Association* 92 (1992): 1223–29.
35. T. D. Lynch, *Public Budgeting in America* (Upper Saddle River, NJ: Prentice-Hall, 1985), p. 4.
36. D. J. Breckon, *Managing Health Promotion Programs: Leadership Skills for the 21st Century,* pp. 141–50.
37. H. H. Schmitz, Decision support: A strategic weapon, *in Healthcare Information Management Systems,* ed. M. J. Ball and coauthors. (New York: Springer, 1992), pp. 42–48.
38. Bartol and Martin, *Management,* pp. 704–5; and W. D. Evers and S. McKinney, Computer tools for the new millennium, *Topics in Clinical Nutrition* 15 (2000): 1–9.
39. A. J. Morrison and K. Roth, Developing global subsidiary mandates, *Business Quarterly* 57 (1993): 104–10.

Chapter 20

1. A. Quinn, Writing a grant proposal that works, *Leader in Action* (Spring 1992): 9–16.
2. M. R. Schiller and J. C. Burge, How to write proposals and obtain funding. In *Research: Successful Approaches,* 2nd ed., ed. E. R. Monsen (Chicago: American Dietetic Association, 2003), pp. 49–68.
3. V. P. White, *Grants: How to Find Out about Them and What to Do Next* (New York: Plenum Press, 1985), p. vii; and Quinn, Writing a grant proposal that works, p. 11.
4. Ibid.
5. B. R. Ferrell and coauthors, Applying for Oncology Nursing Society and Oncology Nursing Foundation grants, *Oncology Nursing Forum* 16 (1989): 723–30.
6. L. E. Miner, J. T. Miner, and J. Griffith, *Proposal Planning & Writing,* 2nd ed. (Phoenix, AZ: Oryx Press, 1998).
7. Ibid.
8. G. N. Eaves, Preparation of the research-grant application: Opportunities and pitfalls, *Grants Magazine* 7 (1984), 151–57.
9. Ibid.
10. Kemp, A practical approach to writing successful grant proposals, *Nurse Practitioner* 16 (1991): 51–56.
11. D. Richards, Ten steps to successful grant writing, *Journal of Nursing Administration* 20 (1990): 20–23.

Appendix A

1. R. Yip and coauthors, Pediatric Nutrition Surveillance System—United States, 1980–1991, *Morbidity and Mortality Weekly Report* 41/No. SS–7 (November 27, 1992): 1–24.
2. R. S. Gibson, *Principles of Nutritional Assessment* (New York: Oxford, 1990), p. 175.
3. Gibson, *Principles of Nutritional Assessment,* pp. 173–74.
4. Ibid., pp. 172–73.
5. Yip and coauthors, Pediatric Nutrition Surveillance System.
6. J. H. Himes and W. H. Dietz, Guidelines for overweight in adolescent preventive services: Recommendations from an expert committee, *American Journal of Clinical Nutrition* 59 (1994): 307–16; S. E. Barlow and W. H. Dietz, Obesity evaluation and treatment: Expert committee recommendations, *Journal of Pediatrics* 102(3) (1998): 29.
7. World Health Organization Expert Committee on Physical Status, The use and interpretation of anthropometry, *Physical Status: Report of a WHO Expert Committee: WHO Technical Report Series 854,* WHO, Geneva, 1996.

Appendix B

1. President's Page, Alternative Medicine, *Journal of the American Dietetic Association* 97 (1997): 1431.

Index

The colors of the pyramid illustrate variety: each color represents one of the five food groups, plus one for oils. Different band widths suggest the proportional contribution of each food group to a healthy diet.

Gradual improvement is encouraged by the slogan. It suggests that individuals can benefit from taking small steps to improve their diet and lifestyle each day.

MyPyramid

STEPS TO A HEALTHIER YOU

MyPyramid.gov

A person climbing steps reminds consumers to be physically active each day.

The narrow slivers of color at the top imply moderation in foods rich in solid fats and added sugars.

Personalization is shown by the person on the steps, the slogan, and the URL. Find the kinds and amounts of food to eat each day at MyPyramid.gov.

The broad bases at the bottom represent nutrient-dense foods that should make up the bulk of the diet.

Greater intakes of grains, vegetables, fruit, and milk are encouraged by the broad bases of orange, green, red, and blue.

GRAINS Make half your grains whole	VEGETABLES Vary your veggies	FRUITS Focus on fruits	MILK Get your calcium-rich foods	MEAT & BEANS Go lean with protein
Eat at least 3 oz. of whole-grain cereals, breads, crackers, rice, or pasta every day 1 oz. is about 1 slice of bread, about 1 cup of breakfast cereal, or $\frac{1}{2}$ cup of cooked rice, cereal, or pasta	Eat more dark-green veggies like broccoli, spinach, and other dark leafy greens Eat more orange vegetables like carrots and sweet potatoes Eat more dry beans and peas like pinto beans, kidney beans, and lentils	Eat a variety of fruit Choose fresh, frozen, canned, or dried fruit Go easy on fruit juices	Go low-fat or fat-free when you choose milk, yogurt, and other milk products If you don't or can't consume milk, choose lactose-free products or other calcium sources such as fortified foods and beverages 1 cup = $1\frac{1}{2}$ oz. natural cheese, or 2 oz. processed cheese	Choose low-fat or lean meats and poultry Bake it, broil it, or grill it Vary your protein routine – choose more fish, beans, peas, nuts, and seeds 1 oz. = 1 oz. meat, poultry, or fish; $\frac{1}{4}$ cup cooked dry beans; 1 egg; 1 tbsp peanut butter; $\frac{1}{2}$ oz. nuts/seeds
For a 2,000-calorie diet, you need the amounts below from each food group.* To find the amounts that are right for you, go to MyPyramid.gov.				
Eat 6 oz. every day	Eat $2\frac{1}{2}$ cups every day	Eat 2 cups every day	Get 3 cups every day; for kids aged 2 to 8, it's 2	Eat $5\frac{1}{2}$ oz. every day

*Make most of your fat choices from fish, nuts, and vegetable oils. Find your allowance for oils at www.MyPyramid.gov.

Source: USDA, 2005